Lasers, Clocks and Drag-Free Control

AF815995

Astrophysics and Space Science Library

EDITORIAL BOARD

Chairman

W. B. BURTON, *National Radio Astronomy Observatory, Charlottesville, Virginia, U.S.A.* (bburton@nrao.edu) and *University of Leiden, The Netherlands* (burton@strw.leidenuniv.nl)

F. BERTOLA, *University of Padua, Italy*
J. P. CASSINELLI, *University of Wisconsin, Madison, U.S.A.*
C. J. CESARSKY, *European Southern Observatory, Garching bei München, Germany*
P. EHRENFREUND, *Leiden University, The Netherlands*
O. ENGVOLD, *University of Oslo, Norway*
A. HECK, *Strasbourg Astronomical Observatory, France*
E. P. J. VAN DEN HEUVEL, *University of Amsterdam, The Netherlands*
V. M. KASPI, *McGill University, Montreal, Canada*
J. M. E. KUIJPERS, *University of Nijmegen, The Netherlands*
H. VAN DER LAAN, *University of Utrecht, The Netherlands*
P. G. MURDIN, *Institute of Astronomy, Cambridge, UK*
F. PACINI, *Istituto Astronomia Arcetri, Firenze, Italy*
V. RADHAKRISHNAN, *Raman Research Institute, Bangalore, India*
B. V. SOMOV, *Astronomical Institute, Moscow State University, Russia*
R. A. SUNYAEV, *Space Research Institute, Moscow, Russia*

Lasers, Clocks and Drag-Free Control

Exploration of Relativistic Gravity in Space

by

Hansjörg Dittus

Claus Lämmerzahl

Slava G. Turyshev

Editors

Springer

Hansjörg Dittus
University of Bremen
Am Fallturm
28359 Bremen, Germany
E-mail: dittus@zarm.uni-bremen.de

Claus Lämmerzahl
University of Bremen
Am Fallturm
28359 Bremen, Germany
E-mail: laemmerzahl@zarm.uni-bremen.de

Slava G. Turyshev
Jet Propulsion Laboratory
California Institute of Technology
4800 Oak Grove Drive
Pasadena, CA 91109, USA
E-mail: turyshev@jpl.nasa.gov

Cover Illustration: Concept for a Deep Space Gravity Explorer, see page 589.

ISSN 0067-0057

ISBN 978-3-642-07073-0 e-ISBN 978-3-540-34377-6

This work is subject to copyright. All rights are reserved, whether the whole or part of the material is concerned, specifically the rights of translation, reprinting, reuse of illustrations, recitation, broadcasting, reproduction on microfilm or in any other way, and storage in data banks. Duplication of this publication or parts thereof is permitted only under the provisions of the German Copyright Law of September 9, 1965, in its current version, and permission for use must always be obtained from Springer-Verlag. Violations are liable to prosecution under the German Copyright Law.

Springer is a part of Springer Science+Business Media.
springer.com
© Springer-Verlag Berlin Heidelberg 2010

The use of general descriptive names, registered names, trademarks, etc. in this publication does not imply, even in the absence of a specific statement, that such names are exempt from the relevant protective laws and regulations and therefore free for general use.

Cover design: F. Steinen, eStudio Calamar, Girona/Spain

Preface

Special and general relativity are the theories describing the physics of space and time. Space and time are explored with clocks and electromagnetic signals. Therefore, special and general relativity are related to precise clocks and the thorough understanding of signal propagation. The ever-increasing accuracy of clocks together with novel methods for precision time transfer and clock synchronization are pivotal for the new generation of experiments probing the validity of Einstein's theories from subatomic distances to cosmic scales.

Such tests are not only motivated by the requirement that fundamental theories like special and general relativity which need the best experimental basis one can obtain, but also by the request to explore as far as possible the range of applicability of these theories, and finally by the search for gravitational waves. The search for quantum gravity and recent progress in astrophysics and cosmology has provided new strong motivation for high-accuracy tests of relativistic gravity. A number of recently proposed experiments will probe the foundations of general relativity by testing the equivalence principle, Lorentz invariances, the universalities of a free fall and gravitational redshift, as well as the constancy of gravitational and fine-structure constants. If detected, a violation of any of these principles will signal the presence of new physics and may show us the way to gravity quantization or/and to a grand unified field theory. As such these experiments have a significant discovery potential and will likely be the focus of the community effort for the next decade.

When conducted in space, these experiments will benefit from well-understood and controlled laboratory environments. A significant advance in the field of experimental gravitational physics can be expected from highly accurate laser ranging paired with new optical and/or microwave frequency standards or based entirely on optical frequency combs together with atomic sensors and drag-free technologies for attitude control. These new technologies allow taking full advantage of the variable gravity potentials, large heliocentric distances, and high velocity and acceleration regimes achievable in the solar system. As a result, the gravity research in the near future can significantly advance knowledge of fundamental physics and will also provide new capabilities to improve our life on Earth.

In the present volume we will discuss the issues that are relevant for future space missions aiming at testing and exploring gravity with much higher accuracy, namely:

- Quest from fundamental physics
- Space conditions
- Space technologies
- Space missions

In particular, we will discuss the present status and expected progress in the laser-enabled technologies (ranging, communication, and interferometry), atomic and optical frequency standards, atomic sensors, and drag-free technologies.

All these issues have been discussed on the 359th WE-Heraeus seminar on "Lasers, Clocks, and Drag-Free: New Technologies for Testing Relativistic Gravity in Space" that took place at the Center for Applied Space Technology and Microgravity (ZARM) at the University of Bremen from 30 May to 1 June 2005. It is our great pleasure to thank all the speakers for their presentations and especially those who were willing to write them up for this volume. We also like to thank the Wilhelm and Else Heraeus Foundation for its generous support without which this seminar could not have been carried through.

Bremen and Pasadena,
May 2007

Hansjörg Dittus
Claus Lämmerzahl
Slava G. Turyshev

Contents

Part I Surveys

Part II Theory

List of Contributors

Neil Ashby
Department of Physics
UCB 390
University of Colorado
Boulder, CO 80309-0390
USA
ashby@boulder.nist.gov

Peter L. Bender
JILA
UCB 440
University of Colorado
Boulder, CO 80309-0440
USA
pbender@jila.colorado.edu

Orfeu Bertolami
Instituto Superior Técnico
Departamento de Física
Av. Rovisco Pais
1049-001 Lisboa
Portugal
orfeu@cosmos.ist.utl.pt

Christian J. Bordé
LNE-SYRTE
UMR8630
Observatoire de Paris
61 avenue de l'Observatoire
75014 Paris
France
chbo@ccr.jussieu.fr

Philippe Bouyer
Laboratoire Charles Fabry de l'Institut d'Optique
Centre National de la Recherche Scientifique et Université Paris Sud 11, Bat. 503,
Campus Universitaire d'Orsay
91403 Orsay Cedex
France
philippe.bouyer@iota.u-psud.fr

Michael H. Brügmann
Theoretisch-Physikalisches Institut
Friedrich-Schiller-Universität Jena
Max-Wien-Platz 1
07743 Jena
Germany
brm@tpi.uni-jena.de

Erwin Brüning
School of Mathematical Sciences
University of KwaZulu-Natal
Durban 4000
South Africa
ebruning@pixie.udw.ac.za

Bernd Dachwald
DLR, Mission Operations Section
Oberpfaffenhofen
82234 Wessling
Germany
bernd.dachwald@dlr.de

Denis Defrère
Faculty of Applied Sciences
University of Liege
chemin des Chevreuils
1 Bât. B52/3 Sart Tilman
4000 Liege
Belgium
denis.defrere@ulg.ac.be

John J. Degnan
Sigma Space Corporation
4801 Forbes Blvd.
Lanham, MD 20706
USA
john.degnan@sigmaspace.com

Hansjörg Dittus
ZARM
University of Bremen
Am Fallturm
28359 Bremen
Germany
dittus@zarm.uni-bremen.de

Walter Fichter
Universität Stuttgart
Institute of Flight Mechanics and Control
70569 Stuttgart
Germany
fichter@ifr.uni-stuttgart.de

Gerhard Heinzel
Max-Planck-Institut für Gravitationsphysik
Albert-Einstein-Institut
Callinstr. 38
30176 Hannover
Germany
gerhard.heinzel@aei.mpg.de

Marc–Thierry Jaekel
Laboratoire de Physique Théorique de l'Ecole Normale Supérieure
CNRS, UPMC
24 rue Lhomond
75231 Paris Cedex 05
France
jaekel@lpt.ens.fr

Ulrich Johann
EADS Astrium GmbH
88039 Friedrichshafen
Germany
ulrich.johann@astrium.eads.net

Sergei Klioner
Lohrmann Observatory
Dresden Technical University
Mommsenstr. 13
01062 Dresden
Germany
klioner@rcs.urz.tu-dresden.de

James M. Kohel
Quantum Sciences and Technology Group
Jet Propulsion Laboratory
California Institute of Technology
Pasadena CA
USA
james.m.kohel@jpl.nasa.gov

Sergei M. Kopeikin
Department of Physics & Astronomy
University of Missouri-Columbia
Columbia, MO 65211
USA
kopeikins@missouri.edu

Claus Lämmerzahl
ZARM
University of Bremen
Am Fallturm
28359 Bremen
Germany
laemmerzahl@zarm.uni-bremen.de

Arnaud Landragin
LNE-SYRTE
UMR8630
Observatoire de Paris
61 avenue de l'Observatoire
75014 Paris
France
arnaud.landragin@obspm.fr

Bernard Linet
Laboratoire de Mathématiques et Physique Théorique
CNRS/UMR 6083
Université François Rabelais
37200 Tours
France
linet@celfi.phys.univ-tours.fr

Nathan E. Lundblad
Quantum Sciences and Technology Group
Jet Propulsion Laboratory
California Institute of Technology
Pasadena CA
USA
nathan.e.lundlbad@jpl.nasa.gov

Lute Maleki
Quantum Sciences and Technology Group
Jet Propulsion Laboratory
California Institute of Technology
Pasadena CA
USA
lute.maleki@jpl.nasa.gov

Jürgen Müller
Institut für Erdmessung (IfE)
University of Hannover
Schneiderberg 50
30167 Hannover
Germany
mueller@ife.uni-hannover.de

Wei–Tou Ni
Purple Mountain Observatory
Chinese Academy of Sciences
Nanjing, 210008
China
wtni@pmo.ac.cn

Kenneth L. Nordtvedt, Jr.
Northwest Analysis
118 Sourdough Ridge Road,
Bozeman, MT 59715
USA
kennordtvedt@bresnan.net

Jorge Páramos
Instituto Superior Técnico
Departamento de Física
Av. Rovisco Pais
1049-001 Lisboa
Portugal
x_jorge@fisica.ist.utl.pt

Franck Pereira dos Santos
LNE-SYRTE
UMR8630
Observatoire de Paris
61 avenue de l'Observatoire
75014 Paris
France
franck.pereira@obspm.fr

Volker Perlick
Technical University Berlin
Institute of Theoretical Physics
Sekr. PN 7-1
Hardenbergstraße 36
10623 Berlin
Germany
vper0433@itp.physik.tu-berlin.de

Achim Peters
Institut für Physik
Humboldt University of Berlin
10117 Berlin
Germany
achim.peters@physik.hu-berlin.de

James D. Phillips
Smithsonian Astrophysical Observatory,
Harvard-Smithsonian Center for Astrophysics
60 Garden Street, MS-63
Cambridge, MA 02138
USA
jphillips@cfa.harvard.edu

Christophe Le Poncin-Lafitte
Département Systèmes de Référence Temps et Espace
CNRS/UMR 8630
Observatoire de Paris
61 avenue de l'Observatoire
75014 Paris
France
leponcin@danof.obspm.fr

John D. Prestage
Quantum Sciences and Technology Group
Jet Propulsion Laboratory
California Institute of Technology
Pasadena CA
USA
john.d.prestage@jpl.nasa.gov

Oliver Preuss
Max–Planck–Institute for Solar System Research
Max-Planck-Str. 2
37191 Katlenburg-Lindau
Germany
opreuss@mps.mpg.de

Antonio Pulido Patón
Purple Mountain Observatory
Chinese Academy of Sciences
Nanjing, 210008
China
antonio@pmo.ac.cn

Andreas Rathke
EADS Astrium GmbH
Dept. AED41
88039 Friedrichshafen
Germany
andreas.rathke@astrium.eads.net

Robert D. Reasenberg
Smithsonian Astrophysical Observatory,
Harvard-Smithsonian Center for Astrophysics
60 Garden Street, MS-63
Cambridge, MA 02138
USA
reasenberg@cfa.harvard.edu

Serge Reynaud
Laboratoire Kastler Brossel
Université Pierre et Marie Curie
case 74, CNRS, ENS
Campus Jussieu
75252 Paris Cedex 05
France
reynaud@spectro.jussieu.fr

Albrecht Rüdiger
Max-Planck-Institut für Gravitationsphysik
Albert-Einstein-Institut
Callinstr. 38
30176 Hannover
Germany
albrecht.ruediger@aei.mpg.de

Gerhard Schäfer
Theoretisch-Physikalisches Institut
Friedrich-Schiller-Universität Jena
Max-Wien-Platz 1
07743 Jena
Germany
g.schaefer@tpi.uni-jena.de

Silvia Scheithauer
Max-Planck-Institut für Astronomie
Königstuhl 17
D-69117 Heidelberg
Germany
scheithauer@mpia.de

Stephan Schiller
Institut für Experimentalphysik
HeinrichHeineUniversitat Dusseldorf
40225 Düsseldorf
Germany
step.schiller@
uni-duesseldorf.de

Alexander Schleicher
EADS Astrium GmbH
88039 Friedrichshafen
Germany
Alexander.Schleicher@
astrium.eads.net

Wolfgang Seboldt
DLR, Institute of Space Simulation
Linder Höhe
51147 Köln
Germany
wolfgang.seboldt@dlr.de

Michael Shao
Jet Propulsion Laboratory
California Institute of Technology
4800 Oak Grove Drive
Pasadena, CA 91109
USA
mshao@huey.jpl.nasa.gov

Pierre Teyssandier
Département Systèmes de Référence
Temps et Espace
CNRS/UMR 8630
Observatoire de Paris
61 avenue de l'Observatoire
75014 Paris
France
Pierre.Teyssandier@obspm.fr

Stephan Theil
ZARM
University of Bremen
Am Fallturm
28359 Bremen
Germany
theil@zarm.uni-bremen.de

Robert J. Thompson
Quantum Sciences and Technology
Group
Jet Propulsion Laboratory
California Institute of Technology
Pasadena CA
USA
robert.j.thompson@jpl.nasa.gov

Massimo Tinto
Jet Propulsion Laboratory
California Institute of Technology,
Pasadena, CA 91109
USA
Massimo.Tinto@jpl.nasa.gov

Michael Tröbs
Max-Planck-Institut für
Gravitationsphysik
Albert-Einstein-Institut
Callinstr. 38
30176 Hannover
Germany
michael.troebs@aei.mpg.de

Slava G. Turyshev
Jet Propulsion Laboratory
California Institute of Technology
4800 Oak Grove Drive
Pasadena, CA 91109
USA
turyshev@jpl.nasa.gov

Stefano Vitale
University of Trento
via Sommarive 14
I-38050 Povo (Trento)
Italy
vitale@science.unitn.it

James G. Williams
Jet Propulsion Laboratory
California Institute of Technology
4800 Oak Grove Drive
Pasadena, CA 91109
USA
james.williams@jpl.nasa.gov

Xuejun Wu
Purple Mountain Observatory
Nanjing 210008
China
and
Department of Physics
Nanjing Normal University
Nanjing 210097
China

Yan Xia
Purple Mountain Observatory
Chinese Academy of Sciences
Nanjing, 210008
China

Chongming Xu
Purple Mountain Observatory
Nanjing 210008
China
and
Department of Physics
Nanjing Normal University
Nanjing 210097
China
cmxu@njnu.edu.cn
cmxu@pmo.ac.cn

Nan Yu
Quantum Sciences and Technology Group
Jet Propulsion Laboratory
California Institute of Technology
Pasadena CA
USA
nan.yu@jpl.nasa.gov

Acronyms

ACES	Atomic clock ensemble in space
AOCS	Attitude and orbit control system
APD	Avalanche photodiode
APOLLO	Apache point observatory lunar laser ranging operation
ASTROD	Astrodynamical space test of relativity using optical devices
BBO	Big bang observer
BCRS	Barycentric celestial reference system
BEC	Bose–Einstein condensation
CASI	Cold atom Sagnac interferometer
CDM	Cold dark matter
CMBR	Cosmic microwave background radiation
CNES	Centre National d'Etudes Spatiales (French Aerospace Center)
CoMRS	Center-of-mass reference system
DE	Dark energy
DECIGO	DECihertz interferometer gravitational wave observatory
DFACS	Drag-free and attitude control system
DM	Dark matter
DLR	Deutsches Zentrum für Luft-und Raumfahrt (German Aerospace Center)
DSGE	Deep-space gravity explorer
DSN	Deep-space network
EOP	Earth orientation parameter
EP	Equivalence principle
ESA	European Space Agency
FEEP	Field emission electric propulsion
FEM	Finite element method
FFCC	Free-falling corner cube
GAIA	Global astrometric interferometer for astrophysics
GCRS	Geocentric celestial reference system
GG	Galileo Galilei

GOCE	Gravity field and steady-state ocean circulation explorer
GOM	Gyromètre à Onde de Matière
GP-A	Gravity Probe A
GP-B	Gravity Probe B
GPS	Global positioning system
GR	General relativity
GTD	Gravitational time delay
GW	Gravitational wave
HIPPARCOS	High-precision parallax collecting satellite
HYPER	Hyper-precision cold atom interferometry in space
ICE	Interférométrie Cohérente pour l'Espace
IERS	International earth rotation and reference systems service
ILRS	International laser ranging service
ISS	International space station
ITRF	International terrestrial reference frame
JASMINE	Japan astrometry satellite mission for infrared exploration
JIMO	Jupiter icy Moons orbiter
JWST	James Webb Space Telescope
LAGEOS	Laser geodynamic satellite
LATOR	Laser astrometric test of relativity
LIGO	Laser interferometer gravitational wave observatory
LISA	Laser interferometer space antenna
LITE	Lidar in-space technology experiment
LITS	Linear ion trap standard
LLI	Local Lorentz invariance
LLR	Lunar laser ranging
LOLA	Lunar orbiter laser altimeter
LPI	Local position invariance
LRE	Laser ranging equipment
LRO	Lunar reconnaissance orbiter
MAGIA	Misura Accurata di G mediante Interferometria Atomica
MCP	Minimal coupling principle
MGS	Mars global surveyor
MICROSCOPE	MICROSatellite a trainée Compensée pour l'Observation du Principe d'Equivalence
MOLA	Mars orbiter laser altimeter
MOND	Modified Newtonian dynamics
MOT	Magneto-optical trap
NASA	National Aeronautics and Space Administration
NEAR	Near Earth asteroid rendezvous
OPTIS	Optical test of the isotropy of space
PARCS	Primary atomic reference clock in space

PHARAO	Projet d'Horloge Atomique par Refroidissement d'Atomes
POEM	Principle of equivalence measurement
POINTS	Precision optical interferometer in space
PPN	Parameterized post-Newtonian
QuITE	Quantum interferometer test of equivalence
RACE	Rubidium atomic clock experiment
RTG	Radioisotope thermoelectric generator
SEE	Satellite energy exchange
SEP	Strong equivalence principle
SLR	Satellite laser ranging
SOREL	Space experiment on gravitational theories
SQUID	Superconducting quantum interference device
STEP	Satellite test of the equivalence principle
STM	Spacetime mission
TAI	Temps atomique international
TCB	Barycentric coordinate time
TDI	Time-delay interferometry
TDIR	Time-delay interferometric ranging
TFG	Tracking frequency laser distance gauge
UFF	Universality of free fall
UGR	Universality of the gravitational redshift
ULE	Ultra-low expansion
VLBI	Very long-baseline interferometer
WEAX	Weak equivalence antimatter experiment
WEP	Weak equivalence principle

Part I

Surveys

Fundamental Physics, Space, Missions and Technologies

Claus Lämmerzahl and Hansjörg Dittus

ZARM, University of Bremen, Am Fallturm, 28359 Bremen, Germany

Summary. We review the relation between fundamental physics, the space conditions which in some cases are of great advantage for carrying through experiments, and already developed and emerging technologies used in fundamental physics space missions.

1 Introduction

Fundamental physics is becoming very exciting these days. The reasons are twofold: on the theoretical side, the unification of general relativity and quantum theory seems to lead to deviations from standard physics. On the experimental side, new developments of high-precision apparatus make new realms of physics accessible, leading to new tests and observations. Consequently, the expectation for "new physics" as well as improvements in experimental devices strongly pushes the efforts for the realization of new experiments and observations. An important aspect in this connection is the quality of the experimental environment. It is clear that most experiments need a noise-free environment with stable thermal, seismic, electric, etc. conditions. Furthermore, some experiments might profit a lot if they will be carried out in a free-fall and non-rotating environment and some experiments necessarily require that environment. This leads one to the conclusion that there are quite a few experiments which will give results that are orders of magnitude better when carried through in space than when carried through on Earth.

In the following survey, which is based on [1], we will shortly review these three aspects, namely the status of the fundamental quests, new experimental developments and the conditions in space. In the case that these three conditions complement one another appropriately, it is reasonable to think about doing these experiments in space, see Fig. 1.

The first dedicated space mission for fundamental physics was GP-A, which measured the gravitational redshift with an until today unrivaled accuracy. It was during the last years that it was recognized that for many

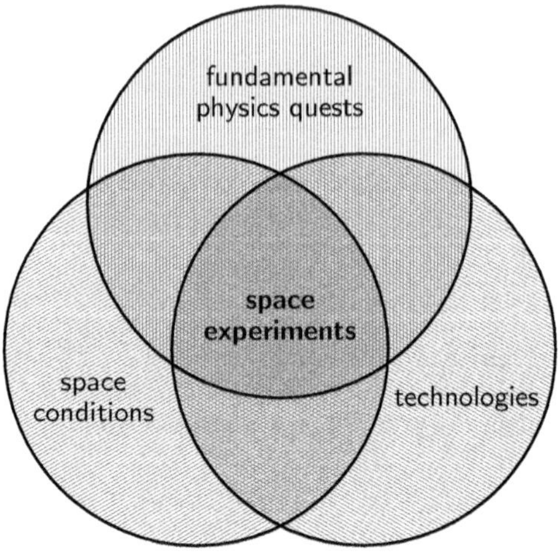

Fig. 1. The conditions for space projects.

experiments space conditions are really indispensable. As a consequence, many space missions have been proposed. Though most of these proposed missions are definitely worth to be carried through, the huge expenses, long planning time and troublesome efforts for space qualification of experimental devices cut down the number of missions expected to fly to a very few. This is a big disadvantage which might be overcome by developing and using more standardized space techniques. Furthermore, in some cases it is more efficient, instead of developing a dedicated mission, to make use of the ISS which is already existing – even at the price of experimental conditions that are not optimal (see Sect. 4)

For the convenience of the reader we give a list of acronyms of the missions mentioned in this short review:

ACES	Atomic clock ensemble in space
ASTROD	Astrodynamical space test of relativity using optical devices
DSGE	Deep-space gravity explorer
GG	Galileo Galilei
GP-A	Gravity probe A
GP-B	Gravity probe B
HYPER	Hyper precision atomic interferometer in space
LAGEOS	Laser geodynamic satellite
LATOR	Laser astrometric test of relativity
LISA	Laser interferometer space antenna
LLR	Lunar laser ranging
MICROSCOPE	Micro-satellite a trainée Compensée pour l'Observation du Principe d'Equivalence
OPTIS	Optical test of the isotropy of space

PHARAO	Projet d'Horloge Atomique par Refroidissement d'Atomes en Orbite
PARCS	Primary atomic reference clock in space
RACE	Rubidium atomic clock experiment
SEE	Satellite energy exchange
STEP	Satellite test of the equivalence principle
STM	Spacetime mission

2 Fundamental Physics

2.1 The General Scheme

Today's fundamental physics is characterized by two schemes [2], see Table 1: by the universal theories and by the four interactions. The universal theories which are applicable to any kind of physical phenomenon are (1) quantum theory, (2) special relativity (SR), (3) general relativity (GR) and (4) many particle physics. The interactions are (1) gravity, (2) electromagnetism, (3) the weak and (4) the strong interaction. Except for gravity, they have been successfully unified. It can be seen that gravity is exceptional since it appears on both sides: it is universal due to its universal couplings and it is a particular interaction.

A big problem in the theoretical description of physics is the incompatibility of quantum theory and GR as relativistic theory of gravity. This can be seen from the fact that in GR from very general assumptions singularities, spatial points, where all the surrounding matter fall into, will occur. Such a localization of matter is forbidden by quantum theory. Such incompatibilities make it necessary to look for a unification of quantum theory and GR, i.e. for a quantum gravity theory (see, e.g. [3], for a recent review). There are several approaches to such a new theory, e.g. string theory, canonical quantum gravity, or non-commutative geometry. In any case, deviations from standard physics given by the theories of Table 1 are predicted. Each of these theories predicts deviations, e.g. from the universality of free fall, modifications

Table 1. Universal theories and the four presently known interactions.

Universal theories	Interactions
– Quantum theory – Special relativity – General relativity – Many particle physics	– Gravity – Electromagnetism – Weak interaction – Strong interaction
Problem: incompatibility between general relativity and quantum theory	**Wish**: unification of all interactions
Possible solution: quantum gravity	

of the electromagnetic and of the gravitational interaction in terms of, e.g. a Yukawa-like potential, etc. All these theoretical considerations are strong reasons to make an effort to get better experimental results. In addition, physics always requires the best experimental foundation of its basic theories being the pillars of modern understanding of physics.

As already mentioned, gravity plays an outstanding role. It plays this role not only because of the reasons described above but also because in most cases a violation of the principles underlying GR can be observed if the description of interactions like the Maxwell's equations or the Dirac equation underlying particle physics is modified. Indeed, only the present form of Maxwell's equations and Dirac equation is compatible with GR, i.e. the structure of the Maxwell and Dirac equations strongly determines the structure of the gravitational field (a modification of Maxwell's equation, for example, leads to a violation of the universality of free fall [4]). Consequently, tests of GR play an outstanding role in this search for new physics.

Behind all these physical structures there are the principles of many particle physics. Today, this field is deeply connected with renormalization group theory. Due to the fact that renormalization group theory not only is a method describing the physics of gases and their phase transitions but, more generally, also is a mathematical method with applications in many parts of physics, from statistical physics to hydrodynamics, solid state physics to elementary particle physics, and even to problems in quantum gravity, it is very important to understand the principles underlying this theory more deeply and to improve the quality of its tests.

2.2 Structure of Gravity

The present theory of gravity, Einstein's general relativity, is based on a set of universality principles [5] (1) the universality of free fall (UFF), (2) the universality of the gravitational redshift (UGR) and (3) a universality with respect to the state of motion of the observer, called local Lorentz invariance (LLI). If these principles are valid, then gravity can be described by a space–time metric as given in the mathematical framework of Riemannian geometry. Further requirements on the structure of the equations the metric has to fulfil then lead to the Einstein field equation, see Fig. 2.

Indeed, most fundamental physics experiments are devoted to tests the principles underlying GR. Owing to the fact that GR deals with the structure of space–time, all tests of GR are tests involving the measurement of time, paths (either paths of light or paths of massive bodies) and directions. Here, clocks play a really fundamental role: they are used for a complete test of the principles underlying SR and also for tests concerning the gravitational redshift, see Fig. 3. That is also the reason why each improvement in the precision of clocks is followed by new tests of SR and GR. The observation of paths, i.e. geodesic paths driven only by the gravitational interaction, is difficult due to many disturbing and competitive effects. Disturbing effects are, e.g., atmospheric drag and radiation from the Earth and the Sun which

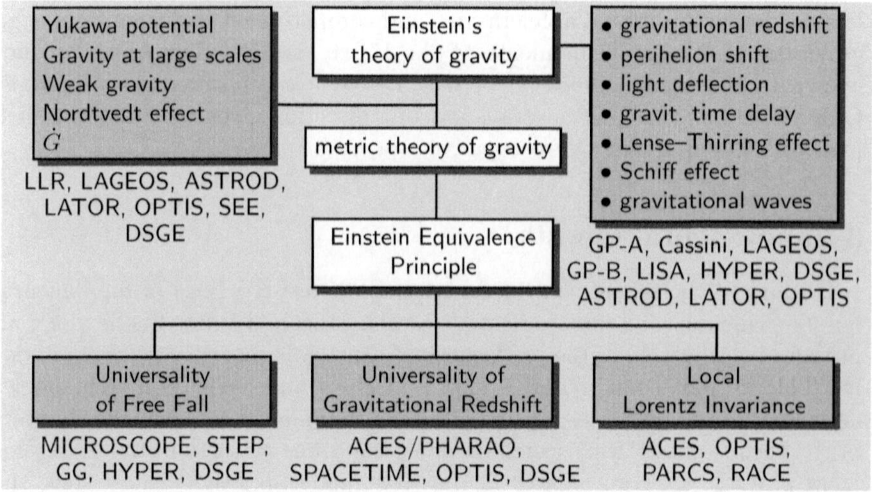

Fig. 2. The structure of experimental exploration of the theory of gravity. Testable issues are described in *grey boxes*, theoretical concepts in *white boxes*. Metric theories of gravity are based on the directly testable principles: universality of free fall, universality of the gravitational redshift and local Lorentz invariance. Particular effects like the Nordtvedt effect, a time-varying G, a deviation from ordinary Newton potential at large distances and for small gravitational acceleration denote deviations from ordinary Einstein's general relativity. The predictions of Einstein's GR are found in the *upper right box*. The missions aiming at the exploration of the various effects are shown below the *grey boxes*.

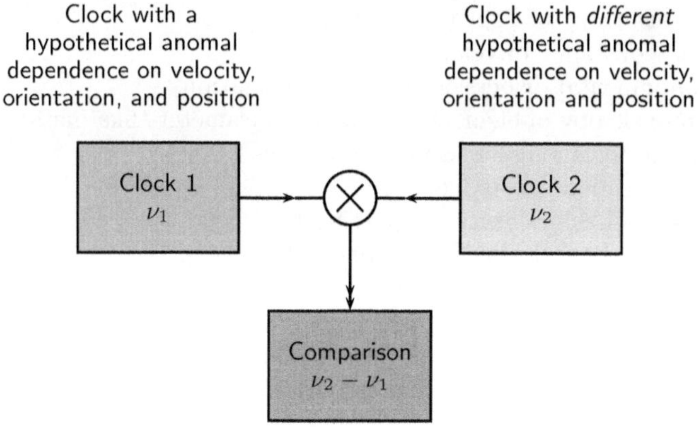

Fig. 3. General scheme for testing SR and GR with clocks. Tests of the constancy of the speed of light and of UGR are essentially carried through with clocks whose frequency may change with orientation, velocity and position. Furthermore, if the clocks are assumed to move with different velocities or to be at different positions, then this scheme also applies to tests of the time dilation and the gravitational redshift.

lead to non-gravitational accelerations and competing effects come from, e.g., gravitational multipole moments of the Earth and the Sun which are not known to the needed precision or which need a complicated data analysis. Only GP-B tests the parallel transport of a direction given by a distinguished physical system, namely gyroscopes.

3 Fundamental Quests

According to the scheme outlined above, there are two lines of fundamental quests (which are, of course, related to each other): the first line is along an improved verification of the predictions of Einstein's theory of general relativity. This, in particular, requires improved clock and navigation technologies as well as an improved description of post-Newtonian effects (see the chapters in the theory part of this volume). The second line consists in the search for "new physics" or, equivalently, for possible quantum gravity effects (see the subsequent contribution of Bertolami, Parámos and Turyshev in this volume).

As already mentioned, all approaches to a quantum gravity theory predict deviations from the principles underlying GR asking for refined tests of all the aspects of GR. More specifically, these experiments look for:

1. Violations of the UFF
2. Violations of the UGR
   - Time and/or position dependence of the fine structure constant and other fundamental constants
   - Time and/or position dependence of the gravitational constant
3. Violations of LLI in many aspects
   - Non-isotropy of light propagation
   - Non-constancy of velocity of light
   - Fundamental dispersion of light propagation
   - Non-isotropy of elementary particle parameters like mass
   - Search for anomalous spin interactions
4. Non-Einsteinian effects like
   - Yukawa-like gravitational potential
   - Modification of gravity at large scales
   - Modification of weak gravity
   - Nordtvedt effect
   - Time variation of the gravitational constant G
   - Different active and passive gravitational mass

The corresponding space missions can be found in Fig. 2. Besides tests of relativity and gravity, there are also issues for testing the fundamentals of the other universal theories like quantum theory and many particle theory:

- Linearity of quantum physics
- Entanglement
- Casimir force
- Physics of Bose–Einstein condensates

- Search for a fundamental decoherence
- Test of renormalization group theory

(See also recent reviews on these topics [3,6–8].)

Since quantum gravity is characterized by the Planck energy $E_P \sim 10^{28}$ eV and laboratory energies are of the order eV for ordinary, e.g. optical laboratory experiments, to GeV for large particle accelerators, quantum gravity effects are too small by many orders of magnitude to be detectable in laboratories. Indeed, expected quantum gravity-induced violations of LLI, for example, are looked for in ultra-high energy cosmic rays. However, all the present predictions from quantum gravity are in fact merely hypotheses; they are not based on complete theories. Furthermore, perhaps some additional mechanism has to be applied which may lead to some enhancement of the expected effect as it is the case for deviations from Newton potential at small distances as predicted from higher dimensional theories. Therefore, there is always a possibility that deviations from standard physics may occur at lower energies than given by the Planck scale. Consequently, *any improvement of the accuracy of experimental results is of great value.*

We also address the question what happens if an experiment shows a violation of one of these basic principles. This not necessarily means that one has found a violation of one of these principles. This effect may also be a result of a new interaction which might well be in accordance with the tested principles. Therefore one first has to search whether this effect can be shielded, or whether one can find a cause of this effect in the sense of a source which creates a field causing this effect. In both cases the effect can be considered to be a new interaction. Only if all these questions are answered appropriately one can speak about a violation of a basic principle.

Particular predictions and models of deviations from standard physics are:

- Violation of the UFF at the 10^{-13} level predicted from dilaton scenarios [9] and at the 10^{-14} level from quintessence theories [10].
- Deviation from GR in terms of the PPN parameters γ and β, again predicted within dilaton scenarios [11].
- Violations of LLI at Planck scale predicted from non-commutative geometry approaches [12].
- The time dependence of constants [10,13–15].
- Additional Yukawa part of the gravitational potential at small distances predicted from higher-dimensional theories [16].
- The gravitational interaction at large distances or for weak gravitational fields (Yukawa for large ranges [17] or MOND for small gravitational accelerations [18]). This may be relevant for the Pioneer anomaly [19].

4 The Space Conditions

In many cases the sensitivity of measuring devices and/or the accuracy of measurements itself will increase if the experiments can be performed under conditions of free fall, i.e. under conditions of weightlessness. The advantages of such conditions are:

1. *The infinitely long, and periodic, free fall.* As an example, long free-fall conditions enable high-precision tests of the UFF for all kinds of structureless (i.e. pointlike) matter.
2. *Long interaction times.* This is, for example, a big advantage in atomic or molecular interferometers, where the laser-cooled atoms or molecules may interact with other external fields for a long time and do not fall out of the interferometer as it happens on Earth. Only in a microgravity environment in space one has the opportunity of a dedicated long exposure to certain interactions.
3. *Large potential differences.* In a large class of experiments (e.g. tests of the gravitational redshift), the magnitude of the signals looked for depends on the difference of the positions of the clocks in the gravitational potential. It is obvious that this can be achieved best by going into space.
4. *Large velocity changes.* For macroscopic devices used for, e.g. testing the dependence of the speed of light with respect to the laboratory velocity (Kennedy–Thorndike tests), the maximum velocity on Earth might be of the order 1,000 $\mathrm{km\,h^{-1}}$. In space this can be increased by about one order of magnitude. For example, the velocity variations along the orbit (e.g. in a high elliptical Earth orbit) is 30 times higher than one can attain using the Earth's rotation.
5. *Availability of long distances.* In space, much longer distances are available than in any laboratory on Earth. This is essential, e.g. in the study of low frequency (10^{-3} Hz) gravity waves using interferometric techniques, where the strain of spacetime is to be measured at or below the 10^{-21} level.
6. *A low noise/vibration environment.* Seismic noise is a limiting factor for many experiments on Earth (e.g. for gravitational wave detectors and for torsion balances) in the frequency range below 10 Hz.
7. For certain interactions, only in space one has the opportunity to search for the corresponding effects. As an example, only in space there are appropriate conditions adapted to the detection of the gravitational time delay, of gravitational waves with very low frequencies, of the Schiff or of the Lense–Thirring effect.
8. Due to the absence of the atmosphere, the true particle content in outer space is directly observable.

As a consequence, there are quite a few instances where it is really indispensable to go to space to achieve improved accuracies as compared to experiments on ground.

However, these advantages have to be compared with some disadvantages. These disadvantages are in many cases:

1. Huge financial effort
2. Long time for preparation and development

3. No direct access to the experiment during operation
4. No possibility of a post-mission analysis of the experimental payload

As an example, the analysis of the Pioneer anomaly heavily rests on properties of materials the spacecraft are built of. These properties may change with time and, thus, in principle may be responsible for the observed effect. A direct access and a post-mission analysis of devices is a big advantage of, e.g. ISS-based experiments.

5 Past and Running Missions

Until now, there have been only very few space missions which carried through dedicated experiments concerning gravitational physics or which could be used for that. Nevertheless, the results of these missions have been widely discussed and led to a much better experimental basis and understanding of gravitational physics. Without these missions some effects like the gravitational redshift or the gravitational time delay would have been confirmed with much lower accuracy, and some effects even would have still remained unobserved like the geodetic precession of the Earth–Moon system or the Lense–Thirring effect. Already from this, one can infer the scientific potential of further space missions.

GP-A. This was the first space mission dedicated to fundamental physics issues. The time given by a H-maser on a rocket was compared with the time of a ground-based H-maser. Due to a three-channel method, one could separate between the Doppler effect from the gravitational redshift. This yielded the up to now best confirmation of the absolute gravitational redshift with an accuracy of $1.4 \cdot 10^{-4}$ [20].

Viking. Due to a microwave link to the Mars explorer one could measure the time needed for a signal to propagate from Earth to the Viking satellite and back. In case of a conjunction, the travel time should be longer according to GR. This could be seen with an accuracy of almost 10^{-3}. Competing and limiting effects came from the solar and Earth's atmosphere.

LLR. Various Apollo Moon missions as well as Russian unmanned Moon missions placed several laser retroreflectors on the Moon. Laser tracking from the Earth yielded an accuracy of about 1 cm after 20 years for the determination of the Earth–Moon distance. Using these data, effects like the geodetic precession of the Earth–Moon system could be verified, the validity of the strong equivalence principle (UFF together with the gravitational self-energy) could be tested, and a comparison of the data with a hypothetical Newtonian Yukawa potential yielded strong estimates on such effects.

LAGEOS. This mission consists of two passive small satellites orbiting the Earth which are laser tracked; it gives information about the gravitational field of the Earth. The orbital data of these satellites have been used to experimentally check for the first time the existence of the gravitomagnetic Lense–Thirring effect with a claimed 20–30% accuracy [21]. The difficulty lies in competing effects from higher gravitational multi-poles of the Earth which are actually much larger than the effect looked for. The basic idea of the data analysis is to combine the data of two satellites (LAGEOS and LAGEOS II) in such a way that the lowest order gravitational multi-poles cancel. Any improvement of Earth gravity models [22] as well as additional data from further satellites [23] will improve the results.

Cassini. The Cassini mission to Saturn is equipped with a multi-frequency radio link to the Earth. Using this technique, the disturbing effects due to the Sun's corona for measurement of the travel time of signals from the Earth to Cassini and back could be removed almost completely. The achieved accuracy of $2 \cdot 10^{-5}$ is almost two orders of magnitude better than the previous Viking result.

6 Possible Future Missions

Here we shortly describe the scientific objectives and other main issues of planned space missions devoted to fundamental physics. For most of these missions, more information can be found in the reviews [2, 24, 25] and the references cited therein.

GP-B. With this mission [26], the gravitomagnetic Schiff or frame-dragging effect, shall be observed with 0.1% accuracy [26]. The main part of the satellite is a huge dewar which maintains a cryogenic environment for 18 months. The main parts of the experimental payload located inside the dewar are gyroscopes made from superconducting rotating spheres and a telescope. The gyroscopes represent the inertial systems which, due to the gravitomagnetic field of the rotating Earth, start to precess. The precession is read out with SQUIDs using a magnetic field attached to the rotating spheres. This direction is then compared with the direction given by distant stars which is observed by the telescope. This mission should have started in 2004.

ACES/PHARAO. In this mission [27] the PHARAO clock, based on a fountain of cold caesium atoms, and a hydrogen maser clock will be brought onboard the ISS, complemented by a microwave link for synchronization with clocks on Earth. With this equipment, better tests of the gravitational redshift can be performed and one can search for a time dependence of the fine structure constant at the 10^{-16} level. Furthermore, establishing such clocks in space represents an enormous improvement over the present level of synchronization that is possible using GPS clocks.

PARCS. This ISS project (see, e.g. the review [24]) consists, similar to PHARAO, of a caesium atomic clock which, together with SUMO, will test the universality of the gravitational redshift, the constancy and isotropy of the speed of light and intends to establish a better time standard by a factor of 20. The accuracy of PARCS will be 10^{-16}.

RACE. For RACE, a double magneto-optical trap (MOT) design is used to multiply launch Rb atoms, which, as compared with Cs, possesses a much lower collision shift error. Furthermore, RACE uses two cavities to interrogate the atomic frequency. Among others, one advantage of that design is the possibility to eliminate the vibrational noise of the ISS. RACE will aim at a clock accuracy of 10^{-17} (see, e.g. [24]).

MICROSCOPE. This already approved mission [28] is devoted to a test of the UFF to an accuracy of 10^{-15} in terms of the Eötvös parameter. The relative accelerations of two pairs of test masses are measured: the first pair consists of a Pt/Rh alloy and special Ti alloy (TA6V), and the second pair consists of two identical Pt/Rh alloys. The second pair is taken for redundancy and drag-free control. The relative acceleration is measured using capacitive sensors. MICROSCOPE will fly in 2011.

LISA. With a huge two-arm interferometer made up of three satellites being 5 million km apart one will detect gravitational waves in the frequency range 10^{-1}–10^{-4} Hz. Each of these satellites carries two phase-locked laser systems and two mirrors. Each mirror is controlled by a drag-free system in order that the satellite very precisely moves on a geodesic path within the timescale of 1,000 s. LISA has the status of an ESA–NASA cornerstone mission. See the contribution of Rüdiger, Heinzel and Tröbs in this volume.

STEP. This mission [29] wants to test the UFF to a precision of 10^{-18}. In contrast to MICROSCOPE it uses cryogenic techniques: SQUIDs are used to determine the relative position between the freely flying test masses. Four pairs of test masses are used, i.e. the pair Pt/Ir–Nb, the pair Nb–Be, and two pairs Pt/Ir–Be, so that there is a redundancy as well as a cyclic condition for which the total acceleration difference between pairs of test masses must add to zero in the case that UFF holds. The high accuracy requires several additional techniques. One of these is that the test masses must have an appropriate design to be insensitive to gravity gradients.

ASTROD. This is a proposed Chinese interplanetary laser ranging mission [30] which aims at (1) an improvement of the determination of the PPN parameters γ and β by three to six orders of magnitude, (2) the detection of gravitational waves below the mHz range and (3) to improve the knowledge of solar system parameters like the angular momentum of the Sun and asteroid masses. The main technique is laser ranging for which new techniques for the coupling of very weak laser light to local oscillators have to be developed.

LATOR. This recently proposed interplanetary ranging mission, see the contribution of Turyshev, Shao and Nordtvedt in this volume, aims at measuring the deflection of light with a precision of 10^{-8}. The main idea is to have two small spacecrafts and a reference on the ISS spanning a triangle and to measure the lengths of the three sides of this triangle and, in addition, the observed angle of the light rays from the satellites to the ISS. From the gravity-induced deformations of ordinary Euclidean geometry of the paths of light one can infer the gravitational influence. The aimed high precision can be achieved by an optical truss provided by a 100-m long multi-channel optical interferometer mounted on the ISS. Besides the deflection of light, the second order effects, the Sun's quadrupole momentum and the Lense–Thirring effect can be measured.

OPTIS. This mission aims at an improvement of quite a few tests of SR and GR, namely (1) the isotropy of the speed of light, (2) the constancy of the speed of light, (3) the Doppler effect (or time dilation), (4) tests of the UGR comparing various atomic clocks, (5) tests of UGR comparing an optical resonator and atomic clocks, (6) measurement of the absolute gravitational redshift, (7) measurement of the Lense–Thirring effect, (8) measurement of the perigee advance and (9) test of Newton's $1/r$ potential. The main components of the experimental payload are clocks (resonators and atomic clocks (H-maser, ion clocks)), a laser link to the Earth and a drag-free control of the satellite. More information can be found in [31].

SPACETIME. With this mission [32], three ion clocks with a stability of the order 10^{-16} will be brought to the Sun as close as about 5 solar radii. From the comparison of these highly accurate clocks during their motion through a strongly changing gravitational potential one gets a huge improvement of tests of the UGR. The errors can be reduced considerably by the possibility to place all three clocks in the same environment.

GG. Here again the UFF is aimed to be tested at an accuracy of 10^{-17}. The main idea of this Italian proposal [33] is that a high frequency modulation of the UFF-violating signal induced by a rotation of the test mass can improve the signal-to-noise ratio.

SEE. In this mission [34], two free-falling interacting test masses are placed in a big container orbiting the Earth in free fall. Using a newly invented and highly precise device for monitoring the positions of the two masses, one can (1) test the validity of the Newton's $1/r$ potential over distances between the two masses and between the Earth and the satellite, (2) make better tests of the UFF, (3) make more precise measurements of the gravitational constant and (4) search for a time dependence of the gravitational constant.

HYPER. This missions aims at (1) measuring the Schiff effect, (2) testing the UFF, (3) searching for a fundamental decoherence in quantum mechanics and (4) making more precise measurement of the fine structure constant [35]. The main feature of this mission is that it wants to employ atomic interferometry in space. A more technical aspect is that with this mission accelerometers (gravity reference sensors) and gyroscopes based on atomic interferometry might be introduced. Working with atomic interferometry, prerequisite techniques are lasers, laser cooling and MOTs.

DSGE. The primary task of the deep-space gravity explorer is to confirm the Pioneer anomaly, the anomalous constant acceleration of the Pioneer 10 and 11 spacecraft of $a_{\text{Pioneer}} = (8.74 \pm 1.33) \cdot 10^{-10}\,\text{m}\,\text{s}^{-2}$ towards the Sun. Further tasks are the precise determination of the direction of the acceleration and of the way how the anomaly turns on. The DSGE may be able to explore the range between solar system physics and galactic physics and cosmology. Furthermore, the exploration of the interplanetary matter content is another science goal of such a mission. See also the article of Johann, Dittus and Lämmerzahl in this volume.

7 Key Technologies

Since gravity is the physics of space and time technologies which are relevant for the determination of position, velocity and time are most important for the exploration of the physics of the gravitational interaction. Furthermore, to provide a force-free motion governed only by the gravitational interaction, other techniques like drag-free control and micro-thrusters are needed. We are describing present and emerging technologies relevant for space missions to explore the gravitational interaction. These technologies nowadays experience very fast improvements.

7.1 Key Space Technologies

Drag-free control. This technique is necessary to assure that the satellite moving as close as possible along a geodesic path is given only by gravity. It needs the interplay between very sensitive inertial sensors and very precise micro-thrusters. For this purpose, algorithms have to be developed which process the signals from the sensors and precisely drive the micro-thrusters, where their corresponding noise properties have to be taken into account. The corresponding control system is called drag-free and attitude control system (DFACS). See the contribution of Theil and of Fichter, Schleicher and Vitale in this volume.

Gravity reference sensors (inertial sensors). In the last years so-called "drag-free sensors" have been developed that offer the opportunity to cancel out all non-gravitational disturbing forces and torques (like air drag, solar pressure, magnetic field disturbances, etc.) on orbiting satellites. The drag-free concept involves centring a free-floating test mass located inside a satellite which is free of external disturbances and follows a purely gravitational orbit. External (non-gravitational) forces and torques will move the satellite relative to the test mass. The change in the relative position is measured and then used to derive the appropriate control force which has to be applied by the DFACS to drive the test mass displacement to zero. Since the satellite is forced to follow the test mass, it follows the same gravitational orbit.

The test mass displacement can be measured electrostatically or magnetically. In most cases (e.g. for MICROSCOPE), a capacitive method is used where the test mass is surrounded by electrodes. One area of the test mass and one electrode form a capacitor and the displacement-induced change of its capacity can be measured, see, e.g. [36].

Micro-thrusters. Micro-thrusters are needed for very precise navigation of satellites. Field electric emission propulsion (FEEP) ion thrusters or colloidal thrusters are used to control the residual acceleration down to $10^{-10}\,\mathrm{m\,s^{-2}}$ in the signal bandwidth. This sets an upper limit for the thrust: linear forces acting on the satellite are less than $50\,\mu\mathrm{N}$ in all three axes and maximum torques are about $10\,\mu\mathrm{N\,m}$. The resolution of thrust control has to be done with an accuracy of about $0.1\,\mu\mathrm{N}$. For a satellite diameter of about 1.5 m, the solar radiation pressure of about $4.4\,\mu\mathrm{N\,m^{-2}}$ and the radiation pressure of the Earth albedo of $1.2\,\mu\mathrm{N\,m^{-2}}$ sum up to a total drag of about $10\,\mu\mathrm{N}$.

Timekeeping. To perform precision positioning and navigation, precise clocks on Earth are needed. Clocks on Earth approach a precision at the 10^{-16} level and may reach in the not too far future an accuracy of 10^{-18}. Such clocks are sensitive to heights at the mm-level. Since the height on Earth is not defined to this accuracy, it is preferable to place such precise clocks in space. A second point to mention is that such precise timekeeping requires a thorough understanding of the basic principles underlying the physics of clocks. In particular, for space navigation one needs a comparison of Earth bound clocks with astrophysical clocks like pulsars and binary systems to exclude a mutual drift of time.

Laser tracking. To measure the distance between a ground station and the satellite (or another object) with very high precision the satellite laser ranging (SLR) technique can be used. This is important for LLR and LAGEOS and will be used in OPTIS, LATOR and ASTROD. Very short laser pulses are transmitted from a telescope in a ground station to a satellite, from which they are retroreflected back to the station by a corner cube reflector. The round trip time is measured and gives the distance. In other words, SLR measures the absolute time of flight of photons so that the geometry of satellite and laser station can be determined precisely as long as the system calibration error is controlled to a negligible level. With the present state-of-the-art one can measure the travel time of signals with an accuracy of 50 ps or better. This is equivalent to an accuracy of 1 cm or less. Currently, NASA is building up the satellite laser ranging 2000 system. SLR2000 is an autonomous and eyesafe photon-counting SLR station with an expected single short-range precision of about 1 cm and a normal point precision better than 3 mm. The system will provide continuous 24-h tracking coverage of artificial satellites at altitudes up to 20,000 km. Approximately 40 laser station systems, distributed all over the world, now contribute to this technique. These stations form a network that is coordinated by the International Laser Ranging Service: ILRS and by a European consortium EUROLAS.

Star trackers. Star trackers are sensors that are used in satellite attitude control to achieve accurate pointing measurements. The sky is scanned by, e.g. a

CCD camera. From the detected star patterns a computer algorithm can determine the pointing direction of the sensor and thus of the spacecraft. Today state-of-the-art high-precision star trackers can provide a single star angular accuracy better than 3 arcsec.

7.2 Key Payload Techniques

Here we describe a few key payload techniques which already played a role in missions or which are planned to be part of future missions.

Gyroscopes. High-precision gyroscopes are based on mechanical gyroscopes with a SQUID-based readout or use laser or cold atom interferometry (see below).

Clocks. All kinds of clocks will play a major role in future fundamental physics missions. The reason is that most of these tests are clock-comparison tests: Tests of LLI consist in comparison of clocks with different internal orientations, of clocks with a different hypothetical velocity dependence (Kennedy–Thorndike tests) and clocks in different states of motion. All redshift experiments compare clocks at different positions in the gravitational field. We shortly list the types of clocks:

- *H-maser.* H-masers are based on a hyperfine transition of the ground state of the hydrogen atom with a life time of about 1 s which is coupled to a resonator. The frequency is 1.420405751 Hz and the Allan deviation is less than 10^{-15}. H-masers are already space qualified.
- *Atomic fountain clocks.* These kinds of clocks are based on the Ramsey interaction scheme. Atoms interact with microwaves during two time intervals. The accuracy of the clock is directly related to the time span between these two interaction intervals. Since the atoms have to be in free motion (free fall) in this time span, only a time span of up to 0.1 s is possible on Earth. In space a much longer time span can be achieved. The present accuracy is better than 10^{-15} [37]. This clock has been used for the best test of UGR [38]. In space one or two orders of magnitude of improvement are feasible.
- *Ion clocks.* Today, ion clocks [39] approach the level of 10^{-16} in their stability (in terms of the Allan variance). Ion clocks are based on hyperfine transitions of trapped ions (e.g. Hg^+, Cd^+, Yb^+).
- *Resonators.* Resonators (cavities) are a realization of so-called *light clocks.* Locking of lasers to optical resonators used for the best Michelson–Morley tests [40] will give highly stabilized frequencies which carry, via $\nu = nc/(2L)$ where L is the length of the cavity, the information about the velocity of light c for propagation along the cavity axis. This information is used to make statements about the isotropy and the constancy of the speed of light. For this, the length of the cavity has to be very stable since otherwise this could mask the searched effect. For cryogenic resonators the

stability is $\delta L/L \leq 7 \cdot 10^{-16}$ [40] which corresponds to the 30th part of the diameter of the proton. The high-dimensional stability requires materials with low thermal expansion coefficients. Another requirement is stable lock of lasers to the cavity which can be obtained. Microwave resonators work in the same way, the only difference is the longer wavelength (in the cm range) of the electromagnetic radiation in the resonator. A particular development in this area are the whispering gallery resonators where the resonators have cylindric geometry and the radiation possesses large angular mode numbers. These have been used for the today's best test [41] of the constancy of c.

Today the most advanced clocks approach an accuracy of 10^{-16}. From the gravitational redshift, it can be seen that these clocks run differently if they are located at a height difference of 90 cm. If the precision of clocks improve by one or two orders, then, due to the fact that the surface of the Earth is not really constant, these clocks cannot define a well-defined time. For that one has to go to space. Only in space the conditions are well defined enough in order that such high-precision apparatus can yield unique and interpretable results.

Lasers. There are already space-proved lasers available. These lasers are diode-pumped Nd:YAG lasers. They possess high intensity and frequency stability. Lasers will be applied in the missions LISA, HYPER, OPTIS, LATOR and ASTROD, indicating the overall importance of this device in space technology.

Frequency combs. Tests of the constancy of c and of the UGR with optical resonators require a high-precision technique for comparing frequencies in the microwave and optical range, differing by more than five orders of magnitude. The recently invented frequency comb is the appropriate technique, see [42] for an overview. This technique is simpler, cheaper, power-saving and more accurate than previous methods. In the laboratory, the comparison of the frequencies with an accuracy of 10^{-15} has already been achieved.

SQUIDs. SQUIDs are based on the flux quantization and on the Josephson effect in superconducting electrical loops. They provide the presently most sensitive magnetic flux detector. Therefore, any low-frequency signal that can be converted into a change of the magnetic flux can be observed with high precision. GP-B [26] and STEP [29] use SQUIDs to measure the position, the linear acceleration and the angular momentum of test masses. For distance measurements the achieved sensitivity is [43] $\delta x \sim 4 \cdot 10^{-14}\,\mathrm{m\,Hz^{-1/2}}$ and for acceleration measurements $\delta a \sim 10^{-14}\,\mathrm{m\,s^{-2}\,Hz^{-1/2}}$, and for the measurement of the angular velocity one gets $\delta\omega \sim 10^{-17}\,\mathrm{deg\,s^{-1}}$ for a year integration time. Therefore SQUIDs are the best gyroscopes available.

Cold atoms. Using laser-cooled atoms it is possible to build up highly precise and sensitive atomic interferometers. These devices can serve as highly precise accelerometers and gyroscopes, as has been demonstrated on Earth.

The achieved sensitivity is $\delta a \sim 10^{-9}\,\mathrm{m\,s^{-2}\,Hz^{-1/2}}$ and $\delta\omega \sim 6 \cdot 10^{-10}\,\mathrm{rad\,s^{-1}}$ $\mathrm{Hz^{-1/2}}$. In the HYPER proposal, atomic interferometry should be used to test the UFF for quantum matter with an accuracy of 10^{-16} and to measure the Schiff effect. There are also ideas to use Bose–Einstein condensates as source for coherent sources of atoms thus enhancing the sensitivity even more, see the article of Bouyer *et al* in this volume.

BEC. Bose–Einstein condensates are a source for coherent atomic beams and, thus, for an atomic laser. Interferometry with coherent atoms will give a much better accuracy.

Machining. In some cases, a high-precision machining of parts of the experimental payload is uttermost important. As examples we mention the gyroscopes for GP-B and the test masses for STEP. In the first case spheres had to be machined with a relative non-sphericity of about 10^{-6} (this means that the radius of a sphere of about 3 cm diameter is precise within 1 μm) and for the tests of the equivalence principle the test masses have to be machined with an absolute accuracy of better than 1 μm.

8 Summary and Outlook

In Fig. 4, we list all the experiments on GR together with completed and planned missions dedicated to such tests. What is not covered by this list are searches for anomalous couplings of spin particles with gravity, the search of a fundamental dispersion of electromagnetic propagation and strong gravity effects like the observation of binary systems and cosmological observations.

8.1 Summary

We outlined the interplay between current quest from fundamental physics, space conditions and the development of new technologies. In some cases, this interplay is very fruitful and can lead to new space missions. We summarize the discussed items in terms of space projects, see Table 2.

8.2 Fundamental vs. Applied Physics

As a final remark, we like to stress that it certainly is a wrong attitude to contrast fundamental research with applied research. It is a matter of experience that fundamental physics and applied physics and technologies are just two sides of the same medal: most new fundamental physics results immediately will be used for applications, and all newly developed technologies will be used for fundamental physics. Furthermore, also technological requirements will trigger fundamental physics research and vice versa.

As one example for the latter we mention that the technological requirement of synchronizing clocks in distant towns for reliable train schedules at the

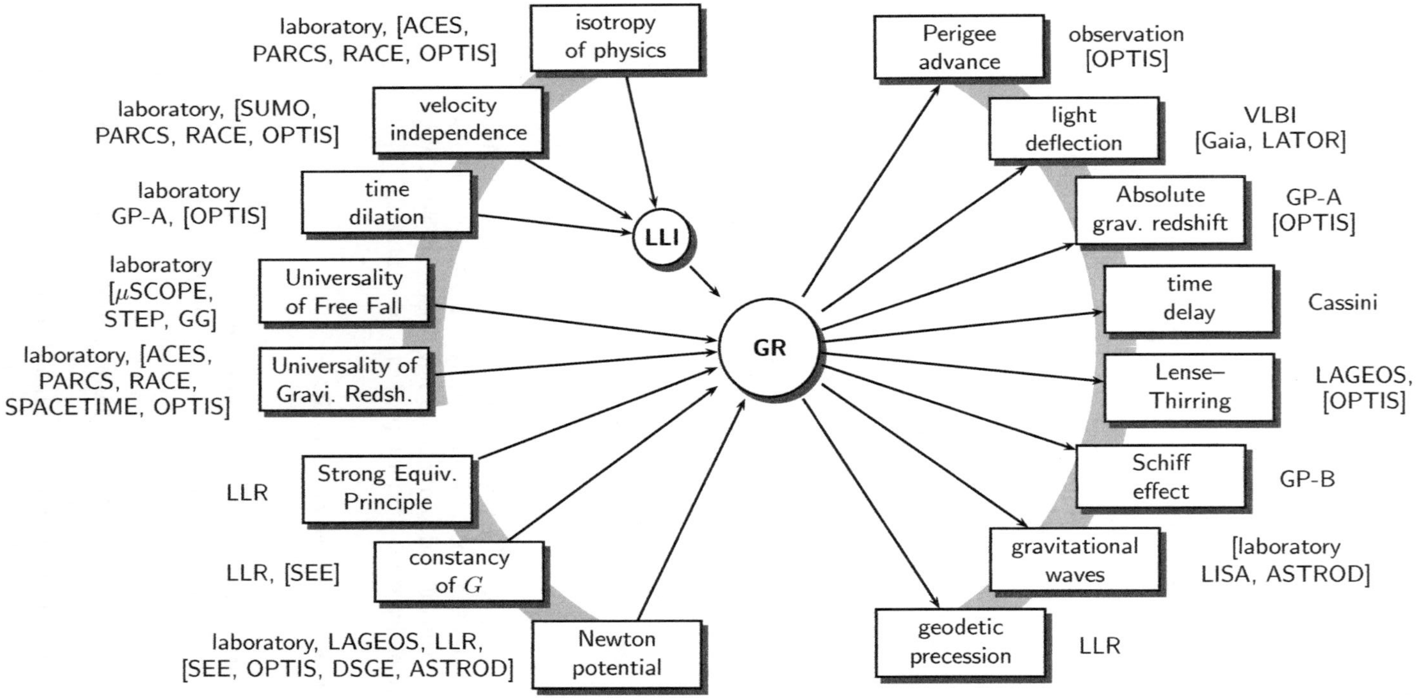

Fig. 4. Tests of GR. On the *left-hand side* tests of the two groups of the foundations of GR are listed, on the *right-hand side* there are predictions of GR. The order of the present accuracy of the results is shown and the methods or missions of these results are also given. Planned missions and their projected order of accuracy are shown in *brackets*. We left out astrophysical observation of strong gravity effects.

Table 2. List of space missions for fundamental physics.

Previous space missions

Mission	Scientific objective	Space condition	Technology
Viking	Grav. time delay	Long distance	Main Technology
GP-A	Grav. redshift	Large difference in grav. potential	Clocks
LLR	Strong EP, $\dot{G}$	Long observation time	Laser
GP-B	Schiff effect	Long free-fall motion	Cryogenic technology
Cassini	Grav. time delay	Long distance	Multi-frequency link, clocks

Space missions under development

Mission	Scientific objective	Space condition	Technology
ACES	Grav. redshift, UGR, $\dot{\alpha}$	Large difference in grav. potential, long free-fall motion of atoms	Clocks
LISA	Grav. waves	Large distance, long free-fall motion	Laser, drag free
MICROSCOPE	UFF	Long free-fall motion	Drag free
Gaja	Light deflection	No atmospheric disturbances	Optical

Proposed space missions

Mission	Scientific objective	Space condition	Technology
LATOR	Light deflection	Large distance	Laser, optics
ASTROD	Light deflection, grav. waves	Large distances, long free-fall motion	Laser, drag free
OPTIS	LLI, UGR, grav. redshift, Lense–Thirring effect	large differences in grav. potential, large velocity	Clocks
HYPER	Schiff effect, UFF, α, decoherence	Long free-fall motion	Cold atoms, lasers
SPACETIME	UGR	Large differences in grav. potential	Clocks
DSGE	Gravity at large distances	Large distances	Drag free, clocks
SEE	UFF, $\dot{G}$	Long free-fall motion	Laser
PARCS	Same as ACES	Same as ACES	Clocks
RACE	Same as ACES	Same as ACES	Clocks

end of the 19th motivated Poincaré and Einstein to revolutionize the notion of time which in turn made possible the development of the theory of SR [44]. SR, on the other side, was necessary to establish a new frame for mechanics and a theory of gravity which then led to the famous insight that mass is related to energy ($E = mc^2$) and to the development of general relativity which is at the basis of today's GPS and Earth science, see, e.g. [45]. Furthermore, even technological requirements in the sense of applications on commercial scales requires fundamental research [46]. Otherwise any technological development is just an ad hoc development without any possibility to systematize the technological achievement or to apply this to other or more general situations. Another example is the development of atomic clocks. Immediately after its development, Hafele and Keating used these clocks to verify the relativistic time dilation in their famous Hafele–Keating experiment [47, 48].

On the other side, new fundamental physics results in most cases lead to the development of new technologies. As examples we may mention superconductivity and the quantum Hall effect. Furthermore, fundamental physics quest motivate the development of new technologies. A recent example for this is GP-B with its gyroscopes, star tracker and drag-free control.

Therefore, fundamental physics and applied technology are two sides of the some medal. Cutting one part away means truncate ones capabilities and possible future developments. The development of technologies based merely on known physics is a dead-end road.

Acknowledgement

Financial support from the German Aerospace Agency DLR and the German Research Foundation DFG is gratefully acknowledged.

References

1. C. Lämmerzahl. General relativity in space and sensitive tests of the equivaence principle. **in** M. Novello, S. Perez Bergliaffa, and R. Ruffini (eds.). *Proceedings of the Tenth Marcel Grossmann Meeting on General Relativity.* World Scientific, Singapore 2006. p. 337.
2. C. Lämmerzahl and H. Dittus. Fundamental physics in space: A guide to present projects. *Ann. Physik.*, 11:95, 2002.
3. C. Kiefer, D. Giulini, and C. Lämmerzahl, editors. *Quantum Gravity – From Theory to Experimental Search*, volume 631 of *Lecture Notes in Physics*, Berlin, 2003. Springer–Verlag.
4. W.-T. Ni. Equivalence principles and electromagnetism. *Phys. Rev. Lett.*, 38:301, 1977.
5. C.M. Will. *Theory and Experiment in Gravitational Physics (Revised Edition).* Cambridge University Press, Cambridge, 1993.

6. G. Amelino-Camelia, C. Lämmerzahl, A. Macias, and H. Müller. The search for quantum gravity signals. In A. Macias, C. Lämmerzahl, and D. Nunez, editors, *Gravitation and Cosmology*, page 30. AIP Conference Proceedings 758, Melville, New York, 2005.
7. D. Mattingly. Modern tests of lorentz invariance. *Living Reviews*, 8:http://www.livingreviews.org/lrr–2005–5, 2005 (cited on April 4, 2006).
8. J. Ehlers and C. Lämmerzahl, editors. *Special Relativity – Will It Survive the Next 101 Years?, Lecture Notes in Physics*, Volume 702, Berlin, 2006. Springer–Verlag.
9. T. Damour, F. Piazza, and G. Veneziano. Runaway dilaton and equivalence principle violations. *Phys. Rev. Lett.*, 89:081601, 2002.
10. C. Wetterich. Crossover quintessence and cosmological history of fundamental constants. *Phys. Lett.*, B 561:10, 2003.
11. T. Damour, F. Piazza, and G. Veneziano. Violations of the equivalence principle in a dilaton–runaway scenario. *Phys. Rev.*, D 66:046007, 2002.
12. S.M. Carroll, J.A. Harvey, V.A. Kostelecký, C.D. Lane, and T. Okamoto. Noncommutative field theory and Lorentz violation. *Phys. Rev. Lett.*, 87:141601, 2001.
13. T. Damour. Gravity, equivalence principle and clocks. Proceedings of the Workshop on the Scientific Applications of Clocks in Space (JPL, Pasadena, November 7–8, 1996).
14. T. Damour. Equivalence principle and clocks. In J. Trân Tanh Vân, J. Dumarchez, J. Reynaud, C. Salomon, S. Thorsett, and J.Y. Vinet, editors, *Gravitational Waves and Experimental Gravity*, page 357. World Publishers, Hanoi, 2000.
15. C. Wetterich. Probing quintessence with time variation of couplings. *Astropart. Phys.*, 10:2, 2003.
16. I. Antoniadis. Physics with large extra dimensions and non–Newtonian gravity at sub–mm distances. In D. Giulini, C. Kiefer, and D. Lämmerzahl, editors, *Quantum Gravity – From Theory to Experimental Search*, page 337. Springer–Verlag, Berlin, 2003.
17. R.H. Sanders. Anti–gravity and galaxy rotation curves. *Astron. Astrophys.*, 136:L21, 1984.
18. R.H. Sanders and S.S. McGough. Modified newtonian dynamics as an alternative to dark matter. *Ann. Rev. Astron. Astrophys.*, 40:263, 2002.
19. J.D. Anderson, P.A. Laing, E.L. Lau, A.S. Liu, M.M. Nieto, and S.G. Turyshev. Study of the anomalous acceleration of Pioneer 10 and 11. *Phys. Rev.*, D 65:082004, 2002.
20. R.F.C. Vessot, M.W. Levine, E.M. Mattison, E.L. Blomberg, T.E. Hoffmann, G.U. Nystrom, B.F. Farrel, R. Decher, P.B. Eby, C.R. Baughter, J.W. Watts, D.L. Teuber, and F.D. Wills. Test of Relativistic Gravitation with a Space–Borne Hydrogen Maser. *Phys. Rev. Lett.* 45:2081, 1980.
21. I. Ciufolini, E. Pavlis, F. Chieppa, E. Fernandes-Vieira, and J. Pérez-Mercader. Test of General Relativity and measurement of the Lense–Thirring effect with two Earth satellites. *Science*, 279:2100, 1998.
22. L. Iorio and A. Morea. The impact of the new Earth gravity models on the measurement of the Lense–Thirring effect. *Gen. Rel. Grav.*, 36:1321, 2004.
23. L. Iorio, I. Ciufolini, S. Schiller, H. Dittus, and C. Lämmerzahl. On the possibility of measuring the Lense–Thirring effect with a LAGEOS–LAGEOS II–OPTIS–mission, 2003. preprint gr-qc/0211013.

24. C. Lämmerzahl, G. Ahlers, N. Ashby, M. Barmatz, P.L. Biermann, H. Dittus, V. Dohm, R. Duncan, K. Gibble, J. Lipa, N.A. Lockerbie, N. Mulders, and C. Salomon. Experiments in Fundamental Physics scheduled and in development for the ISS. *Gen. Rel. Grav.*, 36:615, 2004.
25. H. Dittus and C. Lämmerzahl (eds.). Special Issue on Fundamental Phyiscs on the ISS. *Gen. Rel. Grav.*, 36:471–649, 2004.
26. C.W.F. Everitt, S. Buchman, D.B. DeBra, G.M. Keiser, J.M. Lockhart, B. Muhlfelder, B.W. Parkinson, J.P. Turneaure, and other members of the Gravity Probe B team. Gravity Probe B: Countdown to launch. In C. Lämmerzahl, C.W.F. Everitt, and F.W. Hehl, editors, *Gyros, Clocks, and Interferometers: Testing Relativistic Gravity in Space*, page 52. Springer–Verlag, Berlin, 2001.
27. C. Salomon, N. Dimarcq, M. Abgrall, A. Clairon, P. Laurent, P. Lemonde, G. Santarelli, P. Uhrich, L.G. Bernier, G. Busca, A. Jornod, P. Thomann, E. Samain, P. Wolf, F. Gonzalez, P. Guillemot, S. Leon, F. Nouel, C. Sirmain, and S. Feltham. Cold atoms in space and atomic clocks: Aces. *C.R. Acad. Sci. Paris*, 4:1313, 2004.
28. P. Touboul. MICROSCOPE, testing the equivalence principle in space. *Comptes Rendus de l'Aced. Sci. Série IV: Physique Astrophysique*, 2:1271, 2001.
29. N. Lockerbie, J.C. Mester, R. Torii, S. Vitale, and P.W. Worden. STEP: A status report. In C. Lämmerzahl, C.W.F. Everitt, and F.W. Hehl, editors, *Gyros, Clocks, and Interferometers: Testing Relativistic Gravity in Space*, page 213. Springer–Verlag, Berlin, 2001.
30. W.-T. Ni et al. Mini–astrod mission concept. *Int. J. Mod. Phys.*, D 11:1035, 2002.
31. C. Lämmerzahl, C. Ciufolini, H. Dittus, L. Iorio, H. Müller, A. Peters, E. Samain, S. Scheithauer, and S. Schiller. An Einstein mission for improved tests of special and general relativity. *Gen. Rel. Grav.*, 36:2373, 2004.
32. L. Maleki and J. Prestage. SpaceTime mission: Clock test of relativity at four solar radii. In C. Lämmerzahl, C.W.F. Everitt, and F.W. Hehl, editors, *Gyros, Clocks, and Interferometers: Testing Relativistic Gravity in Space*, page 369. Springer–Verlag, Berlin, 2001.
33. See the informations available on. http://tycho.dm.unipi.it/ nobili/.
34. A.J. Sanders, A.D. Alexeev, S.W. Allison, V. Antonov, K.A. Bronnikov, J.W. Campbell, M.R. Cates, T.A. Corcovilos, D.D. Earl, T. Gadfort, G.T. Gillies, M.J. Harris, N.I. Kolosnitsyn, M.Yu. Konstantinov, V.N. Melnikov, R.J. Newby, R.G. Schunk, and L.L. Smalley. Project SEE (Satellite Energy Exchange): an international effort to develop a space–based mission for precise measurements of gravitation. *Class. Quantum Grav.*, 17:2331, 2000.
35. C. Jentsch, T. Müller, E.M. Rasel, and W. Ertmer. High precision atom interferometry on ground and in space. *Gen. Rel. Grav.*, 36:2197, 2004.
36. P. Touboul. Space accelerometers: Present status. In C. Lämmerzahl, C.W.F. Everitt, and F.W. Hehl, editors, *Gyroscopes, Clock, Interferometers, ...: Testing Relativistic Gravity in Space*, volume LNP 562, page 274. Springer, Berlin, 2001.
37. R. Wynands and S. Weyers. Atomic fountain clocks. *Metrologia*, 42:S64, 2005.
38. A. Bauch and S. Weyers. New experimental limit on the validity of local position invariance. *Phys. Rev.*, D 65:081101(R), 2002.
39. J.D. Prestage, S. Chung, E. Burt, L. Maleki, and R.L. Tjoelker. Stability measurements between Hg^+ LITE 12–pole clocks. Proceedings of the 2001 IEEE International Frequency Control Symposium.

40. H. Müller, S. Herrmann, C. Braxmaier, S. Schiller, and A. Peters. Modern Michelson–Morley experiment using cryogenic optical resonators. *Phys. Rev. Lett.*, 91:020401, 2003.
41. P. Wolf, M.E. Tobar, S. Bize, A. Clairon, A.N. Luiten, and G. Santarelli. Whispering gallery resonators and test of Lorentz invariance. *Gen. Rel. Grav.*, 36:to appear, 2004.
42. S.T. Cundiff, J. Ye, and J.L. Hall. Optical frequency synthesis based on mode–locked lasers. *Rev. Sci. Instr.*, 72:3749, 2001.
43. W. Vodel, H.-J. Dittus, S. Nietzsche, H. Koch, J.v. Zameck Glyscinski, R. Neubert, S. Lochmann, C. Mehls, and D. Lockowandt. High sensitive DC SQUID–based position detectors for application in gravitational experiments at the drop tower Bremen. In C. Lämmerzahl, C.W.F. Everitt, and F.W. Hehl, editors, *Gyroscopes, Clock, Interferometers, ...: Testing Relativistic Gravity in Space*, volume LNP 562, page 248. Springer, Berlin, 2001.
44. P. Gallison. *Einstein's Clocks, Poincaré's Maps*. W.W. Norton, New York, 2003.
45. C. Lämmerzahl. Relativity and techology. *Ann. Phys. (Leipzig)*, 15:5, 2006.
46. M. Carrier. The challenge of practice: Einstein, technological development and conceptual innovation. In J. Ehlers and C. Lämmerzahl, editors, *Special Relativity – Will It Survive the Next 101 Years?*. Springer Verlag, Heidelberg, 2006.
47. J.C. Hafele and R.E. Keating. Around–the–world atomic clocks: Predicted relativistic time gains. *Science*, 177:166, 1972.
48. J.C. Hafele and R.E. Keating. Around–the–world atomic clocks: Observed relativistic time gains. *Science*, 177:168, 1972.

General Theory of Relativity: Will It Survive the Next Decade?

Orfeu Bertolami[1], Jorge Páramos[1], and Slava G. Turyshev[2]

[1] Instituto Superior Técnico, Departamento de Física, Av. Rovisco Pais, 1049-001 Lisboa, Portugal
[2] Jet Propulsion Laboratory, California Institute of Technology, 4800 Oak Grove Drive, Pasadena, CA 91109, USA

Summary. The nature of gravity is fundamental to the understanding of our own solar system, the galaxy, and the structure and evolution of the Universe. Einstein's general theory of relativity is the standard model that is used for almost 90 years to describe gravitational phenomena on various scales. We review the foundations of general relativity, discuss the recent progress in the tests of relativistic gravity, and present motivations for high-accuracy gravitational experiments in space. We also summarize the science objectives and technology needs for the laboratory experiments in space with laboratory being the entire solar system. We discuss the advances in our understanding of fundamental physics anticipated in the near future and evaluate discovery potential for the recently proposed gravitational experiments.

1 Introduction

To understand the Universe in its vast and complex splendor seems a daunting task, yet curiosity and wonder over centuries and civilizations have always led humankind to seek answers to some of the most compelling questions of all: How did the Universe come to be? What is it made of? What forces rule its behavior? Why is it the way it is? What will ultimately become of it? With its prominent influence on natural phenomena at every distance scale, gravitation plays a pivotal role in this intellectual quest.

Gravity was known to humans long before the present-day picture of four fundamental interactions was formed. The nature of gravity is fundamental to the understanding of our solar system, the galaxy, and the structure and evolution of the Universe. It was Newton who first understood that not only gravity dictates the fall of apples and all bodies on Earth, but also planetary motion in our solar system and the Sun itself are governed by the same physical principles. On the larger scales the effects of gravity are even more pronounced, guiding the evolution of the galaxies, galactic clusters, and ultimately determining the fate of the Universe. Presently, Einstein's general theory of relativity is a key to understand a wide range of phenomena, spanning from the dynamics of compact astrophysical objects such as neutron stars and black holes, to cosmology where the Universe itself is the object of study. Its striking predictions include gravitational lensing and waves, and only black holes have not yet been directly confirmed.

The significance that general relativity (GR) plays in our understanding of nature makes the theory a focus of series of experimental efforts performed with ever-increasing accuracy. However, even after more than 90 years since general relativity was born, Einstein's theory has survived every test. Such longevity does not mean that it is absolutely correct, but serves to motivate more precise tests to determine the level of accuracy at which it is violated. This motivates various precision tests of gravity both in laboratories and in space; as a result, we have witnessed an impressive progress in this area over the last two decades. However, there are a number of reasons to question the validity of this theory, both theoretical and experimental.

On the theoretical front, the problems arise from several directions, most dealing with the strong gravitational field regime; this includes the appearance of spacetime singularities and the inability to describe the physics of very strong gravitational fields using the standard of classical description. A way out of this difficulty would be attained through gravity quantization. However, despite the success of modern gauge field theories in describing the electromagnetic, weak, and strong interactions, it is still not understood how gravity should be described at the quantum level. Our two foundational theories of nature, quantum mechanics and GR, are not compatible with each other. In theories that attempt to include gravity, new long-range forces can arise in addition to the Newtonian inverse-square law. Even at the classical level, and assuming the equivalence principle, Einstein's theory does not provide the most general way to establish the spacetime metric. Regardless of whether the cosmological constant should be included, there are also important reasons to consider additional fields, especially scalar fields. Although the latter naturally appear in these modern theories, their inclusion predicts a non-Einsteinian behavior of gravitating systems. These deviations from GR lead to the violation of the equivalence principle, the foundation of general relativity, modification of large-scale gravitational phenomena, and cast doubt upon the constancy of the fundamental "constants." These predictions motivate new searches for very small deviations of relativistic gravity from GR and provide a new theoretical paradigm and guidance for further gravity experiments.

Meanwhile, on the experimental front, recent cosmological observations have forced us to accept the fact that our current understanding of the origin and evolution of the Universe is at best incomplete, and possibly wrong. It turned out that, to our surprise, most of the energy content of the Universe resides in presently unknown dark matter and dark energy that may permeate much, if not all of spacetime. If so, then this dark matter may be accessible to laboratory experimentation. It is likely that the underlying physics that resolve the discord between quantum mechanics and GR will also shed light on cosmological questions addressing the origin and ultimate destiny of the Universe. Recent progress in the development of vastly superior measurement technology placed fundamental physics in a unique position to successfully address these vital questions. Moreover, because of the ever-increasing practical significance of the general theory of relativity (*i.e.*, its use in spacecraft navigation, time transfer, clock synchronization, etalons of time, weight and length, *etc.*), this fundamental theory must be tested to increasing accuracy.

This chapter is organized as follows. Section 2 discusses the foundations of the general theory of relativity and reviews the results of the recent experiments designed to test the foundations of this theory. Section 3 presents motivations for extending the theoretical model of gravity provided by GR; it presents a model arising from string theory, discusses the scalar–tensor theories of gravity, and also highlights phenomenological implications of these proposals. This section also reviews the motivations and the search for new interactions of nature and discusses the hypothesis of gravitational shielding. Section 4 addresses the astrophysical and cosmological phenomena, which led to some recent proposals that modify gravity on large scales; it discusses some of these proposals, and reviews their experimental implications. Section 5 discusses future missions and experiments aiming to expand our knowledge of gravity. Finally, conclusions and an outlook are presented.

2 Testing Foundations of General Relativity

General relativity began its empirical success in 1915, by explaining the anomalous perihelion precession of Mercury's orbit, using no adjustable theoretical parameters. Shortly thereafter, Eddington's 1919 observations of stellar lines-of-sight during a solar eclipse confirmed the doubling of the deflection angles predicted by the Einstein's theory, as compared to Newtonian-like and equivalence principle arguments; this made the theory an instant success. From these beginnings, GR has been extensively tested in the solar system, successfully accounting for all data gathered to date. Thus, microwave ranging to the Viking Lander on Mars yielded accuracy $\sim$0.2 in the tests of GR [1–3]. Spacecraft and planetary radar observations reached an accuracy of $\sim$0.15 [4]. The astrometric observations of quasars on the solar background performed with Very long-baseline interferometry (VLBI) improved the accuracy of the tests of gravity to $\sim$0.045 [5–7]. Lunar laser ranging (LLR) yields a 0.001 verification of GR via precision measurements of the lunar orbit [8–14]. Finally, the recent experiments with the Cassini spacecraft improved the accuracy of the tests to $\sim$0.0023 [15]. As a result, GR became the standard theory of gravity when astrometry and spacecraft navigation are concerned.

To date, GR is also in agreement with the data collected from the binary millisecond pulsars. In fact, a considerable interest has recently developed concerning the physical processes occurring in the strong gravitational field regime with relativistic pulsars providing a promising possibility to test gravity in this qualitatively different dynamical environment. The general theoretical framework for pulsar tests of strong-field gravity was introduced in [16]; the observational data for the initial tests were obtained with PSR1534 [17]. An analysis of strong-field gravitational tests and their theoretical justification was presented in [18–20]. The recent analysis of the pulsar data tested GR to $\sim$0.04 at a 3σ confidence level [21].

In this section, we present the framework used to plan and analyze the data in a weak-field and slow-motion approximation which is appropriate to describe dynamical conditions in the solar system.

2.1 Metric Theories of Gravity and PPN Formalism

Within the accuracy of modern experiments, the weak-field and slow-motion approximation provides a useful starting point for testing the predictions of different metric theories of gravity in the solar system. Following Fock [22,23] and Chandrasekhar [24], a matter distribution in this approximation is often represented by the perfect fluid model with the density of energy–momentum tensor $\widehat{T}^{mn}$ as given by

$$\widehat{T}^{mn} = \sqrt{-g}\left([\rho_0(1+\Pi)+p]\,u^m u^n - p g^{mn}\right) , \tag{1}$$

where ρ_0 is the mass density of the ideal fluid in coordinates of the comoving frame of reference, $u^k = dz^k/ds$ are the components of invariant four-velocity of a fluid element, and $p(\rho)$ is the isentropic pressure connected with ρ by an equation of state. The quantity $\rho\Pi$ is the density of internal energy of an ideal fluid. The definition of Π results from the first law of thermodynamics, through the equation $u^n\left(\Pi_{;n} + p(1/\widehat{\rho})_{;n}\right) = 0$, where the subscript; n denotes a covariant derivative and $\widehat{\rho} = \sqrt{-g}\rho_0 u^0$ is the conserved mass density (see further details in [23–26]). Given the energy–momentum tensor, one finds the solutions of the gravitational field equations for a particular theory of gravity.[1]

Metric theories of gravity have a special position among all the other possible theoretical models. The reason is that, independently of the many different principles at their foundations, the gravitational field in these theories affects the matter directly through the metric tensor g_{mn}, which is determined from the field equations. As a result, in contrast to Newtonian gravity, this tensor expresses the properties of a particular gravitational theory and carries information about the gravitational field of the bodies.

[1] A powerful approach developing a weak-field approximation for GR was presented in [27–29]. It combines an elegant "Maxwell-like" treatise of the spacetime metric in both the *global* and *local* reference frames with the Blanchet–Damour multipole formalism [30]. This approach is applicable for an arbitrary energy–stress tensor and is suitable for addressing problems of strong-field regime. Application of this method to a general N-body problem in a weak-field and slow-motion approximation was developed in [31].

Generalizing on a phenomenological parameterization of the gravitational metric tensor field, which Eddington originally developed for a special case, a method called the parameterized post-Newtonian (PPN) metric has been developed [32–34]. This method represents the gravity tensor's potentials for slowly moving bodies and weak interbody gravity, and is valid for a broad class of metric theories, including GR as a unique case. Several parameters in the PPN metric expansion vary from theory to theory, and they are individually associated with various symmetries and invariance properties of the underlying theory. Gravity experiments can be analyzed in terms of the PPN metric, and an ensemble of experiments will determine the unique value for these parameters, and hence the metric field itself.

As we know it today, observationally, GR is the most successful theory so far as solar system experiments are concerned (see, *e.g.*, [35] for an updated review). The implications of GR for solar system gravitational phenomena are best addressed via the PPN formalism for which the metric tensor of the general Riemannian spacetime is generated by some given distribution of matter in the form of an ideal fluid, given by (1). It is represented by a sum of gravitational potentials with arbitrary coefficients, the PPN parameters. If, for simplicity, one assumes that Lorentz invariance, local position invariance, and total momentum conservation hold, the metric tensor in four dimensions in the so-called PPN-gauge may be written[2] as

$$\begin{aligned} g_{00} &= -1 + 2U - 2\beta U^2 + 2(\gamma+1)\Phi_1 \\ &\quad +2\left[(3\gamma + 1 - 2\beta)\Phi_2 + \Phi_3 + 3\gamma\Phi_4\right] + \mathcal{O}(c^{-5}), \end{aligned} \tag{2}$$

$$\begin{aligned} g_{0i} &= -\frac{1}{2}(4\gamma+3)V_i - \frac{1}{2}W_i + \mathcal{O}(c^{-5}), \\ g_{ij} &= \delta_{ij}(1 + 2\gamma U) + \mathcal{O}(c^{-4}). \end{aligned} \tag{3}$$

The order of magnitude of the various terms is determined according to the rules $U \sim v^2 \sim \Pi \sim p/\rho \sim \epsilon$, $v^i \sim |d/dt|/|d/dx| \sim \epsilon^{1/2}$. The parameter γ represents the measure of the curvature of the spacetime created by the unit rest mass; the parameter β is the measure of the nonlinearity of the law of superposition of the gravitational fields in a theory of gravity or the measure of the metricity. The generalized gravitational potentials, proportional to U^2, result from integrating the energy–stress density (1), are given by

$$U(\mathbf{x},t) = \int d^3\mathbf{x}' \frac{\rho_0(\mathbf{x}',t)}{|\mathbf{x}-\mathbf{x}'|}, \qquad V^\alpha(\mathbf{x},t) = -\int d^3\mathbf{x}' \frac{\rho_0(\mathbf{x}',t)v^\alpha(\mathbf{x}',t)}{|\mathbf{x}-\mathbf{x}'|}, \tag{4}$$

$$W^i(\mathbf{x},t) = \int d^3\mathbf{x}' \rho_0(\mathbf{x}',t) v_j(\mathbf{x}',t) \frac{(x^j - x'^j)(x^i - x'^i)}{|\mathbf{x}-\mathbf{x}'|^3}, \tag{5}$$

---

[2] Note the geometrical units $\hbar = c = G = 1$ are used throughout, as is the metric signature convention $(-+++)$.

$$\Phi_1(\mathbf{x},t) = -\int d^3\mathbf{x}' \frac{\rho_0(\mathbf{x}',t)v^2(\mathbf{x}',t)}{|\mathbf{x}-\mathbf{x}'|}, \qquad \Phi_2(\mathbf{x}',t) = \int d^3\mathbf{x}' \frac{\rho_0(\mathbf{x}',t)U(\mathbf{x}',t)}{|\mathbf{x}-\mathbf{x}'|}, \tag{6}$$

$$\Phi_3(\mathbf{x},t) = \int d^3\mathbf{x}' \frac{\rho_0(\mathbf{x}',t)\Pi(\mathbf{x}',t)}{|\mathbf{x}-\mathbf{x}'|} d^3z'^{\nu}, \qquad \Phi_4(\mathbf{x},t) = \int d^3\mathbf{x}' \frac{p(\mathbf{x}',t)}{|\mathbf{x}-\mathbf{x}'|}. \tag{7}$$

In the complete PPN framework, a particular metric theory of gravity in the PPN formalism might be fully characterized by means of ten PPN parameters [26, 36]. Thus, besides the parameters γ, β, there are eight other parameters $\alpha_1, \alpha_2, \alpha_3, \zeta, \zeta_1, \zeta_2, \zeta_3, \zeta_4$. The formalism uniquely prescribes the values of these parameters for each particular theory under study. In the standard PPN gauge [26] these parameters have clear physical meaning, each quantifying a particular symmetry, conservation law, or fundamental tenant of the structure of spacetime. Thus, in addition to the parameters γ and β discussed above, the group of parameters $\alpha_1, \alpha_2, \alpha_3$ specify the violation of Lorentz invariance (or the presence of the privileged reference frame), the parameter ζ quantifies the violation of the local position invariance, and, finally, the parameters $\zeta_1, \zeta_2, \zeta_3, \zeta_4$ reflect the violation of the law of total momentum conservation for a closed gravitating system. Note that GR, when analyzed in standard PPN gauge, gives $\gamma = \beta = 1$ and all the other eight parameters vanish. The Brans–Dicke theory [37] is the best known of the alternative theories of gravity. It contains, besides the metric tensor, a scalar field and an arbitrary coupling constant ω, which yields the two PPN parameter values, $\beta = 1$, $\gamma = (1+\omega)/(2+\omega)$, where ω is an unknown dimensionless parameter of this theory. More general scalar–tensor theories (see Sect. 3.2) yield values of β different from one [38].

The main properties of the PPN metric tensor given by Eqs. (3)–(7) are well established and widely in use in modern astronomical practice [25, 26, 36, 39–41]. For practical purposes one uses this metric to generate the equations of motion for the bodies of interest. These equations are then used to produce numerical codes in relativistic orbit determination formalisms for planets and satellites [36, 40, 41] as well as for analyzing the gravitational experiments in the solar system [26, 42].

In what follows, we discuss the foundations of general theory of relativity together with our current empirical knowledge on their validity. We take the standard approach to GR according to which the theory is supported by the following basic tenants:

1. Equivalence principle (EP), which states that freely falling bodies do have the same acceleration in the same gravitational field independent of their compositions, which is also known as the principle of universality of the free fall (discussed in Sect. 2.2);
2. Local Lorentz invariance (LLI), which suggests that clock rates are independent of the clock's velocities (discussed in Sect. 2.3);
3. Local position invariance (LPI), which postulates that clock rates are also independent of their spacetime positions (discussed in Sect. 2.4).

2.2 The Equivalence Principle

Since Newton, the question about the equality of inertial and passive gravitational masses has risen in almost every theory of gravitation. Thus, almost 100 years ago, Einstein postulated that not only mechanical laws of motion but also all nongravitational laws should behave in freely falling frames as if gravity were absent. It is this principle that predicts identical accelerations of compositionally different objects in the same gravitational field, and also allows gravity to be viewed as a geometrical property of spacetime, leading to the general relativistic interpretation of gravitation.

Below we shall discuss two different "flavors" of the equivalence principle, the weak and the strong forms of the EP that are currently tested in various experiments performed with laboratory test masses and with bodies of astronomical sizes.

2.2.1 The Weak Equivalence Principle

The weak form of the EP (the WEP) states that the gravitational properties of strong and electroweak interactions obey the EP. In this case, the relevant test-body differences are their fractional nuclear-binding differences, their neutron-to-proton ratios, their atomic charges, *etc.* Furthermore, the equality of gravitational and inertial masses implies that different neutral massive test bodies will have the same free-fall acceleration in an external gravitational field, and therefore in freely falling inertial frames, the external gravitational field appears only in the form of a tidal interaction [43]. Apart from these tidal corrections, freely falling bodies behave as if external gravity is absent [44].

According to GR, the light rays propagating near a gravitating body are achromatically scattered by the curvature of the spacetime generated by the body's gravity field. The entire trajectory of the light ray is bent toward the body by an angle depending on the strength of the body's gravity. In the solar system, the Sun's gravity field produces the largest effect, deflecting the light by as much as $1.75'' \cdot (R_\odot/b)$, where $R_\odot$ is the solar radius and b is the impact parameter. The Eddington's 1919 experiment confirmed the fact that photons obey the laws of free fall in a gravitational field as predicted by GR. The original accuracy was only 10% which was recently improved to 0.0023% by a solar conjunction experiment performed with the Cassini spacecraft [15].

The Pound–Rebka experiment, performed in 1960, further verified effects of gravity on light by testing the universality of gravity-induced frequency shift, $\Delta\nu$, that follows from the WEP

$$\frac{\Delta\nu}{\nu} = \frac{gh}{c^2} = (2.57 \pm 0.26) \times 10^{-15}, \tag{8}$$

where g is the acceleration of gravity and h the height of fall [45].

The WEP can be scrutinized by studying the free fall of antiprotons and antihydrogen, even though the experimental obstacles are considerable; the subject has been extensively reviewed in [46]. This would help in investigating to what extent gravity respects the fundamental CPT symmetry of local quantum field theories, namely if antiparticles fall as particles in a gravitational field. As we shall see later, CPT symmetry may be spontaneously broken in some string/M-theory vacua's; some implications of this will also be mentioned in the context of the validity LLI. The ATHENA (apparatus for high-precision experiments on neutral antimatter) and the ATRAP collaborations at CERN have developed techniques to deal with the difficulties of storing antiprotons and creating an antihydrogen atom (see [47,48] for recent accounts), but no gravitational test has been performed so far. On the other hand, the former CPLEAR collaboration has reported on a test of the WEP involving neutral kaons [49], with limits of 6.5, 4.3, and 1.8×10^{-9}, respectively, for scalar, vector, and tensor potentials originating from the Sun with a range much greater than 1 AU acting on kaons and antikaons. Despite their relevance, these results say nothing about new forces that couple to the baryon number, and therefore are at best complementary to further tests yet to be performed with antiprotons and antihydrogen atoms.

Most extensions to GR are metric in nature, *i.e.*, they assume that the WEP is valid. However, as emphasized by [50,51], almost all extensions to the standard model of particle physics generically predict new forces that would show up as apparent violations of the EP; this occurs specially in theories containing macroscopic-range quantum fields and thus predicting quantum exchange forces that generically violate the WEP, as they couple to generalized "charges," rather than to mass/energy as does gravity [52].

In a laboratory, precise tests of the EP can be made by comparing the free-fall accelerations, a_1 and a_2, of different test bodies. When the bodies are at the same distance from the source of the gravity, the expression for the equivalence principle takes the elegant form

$$\frac{\Delta a}{a} = \frac{2(a_1 - a_2)}{a_1 + a_2} = \left(\frac{M_G}{M_I}\right)_1 - \left(\frac{M_G}{M_I}\right)_2 = \Delta\left(\frac{M_G}{M_I}\right), \tag{9}$$

where M_G and M_I are the gravitational and inertial masses of each body. The sensitivity of the EP test is determined by the precision of the differential acceleration measurement divided by the degree to which the test bodies differ (*e.g.*, composition).

The WEP has been subject to various laboratory tests throughout the years. In 1975, Collela et al. [53] showed with their interferometric experiment that a neutron beam split by a silicon crystal traveling through distinct gravitational paths interferes as predicted by the laws of quantum mechanics, with a gravitational potential given by Newtonian gravity, thus enabling an impressive verification of the WEP applied to an elementary hadron. Present-day technology has achieved impressive limits for the interferometry of atoms rising against gravity, of order 3×10^{-8} [54].

Various experiments have been performed to measure the ratios of gravitational to inertial masses of bodies. Recent experiments on bodies of laboratory dimensions verify the WEP to a fractional precision $\Delta(M_G/M_I) \lesssim 10^{-11}$ by [55], to $\lesssim 10^{-12}$ by [56, 57] and more recently to a precision of $\lesssim 1.4 \times 10^{-13}$ [58]. The accuracy of these experiments is sufficiently high to confirm that the strong, weak, and electromagnetic interactions contribute equally to the passive gravitational and inertial masses of the laboratory bodies. A review of the most recent laboratory tests of gravity can be found in [59].

Quite recently, Nesvizhevsky and collaborators have reported evidence for the existence of gravitational bound states of neutrons [60]; the experiment was, at least conceptually, put forward long ago, in 1978 [61]. Subsequent steps toward the final experiment are described in [62]. This consists in allowing ultra-cold neutrons from a source at the Institute Laue–Langevin reactor in Grenoble to fall toward a horizontal mirror under the influence of the Earth's gravitational field. This potential confines the motion of the neutrons, which do not move continuously vertically, but rather jump from one height to another as predicted by quantum mechanics. It is reported that the minimum measurable energy is of 1.4×10^{-12} eV, corresponding to a vertical velocity of $1.7\,\mathrm{cm\,s^{-1}}$. A more intense beam and an enclosure mirrored on all sides could lead to an energy resolution down to 10^{-18} eV.

We remark that this experiment opens fascinating perspectives, both for testing noncommutative versions of quantum mechanics and for the connection of this theory with gravity [63]. It also enables a new criterion in understanding the conditions for distinguishing quantum from classical behavior in function of the size of an observed system [64].

This impressive evidence of the WEP for laboratory bodies is incomplete for astronomical body scales. The experiments searching for WEP violations are conducted in laboratory environments that utilize test masses with negligible amounts of gravitational self-energy and therefore a large-scale experiment is needed to test the postulated equality of gravitational self-energy contributions to the inertial and passive gravitational masses of the bodies [32]. Recent analysis of the LLR data demonstrated that no composition-dependent acceleration effects [65] are present.

Once the self-gravity of the test bodies is nonnegligible (currently with bodies of astronomical sizes only), the corresponding experiment will be testing the ultimate version of the EP – the strong equivalence principle (SEP) – that is discussed below.

2.2.2 The Strong Equivalence Principle

In its strong form the EP is extended to cover the gravitational properties resulting from gravitational energy itself. In other words, it is an assumption about the way that gravity begets gravity, *i.e.*, about the nonlinear property of gravitation. Although GR assumes that the SEP is exact, alternate metric theories of gravity, such as those involving scalar fields, and other extensions of gravity theory, typically violate the SEP [8,32,66,67]. For the SEP case, the relevant test-body differences are the fractional contributions to their masses by gravitational self-energy. Because of the extreme weakness of gravity, SEP test bodies that differ significantly must have astronomical sizes. Currently, the Earth–Moon–Sun system provides the best solar system arena for testing the SEP.

A wide class of metric theories of gravity are described by the parameterized post-Newtonian formalism [33, 34, 66], which provides a common framework to study the motion of celestial bodies in external gravitational fields. Over the last 35 years, the PPN formalism has become a useful framework for testing the SEP for extended bodies. To facilitate investigation of a possible violation of the SEP, in that formalism the ratio between gravitational and inertial masses, M_G/M_I, is expressed [32, 66] as

$$\left[\frac{M_G}{M_I}\right]_{\mathrm{SEP}} = 1 + \eta\left(\frac{\Omega}{Mc^2}\right), \tag{10}$$

where M is the mass of a body, Ω is the body's (negative) gravitational self-energy, Mc^2 is its total mass energy, and η is a dimensionless parameter that quantifies SEP violation: in fully conservative, Lorentz-invariant theories of gravity [26, 68], the SEP parameter is related to the PPN parameters by $\eta = 4\beta - \gamma - 3$. In GR, $\gamma = \beta = 1$, so that $\eta = 0$.

The self-energy of a body B is given by

$$\left(\frac{\Omega}{Mc^2}\right)_B = -\frac{G}{2M_Bc^2}\int_B d^3\mathbf{x}d^3\mathbf{y}\frac{\rho_B(\mathbf{x})\rho_B(\mathbf{y})}{|\mathbf{x}-\mathbf{y}|}. \tag{11}$$

For a sphere with radius R and uniform density, $\Omega/Mc^2 = -3GM/5Rc^2 = -3v_E^2/10c^2$, where v_E is the escape velocity. Accurate evaluation for solar system bodies requires numerical integration of the expression of (11). Evaluating the standard solar model [69] results in $(\Omega/Mc^2)_\odot \sim -3.52 \times 10^{-6}$. Because gravitational self-energy is proportional to M^2 and also because of the extreme weakness of gravity, the typical values for the ratio (Ω/Mc^2) are $\sim 10^{-25}$ for bodies of laboratory sizes. Therefore, the experimental accuracy of a part in 10^{13} [58], which is so useful for the study of the validity of the WEP, is not sufficient to test on how gravitational self-energy contributes to the inertial and gravitational masses of small bodies. To test the SEP, one must consider planetary-sized extended bodies, where the ratio (11) is considerably higher.

Nordtvedt [8,32,70] suggested several solar system experiments for testing the SEP. One of these was the lunar test. Another, a search for the SEP effect in the motion of the Trojan asteroids, was carried out in [71, 72]. Interplanetary spacecraft tests have been considered in [44] and discussed in [73]. An experiment employing existing binary pulsar data has been proposed [74]. It was pointed out that binary pulsars may provide an excellent possibility for testing the SEP in the new regime of strong self-gravity [18, 19]; however, the corresponding tests have yet to reach competitive accuracy [75, 76].

To date, the Earth–Moon–Sun system has provided the most accurate test of the SEP; recent analysis of LLR data tests the EP to a high precision, yielding $\Delta(M_G/M_I)_{\mathrm{EP}} = (-1.0 \pm 1.4) \times 10^{-13}$ [14]. This result corresponds to a test of the SEP of $\Delta(M_G/M_I)_{\mathrm{SEP}} = (-2.0 \pm 2.0) \times 10^{-13}$ with the SEP violation parameter $\eta = 4\beta - \gamma - 3$ found to be $\eta = (4.4 \pm 4.5) \times 10^{-4}$. Using the recent Cassini result for the PPN parameter γ, PPN parameter β is determined at the level of $\beta - 1 = (1.2 \pm 1.1) \times 10^{-4}$ (see more details in [14]).

2.3 Local Lorentz Invariance

Invariance under Lorentz transformations states that the laws of physics are independent of the frame velocity; this is an underlying symmetry of all current physical theories. However, some evidence recently found in the context of string field theory indicates that this symmetry can be spontaneously broken. Naturally, the experimental verification of this breaking poses a significant challenge. It has already been pointed out that astrophysical observations of distant sources of gamma radiation could hint what is the nature of gravity-induced wave dispersion in vacuum [77, 78] and therefore points toward physics beyond the Standard Model of Particles and Fields (hereafter Standard Model). Limits on Lorentz symmetry violation based on the observations of high-energy cosmic rays with energies beyond 5×10^{19} eV, the so-called Greisen–Zatsepin–Kuzmin (GKZ) cutoff [79], have also been discussed [80–83].

A putative violation of Lorentz symmetry has been a repeated object of interest in the literature. A physical description of the effect of our velocity with respect to a presumably preferred frame of reference relies on a constant background cosmological vector field, as suggested in [84]. Based on the behavior of the renormalization group β-function of non-abelian gauge theories, it has also been argued that Lorentz invariance could be just a low-energy symmetry [85].

Lorentz symmetry breaking due to nontrivial solutions of string field theory was first discussed in [86]. These arise from the string field theory of open strings and may have implications for low-energy physics. For instance, assuming that the contribution of Lorentz-violating interactions to the vacuum energy is about half of the critical density implies that feeble tensor-mediated interactions in the range of $\sim 10^{-4}$ m should exist [87]. Furthermore, Lorentz violation may induce the breaking of conformal symmetry; this, together with inflation, may explain the origin of the primordial magnetic fields required to explain the observed galactic magnetic field [88]. Also, violations of Lorentz invariance may imply in a breaking of the fundamental CPT symmetry of local quantum field theories [89]. Quite remarkably, this can be experimentally verified in neutral-meson [90] experiments,[3] Penning-trap measurements [92], and

[3] These CPT violating effects are unrelated with those due to possible nonlinearities in quantum mechanics, presumably arising from quantum gravity and already investigated by the CPLEAR Collaboration [91].

hydrogen–antihydrogen spectroscopy [93]. This spontaneous breaking of CPT symmetry allows for an explanation of the baryon asymmetry of the Universe: in the early Universe, after the breaking of the Lorentz and CPT symmetries, tensor–fermion interactions in the low-energy limit of string field theories give rise to a chemical potential that creates in equilibrium a baryon–antibaryon asymmetry in the presence of baryon number violating interactions [94].

Limits on the violation of Lorentz symmetry are available from laser interferometric versions of the Michelson–Morley experiment, by comparing the velocity of light, c, and the maximum attainable velocity of massive particles, c_i, up to $\delta \equiv |c^2/c_i^2 - 1| < 10^{-9}$ [95]. More accurate tests can be performed via the Hughes–Drever experiment [96, 97], where one searches for a time dependence of the quadrupole splitting of nuclear Zeeman levels along Earth's orbit. This technique achieves an impressive limit of $\delta < 3 \times 10^{-22}$ [98]. A recent reassessment of these results reveals that more stringent bounds can be reached, up to eight orders of magnitude higher [99]. The parameterized post-Newtonian parameter α_3 can be used to set astrophysical limits on the violation of momentum conservation and the existence of a preferred reference frame. This parameter, which vanishes identically in GR can be accurately determined from the pulse period of pulsars and millisecond pulsars [35, 100]. The most recent results yields a limit on the PPN parameter α_3 of $|\alpha_3| < 2.2 \times 10^{-20}$ [101].

After the cosmic microwave background radiation (CMBR) has been discovered, an analysis of the interaction between the most energetic cosmic-ray particles and the microwave photons was mandatory. As it turns out, the propagation of the ultra-high-energy nucleons is limited by inelastic collisions with photons of the CMBR, preventing nucleons with energies above 5×10^{19} eV from reaching Earth from further than 50–100 Mpc. This is the already mentioned GZK cutoff [79]. However, events where the estimated energy of the cosmic primaries is beyond the GZK cutoff were observed by different collaborations [102–105]. It has been suggested [80, 81] (see also [82]) that slight violations of Lorentz invariance would cause energy-dependent effects that would suppress otherwise inevitable processes such as the resonant scattering reaction, $p + \gamma_{2.73K} \rightarrow \Delta_{1232}$. The study of the kinematics of this process produces a quite stringent constraint on the validity of Lorentz invariance, $\delta < 1.7 \times 10^{-25}$ [83, 106].

Quite recently, the High-Resolution Fly's Eye collaboration suggested that the gathered data show that the GZK cutoff holds for their span of observations [107]. Confirmation of this result is of great importance, and the coming into operation of the Auger collaboration [108] in the near future will undoubtedly provide further insight into this fundamental question. It is also worth mentioning that the breaking of Lorentz invariance can occur in the context of noncommutative field theories [109], even though this symmetry may hold (at least) at first nontrivial order in perturbation theory of the noncommutative parameter [110]. Actually, the idea that the noncommutative parameter may be a Lorentz tensor has been considered in some field theory models [111]. Also, Lorentz symmetry can be broken in loop quantum gravity [112], quantum gravity-inspired spacetime foam scenarios [113], or via the spacetime variation of fundamental coupling constants [114]. For a fairly complete review about Lorentz violation at high energies the reader is directed to [115]. Note that a gravity model where LLI is spontaneously broken was proposed in [116, 117] and solutions where discussed in [118].

2.3.1 Spontaneous Violation of Lorentz Invariance

The impact of a spontaneous violation of Lorentz invariance on theories of gravity is of great interest. In this context, the breaking of Lorentz invariance may be implemented, for instance, by allowing a vector field to roll to its vacuum expectation value. This mechanism requires that the potential that rules the dynamics of the vector field possesses a minimum, in the way similar to the Higgs mechanism [116]. This, so-called, "bumblebee" vector thus acquires an explicit (four-dimensional) orientation, and Lorentz symmetry is spontaneously broken. The action of the bumblebee model is written as

$$S = \int d^4x\sqrt{-g}\left[\frac{1}{2\kappa}\left(R + \xi B^\mu B^\nu R_{\mu\nu}\right) - \frac{1}{4}B^{\mu\nu}B_{\mu\nu} - V(B^\mu B_\mu \pm b^2)\right] , \tag{12}$$

where $B_{\mu\nu} = \partial_\mu B_\nu - \partial_\nu B_\mu$, ξ and b^2 are a real coupling constant and a free real positive constant, respectively. The potential V driving Lorentz and/or CPT violation is supposed to have a minimum at $B^\mu B_\mu \pm b^2 = 0$, $V'(b_\mu b^\mu) = 0$. Since one assumes that the bumblebee field B_μ is frozen at its vacuum expectation value, the particular form of the potential driving its dynamics is irrelevant. The scale of b_μ, which controls the symmetry breaking, must be derived from a fundamental theory, such as string theory or from a low-energy extension to the Standard Model; hence one expects either b of order of the Planck mass, $M_P = 1.2 \times 10^{19}$ GeV, or of order of the electroweak breaking scale, $M_{EW} \simeq 10^2$ GeV.

Previously, efforts to quantify an hypothetical breaking of Lorentz invariance were primarily directed toward the phenomenology of such spontaneous Lorentz symmetry breaking in particle physics. Only recently its implications for gravity have been more thoroughly explored [116, 117]. In that work, the framework for the LSB gravity model is set up, developing the action and using the *vielbein* formalism. A later study [118] considered consequences of such a scenario, assuming three plausible cases (1) the bumblebee field acquires a purely radial vacuum expectation value, (2) a mixed radial and temporal vacuum expectation value, and (3) a mixed axial and temporal vacuum expectation value.

In the first case, an exact black hole solution was found, exhibiting a deviation from the inverse-square law such that the gravitational potential of a point mass at rest depends on the radial coordinate as r^{-1+p} where p is a parameter related to the fundamental physics underlying the breaking of Lorentz invariance. This solution has a removable singularity at a horizon of radius $r_s = (2Mr_0^{-p})^{1/(1-p)}$, slightly perturbed with respect to the usual Scharzschild radius $r_{s0} = 2M$, which protects a real singularity at $r = 0$. Known deviations from Kepler's law yield $p \lesssim 2 \times 10^{-9}$.

In the second case, no exact solution was discovered, although a perturbative method allowed for the characterization of the Lorentz symmetry breaking in terms of the PPN parameters $\beta \approx 1 - (K + K_r)/M$ and $\gamma \approx 1 - (K + 2K_r)/M$, directly proportional to the strength of the induced effect, given by K and $K_r \sim K$, where K and K_r are integration constants arising from the perturbative treatment of the timelike spontaneous LSB superimposed on the vacuum Scharzschild metric. An analogy with Rosen's bimetric theory yields the parameter $\gamma \simeq (A + B)d$, where d is the distance to the central body and A and B rule the temporal and radial components of the vector field vacuum expectation value.

In the third case, a temporal/axial vacuum expectation value for the bumblebee vector field clearly breaks isotropy, thus forbidding a standard PPN analysis. However, for the case of the breaking of Lorentz invariance occurring in the x_1 direction, similar direction-dependent PPN-like parameters were defined as $\gamma_1 \simeq b^2 \cos^2 \theta/2$ and $\gamma_2 \simeq a^2 b^2 \cos^2 \theta/4$, where a and b are proportional, respectively, to the temporal and axial components of the vacuum expectation value acquired by the bumblebee. This enables a crude estimative of the measurable PPN parameter γ, yielding $\gamma \approx b^2(1 - e^2)/4$, where e is the orbit's eccentricity. A comparison with experiments concerning the anisotropy of inertia yields $|b| \leq 2.4 \times 10^{-11}$ [98].

2.4 Local Position Invariance

Given that both the WEP and LLI postulates have been tested with great accuracy, experiments concerning the universality of the gravitational redshift measure the level to which the LPI holds. Therefore, violations of the LPI would imply that the rate of a free-falling clock would be different when compared with a standard one, for instance on the Earth's surface. The accuracy to which the LPI holds as an invariance of Nature can be parameterized through $\Delta\nu/\nu = (1 + \mu)U/c^2$. From the already mentioned Pound–Rebka experiment (cf. (8)) one can infer that $\mu \simeq 10^{-2}$; the most accurate verification of the LPI was performed by Vessot and collaborators, by comparing hydrogen-maser frequencies on Earth and on a rocket flying to altitude of 10^4 km [119]. The resulting bound is $|\mu| < 2 \times 10^{-4}$. Recently, one order-of-magnitude improvement was attained, thus establishing the most stringent result on the LPI so far [120], $|\mu| < 2.1 \times 10^{-5}$ (Table 1).

2.5 Summary of Solar System Tests of Relativistic Gravity

Although, these available experimental data fit quite well with GR, while allowing for the existence of putative extensions of this successful theory,

Table 1. The accuracy of determination of the PPN parameters γ and β [14,35,42].

PPN parameter	Experiment	Result
$\gamma - 1$	Cassini 2003 spacecraft radiotracking	2.3×10^{-5}
	Observations of quasars with astrometric VLBI	3×10^{-4}
$\beta - 1$	Helioseismology bound on perihelion shift	3×10^{-3}
	LLR test of the SEP, assumed: $\eta = 4\beta - \gamma - 3$ and the Cassini result for PPN γ	1.1×10^{-4}

provided any new effects are small at the post-Newtonian scale [26]. We shall here discuss the available phenomenological constraints for alternative theories of gravity.

LLR investigates the SEP by looking for a displacement of the lunar orbit along the direction to the Sun. The equivalence principle can be split into two parts: the WEP tests the sensitivity to composition and the SEP checks the dependence on mass. There are laboratory investigations of the WEP which are about as accurate as LLR [58, 65]. LLR is the dominant test of the SEP with the most accurate testing of the EP at the level of $\Delta(M_G/M_I)_{EP} = (-1.0 \pm 1.4) \times 10^{-13}$ [14]. This result corresponds to a test of the SEP of $\Delta(M_G/M_I)_{SEP} = (-2.0 \pm 2.0) \times 10^{-13}$ with the SEP violation parameter $\eta = 4\beta - \gamma - 3$ found to be $\eta = (4.4 \pm 4.5) \times 10^{-4}$. Using the recent Cassini result for the PPN parameter γ, PPN parameter β is determined at the level of $\beta - 1 = (1.2 \pm 1.1) \times 10^{-4}$ (see Fig. 1).

LLR data yielded the strongest limits to date on variability of the gravitational constant (the way gravity is affected by the expansion of the Universe), the best measurement of the de Sitter precession rate, and is relied upon to generate accurate astronomical ephemerides. The possibility of a time variation of the gravitational constant, G, was first considered by Dirac in 1938 on the basis of his large number hypothesis, and later developed by Brans and Dicke in their theory of gravitation (for more details consult [26,68]). Variation might be related to the expansion of the Universe, in which case $\dot{G}/G = \sigma H_0$, where H_0 is the Hubble constant, and σ is a dimensionless parameter whose value depends on both the gravitational constant and the cosmological model considered. Revival of interest in Brans–Dicke-like theories, with a variable G, was partially motivated by the appearance of string theories where G is considered to be a dynamical quantity [122].

In Brans–Dicke theory, as well as in more general scalar–tensor theories, the gravitational coupling depends on the cosmic time. Observational bounds arising from the timing of the binary pulsar PSR1913+16 yield quite restrictive bounds [123] of $\dot{G}/G = (1.0 \pm 2.3) \times 10^{-11}\,\mathrm{year}^{-1}$, with a magnitude similar to the cosmological bounds available [124–126] (see [127] and references therein for a discussion on a connection with the accelerated expansion of the Universe). Varying-G solar models [128] and measurements of masses and ages of neutron stars yield even more stringent limits [129] of $\dot{G}/G = (-0.6 \pm 2.0) \times 10^{-12}\,\mathrm{year}^{-1}$.

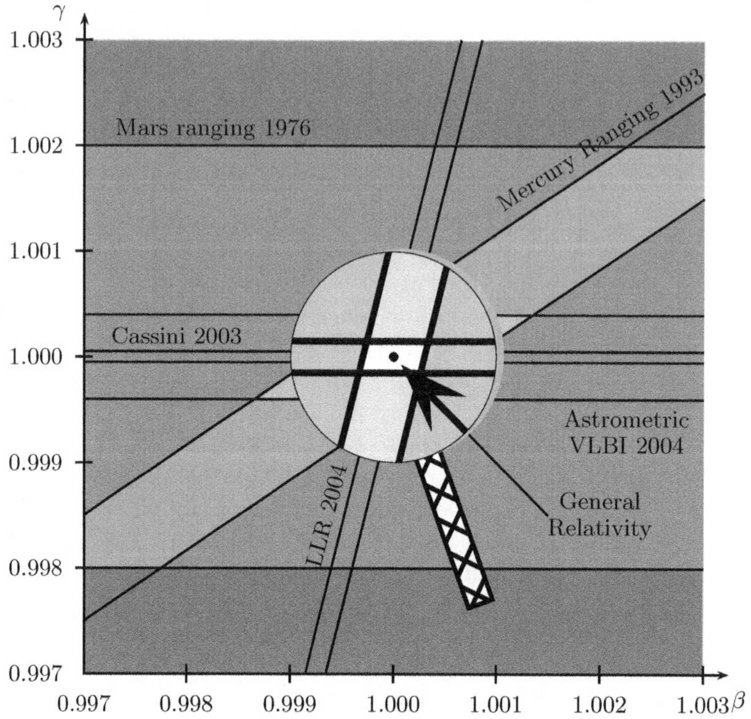

Fig. 1. The progress in determining the PPN parameters γ and β for the last 30 years (adapted from [121]).

The most stringent limit on a change of G comes from LLR, which is one of the important gravitational physics result that LLR provides. GR does not predict a changing G, but some other theories do, thus testing for this effect is important. As we have seen, the most accurate limit published is the current LLR test, yielding $\dot{G}/G = (4 \pm 9) \times 10^{-13}\,\mathrm{year}^{-1}$ [14]. The $\dot{G}/G$ uncertainty is 83 times smaller than the inverse age of the Universe, $t_0 = 13.4\,\mathrm{Gyr}$ with the value for Hubble constant $H_0 = 72\,\mathrm{km\,s^{-1}\,Mpc^{-1}}$ from the WMAP data [130]. The uncertainty for $\dot{G}/G$ is improving rapidly because its sensitivity depends on the square of the data span. This fact puts LLR, with its more than 35 years of history, in a clear advantage as opposed to other experiments.

LLR has also provided the only accurate determination of the geodetic precession. Williams et al. [14] report a test of geodetic precession, which expressed as a relative deviation from GR, is $K_{gp} = -0.0019 \pm 0.0064$. The GP-B satellite should provide improved accuracy over this value, if that mission is successfully completed. LLR also has the capability of determining PPN β and γ directly from the point-mass orbit perturbations. A future possibility is detection of the solar J_2 from LLR data combined with the planetary ranging data. Also possible are dark matter tests, looking for any departure from the inverse-square law of gravity, and checking for a variation of the speed of light. The accurate LLR data have been able to quickly eliminate several suggested alterations of physical laws. The precisely measured lunar motion is a reality that any proposed laws of attraction and motion must satisfy.

3 Search for New Physics Beyond General Relativity

The nature of gravity is fundamental to the understanding of the solar system and the large-scale structure of the Universe. This importance motivates various precision tests of gravity both in laboratories and in space. To date, the experimental evidence for gravitational physics is in agreement with GR; however, there are a number of reasons to question the validity of this theory. Despite the success of modern gauge field theories in describing the electromagnetic, weak, and strong interactions, it is still not understood how gravity should be described at the quantum level. In theories that attempt to include gravity, new long-range forces can arise in addition to the Newtonian inverse-square law. Even at the purely classical level, and assuming the validity of the equivalence principle, Einstein's theory does not provide the most general way to describe the space–time dynamics, as there are reasons to consider additional fields and, in particular, scalar fields.

Although scalar fields naturally appear in the modern theories, their inclusion predicts a non-Einsteinian behavior of gravitating systems. These deviations from GR lead to a violation of the EP, modification of large-scale gravitational phenomena, and imply that the constancy of the "constants" must be reconsidered. These predictions motivate searches for small deviations of relativistic gravity from GR and provide a theoretical paradigm and constructive guidance for further gravity experiments. As a result, this theoretical progress has motivated high-precision tests of relativistic gravity and especially those searching for a possible violation of the equivalence principle. Moreover, because of the ever-increasing practical significance of the general theory of relativity (*i.e.*, its use in spacecraft navigation, time transfer, clock synchronization, standards of time, weight and length, *etc.*) this fundamental theory must be tested to increasing accuracy.

An understanding of gravity at a quantum level will allow one to ascertain whether the gravitational "constant" is a running coupling constant like those of other fundamental interactions of nature. String/M-theory [131] hints a negative answer to this question, given the nonrenormalization theorems of supersymmetry, a symmetry at the core of the underlying principle of string/M-theory, and brane models [132, 133]. However, one-loop higher-derivative quantum gravity models may permit a running gravitational coupling, as these models are asymptotically free, a striking property [134]. In the absence of a screening mechanism for gravity, asymptotic freedom may imply that quantum gravitational corrections take effect on macroscopic and even cosmological scales, which of course has some bearing on the dark matter problem [135] and, in particular, on the subject of the large-scale structure of the Universe [136, 137] (see, however, [124]). Either way, it seems plausible to assume that quantum gravity effects manifest themselves only on cosmological scales.

In this section we review the arguments for high-accuracy experiments motivated by the ideas suggested by proposals of quantization of gravity.

3.1 String/M-Theory

String theory is currently referred to as string/M-theory, given the unification of the existing string theories that is achieved in the context M-theory. Nowadays, it is widely viewed as the most promising scheme to make GR compatible with quantum mechanics (see [131] for an extensive presentation). The closed string theory has a spectrum that contains as zero mass eigenstates the graviton, g_{MN}, the dilaton, Φ, and an antisymmetric second-order tensor, B_{MN}. There are various ways in which to extract the physics of our four-dimensional world, and a major difficulty lies in finding a natural mechanism that fixes the value of the dilaton field, since it does not acquire a potential at any order in string perturbation theory.

Damour and Polyakov [52] have studied a possible mechanism to circumvent the above difficulty by suggesting string loop contributions, which are counted by dilaton interactions, instead of a potential. After dropping the antisymmetric second-order tensor and introducing fermions, $\hat{\psi}$, Yang–Mills fields, $\hat{A}^\mu$, with field strength $\hat{F}_{\mu\nu}$, in a spacetime described by the metric $\hat{g}_{\mu\nu}$, the relevant effective low-energy four-dimensional action is

$$S = \int_M d^4x\sqrt{-\hat{g}}B(\Phi)\left\{\frac{1}{\alpha'}[\hat{R} + 4\hat{\nabla}_\mu\hat{\nabla}^\mu\Phi - 4(\hat{\nabla}\Phi)^2] - \frac{k}{4}\hat{F}_{\mu\nu}\hat{F}^{\mu\nu} - \bar{\hat{\psi}}\gamma^\mu\hat{D}_\mu\hat{\psi} + ...\right\}, \quad (13)$$

where

$$B(\Phi) = e^{-2\Phi} + c_0 + c_1e^{2\Phi} + c_2e^{4\Phi} + ..., \quad (14)$$

α' is the inverse of the string tension and k is a gauge group constant; the constants c_0, c_1, ..., *etc.*, can, in principle, be determined via computation.

To recover Einsteinian gravity, a conformal transformation must be performed with $g_{\mu\nu} = B(\Phi)\hat{g}_{\mu\nu}$, leading to an action where the coupling constants and masses are functions of the rescaled dilaton, ϕ:

$$S = \int_M d^4x\sqrt{-g}\left[\frac{1}{4q}R - \frac{1}{2q}(\nabla\phi)^2 - \frac{k}{4}B(\phi)F_{\mu\nu}F^{\mu\nu} - \bar{\psi}\gamma^\mu D_\mu\psi + ...\right], \quad (15)$$

from which follows that $4q = 16\pi G = \frac{1}{4}\alpha'$ and the coupling constants and masses are now dilaton dependent, through $g^{-2} = kB(\phi)$ and $m_A =$

$m_A(B(\phi))$. Damour and Polyakov proposed the minimal coupling principle (MCP), stating that the dilaton is dynamically driven toward a local minimum of all masses (corresponding to a local maximum of $B(\phi)$). Due to the MCP, the dependence of the masses on the dilaton implies that particles fall differently in a gravitational field, and hence are in violation of the WEP. Although, in the solar system conditions, the effect is rather small being of the order of $\Delta a/a \simeq 10^{-18}$, application of already available technology can potentially test prediction. Verifying this prediction is an interesting prospect, as it would present a distinct experimental signature of string/M-theory. We have no doubts that the experimental search for violations of the WEP, as well as of the fundamental Lorentz and CPT symmetries, present important windows of opportunity to string physics and should be vigorously pursued.

Recent analysis of a potential scalar field's evolution scenario based on action (15) discovered that the present agreement between GR and experiment might be naturally compatible with the existence of a scalar contribution to gravity. In particular, Damour and Nordtvedt [38] (see also [52] for nonmetric versions of this mechanism together with [138] for the recent summary of a dilaton-runaway scenario) have found that a scalar–tensor theory of gravity may contain a "built-in" cosmological attractor mechanism toward GR. These scenarios assume that the scalar coupling parameter $\frac{1}{2}(1-\gamma)$ was of order 1 in the early Universe (say, before inflation), and show that it then evolves to be close to, but not exactly equal to, zero at the present time.

The Eddington parameter γ, whose value in general relativity is unity, is perhaps the most fundamental PPN parameter, in that $\frac{1}{2}(1-\gamma)$ is a measure, for example, of the fractional strength of the scalar gravity interaction in scalar–tensor theories of gravity [18,19]. Within perturbation theory for such theories, all other PPN parameters to all relativistic orders collapse to their general relativistic values in proportion to $\frac{1}{2}(1-\gamma)$. Under some assumptions (see, *e.g.*, [38]), one can even estimate what is the likely order of magnitude of the left-over coupling strength at present time which, depending on the total mass density of the Universe, can be given as $1-\gamma \sim 7.3\times10^{-7}(H_0/\Omega_0^3)^{1/2}$, where Ω_0 is the ratio of the current density to the closure density and H_0 is the Hubble constant in units of $100\,\mathrm{km\,s^{-1}\,Mpc^{-1}}$. Compared to the cosmological constant, these scalar field models are consistent with the supernovae observations for a lower matter density, $\Omega_0 \sim 0.2$, and a higher age, $(H_0t_0) \approx 1$. If this is indeed the case, the level $(1-\gamma) \sim 10^{-6}$–10^{-7} would be the lower bound for the present value of the PPN parameter γ [38]. This is why measuring the parameter γ to accuracy of one part in a billion, as suggested for the laser astrometric test of relativity (LATOR) mission [121], is important.

3.2 Scalar–Tensor Theories of Gravity

In many alternative theories of gravity, the gravitational coupling strength exhibits a dependence on a field of some sort; in scalar–tensor theories, this is a scalar field φ. A general action for these theories can be written as

$$S = \frac{c^3}{4\pi G}\int d^4x\sqrt{-g}\left[\frac{1}{4}f(\varphi)R - \frac{1}{2}g(\varphi)\partial_\mu\varphi\partial^\mu\varphi + V(\varphi)\right] + \sum_i q_i(\varphi)\mathcal{L}_i, \quad (16)$$

where $f(\varphi)$, $g(\varphi)$, $V(\varphi)$ are generic functions, $q_i(\varphi)$ are coupling functions, and $\mathcal{L}_i$ is the Lagrangian density of the matter fields; it is worth mentioning that the graviton-dilaton system in string/M-theory can be viewed as one of such scalar–tensor theories of gravity. An emblematic proposal is the well-known Brans–Dicke theory [37] which corresponds to the specific choice

$$f(\varphi) = \varphi, \qquad g(\varphi) = \frac{\omega}{\varphi}, \quad (17)$$

and a vanishing potential $V(\varphi)$. Note that in the Brans–Dicke theory, the kinetic energy term of the field φ is noncanonical, and the latter has a dimension of energy squared. In this theory, the constant ω marks observational deviations from GR, which is recovered in the limit $\omega \to \infty$. We point out that, in the context of the Brans–Dicke theory, one can operationally introduce the Mach's principle which, we recall, states that the inertia of bodies is due to their interaction with the matter distribution in the Universe. Indeed, in this theory, the gravitational coupling is proportional to φ^{-1}, which depends on the energy–momentum tensor of matter through the field equations. Observational bounds require that $|\omega| \gtrsim 500$ [2, 5], and even higher values $|\omega| \gtrsim 40,000$ are reported in [35]. In the so-called *induced gravity models* [139], the functions of the fields are initially given by $f(\varphi) = \varphi^2$ and $g(\varphi) = 1/2$, and the potential $V(\varphi)$ allows for a spontaneous symmetry breaking, so that the field φ acquires a nonvanishing vacuum expectation value, $f(\langle 0|\varphi|0\rangle) = \langle 0|\varphi^2|0\rangle = M_P^2 = G^{-1}$. Naturally, the cosmological constant is given by the interplay of the value $V(\langle 0|\varphi|0\rangle)$ and all other contributions to the vacuum energy.

Therefore, it is clear that in this setup Newton's constant arises from dynamical or symmetry-breaking considerations. It is mesmerizing to conjecture that the φ field could be locally altered: this would require the coupling of this field with other fields to locally modify its value. This feature can be found in some adjusting mechanisms devised as a solution of the cosmological constant problem (see, *e.g.*, [140] for a list of references). However, Weinberg [140] has shown that these mechanisms are actually unsuitable for this purpose, although they nevertheless contain interesting multifield dynamics. Recent speculations, suggesting that extra dimensions in braneworld scenarios may be rather large [141, 142], bring forth gravitational effects at the much lower scale set by M_5, the five-dimensional Planck mass. Phenomenologically, the existence of extra dimensions should manifest itself through a contribution to Newton's law on small scales, $r \lesssim 10^{-4}$ m, as discussed in Sect. 3.3.

3.3 Search for New Interactions of Nature

The existence of new fundamental forces beyond the already known four fundamental interactions, if confirmed, will have several implications and bring important insights into the physics beyond the Standard Model. A great interest on the subject was sparked after the 1986 claim of evidence for an intermediate range interaction with subgravitational strength [143], both theoretical (see [46] for a review) and experimental, giving rise to a wave of new setups, as well as repetitions of "classical" ones using state-of-the-art technology.

In its simplest versions, a putative new interaction or a fifth force would arise from the exchange of a light boson coupled to matter with a strength comparable to gravity. Planck-scale physics could give origin to such an interaction in a variety of ways, thus yielding a Yukawa-type modification in the interaction energy between point-like masses. This new interaction can be derived, for instance, from extended supergravity theories after dimensional reduction [46, 144], compactification of five-dimensional generalized Kaluza–Klein theories including gauge interactions at higher dimensions [145], and also from string/M-theory. In general, the interaction energy, $V(r)$, between two point masses m_1 and m_2 can be expressed in terms of the gravitational interaction as[4]

$$V(r) = -\frac{G_\infty m_1 m_2}{r}\left(1 + \alpha\, e^{-r/\lambda}\right), \tag{18}$$

where $r = |\mathbf{r}_2 - \mathbf{r}_1|$ is the distance between the masses, G_∞ is the gravitational coupling for $r \to \infty$, and α and λ are, respectively, the strength and range of the new interaction. Naturally, G_∞ has to be identified with Newton's gravitational constant and the gravitational coupling becomes dependent on r. Indeed, the force associated with (18) is given by:

$$\mathbf{F}(\mathbf{r}) = -\nabla V(\mathbf{r}) = -\frac{G(r) m_1 m_2}{r^2}\,\hat{\mathbf{r}}, \tag{19}$$

where

$$G(r) = G_\infty\left[1 + \alpha\,(1 + r/\lambda)e^{-r/\lambda}\right]. \tag{20}$$

The suggestion of existence of a new interaction arose from assuming that the coupling α is not an universal constant, but instead a parameter depending on the chemical composition of the test masses [146]. This comes about if one considers that the new bosonic field couples to the baryon number $B = Z + N$, which is the sum of protons and neutrons. Hence the new interaction between masses with baryon numbers B_1 and B_2 can be expressed through a new fundamental constant, f, as:

$$V(r) = -f^2 \frac{B_1 B_2}{r} e^{-r/\lambda}, \tag{21}$$

such that the constant α can be written as

$$\alpha = -\sigma \left(\frac{B_1}{\mu_1}\right)\left(\frac{B_2}{\mu_2}\right), \tag{22}$$

with $\sigma = f^2/G_\infty m_H^2$ and $\mu_{1,2} = m_{1,2}/m_H$, m_H being the hydrogen mass.

---

[4] We use here the units $c = \hbar = 1$.

The above equations imply that in a Galileo-type experiment a difference in acceleration exists between the masses m_1 and m_2, given by

$$\mathbf{a}_{12} = \sigma \left(\frac{B}{\mu}\right)_{\oplus} \left[\left(\frac{B_1}{\mu_1}\right) - \left(\frac{B_2}{\mu_2}\right)\right] \mathbf{g}, \tag{23}$$

where $\mathbf{g}$ is the field strength of the Earth's gravitational field.

Several experiments (see, for instance, [46, 143] for a list of the most relevant) studied the parameters of a new interaction based on the idea of a composition-dependent differential acceleration, as described in (23), and other composition-independent effect.[5] The current experimental status is essentially compatible with the predictions of Newtonian gravity, in both composition-independent and -dependent setups. The bounds on parameters α and λ are summarized below (Fig. 2):

- Laboratory experiments devised to measure deviations from the inverse-square law are most sensitive in the range $10^{-2}\,\mathrm{m} \lesssim \lambda \lesssim 1\,\mathrm{m}$, constraining α to be smaller than about 10^{-4}.
- Gravimetric experiments sensitive in the range of $10\,\mathrm{m} \lesssim \lambda \lesssim 10^3\,\mathrm{m}$ indicate that $\alpha \lesssim 10^{-3}$.
- Satellite tests probe the ranges of about $10^5\,\mathrm{m} \lesssim \lambda \lesssim 10^7\,\mathrm{m}$ suggest that $\alpha \lesssim 10^{-7}$.

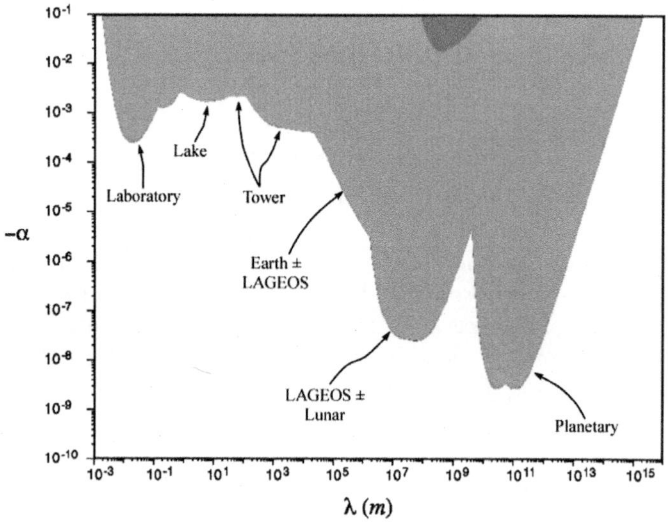

Fig. 2. Experimentally excluded regions for the range and strength of possible new forces, as shown in [148].

---

[5] For instance, neutron interferometry has been suggested to investigate a possible new force that couples to neutron number [147].

– Analysis of the effects of the inclusion of scalar fields into the stellar structure yields a bound in the range $10^8\,\text{m} \lesssim \lambda \lesssim 10^{10}\,\text{m}$, limiting α to be smaller than approximately 10^{-2} [148].

The latter bound, although modest, is derived from a simple computation of the stellar equilibrium configuration in the polytropic gas approximation, when an extra force due to a Yukawa potential is taken into account on the hydrostatic equilibrium equation.

Remarkably, α is so far essentially unconstrained for $\lambda < 10^{-3}\,\text{m}$ and $\lambda > 10^{13}\,\text{m}$. The former range is particularly attractive as a testing ground for new interactions, since forces with submillimetric range seems to be favored from scalar interactions in supersymmetric theories [149]; this is also the case in the recently proposed theories of TeV scale quantum gravity, which stem from the hypothesis that extra dimensions are not necessarily of Planck size [141,142]. The range $\lambda < 10^{-3}$ m also arises if one assumes that scalar [150] or tensor interactions associated with Lorentz symmetry breaking in string theories [87] account for the vacuum energy up to half of the critical density. Putative corrections to Newton's law at millimeter range could have relevant implications, especially taking into account that, in certain models of extra dimensions, these corrections can be as important as the usual Newtonian gravity [142, 151]. From the experimental side, this range has recently been available to experimental verification; state-of-the-art experiments rule out extra dimensions over length scales down to 0.2 mm [152].

3.4 Gravity Shielding: The Majorana Effect

The possibility that matter can shield gravity is not predicted by modern theories of gravity, but it is a recurrent idea and it would cause a violation of the equivalence principle test. In fact, the topic has been more recently reviewed in [125] renewing the legitimacy of this controversial proposal; consequently, a brief discussion is given in this section.

The idea of gravity shielding goes back at least as far as to Majorana's 1920 paper [153]. Since then a number of proposals and studies have been put forward and performed to test the possible absorption of the gravitational force between two bodies when screened from each other by a medium other than vacuum. This effect is a clear gravitational analogue of the magnetic permeability of materials, and Majorana [153] suggested the introduction of a screening or extinction coefficient, h, to measure the shielding of the gravitational force between masses m_1 and m_2 induced by a material with density $\rho(r)$; such an effect can be modeled as

$$F' = \frac{G m_1 m_2}{r^2} \exp\left[-h \int \rho(r)\, dr\right] , \tag{24}$$

which clearly depends on the amount of mass between attracting mass elements and a universal constant h. Naturally, one expects h to be quite small.

Several attempts to derive this parameter from general principles have been made. Majorana gave a closed form expression for a sphere's gravitational to inertial mass ratio. For weak shielding, a simpler expression is given by the linear expansion of the exponential term, $M_G/M_I \approx 1 - hfR\bar{\rho}$, where f is a numerical factor, $\bar{\rho}$ is the mean density, and R is the sphere's radius. For a homogeneous sphere, Majorana and Russell give $f = 3/4$. For a radial density distribution of the form $\rho(r) = \rho(0)(1 - r^2/R^2)^n$, Russell derives $f = (2n + 3)^2/(12n + 12)$. Russell [154] realized that the large masses of the Earth, Moon and planets made the observations of the orbits of these bodies and the asteroid Eros a good test of such a possibility. He made a rough estimate that the equivalence principle was satisfied to a few parts per million, which was much smaller than a numerical prediction based on Majorana's estimate for h. If mass shields gravity, then large bodies such as, for instance, the Moon and the Earth will partly shield their own gravitational attraction. The observable ratio of gravitational mass to inertial mass would not be independent of mass, which would violate the equivalence principle.

Eckhardt [155] showed that LLR can be used to set the limit $h \leq 1.0 \times 10^{-22}\,\mathrm{m^2\,kg^{-1}}$, six orders of magnitude smaller than the geophysical constraint. In [155], an LLR test of the equivalence principle was used to set a modern limit on gravity shielding. That result is updated as follows: the uniform density approximation is sufficient for the Moon and $f\,R\,\bar{\rho} = 4.4 \times 10^9\,\mathrm{kg\,m^{-2}}$. For the Earth we use $n \approx 0.8$ with Russell's expression to get $f\,R\,\bar{\rho} = 3.4 \times 10^{10}\,\mathrm{kg\,m^{-2}}$. Using the difference $-3.0 \times 10^9\,\mathrm{g\,cm^{-2}\,h}$ along with the LLR EP solution for the difference in gravitational to inertial mass ratios gives $h = (3 \pm 5) \times 10^{-24}\,\mathrm{m^2\,kg^{-1}}$ [156]. The value is not significant compared with the uncertainty. To give a sense of scale to the uncertainty, for the gravitational attraction to be diminished by 1/2 would require a column of matter with the density of water stretching at least half way from the solar system to the center of the galaxy. The LLR equivalence principle tests give no evidence that mass shields gravity and the limits are very strong.

For completeness, let us mention that Weber [157] argued that a quasi-static shielding could be predicted from an analysis of relativistic tidal phenomena, concluding that such effect should be extremely small. Finally, the most stringent laboratory limit on the gravitational shielding constant had been obtained during the recent measurement of Newton's constant, resulting in $h \leq 4.3 \times 10^{-15}\,\mathrm{m^2\,kg^{-1}}$ [158].

4 The "Dark Side" of Modern Physics

To a worldwide notice, recent cosmological observations dealt us a challenging puzzle forcing us to accept the fact that our current understanding of the origin and evolution of the Universe is incomplete. Surprisingly, it turns out that most of the energy content of the Universe is in the form of the presently unknown dark matter and dark energy that may likely permeate all of spacetime. It is possible that the underlying physics that resolve the discord between quantum mechanics and GR will also shed light on cosmological questions addressing the origin and ultimate destiny of the Universe.

In this section we consider mechanisms that involve new physics beyond GR to explain the puzzling behavior observed at galactic and cosmological scales.

4.1 Cold Dark Matter

The relative importance of the gravitational interaction increases as one considers large scales, and it is at the largest scales where the observed gravitational phenomena do not agree with our expectations. Thus, based on the motion of the peripheral galaxies of the Coma cluster of galaxies, in 1933 Fritz Zwicky found a discrepancy between the value inferred from the total number of galaxies and brightness of the cluster. Specifically, this estimate of the total amount of mass in the cluster revealed the need for about 400 times more mass than expected. This led Zwicky conclude that there is another form of matter in the cluster which, although unaccounted, contains most of the mass responsible for the gravitational stability of the cluster. This nonluminous matter became known as the "dark matter." The dark matter hypothesis was further supported by related problems, namely the differential rotation of our galaxy, as first discussed by Oort in 1927, and the flatness of galactic rotation curves [159].

The most common approach to these problems is to assume the presence of unseen forms of energy that bring into agreement the observed phenomena with GR. The standard scenario to explain the dynamics of galaxies consists in the introduction of an extra weakly interacting massive particle, the so-called cold dark matter (CDM), that clusters at the scales of galaxies and provides the required gravitational pull to hold them together. The explanation of the observed acceleration of the expansion of the Universe requires however the introduction of a more exotic form of energy, not necessarily associated with any form of matter but associated with the existence of space–time itself – vacuum energy.

Although CDM can be regarded as a natural possibility given our knowledge of elementary particle theory, the existence of a nonvanishing but very small vacuum energy remains an unsolved puzzle for our high-energy understanding of physics. However, the CDM hypothesis finds problems when one begins to look at the details of the observations. Increasingly precise simulations of galaxy formation and evolution, although relatively successful in broad terms, show well-known features that seem at odds with their real counterparts, the most prominent of which might be the "cuspy core" problem and the overabundance of substructure seen in the simulations (see, for instance, [160]).

At the same time, the CDM hypothesis is required to explain the correlations of the relative abundances of dark and luminous matter that seem to hold in a very diverse set of astrophysical objects [161]. These correlations are exemplified in the Tully–Fisher law [162] and can be interpreted as pointing to an underlying acceleration scale, $a_0 \simeq 10^{-10}\,\mathrm{m\,s^{-2}}$, below which the Newtonian potential changes and gravity becomes stronger. This is the basic idea of MOND (modified Newtonian dynamics), a successful phenomenological modification of Newton's potential proposed in 1983 [163] whose predictions for the rotation curves of spiral galaxies have been realized with increasing accuracy as the quality of the data has improved [164]. Interestingly, the critical acceleration required by the data is of order $a_0 \sim cH_0$ where H_0 is today's Hubble constant and c the speed of light (that we will set to 1 from now on). The problem with this idea is that MOND is merely a modification of the classical Poisson equation for the Newtonian potential, and therefore inadequate in any situation in which relativistic effects are important. Efforts have been made to obtain MONDian phenomenology in a relativistic generally covariant theory by including other fields in the action with suitable couplings to the spacetime metric [165].

On the other hand, in what concerns the CDM model, one can state that if matter is purely baryonic, early structure formation does not occur, as its temperature and pressure could not account for the latter. The presence of cold (*i.e.*, nonrelativistic) dark matter allows for gravitational collapse and thus solves this issue. Another hint of the existence of exotic dark matter lies in the observation of gravitational lensing, which may be interpreted as due to the presence of undetected clouds of nonluminous matter between the emitting light source and us, which bends the light path due to its mass. This could also be the cause for the discrepancies in the measured Lyman-alpha forest, the spectra of absorption lines of distant galaxies and quasars. The most likely candidates to account for the dark matter include a linear combination of neutral supersymmetrical particles, the neutralinos (see, *e.g.*, [166]), axions [167], self-interacting scalar particles [168], *etc.*

On a broader sense, one can say that these models do not address in a unified way the dark energy (discussed in Sect. 4.2) and dark matter problems, while a common origin is suggested by the observed coincidence between the critical acceleration scale and the dark energy density. This unification feature is found in the so-called generalized Chaplygin gas [169] (see Sect. 4.2).

4.1.1 Modified Gravity as an Alternative to Dark Matter

There are two types of effects in the dark-matter-inspired gravity theories that are responsible for the infrared modification. First, there is an extra scalar excitation of the spacetime metric besides the massless graviton. The mass of this scalar field is of the order of the Hubble scale in vacuum, but its mass depends crucially on the background over which it propagates. This dependence is such that this excitation becomes more massive near the source, and the extra degree of freedom decouples at short distances in the spacetime of a spherically symmetric mass. This feature makes this excitation behave in a way which resembles the chameleon field of [170–172], however, quite often this "chameleon" field is just a component of the spacetime metric coupled to the curvature.

There is also another effect in these theories: the Planck mass that controls the coupling strength of the massless graviton also undergoes a rescaling or "running" with the distance to the sources (or the background curvature). This phenomenon, although a purely classical one in this context, is reminiscent of the quantum renormalization group running of couplings. So one might wonder if MONDian type actions could be an effective classical description of strong renormalization effects in the infrared that might appear in GR [134, 173], as happens in QCD. A phenomenological approach to structure formation bored on these effects has been attempted in [136]. Other implications, such as lensing, cosmic virial theorem, and nucleosynthesis were analyzed in [124, 174, 175]. Additionally, these models offer a phenomenology that seems well suited to describe an infrared strongly coupled phase of gravity and especially at high energies/curvatures when one may use the GR action or its linearization being a good approximation; however, when one approaches low energies/curvatures one finds a nonperturbative regime. At even lower energies/curvatures perturbation theory is again applicable, but the relevant theory is of scalar–tensor type in a de Sitter space.

Clearly, there are many modifications of the proposed class of actions that would offer a similar phenomenology, such that gravity would be modified below a characteristic acceleration scale of the order of the one required in MOND. Many of these theories also offer the unique possibility of being tested not only through astrophysical observations, but also through well-controlled laboratory experiments where the outcome of an experiment is correlated with parameters that can be determined by means of cosmological and astrophysical measurements.

4.2 Dark Energy as Modern Cosmological Puzzle

In 1998, Perlmutter and collaborators [176] and Riess and collaborators [177] have gathered data of the magnitude-redshift relation of Type Ia supernovae with redshifts $z \geq 0.35$ and concluded that it strongly suggests that we live in an accelerating Universe, with a low matter density with about one third of the energy content of the Universe. Currently, there are about 250 supernovae data points which confirm this interpretation. Dark energy is assumed to be a smooth distribution of nonluminous energy uniformly distributed over the Universe to account for the extra dimming of the light of far away Type Ia supernovae, standard candles for cosmological purposes. If there is a real physical field responsible for dark energy, it may be phenomenologically described in terms of an energy density ρ and pressure p, related instantaneously by the equation-of-state parameter $w = p/\rho$. Furthermore, covariant energy conservation would then imply that ρ dilutes as $a^{-3(1+w)}$, with a being the scale factor. Note that $p = w\rho$ is not necessarily the actual equation of state of the dark energy fluid, meaning that perturbations may not generally obey

$\delta p = w\delta\rho$; however, if one were to have such an equation of state, one can define the speed of sound by $c_s^2 = \partial p/\partial\rho$. The implications of this phenomenology would make much more sense in the context of theories proposed to provide the required microscopic description.

4.2.1 Cosmological Constant and Dark Energy

One of the leading explanations for the accelerated expansion of the Universe is the presence of a nonzero cosmological constant. As can be seen from Einstein's equation, the cosmological term can be viewed not as a geometric prior to the spacetime continuum, but instead interpreted as a energy–momentum tensor proportional to the metric, thus enabling the search for the fundamental physics mechanism behind its value and, possibly, its evolution with cosmic time. An outstanding question in today's physics lies in the discrepancy between the observed value for Λ and the prediction arising from quantum field theory, which yields a vacuum energy density about 120 orders of magnitude larger than the former. To match the observed value, requires a yet unknown cancelation mechanism to circumvent the fine tuning of 120 decimal places necessary to account for the observations. This is so as observations require the cosmological constant to be of order of the critical density $\rho_c = 3H_0^2/8\pi G \simeq 10^{-29}\,\mathrm{g\,cm^{-3}}$:

$$\rho_V \equiv \frac{\Lambda}{8\pi G} \simeq 10^{-29}\mathrm{g\ cm^{-3}} \simeq 10^{-12}\mathrm{eV}^4\,, \tag{25}$$

while the natural number to expect from a quantum gravity theory is $M_P^4 \simeq 10^{76}\,\mathrm{GeV}^4$.

Besides the cosmological constant, a slow-varying vacuum energy[6] of some scalar field, usually referred to as "quintessence" [179], and an exotic fluid like the generalized Chaplygin gas [169] are among other the most discussed candidates to account for this dominating contribution for the energy density. It is worth mentioning that the latter possibility allows for a scenario where dark energy and dark matter are unified.

We mention that the presented bounds result from a variety of sources, of which the most significant are the CMBR, high-z supernovae redshifts and galaxy cluster abundances. These joint constraints establish that the amount of dark energy, dark matter, and baryons are, in terms of the critical density, $\Omega_\Lambda \simeq 0.73$, $\Omega_{\mathtt{DM}} = 0.23$, and $\Omega_{\mathtt{Baryons}} = 0.04$, respectively [180].

Current observational constraints imply that the evolution of dark energy is entirely consistent with $w = -1$, characteristic of a cosmological constant (Λ). The cosmological constant was the first, and remains the simplest, theoretical solution to the dark energy observations. The well-known "cosmological constant problem" – why is the vacuum energy so much smaller than we expect from effective-field-theory considerations? – remains, of course, unsolved.

---

[6] For earlier suggestions see [178].

Recently, an alternative mechanism to explain Λ has arisen out of string theory. It was previously widely perceived that string theory would continue in the path of QED and QCD, wherein the theoretical picture contained few parameters and a uniquely defined ground state. However, recent developments have yielded a theoretical horizon in distinct opposition to this, with a "landscape" of possible vacua generated during the compactification of 11 dimensions down to 3 [181]. Given the complexity of the landscape, anthropic arguments have been put forward to determine whether one vacuum is preferred over another. It is possible that further development of the statistics of the vacua distribution, and characterization of any distinctive observational signatures, such as predictions for the other fundamental coupling constants, might help to distinguish preferred vacua and extend beyond the current vacua counting approach.

Although dark energy is the most obvious and popular possibility to the recently observed acceleration of the Universe, other competing ideas have been investigated, and among them is modifications of gravity on cosmological scales. Indeed, as we discussed earlier, GR is well tested in the solar system, in measurements of the period of binary pulsars, and in the early Universe, via primordial nucleosynthesis. None of these tests, however, probe the ultra-large length scales and low curvatures characteristic of the Hubble radius today. Therefore, one can potentially think that gravity is modified in the very far infrared allowing the Universe to accelerate at late times.

In this section, we discuss some of the gravity modification proposals suggested to provide a description of the observed acceleration of the Universe.

4.2.2 Modified Gravity as an Alternative to Dark Energy

A straightforward possibility is to modify the usual Einstein–Hilbert action by adding terms that blow up as the scalar curvature goes to zero [182,183]. Recently, models involving inverse powers of the curvature have been proposed as an alternative to dark energy [183,184]. In these models, one generically has more propagating degrees of freedom in the gravitational sector than the two contained in the massless graviton in GR. The simplest models of this kind add inverse powers of the scalar curvature to the action ($\Delta\mathcal{L} \propto 1/R^n$), thereby introducing a new scalar excitation in the spectrum. For the values of the parameters required to explain the acceleration of the Universe this scalar field is almost massless in vacuum and one might worry about the presence of a new force contradicting solar system experiments. However, it can be shown that models that involve inverse powers of other invariant, in particular those that diverge for $r \to 0$ in the Schwarzschild solution, generically recover an acceptable weak-field limit at short distances from sources by means of a screening or shielding of the extra degrees of freedom at short distances [185]. Such theories can account for late-time acceleration, but unfortunately typically lead to one of two problems. Either they are in conflict with tests of GR in the solar system, due to the existence of additional dynamical degrees of freedom [186], or they contain ghost-like degrees of freedom that seem difficult to reconcile with fundamental theories. The search is ongoing for versions of this idea that are consistent with experiment.

A more dramatic approach would be to imagine that we live on a brane embedded in a large extra dimension. Although such theories can lead to perfectly conventional gravity on large scales, it is also possible to choose the dynamics in such a way that new effects show up exclusively in the far infrared. An example is the Dvali–Gabadadze–Porrati (DGP) braneworld model, in which the strength of gravity in the bulk is substantially less than that on the brane [187]. Such theories can naturally lead to late-time acceleration [188, 189], but may have difficulties with strong-coupling issues [190]. Furthermore, the DGP model does not properly account for the supernova data, as does its generalization, the Dvali–Turner model, and also other ad hoc modifications of the Friedmann equation, the so-called Cardassian model [191]. Most interestingly, however, DGP gravity and other modifications of GR hold out the possibility of having interesting and testable predictions that distinguish them from models of dynamical dark energy. One outcome of this work is that the physics of the accelerating Universe may be deeply tied to the properties of gravity on relatively short scales, from millimeters to astronomical units.

4.2.3 Scalar Field Models as Candidate for Dark Energy

One of the simplest candidates for dynamical dark energy is a scalar field, φ, with an extremely low-mass and an effective potential, $V(\varphi)$, as shown by (16) [127]. If the field is rolling slowly, its persistent potential energy is responsible for creating the late epoch of inflation we observe today. For the models that include only inverse powers of the curvature, besides the Einstein–Hilbert term, it is however possible that in regions where the curvature is large the scalar has naturally a large mass and this could make the dynamics to be similar to those of GR [192]. At the same time, the scalar curvature, while being larger than its mean cosmological value, is still very small in the solar system (to satisfy the available results of gravitational tests). Although a rigorous quantitative analysis of the predictions of these models for the tests in the solar system is still noticeably missing in the literature, it is not clear whether these models may be regarded as a viable alternative to dark energy.

Effective scalar fields are prevalent in supersymmetric field theories and string/M-theory. For example, string theory predicts that the vacuum expectation value of a scalar field, the dilaton, determines the relationship between the gauge and gravitational couplings. A general, low-energy effective action for the massless modes of the dilaton can be cast as a scalar–tensor theory as (16) with a vanishing potential, where $f(\varphi)$, $g(\varphi)$, and $q_i(\varphi)$ are the dilatonic couplings to gravity, the scalar kinetic term, and gauge and matter fields, respectively, encoding the effects of loop effects and potentially nonperturbative corrections.

A string-scale cosmological constant or exponential dilaton potential in the string frame translates into an exponential potential in the Einstein frame. Such quintessence potentials [193, 194] can have scaling [195], and tracking [196] properties that allow the scalar field energy density to evolve alongside the other matter constituents. A problematic feature of scaling potentials [195] is that they do not lead to accelerative expansion, since the energy density simply scales with that of matter. Alternatively, certain potentials can predict a dark energy density which alternately dominates the Universe and decays away; in such models, the acceleration of the Universe is transient [197–199]. Collectively, quintessence potentials predict that the density of the dark energy dynamically evolves in time, in contrast to the cosmological constant. Similar to a cosmological constant, however, the scalar field is expected to have no significant density perturbations within the causal horizon, so that they contribute little to the evolution of the clustering of matter in large-scale structure [200].

In addition to couplings to ordinary matter, the quintessence field may have nontrivial couplings to dark matter [201, 202]. Nonperturbative string-loop effects do not lead to universal couplings, with the possibility that the dilaton decouples more slowly from dark matter than it does from gravity and fermions. This coupling can provide a mechanism to generate acceleration, with a scaling potential, while also being consistent with equivalence principle tests. It can also explain why acceleration is occurring only recently, through being triggered by the nonminimal coupling to the CDM, rather than a feature in the effective potential [203,204]. Such couplings cannot only generate acceleration, but also modify structure formation through the coupling to CDM density fluctuations [205], in contrast to minimally coupled quintessence models. Dynamical observables, sensitive to the evolution in matter perturbations as well as the expansion of the Universe, such as the matter power spectrum as measured by large-scale surveys and weak lensing convergence spectra, could distinguish nonminimal couplings from theories with minimal effect on clustering. The interaction between dark energy and dark matter is, of course, present in the generalized Chaplygin gas model, as in this proposal the fluid has a dual behavior.

It should be noted that for the runaway-dilaton scenario presented in Sect. 3.1, comparison with the minimally coupled scalar field action,

$$S_\phi = \frac{c^3}{4\pi G}\int d^4x\sqrt{-g}\left[\frac{1}{4}R + \frac{1}{2}\partial_\mu\phi\partial^\mu\phi - V(\phi)\right], \tag{26}$$

reveals that the negative scalar kinetic term leads to an action equivalent to a "ghost" in quantum field theory, and is referred to as "phantom energy" in the cosmological context [206]. Such a scalar field model could in theory generate acceleration by the field evolving *up* the potential toward the maximum. Phantom fields are plagued by catastrophic UV instabilities, as particle excitations have a negative mass [207,208]; the fact that their energy is unbounded from below allows vacuum decay through the production of high-energy real particles and negative energy ghosts that will be in contradiction with the constraints on ultra-high-energy cosmic rays [209].

Such runaway behavior can potentially be avoided by the introduction of higher-order kinetic terms in the action. One implementation of this idea is "ghost condensation" [210]: here, the scalar field has a negative kinetic energy near $\dot{\phi} = 0$, but the quantum instabilities are stabilized by the addition of higher-order corrections to the scalar field Lagrangian of the form $(\partial_\mu \phi \partial^\mu \phi)^2$. The "ghost" energy is then bounded from below, and stable evolution of the dilaton occurs with $w \geq -1$ [211]. The gradient $\partial_\mu \phi$ is nonvanishing in the vacuum, violating Lorentz invariance, and may have interesting consequences in cosmology and laboratory experiments.

In proposing the scalar field as physical and requiring it to replace CDM and DE, one has to also calculate how the scalar field density fluctuations evolve, to compare them with density power spectra from large-scale structure surveys. This is true for the broader set of phenomenological models including the generalized Born–Infeld action, associated to the generalized Chaplygin gas model [169]. Despite being consistent with kinematical observations, it has been pointed that they are disfavored in comparison with the ΛCDM scenario [212, 213], even though solutions have been proposed [214].

5 Gravitational Physics and Experiments in Space

Recent developments in observation astronomy, astrophysics and cosmology have raised important questions related to gravity and other fundamental laws of nature. There are two approaches to physics research in space: one can detect and study signals from remote astrophysical objects, whereas the other relies on a carefully designed experiment. Although the two methods are complementary, the latter has the advantage of utilizing a well-understood and controlled laboratory environment in the solar system. Newly available technologies in conjunction with existing space capabilities offer unique opportunities to take full advantage of the variable gravity potentials, large heliocentric distances, and high velocity and acceleration regimes that are present in the solar system. As a result, solar system experiments can significantly advance our knowledge of fundamental physics and are capable of providing the missing links connecting quarks to the cosmos.

In this section, we discuss theoretical motivation of and innovative ideas for the advanced gravitational space experiments.

5.1 Testable Implications of Recent Theoretical Proposals

The theories that were discussed in the previous section offer a diverse range of characteristic experimental predictions differing from those of GR that would allow their falsification. The most obvious tests would come from the comparison of the predictions of the theory to astrophysical and cosmological observations where the dynamics are dominated by very small gravitational fields. As a result, one might expect that these mechanisms would lead to small effects in the motion of the bodies in the solar system, short- and long-scale modifications of Newton's law, as well as astrophysical phenomena.

In the following sections, we discuss these possible tests and estimate the sizes of the expected effects.

5.1.1 Testing Newton's Law at Short Distances

It was observed that many recent theories predict observable experimental signatures in experiments testing Newton's law at short distances. For instance in the case of MOND-inspired theories discussed in Sect. 4.1.1, there may be an extra scalar excitation of the spacetime metric besides the massless graviton. Thus, in the effective gravitational theory applicable to the terrestrial conditions, besides the massless spin two graviton, one would also have an extra scalar field with gravitational couplings and with a small mass. A peculiar feature of such a local effective theory on a Schwarzschild background is that there will be a preferred direction that will be reflected in an anisotropy of the force that this scalar excitation will mediate. For an experiment conducted in the terrestrial conditions one expects short-range modifications of Newton's law at distances of $\sim$0.1 mm, regime that is close to that already being explored in some laboratory experiments [215, 216].

For an experiment on an Earth-orbiting platform, one explores another interesting regime for which the solar mass and the Sun–Earth distance are the dominant factors in estimating the size of the effects. In this case, the range of interest is $\sim 10^4$ m. However, in measuring the gravitational field of an object one has to measure this field at a distance that is larger than the critical distance for which the self-shielding of the extra scalar excitation induced by the object itself is enough to switch off the modification. This means that, for an experiment in the inner solar system, we could only see significant modifications in the gravitational field of objects whose characteristic distance is smaller than 10^4 m, thus limiting the mass of the body to be below $\sim 10^9$ kg. As an example, one can place an object with mass of 10^3 kg placed on a heliocentric orbit at $\sim$1 AU distance. For this situation, one may expect modifications of the body's gravitational field at distances within the range of ~ 10–10^4 m. Note that at shorter distances the scalar effectively decouples because of the self-gravitational effect of the test object; also, at longer distances the mass induced by solar gravitational field effectively decouples the scalar.

5.1.2 Solar System Tests of Relativistic Gravity

Although many effects expected by gravity modification models are suppressed within the solar system, there are measurable effects induced by some long-distance modifications of gravity (notably the DGP model [187]). For instance, in the case of the precession of the planetary perihelion in the solar system, the anomalous perihelion advance, $\Delta\phi$, induced by a small correction, δV_N, to Newton's potential, V_N, is given in radians per revolution [217] by

$$\Delta\phi \simeq \pi r \frac{d}{dr}\left(r^2 \frac{d}{dr}\left(\frac{\delta V_N}{r V_N}\right)\right). \tag{27}$$

The most reliable data regarding the planetary perihelion advances come from the inner planets of the solar system [218], where most of the corrections are negligible. However, with its excellent 2-cm-level range accuracy [14], LLR offers an interesting possibility to test for these new effects. Evaluating the expected magnitude of the effect to the Earth–Moon system, one predicts an anomalous shift of $\Delta\phi \sim 10^{-12}$, to be compared with the achieved accuracy of 2.4×10^{-11} [217]. Therefore, the theories of gravity modification result in an intriguing possibility of discovering new physics, if one focuses on achieving higher precision in modern astrometrical measurements; this accuracy increase is within the reach and should be attempted in the near future.

The quintessence models discussed in Sect. 4.2.3 offer the possibility of observable couplings to ordinary matter makes these models especially attractive for the tests even on the scales of the solar system. Even if we restrict attention to nonrenormalizable couplings suppressed by the Planck scale, tests from fifth-force experiments and time dependence of the fine-structure constant imply that such interactions must be several orders of magnitude less than expected [219]. Further improvement of existing limits on violations of the equivalence principle in terrestrial experiments and also in space would also provide important constraints on dark energy models.

Another interesting experimental possibility is provided by the "chameleon" effect [170,172]. Thus, by coupling to the baryon energy density, the scalar field value can vary across space from solar system to cosmological scales. Though the small variation of the coupling on Earth satisfies the existing terrestrial experimental bounds, future gravitational experiments in space such as measurements of variations in the gravitational constant or test of equivalence principle may provide critical information for the theory.

There is also a possibility that the dynamics of the quintessence field evolves to a point of minimal coupling to matter. In [52] it was shown that ϕ could be attracted toward a value $\phi_m(x)$ during the matter-dominated era that decoupled the dilaton from matter. For universal coupling, $f(\varphi) = g(\varphi) = q_i(\varphi)$ (see (16)), this would motivate for improving the accuracy of the equivalence principle and other tests of GR. Veneziano [220] suggested that with a large number of non-self-interacting matter species, the coupling constants are determined by the quantum corrections of the matter species, and ϕ would evolve as a runaway dilaton with asymptotic value $\phi_m \to \infty$. More recently [138], the quantity $\frac{1}{2}(1-\gamma)$ has been estimated, within the framework compatible with string theory and modern cosmology, which basically confirms the previous result [38]. This recent analysis discusses a scenario where a composition-independent coupling of dilaton to hadronic matter produces detectable deviations from GR in high-accuracy light deflection experiments in the solar system. This work assumes only some general property of the coupling functions and then only assumes that $(1-\gamma)$ is of order of one at the beginning of the controllably classical part of inflation. It was shown in [138] that one can relate the present value of $\frac{1}{2}(1-\gamma)$ to the cosmological density fluctuations; the level of the expected deviations from GR is $\sim 0.5\times 10^{-7}$ [138]. Note that these predictions are based on the work in scalar–tensor extensions of gravity which are consistent with, and indeed often part of, the present cosmological models and provide a strong motivation for improvement of the accuracy of gravitational tests in the solar system.

5.1.3 Observations on Astrophysical and Cosmological Scales

These new theories also suggest interesting effects on astrophysical and cosmological observations (see for instance [221]). In this respect, one can make unambiguous predictions for the rotation curves of spiral galaxies with the mass-to-light ratio being the only free parameter. Specifically, it has been argued that a skew-symmetric field with a suitable potential could account for galaxy and cluster rotation curves [222]. One can even choose an appropriate potential that would then give rise to flat rotation curves that obey the Tully–Fisher law [162]. But, also other aspects of the observations of galactic dynamics can be used to constrain a MOND-like modification of Newton's potential (see [223]). Also, note that such a theory violates the SEP, as expected for any relativistic theory for MOND [163], since locally physics will intrinsically depend on the background gravitational field. This will be the case if the background curvature dominates the curvature induced by the local system, similar to the "external field effect" in MOND.

At larger scales, where one can use the equivalence with a scalar–tensor theory more reliably, one can then compare the theory against the observations of gravitational lensing in clusters, the growth of large-scale structure, and the fluctuations of the CMBR. In fact, it has been pointed out that if GR was modified at large distances, an inconsistency between the allowed regions of parameter space would allow for dark energy models verification when comparing the bounds on these parameters obtained from CMBR and large-scale structure [224]. This means that although some cosmological observables, like the expansion history of the Universe, can be indistinguishable in modified gravity and dark energy models, this degeneracy is broken when considering other cosmological observations and in particular the growth of large-scale structure and the integrated Sachs–Wolfe effect (ISW) have been shown to be good discriminators for models in which GR is modified [225]. It has been recently pointed out that the fact that in the DGP model the effective Newton's constant increases at late times as the background curvature diminishes causes a suppression of the ISW that brings the theory into better agreement with the CMBR data than the ΛCDM model [226].

5.2 New Experiments and Missions

Theoretical motivations presented above have stimulated development of several highly accurate space experiments. Below we will briefly discuss science objectives and experimental design for several advanced experiments, namely MICROSCOPE, STEP, and HYPER missions, APOLLO LLR facility, and the LATOR mission.

5.2.1 MICROSCOPE, STEP, and HYPER Missions

Ground experiments designed to verify the validity of the WEP are limited by unavoidable microseismic activity of Earth, while the stability of space experiments offers an improvement in the precision of current tests by a factor of 10^6. Most probably, the first test of the WEP in space will be carried out by the MICROSCOPE (MICROSatellite a traine Compensee pour l'Observation du Principe d'Equivalence) mission led by CNES and ESA. The drag-free MICROSCOPE satellite, transporting two pairs of test masses, will be launched into a Sun-synchronous orbit at 600 km altitude. The differential displacements between each test masses of a pair will be measured by capacitive sensors at room temperature, with an expected precision of one part in 10^{15}.

The more ambitious joint ESA/NASA STEP (satellite test of the equivalence principle) mission is proposed to be launched in the near future into a circular, Sun-synchronous orbit with altitude of 600 km. The drag-free STEP spacecraft will carry four pairs of test masses stored in a dewar of superfluid He at a 2 K temperature. Differential displacements between the test masses of a pair will be measured by SQUID (superconducting quantum interference device) sensors, testing the WEP with an expected precision of $\Delta a/a \sim 10^{18}$.

Another quite interesting test of the WEP involves atomic interferometry: high-precision gravimetric measurements can be taken via the interferometry of free-falling cesium atoms, and such a concept has already yielded a precision of seven parts per 10^9 [227]. This can only be dramatically improved in space, through a mission like HYPER (HYPER-precision cold atom interferometry in space). ESA's HYPER spacecraft would be in a Sun-synchronous circular orbit at 700 km altitude. Two atomic Sagnac units are to be accommodated in the spacecraft, comprising four cold atom interferometers that are able to measure rotations and accelerations along two orthogonal planes. By comparing the rates of fall of cesium and rubidium atoms, the resolution of the atom interferometers of the HYPER experiment could, in principle, test the WEP with a precision of one part in 10^{15} or 10^{16} [228].

It is worth mentioning that proposals have been advanced to test the WEP by comparing the rate of fall of protons and antiprotons in a cryogenic vacuum facility that will be available at the ISS [229]. The concept behind this weak equivalence antimatter experiment (WEAX) consists of confining antiprotons for a few weeks in a Penning trap, in a geometry such that gravity would produce a perturbation on the motion of the antiprotons. The expected precision of the experiment is of one part in 10^6, three orders of magnitude better than for a ground experiment.

It is clear that testing the WEP in space requires pushing current technology to the limit; even though no significant violations of this principle are expected, any anomaly would provide significant insight into new and fundamental physical theories. The broad perspectives and the potential impact of testing fundamental physics in space were discussed in [230].

5.2.2 APOLLO: A mm-Class LLR Facility

The Apache point observatory lunar laser-ranging operation is a new LLR effort designed to achieve millimeter-range precision and corresponding order-of-magnitude gains in measurements of fundamental physics parameters. The APOLLO project design and leadership responsibilities are shared between the University of California at San Diego and the University of Washington. In addition to the modeling aspects related to this new LLR facility, a brief description of APOLLO and associated expectations is provided here for reference. A more complete description can be found in [231].

The principal technologies implemented by APOLLO include a robust Nd:YAG laser with 100 ps pulse width, a GPS-slaved 50 MHz frequency standard and clock, a 25-ps-resolution time interval counter, and an integrated avalanche photodiode (APD) array. The APD array, developed at Lincoln Labs, is a new technology that will allow multiple simultaneous photons to be individually time tagged, and provide two-dimensional spatial information for real-time acquisition and tracking capabilities.

The overwhelming advantage APOLLO has over current LLR operations is a 3.5 m astronomical quality telescope at a good site. The site in the Sacramento Mountains of southern New Mexico offers high altitude (2,780 m) and very good atmospheric "seeing" and image quality, with a median image resolution of 1.1 arcsec. Both the image sharpness and large aperture enable the APOLLO instrument to deliver more photons onto the lunar retroreflector and receive more of the photons returning from the reflectors, respectively. Compared with current operations that receive, on average, fewer than 0.01 photons per pulse, APOLLO should be well into the multiphoton regime, with perhaps 5–10 return photons per pulse. With this signal rate, APOLLO will be efficient at finding and tracking the lunar return, yielding hundreds of times more photons in an observation than current operations deliver. In addition to the significant reduction in statistical error ($\sim\sqrt{N}$ reduction), the high-signal rate will allow assessment and elimination of systematic errors in a way not currently possible.

The new LLR capabilities offered by the newly developed APOLLO instrument offer a unique opportunity to improve accuracy of a number of fundamental physics tests. The APOLLO project will push LLR into the regime of millimetric range precision which translates to an order-of-magnitude improvement in the determination of fundamental physics parameters. For the

Earth and Moon orbiting the Sun, the scale of relativistic effects is set by the ratio $(GM/rc^2) \sim v^2/c^2 \sim 10^{-8}$. Relativistic effects are small compared with Newtonian effects. The APOLLO's 1-mm-range accuracy corresponds to 3×10^{-12} of the Earth–Moon distance. The resulting LLR tests of gravitational physics would improve by an order of magnitude: the equivalence principle would give uncertainties approaching 10^{-14}, tests of GR effects would be <0.1%, and estimates of the relative change in the gravitational constant would be 0.1% of the inverse age of the Universe. This last number is impressive considering that the expansion rate of the Universe is approximately one part in $10^{10}\,\text{year}^{-1}$.

Therefore, the gain in our ability to conduct even more precise tests of fundamental physics is enormous, thus this new instrument stimulates development of better and more accurate models for the LLR data analysis at an mm-level [232].

5.2.3 The LATOR Mission

The recently proposed LATOR [121, 233–235] is an experiment designed to test the metric nature of gravitation – a fundamental postulate of Einstein's theory of general relativity. By using a combination of independent time series of highly accurate gravitational deflection of light in the immediate proximity to the Sun, along with measurements of the Shapiro time delay on interplanetary scales (to a precision, respectively, better than 10^{-13} rad and 1 cm), LATOR will significantly improve our knowledge of relativistic gravity. The primary mission objective is to (1) measure the key post-Newtonian Eddington parameter γ with accuracy of a part in 10^9. The quantity $(1-\gamma)$ is a direct measure for presence of a new interaction in gravitational theory, and, in its search, LATOR goes a factor 30,000 beyond the present best result, Cassini's 2003 test. Other mission objectives include (2) first measurement of gravity's nonlinear effects on light to ∼0.01% accuracy; including both the traditional Eddington β parameter via gravity effect on light to ∼0.01% accuracy and also the spatial metric's second-order potential contribution δ (never measured before); (3) direct measurement of the solar quadrupole moment J_2 (currently unavailable) to accuracy of a part in 200 of its expected size; and (4) direct measurement of the "frame-dragging" effect on light due to the Sun's rotational gravitomagnetic field, to 0.1% accuracy. LATOR's primary measurement pushes to unprecedented accuracy the search for cosmologically relevant scalar–tensor theories of gravity by looking for a remnant scalar field in today's solar system. The key element of LATOR is a geometric redundancy provided by the laser ranging and long-baseline optical interferometry.

As a result, LATOR will be able to test the metric nature of the Einstein's general theory of relativity in the most intense gravitational environment available in the solar system – the extreme proximity to the Sun. It will also test alternative theories of gravity and cosmology, notably scalar–tensor theories, by searching for cosmological remnants of scalar field in the solar system.

LATOR will lead to very robust advances in the tests of fundamental physics: this mission could discover a violation or extension of GR, or reveal the presence of an additional long-range interaction in the physical law. There are no analogues to the LATOR experiment; it is unique and is a natural culmination of solar system gravity experiments [121].

LATOR mission is the twenty-first-century version of Michelson–Morley-type experiment searching for a cosmologically evolved scalar field in the solar system. In spite of the previous space missions exploiting radiowaves for tracking the spacecraft, this mission manifests an actual breakthrough in the relativistic gravity experiments as it allows one to take full advantage of the optical techniques that recently became available. LATOR has a number of advantages over techniques that use radiowaves to measure gravitational light deflection. Thus, optical technologies allow low-bandwidth telecommunications with the LATOR spacecraft. The use of the monochromatic light enables the observation of the spacecraft at the limb of the Sun. The use of narrowband filters, coronagraph optics, and heterodyne detection will suppress background light to a level where the solar background is no longer the dominant noise source. The short wavelength allows much more efficient links with smaller apertures, thereby eliminating the need for a deployable antenna. Finally, the use of the ISS enables the test above the Earth's atmosphere – the major source of astrometric noise for any ground-based interferometer. This fact justifies LATOR as a space mission. LATOR is envisaged as a partnership between European and US institutions and with clear areas of responsibility between the space agencies: NASA provides the deep-space mission components, whereas optical infrastructure on the ISS would be an ESA contribution.

6 Conclusions

General theory of relativity is one of the most elegant theories of physics; it is also one of the most empirically verified theories. Thus, almost 90 years of testing have also proved that GR has so far successfully accounted for all encountered phenomena and experiments in the solar system and with binary pulsars. However, the predictions of the theory require still confirmation and detailed analysis, most notably the direct detection of gravitational waves. However, there are new motivations to test the theory to even a higher precisions that already led to a number of experimental proposals to advance the knowledge of fundamental laws of physics.

Recent progress in observational astronomy, astrophysics, and cosmology has raised important questions related to gravity and other fundamental laws of nature. There are two approaches to physics research in space: one can detect and study signals from remote astrophysical objects, whereas the other relies on a carefully designed experiment. Although the two methods are complementary, the latter has the advantage of utilizing a well-understood and controlled laboratory environment in the solar system.

Newly available technologies in conjunction with existing space capabilities offer unique opportunities to take full advantage of the variable gravity potentials, large heliocentric distances, and high velocity and acceleration regimes that are present in the solar system. A common feature of precision gravity experiments is that they must operate in the noise-free environment needed to achieve the ever-increasing accuracy. These requirements are supported by the progress in the technologies, critical for space exploration, namely the highly stable, high-powered, and space-qualified lasers, highly accurate frequency standards, and the drag-free technologies. This progress advances both the science and technology for the laboratory experiments in space with laboratory being the entire solar system. As a result, solar system experiments can significantly advance our knowledge of fundamental physics and are capable of providing the missing links connecting quarks to the cosmos.

To conclude, it is our hope that the recent progress will lead to establishing a more encompassing theory to describe all physical interactions in an unified fashion that harmonizes the spacetime description of GR with quantum mechanics. This unified theory is needed to address many of the standing difficulties we face in theoretical physics: Are singularities an unavoidable property of spacetime? What is the origin of our Universe? How to circumvent the cosmological constant problem and achieve a successful period of inflation and save our Universe from an embarrassing set of initial conditions? The answer to these questions is, of course, closely related to the nature of gravity. It is an exciting prospect to think that experiments carried out in space will be the first to provide the essential insights on the brave new world of the new theories to come.

Acknowledgment

The work of SGT described was carried out at the Jet Propulsion Laboratory, California Institute of Technology, under a contract with the National Aeronautics and Space Administration.

References

1. I.I. Shapiro, C.C. Counselman, III, and R.W. King, *Phys. Rev. Lett.* **36**, 555 (1976).
2. R.D. Reasenberg *et al.*, *Astrophys. J. Lett.* **234**, 4329 (1979) L219.
3. I.I. Shapiro *et al.*, *Journ. Geophys. Res.* **82** (1977).
4. J.D. Anderson, E.L. Lau, S.G. Turyshev, J.G. Williams, and M.M. Nieto, *Bull. Amer. Astron. Soc.* **34**, 833 (2002).
5. D.S. Robertson, W.E. Carter, and W.H. Dillinger, *Nature* **349**, 768 (1991).
6. D.E. Lebach, B.E. Corey, I.I. Shapiro, M.I. Ratner, J.C. Webber, A.E.E. Rogers, J.L. Davis and T.A. Herring, *Phys. Rev. Lett.* **75**, 1439 (1995).

7. S.S. Shapiro, J.L. Davis, D.E. Lebach, and J.S. Gregory, *Phys. Rev. Lett.* **92**, 121101 (2004).
8. K.L. Nordtvedt, Jr., *Phys. Rev.* **170**, 1186 (1968).
9. K.L. Nordtvedt, Jr., *Phys. Rev.* **D 43**, 10 (1991).
10. K.L. Nordtvedt, Jr., *Class. Quantum Gravity* **16**, A101 (1999).
11. K.L. Nordtvedt, Jr., gr-qc/0301024.
12. J.G. Williams, X.X. Newhall, and J.O. Dickey, *Phys. Rev.* **D 53**, 6730 (1996).
13. J.G. Williams, J.D. Anderson, D.H. Boggs, E.L. Lau, and J.O. Dickey, *Bull. Amer. Astron. Soc* **33**, 836 (2001).
14. J.G. Williams, S.G. Turyshev, and D.H. Boggs, *Phys. Rev. Lett.* **93**, 261101 (2004).
15. B. Bertotti, L. Iess, and P. Tortora, *Nature* **425**, 374 (2003).
16. T. Damour, J.H. Taylor, *Phys. Rev.* **D 45**, 1840 (1992).
17. J.H. Taylor, A. Wolszczan, T. Damour, and J.M. Weisberg, *Nature* **355**, 132 (1992).
18. T. Damour, G. Esposito-Farese, *Phys. Rev.* **D 53**, 5541 (1996).
19. T. Damour, G. Esposito-Farese, *Phys. Rev.* **D 54**, 1474 (1996).
20. T. Damour, G. Esposito-Farese, *Phys. Rev.* **D 58**, 042001 (1998).
21. Ch., Lange, F. Camilo, N. Wex, M. Kramer, D.C. Backer, A.G. Lyne, and O. Doroshenko, *Mon. Not. R. Ast. Soc.* **326**, 274 (2001).
22. V.A. Fock, *J. Phys.* **1**, 8 (1939)1; *Rev. Mod. Phys.* **29**, 325 (1957).
23. V.A. Fock, *The Theory of Space, Time and Gravitation.* Fizmatgiz. Moscow (in Russian), (1955). (English translation (1959), Pergamon, Oxford)
24. S. Chandrasekhar, *Astrophys. J.* **142**, 1488 (1965).
25. V.A. Brumberg, *Relativistic Celestial Mechanics.* Nauka: Moscow. (1972) (in Russian); V.A. Brumberg, *Essential Relativistic Celestial Mechanics.* Hilger, Bristol (1991).
26. C.M. Will, *Theory and Experiment in Gravitational Physics*, (Cambridge University Press 1993).
27. T. Damour, M. Soffel and C. Xu, *Phys. Rev.* **D 43**, 3273 (1991); *Phys. Rev.* **D 47**, 3436 (1993); *Phys. Rev.* **D 49**, 618 (1994).
28. L. Blanchet, T. Damour, B.R. Iyer, C.M. Will, and A.G. Wisemann, *Phys. Rev. Lett.* **74**, 3515 (1995).
29. T. Damour and D. Vokrouhlický, *Phys. Rev.* **D 52**, 4455 (1995).
30. L. Blanchet and T. Damour, *Phil. Trans. Roy. Soc. London* **A 320**, 379 (1986); *Ann. Inst. Henri Poincaré* **50**, 77 (1989).
31. S.M. Kopeikin, I. Vlasov, *Phys. Rept.* **400**, 209 (2004) [gr-qc/0403068].
32. K.L. Nordtvedt, Jr., *Phys. Rev.* **169**, 1014 (1968).
33. C.M. Will, *Astrophys. J.* **169**, 125 (1971).
34. C.M. Will and K.L. Nordtvedt, Jr., *Astrophys. J.* **177**, 757 (1972).
35. C.M. Will, *Was Einstein Right? Testing Relativity at the Centenary*, to be published in "100 Years of Relativity: Spacetime Structure - Einstein and Beyond," ed. Abhay Ashtekar (World Scientific, Singapore), gr-qc/0504086.
36. S.G. Turyshev, *Relativistic Navigation: A Theoretical Foundation.* JPL Publication #96-13. Pasadena, CA (1996), gr-qc/9606063.
37. C.H. Brans and R.H. Dicke, *Phys. Rev.* **124**, 925 (1961).
38. T. Damour and K. Nordtvedt, Jr., *Phys. Rev. Lett.* **70**, 2217 (1993); *Phys. Rev.* **D 48**, 3436 (1993).

39. T.D. Moyer, *Mathematical Formulation of the Double Precision Orbit Determination Program (DPODP)*, Jet Propulsion Laboratory Technical Report 32-1527, Pasadena, CA, (1971).
40. T.D. Moyer, *Celest. Mech.* **23**, 33 (1981); *Celest. Mech.* **23**, 57 (1981).
41. E.M. Standish, X.X. Newhall, J.G. Williams, and D.K. Yeomans, *Orbital ephemerides of the Sun, Moon, and planets.* In: "Explanatory Supplement to the Astronomical Almanac," ed. P.K. Seidelmann, (University Science Books, Mill Valley 1992), p. 279.
42. S.G. Turyshev, J.G. Williams, K. Nordtvedt Jr., M. Shao, and T.W. Murphy Jr., *Lect. Notes Phys.* **648**, 301 (2004) [gr-qc/0311039].
43. J.L. Singe, *Relativity: the General Theory,* (Amsterdam: North-Holland, 1960).
44. J.D. Anderson, M. Gross, K.L. Nordtvedt, and S.G. Turyshev, *Astrophys. J.* **459**, 365 (1996) [gr-qc/9510029].
45. R.V. Pound and G.A. Rebka Jr., *Phys. Rev. Lett.* **3**, 439 (1959); R.V. Pound and G.A. Rebka Jr., *Phys. Rev. Lett.* **4**, 337 (1960); R.V. Pound and J.L. Snider, *Phys. Rev. Lett.* **13**, 539 (1964).
46. M.M. Nieto and T. Goldman, *Phys. Rep.* **205**, 221 (1991).
47. M. Amoretti *et al.*, *Nature* **419**, 456 (2002).
48. G. Gabrielse *et al.*, *Phys. Rev. Lett.* **89**, 213401 (2002).
49. A. Apostolakis *et al.* (CPLEAR Collaboration), *Phys. Lett.* **B 452**, 425 (1999).
50. T. Damour, *Class. Quantum Gravity* **13**, A33 (1996).
51. T. Damour, In Proc. of ONERA workshop on "Missions spatiales en physique fondamentale" (Chatillon, 18-19 Jan 2001), a special issue of the Comptes Rendus de l'Academie des Sciences (Paris), edited by C. Bordé and P. Touboul, gr-qc/0109063.
52. T. Damour and A.M. Polyakov, *Gen. Rel. Grav.* **26** (1994) 1171; *Nucl. Phys.* **B 423** (1994) 532.
53. R. Collela, A.W. Overhauser, and S.A. Werner, *Phys. Rev. Lett.* **34** (1975) 1472.
54. M. Kasevich and S. Chu, *Appl. Phys.* **B 44** (1992) 321.
55. P.G. Roll, R. Krotkov, and R.H. Dicke, *Ann. Phys.* (N.Y.) **26**, 442 (1964).
56. V.B. Braginsky and V.I. Panov, *Zh. Eksp. Teor. Fiz.* **61**, 873 (1971); [*Sov. Phys. JETP* **34**, 463 (1972)].
57. Y. Su, R. Heckel, E.G. Adelberger, J.H. Gundlach, M. Harris, G.L. Smith, and H.E. Swanson, *Phys. Rev.* **D 50**, 3614 (1994).
58. E.G. Adelberger, *Class. Quantum Gravity* **18** (2001) 2397.
59. J.H. Gundlach, *New J. Phys.* **7**, 205 (2005).
60. V.V. Nesvizhevsky *et al.*, *Nature* **415**, 297 (2002).
61. V.I. Luschikov and A.I. Frank, *JEPT* **28**, 559 (1978).
62. V.V. Nesvizhevsky *et al.*, *Nucl. Instruments and Methods in Physics Research* **A 440**, 754 (2000).
63. O. Bertolami, J.G. Rosa, C.M.L. Aragão, P. Castorina, and D. Zappalà, *Phys. Rev.* **D 72**, 025010 (2005).
64. O. Bertolami and J.G. Rosa, *Phys. Lett.* **633**, 111 (2006).
65. S. Baessler, B. Heckel, E.G. Adelberger, J. Gundlach, U. Schmidt, and E. Swanson, *Phys. Rev. Lett.* **83**, 3585 (1999).
66. K.L. Nordtvedt, Jr., *Phys. Rev.* **169**, 1017 (1968).
67. K.L. Nordtvedt, Jr., *Phys. Rev.* **D 43**, 3131 (1991).
68. C.M. Will, *Living Rev. Rel.* **4**, 4 (2001), and recent update gr-qc/0510072.

69. R.K. Ulrich, *Astrophys. J.* **258**, 404 (1982).
70. K.L. Nordtvedt, Jr., *Icarus* **12**, 91 (1970).
71. R.B. Orellana and H. Vucetich, *Astron. Astroph.* **200**, 248 (1988).
72. R.B. Orellana and H. Vucetich, *Astron. Astroph.* **273**, 313 (1993).
73. J.D. Anderson and J.G. Williams, *Class. Quantum Gravity* **18**, 2447 (2001).
74. T. Damour and G. Schäfer, *Phys. Rev. Lett.* **66**, 2549 (1991).
75. N. Wex, In Proc. "Gyros, Clocks, and Interferometers: Testing Relativistic Gravity in Space", C. Lämmerzahl *et al.*, eds., *Lecture Notes in Physics* **562** (2001) 381 (Springer, Berlin).
76. D.R. Lorimer and P.C.C. Freire, astro-ph/0404270.
77. G. Amelino-Camelia, J. Ellis, N.E. Mavromatos, D.V. Nanopoulos, and S. Sarkar, *Nature* **393**, 763 (1998).
78. S.D. Biller *et al.*, *Phys. Rev. Lett.* **83**, 2108 (1999).
79. K. Greisen, *Phys. Rev. Lett.* **16**, 748 (1966); G.T. Zatsepin, and V. A. Kuzmin, *JETP Lett.* **41**, 78 (1966).
80. H. Sato and T. Tati, *Progr. Theor. Phys.* **47**, 1788 (1972).
81. S. Coleman and S.L. Glashow, *Phys. Lett.* **B 405**, 249 (1997); *Phys. Rev.* **D 59**, 116008 (1999).
82. L. Gonzales-Mestres, hep-ph/9905430.
83. O. Bertolami, and C.S. Carvalho, *Phys. Rev.* **D 61**, 103002 (2000).
84. P.R. Phillips, *Phys. Rev.* **146**, 966 (1966).
85. H.B. Nielsen and M. Ninomiya, *Nucl. Phys.* **B 141**, 153 (1978).
86. V.A. Kostelecký and S. Samuel, *Phys. Rev.* **D 39**, 683 (1989); *Phys. Rev. Lett.* **63**, 224 (1989).
87. O. Bertolami, *Class. Quantum Gravity* **14**, 2748 (1997).
88. O. Bertolami and D.F. Mota, *Phys. Lett.* **B 455**, 96 (1999).
89. V.A. Kostelecký and R. Potting, *Phys. Rev.* **D 51**, 3923 (1995); *Phys. Lett.* **B 381**, 389 (1996).
90. D. Colladay and V.A. Kostelecký, *Phys. Lett.* **B 344** (1995) 259; *Phys. Rev.* **D 52**, 6224 (1995).
91. R. Adler *et al.* (CPLEAR Collaboration), J. Ellis, J.L. Lopez, N.E. Mavromatos, D.V. Nanopoulos, *Phys. Lett.* **B 364**, 239 (1995).
92. R. Bluhm, V.A. Kostelecký and N. Russell, *Phys. Rev. Lett.* **79**, 1432 (1997); *Phys. Rev.* **D 57**, 3932 (1998).
93. R. Bluhm, V.A. Kostelecký and N. Russell, *Phys. Rev. Lett.* **82** (1999) 2254.
94. O. Bertolami, D. Colladay, V.A. Kostelecký and R. Potting, *Phys. Lett.* **B 395**, 178 (1997).
95. A. Brillet and J.L. Hall, *Phys. Rev. Lett.* **42**, 549 (1979).
96. V.W. Hughes, H.G. Robinson, and V. Beltran-Lopez, *Phys. Rev. Lett.* **4**, 342 (1960).
97. R.W.P. Drever, *Philos. Mag.* **6**, 683 (1961).
98. S.K. Lamoreaux, J.P. Jacobs, B.R. Heckel, F.J. Raab, and E.N. Fortson, *Phys. Rev. Lett.* **57**, 3125 (1986).
99. V.A. Kostelecký and C.D. Lane, *Phys. Rev.* **D 60**, 116010 (1999).
100. J.F. Bell, *Astrophys. J.* **462**, 287 (1996).
101. J.F. Bell and T. Damour, *Class. Quantum Gravity* **13**, 3121 (1996).
102. N. Hayashida *et al.*, (AGASA Collab.), *Phys. Rev. Lett.* **73**, 3491 (1994); M. Takeda *et al.*, (AGASA Collab.), *Phys. Rev. Lett.* **81**, 1163 (1998).
103. D.J. Bird *et al.*, (Fly's Eye Collab.), *Phys. Rev. Lett.* **71**, 3401 (1993); *Astrophys. J.* **424**, 491 (1994); **441**, 144 (1995).

104. M.A. Lawrence, R.J.O. Reid and A.A. Watson (Haverah Park Collab.), *J. Phys.* **G 17**, 733 (1991).
105. N.N. Efimov *et al.*, (Yakutsk Collab.), ICRR Symposium on Astrophysical Aspects of the Most Energetic Cosmic Rays, eds. N. Nagano, F. Takahara (World Scientific, 1991).
106. O. Bertolami, *Gen. Rel. Grav.* **34**, 707 (2002).
107. R.U. Abbasi *et al.*, (High Resolution Fly's Eye Collaboration), *Phys. Lett.* **B 619**, 271 (2005).
108. Pierre Auger Observatory: `http://www.auger.org/icrc2005/`.
109. S.M. Carroll, J.A. Harvey, V.A. Kostelecký, C.D. Lane, and T. Okamoto, *Phys. Rev. Lett.* **87**, 141601 (2001).
110. O. Bertolami and L. Guisado, *Phys. Rev.* **D 67**, 025001 (2003); O. Bertolami, *Mod. Phys. Lett.* **A 29**, 123525 (2005).
111. O. Bertolami and L. Guisado, *JHEP* **0312**, 013 (2003); S. Imai and N. Sasakura, *JHEP* **0009**, 032 (2000); J. Conroy, H.J. Kwee and V. Nazaryan, *Phys. Rev.* **D 68**, 054004 (2003); D. Robbins and S. Sethi, *JHEP* **0307**, 034 (2003).
112. R. Gambini and J. Pullin, *Phys. Rev.* **D 59**, 124021 (1999); J. Alfaro, H.A. Morales-Tecotl and L.F. Urrutia, *Phys. Rev. Lett.* **84**, 2183 (2000).
113. L.J. Garay, *Phys. Rev. Lett.* **80**, 2508 (1998); J.R. Ellis, N.E. Mavromatos and D.V. Nanopoulos, *Gen. Rel. Grav.* **31**, 1257 (1999).
114. V.A. Kostelecký, R. Lehnert and M.J. Perry, *Phys. Rev.* **D 68**, 123511 (2003); O. Bertolami, R. Lehnert, R. Potting and A. Ribeiro, *Phys. Rev.* **D 69**, 083513 (2004).
115. D. Mattingly, *Living Rev. Rel.* **8**, 5 (2005).
116. V.A. Kostelecký, *Phys. Rev.* **D 69**, 105009 (2004).
117. R. Bluhm and V.A. Kostelecký, *Phys. Rev.* **D 71**, 065008 (2005), hep-th/0412320. V.A. Kostelecký and R. Van Kooten, *Phys. Rev.* **D 54**, 5585 (1996).
118. O. Bertolami and J. Páramos *Phys. Rev.* **D 72**, 044001 (2005).
119. R. Vessot *et al.*, *Phys. Rev. Lett.* **45**, 2081 (1980).
120. A. Bauch and S. Weyers, *Phys. Rev.* **D 65**, 081101(R) (2002).
121. S.G. Turyshev, M. Shao and K.L. Nordtvedt, Jr., In proc. "359th WE-Heraeus Seminar: "Lasers, Clock, and Drag-Free: Technologies for Future Exploration in Space and Gravity Tests," University of Bremen, ZARM, Bremen, Germany, 30 May - 1 June 2005, this volume, gr-qc/0601035.
122. W.J. Marciano, *Phys. Rev. Lett.* **52**, 489 (1984).
123. T. Damour, G.W. Gibbons, and J.H. Taylor, *Phys. Rev. Lett.* **61**, 1152 (1988); T. Damour and J.H. Taylor, *Astrophys. J.* **428**, 713 (1994).
124. O. Bertolami and J. Gárcia-Bellido, *Int. J. Mod. Phys.* **D 5**, 97 (1996).
125. G.T. Gillies, *Rep. Prog. Phys.* **60**, 151 (1997).
126. T. Chiba, gr-qc/0110118.
127. O. Bertolami and P.J. Martins, *Phys. Rev.* **D 60**, 064007 (2000).
128. D.B. Guenther, L.M. Krauss, and P. Demarque, *Astrophys. J.* **498**, 871 (1998).
129. S.E. Thorsett, *Phys. Rev. Lett.* **77**, 1432 (1996).
130. D.N. Spergel *et al.*, *Astrophys. J. Suppl.* **148**, 175 (2003); C.L. Bennett *et al.*, *Astrophys. J. Suppl.* **148**, 1 (2003).
131. M. Green, J. Schwarz, and E. Witten, "Superstring Theory," (Cambridge University Press 1987).

132. J. Polchinski, *Phys. Rev. Lett.* **75**, 4724 (1995); P. Horava and E. Witten, *Nucl. Phys.* **B 460**, 506 (1996); A. Lukas, B. A. Ovrut and D. Waldram, *Phys. Rev.* **D 60**, 086001 (1999).
133. L. Randall and R. Sundrum, *Phys. Rev. Lett.* **83**, 3370 (1999).
134. J. Julve and M. Tonin, *Nuovo Cim.* **B 46**, 137 (1978); E.S. Fradkin and A.A. Tseytlin, *Nucl. Phys.* **B 201**, 469 (1982); E.G. Avramidi and A.O. Barvinsky, *Phys. Lett.* **B 159**, 269 (1985).
135. T. Goldman, J. Pérez-Mercader, F. Cooper and M.M. Nieto, *Phys. Lett.* **B 281**, 219 (1992).
136. O. Bertolami, J.M. Mourão and J. Pérez-Mercader, *Phys. Lett.* **B 311**, 27 (1993).
137. O. Bertolami, J.M. Mourão and J. Pérez-Mercader, In. proceedings of the "First Iberian Meeting on Gravity," Eds. M. C. Bento, O. Bertolami, J. M. Mourão, R and F. Picken, (World Scientific Press 1993); O. Bertolami, Proceedings of the International School on Cosmological Dark Matter, Eds. J.W.F. Valle, A. Pérez, (World Scientific Press 1994).
138. T. Damour, F. Piazza, and G. Veneziano, *Phys. Rev. Lett.* **89**, 081601 (2002); *Phys. Rev.* **D 66**, 046007 (2002).
139. Y. Fujii, *Phys. Rev.* **D 9**, 874 (1979); A. Zee, *Phys. Rev. Lett.* **42**, 417 (1979); S.L. Adler, *Rev. Mod. Phys.* **54**, 729 (1982).
140. S. Weinberg, *Rev. Mod. Phys.* **61**, 1 (1989); astro-ph/9610044.
141. I. Antoniadis, *Phys. Lett.* **B 246**, 317 (1990).
142. N. Arkani-Hamed, S. Dimopoulos and G. Dvali, *Phys. Lett.* **B 429**, 263 (1998).
143. E. Fishbach and C. Talmadge, hep-ph/9606249.
144. J. Scherk, *Phys. Lett.* **B 88**, 265 (1979).
145. I. Bars and M. Visser, *Phys. Rev. Lett.* **57**, 25 (1986).
146. T.D. Lee and C.N. Yang, *Phys. Rev.* **98**, 1501 (1955).
147. O. Bertolami, *Mod. Phys. Lett.* **A 6**, 383 (1986).
148. O. Bertolami and J. Páramos *Phys. Rev.* **D 71**, 023501 (2005).
149. S. Dimopoulos and G.F. Guidice, *Phys. Lett.* **B 379**, 105 (1996).
150. S.R. Beane, *Gen. Rel. Grav.* **29**, 945 (1997).
151. E.G. Floratos and G.K. Leontaris, *Phys. Lett.* **B 465**, 95 (1999).
152. C.D. Hoyle *et al.*, *Phys. Rev. Lett.* **86**, 1418 (2001).
153. Q. Majorana, *Philos. Mag.* **39**, 488 (1920).
154. H.N. Russell, *Astrophys. J.* **54**, 334 (1921).
155. D.H. Eckhardt, *Phys. Rev.* **D 42**, 2144 (1990).
156. J.G. Williams, S.G. Turyshev and D.H. Boggs, In proceedings of "Testing the Equivalence Principle on Ground and in Space", Pescara, Italy, September 20 - 23, 2004, C. Lämmerzahl, C.W.F. Everitt and R. Ruffini, editors. To be published by Springer Verlag, Lect. Notes Phys., (2006), gr-qc/0507083.
157. J. Weber, *Phys. Rev.* **146**, 935 (1966).
158. C.S. Unnikrishnan and G.T. Gillies, *Phys. Rev.* **D 61**, 101101(R) (2000).
159. V. Trimble, *Ann. Rev. Astron. Astrophys.* **25**, 425 (1987).
160. J.P. Ostriker and P.J. Steinhardt, *Science* **300**, 1909 (2003).
161. S. McGaugh, *Astrophys. J.* **609**, 652 (2004); *Phys. Rev. Lett.* **95**, 171302 (2005).
162. R.B. Tully and J.R. Fisher, *Astron. Astrophys.* **54**, 661 (1977).
163. M. Milgrom, *Astrophys. J.* **270**, 365 (1983).
164. R.H. Sanders and S.S. McGaugh, *Ann. Rev. Astron. Astrophys.* **40**, 263 (2002).
165. J.D. Bekenstein, *Phys. Rev.* **D 70**, 083509 (2004) [Erratum-ibid. **D 71**, 069901 (2005)].

166. C. Muñoz, *Int. J. Mod. Phys.* **A 19**, 3093 (2004).
167. S. Asztalos *et al.*, *Phys. Rev.* **D 64**, 092003 (2001).
168. M.C. Bento, O. Bertolami, R. Rosenfeld and L. Teodoro, *Phys. Rev.* **D 62**, 041302 (2000); M.C. Bento, O. Bertolami, and R. Rosenfeld, *Phys. Lett.* **B 518**, 276 (2001).
169. M.C. Bento, O. Bertolami, and A.A Sen, *Phys. Rev.* **D 66** (2002) 043507; **D 67**, 063511 (2003); *Phys. Lett.* **B 575**, 172 (2003); *Gen. Rel. Grav.* **35**, 2063 (2003); O. Bertolami, A.A. Sen. S. Sen and P.T. Silva, *Mon. Not. R. Ast. Soc.* **353**, 329 (2004).
170. J. Khoury, A. Weltman, *Phys. Rev. Lett.* **93**, 171104 (2004).
171. J. Khoury and A. Weltman, *Phys. Rev.* **D 69**, 044026 (2004); P. Brax, C.v. de Bruck, A.C. Davis, and A.M. Green, astro-ph/0509878.
172. P. Brax, C. van de Bruck, A.-C. Davis, J. Khoury, A. Weltman, *Phys. Rev.* **D 70**, 123518 (2004).
173. M. Reuter, *Phys. Rev.* **D 57**, 971 (1998); I.L. Shapiro, J. Sola, and H. Stefancic, JCAP **0501**, 012 (2005); M. Reuter and H. Weyer, *Phys. Rev.* **D 70**, 124028 (2004); M. Reuter and H. Weyer, JCAP **0412**, 001 (2004); F. Bauer, gr-qc/0512007.
174. O. Bertolami, *Gen. Rel. Grav.* **29**, 851 (1997).
175. O. Bertolami and F. Nunes, *Phys. Lett.* **B 452**, 108 (1999).
176. S. Perlmutter *et al.*, *Bull. Am. Astron. Soc.* **29**, 1351 (1997) [astro-ph/9812473]; *Astrophys. J.* **517**, 565 (1999) [astro-ph/9812133]; *Astrophys. J.* **507**, 46 (1998).
177. A.G. Riess *et al.*, *Astron. J.* **116**, 1009 (1998).
178. M. Bronstein, *Phys. Z. Sowjetunion* **3** (1938) 73; O. Bertolami, Nuovo Cim. **93 B**, 36 (1986).
179. R.R. Caldwell, R. Dave and P. Steinhardt, *Phys. Lett.* **80** (1998) 1582; L. Wang, R.R. Caldwell, J.P. Ostriker and P. Steinhardt, *Astrophys. J.* **530**, 17 (2000).
180. O. Lahav and A.R. Liddle, astro-ph/0601168.
181. S. Kachru, R. Kallosh, A. Linde, and S.P. Trivedi, *Phys. Rev.* **D 68**, 046005 (2003).
182. S.M. Carroll, V. Duvvuri, M. Trodden and M.S. Turner, *Phys. Rev.* **D 70**, 043528 (2004).
183. S.M. Carroll, A. De Felice, V. Duvvuri, D. A. Easson, M. Trodden, and M.S. Turner, *Phys. Rev.* D**71** (2005) 063513.
184. S. Capozziello, S. Carloni, and A. Troisi, astro-ph/0303041; S.M. Carroll, V. Duvvuri, M. Trodden, and M.S. Turner, *Phys. Rev.* **D 70**, 043528 (2004).
185. I. Navarro and K. Van Acoleyen, gr-qc/0511045.
186. T. Chiba, *Phys. Lett.* **B 575**, 1 (2003).
187. G.R. Dvali, G. Gabadadze, and M. Porrati, *Phys. Lett.* **B 485**, 208 (2000).
188. C. Deffayet, *Phys. Lett.* **B 502**, 199 (2001).
189. C. Deffayet, G.R. Dvali, and G. Gabadadze, *Phys. Rev.* **D 65**, 044023 (2002).
190. M.A. Luty, M. Porrati, and R. Rattazzi, *JHEP* **0309**, 029 (2003).
191. M.C. Bento, O. Bertolami, N.M.C. Santos, and A.A. Sen, *Phys. Rev.* **D 71**, 063501 (2004).
192. J.A.R. Cembranos, gr-qc/0507039; C.G.R. Shao, R.G.R. Cai, B. Wang, and R.K. Su, *Phys. Lett.* **B 633**, 164 (2006) [gr-qc/0511034].
193. C. Wetterich, *Nucl. Phys.* **B 302**, 668 (1988).
194. B. Ratra and J. Peebles, *Phys. Rev.* **D 37**, 321 (1988).
195. P. Ferreira and M. Joyce, *Phys. Rev. Lett.* **79**, 4740 (1997).

196. I. Zlatev, L. Wang, and P. Steinhardt, *Phys. Rev. Lett.* **82** (1999) 896.
197. A. Albrecht, C. Skordis, *Phys. Rev. Lett.* **84** (2000) 2076.
198. S. Dodelson, M. Kaplinghat, and E. Stewart, *Phys. Rev. Lett.* **85**, 5276 (2000).
199. M.C. Bento, O. Bertolami and N.M.C. Santos, *Phys. Rev.* **D 65**, 067301 (2002).
200. P. Ferreira and M. Joyce, *Phys. Rev.* **D 58**, 023503 (1998).
201. G.W. Anderson and S.M. Carroll, astro-ph/9711288.
202. G.R. Farrar and P.J.E. Peebles, *Astrophys. J.* 604 (2004) 1.
203. R. Bean and J. Magueijo, *Phys. Lett.* **B 517**, 177 (2001).
204. M. Gasperini, F. Piazza, G. Veneziano *Phys. Rev.* **D 65** (2002) 023508.
205. R. Bean, *Phys. Rev.* **D 64**, 123516 (2001).
206. R.R. Caldwell, *Phys. Lett.* **B 545**, 23 (2002).
207. S.M. Carroll, M. Hoffman, M. Trodden, *Phys. Rev.* **D 68** (2003) 023509.
208. J.M. Cline, S. Jeon, G.D. Moore, *Phys. Rev.* **D 70** (2004) 043543.
209. P. Sreekumar, et al, *Astrophys. J.* **494** (1998) 523.
210. N. Arkani-Hamed, P. Creminelli, S. Mukohyama, M. Zaldarriaga, *JCAP* **0404** (2004) 001.
211. F. Piazza, S. Tsujikawa, *JCAP* **0407** (2004) 004.
212. H. Sandvik, M. Tegmark, M. Zaldarriaga, I. Waga, *Phys. Rev.* **D 69**, 123524 (2004).
213. R. Bean, O. Dore, *Phys. Rev.* **D 68** (2003) 023515.
214. M.C. Bento, O. Bertolami and A.A. Sen, *Phys. Rev.* **D 71**, 023521 (2004).
215. E.G. Adelberger, B.R. Heckel and A.E. Nelson, *Ann. Rev. Nucl. Part. Sci.* **53**, 77 (2003).
216. E.G. Adelberger, E. Fischbach, D.E. Krause and R.D. Newman, *Phys. Rev.* **D 68**, 062002 (2003).
217. G. Dvali, A. Gruzinov, and M. Zaldarriaga, *Phys. Rev.* **D 68** (2003) 024012; A. Lue and G. Starkman, *Phys. Rev.* **D 67** (2003) 064002.
218. E.V. Pitjeva, Sol. Sys. Res. **39** (2005) 176.
219. S.M. Carroll, *Phys. Rev. Lett.* **81** (1998) 3067.
220. G. Veneziano, *JHEP* **0206**, 051 (2002).
221. A. Aguirre, astro-ph/0310572.
222. J.W. Moffat, gr-qc/0506021.
223. H.S. Zhao and L. Tian, astro-ph/0511754; H.S. Zhao, astro-ph/0511713.
224. M. Ishak, A. Upadhye and D.N. Spergel, astro-ph/0507184.
225. P. Zhang, astro-ph/0511218; K. Koyama and R. Maartens, astro-ph/0511634.
226. I. Sawicki and S.M. Carroll, astro-ph/0510364.
227. A. Peters, K.Y. Chung and S. Chu, *Nature* **400**, 849 (1999).
228. R. Bingham *et al.*, "HYPER – Precision Cold Atom Interferometry in Space", ESA-SCI (2000) 10.
229. R.A. Lewis, G.A. Smith, F.M. Huber and E.W. Messerschmid, "Measuring the gravitational force of anti-protons in space", ESA SP-385 (1996) 439.
230. O. Bertolami, C.J. de Matos, J.C. Grenouilleau, O. Minster and S. Volonte, gr-gc/0405042.
231. T.W. Murphy, Jr., J.D. Strasburg,, C.W. Stubbs, E.G. Adelberger, J. Angle, K. Nordtvedt, J.G. Williams, J.O. Dickey, and B. Gillespie, "The Apache Point Observatory Lunar Laser-Ranging Operation (APOLLO)", Proceedings of 12th International Workshop on Laser, Ranging, Matera, Italy, November 2000, in press, 2002, `http://www.astro.washington.edu/tmurphy/apollo/matera.pdf`

232. J.G. Williams, S.G. Turyshev, and T.W. Murphy, Jr., *Int. J. Mod. Phys.* **D 13**, 567 (2004) [gr-qc/0311021].
233. S.G. Turyshev, M. Shao, and K.L. Nordtvedt, Jr., *Int. J. Mod. Phys.* **D 13**, 2035 (2004) [gr-qc/0410044].
234. S.G. Turyshev, M. Shao, and K.L. Nordtvedt, Jr., *Class. Quantum Grav.* **21**, 2773 (2004) [gr-qc/0311020].
235. S.G. Turyshev, M. Shao, and K.L. Nordtvedt, Jr., In Proc. "The XXII Texas Symposium on Relativistic Astrophysics", Stanford University, December 13-17, 2004, edited by P. Chen, E. Bloom, G. Madejski and V. Petrosian. SLAC-R-752, Stanford e-Conf #C041213, paper #0306, eprint: `http://www.slac.stanford.edu/econf/C041213/`, gr-qc/0502113.

Is the Physics Within the Solar System Really Understood?

Claus Lämmerzahl[1], Oliver Preuss[2], and Hansjörg Dittus[1]

[1] ZARM, University of Bremen, Am Fallturm, 28359 Bremen, Germany
[2] Max-Planck-Institute for Solar System Research, Max-Planck-Str. 2, 37191 Katlenburg-Lindau, Germany

Summary. A collection is made of presently unexplained phenomena within our solar system and in the universe. These phenomena are (1) the Pioneer anomaly, (2) the flyby anomaly, (3) the increase of the astronomical unit, (4) the quadrupole and octupole anomaly, (5) dark energy, and (6) dark matter. A new data analysis of the complete set of Pioneer data is announced to search for systematic effects or to confirm the unexplained acceleration. We also review the mysterious flyby anomaly where the velocities of spacecraft after Earth swingbys are larger than expected. We emphasize the scientific aspects of this anomaly and propose systematic and continuous observations and studies at the occasion of future flybys. Further anomalies within the solar system are the increase of the astronomical unit and the quadrupole and octupole anomaly. We briefly mention dark matter and dark energy as in some cases a relation between them and the solar system anomalies have been speculated.

1 Introduction

Progress in physics has always been stimulated by observations which could not be explained within the presently standard physical theories. In the late nineteenth century the observations and experiments of Bradley as well as Airy who both observed aberration of distant starlight, of Fizeau who observed a dragging of light in moving media not compatible with theory at that time, and finally of Michelson and Morley who showed that the failure of the application of the nonrelativistic mechanical laws to light propagation. All these effects, which could not be made compatible with nonrelativistic physics without introducing various unnatural elements into the theory, culminated into the invention of special relativity. Then, the theoretical incompatibility of Newtonian gravity with special relativity as well as the since long observed perihelion shift of Mercury which has first been attributed to systematic errors or solely to be due to the solar quadrupole moment led to the formulation of general relativity. Later on, the experimental study of atomic spectra which could not be explained using the laws of classical mechanics first led to Bohr's atomic model and, subsequently, to the various formulations of quantum mechanics.

The situation of gravitational physics today bears many similarities. At first, the theoretical inconsistency of quantum mechanics and general relativity makes a new theory combining these two universal theories necessary. Furthermore, there are observations which at least until now and after many years of studies have not yet found any convincing explanation. These observations are (1) dark energy which is necessary – under the assumption of the validity of Einstein's equations – to describe the accelerated expansion of the universe and (2) dark matter which – again under the assumption of general relativity – is necessary to account for the galactic rotation curves, for observed gravitational lensing of light, and for the structure formation in the early universe. Of a slightly weaker observational basis is (3) the Pioneer anomaly, an unexplained constant acceleration of the Pioneer 10 and 11 spacecrafts, (4) the flyby anomaly, an unexplained increase of the velocity of a series of spacecrafts after Earth gravity assists, (5) the recently realized increase of the astronomical unit defined by the distance of the planets from the Sun by approximately 10 m per century, and (6) the quadrupole and octupole anomaly which describes the correlation of the low-l contributions of the cosmic microwave background to the orientation of the solar system.

These six phenomena, including dark energy and dark matter which at this stage are nothing more than a synonym for these observations, had neither found any convincing interpretation or solution nor culminated into a finally convincing theory. Lacking any explanation until now, these phenomena have the potential to be of importance for a new physics.

In this chapter we describe all these unexplained observations, state the open questions, and suggest new observations and new missions to obtain better data for a better analysis of these phenomena.

2 Dark Matter

Dark matter has been introduced to "explain" the gravitational field needed for the galactic rotation curves, the gravitational lensing of galaxies, and the formation of structures in our universe [1]. It also appears in the spectral decomposition of the cosmic microwave background radiation [2]. Dark matter is needed if one assumes Einstein's field equations to be valid. However, there is no single observational hint at particles which could make up this dark matter. As a consequence, there are attempts to describe the same effects by a modification [3] of the gravitational field equations, e.g., of Yukawa form, or by a modification of the dynamics of particles, like the MOND ansatz [4,5],[1] recently formulated in a relativistic frame [7]. Due to the lack of

[1] In a nice short paper Veltman [6] speculates how astronomers may build up laws of gravity by observing gravity on larger scales (scale of galaxies and of the universe) and compares that with a quantum field theory approach.

direct detection of dark matter particles, all those attempts are on the same footing.

3 Dark Energy

Similarly, recent observations of type Ia supernovae indicate that the expansion of the universe is accelerating and that 75% of the total energy density consists of a dark energy component with negative pressure [8, 9]. Furthermore WMAP measurements of the cosmic microwave background [10, 11], the galaxy power spectrum, and the Lyman-alpha forest data lines [12–14] also indicate – when compared with standard cosmological models – the existence of the mysterious dark energy that leads to the acceleration of the universe, rather than a modification of the basic laws of gravitation [15]. However, also in this case there are attempts to give an explanation in terms of modified field equations, see, e.g., [16]. Recently it has been claimed that dark energy or, equivalently, the observed acceleration of the universe can be explained by inhomogeneous cosmological models, such as the spherically symmetric Lemaitre–Tolman–Bondi model, see, e.g., [17–19].

4 The Pioneer Anomaly

The Pioneer anomaly is an anomalous unexplained acceleration of the Pioneer 10 and 11 spacecraft of

$$a_{\text{Pioneer}} = (8.74 \pm 1.33) \cdot 10^{-10}\,\mathrm{m\,s^{-2}} \tag{1}$$

toward the Sun [20, 21]. This acceleration seems to have been turned on after the last flyby at Saturn and stayed constant within a 3% range.

4.1 The Observation

The principle of observation was two-way Doppler tracking: a sender on the Earth emits a signal of frequency ν_0 which is "seen" by the spacecraft as frequency

$$\nu' = \frac{1}{\sqrt{1 - v^2/c^2}} \left(1 - \frac{v}{c}\right) \nu_0 \,. \tag{2}$$

The spacecraft sends this frequency back (with a slight offset what, however, will not affect the principle of measurement), so that the receiver on Earth observes the frequency

$$\nu'' = \frac{1}{\sqrt{1 - v^2/c^2}} \left(1 - \frac{v}{c}\right) \nu' \,. \tag{3}$$

The comparison of the sent and received frequency gives the velocity of the spacecraft

$$y = \frac{\nu'' - \nu_0}{\nu_0} = -2\frac{v/c}{1+v/c} \approx -2\frac{v}{c}. \tag{4}$$

This measured frequency can be compared with the calculated frequency obtained from the calculated orbit given by the gravitational field inside the solar system together with all kinds of modeling needed (see below).

The outcome of the observation was a continuous drift between the observed and calculated frequency shifts

$$\frac{d(y_{\text{obs}} - y_{\text{calc}})}{dt} = (3.84 \pm 0.01) \cdot 10^{-18}\,\text{s}^{-1}. \tag{5}$$

This corresponds to a continuous drift in the velocity of the spacecraft or, equivalently, in the constant acceleration (1).

4.2 Orbit Determination

As the observations are made with tracking stations on the moving Earth observing the frequency of signals, the orbit determination consists of six segments:

- Model of gravitational forces
- Model of external nongravitational forces
- Model of internal (spacecraft) nongravitational forces
- Model of observation stations
- Model of signal propagation
- Codes

$\Longrightarrow$ Orbit and velocity determination

We just mention the main aspects of this scheme. Most of this can be found in [21].

Gravitational forces. The calculation of the orbits has been performed using relativistic equations of motion for celestial bodies including order v^4/c^4:

- The relativistic gravitational accelerations (EIH model) include the Sun, the Moon, and the nine planets as point masses in an isotropic PPN N-body metric
- Newtonian gravity from large asteroids is included
- Terrestrial and lunar figure effects
- Earth tides
- Lunar physical librations have been considered

External nongravitational forces. These forces include:

- Solar radiation and solar wind pressure
- Drag from interplanetary dust

Internal (spacecraft) nongravitational forces. These forces include:

- Thermal radiation
- Attitude-control propulsive maneuvers and propellant (gas) leakage from the spacecraft's propulsion system
- Torques produced from the above forces

Model of observation stations. An orbit determination has to include a model of the ground stations. This is based on:

- Precession, nutation, sidereal rotation, polar motion, tidal effects, and tectonic plates drift. The information on tidal deceleration, nonuniformity of rotation, Love numbers, and Chandler wobble has been obtained from LLR, SLR, and VLBI (from ICRF) measurements.
- Model of DSN antennae and their influence on the tracking data.

Modeling of signal propagation. The propagation of the radio signals includes:

- A relativistic model for light propagation including order v^2/c^2
- Dispersion due to solar wind and interplanetary dust

Codes. Four independent codes have been used for the orbit determination:

- JPL Orbit Determination Program (various generations from 1970 to 2001)
- The Aerospace Corporation code POEAS (during period 1995–2001)
- Goddard Space Flight Center conducted a study in 2003 (data from NSSDC)
- Code of University of Oslo

The definition of these models has to be complemented by a discussion of possible errors. This is tantamount to the search for possible conventional explanations of the effect. We present only a few points.

4.3 Discussion of Some Conventional Effects

In the following, we discuss recent and ongoing work on conventional effects which may contribute to errors or perhaps may also be responsible for the observed acceleration. Not included here is the spin-rotation coupling [22].

Dust

The interplanetary medium consists of (1) interplanetary dust and (2) interstellar dust. The first one consists of hot wind plasma (mainly protons and electrons) distributed within the Kuiper belt (from 30 to 100 AU). The density of this plasma has been modeled to be of the order $\rho_{\rm IPD} \leq 10^{-24}\,{\rm g\,cm^{-3}}$ (Man and Kimura 2000). The second one, interstellar dust, which can be distinguished from the interplanetary dust by its greater impact velocity has been measured by Ulysses to have a density of $\rho_{\rm ISD} \leq 3 \cdot 10^{-26}\,{\rm g\,cm^{-3}}$.

The drag acceleration of a spacecraft moving through dust of density ρ is given by

$$a_{\text{drag}} = -K_{\text{s}}\rho v_{\text{s}}^2 \frac{A_{\text{s}}}{m_{\text{s}}}, \tag{6}$$

where K_{s} is the satellite's drag coefficient which can be taken to be ≈ 2. If we assume the drag acceleration to be the observed anomalous acceleration of the Pioneer spacecraft, then this needs a density which is $3 \cdot 10^5$ larger than the interplanetary dust. Therefore dust cannot be the origin of the Pioneer acceleration [23].

Additional Masses in the Solar System

Additional masses may be present not only in the form of dust but also in the form of larger particles. Irrespective of being dust and of the size of these particles, any additional mass will act as an additional gravitational field which may decelerate the spacecraft when leaving the solar system. Nieto [24] has analytically calculated the gravitational effect of various configurations, i.e., shells, thin rings, and wedges of various density profiles. He obtained that for rings with a density falling off with $1/r$ as well as a wedge with a density falling off like $1/r^2$ yields a nearly constant acceleration (neglecting discontinuities at the sharp boundaries of the matter distributions which are, of course, just results of the mathematical model). However, for the constant acceleration to be of the order of the observed Pioneer acceleration, the mass of the thin ring or the wedge has to be about 10 times the mass of the Earth which is, by far, not compatible with the observations of, e.g., comets.

Accelerated Sun

A nongravitational acceleration of the Sun orthogonal to the ecliptic will also cause an acceleration toward the Sun. Such an accelerated Sun is the consequence of an exact solution of the Einstein equation, the so-called C-metric [25]. In the frame of an accelerated Sun, the equation of motion for test masses reads

$$\ddot{\boldsymbol{r}} + GM_{\odot}\frac{\boldsymbol{r}}{r^3} + \boldsymbol{a}_{\odot} = 0, \tag{7}$$

where $\boldsymbol{r}$ is the distance between the Sun and the test mass and $\boldsymbol{a}_{\odot}$ is the acceleration orthogonal to the eccpliptic. This gives a constant acceleration toward the Sun [25]. However, to obtain an acceleration of the order of the Pioneer anomaly, the acceleration of the Sun orthogonal to the ecliptic has to be larger than what would be obtained if all radiation of the Sun were emitted in one direction.

Cosmic Expansion

Due to the quite good equality $a_{\text{Pioneer}} \approx cH$, where H is the Hubble constant, it has been speculated whether the cosmic expansion has some influence on (1) the signal propagation, (2) the trajectory of the spacecraft, (3) the magnitude of the gravitational field inside the solar system, or (4) the definition of the distance, i.e., the definition of the astronomical unit.

The influence of the expansion of the universe on the procedure of Doppler tracking is negligible. For an expansion described by an Einstein–de Sitter universe

$$ds^2 = -dt^2 + R^2(t)\left(dx^2 + dy^2 + dz^2\right), \tag{8}$$

we obtain the conserved quantities $\nu_u(t)R(t) = const.$, where $\nu = \nu_u = k(u)$ is the measured frequency of a light ray $g(k,k) = 0$; u with $g(u,u) = 1$ is the 4-velocity of an observer at rest in the cosmic substrate. To first order in the expansion,

$$\nu_u(t) = \frac{R(t_0)}{R(t)}\nu_u(t_0) \approx (1 - H(t - t_0))\,\nu_u(t_0), \tag{9}$$

where $H = \dot{R}/R$ is the Hubble constant. This describes the Hubble redshift.

For massive particles we have the conserved quantity

$$R^2(t)\frac{dr(s)}{ds} = const. \quad \Leftrightarrow \quad R(t)\frac{1}{\sqrt{1 - V^2(t)}}V(t) = const., \tag{10}$$

where V, defined by $g(u,v) = 1/\sqrt{1 - V^2/c^2}$, is the measured velocity of an object moving with 4-velocity v along a geodesic, $D_v v = 0$. For small velocities $R(t)V(t) = const.$ implying a slowing down of the velocity

$$V(t) = \frac{R(t_1)}{R(t)}V(t_1) = (1 - H(t - t_1))\,V(t_1). \tag{11}$$

A distance D measured by time-of-flight of light rays is defined by $D = R(r_2 - r_1)$. The measured velocity of a moving object is then

$$\frac{d}{dt}D = \dot{R}(r_2 - r_1) + R\dot{r}_2 = HD + V_2, \tag{12}$$

where V_2 is the velocity of the object measured with respect to the cosmological substrate. As a consequence, the trajectory of an object which has constant distance to an observer, $0 = \dot{D} = HD + V_2$, has to move with velocity

$$V_2 = -HD \tag{13}$$

with respect to the substrate.

Therefore, the two-way Doppler ranging is influenced by three effects (a) the cosmological redshift (9), (b) the slowing down of the velocity of the spacecraft (11), and (c) the velocity of the unit distances with respect to the cosmological substrate (13). Taking all together results in the final two-way Doppler ranging signal

$$\frac{\Delta\nu}{\nu_0} = -2V(t_1)(1 - H(t_2 - t_1)), \tag{14}$$

where t_1 and t_2 are the cosmological time parameters for the emission and reception of the signal. The expansion-induced effect is a chirp of Doppler signal related to an acceleration

$$a = HV = \frac{V}{c} cH, \tag{15}$$

which is by a factor V/c smaller than the observed Pioneer acceleration $a_{\text{Pioneer}} \sim cH$. Therefore, Doppler tracking in an expanding universe cannot account for the observed Pioneer anomaly.

Furthermore, the influence of the expansion of the universe on the gravitational field of the Sun or the planetary orbits is much too small to be of any influence, see Sect. 9.2.

To sum up, the expansion of our universe seems to be of no relevance for the occurrence of the Pioneer acceleration.

Drift of Clocks on Earth

Though a drift of clocks by itself is a nonconventional physics (the drift of atomic clocks with respect to, in our case, gravitational timescales needs some "new physics") it may yield a conventional explanation of the Pioneer anomaly. A quadratic drift of the timekeeping of clocks on Earth

$$t \to t + \frac{1}{2c} a_{\text{P}} t^2 . \tag{16}$$

may simulate the Pioneer anomaly [21, 26]. The numerical value is $\frac{1}{2c} a_{\text{P}} \approx 10^{-18}\,\text{s}^{-1}$. If we assume such a kind of clock drift then the question arises whether this clock drift is consistent with the measurements from other satellites and, in particular, with the ranging of satellites. Another question is whether this time drift is also consistent with the observations of pulsars and binary systems which also define clocks. Pulsars are very stable clocks [27]. Owing to the radiation of gravitational waves the revolution time of binary systems goes down. However, also in this case the stability of this process can be defined [28]. A comparison with a hypothetical drift of atomic clocks on Earth seems not to have been carried through.

4.4 Outlook

New activities are planned for the very near future. During the last months the complete set of Pioneer data has been recovered [31, 32] and brought into a digital form readable by modern computers. These data are now ready for a new data analysis covering all data and the total mission duration, see Table 1. This new data analysis will be carried through at ZARM and at JPL. It is important to find out, e.g., whether the anomalous acceleration was really not present before the last flyby.

Table 1. Pioneer data sets analyzed previously and to be analyzed in the future.

	Previously used data set		Data set to be analyzed	
	Time span	AU Distance	Time span	AU Distance
Pioneer 10	3.1.1987–22.7.1998	40–70.5	8.9.1973–27.4.2002	4.6–80.2
Pioneer 11	5.1.1987–1.10.1990	22.4–31.7	10.4.1974–11.10.1994	1.0–41.7

Table 2. Observed flybys.

Mission	Agency	Year	r_p (km)	v_∞ (km s^{-1})	e	Δv (mm s^{-1})
Galileo	NASA	Dec 1990	959.9	8.949	2.47	3.92 ± 0.08
Galileo	NASA	Dec 1992	303.1	8.877	2.32	No reliable data[a]
NEAR	NASA	Jan 1998	538.8	6.851	1.81	13.46 ± 0.13
Cassini	NASA	Aug 1999	1,173	16.01	5.8	0.11
Stardust	NASA	Jan 2001	5,950	??		No reliable data[b]
Rosetta	ESA	Mar 2005	1,954	3.863	1.327	1.82 ± 0.05
Hayabusa	Japan	May 2004	3,725	??	??	No data available
MESSENGER[c]	Private	Aug 2005	2,347	4.056	1.36	~ 0

r_p pericenter, e eccentricity, v_∞ velocity at infinity, and Δv velocity increase.
[a]Too low orbit with too large atmospheric drag.
[b]Thruster activities.
[c]US spacecraft operated by a private company.

Furthermore, a new Deep-Space Gravity Explorer (DSGE) mission has been proposed [33]. A new DSGE mission has the potentiality to explore the range between solar system physics and the physics of our galaxy and perhaps of the universe which perhaps is characterized by the Pioneer acceleration which is of the order cH and of the characteristic acceleration appearing in MOND theories.

5 The Flyby Anomaly

5.1 The Observations

It has been observed at various occasions that satellites after an Earth swingby possess a significant unexplained velocity increase by a few mm s^{-1}. This unexpected and unexplained velocity increase is called the *flyby anomaly*. According to information from [34, 35, 37], the observed flybys are listed in Table 2. For the actual data for the Galileo and NEAR flyby see Fig. 1.

The data can be put into diagrams where the velocity increase can be plotted as a function of the two orbit parameters eccentricity e and pericenter r_p, see Fig. 2. In general, this is a three-dimensional plot $\Delta v = f(e, r_p)$. For a plot of this surface four data points are far too less. Therefore, in Fig. 2

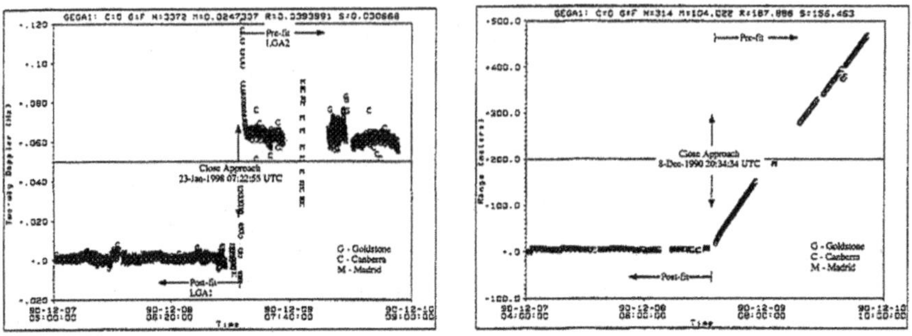

(a) Two-way S-band Doppler residuals and range residuals during the first Galileo flyby

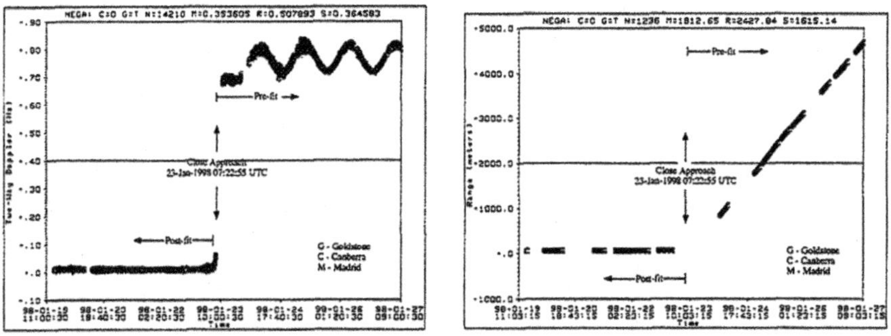

(b) Two-way X-band Doppler residuals and range residuals during NEAR flyby

Fig. 1. Galileo and NEAR flyby data (from [35]).

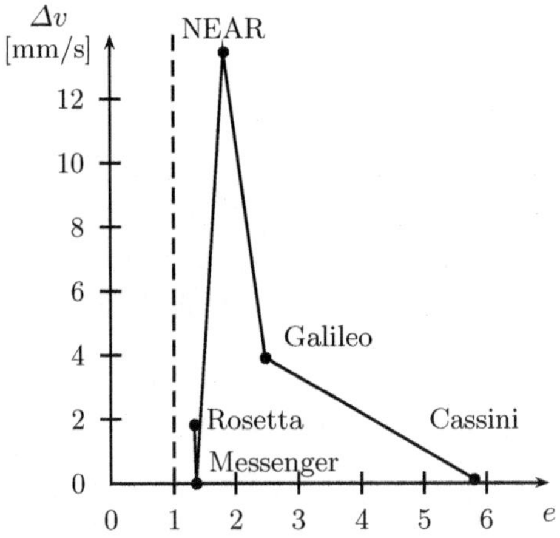

Fig. 2. The velocity increase Δv as function of the eccentricity and of the perigee.

a plot of the velocity increase Δv as function of e and r_p is given.[2] Though from five data points it is much too early to draw any serious conclusion one may speculate the following: if the velocity increase really is due to an unknown gravitational interaction, then it makes sense that (1) the effect goes down with increasing eccentricity, since for larger eccentricity the strength and duration of the interaction with the gravitational field of the Earth goes down and (2) that it also should go down for an eccentricity approaching $e = 1$ because the transition to bound orbits, where no effect has been seen, certainly should show no discontinuity due to many kinds of disturbances like drag and nonideal circumstances like gravitational multipoles, etc. But relying on this poor data base this interpretation is pure speculation only.

The main problem is not just the limited number of flybys for which sufficiently precise data are publicly available so that the anomaly can be seen at all. Even these available data suffer from low cadence (the anomaly often appears between two data points) and so far only allow an anomaly in the speed, but not in the direction of motion, etc. to be identified. Precise data at a much higher cadence of all the motion parameters of the spacecraft prior to, during and after the flyby would allow a qualitatively improved analysis.

5.2 Error Analysis

As a first remark we give the order of the acceleration leading to the velocity increase. This anomalous acceleration, estimated by the velocity increase during the time-of-flight near the Earth, is of the order $10^{-4}\,\mathrm{m\,s^{-2}}$. This is considerably larger than the above-discussed Pioneer anomaly. Below we will use this acceleration as an approximate value to be compared with disturbing influences.

It should be kept in mind that the velocity increase has been observed in the two-way Doppler measurements as well as in the ranging measurements.

Before one starts with possible fundamental explanations of this effect, a serious and reliable error analysis has to be carried through. This analysis has to cover (1) atmospheric mismodeling, (2) ocean tides, (3) if the spacecraft becomes charged, then it may experience an additional force due to the Earth's magnetic field, (4) also the interaction of a hypothetical magnetic moment of the spacecraft with the Earth's magnetic field may give an additional force, (5) ion plasma drag, (6) Earth albedo, and (7) solar wind. Here we give very short first-order estimates on these various effects which certainly have to be improved, and show that even with very rough pessimistic assumptions none of these can be held responsible for the flyby anomaly.

Atmosphere. If a spacecraft of mass m_s and effective area A_s moves with velocity v_s through a medium of density ρ, then it experiences a drag acceleration again given by (6). For a mass of 1 ton, an area of $2\,\mathrm{m}^2$, a

[2] We like to thank J.D. Anderson, J.K. Campbell, and T. Morley for providing us with the relevant data.

velocity of 30 km s^{-1} and an atmospheric density at 1,000 km height of approx. $\rho \approx 10^{-14}$ kg m^{-3} we get an acceleration of $a_{\mathrm{drag}} \approx 4 \cdot 10^{-8}$ m s^{-2} which is far too small to be of any relevance for our problem. Furthermore, this acceleration due to drag has the wrong sign.

Uncertainties in the Earth gravity model. The gravitational field of the Earth is known, combining Grace and Champ data, up to the 360th multipole moment. However, the component of each moment is always equipped with an error bar. This may lead to the idea that the unexplained velocity is due to multipole moments known not precisely enough. However, a study of the trajectories of spacecraft for multipole components within the given error bars shows that this is not possible [36].

Ocean tides. The ocean tides will lead to a change of the Earth's surface of the order of $\delta r \pm 10$ m. This means that the corresponding quadrupole part of the Earth's gravitational potential is of the order $\epsilon = 2\delta r / R_\oplus$ smaller than the monopole part of the Earth, where $R_\oplus$ is the radius of the Earth. Since $\epsilon \approx 10^{-6}$, the corresponding additional acceleration also is factor 10^{-6} smaller than the ordinary acceleration from the monopole part of the Earth's gravitational field. The latter being less than 10 m s^{-2}, the acceleration due to tides is at most 10^{-5} m s^{-2} and, thus, cannot be responsible for the flyby anomaly.

Solid Earth tides. Since Earth solid tides are much smaller than ocean tides, the analysis above shows that this cannot cause the effect.

Charging of the spacecraft. In a recent study of charging of the LISA test masses [38] the charging has been estimated by 10^{-10} C. So, for the whole satellite it might be a conservative assumption that the charge is less than $Q \leq 10^{-7}$ C. A satellite of 1 ton carrying a charge Q and moving with $v = 30$ km s^{-1} in the magnetic field of the Earth which is of the order 0.2 G will experience an acceleration 10^{-8} m s^{-2} far below the observed effect.

Magnetic moment. The force on such a body carrying a magnetic moment $\boldsymbol{m}$ moving in a magnetic field $\boldsymbol{B}$ is $\boldsymbol{F} = \boldsymbol{\nabla}(\boldsymbol{m} \cdot \boldsymbol{B})$. Since the magnetic moment of a spacecraft is not more than 2 A m^2 and the steepness of the magnetic field can be estimated by $|\Delta B / \Delta x| \leq 2 \cdot 10^{-7}$ G m^{-1}, see Fig. 3, the maximum force of a spacecraft is $F \leq 4 \cdot 10^{-11}$ N implying typically a maximum acceleration of $4 \cdot 10^{-15}$ m s^{-2} which safely can be neglected.

Earth albedo. The Earth albedo causes a pressure on the spacecraft of approx. 1 μN m^{-2} which leads, for an effective area of 2 m^2, to a force of 2.4 μN. For a mass of the spacecraft of 1 ton, this will give an acceleration of $a_{\mathrm{albedo}} \approx 2.4 \cdot 10^{-9}$ m s^{-2} which can be neglected compared with the searched for effect of 10^{-4} m s^{-2}.

Solar wind. The solar wind exerts on a spacecraft a pressure of approx. 4 μN m^{-2} which gives an acceleration of max $a_{\mathrm{solar\,wind}} \approx 2.4 \cdot 10^{-9}$ m s^{-2} which again can be safely neglected.

Spin-rotation coupling. A coupling of the helicity of the radiowaves with the rotation of the spacecraft and the rotation of the Earth also leads to an

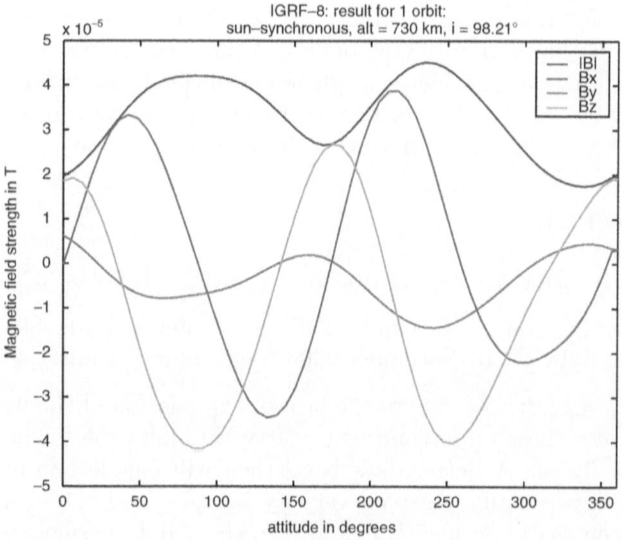

Fig. 3. The magnetic field of the Earth as function of attitude.

effect which simulates a changing velocity [22]. This, however, applies to the two-way Doppler data only. Since simultaneously also ranging, which is independent of the helicity-rotation coupling, indicated an increase of the velocity, spin-rotation cannot be responsible for this observation.

Also estimates of the influence of the Moon including Moon oblateness, the Sun, other planets, relativistic effects, and indirect oblateness of the Earth have been shown to be orders of magnitude smaller than the observed effect [35].

None of these disturbing effects could explain the flyby anomaly.

5.3 Explanations from "New Physics"?

As reported in [35] several nonstandard physical models have been used to explain this velocity increase. Since these considerations have not been published we just mention the discussed models:

- Nonconservative potential energy
- Non-Newtonian gravity (e.g., Moffat, Yukawa, etc.)
- PPN
- Modifications of relativity
- Torsion, i.e., the *eps2 model*. This model is said to fit the data, but is not compatible with the stability of planetary orbits

As reported in [35] none of these could explain the flyby anomaly. Being so large and so near to the Earth, the expansion of the universe should also play no role for this anomaly. In Sect. 9, we sketch a general approach to a description of the motion of test bodies. Among the terms found, there are several which may be considered as phenomenological description of a velocity increase. To be able to pin down a specific interaction term it is necessary to make detailed studies of the nature of the velocity increase.

5.4 Future Flybys

In the near future there will be two flybys, both by Rosetta [37]:

- Rosetta: flyby on 13 November 2007 (pericenter altitude 4,942 km)
- Rosetta: flyby on 13 November 2009 (pericenter altitude 2,483 km)

We strongly suggest that due to the lack of explanation of the flyby anomaly one should use these opportunities to carry through a better observation of the Rosetta flybys. A better data basis then will enable one to establish a correlation between the observed velocity increase and the orbital parameters like eccentricity, perihelion distance to the Earth, perihelion velocity, or inclination. In particular, a continuous observation (Doppler tracking, ranging, positioning, and perhaps other data from the spacecraft like temperature, pressure, etc.) also should give hints to the particular direction of the local acceleration and also on the strength and, thus, to the position dependence of the anomalous force. Furthermore, it would also be of great importance to know whether after a flyby the direction of the motion of the spacecraft also is different compared with the standard result. This information is extremely important for the search for a conventional explanation of this effect or for the modeling of a new force.

As a consequence, a *complete, accurate and continuous observation of spacecraft during flybys is very important to study the nature of this unexplained velocity increase.* It is an advantage of these planned observations that they will in no way whatsoever modify the mission scenario.

Furthermore, since new missions like the Mars Reconnaissance Orbiter will measure very precisely the gravitational field of Mars, also the Mars flyby of Rosetta on 25 February 2007 should be observed as precise as possible to see whether also at Mars a velocity increase will occur. A Mars flyby would provide an excellent augmentation of the Earth flybys. Since Mars possesses other conditions than the Earth (weaker atmosphere, almost no magnetic field, other gravitational field, lower thermal radiation, etc.), many competing effects and also possible explanation schemes based on the fact that the effect as well as the observation occurs at and on the Earth can be ruled out. Therefore *the effect, if it will be observed also at Mars, then will turn out to be universal and beyond any doubt and will become an extremely important science case.* In the case of Mars, it is of course, impossible to have a continuous observation (at least due to a Mars eclipse), but the initial and final velocity should be measured with the best possible precision. Furthermore, if this effect really seems to be existent, then a dedicated mission with a good drag-free control system (which can control the acceleration better than $10\,\mathrm{m\,s^{-2}}$ for 1 s measurement time) for well-defined flybys at Earth for small perigees and, thus, for extreme parameter values might be very helpful for an even better exploration of this effect.

6 The Increase of the Astronomical Unit

6.1 The Observation

From the analysis of radiometric measurements of distances between the Earth and the major planets including observations from Martian orbiters and landers from 1961 to 2003 a secular increase of the astronomical unit of approximately 10 m per century has been reported [39] (see also [40] and the discussion therein).

6.2 Search for Explanation

Time-Dependent Gravitational Constant and Velocity of Light

This increase cannot be explained by a time-dependent gravitational constant G because the $\dot{G}/G$ needed is larger than the restrictions obtained from LLR.

It has also been speculated that a time-dependent change in the velocity of light can be responsible for this effect. Indeed, if the speed of light becomes smaller, then ranging will simulate a drift of distances. However, the inspection of Kepler's third law,

$$\frac{T^2}{a^3} = \frac{4\pi^2}{GM_\odot}, \tag{17}$$

shows that, if one replaces the distance a by a ranging time $a = ct$, then effectively the quotient G/c^3 appears. Only this combination of the gravitational constant and the speed of light governs the ratio between the orbit time, in our case the orbit time of the Earth. Consequently, a time-dependent speed of light is equivalent to a time-dependent gravitational constant. Since the latter has been ruled out to be possibly responsible for an increase of the astronomical unit, also a time-dependent speed of light has to be ruled out.

Cosmic Expansion

The influence of the cosmic expansion on planetary orbits is by many orders of magnitude too small to be in principle responsible for such an observation, see Sect. 9.2. Neither the modification of the gravitational field of the Sun nor the drag of the planetary orbits due to the expansion is big enough to explain this drift.

Clock Drift

An increase of ranged distances might also be due to a drift of the timescale of the form $t \to t + \alpha t^2$ for $\alpha > 0$. This is of the same form as the time drift needed to account for the Pioneer anomaly. From Kepler's third law, one may ask which α is suitable to simulate the increase of the astronomical unit. One obtains $\alpha \approx 3 \cdot 10^{-20}\,\mathrm{s}^{-1}$ what is astonishing close to the clock drift needed for a clock drift simulation of the Pioneer anomaly, see (16) and below.

7 The Quadrupole and Octupule Anomaly

Recently, an anomalous behavior of the low-l contributions to the cosmic microwave background has been reported. It has been shown that (1) there exists an alignment between the quadrupole and octupole with >99.87% C.L. [41] and (2) the quadrupole and octupole are aligned to solar system ecliptic to >99% C.L. [42]. No correlation with the galactic plane has been found.

The reason for this is totally unclear. One may speculate that an unknown gravitational field within the solar system slightly redirects the incoming cosmic microwave radiation (in the similar way as a motion with a certain velocity with respect to the rest frame of the cosmological background redirects the cosmic background radiation and leads to modifications of the dipole and quadrupole parts). Such a redirection should be more pronounced for low-l components of the radiation. It should be possible to calculate the gravitational field needed for such a redirection and then to compare that with the observational data of the solar system and the other observed anomalies.

8 Summary of Anomalies

8.1 Summary

In our opinion, these above-described anomalies split into three groups according to their observational status: (1) the observations related to dark matter and dark energy are beyond any doubt, (2) the Pioneer anomaly and the flyby anomaly are on a good basis, and (3) the increase of the astronomical unit as well as the quadrupole and octupole anomaly is still under debate (Table 3).

This list of anomalies and unexplained phenomena immediately induces a bunch of tasks and questions:

- For each of these phenomena, except for the dark energy and dark matter, one still should try to find a systematic cause.
- One should also try to find conventional explanations of these effects within general relativity leading perhaps to an effect within conventional physics not considered up to now.

Table 3. List of anomalies and their status.

Anomaly	Observational status	Interpretation
Dark energy, dark matter	Well established	Under discussion
Pioneer anomaly, flyby anomaly	Quite well established	Unclear
Quadrupole anomaly, increase of AU	Unclear	Unclear

- A big step in understanding these effects might be to find a relation between two or more of the unexplained phenomena (all of these anomalies very probably are no isolated phenomena).
- This is tantamount to the important question whether all or at least some of these effects have a common cause.
- Motivated by the surprising fact $a_{\text{Pioneer}} \approx cH$ one should analyze the influence of the cosmic expansion on the physics within gravitationally bound systems. There already appeared quite some literature related this topic [43–46], and references therein (see also below).
- Since there seem to exist some strange phenomena within our solar system, one should look whether it might be possible to observe similar things in other gravitating systems like binary systems, binary pulsars, stars moving around the black hole in the center of our galaxy, etc.
- Is something wrong with weak gravity or gravity at large distances?
- Furthermore, one should propose new dedicated missions and space experiments.

8.2 Other Anomalies?

There is one further observation which status is rather unclear bit which perhaps may fit into the other observations. This is the observation of the return time of comets: comets usually come back a few days before they are expected when applying ordinary equations of motion. The delay usually is assigned to the outgassing of these objects. In fact, the delay is used for an estimate of the strength of this outgassing. On the other hand, it has been calculated in [47] that the assumption that starting with 20 AU there is an additional acceleration of the order of the Pioneer anomaly also leads to the effect that comets come back a few days earlier. It is not clear whether this is a serious indications, but a further study of the trajectories of comets is certainly worthwhile.

9 Ways to Describe the Effects

Many approaches have been attempted to explain these anomalies. In most of these attempts a link between one phenomenon and the issues of (1) the influence of the expansion of the universe on the physics within our solar system, (2) dark energy, and (3) dark matter has been tried to establish. We like to emphasize again that indeed it should be the first thing to explore whether there are links between these various observed and unexplained phenomena. It should be strange if all these unexplained phenomena will be really independent of each other, i.e., are not linked by a common (perhaps new) physical principle. (However, it also seems to be a rather difficult task to find similarities between the Pioneer anomaly and the flyby anomaly.)

9.1 Cosmological Constant

One first very simple attempt to generalize general relativity and to incorporate also findings related to dark energy is to describe the physics of gravitating bodies within a theory involving a cosmological constant. The corresponding field equations are

$$R_{\mu\nu} - \frac{1}{2} g_{\mu\nu} R + g_{\mu\nu} \Lambda = \kappa T_{\mu\nu}, \tag{18}$$

where $R_{\mu\nu}$ is the Ricci tensor; R the Ricci scalar; and $T_{\mu\nu}$ the energy–momentum tensor.

In the first step one may use this equation to describe the gravitational field of the Sun and physics within the solar system (Perihelion shift, light deflection, gravitational redshift, gravitational time delay, geodetic precession), see [48] as well as Iorio [49] and Sereno and Jetzer [50]. It comes out that the ordinary cosmological constant $\Lambda = \Lambda_0 = 10^{-52}\,\mathrm{m}^{-2}$ has no observable influence on all the solar system tests. One also can show that if one assumes a constant Λ such large that it may explain the Pioneer acceleration, then this is ruled out by the perihelion shift observation [48]. Therefore, a cosmological constant cannot be an explanation for the Pioneer anomaly. In the article by Jetzer and Sereno [51] also the influence of the cosmological constant on the motion of binary systems has been evaluated with the result that in the future binary systems may be precise enough to "see" the cosmological constant. Some of these effects have also been considered in a Kerr–de Sitter space–time [52].

The calculation of solar system effects in a Schwarzschild–de Sitter space–time is certainly only a simple first step. One may redo this kind of calculation in a more general context like quintessence [15, 53] (for a first step in this direction see [54]) in varying G scenarios [55], in dilaton scenarios [56,57], and in braneworld models [58].

9.2 The Influence of the Expansion of the Universe

We show that the expansion of the universe has no measurable influence whatsoever on the physics within the solar system. This includes the modifications of the gravitational field created by the Sun and the planetary orbits.

Modified Gravitational Field of the Sun

It is conceivable that the cosmic expansion may weaken the gravitational field of the Sun. However, it can be shown [45] that the corresponding effect is far beyond being observable. Starting from an expansion of the local metric around the cosmological background metric $b_{\mu\nu}$

$$g_{\mu\nu} = b_{\mu\nu} + h_{\mu\nu}\,, \quad \text{where} \quad h_{\mu\nu} \ll b_{\mu\nu}, \tag{19}$$

we obtain linearized Einstein equations for $h_{\mu\nu}$ [59]

$$g^{\rho\sigma} D_\rho D_\sigma \bar{h}_{\mu\nu} + 2 g^{\rho\sigma} R^{\kappa}{}_{\mu\rho\nu} \bar{h}_{\kappa\sigma} = 16\pi G T_{\mu\nu}. \tag{20}$$

The static solution for a small spherically symmetric mass distribution is given by

$$h_{00} = \frac{2GM}{R} \frac{\cos(\sqrt{6}|\dot{R}|r)}{r} = \frac{GM}{Rr}\left(1 - 3H^2(Rr)^2 \pm \ldots.\right) \tag{21}$$

At lowest order we obtain the standard Newtonian potential with the measured distance $R(t)r$. Since the modification is quadratic in the Hubble constant, the Newtonian potential practically does not participate in the cosmic expansion. This confirms the findings in [39].

Modified Planetary Orbits

If the planetary orbits may expand due to a drag from the cosmic expansion, then this might be interpreted as if the length unit, astronomical unit, increases so that it might appear that other distances may become smaller. However, from the theorem of adiabatic invariance [60], it can be shown that the parameters of the planetary orbits remain the same with very high stability [45]. This stability is characterized by the factor $\exp(-t/T)$ where t is the timescale of the planetary orbit and T is the timescale of the temporal change of the environment, in our case the Hubble time. Therefore, t/T is of the order 10^{10} so that the planetary orbits are extremely stable.

Another approach is based on the geodesic deviation which relates the relative acceleration of freely moving particles with the curvature of space–time, in our case with the expansion of the universe. This has been taken in [61] as the basis for estimates that this deviation is much too small to be of any influence in the solar system or even within galaxies. For a broader discussion of this approach, see the recent valuable review of Carrera and Giulini [46].

9.3 General Approach to Describe a Modified Particle Dynamics

Gravity describes or, equivalently, can be characterized by the behavior of light and point particles [62]. Light is related to a space–time metric while point particles are related to a geodesic equation. In Riemannian geometry, the mathematical model for general relativity, the equation of motion for particles and the behavior of light are related: The equation of motion for particles is completely determined by the space–time metric. For a general analysis of anomalous gravitational effects, it therefore may be appropriate to start by a general ansatz for the metric and of the equation of motion for particles.[3]

In a weak-field approximation the metric can be written in the form

$$g_{\mu\nu} = \begin{pmatrix} 1 - U & g_{0i} \\ g_{i0} & -\delta_{ij}(1 - V) \end{pmatrix}, \tag{22}$$

where U, V, and $g_{0i} = g_{i0}$ are assumed to be small quantities. U may be identified with the Newtonian potential. We denote $ds^2 = g_{\mu\nu} dx^\mu dx^\nu$.

By definition, the readout of moving clocks is given by

$$T = \int_{\text{worldline of clock}} ds \approx \int \left(1 - U + \dot{x}^2 + V\dot{x}^2\right) dt + \int \boldsymbol{h} \cdot d\boldsymbol{x}, \tag{23}$$

which is the proper time of the clock. A clock at rest will measure the gravitational redshift given by

$$\frac{\nu_1}{\nu_2} = \sqrt{\frac{g_{00}(x_1)}{g_{00}(x_2)}} \approx 1 + U(x_2) - U(x_1). \tag{24}$$

Now we are going to give a general description of the equation of motion for particles. This approach is according to the philosophy of axiomatic approaches to theories of gravity, see, e.g., [62, 63]. The equation of motion for a particle respecting the universality of free fall can be represented as

$$0 = v^\nu \partial_\mu v^\nu + H^\mu(x, v) = \left\{ {\mu \atop \rho\sigma} \right\} v^\mu v^\sigma + \gamma^\mu(x, v), \tag{25}$$

where $\left\{ {\mu \atop \rho\sigma} \right\} = \frac{1}{2} g^{\mu\nu} \left(\partial_\rho g_{\sigma\nu} + \partial_\sigma g_{\rho\nu} - \partial_\nu g_{\mu\nu}\right)$ is the Christoffel connection and $H^\mu(x, v)$ some vector valued function of the position and the velocity. We defined $\gamma^\mu(x, v) = H^\mu(x, v) - \left\{ {\mu \atop \rho\sigma} \right\} v^\rho v^\sigma$. Since no particle parameter enters this equation, the universality of free fall automatically is preserved. However, it is no longer possible to make a transformation to a coordinate system so that

[3] In the axiomatic approach [62], the mathematical relation between the point particle dynamics and the properties of light came in through a compatibility condition. One feature of this condition is a causal dynamics of point particles; particles are not allowed to move faster than light. The requirement of this condition implies a rather restricted geometrical structure, namely a Weylian structure. Since in [62] this condition has been applied to autoparallel curves and since we are here more general in admitting also nonlinear connections as they appear, e.g., in a Finslerian context, the resulting allowed particle dynamics is certainly still more general than a Weylian structure. However, this has to be worked out explicitly.

gravity disappears at one point (Einstein's elevator is not possible). Physically this means that, e.g., the acceleration of a body toward the Earth can depend on the velocity of the body: differently moving bodies feel a different gravitational acceleration which, however, does not depend on the composition or the weight of the (point-like) body.

From this general equation of motion, we can derive the 3-acceleration

$$\begin{aligned}\frac{d^2x^i}{dt^2} &= -\left(\left\{{i \atop \mu\nu}\right\} - \left\{{0 \atop \mu\nu}\right\}\frac{dx^i}{dt}\right)\frac{dx^\mu}{dt}\frac{dx^\nu}{dt} + \frac{1}{\left(\frac{dt}{ds}\right)^2}\left(\gamma^i(v,x) - \frac{dx^i}{dt}\gamma^0(v,x)\right) \\ &= -\left(\left\{{i \atop 00}\right\} - \left\{{0 \atop 00}\right\}\frac{dx^i}{dt}\right) - 2\left(\left\{{i \atop j0}\right\} - \left\{{0 \atop j0}\right\}\frac{dx^i}{dt}\right)\frac{dx^j}{dt} \\ &\quad -\left(\left\{{i \atop jk}\right\} - \left\{{0 \atop jk}\right\}\frac{dx^i}{dt}\right)\frac{dx^j}{dt}\frac{dx^k}{dt} + \frac{1}{\left(\frac{dt}{ds}\right)^2}\left(\gamma^i(v,x) - \frac{dx^i}{dt}\gamma^0(v,x)\right) \\ &\approx \underbrace{\partial_i U}_{\text{Newton}} + \underbrace{(\partial_i h_j - \partial_j h_i)\dot{x}^j}_{\text{Lense–Thirring}} + \dot{x}^2\partial_i V + \dot{x}^i\dot{V} + \Upsilon^i + \Upsilon^i_j\dot{x}^j + \Upsilon^i_{jk}\dot{x}^j\dot{x}^k + \dots,\end{aligned} \tag{26}$$

where we neglected all relativistic corrections since these play no role in the Pioneer and flyby anomalies.

The first term in (26) is the ordinary Newtonian acceleration and the second term the action of the gravitomagnetic field on the orbit of a satellite which has been observed by LAGEOS with a 10% accuracy [64]. This field also acts on spinning objects like gyroscopes and should be confirmed by GP-B with an accuracy better than 1%. The other terms are hypothetical terms beyond ordinary post-Newtonian approximation.

The V term which can be motivated by a running coupling constant to be proportional to the distance $V \sim r^2$ has been introduced by Jaekel and Reynaud [65] to describe the constant anomalous Pioneer acceleration. The other terms, most of them are velocity dependent, have not yet been analyzed. (The influence of an arbitrary force on the trajectories of planets has been analyzed recently in [29] with the main conclusion that any *radial* force that might be considered as being responsible for the Pioneer anomaly is not compatible with the recent analysis of the motion of the outer planets. This indicates that the modification of the equation of motion should include velocity-dependent terms. The same conclusion can be drawn from the fact that any *metric* modification of gravity which leads the Pioneer acceleration is in contradiction to solar system data [30].)

The coefficients $\Upsilon^i_{jk\ldots}$ depend only on the position and may vanish for vanishing gravitating mass. Therefore, the coefficients can contain M, r, r^i, and ∂_i only. Accordingly, these coefficients can be of the form

$$\Upsilon^i = A_{11}\frac{GM}{r^2}\frac{r^i}{r}, \tag{27}$$

$$\Upsilon^i_j = A_{21}\frac{GM}{r^2}\frac{r^i r^j}{r^2} + A_{22}\frac{GM}{r^2}\delta^i_j, \tag{28}$$

$$\Upsilon^i_{jk} = A_{31}\frac{GM}{r^2}\frac{r^i r^j r^k}{r^3} + A_{32}\frac{GM}{r^2}\frac{r^i}{r}\delta_{jk} + A_{33}\frac{GM}{r^2}\frac{r^j}{r}\delta^i_k. \tag{29}$$

Here A_{ij} are constants parametrizing the various contributions. It is understood that the influence of the gravitating body, i.e., all the Υ-coefficients, vanishes at spatial infinity. This is certainly true for a description of the flyby anomaly but may be relaxed for the Pioneer anomaly. Terms which do not vanish at spatial infinity but are of Newtonian form at small distances are polynomials r^l, $l \geq 1$.

The above terms lead to accelerations

$$\ddot{x}^i = A_{11}\frac{GM}{r^2}\frac{r^i}{r}, \tag{30}$$

$$\begin{aligned}\ddot{x}^i &= A_{21}\frac{GM}{r^2}\frac{r^i \boldsymbol{r}\cdot\dot{\boldsymbol{r}}}{cr^2} + A_{22}\frac{GM}{r^2}\frac{\dot{r}^i}{c},\\ &= (A_{21}+A_{22})\frac{GM}{r^2}\frac{r^i \boldsymbol{r}\cdot\dot{\boldsymbol{r}}}{cr^2} + A_{22}\frac{GM}{r^2}\frac{\dot{r}^i_\perp}{c},\end{aligned} \tag{31}$$

$$\ddot{x}^i = A_{31}\frac{GM}{r^2c^2}\frac{r^i(\boldsymbol{r}\cdot\dot{\boldsymbol{r}})^2}{r^3} + A_{32}\frac{GM}{c^2r^2}\frac{r^i}{r}\dot{r}^2 + A_{33}\frac{GM}{c^2r^2}\frac{\dot{r}^i}{r}(\boldsymbol{r}\cdot\dot{\boldsymbol{r}}), \tag{32}$$

where $r^i_\perp = r^i - r^i(\boldsymbol{r}\cdot\dot{\boldsymbol{r}})/r^2$ is the component of the body's velocity orthogonal to the connecting vector $\boldsymbol{r}$.

The first term associated with A_{11} is of Newtonian form and can be combined with the already existing one which amounts to a redefinition of the gravitational constant. The A_{22} term describes an additional acceleration in direction of the velocity. It fades away for large r. The A_{21} term projects the component of the velocity which is parallel to the connecting vector and leads to an acceleration in direction of the connecting vector. This term vanishes at the perigee.

These A_{21} and A_{22} terms may be chosen in such a way that they have the potential to describe an increase of the velocity during a flyby: near the Earth the A_{22} term is dominant since there the connecting vector is more or less orthogonal to the velocity vector. For large r both terms contribute. That is,

$$\ddot{x}^i = \begin{cases} A_{22}\dfrac{GM}{r^2}\dfrac{\dot{r}^i}{c} & \text{for } r \approx r_{\text{perigee}} \\ (A_{21}+A_{22})\dfrac{GM}{r^2}\dfrac{\dot{r}^i}{c} & \text{for } r \text{ large.}\end{cases} \tag{33}$$

Therefore, in principle it is possible to have an acceleration near the perigee (for $A_{22} > 0$) and a deceleration for large distances (for $A_{21} + A_{22} < 0$). Note that for a typical perigee and velocity at perigee the acceleration at perigee for $A_{22} = 1$ is about $10^{-4}\,\mathrm{m\,s^{-2}}$ which is just the value given for a typical Earth flyby, see Sect. 5.2. However, this model does not include the Pioneer deceleration because the acceleration is not constant for large r. It also can be shown that a term of the form (31) leads to unstable planetary orbits.[4]

The A_{32} term just adds to the $\dot{x}^2\partial_i V$ term. Both the A_{31} and the A_{33} terms first project the velocity in direction of the connecting vector and then make out of this an acceleration in direction of the connecting vector and in direction of the velocity. These A_{3i} terms are about five orders of magnitude smaller than the A_{2i} terms and can, thus, play no role in an explanation of the flyby anomaly. The higher-order terms will be of more complicated but similar structure.

In general, this equation of motion does not respect energy conservation: multiplication of (26) with the velocity yields

$$\frac{d}{dt}\left(\tfrac{1}{2}\dot{\boldsymbol{x}}^2 - U\right) = 2\dot{x}^2\dot{V} + \dot{\boldsymbol{x}}\cdot\boldsymbol{\Upsilon} + \Upsilon^i_j\dot{x}^j\dot{x}^i + \Upsilon^i_{jk}\dot{x}^j\dot{x}^k\dot{x}^i + \ldots . \tag{34}$$

Therefore, the terms on the right-hand side might be candidates for effects reducing or enlarging the kinetic energy of moving bodies and, thus, may play a role in the description of the flyby or the Pioneer anomaly.

It should be clear from the independence of the metric from the equation of motion for point particles that it is necessary both to track position and velocity of the satellite and to have a clock onboard to determine all components of the space–time metric.

10 Summary and Outlook

We collected the anomalies related to the physics of the solar system and discussed to some extent the error sources and possibilities to explain these anomalies. In particular, we tried to find similarities or fundamental differences between these anomalies.

As final statement we like to stress that there are at least three important science cases related to the exploration of these anomalies which we strongly suggest to be tackled in the near future:

1. Analysis of the complete set of Pioneer data
2. Continuous and complete (velocity, distance, time onboard, and direction) observations of future flybys

[4] Nearly all experiences with the gravitational interaction come from the observation of bodies in *bound* orbits: moons are in bound orbits around planets, planets around stars, stars are bound within the galaxy, galaxies are bound in galaxy clusters. We know only a very few examples of escape orbits mainly form artificial satellites. Therefore one may speculate that escape orbits might behave differently than bound orbits, an idea which also came up in discussions of the Pioneer anomaly. However, it is at least not obvious how this property of an orbit can be built into the equation of motion.

3. Search for clock drifts by comparison of clock rates on Earth with clocks defined by astrophysical systems

Acknowledgments

We like to thank J. Anderson, J.K. Campbell, D. Giulini, S. Grotian, P. Jetzer, G. Krasinsky, B. Mashhoon, T. Morley, E. Pitjeva, G. Schäfer, S. Solanki, and S. Turyshev for support, information, and various discussions. Financial support from the German Research Foundation DFG and the German Aerospace Agency DLR is gratefully acknowledged.

References

1. T. Sumner. Experimental searches for dark matter. *Living Rev. Relativity*, 5:`http://www.livingreviews.org/lrr-2002-4` (cited on 20.11.2005), 2002.
2. W. Hu. Mapping the dark matter through the cosmic microwave background damping tail. *Astrophys. J.*, 557:L79, 2001.
3. R.H. Sanders. Anti–gravity and galaxy rotation curves. *Astron. Astrophys.*, 136:L21, 1984.
4. M. Milgrom. MOND - theoretical aspects. *New Astr. Rev.*, 46:741, 2002.
5. R.H. Sanders and S.S. McGough. Modified newtonian dynamics as an alternative to dark matter. *Ann. Rev. Astron. Astrophys.*, 40:263, 2002.
6. M. Veltman. Problems and difficulties in standard model and gravitation. In T. Curtright, A. Perlmutter, and S. Mintz, editors, *The Launching of a Belle Epoche of High Energy Physics and Cosmology*, page 75. World Scientific, Singapore, 2003.
7. J.D. Bekenstein. Relativistic gravitation theory for the MOND paradigm. astro-ph/0403694.
8. A.G. Riess, A.V. Filippenko, and P. Challis et al. Measurements of Omega and Lambda from 42 high–redshift supernovae. *Astron. J.*, 116:1009, 1998.
9. S. Perlmutter, G. Aldering, and G. Goldhaber et al. Measurements of Omega and Lambda from 42 high–redshift supernovae. *Astroph. J.*, 517:565, 1999.
10. C.L. Bennet et al. First-year Wilkinson Microwave Anisotropy Probe (WMAP) observations: Preliminary maps and basic results. *Astrophys. J. Suppl. Ser.*, 148:1, 2003.
11. N.N. Spergel et al. First-year Wilkinson Microwave Anisotropy Probe (WMAP) observations: Determination of cosmological parameters. *Astrophys. J. Suppl. Ser.*, 148:175, 2003.
12. C. van de Bruck and W. Priester. The cosmological constant λ, the age of the universe and Dark Matter: Clues from the Lyα–forest. In H.V. Klapdor-Kleingrothaus, editor, *Dark Matter in Astrophysics and Particle Physics 1998: Proceedings of the Second International Conference on Dark Matter in Astrophysics and Particle*, page . Inst. of Physics, London, 1998.
13. J.O. Overduin and W. Priester. How dominant is the vacuum? *Naturwiss.*, 88:229, 2001.

14. M. Tegmark et al. Cosmological parameters from SDSS and WMAP. *Phys. Rev.*, D 69:103501, 2004.
15. P.J.E. Peebles and B. Ratra. The cosmological constant and dark energy. *Rev. Mod. Phys.*, 75:559, 2003.
16. S. Nojiri and S.D. Odintsov. Introduction to modified gravity and gravitational alternative for dark energy, 2006. hep-th/0601213.
17. M.N. Celerier. Do we really see a cosmological constant in the supernovae data? *Astron. Astroph.*, 353:63, 2000.
18. R.A. Vanderveld, E.E. Flanagan, and I. Wasserman. Mimicking dark energy with lemaitre-tolman-bondi models: Weak central singularities and critical points, 2006. astro-ph/0602476.
19. P.S. Apostolopoulos, N. Broudzakis, N. Tetradis, and E. Tsavara. Cosmological acceleration and gravitational collapse, 2006. astro-ph/0603234.
20. J.D Anderson, P.A. Laing, E.L. Lau, A.S. Liu, M.M. Nieto, and S.G. Turyshev. Indication, from Pioneer 10/11, Galileo, and Ulysses Data, of an Apparent Anomalous, Weak, Long–Range Acceleration. *Phys. Rev. Lett.*, 81:2858, 1998.
21. J.D. Anderson, P.A. Laing, E.L. Lau, A.S. Liu, M.M. Nieto, and S.G. Turyshev. Study of the anomalous acceleration of Pioneer 10 and 11. *Phys. Rev.*, D 65:082004, 2002.
22. J.D. Anderson and B. Mashhoon. Pioneer anomaly and the helicity-rotation coupling. *Phys. Lett.*, A 315:199, 2003.
23. M.M. Nieto, S. Turyshev, and J.D. Anderson. Directly measured limit on the interplanetary matter density from Pioneer 10 and 11. *Phys. Lett. B*, 613:11, 2005.
24. M.M. Nieto. Analytic gravitational–force calculations for models of the kuiper belt, with application to the Pioneer anomaly. *Phys. Rev.*, D 72:083004, 2005.
25. D. Bini, C. Cherubini, and B. Mashhoon. Vacuum c-metric and the gravitational stark effect. *Phys. Rev.*, D 70:044020, 2004.
26. A.F. Ranada. The Pioneer anomaly as acceleration of the clocks. *Found. Phys.*, 34:1955, 2005.
27. N. Wex. Pulsar timing – strong gravity clock experiments. In C. Lämmerzahl, C.W.F. Everitt, and F.W. Hehl, editors, *Gyros, Clocks, and Interferometers: Testing Relativistic Gravity in Space*, page 381. Springer–Verlag, Berlin, 2001.
28. D.M. Matsakis, J.H. Taylor, and T.M. Eubanks. A statistic for describing pulsar and clock stabilities. *Astron. Astrophys.*, 326:924, 1997.
29. L. Iorio. What do the orbital motions of the outer planets of the solar system tell us about the pioneer anomaly? gr.qc/0601055.
30. K. Tangen. Could the pioneer anomaly have a gravitational origin? gr-qc/0602089.
31. S.G. Turyshev, V.T. Toth, L.R. Kellog, E.L. Lau, and K.J. Lee. The study of the Pioneer anomaly: New data and objectives for new investigation. *Int. J. Mod. Phys.*, D 15:1, 2006.
32. V.T. Toth and S.G. Turyshev. The Pioneer anomaly: Seeking an explanation in newly recovered data, 2006. gr-qc/0603016.
33. H. Dittus and the Pioneer Explorer Collaboration. A mission to explore the Pioneer anomaly. In Á. Giménez et al, editor, *Trends in Space Science and Cosmic Vision 2020*, page 3. ESA, Noordwijk, 2005; gr-qc/0506139.
34. J.D. Anderson and J.G. Williams. Long-range tests of the equivalence principle. *Class. Quantum Grav.*, 18:2447, 2001.

35. P.G. Antreasian and J.R. Guinn. Investigations into the unexpected delta-v increase during the Earth Gravity Assist of GALILEO and NEAR. In., editor, *Astrodynamics Specialist Conf. and Exhibition*, pages paper no 98–4287. American Institute of Aeronautics and Astronautics, Washington, 1998.
36. J.K. Campbell. private communication.
37. T. Morley. Private communication.
38. T. Sumner, H. Araujo, D. Davidge, A. Howard, C. Lee, G. Rochester, D. Shaul, and P. Wass. Description of charging/discharging process of the LISA sensors. *Class. Quantum Grav*, 21:S597, 2004.
39. G.A. Krasinsky and V.A. Brumberg. Secular increase of astronomical unit from analysis of the major planets motions, and its interpretation. *Celest. Mech. & Dyn. Astron.*, 90:267, 2004.
40. E.M. Standish. The Astronomical Unit now. In D.W. Kurtz, editor, *Transits of Venus: New Views of the Solar System and Galaxy, , Proceedings IAU Colloquium No. 196*, page 163. Cambridge University Press, Cambridge, 2005.
41. A. de Oliveira-Costa, M. Tegmark, M.J. Devlin, L. Page, A.D. Miller, C.B. Netterfield, and Y. Xu. Solar and stellar system tests of the cosmological constant. *Phys. Rev.*, D 71:043004, 2005.
42. D.J. Schwarz, G.D. Starkman, D. Huterer, and C.J. Copi. Is the low–ℓ microwave background cosmic? *Phys. Rev. Lett.*, 93:221301, 2004.
43. A. Krasiński. *Inhomogenous Cosmological Models.* Cambridge University Press, Cambridge, 1997.
44. W.B. Bonnor. A generalization of the Einstein–Straus vacuole. *Class. Quantum Grav.*, 17:2739, 2000.
45. C. Lämmerzahl and H. Dittus. Doppler tracking in the expanding universe, 2005. preprint, ZARM, University of Bremen.
46. M. Carrera and D. Giulini. On the influence of the global cosmological expansion on the local dynamics in the solar system, 2005. gr-qc/0602098.
47. G.L. Page, D.S. Dixon, and J.F. Wallin. Can minor planets be used to assess gravity in the outer solar system?, 2005. astro-ph/0504367, to appear in Astrophys. J.
48. V. Kagramanova, J. Kunz, and C. Lämmerzahl. Solar system effects in Schwarzschild–de Sitter space–time. *Phys. Lett.*, 634:465, 2006.
49. L. Iorio. Can solar system observations tell us something about the cosmological constant? gr-qc/0511137.
50. M. Sereno and M. Jetzer. Solar and stellar system tests of the cosmological constant. *Phys. Rev.*, D 73:063004, 2006.
51. P. Jetzer and M. Sereno. Two-body problem with the cosmological constant and observational constraints. *Phys. Rev.*, D 73:044015, 2006.
52. A.W. Kerr, J.C. Hauck, and B. Mashhoon. Standard clocks, orbital precession and the cosmological constant. *Class. Qauntum Grav.*, 20:2727, 2003.
53. C. Wetterich. Cosmologies with variable Newton's "constant". *Nucl. Phys.*, B 302:645, 1988.
54. J.P. Mbelek. General relativity and quintessence explain the Pioneer anomaly. gr-qc/0402088.
55. M. Reuter and H. Weyer. Running newton constant, improved gravitational actions, and galaxy rotation curves. *Phys. Rev.*, D 70:124028, 2004.
56. T. Damour, F. Piazza, and G. Veneziano. Runaway dilaton and equivalence principle violations. *Phys. Rev. Lett.*, 89:081601, 2002.

57. T. Damour, F. Piazza, and G. Veneziano. Violations of the equivalence principle in a dilaton–runaway scenario. *Phys. Rev.*, D 66:046007, 2002.
58. D. Dvali, G. Gabadadze, and M. Porrati. 4D gravity on a brane in 5D Minkowski space. *Phys. Lett.*, B 485:208, 2000.
59. C.W. Misner, K. Thorne, and J.A. Wheeler. *Gravitation.* Freeman, San Francisco, 1973.
60. L.D. Landau and E.M. Lifshitz. *Mechanics.* Pergamon Press, Oxford, 1976.
61. F.I. Cooperstock, V. Faraoni, and D.N. Vollik. The influence of the cosmological expansion on local systems. *Astroph. J.*, 503:61, 1998.
62. J. Ehlers, F.A.E. Pirani, and A. Schild. The geometry of free fall and light propagation. In L. O'Raifeartaigh, editor, *General Relativity, Papers in Honour of J.L. Synge*, page 63. Clarendon Press, Oxford, 1972.
63. C. Lämmerzahl. On the equivalence principle in quantum mechanics. *Gen. Rel. Grav.*, 28:1043, 1996.
64. I. Ciufolini. Frame-dragging and Lense–Thirring effect. *Gen. Rel. Grav.*, 36: 2257, 2004.
65. M.-T. Jaekel and S. Reynaud. Gravity tests in the solar system and the pioneer anomaly. *Mod. Phys. Lett.*, A 20:to appear, 2005.
66. I. Mann and H. Kimura, Interstellar dust properties derived from mass density, mass distribution and flux rates in the heliosphere, *J. Geophys. Res.* A 105:10317, 2000.

Part II

Theory

Propagation of Light in the Gravitational Field of Binary Systems to Quadratic Order in Newton's Gravitational Constant

G. Schäfer and Michael H. Brügmann

Theoretisch-Physikalisches Institut, Friedrich-Schiller-Universität Jena, Max-Wien-Platz 1, 07743 Jena, Germany

Summary. The propagation of light is treated in the postlinear gravitational field of binary systems. The light deflection is calculated to quadratic order in Newton's gravitational constant and fourth order in the inverse power of the speed of light. Similarities and dissimilarities of linearized gravity and electrodynamics are discussed. A recent speed-of-gravity controversy is investigated.

1 Introduction

Today, technology has achieved a level at which the extremely high precision of current ground-based radio interferometric observations are approaching an accuracy of 1 μarcsec. Moreover the planned space-based astrometric telescope (GAIA)[1] and the space interferometric mission (SIM) are going to measure the positions and/or the parallaxes of celestial objects with uncertainties in the range 10^{-5}–10^{-6} arcsec. Furthermore the interferometer for the planned laser astrometric test of relativity mission (LATOR) will be able to measure light deflection angles of the order 10^{-8} arcsec.

To reach the desired accuracies of 10^{-6}–10^{-8} arcsec in the computation of light deflection in gravitational fields, corrections arising from the lack of spherical symmetry of the gravitating system, the motion of the gravitating masses, and the relativistic definition of the center of mass must be taken into account.

In this chapter, the light deflection in the postlinear gravitational field of two-bounded point-like masses is treated. However, to gain more insight into the interrelation between the Einstein field equations and the Maxwell equations, in the first part of the paper, the linear gravitational field is treated and its structural similarity with the electromagnetic field is discussed. Emphasis is put on the difference between linearized gravity and electrodynamics.

[1] The name GAIA derives from global astrometric interferometer for astrophysics, since GAIA was originally planned as a space-based interferometer.

In the second part of the paper, but still within the context of linearized gravity, a recent well-known speed-of-gravity controversy is discussed using a clear-cut approach for clarification. Finally, in the third part of the paper, in going over to the postlinear gravitational field, the light deflection is treated.

Notation

Let us summarize the notation and symbols used in this paper:

1. G is the Newtonian constant of gravitation
2. c is the speed of light
3. in Sect. 3, by c_{g} we denote the speed of gravity
4. The Greek indices α, β, γ, etc. are space–time indices and run from 0 to 3
5. The Latin indices i, j, k, etc. are spatial indices and run from 1 to 3
6. $\eta_{\mu\nu} = \eta^{\mu\nu} = \mathrm{diag}(-1, 1, 1, 1)$ is the Minkowski metric
7. $g_{\mu\nu}$ is a metric tensor of curved, four-dimensional space–time, depending on spatial coordinates and time
8. We suppose that space–time is covered by a harmonic coordinate system $(x^\mu) = (x^0, x^i)$, where $x^0 = ct$, t being the time coordinate
9. The three-dimensional quantities (3-vectors) are denoted by $\boldsymbol{a} = a^i$
10. The three-dimensional unit vector in the direction of $\boldsymbol{a}$ is denoted by $\boldsymbol{e}_a = e_a^i$
11. The Latin indices are lowered and raised by means of the unit matrix $\delta_{ij} = \delta^{ij} = \mathrm{diag}(1, 1, 1)$
12. By $_{,\sigma} = \partial_\sigma$ we denote the partial derivative with respect to the coordinate x^σ
13. The scalar product of any two 3-vectors $\boldsymbol{a}$ and $\boldsymbol{b}$ with respect to the Euclidean metric δ_{ij} is denoted by $\boldsymbol{a} \cdot \boldsymbol{b}$ and can be computed as $\boldsymbol{a} \cdot \boldsymbol{b} = \delta_{ij} a^i b^j = a^i b^i$
14. The Euclidean norm of a 3-vector $\boldsymbol{a}$ is denoted by $a = |\boldsymbol{a}|$ and can be computed as $a = (\delta_{mn} a^m a^n)^{1/2}$
15. By $\boldsymbol{l}_{(0)}$ we denote the vector tangent to the unperturbed light ray and the unit vector $\boldsymbol{e}_{(0)}$ is defined by $\boldsymbol{e}_{(0)} = \boldsymbol{l}_{(0)} / |\boldsymbol{l}_{(0)}|$
16. ∇ denotes the vector operator $\boldsymbol{e}_x \partial_x + \boldsymbol{e}_y \partial_y + \boldsymbol{e}_z \partial_z$
17. Δ denotes the usual Laplace operator in flat space
18. By $\Box \equiv \eta^{\mu\nu} \partial_\mu \partial_\nu = -\partial_0^2 + \Delta$ we denote the flat d'Alembertian operator

2 Analogies Between Electrodynamics and Einsteinian Gravity

In linearized approximation, the complicated Einstein theory with the group of general coordinate transformations as symmetry group simplifies to an abelian gauge theory. Electrodynamics is an abelian gauge theory too, if also with a single group parameter in contrast to linearized gravity theory which has four group parameters, so there are analogies between both theories to be expected.

2.1 Gauge-Invariant Electrodynamics

In vacuum space–time, the Maxwell equations have the form (Gaussian units)

$$\boldsymbol{\nabla}\cdot\boldsymbol{B}=0, \qquad \boldsymbol{\nabla}\times\boldsymbol{E}+\frac{1}{c}\frac{\partial}{\partial t}\boldsymbol{B}=0, \tag{1}$$

$$\boldsymbol{\nabla}\cdot\boldsymbol{E}=4\pi\varrho, \qquad \boldsymbol{\nabla}\times\boldsymbol{B}-\frac{1}{c}\frac{\partial}{\partial t}\boldsymbol{E}=\frac{4\pi}{c}\boldsymbol{j}. \tag{2}$$

Hereof the conservation equation for the charge follows:

$$\frac{\partial}{\partial t}\varrho+\boldsymbol{\nabla}\cdot\boldsymbol{j}=0. \tag{3}$$

In covariant form, the Maxwell equations read,

$$\partial_\nu F^{\mu\nu}=\frac{4\pi}{c}j^\mu, \qquad \partial_\sigma F_{\mu\nu}+\partial_\mu F_{\nu\sigma}+\partial_\nu F_{\sigma\mu}=0, \tag{4}$$

and the conservation equation takes the form

$$\partial_\mu j^\mu=0. \tag{5}$$

Here, the definitions hold,

$$F_{\mu\nu}=(\boldsymbol{E},\boldsymbol{B}), \qquad j^\mu=(c\varrho,\boldsymbol{j}), \qquad F_{\mu\nu}=-F_{\nu\mu}. \tag{6}$$

The Lorentz force and power expressions are, respectively,

$$\boldsymbol{k}=\varrho\boldsymbol{E}+\frac{1}{c}\boldsymbol{j}\times\boldsymbol{B}, \qquad \boldsymbol{k}\cdot\boldsymbol{v}=\boldsymbol{E}\cdot\boldsymbol{j}, \tag{7}$$

where $\boldsymbol{j}=\varrho\boldsymbol{v}$. In covariant notation, the four-dimensional force density reads

$$k_\mu=\frac{1}{c}F_{\mu\nu}j^\nu=\left(-\frac{1}{c}\boldsymbol{E}\cdot\boldsymbol{j},\boldsymbol{k}\right). \tag{8}$$

All the given expressions in this section have physical meaning, locally.

2.2 Electrodynamics in Gauge-Field Form

Introducing the gauge-field A_μ according to

$$\boldsymbol{E}=-\nabla\phi-\frac{1}{c}\frac{\partial}{\partial t}\mathbf{A}, \qquad \boldsymbol{B}=\boldsymbol{\nabla}\times\boldsymbol{A} \tag{9}$$

or, in four-dimensional form,

$$F_{\mu\nu}=\partial_\mu A_\nu-\partial_\nu A_\mu, \qquad A_\mu=(-\phi,\boldsymbol{A}), \tag{10}$$

the field equations (1) and (2) transform into the equations

$$-\partial^{\nu}\partial_{\nu}A^{\mu} + \partial^{\mu}\partial_{\nu}A^{\nu} = \frac{4\pi}{c}j^{\mu}. \tag{11}$$

A gauge transformation is given by,

$$A'_{\mu} = A_{\mu} + \partial_{\mu}\Lambda, \qquad F'_{\mu\nu} = F_{\mu\nu}. \tag{12}$$

It includes one arbitrary function Λ. The Lorentz gauge condition is defined by

$$\frac{1}{c}\frac{\partial}{\partial t}\phi + \nabla \cdot \boldsymbol{A} = \partial_{\mu}A^{\mu} = 0. \tag{13}$$

Herewith, the field equations (11) result in

$$\partial^{\nu}\partial_{\nu}A^{\mu} = -\frac{4\pi}{c}j^{\mu}. \tag{14}$$

In gauge-field form, the Maxwell equations were put onto a footing which is close in form to the Einstein field equations in linearized approximation.

2.3 The Linearized Einstein Theory

In linearized approximation, applying the harmonic or Hilbert–Lorentz gauge condition, the Einstein field equations read, e.g., see [1]

$$\partial^{\lambda}\partial_{\lambda}\bar{h}^{\mu\nu} = -\frac{16\pi G}{c^4}T^{\mu\nu}, \tag{15}$$

where the harmonic coordinate condition reads

$$\partial_{\mu}\bar{h}^{\mu\nu} = 0. \tag{16}$$

The field equations (15) together with the harmonic coordinate condition (16) imply the conservation law for the matter stress–energy tensor

$$\partial_{\mu}T^{\mu\nu} = 0. \tag{17}$$

The barred field $\bar{h}^{\mu\nu}$ is connected with the metric tensor $g_{\mu\nu}$ as follows

$$h_{\mu\nu} = \bar{h}_{\mu\nu} - \frac{1}{2}\eta_{\mu\nu}\eta^{\alpha\beta}\bar{h}_{\alpha\beta}, \qquad g_{\mu\nu} = \eta_{\mu\nu} + h_{\mu\nu}, \tag{18}$$

where raising and lowering of indices are with the Minkowski metric. Introducing the notations, cf. [2]

$$T^{00} = \varrho c^2, \qquad T^{0i} = cj^i, \tag{19}$$

$$\bar{h}^{00} = 4\varphi/c^2, \qquad \bar{h}^{0i} = 4a^i/c^2, \qquad \bar{h}^{ij} = O(1/c^4), \tag{20}$$

and

$$\boldsymbol{E} = -\nabla\varphi - \frac{1}{c}\frac{\partial}{\partial t}\boldsymbol{a}, \qquad \boldsymbol{B} = \nabla \times \boldsymbol{a}, \tag{21}$$

the field equations (15) take the form,

$$\boldsymbol{\nabla} \cdot \boldsymbol{B} = 0, \qquad \boldsymbol{\nabla} \times \boldsymbol{E} + \frac{1}{c}\frac{\partial}{\partial t}\boldsymbol{B} = 0, \tag{22}$$

$$\boldsymbol{\nabla} \cdot \boldsymbol{E} = 4\pi\varrho, \qquad \boldsymbol{\nabla} \times \boldsymbol{B} - \frac{1}{c}\frac{\partial}{\partial t}\boldsymbol{E} = \frac{4\pi}{c}\boldsymbol{j}. \tag{23}$$

Hereof the time component of the conservation equation (17) follows

$$\frac{\partial}{\partial t}\varrho + \boldsymbol{\nabla} \cdot \boldsymbol{j} = 0, \qquad (\partial_\nu T^{0\nu} = 0). \tag{24}$$

Obviously, whereas from the field equations (15), together with the gauge condition (16), four conservation equations follow, namely (17), from the field equations (22) and (23), only one conservation equation results (24).

The force and power expressions have to be added to (22) and (23) from outside because $(c\varrho, j^i)$ is treated therein as 4-vector and not as components of a tensor

$$\boldsymbol{k} = -\left(\varrho\boldsymbol{E} + \frac{4}{c}\boldsymbol{j} \times \boldsymbol{B}\right), \qquad \boldsymbol{k} \cdot \boldsymbol{v} = -\boldsymbol{E} \cdot \boldsymbol{j}, \tag{25}$$

where $\boldsymbol{j} = \varrho\boldsymbol{v}$ and where a point-mass model has been assumed for the matter.

For point masses, some analogy between electrodynamics and the linearized Einstein theory has been achieved apart from a minus sign and a factor of 4. The first difference relates to the attraction of gravity for all masses and the second one to the tensorial structure of gravity. However, there is a much bigger difference present which also relates to the treatment of $(c\varrho, j^i)$ as 4-vector. The electromagnetic field equations (4) are gauge invariant against the transformation

$$A'^0 = A^0 - \partial_0\Lambda, \tag{26}$$

$$A'^i = A^i + \partial_i\Lambda. \tag{27}$$

The linearized Einstein field equations in electrodynamic form, (22) and (23), however, are not invariant against the gauge transformations of linearized gravity which are given by, containing four arbitrary functions ϵ^μ,

$$\bar{h}'^{00} = \bar{h}^{00} - \partial_0\epsilon^0 + \partial_j\epsilon^j, \tag{28}$$

$$\bar{h}'^{0i} = \bar{h}^{0i} + \partial_i\epsilon^0 - \partial_0\epsilon^i, \tag{29}$$

$$\bar{h}'^{ij} = \bar{h}^{ij} + \partial_i\epsilon^j + \partial_j\epsilon^i - \delta_{ij}\partial_\mu\epsilon^\mu. \tag{30}$$

Only in the case of vanishing ϵ^i, the above field equations (22) and (23) are invariant. This means that the linearized Einstein field equations in the electrodynamic form, (22) and (23), have no physical meaning, locally, in contrast to the Maxwell equations (1) and (2).

2.4 The Linearized Einstein Theory in Gauge-Invariant Form

A locally gauge-invariant representation of the linearized Einstein theory can be achieved with the aid of the Riemann curvature tensor

$$R_{\mu\nu\sigma\tau} = \frac{1}{2}(\partial_\nu\partial_\sigma h_{\mu\tau} + \partial_\mu\partial_\tau h_{\nu\sigma} - \partial_\nu\partial_\tau h_{\mu\sigma} - \partial_\mu\partial_\sigma h_{\nu\tau}) \tag{31}$$

which is an invariant object under the gauge transformations (28)–(30)

$$R'_{\mu\nu\sigma\tau} = R_{\mu\nu\sigma\tau}. \tag{32}$$

Calling, respectively, e.g., see [3],

$$E_{ij} = R_{i0j0}, \qquad H_{ij} = \frac{1}{2}\epsilon_{ikl}R_{klj0}, \tag{33}$$

the electric and magnetic components of the curvature tensor, all its components can be recovered in the form:

$$R_{i0j0} = E_{ij}, \qquad R_{ijk0} = \epsilon_{ijl}H_{lk}, \tag{34}$$

$$R_{ijkl} = \epsilon_{ijm}\epsilon_{kln}\left(-E_{mn} + \frac{1}{2}J_{mn}\right), \tag{35}$$

$$J_{ij} = \frac{8\pi G}{c^4}\left(-T_{ij} + \frac{1}{2}\delta_{ij}(T_{00} + T_{kk})\right). \tag{36}$$

The fully gauge-invariant field equations for linearized Einstein theory read

$$\nabla \cdot \mathbf{H} = 0, \qquad \nabla \times \mathbf{E} + \frac{1}{c}\frac{\partial}{\partial t}\mathbf{H} = 0, \tag{37}$$

$$\nabla \cdot \mathbf{E} = \nabla \cdot \mathbf{J}, \qquad \nabla \times \mathbf{H}^T - \frac{1}{c}\frac{\partial}{\partial t}\mathbf{E} = -\frac{1}{c}\frac{\partial}{\partial t}\mathbf{J}, \tag{38}$$

where $\mathbf{H}^T$ denotes the transposed of the dyadic $\mathbf{H}$; $\mathbf{E}^T = \mathbf{E}$, $\mathbf{J}^T = \mathbf{J}$. (Notice the similarity of the inhomogeneous equations with the macroscopic Maxwell equations with polarization, i.e., dipole sources.) These equations do have local meaning as the expression

$$K_i = -\left(c^2E_{ij}X^j + 2c\epsilon_{ikl}V^kH_{lj}X^j\right) = -\left(E_i + \frac{2}{c}(\mathbf{V} \times \mathbf{H})_i\right) \tag{39}$$

does which describes the tidal force on two particles with unit mass, separated by the vector X^i ($V^i = \frac{dX^i}{dt}$), where

$$E_i = c^2E_{ij}X^j, \qquad H_i = c^2H_{ij}X^j. \tag{40}$$

The second-order field equations for components of the Riemann tensor read

$$\Box E_{ij} = \frac{8\pi G}{c^4}\left[\partial_0^2(T_{ij} - \frac{1}{2}\delta_{ij}T) + \partial_i\partial_j(T_{00} + \frac{1}{2}T) - \partial_0(\partial_iT_{j0} + \partial_jT_{i0})\right], \tag{41}$$

$$\Box H_{ij} = \frac{8\pi G}{c^4}\epsilon_{ilk}\left[\partial_0\partial_k(T_{jl} - \frac{1}{2}\delta_{jl}T) - \partial_k\partial_j T_{l0}\right]. \tag{42}$$

Under stationarity conditions, the field equations become,

$$\Delta E_{ij} = \frac{4\pi G}{c^4}\partial_i\partial_j(T_{00} + T_{kk}), \tag{43}$$

$$\Delta H_{ij} = \frac{8\pi G}{c^4}\epsilon_{ilk}\partial_l\partial_j T_{0k}. \tag{44}$$

In the Newtonian limit, the well-known tidal-force potential results,

$$E_{ij} = -\frac{1}{c^2}\partial_i\partial_j\varphi, \qquad \varphi = G\int d^3x'\frac{\varrho(\boldsymbol{x}')}{|\boldsymbol{x} - \boldsymbol{x}'|}. \tag{45}$$

3 On the Speed-of-Gravity Controversy

Recently it has been claimed that the speed of gravity should be measurable by radio observations of a bright radio quasar J0842+1835, during the time of its line-of-sight close angular encounter with Jupiter by very long baseline interferometer (VLBI), predicted to occur on 8 September 2002 [4]. The theoretical basis of above erroneous conclusion rests upon interpreting relativistic corrections to the famous Shapiro delay. The first criticism of [4], raised by Asada, points out that the excess time delay is a light-cone effect only, hence should not involve the speed of gravity [5]. Recently, inaugurated by a new paper which supports Kopeikin's earlier interpretation of "the higher-order Shapiro delay" [6], a strong criticism was raised by Will, who presented a detailed calculation for the relativistic corrections to the Shapiro delay in the parametrized post-Newtonian framework and showed that the above-mentioned VLBI measurements are insensitive to the speed of propagation of gravity [7]. In the final version of his recent publication, Kopeikin strongly criticized the conclusions reached by Asada and Will by pointing out unsatisfactory aspects, both conceptual and calculational, of their treatments [6]. In this article, we will provide a firm mathematical footing to the analysis of Asada and Will, and convincingly show that the speed of gravity is not sensitive to the measurements of radio waves, emitted by the quasar J0842+1835 and deflected by the moving Jupiter, performed by VLBI. We shall also point out the conceptual error committed by Kopeikin which allowed him to interpret erroneously the above-mentioned VLBI observations [8].

3.1 The Approach by C.M. Will

Though the final expression for the relativistic time delay presented by Asada is consistent with that obtained by Kopeikin (compare (10) and (12) in [5] and [4], respectively), he pointed out that Asada's derivation assumed that the position of Jupiter be fixed at retarded light-cone time, which makes his derivation somewhat ad hoc. Kopeikin also raised few concerns over the higher-order time-delay computations in [7], especially the way time-delay integral was evaluated (refer Sect. B in [6]). Below, we will present an elegant integration of the relativistic time-delay equation, which is free of blemishes associated with Will's treatment, as indicated by Kopeikin. This will help us to justify mathematically Asada's result too.

We start with the time-delay equation, (16) of [7], but dropping the summation symbol there,

$$\Delta(t_r, t_e) = (1+\gamma)\frac{Gm_a}{c^2}\int_{t_e}^{t_r} \frac{(1-(2+\zeta)\boldsymbol{e}_{(0)}\cdot\boldsymbol{v}_a(s_a)/c)dt}{|\boldsymbol{z}-\boldsymbol{x}_a(s_a)| - \boldsymbol{v}_a(s_a)\cdot(\boldsymbol{z}-\boldsymbol{x}_a(s_a))/c_g}, \quad (46)$$

where $\boldsymbol{x}_a$, $\boldsymbol{v}_a$, and m_a are the position vector, the velocity vector, and the mass of the gravitational source, respectively. t_e and t_r denote the light ray (photon) emission and reception instances. The Newtonian gravitational constant and the speed of gravity are denoted by G and c_g. The constant unit vector along the incoming light ray $\boldsymbol{e}_{(0)}$ helps us to define the unperturbed photon trajectory as

$$\boldsymbol{z} \equiv \boldsymbol{z}(t) = \boldsymbol{e}_{(0)}c(t-t_e) + \boldsymbol{z}_e, \quad (47)$$

and the retarded time s_a is given by $s_a = t - |\boldsymbol{z}-\boldsymbol{x}_a(s_a)|/c_g$. The underlying reference frame is an inertial frame where as well the observer as the source of the light ray is treated to be at rest. The time-delay expression in the Einstein theory results from putting $\gamma = 1, \zeta = 0$, and $c_g = c$. The advantage of working within a well-posed generalized framework is the natural difference therein between the speed of gravity and the speed of light.

It may already be noted here that terms of the type $\boldsymbol{e}_{(0)}\cdot\mathbf{v}_a(s_a)/c$ in the numerator of (46) can be neglected for the interpretation of the data from the Jupiter VLBI experiment. Only the denominator in (46) is relevant. Throughout the rest of the chapter, we will assume that the source of the gravitational field is uniformly moving, making $\boldsymbol{v}_a$ a constant.

Following techniques used in the computation of electric and magnetic fields, using Liénard–Wiechert potentials (see [10], Sect. 63), we write the denominator in the integrand of (46) as

$$|\boldsymbol{z}-\boldsymbol{x}_a(s_a)| - \boldsymbol{v}_a\cdot\frac{\boldsymbol{z}-\boldsymbol{x}_a(s_a)}{c_g} = |\boldsymbol{z}-\boldsymbol{x}_a(t)|\left(1-\left(\frac{v_a}{c_g}\right)^2\sin^2\theta_t\right)^{1/2}, \quad (48)$$

where θ_t is the angle between $\boldsymbol{z}-\boldsymbol{x}_a(t)$ and $\boldsymbol{v}_a$. To elegantly integrate (46), we introduce the following expression, where the retardation is with respect to the speed of light c,

$$|\boldsymbol{z}-\boldsymbol{x}_a(u_a)| - \boldsymbol{v}_a\cdot\frac{\boldsymbol{z}-\boldsymbol{x}_a(u_a)}{c} = |\boldsymbol{z}-\boldsymbol{x}_a(t)|\left(1-\left(\frac{v_a}{c}\right)^2\sin^2\theta_t\right)^{1/2} \quad (49)$$

where $u_a = t - |\boldsymbol{z} - \boldsymbol{x}_a(u_a)|/c$. Using the above expression, we may write (48) as

$$|\boldsymbol{z} - \boldsymbol{x}_a(s_a)| - \boldsymbol{v}_a \cdot \frac{\boldsymbol{z} - \boldsymbol{x}_a(s_a)}{c_g}$$
$$= \left[|\boldsymbol{z} - \boldsymbol{x}_a(u_a)| - \boldsymbol{v}_a \cdot \frac{\boldsymbol{z} - \boldsymbol{x}_a(u_a)}{c}\right] \left(\frac{1 - (v_a/c_g)^2 \sin^2\theta_t}{1 - (v_a/c)^2 \sin^2\theta_t}\right)^{1/2}. \quad (50)$$

Restricting right-hand side of above equation to $O(v_a^2)$ and plugging it in (46) we obtain,

$$\Delta(t_r, t_e) = (1+\gamma)\frac{Gm_a}{c^2}\int_{t_e}^{t_r} \frac{(1 - (2+\zeta)\boldsymbol{e}_{(0)} \cdot \mathbf{v}_a/c)dt}{|\boldsymbol{z} - \boldsymbol{x}_a(u_a)| - \boldsymbol{v}_a \cdot (\boldsymbol{z} - \boldsymbol{x}_a(u_a))/c}. \quad (51)$$

Using (25), (28), (45), and (50) in [9], which are quite the merit equations of that paper, it is straightforward to obtain, without any further approximation, the relativistic time-delay expression in the following form

$$\Delta(t_r, t_e) = -(1+\gamma)\frac{Gm_a}{c^3}\left(1 - (1+\zeta)\boldsymbol{e}_{(0)} \cdot \frac{\boldsymbol{v}_a}{c}\right) \ln \frac{r_{ra}(u_r) - \boldsymbol{e}_{(0)} \cdot \boldsymbol{r}_{ra}(u_r)}{r_{ea}(u_e) - \boldsymbol{e}_{(0)} \cdot \boldsymbol{r}_{ea}(u_e)}, \quad (52)$$

where the retarded times u_r and u_e are given by

$$u_i = t_i - \frac{r_{ia}(u_i)}{c}, \qquad i = r, e \quad (53)$$

and $\boldsymbol{r}_{ia}(u_i) = \boldsymbol{z}_i - \boldsymbol{x}_a(u_i)$ with $r_{ia}(u_i) = |\boldsymbol{r}_{ia}(u_i)|$.

It is clear that above equation is not very useful, as it involves unknown constants like t_e, the instant of time when the photon was ejected and $\boldsymbol{z}_e$, a vector associated with its origin. To eliminate t_e and $\boldsymbol{z}_e$, we introduce a second observer and let t_1 and t_2 be the reception times at these receivers for a photon characterized by t_e and $\boldsymbol{z}_e$. The relativistic time delay, given by (52), becomes

$$\Delta(t_2, t_1) = -(1+\gamma)\frac{Gm_J}{c^3}\left(1 - (1+\zeta)\boldsymbol{e}_{(0)} \cdot \frac{\boldsymbol{v}_J}{c}\right) \ln \frac{r_{2J}(u_2) - \boldsymbol{e}_{(0)} \cdot \boldsymbol{r}_{2J}(u_2)}{r_{1J}(u_1) - \boldsymbol{e}_{(0)} \cdot \boldsymbol{r}_{1J}(u_1)}, \quad (54)$$

where the index J stands for the Jupiter. The new retarded instances are

$$u_i = t_i - \frac{r_{iJ}(u_i)}{c}, \qquad i = 1, 2 \quad (55)$$

along with

$$\boldsymbol{r}_{iJ}(u_i) = \boldsymbol{z}_i - \boldsymbol{x}_J(u_i), \quad r_{iJ} = |\boldsymbol{r}_{iJ}(u_i)|, \qquad i = 1, 2. \quad (56)$$

It is clear that in the expression for the relativistic time delay, given by (54), the speed of gravity plays absolutely no role. The expression, which should be used to interpret astronomical observations like that made by VLBI on the fall of 2002, may be obtained by simply replacing $(1-(1+\zeta)\boldsymbol{e}_{(0)}\cdot\boldsymbol{v}_J/c)$ by 1 in (54). This is so as the effects associated with the g_{0i}-component of the gravitational field (see [7]), the so-called *gravitomagnetic field*, may be neglected during such observations. The final expression for the relativistic Shapiro delay reads

$$\Delta(t_2,t_1) = -(1+\gamma)\frac{Gm_J}{c^3}\ln\frac{r_{2J}(u_2)-\boldsymbol{e}_{(0)}\cdot\boldsymbol{r}_{2J}(u_2)}{r_{1J}(u_1)-\boldsymbol{e}_{(0)}\cdot\boldsymbol{r}_{1J}(u_1)}. \tag{57}$$

This proves the correctness of the ansatz used in [5], when $\gamma = 1$. The above equation also agrees with (34) and (35) of [7]. We feel that it is important to stress again what (54) or (57) really implies. They demonstrate that, whenever measurements of the gravitational time delay for electromagnetic radiation passing by a moving massive object, similar to the VLBI observations of 8 September 2002, are interpreted, the only field velocity that enters the analysis is that of the light.

3.2 The Treatment by S.M. Kopeikin

In this section, we closely scrutinize Kopeikin's arguments to see how he reached his erroneous conclusion that the above-mentioned VLBI observations measure the speed of gravity. The time-delay equation employed by Kopeikin reads

$$\Delta(t_r,t_e) = \frac{2Gm_a}{c^2}\int_{t_e}^{t_r}\frac{\left(1-2\boldsymbol{e}_{(0)}\cdot\boldsymbol{v}_a(s_a)/c_g\right)dt}{|\boldsymbol{z}-\boldsymbol{x}_a(s_a)|-\boldsymbol{v}_a(s_a)\cdot(\boldsymbol{z}-\boldsymbol{x}_a(s_a))/c_g}, \tag{58}$$

where

$$s_a = \tau - \frac{|\boldsymbol{z}-\boldsymbol{x}_a(s_a)|}{c_g} \quad \text{with} \quad \tau \equiv \frac{ct}{c_g}. \tag{59}$$

The velocities $\boldsymbol{v}_a(s_a)$ are also defined with respect to the new time variable τ. However, for the light propagation he still uses (47), which is

$$\boldsymbol{z} \equiv \boldsymbol{z}(t) = \boldsymbol{e}_{(0)}c(t-t_e)+\boldsymbol{z}_e, \tag{60}$$

Note that (58) is quite similar in form to (46), we employed in Sect. 3.1. We integrate (58) in the same manner as the time-delay integral was performed in Sect. 3.1. The final result, expressed in terms of τ, reads

$$\Delta(\tau_r,\tau_e) = -\frac{2Gm_a}{c^3}\left(1-\boldsymbol{e}_{(0)}\cdot\frac{\boldsymbol{v}_a}{c_g}\right)\ln\frac{r_{ra}(s_r)-\boldsymbol{e}_{(0)}\cdot\boldsymbol{r}_{ra}(s_r)}{r_{ea}(s_e)-\boldsymbol{e}_{(0)}\cdot\boldsymbol{r}_{ea}(s_e)}, \tag{61}$$

where the retarded times s_e and s_r, associated with the positions of emission and reception of the photon, are

$$s_i = \tau_i - \frac{r_{ia}(s_i)}{c}\,, \qquad i = e, r \tag{62}$$

In above equation, $r_{ea}(s_e)$ and $r_{ra}(s_r)$ are given by

$$r_{ia}(s_i) = |\boldsymbol{z}_i - \boldsymbol{x}_a(s_i)|\,, \quad \boldsymbol{r}_{ia}(s_i) = \boldsymbol{z}_i - \boldsymbol{x}_a(s_i)\,, \quad i = e, r. \tag{63}$$

Introducing a second observer and following exactly what have been done after (52) to get (57), we obtain, for $\gamma = 1$,

$$\Delta(\tau_2, \tau_1) = -\frac{2Gm_J}{c^3} \ln \frac{r_{2J}(s_2) - \boldsymbol{e}_{(0)} \cdot \boldsymbol{r}_{2J}(s_2)}{r_{1J}(s_1) - \boldsymbol{e}_{(0)} \cdot \boldsymbol{r}_{1J}(s_1)}, \tag{64}$$

where τ_1 and τ_2 are the fiducial reception times for the deflected photon at the positions of two VLBI observers.

Since the time t, associated with the photon propagation, is related to the fiducial time τ by $t = (c_g/c)\,\tau$, we are free to introduce another retardation $u = (c_g/c)\,s$. This indicates that we have the freedom to replace Kopeikin's arbitrarily defined retardations s_1 and s_2 with u_1 and u_2, where

$$u_a = t - \frac{|\boldsymbol{z} - \boldsymbol{x}_a(u_a)|}{c}, \quad a = 1, 2. \tag{65}$$

In terms of u_1 and u_2, (64) completely agrees with (57) when $\gamma = 1$. We emphasize that it is the time t, associated with the propagation of light, that is involved in the true measurements of velocities and hence to be used in the interpretation of astronomical observations. Kopeikin, however, used fiducial τ to interpret the VLBI observations of 8 September 2002.

4 Light Deflection in the Gravitational Field of a Compact Binary System

In this section, we shall recapitulate the computations of the angle of light deflection in the gravitational field of a compact binary system in the linear and postlinear approximations, which were presented in [9,11]. Both the light source and the observer are assumed to be located at infinity in an asymptotically flat space. The equations of light propagation are explicitly integrated to the second order in G/c^2. We assume that the impact parameter $|\boldsymbol{\xi}|$ is much larger (five times or more) than the distance r_{12} between the two components of the binary system.

4.1 Light Propagation and Light Deflection in the Gravitational Field of Compact Binary System

Since the light ray is propagating in a weak gravitational field, we can assume that the light propagation is very well governed by the laws of geometric optics, whereby light rays (photons) move in curved space–time along null geodesics. The equations of null geodesics with the time coordinate as parameter are given by (e.g., see [12])

$$\frac{dl^i}{dt} + \Gamma^i_{\alpha\beta} l^\alpha l^\beta = c^{-1} \Gamma^0_{\nu\sigma} l^\nu l^\sigma l^i, \tag{66}$$

where $\Gamma^\mu_{\rho\sigma}$ are the Christoffel symbols of the second kind and $l^\mu = \frac{dz^\mu}{dt}$ denotes the 4-vector $l^\mu = (c, l^i)$. Here, it is important to point out that l^μ is not exactly a 4-vector since we differentiate with respect to the time coordinate t. So l^μ is a 4-vector up to a factor. The spatial part of l^μ given by $l^i = dz^i/dt$ is the 3-vector tangent to the light ray $z^i(t)$. In the present case of null geodesics, l^μ has to fulfill the condition

$$l^2 \equiv g_{\mu\nu}[z^0, z^i(t), G] l^\mu l^\nu = 0. \tag{67}$$

Now we consider a light ray $z^i(t)$ that is propagating in a curved space–time $g_{\mu\nu}[z^0, z^i(t), G]$. If the gravitational field is weak, we can write the fundamental metric tensor as a power series in the gravitational constant G

$$g_{\mu\nu}[z^0, z^i(t), G] \equiv \eta_{\mu\nu} + \sum_{n=1}^{\infty} h^{(n)}_{\mu\nu}[z^0, z^i(t), G], \tag{68}$$

where $\eta_{\mu\nu}$ is the Minkowski metric and $h^{(n)}_{\mu\nu}[z^0, z^i(t), G]$ is a perturbation of the order n in the gravitational constant G (physically, this means an expansion in the dimensionless parameter Gm/c^2d which is very small, d being the characteristic length of the problem and m a characteristic mass).

To obtain from (66) the equations of light propagation for the metric given in (68), we substitute the Christoffel symbols into (66). To save writing we denote the metric coefficients $h^{(1)}_{pq}[z^0, z^i(t), G]$ and $h^{(2)}_{pq}[z^0, z^i(t), G]$ by $h^{(1)}_{pq}$ and $h^{(2)}_{pq}$. Then the resulting equation of light propagation to the second order in G/c^2 is

$$\begin{aligned}
\frac{dl^i}{dt} = &\frac{1}{2} c^2 h^{(1)}_{00,i} - c^2 h^{(1)}_{0i,0} - c\, h^{(1)}_{0i,m} l^m + c\, h^{(1)}_{0m,i} l^m - c\, h^{(1)}_{mi,0} l^m \\
&- h^{(1)}_{mi,n} l^m l^n + \frac{1}{2} h^{(1)}_{mn,i} l^m l^n - \frac{1}{2} c\, h^{(1)}_{00,0} l^i - h^{(1)}_{00,k} l^k l^i \\
&+ \left(\frac{1}{2} c^{-1} h^{(1)}_{mp,0} - c^{-1} h^{(1)}_{0p,m} \right) l^m l^p l^i + \frac{1}{2} c^2 h^{(2)}_{00,i} - \frac{1}{2} c^2 h^{(1)ik} h^{(1)}_{00,k} \\
&- h^{(2)}_{00,k} l^k l^i - \left(h^{(2)}_{mi,n} - \frac{1}{2} h^{(2)}_{mn,i} \right) l^m l^n + h^{(1)ik} \left(h^{(1)}_{mk,n} - \frac{1}{2} h^{(1)}_{mn,k} \right) l^m l^n \\
&- h^{(1)}_{00} h^{(1)}_{00,k} l^k l^i,
\end{aligned} \tag{69}$$

where by $,0$ and $,i$ we denote $\partial/\partial z^0$ and $\partial/\partial z^i$, respectively. To calculate the light deflection we need to solve (69) for l^i. To solve this complicated nonlinear differential equation, we turn to approximation techniques.

The Approximation Scheme

We can write the 3-vector $l^i(t)$ as

$$l^i(t) = l^i_{(0)} + \sum_{n=1}^{\infty} \delta l^i_{(n)}(t), \tag{70}$$

where $l^i_{(0)}$ denotes the constant incoming tangent vector $l^i(-\infty)$ and $\delta l^i_{(n)}(t)$ the perturbation of the constant tangent vector $l^i_{(0)}$ of order n in G. After introducing the expression for $l^i(t)$ given by (70) into (69), we obtain differential equations for the perturbations $\delta l^i_{(1)}$ and $\delta l^i_{(2)}$

$$\begin{aligned}\frac{d\delta l^i_{(1)}}{dt} = &\frac{1}{2}c^2 h^{(1)}_{00,i} - c^2 h^{(1)}_{0i,0} - c\, h^{(1)}_{0i,m} l^m_{(0)} + c\, h^{(1)}_{0m,i} l^m_{(0)} - c\, h^{(1)}_{mi,0} l^m_{(0)} \\ &- h^{(1)}_{mi,n} l^m_{(0)} l^n_{(0)} + \frac{1}{2} h^{(1)}_{mn,i} l^m_{(0)} l^n_{(0)} - \frac{1}{2} c\, h^{(1)}_{00,0} l^i_{(0)} - h^{(1)}_{00,k} l^k_{(0)} l^i_{(0)} \\ &+ \left(\frac{1}{2}c^{-1} h^{(1)}_{mp,0} - c^{-1} h^{(1)}_{0p,m}\right) l^m_{(0)} l^p_{(0)} l^i_{(0)}\end{aligned} \tag{71}$$

and

$$\begin{aligned}\frac{d\delta l^i_{(2)}}{dt} = &\frac{1}{2}c^2 h^{(2)}_{00,i} - \frac{1}{2}c^2 h^{(1)ik} h^{(1)}_{00,k} - h^{(2)}_{00,k} l^k_{(0)} l^i_{(0)} - \left(h^{(2)}_{mi,n} - \frac{1}{2} h^{(2)}_{mn,i}\right) l^m_{(0)} l^n_{(0)} \\ &+ h^{(1)ik}\left(h^{(1)}_{mk,n} - \frac{1}{2} h^{(1)}_{mn,k}\right) l^m_{(0)} l^n_{(0)} - h^{(1)}_{00} h^{(1)}_{00,k} l^k_{(0)} l^i_{(0)} \\ &- c\, h^{(1)}_{0i,m} \delta l^m_{(1)} + c\, h^{(1)}_{0m,i} \delta l^m_{(1)} - c\, h^{(1)}_{mi,0} \delta l^m_{(1)} \\ &- h^{(1)}_{mi,n} \delta l^m_{(1)} l^n_{(0)} - h^{(1)}_{mi,n} l^m_{(0)} \delta l^n_{(1)} + h^{(1)}_{mn,i} \delta l^m_{(1)} l^n_{(0)} \\ &- \frac{1}{2} c h^{(1)}_{00,0} \delta l^i_{(1)} - h^{(1)}_{00,k} \delta l^k_{(1)} l^i_{(0)} - h^{(1)}_{00,k} l^k_{(0)} \delta l^i_{(1)} \\ &+ c^{-1} h^{(1)}_{mp,0} \delta l^m_{(1)} l^p_{(0)} l^i_{(0)} - c^{-1} h^{(1)}_{0p,m} \delta l^m_{(1)} l^p_{(0)} l^i_{(0)} - c^{-1} h^{(1)}_{0p,m} l^m_{(0)} \delta l^p_{(1)} l^i_{(0)} \\ &+ \left(\frac{1}{2}c^{-1} h^{(1)}_{mp,0} - c^{-1} h^{(1)}_{0p,m}\right) l^m_{(0)} l^p_{(0)} \delta l^i_{(1)}.\end{aligned} \tag{72}$$

To calculate the perturbations $\delta l^i_{(1)}(t)$ and $\delta l^i_{(2)}(t)$, we have to integrate (71) and (72) along the light ray trajectory to the appropriate order.

Angle of Light Deflection

The dimensionless vector $\alpha^i_{(n)}$ of order n in G, describing the angle of total deflection of the light ray measured at the point of observation and computed with respect to the vector $l^i_{(0)}$, is given by

$$\alpha^i_{(n)}(t) = P^i_q \frac{\delta l^q_{(n)}(t)}{|\boldsymbol{l}_{(0)}|}, \tag{73}$$

where $\delta l^i_{(n)}$ is the perturbation of the constant tangent vector of order n in G. Here,

$$P^i_q = \delta^i_q - e^i_{(0)} e_{(0)q}, \tag{74}$$

with $e^i_{(0)} = l^i_{(0)}/|\boldsymbol{l}_{(0)}|$, is the projection tensor onto the plane orthogonal to the vector $l^i_{(0)}$. In the case of light rays (photons) $|\boldsymbol{l}_{(0)}| = c$.

4.2 The Gravitational Field of a Compact Binary in the Linear Approximation

In the linear approximation (68) reduces to

$$g_{\mu\nu}(t, \boldsymbol{x}) = \eta_{\mu\nu} + h^{(1)}_{\mu\nu}(t, \boldsymbol{x}). \tag{75}$$

The metric perturbation $h^{(1)}_{\mu\nu}(t, \boldsymbol{x})$ can be found by solving the Einstein field equations which in the linear approximation and in the harmonic gauge (see [12]) are given by

$$\Box h^{(1)}_{\mu\nu}(t, \boldsymbol{x}) = -16\pi \frac{G}{c^4} S_{\mu\nu}(t, \boldsymbol{x}), \tag{76}$$

where

$$S_{\mu\nu}(t, \boldsymbol{x}) = T_{\mu\nu}(t, \boldsymbol{x}) - \tfrac{1}{2}\eta_{\mu\nu} T^{\lambda}{}_{\lambda}(t, \boldsymbol{x}). \tag{77}$$

As is well known, the solution of these equations has the form of a Liénard–Wiechert potential (e.g., see [13]).

For a binary system the matter stress–energy tensor reads

$$T^{\mu\nu}(t, \boldsymbol{x}) = \sum_{a=1}^{2} \mu_a(t) v^\mu_a v^\nu_a \delta(\boldsymbol{x} - \boldsymbol{x}_a), \tag{78}$$

where the trajectory of the mass m_a (in harmonic coordinates) is denoted by $\boldsymbol{x}_a(t)$; the coordinate velocity is $\boldsymbol{v}_a(t) = d\boldsymbol{x}_a(t)/dt$ and $v^\mu \equiv (c, \boldsymbol{v}_a)$; $\mu_a(t)$ represents a time-dependent mass of the body a defined by

$$\mu_a(t) = \frac{m_a}{\sqrt{1 - v_a^2(t)/c^2}}, \tag{79}$$

where m_a is the (constant) Schwarzschild mass.

After performing the integration of (76) with the help of the flat-retarded propagator, we finally get

$$h^{(1)}_{\mu\nu}(t,\boldsymbol{x}) = 4\frac{G}{c^4}\sum_{a=1}^{2}\frac{\mu_a(s_a)v_{a\mu}(s_a)v_{a\nu}(s_a) - (1/2)\eta_{\mu\nu}\mu_a(s_a)v^{\lambda}_a(s_a)v_{a\lambda}(s_a)}{r_a(s_a) - (1/c)(\boldsymbol{v}_a(s_a)\cdot\boldsymbol{r}_a(s_a))}, \tag{80}$$

where $\boldsymbol{r}_a(s_a)$ is given by $\boldsymbol{r}_a(s_a) = \boldsymbol{x} - \boldsymbol{x}_a(s_a)$, and $r_a(s_a)$ is the Euclidean norm of $\boldsymbol{r}_a(s_a)$. In (80) s_a denotes the retarded time $s_a = s_a(t,\boldsymbol{x})$ for the ath body which is a solution of the light-cone equation

$$s_a + \frac{1}{c}r_a(s_a) = t. \tag{81}$$

4.3 The Angle of Light Deflection in the Linear Approximation

By virtue of (71), (73), and considering that the metric coefficients $h^{(1)}_{\mu\nu}$ in (71) are smooth functions of t and $\boldsymbol{z}$, it can be shown that the expression for the angle of light deflection is given by (e.g., see [9])

$$\alpha^i_{(1)}(\tau) = \frac{1}{2c}\int_{-\infty}^{\tau} d\sigma l^{\alpha}_{(0)}l^{\beta}_{(0)}\hat{\partial}_i h^{(1)}_{\alpha\beta}(\tau,\boldsymbol{z}(\tau)) - \frac{1}{c}P^i_q l_{(0)\delta}h^{(1)\delta q}(\tau,\boldsymbol{z}(\tau)), \tag{82}$$

where $\hat{\partial}_i \equiv P^q_i\partial/\partial\xi^q$. Here, τ is an independent parameter defined by

$$\tau = t - t^*, \tag{83}$$

where t^* is the time of closest approach of the unperturbed light ray to the origin of an asymptotically flat harmonic coordinate system. Then the equation of the unperturbed light ray can be represented by

$$\boldsymbol{z}(\tau)_{\text{unpert.}} = \tau\boldsymbol{l}_{(0)} + \boldsymbol{\xi}, \tag{84}$$

where $\boldsymbol{\xi}$ is a vector directed from the origin of the coordinate system toward the point of closest approach (i.e., the impact parameter).

The integral in (82) can be calculated by applying the method developed by Kopeikin and Schäfer in [9]. After inserting the metric coefficients (80) into (82) and computing the integral, we finally obtain

$$\alpha^i_{(1)}(\tau) = \sum_a \frac{4(G/c^3)m_a\left[1 - \frac{\boldsymbol{e}_{(0)}\cdot\boldsymbol{v}_a(s_a)}{c}\right]}{\sqrt{1-\frac{v_a^2(s_a)}{c^2}}\left[r_a(\tau,s_a) - \frac{\boldsymbol{v}_a(s_a)\cdot\boldsymbol{r}_a(\tau,s_a)}{c}\right]}P^i_q v^q_a(s_a)$$
$$-\sum_a \frac{2(G/c^2)m_a\left[1 - \frac{\boldsymbol{e}_{(0)}\cdot\boldsymbol{v}_a(s_a)}{c}\right]^2\left[r_a(\tau,s_a) + (\boldsymbol{e}_{(0)}\cdot\boldsymbol{r}_a(\tau,s_a))\right]P^i_q r^q_a(\tau,s_a)}{\sqrt{1-\frac{v_a^2(s_a)}{c^2}}\left[r_a^2(\tau,s_a) - (\boldsymbol{e}_{(0)}\cdot\boldsymbol{r}_a(\tau,s_a))^2\right]\left[r_a(\tau,s_a) - \frac{\boldsymbol{v}_a(s_a)\cdot\boldsymbol{r}_a(\tau,s_a)}{c}\right]}. \tag{85}$$

For an observer located at infinity, we find

$$\alpha^i_{(1)} = \lim_{\tau\to\infty} \alpha^i_{(1)}(\tau)$$
$$= -4\frac{G}{c^2}\sum_{a=1}^{2} \frac{m_a\left[1-\frac{\boldsymbol{e}_{(0)}\cdot\boldsymbol{v}_a(s_a)}{c}\right]}{\sqrt{1-\frac{v_a^2(s_a)}{c^2}}R_a(s_a)}\left[\xi^i - P^i_q x^q_a(s_a)\right], \tag{86}$$

where the quantity $R_a(s_a)$ is defined by

$$R_a(s_a) = r_a^2(0, s_a) - (\boldsymbol{e}_{(0)}\cdot\boldsymbol{x}_a(s_a))^2. \tag{87}$$

4.4 The Postlinear Gravitational Field of a Compact Binary

In [14, 15], it was shown that leading order terms for the effect of light deflection in the linear gravitational field in the case of a small impact parameter $|\boldsymbol{\xi}|$ (i.e., an impact parameter small with respect to the distance between the deflector and the observer) depend neither on the radiative part ($\sim 1/|\boldsymbol{\xi}|$) of the gravitational field nor on the intermediate ($\sim 1/|\boldsymbol{\xi}|^2$) zone terms. The main effect rather comes from the near zone ($\sim 1/|\boldsymbol{\xi}|^3$) terms. Taking into account this property of strong suppression of the influence of gravitational waves on the light propagation, we can assume that the light deflection in the postlinear gravitational field of a compact binary is mainly determined by the near zone metric.

The Metric in the Near Zone

In [16], Blanchet et al. calculated the conservative 2PN harmonic coordinate metric for the near zone of a system of two-bounded point-like masses as function of the distance $\boldsymbol{z}$ and of the positions and velocities of the masses $\boldsymbol{x}_a(t)$ and $\boldsymbol{v}_a(t)$, respectively, with $a = 1, 2$. For the sake of simplicity we split the 2PN metric into two parts: the G-2PN and GG-2PN parts.

G-2PN Metric

The G-2PN part is given by

$$h^{(1)}_{00} = 2\frac{G}{c^2}\sum_{a=1}^{2}\frac{m_a}{r_a} + \frac{G}{c^4}\sum_{a=1}^{2}\frac{m_a}{r_a}\left[-(\boldsymbol{n}_a\cdot\boldsymbol{v}_a)^2 + 4v_a^2\right],$$
$$h^{(1)}_{0p} = -4\frac{G}{c^3}\sum_{a=1}^{2}\frac{m_a}{r_a}v_a^p,$$
$$h^{(1)}_{pq} = 2\frac{G}{c^2}\sum_{a=1}^{2}\frac{m_a}{r_a}\delta^{pq} + \frac{G}{c^4}\sum_{a=1}^{2}\frac{m_a}{r_a}\left[-(\boldsymbol{n}_a\cdot\boldsymbol{v}_a)^2\delta^{pq} + 4v_a^p v_a^q\right], \tag{88}$$

where v_a^p denotes the velocity of the mass m_a, and n_a^p is the unit vector defined by $n_a^p = r_a^p/r_a$. By r_a^p we denote the vector $r_a^p = z^p - x_a^p(t)$ and by r_a we denote its Euclidean norm $r_a = |\boldsymbol{z} - \boldsymbol{x}_a(t)|$.

Here, it is worthwhile to point out that the parts of the G-2PN metric which contain the accelerations of the masses were introduced into the part of the GG-2PN metric after substituting the accelerations by explicit functions of the coordinate positions of the masses by means of the Newtonian equations of motion.

GG-2PN Metric

The GG-2PN part is given by

$$h_{00}^{(2)} = \frac{G^2}{c^4}\left\{-2\frac{m_1^2}{r_1^2} + m_1 m_2\left(-\frac{2}{r_1 r_2} - \frac{r_1}{2r_{12}^3} + \frac{r_1^2}{2r_2 r_{12}^3} - \frac{5}{2r_2 r_{12}}\right)\right\}$$
$$+\frac{G^2}{c^4}(1 \leftrightarrow 2),$$
$$h_{pq}^{(2)} = \frac{G^2}{c^4}\left\{\delta^{pq}\left[\frac{m_1^2}{r_1^2} + m_1 m_2\left(\frac{2}{r_1 r_2} - \frac{r_1}{2r_{12}^3} + \frac{r_1^2}{2r_2 r_{12}^3} - \frac{5}{2r_1 r_{12}} + \frac{4}{r_{12}S}\right)\right]\right.$$
$$\left.+\frac{m_1^2}{r_1^2}n_1^p n_1^q - 4m_1 m_2 n_{12}^p n_{12}^q\left(\frac{1}{S^2} + \frac{1}{r_{12}S}\right)\right\}$$
$$+\frac{4G^2 m_1 m_2}{c^4 S^2}\left(n_1^{(p}n_2^{q)} + 2n_1^{(p}n_{12}^{q)}\right) + \frac{G^2}{c^4}(1 \leftrightarrow 2), \tag{89}$$

where the symbol $(1 \leftrightarrow 2)$ refers to the preceding term in braces but with the labels 1 and 2 exchanged; by S we denote $S = r_1 + r_2 + r_{12}$, where $r_1 = |\boldsymbol{z} - \boldsymbol{x}_1(t)|$, $r_2 = |\boldsymbol{z} - \boldsymbol{x}_2(t)|$, and $r_{12} = |\boldsymbol{x}_1(t) - \boldsymbol{x}_2(t)|$. The vectors n_1^p, n_2^p, and n_{12}^p are unit vectors defined by $n_1^p = r_1^p/r_1$, $n_2^p = r_2^p/r_2$, and $n_{12}^p = r_{12}^p/r_{12}$.

The Barycentric Coordinate System

We use a harmonic coordinate system, the origin of which coincides with the 1PN-center of mass. Using the 1PN-accurate center of mass theorem of [17], we can express the individual center of mass frame positions of the two masses in terms of the relative position $\boldsymbol{r}_{12} \equiv \boldsymbol{x}_1 - \boldsymbol{x}_2$ and the relative velocity $\boldsymbol{v}_{12} \equiv \boldsymbol{v}_1 - \boldsymbol{v}_2$ as

$$\boldsymbol{x}_1 = \left[X_2 + \frac{1}{c^2}\epsilon_{1\mathrm{PN}}\right]\boldsymbol{r}_{12}, \tag{90}$$

$$\boldsymbol{x}_2 = \left[-X_1 + \frac{1}{c^2}\epsilon_{1\mathrm{PN}}\right]\boldsymbol{r}_{12}, \tag{91}$$

where X_1, X_2, and ϵ_{1PN} are given by

$$X_1 \equiv \frac{m_1}{M}, \tag{92}$$

$$X_2 \equiv \frac{m_2}{M}, \tag{93}$$

$$\epsilon_{1\mathrm{PN}} = \frac{\nu(m_1 - m_2)}{2M}\left[v_{12}^2 - \frac{GM}{r_{12}}\right]. \tag{94}$$

Here, we have introduced

$$M \equiv m_1 + m_2, \; v_{12} = |\boldsymbol{v}_{12}| \tag{95}$$

and

$$\nu \equiv \frac{m_1 m_2}{M^2}. \tag{96}$$

It is important to remark that, in our computation of the postlinear light deflection up to the order G^2/c^4, we need only to consider the 1PN corrections to the Newtonian center of mass, because, as we shall see in Sect. 4.5, the 2PN corrections to the Newtonian center of mass are related to postlinear light deflection terms of order higher than G^2/c^4.

4.5 The Postlinear Angle of Light Deflection

From (72) and (73), we see that the postlinear angle of light deflection $\alpha^i_{(2)}$ is a function of the GG-2PN metric coefficients $h^{(2)}_{\mu\nu}$, the G-2PN metric coefficients $h^{(1)}_{\mu\nu}$, and the linear perturbation $\delta l^i_{(1)}(\tau)$. To facilitate the computations, we separate the light deflection terms that are functions of the GG-2PN metric coefficients from the terms that are functions of the G-2PN metric coefficients and the linear perturbations.

The Linear Perturbation $\delta l^i_{(1)}(\tau)$

From (71) it follows that the perturbation $\delta l^i_{(1)}(\tau)$ is given by

$$\begin{aligned}\delta l^i_{(1)}(\tau) = &\frac{1}{2}\int_{-\infty}^{\tau} d\sigma\, l^\alpha_{(0)} l^\beta_{(0)} h^{(1)}_{\alpha\beta,i}\Big|_{(\rightarrow)} - c\, h^{(1)}_{0i} - h^{(1)}_{mi} l^m_{(0)} - h^{(1)}_{00} l^i_{(0)}\\ &+\frac{1}{2}c\int_{-\infty}^{\tau} d\sigma\, h^{(1)}_{00,0} l^i_{(0)}\Big|_{(\rightarrow)}\\ &+\int_{-\infty}^{\tau} d\sigma\, l^m_{(0)} l^p_{(0)}\left[\frac{1}{2}c^{-1}h^{(1)}_{mp,0} - c^{-1}h^{(1)}_{0p,m}\right] l^i_{(0)}\Big|_{(\rightarrow)}.\end{aligned} \tag{97}$$

On the right-hand side of (97) after evaluating the partial derivatives of the metric coefficients with respect to the photon's coordinates (i.e., $(z^0, z^i(t))$), we replace in the integrals the photon trajectory by its unperturbed approximation $z^i(\sigma)_{\rm unpert.} = \sigma l^i_{(0)} + \xi^i$ and the time coordinate z^0 by $\sigma + t^*$. In this chapter, we denote this operation by the symbol $|_{(\rightarrow)}$. After introducing the G-2PN metric coefficients (88) into (97), we obtain the explicit expression for $\delta l^i_{(1)}(\tau)$ which we have to integrate. Since the G-2PN metric coefficients are functions of the positions and velocities of the masses $\boldsymbol{x}_a(t)$ and $\boldsymbol{v}_a(t)$, respectively, the expression for $\delta l^i_{(1)}(\tau)$ is a function of these quantities. This means that we have to take into account the motion of the masses when we are going to compute the integrals. Considering that the influence of the gravitational field on the light propagation is strongest near the barycenter of the binary and that the velocities of the masses are small with respect to the velocity of light, we are allowed to make the following approximations:

1. We may assume that the linear gravitational field is determined by the positions and velocities of the masses taken at the time of closest approach ($t = t^*$) of the unperturbed light ray to the barycenter of the binary (i.e., to the origin of the asymptotically flat harmonic coordinate system). The expression, resulting from (97) after introducing the G-2PN metric coefficients and setting $t = t^*$ for the positions and velocities and computing the integrals, is denoted by $\delta l^i_{(1)\rm I}(\tau)$.
2. We treat the effect of the motion of the masses on light propagation as a correction to the expression of $\delta l^i_{(1)\rm I}(\tau)$, which we denote by $\delta l^i_{(1)\rm II}(\tau)$. We shall compute this correction in Sect. 4.6.

The total linear perturbation $\delta l^i_{(1)}(\tau)$ is then given by

$$\delta l^i_{(1)}(\tau) = \delta l^i_{(1)\rm I}(\tau) + \delta l^i_{(1)\rm II}(\tau). \tag{98}$$

Consequently, the corresponding angle of light deflection reads

$$\alpha^i_{(1)}(\tau) = \frac{1}{c} P^i_q \left[\delta l^i_{(1)\rm I}(\tau) + \delta l^i_{(1)\rm II}(\tau)\right], \tag{99}$$

where P^i_q is defined by (74).

Here, it is important to remark that to obtain the total linear light deflection we have to add to (99) terms arising from the 1PN corrections in the positions of the masses, which we shall compute in Sect. 4.6. Since these terms are proportional to v_{12}^2/c^2, it is easy to see by virtue of the virial theorem that they are of the same order as the terms in G^2/c^4.

The Postlinear Light Deflection Terms That Depend on the GG-2PN Metric

It follows from (72) and (73) that a part of the postlinear light deflection is given by:

$$\alpha^i_{(2)\rm I} = \frac{1}{c} P^i_q \left[\frac{1}{2} c^2 \int_{-\infty}^{\infty} d\tau h^{(2)}_{00,q}\Big|_{(\rightarrow)} + \int_{-\infty}^{\infty} d\tau \left[\frac{1}{2} h^{(2)}_{mn,q} - h^{(2)}_{qm,n}\right] l^m_{(0)} l^n_{(0)}\Big|_{(\rightarrow)}\right]. \tag{100}$$

Upon introducing the GG-2PN metric coefficients given by (89) into (100), we obtain integrals whose integrands are functions of the distances r_1, r_2, S, and their inverses. Through the distances r_1, r_2, and S, the resulting integrals from (100) are functions of the positions of the masses $\boldsymbol{x}_a(t)$.

For the same reason as in the case of the linear perturbation, we are here allowed to fix the values of the positions of the masses $\boldsymbol{x}_a(t)$ to their values at the time t^* before performing the integration.

To evaluate the integrals that cannot be represented by elementary functions, we resort as usual to a series expansion of the integrands. The order of the expansion should be chosen in a consistent manner with the expansion in terms of G/c^2.

The Postlinear Light Deflection Terms That Depend on the G-2PN Metric

We denote the postlinear light deflection terms, which are functions of the G-2PN metric coefficients and the linear perturbations $\delta l^i_{(1)}(\tau)$, by $\alpha^i_{(2)\mathrm{II}}$. It follows from (72) and (73) that the resulting expression for the postlinear light deflection $\alpha^i_{(2)\mathrm{II}}$ is given by

$$\begin{aligned}
\alpha^i_{(2)\mathrm{II}} = \frac{1}{c} P^i_q \Bigg[& -\frac{1}{2} c^2 \int_{-\infty}^{\infty} d\tau\, h^{(1)qm} h^{(1)}_{00,m}|_{(\rightarrow)} \\
& + \int_{-\infty}^{\infty} d\tau \left[h^{(1)qp} \left(h^{(1)}_{mp,n} - \frac{1}{2} h^{(1)}_{mn,p} \right) \right] l^m_{(0)} l^n_{(0)} \Big|_{(\rightarrow)} \\
& + c \int_{-\infty}^{\infty} d\tau \left[h^{(1)}_{0m,q} - h^{(1)}_{0q,m} \right] \delta l^m_{(1)}(\tau) \Big|_{(\rightarrow)} \\
& + \int_{-\infty}^{\infty} d\tau \left[h^{(1)}_{mn,q} \delta l^m_{(1)}(\tau) l^n_{(0)} - h^{(1)}_{mq,n} \delta l^m_{(1)}(\tau) l^n_{(0)} - h^{(1)}_{mq,n} l^m_{(0)} \delta l^n_{(1)}(\tau) \right] \Big|_{(\rightarrow)} \\
& - \int_{-\infty}^{\infty} d\tau\, h^{(1)}_{00,k} l^k_{(0)} \delta l^q_{(1)}(\tau) \Big|_{(\rightarrow)} \\
& - \frac{1}{c} \int_{-\infty}^{\infty} d\tau\, h^{(1)}_{0p,m} l^m_{(0)} l^p_{(0)} \delta l^q_{(1)}(\tau) \Big|_{(\rightarrow)} \Bigg] . \qquad (101)
\end{aligned}$$

To compute $\alpha^i_{(2)\mathrm{II}}$, we introduce the expressions for the perturbations $\delta l^i_{(1)}(\tau)$ given by (98) and the G-2PN metric coefficients given by (88) into the expression for $\alpha^i_{(2)\mathrm{II}}$. Here, we may use the same approximations as before, i.e., we can fix the values of the positions and velocities of the masses to their values at the time t^* before performing the integrals. As explained in the preceding section, with the help of a Taylor expansion of the integrands we can evaluate the integrals, which cannot be represented by elementary functions.

4.6 Relativistic Corrections

In this section, we give a brief account of the corrections that we have to consider in the calculation of the linear and postlinear light deflection. Further details are given in [11]:

- *Light deflection and the motion of the masses*
  As we mentioned before, the general expression for the linear perturbation $\delta l^i_{(1)}(\tau)$ is through the G-2PN metric coefficients, a function of the positions and velocities of the components of the binary.
  To find the correction terms to the linear perturbation $\delta l^i_{(1)\mathrm{I}}(\tau)$ and postlinear light deflection, we perform the Taylor expansion of the general expression for $\delta l^i_{(1)}(\tau)$ (i.e., of the expression resulting from the introduction of the G-2PN metric coefficients (88) into (97)) in which the coefficients depend on the sources' coordinates and their successive derivatives with respect to t, namely

$$\frac{dx^i_a}{dt} = v^i_a(t); \quad \frac{d^2x^i_a}{dt^2} = \frac{dv^i_a}{dt} = a^i_a(t); \quad \dots$$

  The corrections arising from the motion of the binary's components are denoted by $\delta l^i_{(1)\mathrm{II}}$ and $\alpha^i_{(2)\mathrm{III}}$.
- *The postlinear light deflection and the perturbed light trajectory*
  The linear perturbation of the light trajectory reads

$$\delta z^i_{(1)}(\tau) = \int d\tau \left[\delta l^i_{(1)\mathrm{I}}(\tau) + \delta l^i_{(1)\mathrm{II}}(\tau)\right] + K^i, \tag{102}$$

  where K^i is a vectorial integration constant. After introducing the perturbation $\delta z^i_{(1)}(\tau)$ into (99), we obtain additional postlinear light deflection terms, which are denoted by $\alpha^i_{(2)\mathrm{IV}}$.
- *Light deflection and the center of mass*
  After introducing the 1PN corrections in the positions given by

$$\delta \boldsymbol{x}_1 = \delta \boldsymbol{x}_2 = \frac{1}{c^2}\epsilon_{1\mathrm{PN}}\boldsymbol{r}_{12} \tag{103}$$

  into (99), we get additional terms to the linear and postlinear light deflection, which we denote by $\tilde{\alpha}^i_{(1)(2)}$. From (94) and (103), it is easy to see that the corrections vanish for equal masses and circular orbits.

4.7 The Total Linear and Postlinear Light Deflection

The total linear light deflection results from summing up (99) with the correction terms arising from the part of $\tilde{\alpha}^i_{(1)(2)}$ that is linear in G. Consequently the total linear light deflection reads

$$\alpha^i_{(1)}(\tau)_{\mathrm{tot.}} = \frac{1}{c}P^i_q\left[\delta l^i_{(1)\mathrm{I}}(\tau) + \delta l^i_{(1)\mathrm{II}}(\tau)\right] + \tilde{\alpha}^i_{(1)(2)}(G). \tag{104}$$

From Sects. 4.5 and 4.6, it follows that the total postlinear light deflection up to the order G^2/c^4 is given by

$$\alpha^i_{(2)} = \alpha^i_{(2)\mathrm{I}} + \alpha^i_{(2)\mathrm{II}} + \alpha^i_{(2)\mathrm{III}} + \alpha^i_{(2)\mathrm{IV}} + \tilde{\alpha}^i_{(1)(2)}(G^2), \tag{105}$$

where $\tilde{\alpha}^i_{(1)(2)}(G^2)$ denotes the part of $\tilde{\alpha}^i_{(1)(2)}$ that is quadratic in G.

4.8 Results

From (104) and (105), we obtain the general formulas for the angle of light deflection linear and quadratic in G. These formulas are given in an explicit form in [11]. Here, to study the important features of the derived formulas and in view of an application of the obtained formulas to the double pulsar PSR J0737-3039, we shall consider only the special case when the light ray is originally parallel to the orbital plane of a binary with equal masses (see [18]). In this case the resulting expressions for the angle of light deflection linear and quadratic in G (see [11]) read

$$\begin{aligned}
\alpha^i_{(1)\|} = \frac{G M_{\mathrm{ADM}}}{c^2\,\xi} \Bigg\{ & -4 + \left[1 - (\boldsymbol{e}_{(0)}\cdot\boldsymbol{n}_{12})^2\right]\left(\frac{r_{12}}{\xi}\right)^2 \\
& + \left[-\frac{1}{4} + \frac{1}{2}(\boldsymbol{e}_{(0)}\cdot\boldsymbol{n}_{12})^2 - \frac{1}{4}(\boldsymbol{e}_{(0)}\cdot\boldsymbol{n}_{12})^4\right]\left(\frac{r_{12}}{\xi}\right)^4 \\
& + \left[\frac{1}{16} - \frac{3}{16}(\boldsymbol{e}_{(0)}\cdot\boldsymbol{n}_{12})^2 + \frac{3}{16}(\boldsymbol{e}_{(0)}\cdot\boldsymbol{n}_{12})^4 - \frac{1}{16}(\boldsymbol{e}_{(0)}\cdot\boldsymbol{n}_{12})^6\right]\left(\frac{r_{12}}{\xi}\right)^6 \\
& + \Bigg[-\frac{1}{64} + \frac{1}{16}(\boldsymbol{e}_{(0)}\cdot\boldsymbol{n}_{12})^2 - \frac{3}{32}(\boldsymbol{e}_{(0)}\cdot\boldsymbol{n}_{12})^4 + \frac{1}{16}(\boldsymbol{e}_{(0)}\cdot\boldsymbol{n}_{12})^6 \\
& \quad - \frac{1}{64}(\boldsymbol{e}_{(0)}\cdot\boldsymbol{n}_{12})^8\Bigg]\left(\frac{r_{12}}{\xi}\right)^8 \\
& + \Bigg[\frac{1}{256} - \frac{5}{256}(\boldsymbol{e}_{(0)}\cdot\boldsymbol{n}_{12})^2 + \frac{5}{128}(\boldsymbol{e}_{(0)}\cdot\boldsymbol{n}_{12})^4 - \frac{5}{128}(\boldsymbol{e}_{(0)}\cdot\boldsymbol{n}_{12})^6 \\
& \quad + \frac{5}{256}(\boldsymbol{e}_{(0)}\cdot\boldsymbol{n}_{12})^8 - \frac{1}{256}(\boldsymbol{e}_{(0)}\cdot\boldsymbol{n}_{12})^{10}\Bigg]\left(\frac{r_{12}}{\xi}\right)^{10} \\
& + \Bigg[-\frac{1}{1024} + \frac{3}{512}(\boldsymbol{e}_{(0)}\cdot\boldsymbol{n}_{12})^2 - \frac{15}{1024}(\boldsymbol{e}_{(0)}\cdot\boldsymbol{n}_{12})^4 + \frac{5}{256}(\boldsymbol{e}_{(0)}\cdot\boldsymbol{n}_{12})^6 \\
& \quad - \frac{15}{1024}(\boldsymbol{e}_{(0)}\cdot\boldsymbol{n}_{12})^8 + \frac{3}{512}(\boldsymbol{e}_{(0)}\cdot\boldsymbol{n}_{12})^{10} \\
& \quad - \frac{1}{1024}(\boldsymbol{e}_{(0)}\cdot\boldsymbol{n}_{12})^{12}\Bigg]\left(\frac{r_{12}}{\xi}\right)^{12} + \mathcal{O}\left[\left(\frac{r_{12}}{\xi}\right)^{14}\right]\Bigg\} e^i_\xi
\end{aligned}$$

$$
\begin{aligned}
&+\frac{GM_{\rm ADM}}{c^3\,\xi}\Bigg\{-\left(\boldsymbol{e}_{(0)}\cdot\boldsymbol{v}_{12}\right)\left(\frac{r_{12}}{\xi}\right)\\
&+\left[\frac{1}{4}\left(\boldsymbol{e}_{(0)}\cdot\boldsymbol{v}_{12}\right)\left[1+\left(\boldsymbol{e}_{(0)}\cdot\boldsymbol{n}_{12}\right)^2\right]-\frac{1}{2}\left(\boldsymbol{n}_{12}\cdot\boldsymbol{v}_{12}\right)\left(\boldsymbol{e}_{(0)}\cdot\boldsymbol{n}_{12}\right)\right]\left(\frac{r_{12}}{\xi}\right)^3\\
&+\Bigg[-\frac{1}{16}\left(\boldsymbol{e}_{(0)}\cdot\boldsymbol{v}_{12}\right)\left[1+2\left(\boldsymbol{e}_{(0)}\cdot\boldsymbol{n}_{12}\right)^2-3\left(\boldsymbol{e}_{(0)}\cdot\boldsymbol{n}_{12}\right)^4\right]\\
&\quad+\frac{1}{4}\left(\boldsymbol{n}_{12}\cdot\boldsymbol{v}_{12}\right)\left(\boldsymbol{e}_{(0)}\cdot\boldsymbol{n}_{12}\right)\left[1-\left(\boldsymbol{e}_{(0)}\cdot\boldsymbol{n}_{12}\right)^2\right]\Bigg]\left(\frac{r_{12}}{\xi}\right)^5\\
&+\Bigg[\frac{1}{64}\left(\boldsymbol{e}_{(0)}\cdot\boldsymbol{v}_{12}\right)\left[1+3\left(\boldsymbol{e}_{(0)}\cdot\boldsymbol{n}_{12}\right)^2-9\left(\boldsymbol{e}_{(0)}\cdot\boldsymbol{n}_{12}\right)^4+5\left(\boldsymbol{e}_{(0)}\cdot\boldsymbol{n}_{12}\right)^6\right]\\
&\quad-\frac{3}{32}(\boldsymbol{n}_{12}\cdot\boldsymbol{v}_{12})(\boldsymbol{e}_{(0)}\cdot\boldsymbol{n}_{12})\left[1-2(\boldsymbol{e}_{(0)}\cdot\boldsymbol{n}_{12})^2+(\boldsymbol{e}_{(0)}\cdot\boldsymbol{n}_{12})^4\right]\Bigg]\left(\frac{r_{12}}{\xi}\right)^7\\
&\quad+\mathcal{O}\left[\left(\frac{r_{12}}{\xi}\right)^9\right]\Bigg\}P^i_q n^q_{12}\\
&+\frac{GM_{\rm ADM}}{c^3\,\xi}\left(\boldsymbol{e}_{(0)}\cdot\boldsymbol{n}_{12}\right)\Bigg\{\frac{1}{2}\left(\frac{r_{12}}{\xi}\right)-\frac{1}{8}\left[1-\left(\boldsymbol{e}_{(0)}\cdot\boldsymbol{n}_{12}\right)^2\right]\left(\frac{r_{12}}{\xi}\right)^3\\
&\quad+\frac{1}{32}\left[1-2\left(\boldsymbol{e}_{(0)}\cdot\boldsymbol{n}_{12}\right)^2+\left(\boldsymbol{e}_{(0)}\cdot\boldsymbol{n}_{12}\right)^4\right]\left(\frac{r_{12}}{\xi}\right)^5\\
&\quad-\frac{1}{128}\left[1-3\left(\boldsymbol{e}_{(0)}\cdot\boldsymbol{n}_{12}\right)^2+3\left(\boldsymbol{e}_{(0)}\cdot\boldsymbol{n}_{12}\right)^4-\left(\boldsymbol{e}_{(0)}\cdot\boldsymbol{n}_{12}\right)^6\right]\left(\frac{r_{12}}{\xi}\right)^7\\
&\quad+\mathcal{O}\left[\left(\frac{r_{12}}{\xi}\right)^9\right]\Bigg\}P^i_q v^q_{12}\\
&+\frac{GM_{\rm ADM}}{c^4\,\xi}\left(\boldsymbol{e}_{(0)}\cdot\boldsymbol{n}_{12}\right)^2\left\{\frac{1}{4}\left[v_{12}^2-\left(\boldsymbol{e}_{(0)}\cdot\boldsymbol{v}_{12}\right)^2\right]\left(\frac{r_{12}}{\xi}\right)^2+\mathcal{O}\left[\left(\frac{r_{12}}{\xi}\right)^4\right]\right\}e^i_\xi
\end{aligned}
\tag{106}
$$

and

$$
\begin{aligned}
\alpha^i_{(2)\|}&=\frac{G^2M^2_{\rm ADM}}{c^4\,\xi^2}\Bigg\{-\frac{15}{4}\pi-\frac{1}{24}\left[4+\left(\boldsymbol{e}_{(0)}\cdot\boldsymbol{n}_{12}\right)^2+\left(\boldsymbol{e}_{(0)}\cdot\boldsymbol{n}_{12}\right)^4\right]\left(\frac{r_{12}}{\xi}\right)\\
&\quad+\frac{3}{256}\pi\left[250-797\left(\boldsymbol{e}_{(0)}\cdot\boldsymbol{n}_{12}\right)^2\right]\left(\frac{r_{12}}{\xi}\right)^2+\mathcal{O}\left[\left(\frac{r_{12}}{\xi}\right)^3\right]\Bigg\}e^i_\xi\\
&\quad+\frac{G^2M^2_{\rm ADM}}{c^4\,\xi^2}\left\{4\left(\boldsymbol{e}_{(0)}\cdot\boldsymbol{n}_{12}\right)\left(\frac{r_{12}}{\xi}\right)^2+\mathcal{O}\left[\left(\frac{r_{12}}{\xi}\right)^3\right]\right\}P^i_q n^q_{12},
\end{aligned}
\tag{107}
$$

where in this case the ADM mass is given by

$$M_{\mathrm{ADM}} = M\left[1 + \frac{1}{4}\left(\frac{v_{12}^2}{2c^2} - \frac{GM}{c^2 r_{12}}\right)\right]. \tag{108}$$

Here, we have assumed that the mass of each component of the binary is equal to $M/2$. In (106) the components e_ξ^i, $P_q^i n_{12}^q$, and $P_q^i v_{12}^q$ of the linear light deflection were expanded to the order $(r_{12}/\xi)^{12}$, $(r_{12}/\xi)^7$, and $(r_{12}/\xi)^7$, respectively, to reach the accuracy of the postlinear light deflection (107).

In (106)–(108), the quantities $\boldsymbol{n}_{12}$, v_{12}, and r_{12} are taken at the time t^*. Note that in this case the correction arising from the shift of the 1PN-center of mass with respect to the Newtonian center of mass (see (103)) vanishes.

In the limit $r_{12} \to 0$ (106) and (107) reduce to

$$\alpha^i_{(1)(\mathrm{E})} = -4\frac{GM_{ADM}}{c^2\xi}e^i_\xi \tag{109}$$

and

$$\alpha^i_{(2)(\mathrm{E-S})} = -\frac{15}{4}\pi\frac{G^2M^2_{ADM}}{c^4\xi^2}e^i_\xi, \tag{110}$$

which are the Einstein and Epstein–Shapiro light deflection angles, respectively [19].

Application of the formulas for the deflection angle given by (106) and (107) to the double pulsar PSR J0737-3039 for an impact parameter five times greater than the relative separation distance of the binary's components shows that the absolute corrections to an Epstein–Shapiro angle of about 10^{-6} arcsec lie between 10^{-7} and 10^{-8} arcsec.

5 Concluding Remarks

The main steps in the computations of the angle of light deflection in the gravitational field of a compact binary in the linear and postlinear approximations were recapitulated.

The equations of light propagation were explicitly integrated to the second order in G/c^2.

The expressions for the angle of light deflection in the event that the light ray is originally parallel to the orbital of a binary with equal masses were given in an explicit form. In the limit $r_{12} \to 0$ the Einstein angle and the Epstein–Shapiro light deflection angle were obtained from the expressions for the linear and postlinear light deflection, respectively.

Application of the derived formulas for the deflection angle to the double pulsar PSR J0737-3039 for an impact parameter five times greater than the relative separation distance of the binary's components has shown that the absolute corrections to an Epstein–Shapiro angle of about 10^{-6} arcsec lie between 10^{-7} and 10^{-8} arcsec.

We conclude that the corrections to the Epstein–Shapiro light deflection angle are beyond the sensitivity of the current astronomical interferometers. Nevertheless, taking into account that the interferometer for the planned mission LATOR (see [20]) will be able to measure light deflection angles of the order 10^{-8} arcsec, we believe that the corrections to the Epstein–Shapiro light deflection could well be measured by space-borne interferometers in the foreseeable future.

On the level of the light propagation in linear gravitational fields, the controversy on the speed-of-gravity measurement by the radio observations of the bright radio quasar J0842+1835 has been investigated. The conclusion has been drawn that, in that measurement, no speed-of-gravity effect was included.

Finally, a comparison of linearized Einstein's field equations with electrodynamics has been undertaken to clearly show the similarities and dissimilarities between both theories. We feel that this comparison should be useful for those researchers who like to think about linearized Einstein theory in terms of electrodynamics because on this route errors may enter easily when ignoring the different invariance groups of the both theories.

References

1. C.W. Misner, K.S. Thorne, and J.A. Wheeler, *Gravitation* (W.H. Freeman, San Francisco, 1973)
2. B. Mashhoon, in *Reference Frames and Gravitomagnetism*, Eds. J.F. Pascual-Sánchez, L. Floría, A. San Miguel, and F. Vicente (World Scientific, Singapore, 2001), p. 121; also see, *Gravitoelectromagnetism: A Brief Review*, gr-qc/0311030
3. B.S. DeWitt, in *Gravitation: An Introduction to Current Research*, Ed. L. Witten (Wiley, New York, 1962), p. 266; in *Relativity, Groups and Topology*, Eds. C. DeWitt and B. DeWitt (Gordon and Breach, New York, 1965), p. 585
4. S.M. Kopeikin, ApJ Letters **556**, L1 (2001)
5. H. Asada, ApJ Letters **572**, L69 (2002)
6. S.M. Kopeikin, Phys. Lett. **A312**, 147 (2003)
7. C.M. Will, ApJ **590**, 683 (2003)
8. E.B. Fomalont and S.M. Kopeikin, ApJ **598**, 704 (2003)
9. S.M. Kopeikin and G. Schäfer, Phys. Rev. D **60**, 124002 (1999)
10. L.D. Landau and E.M. Lifshitz, *The Classical Theory of Fields* (Pergamon Press, Oxford, 1971)
11. M.H. Brügmann, Phys. Rev. D **72**, 024012 (2005)
12. S. Weinberg, *Gravitation and Cosmology* (Wiley, New York, 1972)
13. J.D. Jackson, *Classical Electrodynamics* (Wiley, New York, 1999)
14. T. Damour and G. Esposito-Farèse, Phys. Rev. D **58**, 044003 (1998)
15. S.M. Kopeikin, G. Schäfer, C.R. Gwinn, and T.M. Eubanks, Phys. Rev. D **59**, 084023 (1999)
16. L. Blanchet, G. Faye, and B. Ponsot, Phys. Rev. D **58**, 124002 (1998)

17. T. Damour and N. Deruelle, C. R. Acad. Sci. Paris **293**, 877 (1981)
18. A.G. Lyne, M. Burgay, M. Kramer, A. Possenti, R.N. Manchester, F. Camilo, M.A. McLaughlin, D.R. Lorimer, N. D'Amico, B.C. Joshi, J. Reynolds, and P.C.C. Freire, Science **303**, 1153 (2004)
19. R. Epstein and I.I. Shapiro, Phys. Rev. D **22**, 2947 (1980)
20. S.G. Turyshev, M. Shao, and K. Nordtvedt, Jr., Class. Quantum Grav, **21**, 2773 (2004); Int. J. Mod. Phys. D **13**, 2035 (2004)

On the Radar Method in General-Relativistic Spacetimes

V. Perlick

TU Berlin, Institute of Theoretical Physics, Sekr. PN 7-1,
Hardenbergstrasse 36, 10623 Berlin, Germany

Summary. If a clock, mathematically modeled by a parametrized timelike curve in a general-relativistic spacetime, is given, the radar method assigns a time and a distance to every event which is sufficiently close to the clock. Several geometric aspects of this method are reviewed and their physical interpretation is discussed.

1 Introduction

When Einstein was asked about the meaning of time he used to say: "Time is the reading of a clock." Taking this answer seriously, one is forced to accept that time is directly defined only at the position of a clock; if one wants to assign a time to events at some distance from the clock, one needs an additional prescription. As such prescription, Einstein suggested the radar method with light rays.

Although originally designed for special relativity, the radar method works equally well in general relativity. What one needs is a clock in an arbitrary general-relativistic spacetime. Here and in the following, our terminology is as follows. A *general-relativistic spacetime* is a four-dimensional manifold M with a smooth metric tensor field g of Lorentzian signature and a time orientation; the latter means that a globally consistent distinction between future and past has been made. A *clock* is a smooth embedding $\gamma : t \mapsto \gamma(t)$ from a real interval into M such that the tangent vector $\dot{\gamma}(t)$ is everywhere timelike with respect to g and future-pointing. This terminology is justified because we can interpret the value of the parameter t as the reading of a clock. Note that our definition of a clock does not demand that "its ticking be uniform" in any sense. Only smoothness and monotonicity are required.

The radar method assigns a time and a distance to an event q in the following way. One has to send a light ray from an event on the curve γ, say $\gamma(t_1)$, to q and receive the reflected light ray at another event on γ, say $\gamma(t_2)$

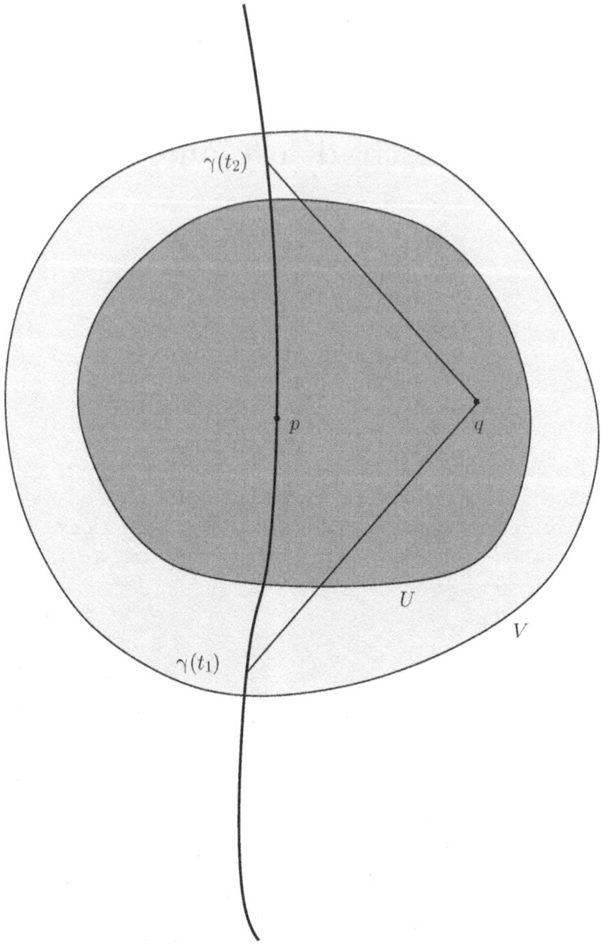

Fig. 1. The radar method.

(see Fig. 1). The *radar time* T and the *radar distance* R of the event q with respect to γ are then defined by

$$T = \frac{1}{2}\big(t_2 + t_1\big), \tag{1}$$

$$R = \frac{1}{2}\big(t_2 - t_1\big). \tag{2}$$

Here and in the following, "light ray" tacitly means "freely propagating light ray," i.e., it is understood that there is no optical medium and that mirrors or other appliances that deviate a light ray are not used. Adopting the standard formalism of general relativity, "light ray" is then just another word for "lightlike geodesic of the spacetime metric g."

In the following, we discuss the radar method from a geometrical point of view, reviewing some known results and formulating a few new ones. The radar method has obvious relevance for the communication with satellites in the solar system, because all such communication is made with the help of electromagnetic radiation that can be modeled, in almost all cases, in terms of light rays. By sending a light ray to a satellite and receiving the reflected signal, the radar time T and the radar distance R of events at the satellite are directly measurable quantities. Note that we do not need an experimentalist at the event q where the light ray is reflected; a passive reflecting body, such as the LAGEOS satellites, would do.

2 Radar Neighborhoods

The radar time T and the radar distance R of an event q with respect to a clock γ are well defined if there is precisely one future-pointing and precisely one past-pointing light ray from q to γ. Neither existence nor uniqueness of such light rays is guaranteed.

It is possible that an event q cannot be connected to γ by any future-pointing (or any past-pointing) light ray. There are two physically different situations in which this occurs: first, q may be in a "shadow" of some obstacle that lies in the direction to γ; second, q may be behind an "event horizon" of γ (see Fig. 2).

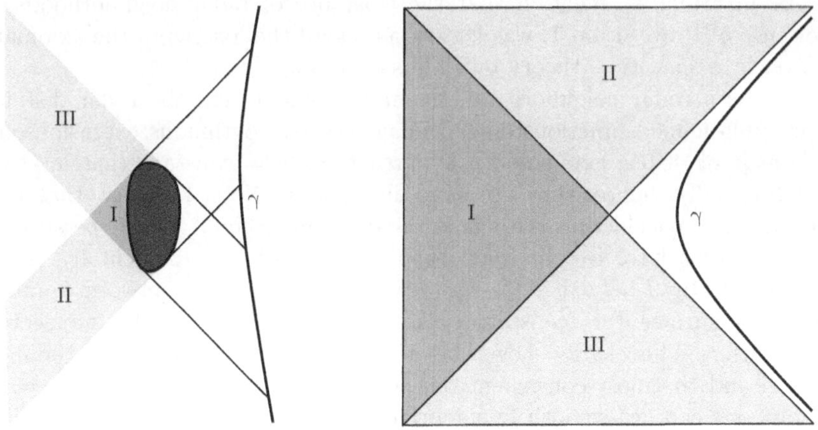

Fig. 2. Shadows (*left*) and horizons (*right*) are obstacles for the radar method. The example on the *left* shows a clock γ in Minkowski spacetime with a subset removed. The example on the *right* shows a clock γ with uniform acceleration in Minkowski spacetime. In both cases, events in the region II cannot be connected to γ by a future-pointing light ray, events in the region III cannot be connected to γ by a past-pointing light ray, and events in the region I cannot be connected to γ by any light ray.

It is also possible that an event q can be connected to γ by two or more future-pointing (or past-pointing) light rays. Whenever the future light cone (or the past light cone) of q has a caustic or a transverse self-intersection, it meets some timelike curves at least twice, see [26] or [27] for a detailed discussion. If the past light cone of q intersects γ at least twice, an observer at q sees two or more images of γ, i.e., we are in a gravitational lensing situation. Figure 5 shows an example of a past light cone that has two intersections with appropriately chosen timelike curves, as is geometrically evident from the picture.

These observations clearly show that, in an arbitrary general-relativistic spacetime, the radar method does not work globally. However, it always works locally. This is demonstrated by the following simple proposition.

Proposition 1. *Let γ be a clock in an arbitrary general-relativistic spacetime and $p = \gamma(t_0)$ be some point on γ. Then there are open subsets U and V of the spacetime with $p \in U \subset V$ such that every point q in $U \setminus \mathrm{image}(\gamma)$ can be connected to the world line of γ by precisely one future-pointing and precisely one past-pointing light ray that stays within V (see Fig. 1). In this case, U is called a* radar neighborhood *of p with respect to γ.*

To prove this, we just have to recall that every point in a general-relativistic spacetime admits a convex normal neighborhood, i.e., a neighborhood V such that any two points in V can be connected by precisely one geodesic that stays within V. Having chosen such a V, it is easy to verify that every sufficiently small neighborhood U of p satisfies the desired property.

As an aside, we mention that the existence of radar neighborhoods, in the sense of Proposition 1, was chosen as one of the axioms in the axiomatic approach to spacetime theory by Ehlers et al. [5].

If U is a radar neighborhood, the radar time T and the radar distance R are well-defined functions on $U \setminus \mathrm{image}(\gamma)$. By continuous extension onto the image of γ, one gets smooth hypersurfaces $T = $ constant that intersect γ orthogonally; hence, they are spacelike near γ. Note, however, that they need not be spacelike on the whole radar neighborhood. The hypersurfaces $R = $ constant have a cylindrical topology (see Fig. 3). Incidentally, if one replaces (1) by $T = p t_1 + (1 - p) t_2$ with any number p between 0 and 1, each hypersurface $T = $ constant gets a conic singularity at the intersection point with γ. This clearly shows that the choice of the factor $1/2$ is the most natural and the most convenient one. (If one allows for *direction-dependent* factors, one can get smooth hypersurfaces with factors other than $1/2$. This idea, which however seems a little bit contrived, was worked out by Havas [13] where the reader can find more on the "conventionalism debate" around the factor $1/2$.)

By covering γ with radar neighborhoods U (and the pertaining convex normal neighborhoods V), it is easy to verify that T and R coincide on the intersection of any two radar neighborhoods. Hence, T and R are well defined on some tubular neighborhood of γ. We will now investigate how large this

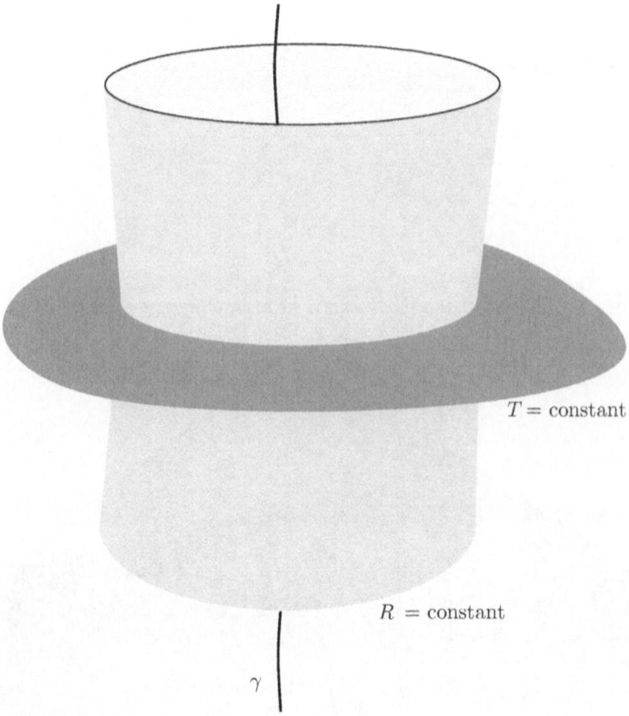

Fig. 3. Hypersurfaces T = constant and hypercylinders R = constant defined by the radar method.

neighborhood can be for the case of a clock moving in the solar system, the latter being modeled by the Schwarzschild spacetime around the Sun.

To that end we consider the Schwarzschild spacetime around a nontransparent spherical body of radius r_* and mass m. (The radius is measured in terms of the radial Schwarzschild coordinate and for the mass we use geometrical units, i.e., the Schwarzschild radius is $2m$.) Using the standard deflection formula for light rays in the Schwarzschild spacetime, the following result can be easily verified. If a bundle of light rays comes in initially parallel from infinity, the rays that graze the surface of the central body will meet the axis of symmetry of the bundle at radius

$$r_f = \frac{r_*}{4\frac{m}{r_*} + O\big((\frac{m}{r_*})^2\big)} \approx \frac{r_*^2}{4m} \tag{3}$$

see Fig. 4. This radius r_f is sometimes called the *focal length* of a nontransparent body of radius r_* and mass m. If we insert the values of our Sun, we find

$$r_f \approx 550 \text{ a.u.} \tag{4}$$

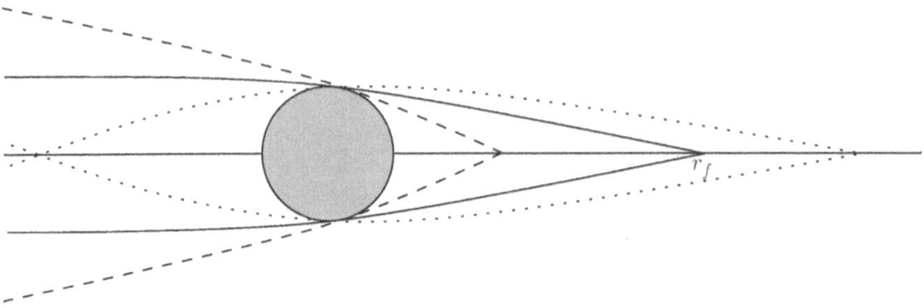

Fig. 4. The focal length r_f of a nontransparent spherical body.

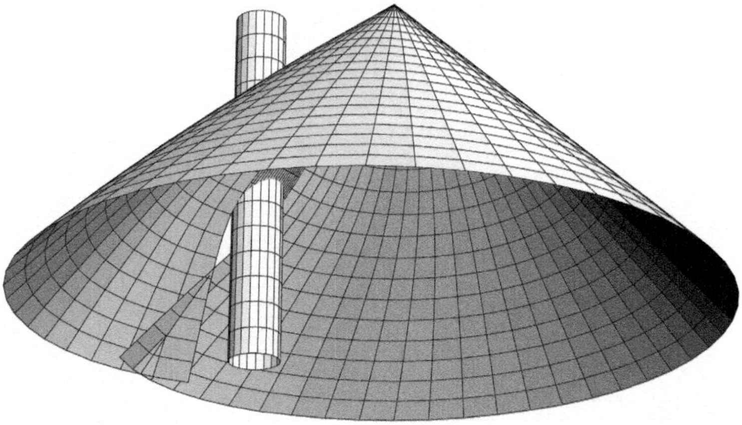

Fig. 5. Past light cone of an event at radius $r > r_f$ in the field of a nontransparent gravitating body. The "chimney" is the world tube of the gravitating body. The "shadow" is clearly seen as a gap in the light cone. In this $2+1$-dimensional picture the light cone forms a transverse self-intersection. If the gravitating body is spherical symmetric, taking the missing spatial dimension into account shows that actually a sphere's worth of light rays is focused at each point of the intersection set.

where 1 a.u. = 1 astronomical unit is the average distance from the Earth to the Sun. From any event at $r < r_f$, the future-pointing and past-pointing light rays spread out without intersecting each other. They cover the whole space $r > r_*$ with the exception of those points that lie in the "shadow" cast by the central body (see Fig. 4). By contrast, light rays from an event at $r > r_f$ do intersect; the past light cone of such an event is shown in Fig. 5.

As a consequence, for a clock γ moving arbitrarily in the region $r > r_*$, an event q at a radius r with $r_* < r < r_f$ can be connected to the world line of γ by at most one future-pointing and at most one past-pointing light ray. We shall make the additional assumption that γ is inextendible and approaches neither the surface of the central body nor infinity in the future or in the past. This assures that there are no event horizons for γ. As a consequence, any event q at radius r with $r_* < r < r_f$ can be connected to γ by precisely one future-pointing and precisely one past-pointing light ray unless γ moves through the shadow cast by the central body for light rays issuing from q.

An event q at radius $r > r_f$, on the other hand, may be connected to the world line of a clock by several future-pointing (or past-pointing) light rays. This is geometrically evident from Fig. 5.

So for any clock in the solar system, the radar method assigns a unique time T and a unique distance R to any event at radius $r < r_f$, with the exception of those events for which the clock lies in the shadow of the central body. Note that for all existing spacecraft the distance from the Sun is considerably smaller than $r_f = 550\,\text{a.u.}$ (In October 2005, the spacecraft farthest away from the Sun was Pioneer 10 with a distance of 89 a.u.)

The idea of sending a spacecraft to $r > 550\,\text{a.u.}$ was brought forward by Eshleman [8] in 1979. What makes this idea attractive is the possibility of observing distant light sources strongly magnified by the focusing effect of the gravitational field of the Sun (see again Fig. 4). For a detailed discussion of the perspectives of such a mission, see Turyshev and Andersson [37].

It should be emphasized that our consideration applies only to a nontransparent body. If the central body is transparent, light rays passing through the central region of the body are focussed at a radius that is much smaller than the r_f given above. If the interior is modeled by a perfect fluid with constant density, one finds for the Sun a focal length of 30 a.u., in comparison to the 550 a.u. for the nontransparent case, see Nemiroff and Ftaclas [22]. A transparent Sun is a reasonable model for neutrino radiation (which travels approximately, though not precisely, on lightlike geodesics) and for gravitational radiation (which travels along lightlike geodesics if modeled as a linear perturbation of the Schwarzschild background). So, the focusing at 30 a.u. might have some futuristic perspective in view of neutrino astronomy and gravitational wave astronomy.

3 Characterization of Standard Clocks with the Radar Method

If we reparametrize the curve γ, the hypersurfaces $T = \text{constant}$ and $R = \text{constant}$ change. Therefore, the radar method can be used to characterize distinguished parametrizations of world lines, i.e., distinguished clocks. In a general-relativistic spacetime, the standard clock parametrization is defined by the condition

$$\frac{d}{dt} g\big(\dot{\gamma}(t), \dot{\gamma}(t)\big) = 0 \tag{5}$$

where g is the spacetime metric. This defines a parametrization along any timelike curve that is unique up to affine transformations, $t \mapsto at + b$ with real constants a and b. As we restrict to future-pointing parametrizations, a must be positive. Then the choice of a determines the unit and the choice of b determines the zero on the dial. By choosing a appropriately, we can fix the unit of a standard clock such that $g\big(\dot{\gamma}(t),\dot{\gamma}(t)\big) = -1$. Then the parameter of the clock is called *proper time*. Note that under an affine reparametrization $t \mapsto at + b$ the radar time and the radar distance transform according to $T \mapsto aT + b$ and $R \mapsto aR$, i.e., the hypersurfaces $T =$ constant and $R =$ constant are relabeled but remain unchanged.

With the help of the radar method, one can formulate an operational prescription that allows to test whether a clock is a standard clock. This prescription is now briefly reviewed, for details and proofs see [24]. Here we assume that the test is made in a general-relativistic spacetime; in [24] the more general case of a Weylian spacetime is considered.

To test whether a clock γ behaves like a standard clock in a particular event $\gamma(t_0)$, we emit at this event two freely falling particles in spatially opposite directions. These two freely falling particles are mathematically modeled by timelike geodesics μ and $\overline{\mu}$, and the condition that they are emitted in spatially opposite directions means that the future-oriented tangent vector to γ is a convex linear combination of the future-oriented tangent vectors to μ and $\overline{\mu}$. If we restrict to a radar neighborhood of $\gamma(t_0)$, the radar method assigns a time T and a distance R to each event on μ, and a time $\overline{T}$ and a distance $\overline{R}$ to each event on $\overline{\mu}$ (see Fig. 6). These quantities can be actually measured provided that the two freely falling particles are reflecting objects. From these

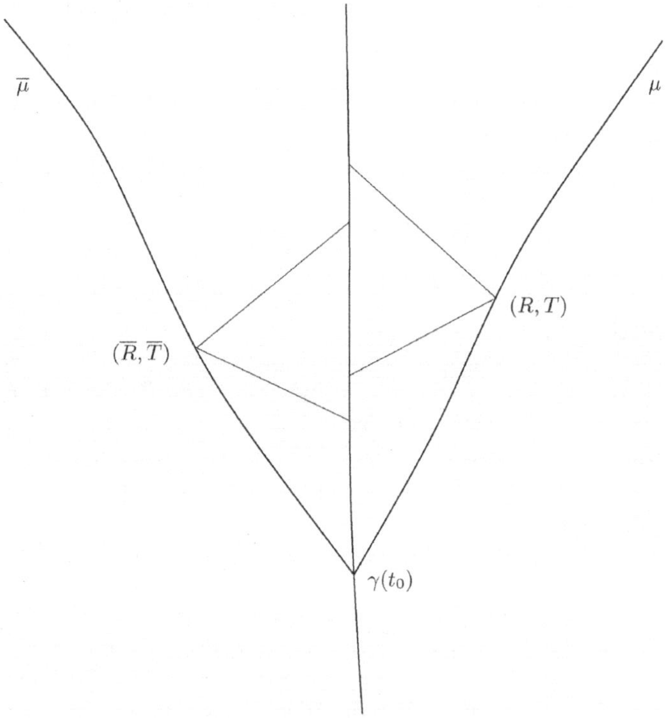

Fig. 6. Testing a clock as a standard clock with the radar method.

measured quantities we can calculate the differential quotients dR/dT and d^2R/dT^2 along μ and the differential quotients $d\overline{R}/d\overline{T}$ and $d^2\overline{R}/d\overline{T}^2$ along $\overline{\mu}$, i.e., the *radar velocity* and the *radar acceleration* of the two freely falling particles. It is shown in [24] that the standard clock condition (5) holds at $t = t_0$ (which corresponds to $T = \overline{T} = t_0$) if and only if

$$\frac{\dfrac{d^2R}{dT^2}}{1-\left(\dfrac{dR}{dT}\right)^2}\Bigg|_{T=t_0} = -\frac{\dfrac{d^2\overline{R}}{d\overline{T}^2}}{1-\left(\dfrac{d\overline{R}}{d\overline{T}}\right)^2}\Bigg|_{\overline{T}=t_0}. \tag{6}$$

This prescription can be used, in particular, to directly test whether atomic clocks are standard clocks. All experiments so far are in agreement with this hypothesis, but a direct test has not been made.

There are alternative characterizations of standard clocks by Marzke and Wheeler [19] and Kundt and Hoffman [18] which also work with light rays and freely falling particles. The advantages of the method reviewed here in comparison to these two older methods are outlined in [25].

4 Radar Coordinates, Optical Coordinates, and Fermi Coordinates

Given any clock γ in any general-relativistic spacetime, the radar method assigns, as outlined above, to each event q in some tubular neighborhood of γ a radar time T and a radar distance R. To get a coordinate system (*radar coordinates*) on this tubular neighborhood, we may add two angular coordinates ϑ and φ in the following way. Choose at each point $\gamma(t)$ an orthonormal tetrad $(E_0(t), E_1(t), E_2(t), E_3(t))$, smoothly dependent on t, such that $E_0(t)$ is future-pointing and tangent to γ. To each event q consider the past-oriented light ray, in the notation of Fig. 1, from $\gamma(t_2)$ to q. The tangent vector to this light ray at $\gamma(t_2)$ must be proportional to a vector of the form $-E_0(t_2) + \cos\varphi \sin\vartheta\, E_1(t_2) + \sin\varphi \sin\vartheta\, E_2(t_2) + \cos\vartheta\, E_3(t_2)$ which defines ϑ and φ. Thus, ϑ and φ indicate at which point on the sky of γ the event q is seen. Just as with ordinary spherical coordinates, there are coordinate singularities at $R = 0$ and at $\sin\vartheta = 0$, and φ has to be identified with $\varphi + 2\pi$. Apart from these obvious pathologies, the radar coordinates $(T, R, \vartheta, \varphi)$ form a well-defined coordinate system on some tubular neighborhood of γ. There are two possibilities of modifying the radar coordinates without changing the information contained in them. First, one may replace T and R by t_1 and t_2, according to (1) and (2), and use the modified radar coordinates $(t_1, t_2, \vartheta, \varphi)$. Second, one may switch to Cartesian-like coordinates (T, x, y, z) by introducing $x = R\cos\varphi\sin\vartheta$, $y = R\sin\varphi\sin\vartheta$, and $z = R\cos\vartheta$ to remove the coordinate singularities at $R = 0$ and $\sin\vartheta = 0$. Radar coordinates have been used as a tool, e.g., in the axiomatic approach to spacetime theory of Schröter and Schelb [32–34].

We will now compare radar coordinates with two other kinds of coordinate systems that can be introduced near the world line of any clock γ: "optical coordinates" and "Fermi coordinates." We will see that there are some similarities but also major differences between these three types of coordinate systems. For an alternative discussion of optical coordinates and Fermi coordinates, see Synge [35].

Optical coordinates were introduced by Temple [36]. The alternative name *observational coordinates* is also common, see Ellis et al. [6, 7]. They assign to the event q the four-tuple $(t_2, s, \vartheta, \varphi)$, where t_2, ϑ, and φ have the same meaning as above and s is the "affine length" (or "projected length") along the past-oriented light ray from $\gamma(t_2)$ to q. Using the exponential map exp determined by the spacetime metric, s can be defined by the equation

$$q = \exp_{\gamma(t_2)}\Big(s\big(-E_0(t_2) + \cos\varphi \sin\vartheta\, E_1(t_2) + \sin\varphi \sin\vartheta\, E_2(t_2) + \cos\vartheta\, E_3(t_2)\big)\Big). \tag{7}$$

Just as radar coordinates, optical coordinates are well defined, apart from the obvious coordinate singularities at $s = 0$ and $\sin\vartheta = 0$ on some tubular neighborhood of γ. The boundary of this neighborhood is reached when the past light cone of an event on γ develops a caustic or a transverse self-intersection. (Beyond such points, the optical coordinates are multivalued. This does not mean that they are useless there; however, they do not define a coordinate system in the usual sense.) As radar coordinates require a similar condition not only on past light cones but also on future light cones, the domain of radar coordinates is always contained in the domain of optical coordinates. Also, there is an important advantage of optical coordinates in view of calculations: optical coordinates only require to calculate the past-pointing lightlike geodesics issuing from points on γ; radar coordinates require to calculate past-pointing and future-pointing lightlike geodesics from points on γ, and to determine their intersections. Nonetheless, for applications in the solar system radar coordinates are advantageous because they have an operational meaning. In principle, optical coordinates also have an operational meaning: $(t_2, \vartheta, \varphi)$ are the same as in radar coordinates, and for the affine (or projected) length s a prescription of measurement was worked out by Ruse [29] after this length measure had been introduced mathematically by Kermack, McCrea, and Whittaker [16]. However, this prescription requires the distribution of assistants with rigid rods along each light ray issuing from γ into the past which is, of course, totally unrealistic in an astronomical situation. In this sense, optical coordinates have an operational meaning only in principle but not in practice, whereas radar coordinates have an operational meaning both in principle and in practice, at least in the solar system. In cosmology, however, this is no longer true. Then the radar coordinates, just as the optical coordinates, have an operational meaning only in principle but not in practice: sending a light ray to a distant galaxy and waiting for the reflected ray is a ridiculous idea. As a matter of fact, optical coordinates are much more useful in cosmology than radar coordinates. Although s is not directly measurable, it is related in some classes of spacetimes to other distance measures, such as the redshift or the angular diameter distance, which can be used to replace s. For applications of optical coordinates in cosmology, see [7]. As the simplest example, one may consider optical coordinates and radar coordinates in Robertson–Walker spacetimes, cf. Jennison and McVittie [15] and Fletcher [10].

We now turn to *Fermi coordinates* which were introduced by Enrico Fermi [9]. Let us recall how they are defined. As above, we have to choose along γ an orthonormal tetrad $(E_0(t), E_1(t), E_2(t), E_3(t))$ with E_0 tangent to γ. Following Fermi, we require that the covariant derivative of each spatial axis E_μ $(\mu = 1, 2, 3)$ is parallel to the tangent of γ. This *Fermi transport* law can be operationally realized by means of gyroscope axes [20] or Synge's *bouncing photon* method [28, 35]. (Actually, the construction below can be carried through equally well if the spatial axes are not Fermi parallel. What is needed is only smooth dependence on the foot-point, just as with radar coordinates and optical coordinates.) Then every event q in a sufficiently small tubular neighborhood of γ can be written in the form

$$q = \exp_{\gamma(\tau)}\Big(\rho\left(\cos\phi\sin\theta\, E_1(\tau) + \sin\phi\sin\theta\, E_2(\tau) + \cos\theta\, E_3(t_2)\right)\Big). \quad (8)$$

The Fermi coordinates of the point q are the four numbers $(\tau, \rho, \theta, \phi)$. Thus, each surface $\tau =$ constant is generated by the geodesics issuing orthogonally from the point $\gamma(\tau)$. The distance ρ is defined analogously to the affine length in the optical coordinates, but now along spacelike rather than lightlike geodesics. Also, the angular coordinates θ and ϕ are analogous to the angular coordinates ϑ and φ in the radar and optical coordinates, but now they indicate the direction of a spacelike vector, rather than the direction of the spatial part of a lightlike vector. Just as the other two coordinate systems, Fermi coordinates are well defined only on some tubular neighborhood of γ. There are two reasons that limit this neighborhood. First, a hypersurface $\tau =$ constant might develop caustics or self-intersections. Second, two hypersurfaces $\tau =$ constant might intersect. In contrast to radar coordinates, Fermi coordinates are insensitive to reparametrizations of γ (apart from the fact that the surfaces $\tau =$ constant are relabeled). The difficulty involved in their calculation is the same as for optical coordinates which is considerably less than for radar coordinates, as already mentioned above. The essential drawback of Fermi coordinates is in the fact that they have absolutely no operational meaning: none of the four coordinates τ, ρ, θ, and ϕ can be measured because there is no prescription for physically realizing a spacelike geodesic orthogonal to a world line.

In spite of this fact, Fermi coordinates have found many applications because sometimes physically relevant effects can be conveniently calculated in terms of Fermi coordinates. For a plea in favor of Fermi coordinates, in comparison to radar coordinates, see Bini et al. [1]. In Minkowski spacetime, e.g., it is fairly difficult to calculate the radar time hypersurfaces T = constant for an accelerating clock. By contrast, the Fermi time hypersurfaces τ = constant are just the hyperplanes perpendicular to the world line which are quite easy to determine. (Of course, for an accelerating clock these hyperplanes necessarily intersect, so they cannot form a smooth foliation on all of Minkowski spacetime.) It is an interesting question to ask for which clocks the radar time hypersurfaces T = constant coincide with the Fermi time hypersurfaces τ = constant. For standard clocks (recall Sect. 3) in Minkowski spacetime, Dombrowski et al. [4] have found the following answer.

Proposition 2. *Let γ be a standard clock in Minkowski spacetime. Then the following two statements are equivalent:*

(a) *The radar time hypersurfaces T =* constant *are hyperplanes, i.e., they coincide with the Fermi time hypersurfaces τ =* constant.
(b) *The 4-acceleration of γ is constant (i.e., a Fermi-transported vector along γ).*

A world line with constant 4-acceleration in Minkowski spacetime is either a straight line ("inertial observer," for which the 4-acceleration is zero) or a hyperbola ("Rindler observer," for which the 4-acceleration is a nonzero Fermi-transported vector, see Fig. 2). It is easy to check that, indeed, in both cases the radar time hypersurfaces with respect to proper time parametrization are hyperplanes. The nontrivial statement of Proposition 2 is in the fact that these are the *only* cases for which the radar time hypersurfaces are hyperplanes.

We end this section with a remark on the fact that the term "radar coordinates" has been used in the literature also in another way. Instead of supplementing the radar time T and the radar distance R with two angular coordinates, one could choose a second clock $\tilde{\gamma}$ which defines a radar time $\tilde{T}$ and a radar distance $\tilde{R}$. If the two clocks are sufficiently close, $(T, R, \tilde{T}, \tilde{R})$ can be used as coordinates on some open subset which is not a tubular neighborhood of either clock. Of course, one can replace (T, R) by (t_1, t_2) according to (1) and (2), and analogously $(\tilde{T}, \tilde{R})$ by $(\tilde{t}_1, \tilde{t}_2)$. In the coordinates $(t_1, t_2, \tilde{t}_1, \tilde{t}_2)$, which are used, e.g., by Ehlers et al. [5], the coordinate hypersurfaces are light cones. Thus, the construction makes use of the fact that four light cones generically intersect in a point. (Two light cones generically intersect in a two-dimensional manifold, where "generically" means that we have to exclude points where one of the light cones fails to be a submanifold and points where the two light cones are tangent. Similarly, three light cones generically intersect in a one-dimensional manifold.) In this sense the radar coordinates of Ehlers et al. are similar to the GPS type coordinates of Blagojević et al. [2]. The only difference is that the latter characterize each point as intersection of four future light cones that issue from four given world lines ("GPS satellites"), whereas the former characterize each point as intersection of two future and two past light cones that issue from two given world lines.

5 Synchronization of Clocks

Let γ be a clock in an arbitrary general-relativistic spacetime, and consider a second clock $\tilde{\gamma}$. If $\tilde{\gamma}$ is sufficiently close to γ, the radar method, carried through with respect to the clock γ, assigns a unique time $T(\tilde{t})$ and a unique distance $R(\tilde{t})$ to each event $\tilde{\gamma}(\tilde{t})$. We say that $\tilde{\gamma}$ is *synchroneous* to γ if $T(\tilde{t}) = \tilde{t}$ for all $\tilde{t}$ in the considered time interval. (Instead of synchroneous one may say *Einstein synchroneous* or *radar synchroneous* to be more specific.) Clearly, for every world line sufficiently close to γ, there is a unique parametrization that is synchroneous to γ. Selecting this particular parametrization is called *synchronization* with γ. Note that the relation of being synchroneous is not symmetric: $\tilde{\gamma}$ may be synchroneous to γ without γ being synchroneous to $\tilde{\gamma}$. As an example, we may choose two affinely parametrized straight timelike lines γ and $\tilde{\gamma}$ in Minkowski spacetime that are not parallel. If we arrange the parameters such that $\tilde{\gamma}$ is synchroneous to γ, the converse is not true. Also, the relation of being synchroneous is not transitive: if $\tilde{\gamma}$ is synchroneous to γ and $\hat{\gamma}$ is synchroneous to $\tilde{\gamma}$, it is not guaranteed that $\hat{\gamma}$ is synchroneous to γ. This nontransitivity is best illustrated with the *Sagnac effect*: consider a family of clocks along the rim of a rotating circular platform in Minkowski spacetime. Starting with any one of these clocks, synchronize each clock with its neighbor on the right. Then there is a deficit time interval after completing the full circle.

Proposition 3 characterizes the special situation that two clocks are mutually synchroneous.

Proposition 3. *Let $\gamma : \mathbb{R} \longrightarrow M$ and $\tilde{\gamma} : \mathbb{R} \longrightarrow M$ be two clocks, in an arbitrary spacetime, for which the parameter extends from $-\infty$ to $+\infty$. Assume that the world lines of the two clocks have no intersection but are sufficiently close to each other such that the radar method can be carried through in both directions. If $\tilde{\gamma}$ is synchroneous to γ and γ is synchroneous to $\tilde{\gamma}$, then the radar distance R of $\tilde{\gamma}$ with respect γ is a constant R_0, and the radar distance $\tilde{R}$ of γ with respect to $\tilde{\gamma}$ is the same constant R_0.*

Proof. The radar method carried through with γ assigns to each event $\tilde{\gamma}(\tilde{t})$ a time $T(\tilde{t})$ and a distance $R(\tilde{t})$. Analogously, the radar method carried through with respect to $\tilde{\gamma}$ assigns to each event $\gamma(t)$ a time $\tilde{T}(t)$ and a distance $\tilde{R}(t)$. This implies the following identities (see Fig. 7)

$$\begin{aligned} t &= T\big(\tilde{T}(t) - \tilde{R}(t)\big) + R\big(\tilde{T}(t) - \tilde{R}(t)\big) \, , \\ \tilde{t} &= \tilde{T}\big(T(\tilde{t}) - R(\tilde{t})\big) + \tilde{R}\big(T(\tilde{t}) - R(\tilde{t})\big) \, . \end{aligned} \tag{9}$$

If the clocks are mutually synchroneous, T and $\tilde{T}$ are the identity maps, so (9) simplifies to

$$\begin{aligned} \tilde{R}(t) &= R\big(t - \tilde{R}(t)\big) \, , \\ R(\tilde{t}) &= \tilde{R}\big(\tilde{t} - R(\tilde{t})\big) \, . \end{aligned} \tag{10}$$

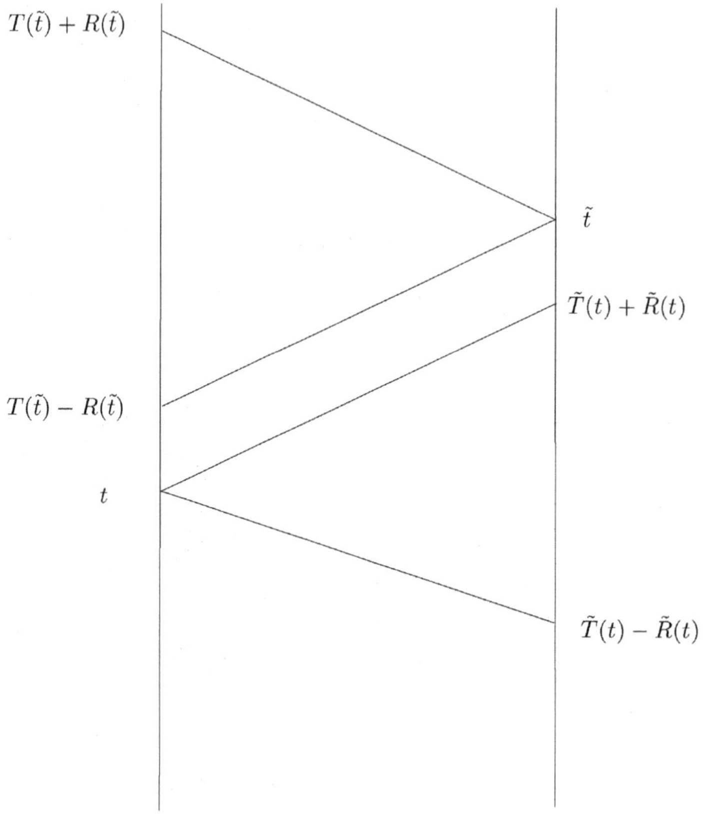

Fig. 7. Illustration of the proof of Proposition 3.

These equations hold for all t and for all $\tilde{t}$ in $\mathbb{R}$. By considering the special case $t = \tilde{t} - R(\tilde{t})$ we find

$$R(\tilde{t}) = R\big(\tilde{t} - 2R(\tilde{t})\big) \tag{11}$$

for all $\tilde{t}$ in $\mathbb{R}$. To ease notation, we drop the tilde in the following. By induction, (11) yields

$$R(t) = R\big(t - 2nR(t)\big) \qquad \text{for all } n \in \mathbb{N}. \tag{12}$$

It is now our goal to prove that (12) implies that R is a constant. By contradiction, assume there is a point where R has negative derivative, $R'(t_*) < 0$. Then we must have

$$\frac{t_* \, - \, (t_* + \varepsilon) \, + \, 2\, R(t_* + \varepsilon)}{2\big(R(t_*) - R(t_* + \varepsilon)\big)} \longrightarrow \infty \qquad \text{for } \varepsilon \to +0, \tag{13}$$

because, by our assumption that the world lines of the two clocks do not intersect, $R(t_*) > 0$. Thus, there is an infinite sequence t_n that converges toward t_* from above, such that

$$n = \frac{t_* - t_n + 2\,R(t_n)}{2\big(R(t_*) - R(t_n)\big)} \qquad \text{for all sufficiently large } n \in \mathbb{N}\,. \tag{14}$$

As (14) can be rewritten as

$$t_n - 2\,(n+1)\,R(t_n) = t_* - 2\,n\,R(t_*)\,, \tag{15}$$

our earlier result (12) yields $R(t_n) = R(t_*)$ for all members t_n of our sequence, which obviously contradicts the assumption $R'(t_*) < 0$. We have thus proven that $R'(t) \geq 0$ for all t. But then we must have $R'(t) = 0$ for all t, because, again by (12), to every t there is a smaller parameter value at which the function R takes the same value. Hence, R must be a constant, $R(t) = R_0$ for all t. It is obvious from (10) that then $\tilde{R}$ must take the same constant value. □

We illustrate this result with an example in Minkowski spacetime, using standard coordinates (x^0, x^1, x^2, x^3) such that the metric takes the form

$$g = -(dx^0)^2 + (dx^1)^2 + (dx^2)^2 + (dx^3)^2\,. \tag{16}$$

We consider the two clocks

$$\gamma(t) = (\,t\,,0\,,0\,,0\,)\,, \tag{17}$$

$$\tilde{\gamma}(\tilde{t}) = \big(\sqrt{1-\omega^2 R_0^2}\,\tilde{t}\,,\,R_0 \cos\omega\tilde{t}\,,\,R_0 \sin\omega\tilde{t}\,,\,0\big)\,, \tag{18}$$

where ω and R_0 are constants such that $\omega^2 R_0^2 < 1$. In both cases the parameter is proper time, i.e., both clocks are standard clocks with the usual choice of the time unit. The first clock is at rest at the origin of the coordinate system, the other clock moves with constant angular velocity ω on a circle with radius R_0 around the origin. An elementary exercise shows that the radar method carried through with respect to γ assigns to each event $\tilde{\gamma}(\tilde{t})$ the time $T(\tilde{t}) = \sqrt{1-\omega^2 R_0^2}\,\tilde{t}$ and the distance $R(\tilde{t}) = R_0$. On the other hand, the radar method carried through with respect to $\tilde{\gamma}$ assigns to each event $\gamma(t)$ the time $\tilde{T}(t) = t/\sqrt{1-\omega^2 R_0^2}$ and the distance $\tilde{R}(t) = R_0/\sqrt{1-\omega^2 R_0^2}$. Thus, neither clock is synchroneous to the other, and they assign to each other constant but different distances. Now let us modify this example by changing the time unit for $\tilde{\gamma}$ according to the affine transformation $\tilde{t} \mapsto \hat{t} = \sqrt{1-\omega^2 R_0^2}\,\tilde{t}$. This transformation replaces $\tilde{\gamma}$ with a new clock $\hat{\gamma}$,

$$\hat{\gamma}(\hat{t}) = \Big(\,\hat{t}\,,\,R_0 \cos\frac{\omega\,\hat{t}}{\sqrt{1-\omega^2 R_0^2}}\,,\,R_0 \sin\frac{\omega\,\hat{t}}{\sqrt{1-\omega^2 R_0^2}}\,,\,0\Big)\,. \tag{19}$$

Note that $\hat{\gamma}$ is still a standard clock, but not with the usual time unit. We now find that the radar method carried through with respect to γ assigns to each event $\hat{\gamma}(\hat{t})$ the time $T(\hat{t}) = \hat{t}$ and the distance $R(\tilde{t}) = R_0$. On the other hand, the radar method carried through with respect to $\hat{\gamma}$ assigns to each event $\gamma(t)$ the time $\hat{T}(t) = t$ and the distance $\hat{R}(t) = R_0$. This modified example illustrates that Proposition 3 may apply to situations where there is no symmetry between the two clocks.

6 Observer Fields

By an *observer field* on a general-relativistic spacetime, we mean a smooth vector field V which is everywhere timelike and future-pointing. An observer field V is called a *standard observer field* if $g(V, V) = -1$. According to our earlier terminology, integral curves of observer fields are clocks, and integral curves of standard observer fields are standard clocks with the usual choice of time unit. For the sake of brevity, we will refer to the integral curves of an observer field V as to "clocks in V." Note that V fixes the parametrization for each of its integral curves uniquely up to an additive constant, i.e., for each clock in V there is still the freedom of "choosing the zero point on the clock's dial."

In this section, we consider the following four properties of an observer field V, and for each of them we give necessary and sufficient conditions on V under which it is satisfied:

Property A. For each clock γ in V, any other clock in V that is sufficiently close to γ such that the radar method can be carried through is synchroneous with γ, provided that the additive constant has been chosen appropriately.

Property B. For each clock γ in V, any other clock in V that is sufficiently close to γ such that the radar method can be carried through has temporally constant radar distance from γ.

Property C. For any three clocks γ_1, γ_2, and γ_3 in V which are sufficiently close to each other, the following is true: if one light ray from γ_1 to γ_3 intersects the world line of γ_2, then all light rays from γ_1 to γ_3 intersect the world line of γ_2.

Property D. For any two clocks γ_1 and γ_2 in V that are sufficiently close to each other, the light rays from γ_1 to γ_2 and the light rays from γ_2 to γ_1 span the same 2-surface.

All four properties are obviously closely related to the radar method, and we will discuss them one by one. In the following we have to assume that the reader is familiar with the standard textbook decomposition of the covariant derivative of an observer field into acceleration, rotation, shear, and expansion, and with the related physical interpretation.

We begin with Property A. We emphasize that, in the formulation of this property, we restricted to clocks that are sufficiently close to each other such that the radar method can be carried through, but *not* to clocks that are infinitesimally close. The synchronizability condition for infinitesimally close clocks is a standard textbook matter, see, e.g., Sachs and Wu [30], Sects. 2.3 and 5.3. One finds that this condition is satisfied, for an appropriately rescaled observer field $e^f V$, if and only if V is irrotational, i.e., locally hypersurface orthogonal. The rescaling means that the clocks of the observers have to be changed appropriately. The synchronization condition for clocks that are not infinitesimally close to each other is less known. It is given in Proposition 4.

Proposition 4. (i) *A standard observer field V satisfies Property A if and only if V is an irrotational Killing vector field.*
(ii) *An arbitrary (not necessarily standard) observer field V satisfies Property A if and only if V is an irrotational conformal Killing vector field.*

Proof. The hard part of the proof is in a paper by Kuang and Liang [17] who proved the following. If V is a standard observer field, any point admits a neighborhood that can be sliced into hypersurfaces that are synchronization hypersurfaces for all clocks in V if and only if V is proportional to an irrotational Killing vector field. In this case, the flow of the Killing vector field maps synchronization hypersurfaces onto synchronization hypersurfaces. Clearly, Property A requires in addition that the hypersurfaces can be labeled such that along each integral curve of V the labeling coincides with proper time. Thus, the flow of V itself must map synchronization hypersurfaces onto synchronization hypersurfaces. This completes the proof of Proposition 4(i). Now let V be an arbitrary observer field on the spacetime (M, g). Then it is a standard observer field on the conformally rescaled spacetime $(M, -g(V, V)^{-1}g)$. Clearly, as a conformal factor does not affect the paths of lightlike geodesics, V satisfies Property A on the original spacetime if and only if it satisfies Property A on the conformally rescaled spacetime. By Proposition 4(i), the latter is true if and only if V is a normalized irrotational Killing vector field of the metric $-g(V, V)^{-1}g$ and, thus, if and only if V is an irrotational conformal Killing vector field of the original metric g. This completes the proof of Proposition 4(ii). □

A spacetime that admits an irrotational Killing vector field normalized to -1 is called *ultrastatic*, and a spacetime that admits an irrotational conformal Killing vector field is called *conformally static*. Hence, we can summarize that ultrastaticity is necessary and sufficient for the existence of a standard observer field that satisfies Property A, and conformal staticity is necessary and sufficient for the existence of a (not necessarily standard) observer field that satisfies Property A. A simple and instructive example is an expanding Robertson–Walker spacetime. Such a spacetime admits a timelike conformal Killing vector field W orthogonal to hypersurfaces such that $g(W, W)$ is nonconstant along the integral curves of W. The flow lines of W are often referred to as the "Hubble flow." By Proposition 4(ii), the observer field W satisfies Property A, i.e., if we use on the Hubble flow lines a parametrization adapted to W (often called "conformal time"), then the clocks are synchroneous. However, the standard observer field V that results by normalizing W does not satisfy Property A, i.e., if we use on the Hubble flow lines the parametrization by proper time, the clocks are not synchroneous (unless they are infinitesimally close to each other). This example demonstrates that it is sometimes mathematically convenient to use nonstandard observer fields.

We now turn to Property B which may be viewed as a rigidity condition. Again, there is a well-known textbook result on the situation where only clocks that are infinitesimally close are considered: for a standard observer field, any two clocks that are infinitesimally close to each other have temporally constant radar distance if and only if V has vanishing shear and vanishing expansion. This is known as the *Born rigidity* condition, referring to a classical paper by Born [3] who introduced this rigidity notion in *special relativity*. The differential equations for Born rigid observer fields in *general relativity* were first written by Salzmann and Taub [31]. They have nontrivial integrability conditions, i.e., Born rigid observer fields do not exist on arbitrary spacetimes. The following important result is known as the *generalized Herglotz–Noether theorem*: if V is a Born-rigid, not hypersurface orthogonal standard observer field on a spacetime with constant curvature, then V is proportional to a Killing vector field. This was proven by Herglotz [14] and Noether [23] for the case of vanishing curvature (Minkowski spacetime) and generalized by Williams [38] to the case of positive or negative curvature (deSitter or anti-deSitter spacetime). As in the case of the synchronization condition, the rigidity condition for clocks that are not infinitesimally close to each other is less well known. It is given in Proposition 5.

Proposition 5. (i) *A standard observer field V satisfies Property B if and only if V is proportional to a Killing vector field.*
(ii) *An arbitrary (not necessarily standard) observer field V satisfies Property B if and only if $V = e^f W$, where W is a conformal Killing vector field and f is a scalar function that is constant along each integral curve of V.*

Proof. For the proof of Proposition 5(i) we refer to Müller zum Hagen [21]. To prove Proposition 5(ii), let V be an arbitrary observer field. As in the proof of Proposition 4(ii), we make use of the fact that V satisfies Property B with respect to the metric g if and only if V satisfies Property B with respect to the conformally rescaled metric $-g(V,V)^{-1}g$. The latter is true, by Proposition 5(i), if and only if $V = e^f W$ where W is a Killing vector field of the metric $-g(V,V)^{-1}g$. The latter condition is true if W is a conformal Killing vector field of the original metric g and $-g(V,V)^{-1}g(W,W) = e^{-2f}$ is constant along each integral curve of V. This completes the proof. □

A spacetime that admits a timelike Killing vector field is called *stationary* and a spacetime that admits a timelike conformal vector field is called *conformally stationary*. Hence, we can summarize that stationarity is necessary and sufficient for the existence of a standard observer field with Property B, and that conformal stationarity is necessary and sufficient for the existence of a (not necessarily standard) observer field with Property B. As an example, we may again consider the Hubble flow in an expanding Robertson–Walker spacetime. As with Property A, Property B is satisfied if we use conformal time but not if we use proper time.

We now turn to Property C. This property can be rephrased in the following way. If, from the position of one clock in V, two other clocks in V are seen at the same spot in the sky (i.e., one behind the other), then this will be true for all times. In a more geometric wording, Property C requires that the light rays issuing from any one integral curve of V into the past together with the integral curves of V are surface forming. In Hasse and Perlick [12], observer fields with this property were called *parallax free*, and Proposition 6 was proven.

Proposition 6. *An observer field V satisfies Property C if and only if V is proportional to a conformal Killing vector field.*

The "if" part follows from the well-known fact that the flow of a conformal Killing vector field maps light rays onto light rays. The proof of the "only if" part is more involved, see [12]. Clearly, Property C refers only to the motion of the clocks, but not to their "ticking" (i.e., not to the parametrization). Hence, it is irrelevant whether we consider standard observer fields or nonstandard observer fields.

Finally we turn to Property D which is a way of saying that light rays from γ_1 to γ_2 take the same spatial paths as light rays from γ_2 to γ_1. If this property is satisfied, there is a timelike 2-surface between γ_1 and γ_2 that is ruled by two families of lightlike geodesics. Note that *any* timelike 2-surface is ruled by two families of lightlike curves; in general, however, these will not be geodesics. Foertsch et al. [11] have shown that a timelike 2-surface is ruled by two families of lightlike geodesics if and only if its second fundamental form is a multiple of its first fundamental form. In the mathematical literature, such surfaces are called *totally umbilic*. Some construction methods and examples of timelike totally umbilic 2-surfaces are discussed in Foertsch et al. [11]. Note that in an arbitrary spacetime totally umbilic 2-surfaces need not exist. This shows that Property D, which requires such a 2-surface between any two sufficiently close integral curves of some observer field, is quite restrictive. A criterion is given in Proposition 7.

Proposition 7. *An observer field V satisfies Property D if and only if V is proportional to an irrotational conformal Killing vector field.*

Proof. The proof of the "if" part follows from Foertsch et al. [11], Proposition 3. To prove the "only if" part, fix any event p and let γ be the clock in V that passes through p. On a neighborhood of p, with the world line of γ omitted, consider two vector fields X and Y such that the integral curves of X are future-pointing light rays and the integral curves of Y are past-pointing light rays issuing from the world line of γ. This condition fixes X and Y uniquely up to nowhere vanishing scalar factors. Property D requires X and Y to be surface forming and V to be tangent to these surfaces. The first condition is true, by the well-known Frobenius theorem, if and only if the Lie bracket of X and Y is a linear combination of X and Y, and the second condition is true if and only if V is a linear combination of X and Y. As a consequence, the Lie bracket of Y and V must be a linear combination of Y and V, i.e., Y and V must be surface forming. This proves that V must satisfy Property C. Hence, by Proposition 6, V must be proportional to a conformal Killing vector field. What remains to be shown is that this conformal Killing vector field is irrotational, i.e., hypersurface orthogonal. To that end we come back to the observation that V is a linear combination of X and Y. This means that, for any integral curve of V in the considered neighborhood, light rays from γ are seen in the same spatial direction in which light rays to γ are emitted. This is true, in particular, for integral curves of V that are infinitesimally close to γ. Synge [35] and, in a simplified way, Pirani [28] have shown that this "bouncing photon construction" implies that the connecting vector between the two world lines is Fermi transported. This is true for *all* pairs of infinitesimally neighboring world lines of V if and only if V is irrotational. This completes the proof of Proposition 7. □

The important result to be kept in mind is that, in a spacetime that is not conformally stationary, it is impossible to find an observer field that satisfies any of the four Properties A, B, C, and D. This demonstrates that several features of the radar method, which intuitively might be taken for granted, are actually not satisfied in many cases of interest.

References

1. D. Bini, L. Lusanna, and B. Mashhoon. Limitations of radar coordinates. *Int. J. Modern Phys.*, D 14:1413–1429, 2005.
2. M. Blagojević, J. Garecki, F. W. Hehl, and Y. N. Obukhov. Real null coframes in general relativity and GPS type coordinates. *Phys. Rev.*, D 65:044018, 2002.
3. M. Born. Die Theorie des starren Elektrons in der Kinematik des Relativitätsprinzips. *Annalen der Physik*, 30:1–56, 1909.
4. P. Dombrowski, J. Kuhlmann, and U. Proff. On the spatial geometry of a non-inertial observer in special relativity. In T. J. Willmore and N. J. Hitchin, editors, *Global Riemannian Geometry*, pages 177–193, Chichester, UK, 1984. Horwood.
5. J. Ehlers, F. Pirani, and A. Schild. The geometry of free fall and light propagation. In L. O'Raifeartaigh, editor, *General Relativity*, pages 63–84, Oxford, UK, 1972. Clarendon.
6. G. F. R. Ellis. Limits to verification in cosmology. In J. Ehlers, editor, *Proceedings of the 9th Texas Symposium on Relativistic Astrophysics*, volume 336 of *Annals of the New York Academy of Sciences*, pages 130–160, New York, U.S.A., 1980. New York Academy of Sciences.
7. G. F. R. Ellis, S. D. Nel, R. Maartens, W. R. Stoeger, and A. P. Whitman. Ideal observational cosmology. *Phys. Rep.*, 124:315–417, 1985.
8. R. v. Eshleman. Gravitational lens of the sun – Its potential for observations and communications over interstellar distances. *Science*, 205:1133–1135, 1979.
9. E. Fermi. Sopra i fenomeni che avvengono in vicinanza di una linea ovaria. *Rendiconti Accad. Linc. Roma*, 31(1):21–23, 1922.
10. R. C. Fletcher. Light exchange in an expanding universe in fixed coordinates. *Amer. J. Phys.*, 62:648–656, 1994.
11. T. Foertsch, W. Hasse, and V. Perlick. Inertial forces and photon surfaces in arbitrary spacetimes. *Class. Quant. Grav.*, 20:4635–4651, 2003.

12. W. Hasse and V. Perlick. Geometrical and kinematical characterization of parallax-free world models. *J. Math. Phys.*, 29:2064–2068, 1988.
13. P. Havas. Simultaneity, conventionalism, general covariance and the special theory of relativity. *Gen. Relativ. Gravit.*, 19:435–453, 1987.
14. G. Herglotz. Über den vom Standpunkt des Relativitätsprinzips als "starr" zu bezeichnenden Körper. *Annalen der Physik*, 31:393–415, 1910.
15. R. C. Jennison and G. C. McVittie. Radar distance in Robertson–Walker spacetimes. *Royal Society Edinburgh Communications (Physical Sciences)*, 1:1–17, 1975.
16. W. O. Kermack, W. H. McCrea, and E. T. Whittacker. On properties of null geodesics and their application to the theory of radiation. *Proc. Roy. Soc. Edinburgh*, 53:31–47, 1932.
17. Z. Kuang and C. Liang. All space-times admitting strongly synchronizable reference frames are static. *J. Math. Phys.*, 34:1016–1021, 1993.
18. W. Kundt and B. Hoffmann. Determination of gravitational standard time. In Anonymous, editor, *Recent developments in general relativity*, pages 303–306, Oxford, UK, 1962. Pergamon.
19. R. F. Marzke and J. A. Wheeler. Gravitation as geometry. I: The geometry of space-time and the geometrodynamical standard meter. In H. Y. Chiu and W. F. Hoffmann, editors, *Gravitation and relativity*, pages 40–64, New York, USA, 1964. Benjamin.
20. C. Misner, K. Thorne, and J. A. Wheeler. *Gravitation*. Freeman, San Francisco, 1973.
21. H. Müller zum Hagen. A new physical characterization of stationary and static space-times. *Proc. Cambridge Philos. Soc.*, 71:381–389, 1972.
22. R. J. Nemiroff and C. Ftaclas. Our sun as a gravitational lens. *Bull. Amer. Astron. Soc.*, 29:827, 1997.
23. F. Noether. Zur Kinematik des starren Körpers in der Relativtheorie. *Annalen der Physik*, 31:919–944, 1910.
24. V. Perlick. Characterization of standard clocks by means of light rays and freely falling particles. *Gen. Relativ. Gravit.*, 19:1059–1073, 1987.
25. V. Perlick. Characterization of standard clocks in general relativity. In U. Majer and H.-J. Schmidt, editors, *Semantical aspects of spacetime theories*, pages 169–180, Mannheim, Germany, 1994. BI Wissenschaftsverlag.
26. V. Perlick. Criteria for multiple imaging in Lorentzian manifolds. *Class. Quant. Grav.*, 13:529–537, 1996.
27. V. Perlick. Gravitational lensing from a spacetime perspective. *Living Rev. Relativity*, 7(9), 2004. `http://www.livingreviews.org/lrr-2004-9`.
28. F. Pirani. A note on bouncing photons. *Bull. Acad. Polon. Sci., Ser. Sci. Mat. Astr. Phys.*, 13:239–242, 1965.
29. H. S. Ruse. On the measurement of spatial distance in a curved space-time. *Proc. Roy. Soc. Edinburgh*, 53:79–88, 1933.
30. R. K. Sachs and H.-H. Wu. *General relativity for mathematicians*. Springer, Berlin Heidelberg New York, 1977.
31. G. Salzmann and A. H. Taub. Born-type rigid motion in relativity. *Phys. Rev.*, 95:1659–1669, 1954.
32. U. Schelb. An axiomatic basis of space-time theory. III: Construction of a differentiable manifold. *Rep. Math. Phys.*, 31:297–309, 1992.
33. J. Schröter. An axiomatic basis of space-time theory. I: Construction of a causal space with coordinates. *Rep. Math. Phys.*, 26:303–333, 1988.

34. J. Schröter and U. Schelb. An axiomatic basis of space-time theory. II: Construction of a c^0- manifold. *Rep. Math. Phys.*, 31:5–27, 1992.
35. J. L. Synge. *Relativity. The general theory.* North-Holland, Amsterdam, 1960.
36. G. Temple. New systems of normal co-ordinates for relativistic optics. *Proc. Roy. Soc. London*, A 168:122–148, 1938.
37. S. G. Turyshev and B. G. Andersson. The 550-au mission: a critical discussion. *Mon. Not. Roy. Astron. Soc.*, 341:577–582, 2003.
38. G. Williams. Motions in relativistic spaces. *J. Math. Anal. Appl.*, 22:646–657, 1968.

A Universal Tool for Determining the Time Delay and the Frequency Shift of Light: Synge's World Function

Pierre Teyssandier[1], Christophe Le Poncin-Lafitte[1], and Bernard Linet[2]

[1] Département Systèmes de Référence Temps et Espace, CNRS/UMR 8630, Observatoire de Paris, 61 avenue de l'Observatoire, F-75014 Paris, France
[2] Laboratoire de Mathématiques et Physique Théorique, CNRS/UMR 6083, Université François Rabelais, F-37200 Tours, France

Summary. In almost all of the studies devoted to the time delay and the frequency shift of light, the calculations are based on the integration of the null geodesic equations. However, the above-mentioned effects can be calculated without integrating the geodesic equations if one is able to determine the bifunction $\Omega(x_A, x_B)$ giving half the squared geodesic distance between two points x_A and x_B (this bifunction may be called *Synge's world function*). In this chapter, $\Omega(x_A, x_B)$ is determined up to the order $1/c^3$ within the framework of the PPN formalism. The case of a stationary gravitational field generated by an isolated, slowly rotating axisymmetric body is studied in detail. The calculation of the time delay and the frequency shift is carried out up to the order $1/c^4$. Explicit formulae are obtained for the contributions of the mass, of the quadrupole moment, and of the internal angular momentum when the only post-Newtonian parameters different from zero are β and γ. It is shown that the relative frequency shift induced by the mass quadrupole moment of the Earth at the order $1/c^3$ will be bounded by 10^{-16} in space experiments like ESA's Atomic Clock Ensemble in Space (ACES) mission. Other contributions are briefly discussed.

1 Introduction

A lot of fundamental tests of gravitational theories rest on highly precise measurements of the travel time and/or the frequency shift of electromagnetic signals propagating through the gravitational field of the solar system. In practically all of the previous studies, the explicit expressions of such travel times and frequency shifts as predicted by various metric theories of gravity are derived from an integration of the null geodesic differential equations. This method works quite well within the first post-Minkowskian approximation, as it is shown by the results obtained, e.g., in [1–5]. Of course, it works also within the post-Newtonian approximation, especially in the case of a static, spherically symmetric space–time treated up to order $1/c^3$ [6, 7]. However, the solution of the geodesic equations requires heavy calculations when one has to take into account the presence of mass multipoles in the field or the tidal effects due to the planetary motions, and the calculations become quite complicated in the post-post-Minkowskian approximation [8], especially in the dynamical case [9].

The aim of this chapter is to present a quite different procedure recently developed by two of us. Based on Synge's world function [10], this procedure avoids the integration of the null geodesic equations and is particularly convenient for determining the light rays which connect an emitter and a receiver having specified spatial locations at a finite distance. Thus, we are able to extend the previous calculations of the time delay and of the frequency shift up to the order $1/c^4$. As a consequence, it is now possible to predict the time/frequency transfers in the vicinity of the Earth at a level of accuracy which amounts to 10^{-18} in fractional frequency. This level of accuracy is expected to be reached in the foreseeable future with optical atomic clocks [11].

The plan of the chapter is as follows. First, in Sect. 2, the definition of the time transfer function is given and the invariant expression of the frequency shift is recalled. It is shown that explicit expressions of the frequency shift can be derived when the time transfer functions are known. In Sect. 3, the relevant properties of Synge's world function are recalled. In Sect. 4, the general expressions of the world function and of the time transfer function are obtained within the Nordtvedt–Will parametrized post-Newtonian (PPN) formalism. In Sect. 5, the case of a stationary field generated by an isolated, slowly rotating axisymmetric body is analyzed in detail. It is shown that the contributions of the mass and spin multipoles can be obtained by straightforward derivations of a single function. Retaining only the terms due to the mass M, to the quadrupole moment J_2, and to the intrinsic angular momentum $\boldsymbol{S}$ of the rotating body, explicit expansions of the world function and of the time transfer function are derived up to the order $1/c^3$ and $1/c^4$, respectively. The same formalism yields the vectors tangent to the light ray at the emitter and at the receiver up to the order $1/c^3$. In Sect. 6, the frequency shift is developed up to the order $1/c^4$ on the assumption that β and γ are the only nonvanishing post-Newtonian parameters. Explicit expressions are obtained for the contributions of J_2 and $\boldsymbol{S}$. Numerical estimates are given for ESA's Atomic Clock Ensemble in Space (ACES) mission [12, 13]. Concluding remarks are given in Sect. 7.

Equivalent results formulated with slightly different notations may be found in [14] and an extension of the method to the general post-Minkowskian approximation is given in [15].

Notations

In this work, G is the Newtonian gravitational constant and c is the speed of light in a vacuum. The Lorentzian metric of space–time is denoted by g. The signature adopted for g is $(+\,-\,-\,-)$. We suppose that the space–time is covered by one global coordinate system $(x^\mu) = (x^0, \boldsymbol{x})$, where $x^0 = ct$, t being a time coordinate, and $\boldsymbol{x} = (x^i)$, the x^i being quasi-Cartesian coordinates. We choose coordinates x^i so that the curves of equations $x^i =$ const are timelike. This choice means that $g_{00} > 0$ everywhere. We employ the vector notation $\boldsymbol{a}$ to denote either $\{a^1, a^2, a^3\} = \{a^i\}$ or $\{a_1, a_2, a_3\} = \{a_i\}$. Considering two such quantities $\boldsymbol{a}$ and $\boldsymbol{b}$ with for instance $\boldsymbol{a} = \{a^i\}$, we use $\boldsymbol{a} \cdot \boldsymbol{b}$ to denote $a^i b^i$ if $\boldsymbol{b} = \{b^i\}$ or $a^i b_i$ if $\boldsymbol{b} = \{b_i\}$ (the Einstein convention on the repeated indices is used). The quantity $|\boldsymbol{a}|$ stands for the ordinary Euclidean norm of $\boldsymbol{a}$.

2 Time Transfer Functions, Time Delay, and Frequency Shift

We consider here electromagnetic signals propagating through a vacuum between an emitter A and a receiver B. We suppose that these signals may be assimilated to light rays traveling along null geodesics of the metric (geometric optics approximation). We call x_A the point of emission by A and x_B the point of reception by B. We put $x_A = (ct_A, \boldsymbol{x}_A)$ and $x_B = (ct_B, \boldsymbol{x}_B)$. We assume that there do not exist two distinct null geodesics starting from x_A and intersecting the world line of B. These assumptions are clearly satisfied in all experiments currently envisaged in the solar system.

2.1 Time Transfer Functions and Time Delay

The quantity $t_B - t_A$ is the (coordinate) travel time of the signal. Upon the above-mentioned assumptions, $t_B - t_A$ may be considered either as a function of the instant of emission t_A and of $\boldsymbol{x}_A$, $\boldsymbol{x}_B$, or as a function of the instant of reception t_B and of $\boldsymbol{x}_A$ and $\boldsymbol{x}_B$. So, we can in general define two distinct (coordinate) time transfer functions, $\mathcal{T}_e$ and $\mathcal{T}_r$ by putting:

$$t_B - t_A = \mathcal{T}_e(t_A, \boldsymbol{x}_A, \boldsymbol{x}_B) = \mathcal{T}_r(t_B, \boldsymbol{x}_A, \boldsymbol{x}_B) \,. \tag{1}$$

We call $\mathcal{T}_e$ the emission time transfer function and $\mathcal{T}_r$ the reception time transfer function. As we shall see below, the main problem will consist in determining explicitly these functions when the metric is given. Of course, it is, in principle, sufficient to determine one of these functions.

We shall put

$$R_{AB} = |\boldsymbol{x}_B - \boldsymbol{x}_A| \tag{2}$$

throughout this work. The time delay is then defined as $t_B - t_A - R_{AB}/c$. It is well known that this quantity is > 0 in Schwarzschild space–time, which explains its designation [16].

2.2 Frequency Shift

Denote by u_A^α and u_B^α the unit 4-velocity vectors of the emitter at x_A and of the receiver at x_B, respectively. Let Γ_{AB} be the null geodesic path connecting

x_A and x_B, described by parametric equations $x^\alpha = x^\alpha(\zeta)$, ζ being an affine parameter. Denote by l^μ the vector tangent to Γ_{AB} defined as

$$l^\mu = \frac{dx^\mu}{d\zeta} . \tag{3}$$

Let ν_A be the frequency of the signal emitted at x_A as measured by a clock comoving with A, and ν_B be the frequency of the same signal received at x_B as measured by a clock comoving with B. The ratio ν_A/ν_B is given by the well-known formula [10]

$$\frac{\nu_A}{\nu_B} = \frac{u_A^\mu (l_\mu)_A}{u_B^\mu (l_\mu)_B} . \tag{4}$$

Since it is assumed that the emission and reception points are connected by a single null geodesic, it is clear that $(l_\mu)_A$ and $(l_\mu)_B$ may be considered either as functions of the instant of emission t_A and of $\boldsymbol{x}_A$, $\boldsymbol{x}_B$, or as functions of the instant of reception t_B and of $\boldsymbol{x}_A$ and $\boldsymbol{x}_B$. Therefore, we may write

$$\frac{\nu_A}{\nu_B} = \mathcal{N}_e(u_A, u_B; t_A, \boldsymbol{x}_A, \boldsymbol{x}_B) = \mathcal{N}_r(u_A, u_B; t_B, \boldsymbol{x}_A, \boldsymbol{x}_B) . \tag{5}$$

Denote by $\boldsymbol{v}_A = (d\boldsymbol{x}/dt)_A$ and $\boldsymbol{v}_B = (d\boldsymbol{x}/dt)_B$ the coordinate velocities of the observers at x_A and x_B, respectively:

$$\boldsymbol{v}_A = \left(\frac{d\boldsymbol{x}}{dt}\right)_A , \quad \boldsymbol{v}_B = \left(\frac{d\boldsymbol{x}}{dt}\right)_B . \tag{6}$$

It is easy to see that the formula (4) may be written as

$$\frac{\nu_A}{\nu_B} = \frac{u_A^0}{u_B^0} \frac{(l_0)_A}{(l_0)_B} \frac{q_A}{q_B} , \quad q_A = 1 + \frac{1}{c} \widehat{\boldsymbol{l}}_A \cdot \boldsymbol{v}_A , \quad q_B = 1 + \frac{1}{c} \widehat{\boldsymbol{l}}_B \cdot \boldsymbol{v}_B , \tag{7}$$

where $\widehat{\boldsymbol{l}}_A$ and $\widehat{\boldsymbol{l}}_B$ are the quantities defined as

$$\widehat{\boldsymbol{l}}_A = \left\{ \left(\frac{l_i}{l_0} \right)_A \right\} , \quad \widehat{\boldsymbol{l}}_B = \left\{ \left(\frac{l_i}{l_0} \right)_B \right\} . \tag{8}$$

It is immediately deduced from (7) that an explicit expression of $\mathcal{N}_e$ (resp., $\mathcal{N}_r$) can be derived when the time transfer function $\mathcal{T}_e$ (resp., $\mathcal{T}_r$) is known. Indeed, one has Theorem 1 [15].

Theorem 1. *Consider a signal emitted at point $x_A = (ct_A, \boldsymbol{x}_A)$ and received at point $x_B = (ct_B, \boldsymbol{x}_B)$. Denote by l^μ the vector $dx^\mu/d\zeta$ tangent to the null geodesic at point $x(\zeta)$, ζ being any affine parameter, and put*

$$\widehat{l}_i = \left(\frac{l_i}{l_0} \right) . \tag{9}$$

Then, one has relations as follow at x_A *and at* x_B

$$\left(\widehat{l}_i\right)_A = c\frac{\partial \mathcal{T}_e}{\partial x_A^i}\left[1+\frac{\partial \mathcal{T}_e}{\partial t_A}\right]^{-1} = c\frac{\partial \mathcal{T}_r}{\partial x_A^i}, \tag{10}$$

$$\left(\widehat{l}_i\right)_B = -c\frac{\partial \mathcal{T}_e}{\partial x_B^i} = -c\frac{\partial \mathcal{T}_r}{\partial x_B^i}\left[1-\frac{\partial \mathcal{T}_r}{\partial t_B}\right]^{-1}, \tag{11}$$

$$\frac{(l_0)_A}{(l_0)_B} = 1+\frac{\partial \mathcal{T}_e}{\partial t_A} = \left[1-\frac{\partial \mathcal{T}_r}{\partial t_B}\right]^{-1}, \tag{12}$$

where $\mathcal{T}_e$ *and* $\mathcal{T}_r$ *are taken at* $(t_A, \boldsymbol{x}_A, \boldsymbol{x}_B)$ *and* $(t_B, \boldsymbol{x}_A, \boldsymbol{x}_B)$*, respectively.*

This theorem may be straightforwardly deduced from a fundamental property of the world function that we introduce in Sect. 3.

Case of a stationary space–time. In a stationary space–time, we can choose coordinates (x^μ) such that the metric does not depend on x^0. Then, the travel time of the signal only depends on $\boldsymbol{x}_A, \boldsymbol{x}_B$. This means that (1) reduces to a single relation of the form

$$t_B - t_A = \mathcal{T}(\boldsymbol{x}_A, \boldsymbol{x}_B)\,. \tag{13}$$

It immediately follows from (10) and (11) that

$$(\widehat{l}_i)_A = c\frac{\partial}{\partial x_A^i}\mathcal{T}(\boldsymbol{x}_A, \boldsymbol{x}_B)\,, \tag{14}$$

$$(\widehat{l}_i)_B = -c\frac{\partial}{\partial x_B^i}\mathcal{T}(\boldsymbol{x}_A, \boldsymbol{x}_B)\,, \tag{15}$$

$$\frac{(l_0)_A}{(l_0)_B} = 1\,. \tag{16}$$

As a consequence, the formula (7) reduces now to

$$\frac{\nu_A}{\nu_B} = \frac{u_A^0}{u_B^0}\,\frac{1+\boldsymbol{v}_A\cdot\boldsymbol{\nabla}_{\boldsymbol{x}_A}\mathcal{T}}{1-\boldsymbol{v}_B\cdot\boldsymbol{\nabla}_{\boldsymbol{x}_B}\mathcal{T}}, \tag{17}$$

where $\boldsymbol{\nabla}_{\boldsymbol{x}} f$ denotes the usual gradient operator acting on $f(\boldsymbol{x})$.

It is worthy of note that $(1, \{(\widehat{l}_i)_A\})$ and $(1, \{(\widehat{l}_i)_B\})$ constitute a set of covariant components of the vector tangent to the light ray at $\boldsymbol{x}_A$ and $\boldsymbol{x}_B$, respectively. This tangent vector corresponds to the affine parameter chosen so that $(l_0)_A = (l_0)_B = 1$.

3 The World Function and Its Post-Newtonian Limit

3.1 Definition and Fundamental Properties

For a moment, consider x_A and x_B as arbitrary points. We assume that there exists one and only one geodesic path, say $\varGamma_{AB}$, which links these two points.

This assumption means that point x_B belongs to the normal convex neighborhood [17] of point x_A (and conversely that x_A belongs to the normal convex neighborhood of point x_B). The world function is the two-point function $\Omega(x_A, x_B)$ defined by

$$\Omega(x_A, x_B) = \frac{1}{2}\epsilon_{AB}[s_{AB}]^2 , \tag{18}$$

where s_{AB} is the geodesic distance between x_A and x_B, namely

$$s_{AB} = \int_{\Gamma_{AB}} \sqrt{g_{\mu\nu}dx^\mu dx^\nu} \tag{19}$$

and $\epsilon_{AB} = 1, 0, -1$ according as Γ_{AB} is a timelike, a null, or a spacelike geodesic. An elementary calculation shows that $\Omega(x_A, x_B)$ may be written in any case as [10]

$$\Omega(x_A, x_B) = \frac{1}{2}\int_0^1 g_{\mu\nu}(x^\alpha(\lambda))\frac{dx^\mu}{d\lambda}\frac{dx^\nu}{d\lambda}d\lambda , \tag{20}$$

where the integral is taken along Γ_{AB}, λ denoting the unique affine parameter along Γ_{AB} which fulfills the boundary conditions $\lambda_A = 0$ and $\lambda_B = 1$.

It follows from (16) or (18) that the world function $\Omega(x_A, x_B)$ is unchanged if we perform any admissible coordinate transformation.

The utility of the world function for our purpose comes from the following properties [10, 15]:

1. The vectors $(dx^\alpha/d\lambda)_A$ and $(dx^\alpha/d\lambda)_B$ tangent to the geodesic Γ_{AB}, respectively, at x_A and x_B are given by

$$\left(g_{\alpha\beta}\frac{dx^\beta}{d\lambda}\right)_A = -\frac{\partial\Omega}{\partial x_A^\alpha} , \quad \left(g_{\alpha\beta}\frac{dx^\beta}{d\lambda}\right)_B = \frac{\partial\Omega}{\partial x_B^\alpha} . \tag{21}$$

   As a consequence, if $\Omega(x_A, x_B)$ is explicitly known, the determination of these vectors does not require the integration of the differential equations of the geodesic.
2. Two points x_A and x_B are linked by a null geodesic if and only if the condition

$$\Omega(x_A, x_B) = 0 \tag{22}$$

   is fulfilled. Thus, $\Omega(x_A, x) = 0$ is the equation of the null cone $\mathcal{C}(x_A)$ at x_A.

Consequently, if the bifunction $\Omega(x_A, x_B)$ is explicitly known, it is, in principle, possible to determine the emission time transfer function $\mathcal{T}_e$ by solving the equation

$$\Omega(ct_A, \boldsymbol{x}_A, ct_B, \boldsymbol{x}_B) = 0 \tag{23}$$

for t_B. It must be pointed out, however, that solving (23) for t_B yields two distinct solutions t_B^+ and t_B^- since the timelike curve $x^i = x_B^i$ cuts the light cone

$\mathcal{C}(x_A)$ at two points x_B^+ and x_B^-; x_B^+ being in the future of x_B^-. Since we regard x_A as the point of emission of the signal and x_B as the point of reception, we shall exclusively focus our attention on the determination of $t_B^+ - t_A$ (clearly, the determination of $t_B^- - t_A$ comes within the same methodology). For the sake of brevity, we shall henceforth write t_B instead of t_B^+.

Of course, solving (23) for t_A yields the reception time transfer function $\mathcal{T}_r$.

Generally, extracting the time transfer functions from (23), next using (10) or (11) will be more straightforward than deriving the vectors tangent at x_A and x_B from (21), next imposing the constraint (22).

To finish, note that Theorem 1 is easily deduced from the identities

$$\Omega(ct_A, \boldsymbol{x}_A, ct_A + c\mathcal{T}_e(t_A, \boldsymbol{x}_A, \boldsymbol{x}_B), \boldsymbol{x}_B) \equiv 0$$

and

$$\Omega(ct_B - c\mathcal{T}_r(t_B, \boldsymbol{x}_A, \boldsymbol{x}_B), \boldsymbol{x}_A, ct_B, \boldsymbol{x}_B) \equiv 0 .$$

3.2 General Expression of the World Function in the Post-Newtonian Limit

We assume that the metric may be written as

$$g_{\mu\nu} = \eta_{\mu\nu} + h_{\mu\nu} \tag{24}$$

throughout space–time, with $\eta_{\mu\nu} = \mathrm{diag}(1, -1, -1, -1)$. Let $\Gamma_{AB}^{(0)}$ be the straight line defined by the parametric equations $x^\alpha = x^\alpha_{(0)}(\lambda)$, with

$$x^\alpha_{(0)}(\lambda) = (x^\alpha_B - x^\alpha_A)\lambda + x^\alpha_A , \quad 0 \le \lambda \le 1 . \tag{25}$$

With this definition, the parametric equations of the geodesic Γ_{AB} connecting x_A and x_B may be written in the form

$$x^\alpha(\lambda) = x^\alpha_{(0)}(\lambda) + X^\alpha(\lambda) , \quad 0 \le \lambda \le 1 , \tag{26}$$

where the quantities $X^\alpha(\lambda)$ satisfy the boundary conditions

$$X^\alpha(0) = 0 , \quad X^\alpha(1) = 0 . \tag{27}$$

Inserting (24) and $dx^\mu(\lambda)/d\lambda = x^\mu_B - x^\mu_A + dX^\mu(\lambda)/d\lambda$ in (16), then developing and noting that

$$\int_0^1 \eta_{\mu\nu}(x^\mu_B - x^\mu_A)\frac{dX^\nu}{d\lambda}d\lambda = 0$$

by virtue of (27), we find the rigorous formula

$$\begin{aligned}\Omega(x_A, x_B) = \Omega^{(0)}(x_A, x_B) &+ \frac{1}{2}(x^\mu_B - x^\mu_A)(x^\nu_B - x^\nu_A)\int_0^1 h_{\mu\nu}(x^\alpha(\lambda))d\lambda \\ &+ \frac{1}{2}\int_0^1 \left[g_{\mu\nu}(x^\alpha(\lambda))\frac{dX^\mu}{d\lambda}\frac{dX^\nu}{d\lambda} + 2(x^\mu_B - x^\mu_A)h_{\mu\nu}(x^\alpha(\lambda))\frac{dX^\nu}{d\lambda}\right] d\lambda,\end{aligned} \tag{28}$$

where the integrals are taken over Γ_{AB} and $\Omega^{(0)}(x_A, x_B)$ is the world function in Minkowski space–time

$$\Omega^{(0)}(x_A, x_B) = \frac{1}{2}\eta_{\mu\nu}(x_B^\mu - x_A^\mu)(x_B^\nu - x_A^\nu) . \tag{29}$$

Henceforth, we shall consider only weak gravitational fields generated by self-gravitating extended bodies within the slow-motion, post-Newtonian approximation. So, we assume that the potentials $h_{\mu\nu}$ may be expanded as follows

$$\begin{aligned} h_{00} &= \frac{1}{c^2}h_{00}^{(2)} + \frac{1}{c^4}h_{00}^{(4)} + O(6) , \\ h_{0i} &= \frac{1}{c^3}h_{0i}^{(3)} + O(5) , \\ h_{ij} &= \frac{1}{c^2}h_{ij}^{(2)} + O(4) . \end{aligned} \tag{30}$$

From these expansions and from the Euler–Lagrange equations satisfied by any geodesic curve, namely

$$\frac{d}{d\lambda}\left(g_{\alpha\beta}\frac{dx^\beta}{d\lambda}\right) = \frac{1}{2}\partial_\alpha h_{\mu\nu}\frac{dx^\mu}{d\lambda}\frac{dx^\nu}{d\lambda} , \tag{31}$$

it results that $X^\mu(\lambda) = O(2)$ and that $dx^\mu/d\lambda = x_B^\mu - x_A^\mu + O(2)$. As a consequence, $h_{\mu\nu}(x^\alpha(\lambda)) = h_{\mu\nu}(x^\alpha_{(0)}(\lambda)) + O(4)$ and the third and fourth terms in the RHS of (28) are of order $1/c^4$. These features result in an expression for $\Omega(x_A, x_B)$ as follows

$$\Omega(x_A, x_B) = \Omega^{(0)}(x_A, x_B) + \Omega^{(PN)}(x_A, x_B) + O(4) , \tag{32}$$

where

$$\begin{aligned} \Omega^{(PN)}(x_A, x_B) = &\frac{1}{2c^2}(x_B^0 - x_A^0)^2 \int_0^1 h_{00}^{(2)}(x^\alpha_{(0)}(\lambda))d\lambda \\ &+\frac{1}{2c^2}(x_B^i - x_A^i)(x_B^j - x_A^j) \int_0^1 h_{ij}^{(2)}(x^\alpha_{(0)}(\lambda))d\lambda \\ &+\frac{1}{c^3}(x_B^0 - x_A^0)(x_B^i - x_A^i) \int_0^1 h_{0i}^{(3)}(x^\alpha_{(0)}(\lambda))d\lambda, \end{aligned} \tag{33}$$

the integral being now taken over the line $\Gamma_{AB}^{(0)}$ defined by (25).

The formulae (32) and (33) yield the general expression of the world function up to the order $1/c^3$ within the framework of the 1 PN approximation. We shall see in Sect. 3.3 that this approximation is sufficient to determine the time transfer functions up to the order $1/c^4$. It is worthy of note that the method used above would as well lead to the expression of the world function in the linearized weak-field limit previously found by Synge [10].

3.3 Time Transfer Functions at the Order $1/c^4$

Suppose that x_B is the point of reception of a signal emitted at x_A. Taking (32) into account, (22) may be written in the form

$$\Omega^{(0)}(x_A, x_B) + \Omega^{(PN)}(x_A, x_B) = O(4)\,, \tag{34}$$

which implies the relation

$$t_B - t_A = \frac{1}{c}R_{AB} - \frac{\Omega^{(PN)}(ct_A, \boldsymbol{x}_A, ct_B, \boldsymbol{x}_B)}{cR_{AB}} + O(4)\,. \tag{35}$$

Using iteratively this relation, we find for the emission time transfer function

$$\mathcal{T}_e(t_A, \boldsymbol{x}_A, \boldsymbol{x}_B) = \frac{1}{c}R_{AB} - \frac{\Omega^{(PN)}(ct_A, \boldsymbol{x}_A, ct_A + R_{AB}, \boldsymbol{x}_B)}{cR_{AB}} + O(5) \tag{36}$$

and for the reception time transfer function

$$\mathcal{T}_r(t_B, \boldsymbol{x}_A, \boldsymbol{x}_B) = \frac{1}{c}R_{AB} - \frac{\Omega^{(PN)}(ct_B - R_{AB}, \boldsymbol{x}_A, ct_B, \boldsymbol{x}_B)}{cR_{AB}} + O(5)\,. \tag{37}$$

These last formulae show that the time transfer functions can be explicitly calculated up to the order $1/c^4$ when $\Omega^{(PN)}(x_A, x_B)$ is known. This fundamental result will be exploited in the following sections.

It is worthy of note that a comparison of (36) and (37) immediately gives the following relations:

$$\mathcal{T}_r(t_B, \boldsymbol{x}_A, \boldsymbol{x}_B) = \mathcal{T}_e\left(t_B - \frac{R_{AB}}{c}, \boldsymbol{x}_A, \boldsymbol{x}_B\right) + O(5) \tag{38}$$

and conversely

$$\mathcal{T}_e(t_A, \boldsymbol{x}_A, \boldsymbol{x}_B) = \mathcal{T}_r\left(t_A + \frac{R_{AB}}{c}, \boldsymbol{x}_A, \boldsymbol{x}_B\right) + O(5)\,. \tag{39}$$

The quantity $\Omega^{(PN)}(ct_A, \boldsymbol{x}_A, ct_A + R_{AB}, \boldsymbol{x}_B)$ in (36) may be written in an integral form by using (33), in which R_{AB} and $R_{AB}\lambda + ct_A$ are substituted for $x^0_B - x^0_A$ and for $x^0_{(0)}(\lambda)$, respectively. As a consequence

$$\mathcal{T}_e(t_A, \boldsymbol{x}_A, \boldsymbol{x}_B) = \frac{1}{c}R_{AB}\left\{1 - \frac{1}{2c^2}\int_0^1 \left[h^{(2)}_{00}(z^\alpha_+(\lambda)) + h^{(2)}_{ij}(z^\alpha_+(\lambda))N^iN^j + \frac{2}{c}h^{(3)}_{0i}(z^\alpha_+(\lambda))N^i\right]d\lambda\right\} + O(5), \tag{40}$$

the integral being taken over curve $\Gamma_{AB}^{(0)+}$ defined by the parametric equations $x^\alpha = z_+^\alpha(\lambda)$, where

$$z_+^0(\lambda) = R_{AB}\lambda + ct_A\,, \quad z_+^i(\lambda) = R_{AB}N^i\lambda + x_A^i\,, \quad 0 \leq \lambda \leq 1\,, \tag{41}$$

with

$$R_{AB} = \mid \boldsymbol{R}_{AB} \mid, \qquad N^i = \frac{x_B^i - x_A^i}{R_{AB}}\,. \tag{42}$$

We note that $\Gamma_{AB}^{(0)+}$ is a null geodesic path of Minkowski metric from x_A, having the above-defined quantities N^i as direction cosines.

A similar reasoning leads to an expression as follows for $\mathcal{T}_r$

$$\mathcal{T}_r(t_B, \boldsymbol{x}_A, \boldsymbol{x}_B) = \frac{1}{c} R_{AB} \Bigg\{ 1 - \frac{1}{2c^2} \int_0^1 \Big[h_{00}^{(2)}(z_-^\alpha(\lambda)) + h_{ij}^{(2)}(z_-^\alpha(\lambda)) N^i N^j + \frac{2}{c} h_{0i}^{(3)}(z_-^\alpha(\lambda)) N^i \Big] d\lambda \Bigg\} + O(5), \tag{43}$$

the integral being now taken over curve $\Gamma_{AB}^{(0)-}$ defined by the parametric equations $x^\alpha = z_-^\alpha(\lambda)$, where

$$z_-^0(\lambda) = -R_{AB}\lambda + ct_B\,, \quad z_-^i(\lambda) = -R_{AB}N^i\lambda + x_B^i\,, \quad 0 \leq \lambda \leq 1\,. \tag{44}$$

Curve $\Gamma_{AB}^{(0)-}$ is a null geodesic path of Minkowski metric arriving at x_B and having N^i as direction cosines.

4 World Function and Time Transfer Functions Within the Nordtvedt–Will PPN Formalism

4.1 Metric in the 1 PN Approximation

In this section, we use the Nordtvedt–Will post-Newtonian formalism involving ten parameters β, γ, ξ, α_1, ..., ζ_4 [18]. We introduce slightly modified notations to be closed of the formalism recently proposed by Klioner and Soffel [20] as an extension of the post-Newtonian framework elaborated by Damour et al. [21] for general relativity. In particular, we denote by $\boldsymbol{v}_r$ the velocity of the center of mass O relative to the universe rest frame.[1]

Although our method is not confined to any particular assumption on the matter, we suppose here that each source of the field is described by the energy–momentum tensor of a perfect fluid

$$T^{\mu\nu} = \rho c^2 \left[1 + \frac{1}{c^2}\left(\Pi + \frac{p}{\rho} \right) \right] u^\mu u^\nu - p g^{\mu\nu}\,, \tag{45}$$

[1] This velocity is noted $\boldsymbol{w}$ in [18].

where ρ is the rest mass density, Π is the specific energy density (ratio of internal energy density to rest mass density), p is the pressure, and u^μ is the unit 4-velocity of the fluid. In this section and in the following one, $\boldsymbol{v}$ is the coordinate velocity $d\boldsymbol{x}/dt$ of an element of the fluid. We introduce the conserved mass density ρ^* given by

$$\rho^* = \rho\sqrt{-g}u^0 = \rho\left[1 + \frac{1}{c^2}\left(\frac{1}{2}v^2 + 3\gamma U\right) + O(4)\right], \tag{46}$$

where $g = \det(g_{\mu\nu})$ and U is the Newtonian-like potential

$$U(x^0, \boldsymbol{x}) = G\int \frac{\rho^*(x^0, \boldsymbol{x}')}{|\boldsymbol{x} - \boldsymbol{x}'|} d^3\boldsymbol{x}'. \tag{47}$$

To obtain a more simple form than the usual one for the potentials h_{0i}, we suppose that the chosen (x^μ) are related to a standard post-Newtonian gauge $(\overline{x}^\mu)$ by the transformation

$$x^0 = \overline{x}^0 + \frac{1}{c^3}\left[(1 + 2\xi + \alpha_2 - \zeta_1)\partial_t\chi - 2\alpha_2\boldsymbol{v}_r \cdot \boldsymbol{\nabla}\chi\right], \quad x^i = \overline{x}^i, \tag{48}$$

where χ is the superpotential defined by

$$\chi(x^0, \boldsymbol{x}) = \frac{1}{2}G\int \rho^*(x^0, \boldsymbol{x}')\,|\boldsymbol{x} - \boldsymbol{x}'|\, d^3\boldsymbol{x}'. \tag{49}$$

Moreover, we define $\widehat{\rho}$ by

$$\widehat{\rho} = \rho^*\left[1 + \frac{1}{2}(2\gamma + 1 - 2\xi + \alpha_3 + \zeta_1)\frac{v^2}{c^2} + (1 - 2\beta + \xi + \zeta_2)\frac{U}{c^2} + (1 + \zeta_3)\frac{\Pi}{c^2}\right.$$
$$\left. + (3\gamma - 2\xi + 3\zeta_4)\frac{p}{\rho^* c^2} - \frac{1}{2}(\alpha_1 - \alpha_3)\frac{v_r^2}{c^2} - \frac{1}{2}(\alpha_1 - 2\alpha_3)\frac{\boldsymbol{v}_r \cdot \boldsymbol{v}}{c^2} + O(4)\right]. \tag{50}$$

Then, the post-Newtonian potentials read

$$h_{00} = -\frac{2}{c^2}w + \frac{2\beta}{c^4}w^2 + \frac{2\xi}{c^4}\phi_W + \frac{1}{c^4}(\zeta_1 - 2\xi)\phi_v - \frac{2\alpha_2}{c^4}v_r^i v_r^j \partial_{ij}\chi + O(6), \tag{51}$$

$$\boldsymbol{h} \equiv \{h_{0i}\} = \frac{2}{c^3}\left[\left(\gamma + 1 + \frac{1}{4}\alpha_1\right)\boldsymbol{w} + \frac{1}{4}\alpha_1 w\, \boldsymbol{v}_r\right] + O(5), \tag{52}$$

$$h_{ij} = -\frac{2\gamma}{c^2}w\delta_{ij} + O(4), \tag{53}$$

where

$$w(x^0, \boldsymbol{x}) = G\int \frac{\widehat{\rho}(x^0, \boldsymbol{x}')}{|\boldsymbol{x} - \boldsymbol{x}'|} d^3\boldsymbol{x}'$$
$$+ \frac{1}{c^2}\left[(1 + 2\xi + \alpha_2 - \zeta_1)\partial_{tt}\chi - 2\alpha_2\boldsymbol{v}_r \cdot \boldsymbol{\nabla}(\partial_t\chi)\right], \tag{54}$$

$$\phi_W(x^0, \boldsymbol{x}) = G^2 \int \frac{\rho^*(x^0, \boldsymbol{x}')\rho^*(x^0, \boldsymbol{x}'')(\boldsymbol{x} - \boldsymbol{x}')}{|\boldsymbol{x} - \boldsymbol{x}'|^3} \times \left(\frac{\boldsymbol{x}' - \boldsymbol{x}''}{|\boldsymbol{x} - \boldsymbol{x}''|} - \frac{\boldsymbol{x} - \boldsymbol{x}''}{|\boldsymbol{x}' - \boldsymbol{x}''|} \right) d^3\boldsymbol{x}' d^3\boldsymbol{x}'', \tag{55}$$

$$\phi_v(x^0, \boldsymbol{x}) = G \int \frac{\rho^*(x^0, \boldsymbol{x}')[\boldsymbol{v}(x^0, \boldsymbol{x}') \cdot (\boldsymbol{x} - \boldsymbol{x}')]^2}{|\boldsymbol{x} - \boldsymbol{x}'|^3} d^3\boldsymbol{x}', \tag{56}$$

$$\boldsymbol{w}(x^0, \boldsymbol{x}) = G \int \frac{\rho^*(x^0, \boldsymbol{x}')\boldsymbol{v}(x^0, \boldsymbol{x}')}{|\boldsymbol{x} - \boldsymbol{x}'|} d^3\boldsymbol{x}' . \tag{57}$$

4.2 Determination of the World Function and of the Time Transfer Functions

For the post-Newtonian metric given by (51–57), it follows from (33) that $\Omega(x_A, x_B)$ may be written up to the order $1/c^3$ in the form given by (32) with

$$\Omega^{(PN)}(x_A, x_B) = \Omega_w^{(PN)}(x_A, x_B) + \Omega_{\boldsymbol{w}}^{(PN)}(x_A, x_B) + \Omega_{\boldsymbol{v}_r}^{(PN)}(x_A, x_B) , \tag{58}$$

where

$$\Omega_w^{(PN)}(x_A, x_B) = -\frac{1}{c^2} \left[(x_B^0 - x_A^0)^2 + \gamma R_{AB}^2 \right] \int_0^1 w(x_{(0)}^\alpha(\lambda)) d\lambda , \tag{59}$$

$$\Omega_{\boldsymbol{w}}^{(PN)}(x_A, x_B) = \frac{2}{c^3} \left(\gamma + 1 + \frac{1}{4}\alpha_1 \right) (x_B^0 - x_A^0) \times \boldsymbol{R}_{AB} \cdot \int_0^1 \boldsymbol{w}(x_{(0)}^\alpha(\lambda)) d\lambda , \tag{60}$$

$$\Omega_{\boldsymbol{v}_r}^{(PN)}(x_A, x_B) = \frac{1}{2c^3} \alpha_1 (x_B^0 - x_A^0)(\boldsymbol{R}_{AB} \cdot \boldsymbol{v}_r) \int_0^1 w(x_{(0)}^\alpha(\lambda)) d\lambda , \tag{61}$$

the integrals being calculated along the curve defined by (25).

The emission time transfer function is easily obtained by using (36) or (40). We get

$$\mathcal{T}_e(t_A, \boldsymbol{x}_A, \boldsymbol{x}_B) = \frac{1}{c} R_{AB} + \frac{1}{c^3} (\gamma + 1) R_{AB} \int_0^1 w(z_+^\alpha(\lambda)) d\lambda - \frac{2}{c^4} \boldsymbol{R}_{AB} \cdot \left[(\gamma + 1 + \frac{1}{4}\alpha_1) \int_0^1 \boldsymbol{w}(z_+^\alpha(\lambda)) d\lambda + \frac{1}{4}\alpha_1 \boldsymbol{v}_r \int_0^1 w(z_+^\alpha(\lambda)) d\lambda \right] + O(5) , \tag{62}$$

the integral being evaluated along the curve $\Gamma_{AB}^{(0)+}$ defined by (41).

The reception time transfer function is given by

$$\mathcal{T}_r(t_B, \boldsymbol{x}_A, \boldsymbol{x}_B) = \frac{1}{c} R_{AB} + \frac{1}{c^3}(\gamma + 1) R_{AB} \int_0^1 w(z_-^\alpha(\lambda)) d\lambda$$
$$-\frac{2}{c^4} \boldsymbol{R}_{AB} \cdot \left[(\gamma + 1 + \frac{1}{4}\alpha_1) \int_0^1 \boldsymbol{w}(z_-^\alpha(\lambda)) d\lambda \right.$$
$$\left. + \frac{1}{4}\alpha_1 \boldsymbol{v}_r \int_0^1 w(z_-^\alpha(\lambda)) d\lambda \right] + O(5) \,, \quad (63)$$

the integral being evaluated along the curve $\Gamma_{AB}^{(0)-}$ defined by (44).

Let us emphasize that, since $w = U + O(2)$, w may be replaced by the Newtonian-like potential U in (59–62).

4.3 Case of a Stationary Source

In what follows, we suppose that the gravitational field is generated by a single stationary source. Then, $\partial_t \chi = 0$ and the potentials w and $\boldsymbol{w}$ do not depend on time. In this case, the integration involved in (59–61) can be performed by a method due to Buchdahl [19]. Introducing the auxiliary variables $\boldsymbol{y}_A = \boldsymbol{x}_A - \boldsymbol{x}'$ and $\boldsymbol{y}_B = \boldsymbol{x}_B - \boldsymbol{x}'$, and replacing in (25) the parameter λ by $u = \lambda - 1/2$, a straightforward calculation yields

$$\int_0^1 w(\boldsymbol{x}_{(0)}(\lambda)) d\lambda = G \int \widehat{\rho}(\boldsymbol{x}') F(\boldsymbol{x}', \boldsymbol{x}_A, \boldsymbol{x}_B) d^3\boldsymbol{x}', \quad (64)$$

$$\int_0^1 \boldsymbol{w}(\boldsymbol{x}_{(0)}(\lambda)) d\lambda = G \int \rho^*(\boldsymbol{x}') \boldsymbol{v}(\boldsymbol{x}') F(\boldsymbol{x}', \boldsymbol{x}_A, \boldsymbol{x}_B) d^3\boldsymbol{x}' \,, \quad (65)$$

where the kernel function $F(\boldsymbol{x}', \boldsymbol{x}_A, \boldsymbol{x}_B)$ has the expression

$$F(\boldsymbol{x}', \boldsymbol{x}_A, \boldsymbol{x}_B) = \int_{-1/2}^{1/2} \frac{du}{|(\boldsymbol{y}_B - \boldsymbol{y}_A)u + \frac{1}{2}(\boldsymbol{y}_B + \boldsymbol{y}_A)|} \,. \quad (66)$$

Noting that $\boldsymbol{y}_B - \boldsymbol{y}_A = \boldsymbol{R}_{AB}$, which implies that $|\boldsymbol{y}_B - \boldsymbol{y}_A| = R_{AB}$, we find

$$F(\boldsymbol{x}, \boldsymbol{x}_A, \boldsymbol{x}_B) = \frac{1}{R_{AB}} \ln \left(\frac{|\boldsymbol{x} - \boldsymbol{x}_A| + |\boldsymbol{x} - \boldsymbol{x}_B| + R_{AB}}{|\boldsymbol{x} - \boldsymbol{x}_A| + |\boldsymbol{x} - \boldsymbol{x}_B| - R_{AB}} \right) . \quad (67)$$

Inserting (64), (65), and (67) in (59–61) and (62) will enable one to obtain quite elegant expressions for $\Omega^{(PN)}(x_A, x_B)$ and for $\mathcal{T}(\boldsymbol{x}_A, \boldsymbol{x}_B)$.

5 Isolated, Slowly Rotating Axisymmetric Body

Henceforth, we suppose that the light is propagating in the gravitational field of an isolated, slowly rotating axisymmetric body. The gravitational field is assumed to be stationary. The main purpose of this section is to determine the influence of the mass and spin multipole moments of the rotating body on the coordinate time transfer and on the direction of light rays. From these results, it will be possible to obtain a relativistic modeling of the one-way time transfers and frequency shifts up to the order $1/c^4$ in a geocentric nonrotating frame.

Since we treat the case of a body located very far from the other bodies of the universe, the global coordinate system (x^μ) used until now can be considered as a local (i.e., geocentric) one. So, in agreement with the UAI/UGG Resolution B1 (2000) [22], we shall henceforth denote by W and $\boldsymbol{W}$ the quantities w and $\boldsymbol{w}$, respectively, defined by (54) and (57) and we shall denote by $G_{\mu\nu}$ the components of the metric. However, we shall continue here with using lower case letters for the geocentric coordinates to avoid too heavy notations.

The center of mass O of the rotating body is taken as the origin of the quasi-Cartesian coordinates $(\boldsymbol{x})$; we choose the axis of symmetry as the x^3-axis. We assume that the body is rotating about Ox^3 with a constant angular velocity $\boldsymbol{\omega}$, so that

$$\boldsymbol{v}(\boldsymbol{x}) = \boldsymbol{\omega} \times \boldsymbol{x} \,. \tag{68}$$

In what follows, we put $r = \mid \boldsymbol{x} \mid$, $r_A = \mid \boldsymbol{x}_A \mid$, and $r_B = \mid \boldsymbol{x}_A \mid$. We call θ the angle between $\boldsymbol{x}$ and Ox^3. We consider only the case where all points of the segment joining $\boldsymbol{x}_A$ and $\boldsymbol{x}_B$ are outside the body. We denote by r_e the radius of the smallest sphere centered on O and containing the body (for celestial bodies, r_e is the equatorial radius). In this section, we assume the convergence of the multipole expansions formally derived below at any point outside the body, even if $r < r_e$.

5.1 Multipole Expansions of the Potentials

According to (54), (57), and (68), the gravitational potentials W and $\boldsymbol{W}$ obey the equations

$$\nabla^2 W = -4\pi G \widehat{\rho} \,, \quad \nabla^2 \boldsymbol{W} = -4\pi G \rho^* \boldsymbol{\omega} \times \boldsymbol{x} \,. \tag{69}$$

It follows from (69) that the potential W is a harmonic function outside the rotating body. As a consequence, W may be expanded in a multipole series of the form

$$W(\boldsymbol{x}) = \frac{GM}{r} \left[1 - \sum_{n=2}^{\infty} J_n \left(\frac{r_e}{r} \right)^n P_n(\cos\theta) \right] . \tag{70}$$

In this expansion, P_n is the Legendre polynomial of degree n and the quantities M, $J_2, \ldots, J_n, \ldots$ correspond to the generalized Blanchet–Damour mass multipole moments in general relativity [23].

For the sake of simplicity, put

$$z = x^3 \,. \tag{71}$$

Taking into account the identity

$$\frac{\partial^n}{\partial z^n}\left(\frac{1}{r}\right) = \frac{(-1)^n n!}{r^{1+n}} P_n(z/r)\,, \quad z = x^3\,, \tag{72}$$

it may be seen that

$$W(\boldsymbol{x}) = GM\left[\frac{1}{r} - \sum_{n=2}^{\infty}\frac{(-1)^n}{n!} J_n r_e^n \frac{\partial^n}{\partial z^n}\left(\frac{1}{r}\right)\right]. \tag{73}$$

Substituting for W from (73) into (69) yields an expansion for $\widehat{\rho}$ as follows

$$\widehat{\rho}(\boldsymbol{x}) = M\left[\delta^{(3)}(\boldsymbol{x}) - \sum_{n=2}^{\infty}\frac{(-1)^n}{n!} J_n r_e^n \frac{\partial^n}{\partial z^n}\delta^{(3)}(\boldsymbol{x})\right], \tag{74}$$

where $\delta^{(3)}(\boldsymbol{x})$ is the Dirac distribution supported by the origin O. This expansion of $\widehat{\rho}$ in a multipole series will be exploited in Sect. 5.2.

Now, substituting (68) into (57) yields for the vector potential $\boldsymbol{W}$

$$\boldsymbol{W}(\boldsymbol{x}) = G\int \frac{\rho^*(\boldsymbol{x}')\boldsymbol{\omega}\times\boldsymbol{x}'}{|\boldsymbol{x}-\boldsymbol{x}'|} d^3\boldsymbol{x}'\,. \tag{75}$$

It is possible to show that this vector may be written as

$$\boldsymbol{W} = -\frac{1}{2}\boldsymbol{\omega}\times\boldsymbol{\nabla}\mathcal{V}\,, \tag{76}$$

where $\mathcal{V}$ is an axisymmetric function satisfying the Laplace equation $\boldsymbol{\nabla}^2\mathcal{V} = 0$ outside the body. Consequently, we can expand $\mathcal{V}$ in a series of the form

$$\mathcal{V}(\boldsymbol{x}) = \frac{GI}{r}\left[1 - \sum_{n=1}^{\infty} K_n\left(\frac{r_e}{r}\right)^n P_n(\cos\theta)\right], \tag{77}$$

where I and each K_n are constants. Substituting for $\mathcal{V}$ from (77) into (76), and then using the identity

$$(n+1)P_n(z/r) + (z/r)P_n'(z/r) = P_{n+1}'(z/r)\,, \tag{78}$$

we find an expansion for $\boldsymbol{W}$ as follows

$$\boldsymbol{W}(\boldsymbol{x}) = \frac{GI\boldsymbol{\omega}\times\boldsymbol{x}}{2r^3}\left[1 - \sum_{n=1}^{\infty} K_n\left(\frac{r_e}{r}\right)^n P_{n+1}'(\cos\theta)\right], \tag{79}$$

which coincides with a result previously obtained by one of us [24]. This coincidence shows that I is the moment of inertia of the body about the z-axis. Thus, the quantity $\boldsymbol{S} = I\boldsymbol{\omega}$ is the intrinsic angular momentum of the rotating body. The coefficients K_n are completely determined by the density distribution ρ^*

and by the shape of the body [24, 25]. The quantities $I, K_1, K_2, \dots K_n, \dots$ correspond to the Blanchet–Damour spin multipoles in the special case of a stationary axisymmetric gravitational field.

Equation (79) may also be written as

$$\boldsymbol{W}(\boldsymbol{x}) = -\frac{1}{2} G \boldsymbol{S} \times \boldsymbol{\nabla} \left[\frac{1}{r} - \sum_{n=1}^{\infty} \frac{(-1)^n}{n!} K_n r_e^n \frac{\partial^n}{\partial z^n} \left(\frac{1}{r} \right) \right] . \tag{80}$$

Consequently, the density of mass current can be expanded in the multipole series

$$\rho^*(\boldsymbol{x})(\boldsymbol{\omega} \times \boldsymbol{x}) = -\frac{1}{2} \boldsymbol{S} \times \boldsymbol{\nabla} \left[\delta^{(3)}(\boldsymbol{x}) - \sum_{n=1}^{\infty} \frac{(-1)^n}{n!} K_n r_e^n \frac{\partial^n}{\partial z^n} \delta^{(3)}(\boldsymbol{x}) \right] . \tag{81}$$

This expansion may be compared with the expansion of $\widehat{\rho}$ given by (74).

5.2 Multipole Structure of the World Function

The function $\Omega^{(PN)}(x_A, x_B)$ is determined by (58–61) where w and $\boldsymbol{w}$ are, respectively, replaced by W and $\boldsymbol{W}$. The integrals involved in the RHS of (58–61) are given by (64) and (65). Substituting (74) into (64) and using the properties of the Dirac distribution, we obtain

$$\int_0^1 W\left(\boldsymbol{x}_{(0)}(\lambda)\right) d\lambda = GM \left[1 - \sum_{n=2}^{\infty} \frac{1}{n!} J_n r_e^n \frac{\partial^n}{\partial z^n} \right] F(\boldsymbol{x}, \boldsymbol{x}_A, \boldsymbol{x}_B) \Big|_{\boldsymbol{x}=0} . \tag{82}$$

Similarly, substituting (81) into (65), we get[2]

$$\int_0^1 \boldsymbol{W}\left(\boldsymbol{x}_{(0)}(\lambda)\right) d\lambda = \frac{1}{2} G \boldsymbol{S} \times \boldsymbol{\nabla} \left[1 - \sum_{n=1}^{\infty} \frac{1}{n!} K_n r_e^n \frac{\partial^n}{\partial z^n} \right] F(\boldsymbol{x}, \boldsymbol{x}_A, \boldsymbol{x}_B) \Big|_{\boldsymbol{x}=0} . \tag{83}$$

These formulae show that the multipole expansion of $\Omega^{(PN)}(x_A, x_B)$ can be thoroughly calculated by straightforward differentiations of the kernel function $F(\boldsymbol{x}, \boldsymbol{x}_A, \boldsymbol{x}_B)$ given by (67). They constitute an essential result, since they give an algorithmic procedure for determining the multipole expansions of the time transfer function and of the frequency shift in a stationary axisymmetric field (see also [2]).

To obtain explicit formulae, we shall only retain the contributions due to M, J_2, and $\boldsymbol{S}$ in the expansion yielding $\Omega_W^{(PN)}$ and $\Omega_{\boldsymbol{W}}^{(PN)}$. Then, denoting the unit vector along the z-axis by $\boldsymbol{k}$ and noting that $\boldsymbol{S} = S\boldsymbol{k}$, we get for $\Omega_W^{(1)}(x_A, x_B)$

[2] Note that the sign of (55) in [14] is erroneous.

$$\Omega_W^{(PN)}(x_A, x_B)$$

$$= -\frac{GM}{c^2}\frac{(x_B^0 - x_A^0)^2 + \gamma R_{AB}^2}{R_{AB}} \ln\left(\frac{r_A + r_B + R_{AB}}{r_A + r_B - R_{AB}}\right)$$

$$+\frac{2GM}{c^2} J_2 r_e^2 \frac{(x_B^0 - x_A^0)^2 + \gamma R_{AB}^2}{[(r_A + r_B)^2 - R_{AB}^2]^2}(r_A + r_B)\left(\frac{\boldsymbol{k}\cdot\boldsymbol{x}_A}{r_A} + \frac{\boldsymbol{k}\cdot\boldsymbol{x}_B}{r_B}\right)^2$$

$$-\frac{GM}{c^2} J_2 r_e^2 \frac{(x_B^0 - x_A^0)^2 + \gamma R_{AB}^2}{(r_A + r_B)^2 - R_{AB}^2}\left[\frac{(\boldsymbol{k}\times\boldsymbol{x}_A)^2}{r_A^3} + \frac{(\boldsymbol{k}\times\boldsymbol{x}_B)^2}{r_B^3}\right] + \ldots \quad (84)$$

and for $\Omega_{\boldsymbol{W}}^{(PN)}(x_A, x_B)$

$$\Omega_{\boldsymbol{W}}^{(PN)}(x_A, x_B) = \left(\gamma + 1 + \frac{1}{4}\alpha_1\right)\frac{2GS}{c^3}$$

$$\times (x_B^0 - x_A^0)\frac{r_A + r_B}{r_A r_B}\frac{\boldsymbol{k}\cdot(\boldsymbol{x}_A \times \boldsymbol{x}_B)}{(r_A + r_B)^2 - R_{AB}^2} + \ldots . \quad (85)$$

Finally, owing to the limit $|\alpha_1| < 4\times10^{-4}$ furnished in [18], we shall henceforth neglect all the multipole contributions in $\Omega_{\boldsymbol{v}_r}^{(PN)}(x_A, x_B)$. Thus, we get

$$\Omega_{\boldsymbol{v}_r}^{(PN)}(x_A, x_B) = \alpha_1 \frac{GM}{2c^3}(x_B^0 - x_A^0)\frac{\boldsymbol{R}_{AB}\cdot\boldsymbol{v}_r}{R_{AB}} \ln\left(\frac{r_A + r_B + R_{AB}}{r_A + r_B - R_{AB}}\right) + \cdots . \quad (86)$$

In this section and in the following one, the symbol $+\ldots$ stands for the contributions of higher multipole moments which are neglected. For the sake of brevity, when $+\cdots$ is used, we systematically omit to mention the symbol $O(n)$ which stands for the neglected post-Newtonian terms.

5.3 Time Transfer Function up to the Order $1/c^4$

In what follows, we put

$$\boldsymbol{n}_A = \frac{\boldsymbol{x}_A}{r_A}, \qquad \boldsymbol{n}_B = \frac{\boldsymbol{x}_B}{r_B}, \quad (87)$$

and

$$\boldsymbol{N}_{AB} = \{N^i\} = \frac{\boldsymbol{x}_B - \boldsymbol{x}_A}{R_{AB}} . \quad (88)$$

Furthermore, we use systematically the identity

$$(r_A + r_B)^2 - R_{AB}^2 = 2 r_A r_B (1 + \boldsymbol{n}_A \cdot \boldsymbol{n}_B) . \quad (89)$$

By substituting R_{AB} for $x_B^0 - x_A^0$ into (84–86) and inserting the corresponding expression of $\Omega^{(PN)}$ into (36), we get an expression for the time transfer function as follows

$$\mathcal{T}(\boldsymbol{x}_A, \boldsymbol{x}_B) = \frac{1}{c}R_{AB} + \mathcal{T}_M(\boldsymbol{x}_A, \boldsymbol{x}_B)$$

$$+\mathcal{T}_{J_2}(\boldsymbol{x}_A, \boldsymbol{x}_B) + \mathcal{T}_{\boldsymbol{S}}(\boldsymbol{x}_A, \boldsymbol{x}_B) + \mathcal{T}_{\boldsymbol{v}_r}(\boldsymbol{x}_A, \boldsymbol{x}_B) + \ldots , \quad (90)$$

where

$$\mathcal{T}_M(\boldsymbol{x}_A, \boldsymbol{x}_B) = (\gamma+1)\frac{GM}{c^3}\ln\left(\frac{r_A + r_B + R_{AB}}{r_A + r_B - R_{AB}}\right), \tag{91}$$

$$\begin{aligned}\mathcal{T}_{J_2}(\boldsymbol{x}_A, \boldsymbol{x}_B) = & -\frac{\gamma+1}{2}\frac{GM}{c^3}J_2\frac{r_e^2}{r_A r_B}\frac{R_{AB}}{1+\boldsymbol{n}_A\cdot\boldsymbol{n}_B} \\ & \times\left[\left(\frac{1}{r_A}+\frac{1}{r_B}\right)\frac{(\boldsymbol{k}\cdot\boldsymbol{n}_A + \boldsymbol{k}\cdot\boldsymbol{n}_B)^2}{1+\boldsymbol{n}_A\cdot\boldsymbol{n}_B}\right. \\ & \left.-\frac{1-(\boldsymbol{k}\cdot\boldsymbol{n}_A)^2}{r_A} - \frac{1-(\boldsymbol{k}\cdot\boldsymbol{n}_B)^2}{r_B}\right],\end{aligned} \tag{92}$$

$$\mathcal{T}_{\boldsymbol{S}}(\boldsymbol{x}_A, \boldsymbol{x}_B) = -\left(\gamma+1+\frac{1}{4}\alpha_1\right)\frac{GS}{c^4}\left(\frac{1}{r_A}+\frac{1}{r_B}\right)\frac{\boldsymbol{k}\cdot(\boldsymbol{n}_A\times\boldsymbol{n}_B)}{1+\boldsymbol{n}_A\cdot\boldsymbol{n}_B}, \tag{93}$$

$$\mathcal{T}_{\boldsymbol{v}_r}(\boldsymbol{x}_A, \boldsymbol{x}_B) = -\alpha_1\frac{GM}{2c^4}(\boldsymbol{N}_{AB}\cdot\boldsymbol{v}_r)\ln\left(\frac{r_A + r_B + R_{AB}}{r_A + r_B - R_{AB}}\right). \tag{94}$$

The time transfer is thus explicitly determined up to the order $1/c^4$. The term of order $1/c^3$ given by (91) is the well-known Shapiro time delay [16]. Equations (92) and (93) extend results previously found for $\gamma = 1$ and $\alpha_1 = 0$ [1]. However, our derivation is more straightforward and yields formulae which are more convenient to calculate the frequency shifts. As a final remark, it is worthy of note that $\mathcal{T}_M$ and $\mathcal{T}_{J_2}$ are symmetric in $(\boldsymbol{x}_A, \boldsymbol{x}_B)$, while $\mathcal{T}_{\boldsymbol{S}}$ and $\mathcal{T}_{\boldsymbol{v}_r}$ are antisymmetric in $(\boldsymbol{x}_A, \boldsymbol{x}_B)$.

5.4 Directions of Light Rays at $\boldsymbol{x}_A$ and $\boldsymbol{x}_B$ up to the Order $1/c^3$

To determine the vectors tangent to the ray path at x_A and x_B, we use (14) and (15) where $\mathcal{T}$ is replaced by the expression given by (90–94). It is clear that $\widehat{\boldsymbol{l}}_A$ and $\widehat{\boldsymbol{l}}_B$ may be written as

$$\widehat{\boldsymbol{l}}_A = -\boldsymbol{N}_{AB} + \boldsymbol{\lambda}_e(\boldsymbol{x}_A, \boldsymbol{x}_B), \tag{95}$$

$$\widehat{\boldsymbol{l}}_B = -\boldsymbol{N}_{AB} + \boldsymbol{\lambda}_r(\boldsymbol{x}_A, \boldsymbol{x}_B), \tag{96}$$

where $\boldsymbol{\lambda}_e$ and $\boldsymbol{\lambda}_r$ are perturbation terms due to $\mathcal{T}_M$, $\mathcal{T}_{J_n}$, $\mathcal{T}_{\boldsymbol{S}}$, $\mathcal{T}_{K_n}$, ... For the expansion of $\mathcal{T}$ given by (90–94), we find

$$\boldsymbol{\lambda}_e(\boldsymbol{x}_A, \boldsymbol{x}_B) = -\boldsymbol{\lambda}_M(\boldsymbol{x}_B, \boldsymbol{x}_A) - \boldsymbol{\lambda}_{J_2}(\boldsymbol{x}_B, \boldsymbol{x}_A) + \boldsymbol{\lambda}_{\boldsymbol{S}}(\boldsymbol{x}_B, \boldsymbol{x}_A) + \boldsymbol{\lambda}_{\boldsymbol{v}_r}(\boldsymbol{x}_B, \boldsymbol{x}_A) + \cdots, \tag{97}$$

$$\boldsymbol{\lambda}_r(\boldsymbol{x}_A, \boldsymbol{x}_B) = \boldsymbol{\lambda}_M(\boldsymbol{x}_A, \boldsymbol{x}_B) + \boldsymbol{\lambda}_{J_2}(\boldsymbol{x}_A, \boldsymbol{x}_B) + \boldsymbol{\lambda}_{\boldsymbol{S}}(\boldsymbol{x}_A, \boldsymbol{x}_B) + \boldsymbol{\lambda}_{\boldsymbol{v}_r}(\boldsymbol{x}_A, \boldsymbol{x}_B) + \cdots, \tag{98}$$

where $\boldsymbol{\lambda}_M$, $\boldsymbol{\lambda}_{J_2}$, $\boldsymbol{\lambda}_{\boldsymbol{S}}$, and $\boldsymbol{\lambda}_{\boldsymbol{v}_r}$ stand for the contributions of $\mathcal{T}_M$, $\mathcal{T}_{J_2}$, $\mathcal{T}_{\boldsymbol{S}}$, and $\mathcal{T}_{\boldsymbol{v}_r}$, respectively. We get from (91)

$$\boldsymbol{\lambda}_M(\boldsymbol{x}_A,\boldsymbol{x}_B) = -(\gamma+1)\frac{GM}{c^2}\left(\frac{1}{r_A}+\frac{1}{r_B}\right)\frac{1}{1+\boldsymbol{n}_A\cdot\boldsymbol{n}_B}\left(\boldsymbol{N}_{AB}-\frac{R_{AB}}{r_A+r_B}\boldsymbol{n}_B\right). \tag{99}$$

From (92), we get

$$\begin{aligned}
&\boldsymbol{\lambda}_{J_2}(\boldsymbol{x}_A,\boldsymbol{x}_B)\\
&= (\gamma+1)\frac{GM}{c^2}\left(\frac{1}{r_A}+\frac{1}{r_B}\right)J_2\frac{r_e^2}{r_Ar_B}\frac{1}{(1+\boldsymbol{n}_A\cdot\boldsymbol{n}_B)^2}\\
&\times\left\{\boldsymbol{N}_{AB}\left[\frac{(\boldsymbol{k}\cdot\boldsymbol{n}_A+\boldsymbol{k}\cdot\boldsymbol{n}_B)^2}{1+\boldsymbol{n}_A\cdot\boldsymbol{n}_B}\left(\frac{r_A}{r_B}+\frac{r_B}{r_A}+\frac{1}{2}-\frac{3}{2}\boldsymbol{n}_A\cdot\boldsymbol{n}_B\right)\right.\right.\\
&\quad\left.-\frac{1}{2}\frac{r_Ar_B}{r_A+r_B}\left(\frac{1-(\boldsymbol{k}\cdot\boldsymbol{n}_A)^2}{r_A}+\frac{1-(\boldsymbol{k}\cdot\boldsymbol{n}_B)^2}{r_B}\right)\left(\frac{r_A}{r_B}+\frac{r_B}{r_A}+1-\boldsymbol{n}_A\cdot\boldsymbol{n}_B\right)\right]\\
&\quad-\boldsymbol{n}_B\frac{R_{AB}}{r_A+r_B}\left[\frac{(\boldsymbol{k}\cdot\boldsymbol{n}_A+\boldsymbol{k}\cdot\boldsymbol{n}_B)^2}{1+\boldsymbol{n}_A\cdot\boldsymbol{n}_B}\left(\frac{r_A}{r_B}+\frac{r_B}{r_A}+\frac{3}{2}-\frac{1}{2}\boldsymbol{n}_A\cdot\boldsymbol{n}_B\right)\right.\\
&\quad-\frac{1}{2}\left[1-3(\boldsymbol{k}\cdot\boldsymbol{n}_B)^2\right]\frac{r_A(2+\boldsymbol{n}_A\cdot\boldsymbol{n}_B)+r_B}{r_B}\\
&\quad\left.-\frac{1}{2}(r_A+r_B)\left(\frac{1-(\boldsymbol{k}\cdot\boldsymbol{n}_A)^2}{r_A}-\frac{2(\boldsymbol{k}\cdot\boldsymbol{n}_A)(\boldsymbol{k}\cdot\boldsymbol{n}_B)}{r_B}\right)\right]\\
&\quad\left.+\boldsymbol{k}\frac{R_{AB}}{r_B}\left[(\boldsymbol{k}\cdot\boldsymbol{n}_A)+(\boldsymbol{k}\cdot\boldsymbol{n}_B)\frac{r_A(2+\boldsymbol{n}_A\cdot\boldsymbol{n}_B)+r_B}{r_A+r_B}\right]\right\}.
\end{aligned} \tag{100}$$

From (93) and (94), we derive the other contributions that are not neglected here:

$$\begin{aligned}
\boldsymbol{\lambda}_{\boldsymbol{S}}(\boldsymbol{x}_A,\boldsymbol{x}_B) &= \left(\gamma+1+\frac{1}{4}\alpha_1\right)\frac{GS}{c^3r_B}\left(\frac{1}{r_A}+\frac{1}{r_B}\right)\frac{1}{1+\boldsymbol{n}_A\cdot\boldsymbol{n}_B}\\
&\times\left\{\boldsymbol{k}\times\boldsymbol{n}_A-\frac{\boldsymbol{k}\cdot(\boldsymbol{n}_A\times\boldsymbol{n}_B)}{1+\boldsymbol{n}_A\cdot\boldsymbol{n}_B}\left[\boldsymbol{n}_A+\frac{r_A(2+\boldsymbol{n}_A\cdot\boldsymbol{n}_B)+r_B}{r_A+r_B}\boldsymbol{n}_B\right]\right\},
\end{aligned} \tag{101}$$

$$\begin{aligned}
\boldsymbol{\lambda}_{\boldsymbol{v}_r}(\boldsymbol{x}_A,\boldsymbol{x}_B) &= \alpha_1\frac{GM}{2c^3}\left[\frac{\boldsymbol{v}_r-(\boldsymbol{v}_r\cdot\boldsymbol{N}_{AB})\boldsymbol{N}_{AB}}{R_{AB}}\ln\left(\frac{r_A+r_B+R_{AB}}{r_A+r_B-R_{AB}}\right)\right.\\
&\left.+\frac{(\boldsymbol{v}_r\cdot\boldsymbol{N}_{AB})}{1+\boldsymbol{n}_A\cdot\boldsymbol{n}_B}\left(\frac{1}{r_A}+\frac{1}{r_B}\right)\left(\boldsymbol{N}_{AB}-\frac{R_{AB}}{r_A+r_B}\boldsymbol{n}_B\right)\right].
\end{aligned} \tag{102}$$

We note that the mass and the quadrupole moment yield contributions of order $1/c^2$, while the intrinsic angular momentum and the velocity relative to the universe rest frame yield contributions of order $1/c^3$.

5.5 Sagnac Terms in the Time Transfer Function

In experiments like ACES Mission, recording the time of emission t_A will be more practical than recording the time of reception t_B. So, it will be very convenient to form the expression of the time transfer $\mathcal{T}(\boldsymbol{x}_A, \boldsymbol{x}_B)$ from $\boldsymbol{x}_A(t_A)$ to $\boldsymbol{x}_B(t_B)$ in terms of the position of the receiver B at the time of emission t_A. For any quantity $Q_B(t)$ defined along the world line of the station B, let us put $\widetilde{Q}_B = Q(t_A)$. Thus we may write $\widetilde{\boldsymbol{x}}_B(t_A)$, $\widetilde{r}_B(t_A)$, $\widetilde{\boldsymbol{v}}_B(t_A)$, $\widetilde{v}_B = \mid \widetilde{\boldsymbol{v}}_B \mid$, etc.

Now, let us introduce the instantaneous coordinate distance $\boldsymbol{D}_{AB} = \widetilde{\boldsymbol{x}}_B - \boldsymbol{x}_A$ and its norm D_{AB}. Since we want to know $t_B - t_A$ up to the order $1/c^4$, we can use the Taylor expansion of $\boldsymbol{R}_{AB}$

$$\boldsymbol{R}_{AB} = \boldsymbol{D}_{AB} + (t_B - t_A)\widetilde{\boldsymbol{v}}_B + \frac{1}{2}(t_B - t_A)^2\, \widetilde{\boldsymbol{a}}_B + \frac{1}{6}(t_B - t_A)^3\, \widetilde{\boldsymbol{b}}_B + \cdots ,$$

where $\boldsymbol{a}_B$ is the acceleration of B and $\boldsymbol{b}_B = d\boldsymbol{a}_B/dt$. Using iteratively this expansion together with (90), we get

$$\begin{aligned}\mathcal{T}(\boldsymbol{x}_A, \boldsymbol{x}_B) = \mathcal{T}(\boldsymbol{x}_A, \widetilde{\boldsymbol{x}}_B) &+ \frac{1}{c^2}\boldsymbol{D}_{AB} \cdot \widetilde{\boldsymbol{v}}_B \\ &+ \frac{1}{2c^3} D_{AB} \left[\frac{(\boldsymbol{D}_{AB} \cdot \widetilde{\boldsymbol{v}}_B)^2}{D_{AB}^2} + \widetilde{v}_B^2 + \boldsymbol{D}_{AB} \cdot \widetilde{\boldsymbol{a}}_B \right] \\ &+ \frac{1}{c^4} \Bigg[(\boldsymbol{D}_{AB} \cdot \widetilde{\boldsymbol{v}}_B) \left(\widetilde{v}_B^2 + \boldsymbol{D}_{AB} \cdot \widetilde{\boldsymbol{a}}_B \right) \\ &+ \frac{1}{2} D_{AB}^2 \left(\widetilde{\boldsymbol{v}}_B \cdot \widetilde{\boldsymbol{a}}_B + \frac{1}{3} \boldsymbol{D}_{AB} \cdot \widetilde{\boldsymbol{b}}_B \right) \Bigg] \\ &+ \frac{1}{c} \frac{\boldsymbol{D}_{AB}}{D_{AB}} \cdot \widetilde{\boldsymbol{v}}_B \left[\mathcal{T}_M(\boldsymbol{x}_A, \widetilde{\boldsymbol{x}}_B) + \mathcal{T}_{J_2}(\boldsymbol{x}_A, \widetilde{\boldsymbol{x}}_B) \right] \\ &- \frac{1}{c^2} D_{AB} \widetilde{\boldsymbol{v}}_B \cdot \left[\boldsymbol{\lambda}_M(\boldsymbol{x}_A, \widetilde{\boldsymbol{x}}_B) + \boldsymbol{\lambda}_{J_2}(\boldsymbol{x}_A, \widetilde{\boldsymbol{x}}_B) \right] + \cdots , \end{aligned} \qquad (103)$$

where $\mathcal{T}(\boldsymbol{x}_A, \widetilde{\boldsymbol{x}}_B)$ is obtained by substituting $\widetilde{\boldsymbol{x}}_B$, $\widetilde{r}_B$, and $\boldsymbol{D}_{AB}$, respectively, for $\boldsymbol{x}_B$, r_B, and $\boldsymbol{R}_{AB}$ into the time transfer function defined by (90–94). This expression extends the previous formula [6] to the next order $1/c^4$. The second, the third, and the fourth terms in (103) represent pure Sagnac terms of order $1/c^2$, $1/c^3$, and $1/c^4$, respectively. The fifth and the sixth terms are contributions of the gravitational field mixed with the coordinate velocity of the receiving station. Since these last two terms are of order $1/c^4$, they might be calculated for the arguments $(\boldsymbol{x}_A, \boldsymbol{x}_B)$.

6 Frequency Shift in the Field of a Rotating Axisymmetric Body

6.1 General Formulae up to the Fourth Order

It is possible to derive the ratio q_A/q_B up to the order $1/c^4$ from our results in Sect. 4 since $\widehat{\boldsymbol{l}}_A$ and $\widehat{\boldsymbol{l}}_B$ are given up to the order $1/c^3$ by (95–98). Denoting by $\widehat{\boldsymbol{l}}^{(n)}/c^n$ the $O(n)$ terms in $\widehat{\boldsymbol{l}}$, q_A/q_B may be expanded as

$$\frac{q_A}{q_B} = 1 - \frac{1}{c}\frac{\boldsymbol{N}_{AB}\cdot(\boldsymbol{v}_A - \boldsymbol{v}_B)}{1 - \boldsymbol{N}_{AB}\cdot\dfrac{\boldsymbol{v}_B}{c}} + \frac{1}{c^3}\left[\widehat{\boldsymbol{l}}^{(2)}_A\cdot\boldsymbol{v}_A - \widehat{\boldsymbol{l}}^{(2)}_B\cdot\boldsymbol{v}_B\right]$$
$$+\frac{1}{c^4}\left[\widehat{\boldsymbol{l}}^{(3)}_A\cdot\boldsymbol{v}_A - \widehat{\boldsymbol{l}}^{(3)}_B\cdot\boldsymbol{v}_B\right]$$
$$+\frac{1}{c^4}\boldsymbol{N}_{AB}\cdot\left[\left(\widehat{\boldsymbol{l}}^{(2)}_B\cdot\boldsymbol{v}_B\right)(\boldsymbol{v}_A - 2\boldsymbol{v}_B) + \left(\widehat{\boldsymbol{l}}^{(2)}_A\cdot\boldsymbol{v}_A\right)\boldsymbol{v}_B\right] + O(5)\,. \quad (104)$$

To be consistent with this expansion, we have to perform the calculation of u^0_A/u^0_B at the same level of approximation. For a clock delivering a proper time τ, $1/u^0$ is the ratio of the proper time $d\tau$ to the coordinate time dt. To reach the suitable accuracy, it is therefore necessary to take into account the terms of order $1/c^4$ in g_{00}. For the sake of simplicity, we shall henceforth confine ourselves to the fully conservative metric theories of gravity without preferred location effects, in which all the PPN parameters vanish except β and γ. Since the gravitational field is assumed to be stationary, the chosen coordinate system is then a standard post-Newtonian gauge and the metric reduces to its usual form

$$G_{00} = 1 - \frac{2}{c^2}W + \frac{2\beta}{c^4}W^2 + O(6),$$
$$\{G_{0i}\} = \frac{2(\gamma+1)}{c^3}\boldsymbol{W} + O(5), \quad (105)$$
$$G_{ij} = -\left(1 + \frac{2\gamma}{c^2}W\right)\delta_{ij} + O(4)\,, \quad (106)$$

where W given by (54) reduces to

$$W(\boldsymbol{x}) = U(\boldsymbol{x}) + \frac{G}{c^2}\int\frac{\rho^*(\boldsymbol{x}')}{|\boldsymbol{x}-\boldsymbol{x}'|}\left[\left(\gamma+\frac{1}{2}\right)v^2 + (1-2\beta)U + \Pi + 3\gamma\frac{p}{\rho^*}\right]d^3\boldsymbol{x}'\,, \quad (107)$$

and $\boldsymbol{W}$ is given by (75). As a consequence, for a clock moving with the coordinate velocity $\boldsymbol{v}$, the quantity $1/u^0$ is given by the formula

$$\frac{1}{u^0} \equiv \frac{d\tau}{dt} = 1 - \frac{1}{c^2}\left(W + \frac{1}{2}v^2\right) + \frac{1}{c^4}\left[\left(\beta - \frac{1}{2}\right)W^2 - \left(\gamma + \frac{1}{2}\right)Wv^2\right.$$
$$\left. - \frac{1}{8}v^4 + 2(\gamma+1)\boldsymbol{W}\cdot\boldsymbol{v}\right] + O(6), \quad (108)$$

from which it is easily deduced that

$$\frac{u_A^0}{u_B^0} = 1 + \frac{1}{c^2}\left(W_A - W_B + \frac{1}{2}v_A^2 - \frac{1}{2}v_B^2\right) + \frac{1}{c^4}\Big\{(\gamma+1)(W_A v_A^2 - W_B v_B^2)$$
$$+\frac{1}{2}(W_A - W_B)\left[W_A - W_B + 2(1-\beta)(W_A + W_B) + v_A^2 - v_B^2\right]$$
$$- 2(\gamma+1)(\boldsymbol{W}_A \cdot \boldsymbol{v}_A - \boldsymbol{W}_B \cdot \boldsymbol{v}_B) + \frac{3}{8}v_A^4 - \frac{1}{4}v_A^2 v_B^2 - \frac{1}{8}v_B^4\Big\} + O(6)\,. \tag{109}$$

It follows from (104) and (109) that the frequency shift $\delta\nu/\nu$ is given by

$$\frac{\delta\nu}{\nu} \equiv \frac{\nu_A}{\nu_B} - 1 = \left(\frac{\delta\nu}{\nu}\right)_c + \left(\frac{\delta\nu}{\nu}\right)_g , \tag{110}$$

where $(\delta\nu/\nu)_c$ is the special-relativistic Doppler effect

$$\left(\frac{\delta\nu}{\nu}\right)_c = -\frac{1}{c}\boldsymbol{N}_{AB}\cdot(\boldsymbol{v}_A - \boldsymbol{v}_B)$$
$$+\frac{1}{c^2}\left[\frac{1}{2}v_A^2 - \frac{1}{2}v_B^2 - (\boldsymbol{N}_{AB}\cdot(\boldsymbol{v}_A - \boldsymbol{v}_B))(\boldsymbol{N}_{AB}\cdot\boldsymbol{v}_B)\right]$$
$$-\frac{1}{c^3}\left[(\boldsymbol{N}_{AB}\cdot(\boldsymbol{v}_A - \boldsymbol{v}_B))\left(\frac{1}{2}v_A^2 - \frac{1}{2}v_B^2 + (\boldsymbol{N}_{AB}\cdot\boldsymbol{v}_B)^2\right)\right]$$
$$+\frac{1}{c^4}\left[\frac{3}{8}v_A^4 - \frac{1}{4}v_A^2 v_B^2 - \frac{1}{8}v_B^4\right.$$
$$\left. - (\boldsymbol{N}_{AB}\cdot(\boldsymbol{v}_A - \boldsymbol{v}_B))(\boldsymbol{N}_{AB}\cdot\boldsymbol{v}_B)\left(\frac{1}{2}v_A^2 - \frac{1}{2}v_B^2 + (\boldsymbol{N}_{AB}\cdot\boldsymbol{v}_B)^2\right)\right]$$
$$+O(5) \tag{111}$$

and $(\delta\nu)/\nu)_g$ contains all the contribution of the gravitational field, eventually mixed with kinetic terms

$$\left(\frac{\delta\nu}{\nu}\right)_g = \frac{1}{c^2}(W_A - W_B)$$
$$-\frac{1}{c^3}\left[(W_A - W_B)(\boldsymbol{N}_{AB}\cdot(\boldsymbol{v}_A - \boldsymbol{v}_B)) - \widehat{\boldsymbol{l}}_A^{(2)}\cdot\boldsymbol{v}_A + \widehat{\boldsymbol{l}}_B^{(2)}\cdot\boldsymbol{v}_B\right]$$
$$+\frac{1}{c^4}\Big\{(\gamma+1)(W_A v_A^2 - W_B v_B^2)$$
$$+\frac{1}{2}(W_A - W_B)\left[W_A - W_B + 2(1-\beta)(W_A + W_B) + v_A^2 - v_B^2\right.$$
$$\left. -2(\boldsymbol{N}_{AB}\cdot(\boldsymbol{v}_A - \boldsymbol{v}_B))(\boldsymbol{N}_{AB}\cdot\boldsymbol{v}_B)\right]$$
$$+\boldsymbol{N}_{AB}\cdot\left[\left(\widehat{\boldsymbol{l}}_B^{(2)}\cdot\boldsymbol{v}_B\right)(\boldsymbol{v}_A - 2\boldsymbol{v}_B) + \left(\widehat{\boldsymbol{l}}_A^{(2)}\cdot\boldsymbol{v}_A\right)\boldsymbol{v}_B\right]$$
$$+\left(\widehat{\boldsymbol{l}}_A^{(3)} - 2(\gamma+1)\boldsymbol{W}_A\right)\cdot\boldsymbol{v}_A - \left(\widehat{\boldsymbol{l}}_B^{(3)} - 2(\gamma+1)\boldsymbol{W}_B\right)\cdot\boldsymbol{v}_B\Big\}$$
$$+O(5)\,. \tag{112}$$

It must be emphasized that the formulae (108) and (109) are valid within the PPN framework without adding special assumption, provided that β and γ are the only nonvanishing post-Newtonian parameters. On the other hand, (112) is valid only for stationary gravitational fields. In the case of an axisymmetric rotating body, we shall obtain an approximate expression of the frequency shift by inserting the following developments in (112), yielded by (97–102):

$$\frac{\widehat{\boldsymbol{l}}_A^{(2)}}{c^2} = -\boldsymbol{\lambda}_M(\boldsymbol{x}_B, \boldsymbol{x}_A) - \boldsymbol{\lambda}_{J_2}(\boldsymbol{x}_B, \boldsymbol{x}_A) + \dots , \qquad \frac{\widehat{\boldsymbol{l}}_A^{(3)}}{c^3} = \boldsymbol{\lambda}_S(\boldsymbol{x}_B, \boldsymbol{x}_A) + \dots ,$$

$$\frac{\widehat{\boldsymbol{l}}_B^{(2)}}{c^2} = \boldsymbol{\lambda}_M(\boldsymbol{x}_A, \boldsymbol{x}_B) + \boldsymbol{\lambda}_{J_2}(\boldsymbol{x}_A, \boldsymbol{x}_B) + \dots , \qquad \frac{\widehat{\boldsymbol{l}}_B^{(3)}}{c^3} = \boldsymbol{\lambda}_S(\boldsymbol{x}_A, \boldsymbol{x}_B) + \dots ,$$

the function $\boldsymbol{\lambda}_S$ being now given by (101) written with $\alpha_1 = 0$. Let us recall that the symbol $+\cdots$ stands for the contributions of the higher multipole moments which are neglected.

6.2 Application in the Vicinity of the Earth

To perform numerical estimates of the frequency shifts in the vicinity of the Earth, we suppose now that A is onboard the International Space Station (ISS) orbiting at the altitude $H = 400\,\mathrm{km}$ and that B is a terrestrial station. It will be the case for the ACES mission. We use $r_B = 6.37 \times 10^6\,\mathrm{m}$ and $r_A - r_B = 400\,\mathrm{km}$. For the velocity of ISS, we take $v_A = 7.7 \times 10^3\,\mathrm{m\,s^{-1}}$ and for the terrestrial station, we have $v_B \leq 465\,\mathrm{m\,s^{-1}}$. The other useful parameters concerning the Earth are $GM = 3.986 \times 10^{14}\,\mathrm{m^3\,s^{-2}}$, $r_e = 6.378 \times 10^6\,\mathrm{m}$, $J_2 = 1.083 \times 10^{-3}$; for $n \geq 3$, the multipole moments J_n are in the order of 10^{-6}. With these values, we get $W_B/c^2 \approx GM/c^2 r_B = 6.95 \times 10^{-10}$ and $W_A/c^2 \approx GM/c^2 r_A = 6.54 \times 10^{-10}$. From these data, it is easy to deduce the following upper bounds: $| \boldsymbol{N}_{AB} \cdot \boldsymbol{v}_A/c | \leq 2.6 \times 10^{-5}$ for the satellite, $| \boldsymbol{N}_{AB} \cdot \boldsymbol{v}_B/c | \leq 1.6 \times 10^{-6}$ for the ground station, and $| \boldsymbol{N}_{AB} \cdot (\boldsymbol{v}_A - \boldsymbol{v}_B)/c | \leq 2.76 \times 10^{-5}$ for the first-order Doppler term.

Our purpose is to obtain correct estimates of the effects in (112) which are greater than or equal to 10^{-18} for an axisymmetric model of the Earth. At this level of approximation, it is not sufficient to take into account the J_2-terms in $(W_A - W_B)/c^2$. First, the higher-multipole moments J_3, J_4, ... yield contribution of order 10^{-15} in W_A/c^2. Second, owing to the irregularities in the distribution of masses, the expansion of the geopotential in a series of spherical harmonics is probably not convergent at the surface of the Earth. For these reasons, we do not expand $(W_A - W_B)/c^2$ in (112).

However, for the higher-order terms in (112), we can apply the explicit formulae obtained in Sect. 5. Indeed, since the difference between the geoid and the reference ellipsoid is less than 100 m, W_B/c^2 may be written as [26]

$$\frac{1}{c^2} W_B = \frac{GM}{c^2 r_B} + \frac{GM r_e^2 J_2}{2c^2 r_B^3}(1 - 3\cos^2\theta) + \frac{1}{c^2}\triangle W_B \,, \qquad (113)$$

where the residual term $\triangle W_B/c^2$ is such that $|\triangle W_B/c^2| \leq 10^{-14}$. At a level of experimental uncertainty about 10^{-18}, this inequality allows to retain only the contributions due to M, J_2, and $\boldsymbol{S}$ in the terms of orders $1/c^3$ and $1/c^4$. As a consequence, the formula (112) reduces to

$$\left(\frac{\delta\nu}{\nu}\right)_g = \frac{1}{c^2}(W_A - W_B) + \frac{1}{c^3}\left(\frac{\delta\nu}{\nu}\right)_M^{(3)} + \frac{1}{c^3}\left(\frac{\delta\nu}{\nu}\right)_{J_2}^{(3)} + \dots$$
$$+\frac{1}{c^4}\left(\frac{\delta\nu}{\nu}\right)_M^{(4)} + \frac{1}{c^4}\left(\frac{\delta\nu}{\nu}\right)_{\boldsymbol{S}}^{(4)} + \dots, \qquad (114)$$

where the different terms involved in the RHS are separately explicited and discussed in what follows.

By using (89), it is easy to see that $(\delta\nu/\nu)_M^{(3)}$ is given by

$$\left(\frac{\delta\nu}{\nu}\right)_M^{(3)} = -\frac{GM(r_A + r_B)}{r_A r_B}\left[\left(\frac{\gamma+1}{1+\boldsymbol{n}_A\cdot\boldsymbol{n}_B} - \frac{r_A - r_B}{r_A + r_B}\right)[\boldsymbol{N}_{AB}\cdot(\boldsymbol{v}_A - \boldsymbol{v}_B)]\right.$$
$$\left.+(\gamma+1)\frac{R_{AB}}{r_A + r_B}\,\frac{\boldsymbol{n}_A\cdot\boldsymbol{v}_A + \boldsymbol{n}_B\cdot\boldsymbol{v}_B}{1+\boldsymbol{n}_A\cdot\boldsymbol{n}_B}\right]. \qquad (115)$$

The contribution of this term is bounded by 5×10^{-14} for $\gamma = 1$, in accordance with a previous analysis [6].

6.3 Influence of the Quadrupole Moment at the Order $1/c^3$

It follows from (100) and (112) that the term $(\delta\nu/\nu)_{J_2}^{(3)}$ in (114) is given by

$$\left(\frac{\delta\nu}{\nu}\right)_{J_2}^{(3)}$$
$$= \frac{GM}{2r_e}J_2\left(\boldsymbol{N}_{AB}\cdot(\boldsymbol{v}_A - \boldsymbol{v}_B)\right)\left[\left(\frac{r_e}{r_A}\right)^3\left[3(\boldsymbol{k}\cdot\boldsymbol{n}_A)^2 - 1\right] - \left(\frac{r_e}{r_B}\right)^3\left[3(\boldsymbol{k}\cdot\boldsymbol{n}_B)^2 - 1\right]\right]$$
$$+(\gamma+1)GM\left(\frac{1}{r_A} + \frac{1}{r_B}\right)J_2\,\frac{r_e^2}{r_A r_B}\,\frac{1}{(1+\boldsymbol{n}_A\cdot\boldsymbol{n}_B)^2}$$
$$\times\left\{[\boldsymbol{N}_{AB}\cdot(\boldsymbol{v}_A - \boldsymbol{v}_B)]\left[\frac{(\boldsymbol{k}\cdot\boldsymbol{n}_A + \boldsymbol{k}\cdot\boldsymbol{n}_B)^2}{1+\boldsymbol{n}_A\cdot\boldsymbol{n}_B}\left(\frac{r_A}{r_B} + \frac{r_B}{r_A} + \frac{1}{2} - \frac{3}{2}\boldsymbol{n}_A\cdot\boldsymbol{n}_B\right)\right.\right.$$
$$-\frac{1}{2}\left(1 - \frac{r_A(\boldsymbol{k}\cdot\boldsymbol{n}_B)^2 + r_B(\boldsymbol{k}\cdot\boldsymbol{n}_A)^2}{r_A + r_B}\right)$$
$$\left.\times\left(\frac{r_A}{r_B} + \frac{r_B}{r_A} + 1 - \boldsymbol{n}_A\cdot\boldsymbol{n}_B\right)\right]$$
$$+\frac{R_{AB}}{r_A + r_B}(\boldsymbol{n}_A\cdot\boldsymbol{v}_A + \boldsymbol{n}_B\cdot\boldsymbol{v}_B)\frac{(\boldsymbol{k}\cdot\boldsymbol{n}_A + \boldsymbol{k}\cdot\boldsymbol{n}_B)^2}{1+\boldsymbol{n}_A\cdot\boldsymbol{n}_B}\left(\frac{r_A}{r_B} + \frac{r_B}{r_A} + \frac{3}{2} - \frac{1}{2}\boldsymbol{n}_A\cdot\boldsymbol{n}_B\right)$$
$$-\frac{1}{2}\frac{R_{AB}}{r_A}(\boldsymbol{n}_A\cdot\boldsymbol{v}_A)\left[1 - 3(\boldsymbol{k}\cdot\boldsymbol{n}_A)^2\right]\frac{r_A + r_B(2 + \boldsymbol{n}_A\cdot\boldsymbol{n}_B)}{r_A + r_B}$$

$$
\begin{aligned}
&-\frac{1}{2}\frac{R_{AB}}{r_B}(\boldsymbol{n}_B\cdot\boldsymbol{v}_B)\left[1-3(\boldsymbol{k}\cdot\boldsymbol{n}_B)^2\right]\frac{r_A(2+\boldsymbol{n}_A\cdot\boldsymbol{n}_B)+r_B}{r_A+r_B}\\
&+R_{AB}\left[\left(\frac{\boldsymbol{n}_A\cdot\boldsymbol{v}_A}{r_A}+\frac{\boldsymbol{n}_B\cdot\boldsymbol{v}_B}{r_B}\right)(\boldsymbol{k}\cdot\boldsymbol{n}_A)(\boldsymbol{k}\cdot\boldsymbol{n}_B)\right.\\
&\qquad\left.-\frac{1}{2}(\boldsymbol{n}_A\cdot\boldsymbol{v}_A)\frac{1-(\boldsymbol{k}\cdot\boldsymbol{n}_B)^2}{r_B}-\frac{1}{2}(\boldsymbol{n}_B\cdot\boldsymbol{v}_B)\frac{1-(\boldsymbol{k}\cdot\boldsymbol{n}_A)^2}{r_A}\right]\\
&-\frac{R_{AB}}{r_A}(\boldsymbol{k}\cdot\boldsymbol{v}_A)\left[\boldsymbol{k}\cdot\boldsymbol{n}_A\frac{r_A+r_B(2+\boldsymbol{n}_A\cdot\boldsymbol{n}_B)}{r_A+r_B}+\boldsymbol{k}\cdot\boldsymbol{n}_B\right]\\
&\left.-\frac{R_{AB}}{r_B}(\boldsymbol{k}\cdot\boldsymbol{v}_B)\left[\boldsymbol{k}\cdot\boldsymbol{n}_A+\boldsymbol{k}\cdot\boldsymbol{n}_B\frac{r_A(2+\boldsymbol{n}_A\cdot\boldsymbol{n}_B)+r_B}{r_A+r_B}\right]\right\}. \qquad (116)
\end{aligned}
$$

One has $|\boldsymbol{v}_A/c|=2.6\times10^{-5}$, $|\boldsymbol{v}_B/c|\leq1.6\times10^{-6}$, and $K_{AB}=3.77\times10^{-3}$. A crude estimate can be obtained by neglecting in (116) the terms involving the scalar products $\boldsymbol{n}_B\cdot\boldsymbol{v}_B$ and $\boldsymbol{k}\cdot\boldsymbol{v}_B$. Since the orbit of ISS is almost circular, the scalar product $\boldsymbol{n}_A\cdot\boldsymbol{v}_A$ can also be neglected. On these assumptions, we find for $\gamma=1$

$$
\left|\frac{1}{c^3}\left(\frac{\delta\nu}{\nu}\right)^{(3)}_{J_2}\right|\leq1.3\times10^{-16}. \qquad (117)
$$

As a consequence, it will perhaps be necessary to take into account the $O(3)$ contributions of J_2 in the ACES mission. This conclusion is to be compared with the order of magnitude given in [6] without a detailed calculation. Of course, a better estimate might be found if the inclination $i=51.6\,\mathrm{deg}$ of the orbit with respect to the terrestrial equatorial plane and the latitude $\pi/2-\theta_B$ of the ground station was taken into account.

6.4 Frequency Shifts of Order $1/c^4$

The term $(\delta\nu/\nu)^{(4)}_M$ in (114) is given by

$$
\begin{aligned}
\left(\frac{\delta\nu}{\nu}\right)^{(4)}_M &= (\gamma+1)\left(\frac{GM}{r_A}v_A^2-\frac{GM}{r_B}v_B^2\right)-\frac{GM(r_A-r_B)}{2r_Ar_B}(v_A^2-v_B^2)\\
&+\frac{1}{2}\left(\frac{GM}{r_Ar_B}\right)^2\left[(r_A-r_B)^2+2(\beta-1)(r_A^2-r_B^2)\right]-\frac{GM(r_A+r_B)}{r_Ar_B}\\
&\times\left[\left(\frac{2(\gamma+1)}{1+\boldsymbol{n}_A\cdot\boldsymbol{n}_B}-\frac{r_A-r_B}{r_A+r_B}\right)\left[\boldsymbol{N}_{AB}\cdot(\boldsymbol{v}_A-\boldsymbol{v}_B)\right](\boldsymbol{N}_{AB}\cdot\boldsymbol{v}_B)\right.\\
&\qquad+\frac{\gamma+1}{1+\boldsymbol{n}_A\cdot\boldsymbol{n}_B}\frac{R_{AB}}{r_A+r_B}\Big\{(\boldsymbol{n}_A\cdot\boldsymbol{v}_A)(\boldsymbol{N}_{AB}\cdot\boldsymbol{v}_B)\\
&\qquad\left.-\left[\boldsymbol{N}_{AB}\cdot(\boldsymbol{v}_A-2\boldsymbol{v}_B)\right](\boldsymbol{n}_B\cdot\boldsymbol{v}_B)\Big\}\right]. \qquad (118)
\end{aligned}
$$

The dominant term $(\gamma + 1)GMv_A^2/r_A$ in (118) induces a correction to the frequency shift which amounts to 10^{-18}. So, it will certainly be necessary to take this correction into account in experiments performed in the foreseeable future.

The term $(\delta\nu/\nu)_{\boldsymbol{S}}^{(4)}$ is the contribution of the intrinsic angular momentum to the frequency shift. Substituting (79) and (101) into (112), it may be seen that

$$\left(\frac{\delta\nu}{\nu}\right)_{\boldsymbol{S}}^{(4)} = (\mathcal{F}_{\boldsymbol{S}})_A - (\mathcal{F}_{\boldsymbol{S}})_B\,, \tag{119}$$

where

$$(\mathcal{F}_{\boldsymbol{S}})_A = (\gamma+1)\,\frac{GS}{r_A^2}\left(1+\frac{r_A}{r_B}\right)\boldsymbol{v}_A\cdot\left\{\frac{\boldsymbol{k}\times\boldsymbol{n}_B}{1+\boldsymbol{n}_A\cdot\boldsymbol{n}_B} - \frac{r_B}{r_A+r_B}\,\boldsymbol{k}\times\boldsymbol{n}_A\right.$$
$$\left. + \frac{\boldsymbol{k}\cdot(\boldsymbol{n}_A\times\boldsymbol{n}_B)}{(1+\boldsymbol{n}_A\cdot\boldsymbol{n}_B)^2}\left[\frac{r_A+r_B(2+\boldsymbol{n}_A\cdot\boldsymbol{n}_B)}{r_A+r_B}\boldsymbol{n}_A+\boldsymbol{n}_B\right]\right\}, \tag{120}$$

$$(\mathcal{F}_{\boldsymbol{S}})_B = (\gamma+1)\,\frac{GS}{r_B^2}\left(1+\frac{r_B}{r_A}\right)\boldsymbol{v}_B\cdot\left\{\frac{\boldsymbol{k}\times\boldsymbol{n}_A}{1+\boldsymbol{n}_A\cdot\boldsymbol{n}_B} - \frac{r_A}{r_A+r_B}\,\boldsymbol{k}\times\boldsymbol{n}_B\right.$$
$$\left. - \frac{\boldsymbol{k}\cdot(\boldsymbol{n}_A\times\boldsymbol{n}_B)}{(1+\boldsymbol{n}_A\cdot\boldsymbol{n}_B)^2}\left[\boldsymbol{n}_A+\frac{r_A(2+\boldsymbol{n}_A\cdot\boldsymbol{n}_B)+r_B}{r_A+r_B}\boldsymbol{n}_B\right]\right\}. \tag{121}$$

To make easier the discussion, it is useful to introduce the angle ψ between $\boldsymbol{x}_A$ and $\boldsymbol{x}_B$, and the angle i_p between the plane of the photon path and the equatorial plane. These angles are defined by

$$\cos\psi = \boldsymbol{n}_A\cdot\boldsymbol{n}_B\,,\quad 0\leq\psi<\pi\,,\quad \boldsymbol{k}\cdot(\boldsymbol{n}_A\times\boldsymbol{n}_B) = \sin\psi\cos i_p\,,\quad 0\leq i_p<\pi\,.$$

With these definitions, it is easily seen that

$$\frac{\boldsymbol{k}\cdot(\boldsymbol{n}_A\times\boldsymbol{n}_B)}{1+\boldsymbol{n}_A\cdot\boldsymbol{n}_B} = \cos i_p \tan\frac{\psi}{2}\,.$$

Let us apply our formulas to ISS. Due to the inequality $v_B/v_A \leq 6\times10^{-2}$, the term $(\mathcal{F}_{\boldsymbol{S}})_B$ in (119) may be neglected. From (120), it is easily deduced that

$$\mid(\mathcal{F}_{\boldsymbol{S}})_A\mid\leq(\gamma+1)\,\frac{GS}{r_A^2}\left(1+\frac{r_A}{r_B}\right)\frac{2+3\,|\tan\psi/2|}{|1+\cos\psi|}\,v_A\,.$$

Assuming $0\leq\psi\leq\pi/2$, we have $(2+3\,|\tan\psi/2\,|)/\,|1+\cos\psi\,|\leq 5$. Inserting this inequality in the previous one and taking for the Earth $GS/c^3r_A^2 = 3.15\times10^{-16}$, we find

$$\left|\frac{1}{c^4}\left(\frac{\delta\nu}{\nu}\right)_{\boldsymbol{S}}^{(4)}\right| \leq (\gamma+1)\times10^{-19}\,. \tag{122}$$

Thus, we get an upper bound which is slightly greater than the one estimated by retaining only the term $h_{0i}v^i/c$ in (109). However, our formula confirms that the intrinsic angular momentum of the Earth will not affect the ACES experiment.

7 Conclusion

It is clear that the world function $\Omega(x_A, x_B)$ constitutes a powerful tool for determining the time delay and the frequency shift of electromagnetic signals in a weak gravitational field. The analytical derivations given here are obtained within the Nordtvedt–Will PPN formalism. We have found the general expression of $\Omega(x_A, x_B)$ up to the order $1/c^3$. This result yields the expression of the time transfer functions $\mathcal{T}_e(t_A, \boldsymbol{x}_A, \boldsymbol{x}_B)$ and $\mathcal{T}_r(t_B, \boldsymbol{x}_A, \boldsymbol{x}_B)$ up to the order $1/c^4$. We point out that γ and α_1 are the only post-Newtonian parameters involved in the expressions of the world function and of the time transfer functions within the limit of the considered approximation.

We have treated in detail the case of an isolated, axisymmetric rotating body, assuming that the gravitational field is stationary and that the body is moving with a constant velocity $\boldsymbol{v}_r$ relative to the universe rest frame. We have given a systematic procedure for calculating the terms due to the multipole moments in the world function $\Omega(x_A, x_B)$ and in the single time transfer function $\mathcal{T}(\boldsymbol{x}_A, \boldsymbol{x}_B)$. These terms are obtained by straightforward differentiations of a kernel function. We have explicitly derived the contributions due to the mass M, to the quadrupole moment J_2, and to the intrinsic angular momentum $\boldsymbol{S}$ of the rotating body.

Assuming for the sake of simplicity that only β and γ are different from zero, we have determined the general expression of the frequency shift up to the order $1/c^4$. We have derived an explicit formula for the contributions of J_2 at the order $1/c^3$. Our method would give as well the quadrupole contribution at the order $1/c^4$ in case of necessity. Furthermore, we have obtained a thorough expression for the contribution of the mass monopole at the fourth order, as well as the contribution of the intrinsic angular momentum $\boldsymbol{S}$, which is also of order $1/c^4$. It must be pointed out that our calculations give also the vectors tangent to the light ray at the emission and reception points. So, our results could be used for determining the contributions of J_2 and $\boldsymbol{S}$ to the deflection of light.

On the assumption that the gravitational field is stationary, our formulae yield all the gravitational corrections to the frequency shifts up to 10^{-18} in the vicinity of the Earth. Numerically, the influence of the Earth quadrupole moment at the order $1/c^3$ is in the region of 10^{-16} for a clock installed onboard ISS and compared with a ground clock. As a consequence, this effect will probably be observable during the ACES mission. We also note that the leading term in the fourth-order frequency shift due to the mass monopole is equal to 10^{-18} for a clock installed onboard ISS and compared with a ground clock. As a consequence, this effect could be observable in the foreseeable future with atomic clocks using optical transitions.

References

1. S. A. Klioner, Sov. Astron. **35**, 523 (1991); S. A. Klioner, and S. M. Kopeikin, Astron. J. **104**, 897 (1992); see references therein.
2. S. M. Kopeikin, J. Math. Phys. **38**, 2587 (1997).
3. S. M. Kopeikin, and G. Schäfer, Phys. Rev. D **60**, 124002 (1999).
4. S. A. Klioner, *preprint* astro-ph/0107457.
5. S. M. Kopeikin, and B. Mashhoon, Phys. Rev. D **65**, 064025 (2002).
6. L. Blanchet, C. Salomon, P. Teyssandier, and P. Wolf, Astron. Astrophys. **370**, 320 (2001).
7. N. Ashby, IEEE International Frequency Control Symposium, 320 (1998).
8. G. W. Richter and R. A. Matzner, Phys. Rev. D **28**, 3007 (1983).
9. M. H. Brügmann, Phys. Rev. D **72**, 024012 (2005).
10. J. L. Synge, *Relativity: The General Theory* (North-Holland, 1964); see also H. A. Buchdahl, Int. J. Theor. Phys. **24**, 457 (1985); *ibid.* **29**, 209 (1990).
11. R. Holzwarth *et al.*, Phys. Rev. Lett. **85**, 2264 (2000). Th. Udem *et al.*, *ibid.* **86**, 4996 (2001).
12. C. Salomon, and C. Veillet, in Proc. Symp. on Space Station Utilisation, ESA-SP 385, 295 (1996).
13. A. Spallicci *et al.*, Class. Quant. Grav. **14**, 2971 (1997).
14. B. Linet and P. Teyssandier, Phys. Rev. D **66**, 024045 (2002).
15. C. Le Poncin-Lafitte, B. Linet, and P. Teyssandier, Class. Quantum Grav. **21**, 4463 (2004).
16. I. I. Shapiro, Phys. Rev. Lett. **13**, 789 (1964).
17. E. Poisson, Class. Quantum Grav. **21** R 153 (2004).
18. C. M. Will, *Theory and Experiment in Gravitational Physics* (Cambridge University Press, revised edition, 1993).
19. H. A. Buchdahl, Austral. J. Phys. **32**, 405 (1979).
20. S. A. Klioner, and M. H. Soffel, Phys. Rev. D **62**, 024019 (2000).
21. T. Damour, M. Soffel, and C. Xu, Phys. Rev. D **43**, 3273 (1991); *ibid.* **45**, 1017 (1992); *ibid.* **47**, 3124 (1993).
22. UAI Resolutions Adopted at the 24th General Assembly (Manchester, August 2000), to be published in Transactions of the UAI **24** B.
23. L. Blanchet, and T. Damour, Phil. Trans. Roy. Soc. Lond. A **320**, 379 (1986), Annales Poincaré Phys. Theor. **50**, 377 (1989).
24. P. Teyssandier, Phys. Rev. D **16**, 946 (1977); *ibid.* **18**, 1037 (1978).
25. R. J. Adler, and A. S. Silbergleit, Int. J. Theor. Phys. **39**, 1291 (2000).
26. P. Wolf, and G. Petit, Astron. Astrophys. **304**, 653 (1995); G. Petit, and P. Wolf, IEEE Trans. Instrum. Measur. **46**, 201 (1997).

Unified Formula for Comparison of Clock Rates and Its Applications

C. Xu[1], X. Wu[1], and E. Brüning[2]

[1] Purple Mountain Observatory, Nanjing 210008, China
Department of Physics, Nanjing Normal University, Nanjing 210097, China
[2] School of Mathematical Sciences, University of KwaZulu-Natal, Durban 4000, South Africa

Summary. In this chapter, we deduce a unified formula which allows to discuss the comparison of clock rates at two different space–time points. In the case of a perturbed Robertson–Walker metric, our formula returns to an equation for the comparison of clock rates at different cosmic space–time points, which includes the Hubble redshift, the Doppler effect, the gravitational redshift, and the Rees–Sciama effects. In the case of the solar system, when the 2PN metric is substituted into the unified formula, the comparison of the clock rates both on the earth and a space station could be made. It might be useful for the discussion on the precise measurement on future ACES and ASTROD.

1 Introduction

One of the most basic experiments in physics is the measurement of times. Recently, atomic clocks with a time-keeping accuracy of the order of 10^{-18} in fractional frequency have been considered [1,2]. Also a spatial experiment named Atomic Clock Ensemble in Space (ACES) mission [3,4] is scheduled to be launched in 2006 by European Space Agency (ESA). The purpose of ACES is to obtain an accuracy of order 10^{-16} in fractional frequency. In such a situation (10^{-16}–10^{-18} level), 2PN (second post-Newtonian) approximate framework has to be carried out before hand. Also, Astrodynamical Space Test of Relativity using Optical Devices (ASTROD) [5,6] is planned. The accuracy of measuring γ (about 10^{-9}) and other parameters will depend on the stability of the lasers or clocks. This plan also needs a 2PN level on the comparison of clock rates and equations of motion for planets. The precision of 2PN level on the comparison of clock rates (or time transfer) has been discussed in [7,8] by means of world function. But as we know, the calculation of the world function is not easy. Therefore, we deduce a unified formula in a different way. Our unified formula can also be applied to cosmos and easily extended to an even higher order (higher than 2PN level).

Many formulae have been suggested for the comparison of clock rates at different positions, based on certain simplifying assumptions about which effects are dominant. The change of the clock rates can be related to the relativistic Doppler effect, the gravitational redshift, the Hubble redshift, the Rees–Sciama effect, and so on. The physical conditions causing all these effects may prevail at the same time. In early 1990s, the Hubble redshift, the gravitational redshift, the Doppler effect, and the Rees-Sciama effect have been combined into one equation (to see (6) in [9]) in first-order approximation

$$1+z=\frac{R(\tau_o)}{R(\tau_e)}\left\{1+\frac{5}{3}(\phi_e-\phi_o)+2\int_{\tau_e}^{\tau_o} d\tau \mathbf{l}\cdot\nabla\phi+\mathbf{n}\cdot(\mathbf{v}_e-\mathbf{v}_o)\right\}, \tag{1}$$

where z is redshift, ϕ is gravitational potential, the subscript $e(o)$ denote the emitting (observer) point, $\mathbf{l}=\mathbf{k}/k_o$ (k^α is the tangent vector to the null geodesic connecting the emitting point and observer), and $\mathbf{v}$ is 3-velocity. Since all of terms are the level of the first-order approximation, the coupling terms do not exist. Also they do not deduce (1) through an exact method, it is difficult for us to extend the formula to higher-order precision.

Accordingly, a comprehensive approach, starting from first principles, is needed in which the physical conditions for all these effects are taken into account at the same time. Such an approach should lead us to a synthetic formula which reflects all these effects in a compact way and which should provide additional information, due to possible interactions which could not be incorporated in the isolated approaches for the individual effects.

In general, a comparison of the clock rates between $\Delta\tau_A$ and $\Delta\tau_B$ by means of differential coordinate time Δt_A and Δt_B in global coordinate can be achieved. The relation of the coordinate time between A and B is established by null geodesic line (light ray) [10]

$$cdt=\frac{-g_{0i}dx^i\pm\sqrt{(g_{0i}g_{oj}-g_{00}g_{ij})dx^idx^j}}{g_{00}}. \tag{2}$$

The minus and plus sign are taken in I and III quadrants (in $x-t$ coordinates) and II and IV quadrants, respectively. Normally we take the minus sign. Using these ideal we first time deduce a unified formula for the comparison of clock rates by means of "calculus of differences." Substituting the simplest perturbed Robertson–Walker metric into the unified formula, we obtain a formula for the comparison of clock rates at different cosmic space–time points, which includes the Hubble redshift, the Doppler effect, the gravitational redshift, and the Rees–Sciama effects. By using the 2PN metric in multiple coordinates [11], the 2PN comparison of clock rates both on the earth and a space station in the solar system is made, it may be useful for the precise measurement of ACES and ASTROD in future.

2 General Formula

In a global coordinates (ct, x^i), a source A moves with a velocity v_A^i and a receiver B with a velocity v_B^i. The clock rates in A and B are directly related with their own proper time $\Delta\tau_A$ and $\Delta\tau_B$. To compare them, we need to know the relation between the time interval Δt_A and Δt_B, because

$$\frac{\Delta\tau_A}{\Delta\tau_B} = \frac{\Delta\tau_A}{\Delta t_A}\frac{\Delta t_A}{\Delta t_B}\frac{\Delta t_B}{\Delta\tau_B}\,. \tag{3}$$

Since $-c^2 d\tau^2 = ds^2$, therefore if the velocity of a standard clock (A or B) in the global coordinates is $\mathbf{v}$, we have

$$\Delta t = \frac{\Delta\tau}{\sqrt{-[g_{00} + 2g_{0i}v^i/c + g_{ij}v^i v^j/c^2]}}\,, \tag{4}$$

where $g_{\mu\nu}$ (g_{00}, g_{0i}, and g_{ij}) are the global metric. As abbreviation we introduce

$$G_A = -(g_{00}(A) + 2g_{0i}(A)v_A^i/c + g_{ij}(A)v_A^i v_A^j/c^2)\,,$$
$$G_B = -(g_{00}(B) + 2g_{0i}(B)v_B^i/c + g_{ij}(B)v_B^i v_B^j/c^2)\,.$$

One of the main purpose of our chapter is to calculate the relation between Δt_A and Δt_B by means of a "calculus of differences." Assuming that, at t_{A_1} (coordinate time) a source A emits a first pulse at position $A_1(x_{A_1}^i)$, then a receiver B received the first pulse at position $B_1(x_{B_1}^i)$ at time t_{B_1}. A second pulse is emitted from A at position $A_2(x_{A_2}^i)$ at t_{A_2}, which is received by B at position $B_2(x_{B_2}^i)$ at time t_{B_2}. Then the relation between the emission time and reception time can be rewritten as

$$t_B = t_A + \frac{1}{c}\int_A^B \frac{-g_{0i}\frac{dx^i}{dx} - \sqrt{(g_{0i}g_{0j} - g_{00}g_{ij})\frac{dx^i}{dx}\frac{dx^j}{dx}}}{g_{00}}dx\,, \tag{5}$$

where we define $dx^2 \equiv \delta_{ij}dx^i dx^j$, the geometric meaning of dx is the spatial differential length of the line in the flat space.

In a weak field, $g_{0i}g_{0j}$ is a small quantities ($\sim O(6)$) and the spatial conformal isotropic condition [11, 12] is

$$g_{00}g_{ij} = -\delta_{ij} - \frac{q_{ij}}{c^4} + O(6)\,, \tag{6}$$

where q_{ij} is a spatial anisotropic contribution in the second order, $O(6)$ is the abbreviation symbol for $O(c^{-6})$ as well as $O(n)$ for $O(c^{-n})$. Then (5) simplifies to

$$t_B = t_A - \frac{1}{c}\int_A^B \frac{1}{g_{00}}\left[1 + g_{0i}\frac{dx^i}{dx} + \frac{q_{ij}}{2c^4}\frac{dx^i}{dx}\frac{dx^j}{dx}\right]dx\,. \tag{7}$$

Briefly, we define

$$F(t,x^i) \equiv \frac{-g_{0i}\frac{dx^i}{dx} - \sqrt{(g_{0i}g_{0j} - g_{00}g_{ij})\frac{dx^i}{dx}\frac{dx^j}{dx}}}{cg_{00}}, \tag{8}$$

Equation (5) becomes $t_B = t_A + \int_A^B F(t,x^i)dx$. According to the "calculus of differences," we have

$$\Delta t_B = \Delta t_A + \Delta\left[\int_A^B F(t,x^i)dx\right]. \tag{9}$$

Note that in this formula we are dealing with finite differences, not with infinitesimals ones as in the calculus of variations. In particular in (6) one would have to use $x(A)$ and $x(B)$, respectively, as integration boundaries.

If we only consider a linear approximation in "calculus of differences," the difference of the integral in (6) can be divided into three parts

$$\Delta\int_A^B F dx = \int_{x(B)}^{x(B+\Delta B)} F dx - \int_{x(A)}^{x(A+\Delta A)} F dx + \int_{x(A)}^{x(B)} \Delta F dx, \tag{10}$$

where $A + \Delta A$ and $B + \Delta B$ are corresponding to A_2 and B_2. $\Delta x(A)$ and $\Delta x(B)$ are given by

$$\Delta x(A) \equiv x(A + \Delta A) - x(A) = \frac{\mathbf{k}_A}{|k_A|}\cdot\left.\frac{d\mathbf{x}}{dx}\right|_A \Delta x = \frac{\mathbf{k}_A}{|k_A|}\cdot\mathbf{v}_A\Delta t_A, \tag{11}$$

$$\Delta x(B) \equiv x(B + \Delta B) - x(B) = \frac{\mathbf{k}_B}{|k_B|}\cdot\left.\frac{d\mathbf{x}}{dx}\right|_B \Delta x = \frac{\mathbf{k}_B}{|k_B|}\cdot\mathbf{v}_B\Delta t_B. \tag{12}$$

Here $\mathbf{k}_A$ is the wave vector at point A of the light signal emitted from A and received at B. $\mathbf{k}_B$ is the value of this wave vector at the point B. In (11) and (12) we can replace $\frac{dx^i}{dx}\Delta x$ by $\frac{dx^i}{dt}\Delta t$. The first and second terms in (10) can be written as

$$\int_{x(B)}^{x(B)+x(\Delta B)} F dx = F(B)\frac{\mathbf{k}_B\cdot\mathbf{v}_B}{|k_B|}\Delta t_B, \tag{13}$$

$$-\int_{x(A)}^{x(A)+x(\Delta A)} F dx = -F(A)\frac{\mathbf{k}_A\cdot\mathbf{v}_A}{|k_A|}\Delta t_A. \tag{14}$$

The last term of (10) is the difference of the integral between line 2 and line 1 when $\mathbf{v}_A = \mathbf{v}_B = 0$ (i.e., the boundary of integral is fixed), which can be expanded as

$$\int_{x(A)}^{x(B)} \left(F(t+\Delta t, x^i + \Delta x^i) - F(t,x^i)\right) dx = \int_{x(A)}^{x(B)} \left(\frac{\partial F}{\partial t}\Delta t + \frac{\partial F}{\partial x^i}\Delta x^i\right) dx. \tag{15}$$

As we know, if F is independent of time, then $\int_{x(A)}^{x(B)} \Delta F dx = \int_{x(A)}^{x(B)} \frac{\partial F}{\partial x^i}\Delta x^i dx = 0$, since for fixed boundaries the light ray is unique (no deviation). When F is dependent on time, there are two curves. The second term $\int_{x(A)}^{x(B)} \frac{\partial F}{\partial x^i} \Delta x^i dx$ caused by time-dependent metric is a higher-order term compared with $\int_{x(A)}^{x(B)} \frac{\partial F}{\partial t} \Delta t dx$, i.e.,

$$\int_{x(A)}^{x(B)} \frac{\partial F}{\partial x^i} \Delta x^i dx \ll \int_{x(A)}^{x(B)} \frac{\partial F}{\partial t} \Delta t dx \,. \tag{16}$$

Finally we substitute (13–15) into (6), and use (4). The unified formula is obtained as

$$\frac{\Delta\tau_B}{\Delta\tau_A} = \sqrt{\frac{G_B}{G_A}} \left(\frac{1 - F(A) \frac{\mathbf{k}_A \cdot \mathbf{v}_A}{|k_A|}}{1 - F(B) \frac{\mathbf{k}_B \cdot \mathbf{v}_B}{|k_B|}} \right)$$
$$+ \frac{\sqrt{G_B}}{\Delta\tau_A \left(1 - F(B) \frac{\mathbf{k}_B \cdot \mathbf{v}_B}{|k_B|}\right)} \int_{x(A)}^{x(B)} \left(\frac{\partial F}{\partial t} \Delta t + \frac{\partial F}{\partial x^i} \Delta x^i \right) dx \,. \tag{17}$$

As an example, we consider the Doppler effect of a moving source in Minkowski metric

$$g_{\mu\nu} = \eta_{\mu\nu} = \begin{pmatrix} -1 & 0 \\ 0 & \delta_{ij} \end{pmatrix} , \tag{18}$$

since $\mathbf{v}_B = 0$, $G_B = 1$, $G_A = 1 - \frac{v_A^2}{c^2}$, and $\frac{\partial F}{\partial t} = \frac{\partial F}{\partial x^i} = 0$, $F(A) = \frac{1}{c}$, so that

$$\frac{\Delta\tau_B}{\Delta\tau_A} = \frac{1 - \frac{\mathbf{v}}{c} \cdot \frac{\mathbf{k}_A}{|k_A|}}{\sqrt{1 - \frac{v_A^2}{c^2}}} \,. \tag{19}$$

This is just the formula of the Doppler effect in the special relativity.

The other simple example is the gravitational redshift. Considering a static gravitational field (e.g., Schwarzschild metric), in which both source and receiver without moving ($\mathbf{v}_A = \mathbf{v}_B = 0$), the unified form then becomes

$$\frac{\Delta\tau_B}{\Delta\tau_A} = \sqrt{\frac{-g_{00}(B)}{-g_{00}(A)}} \simeq 1 - \frac{w(B)}{c^2} + \frac{w(A)}{c^2} , \tag{20}$$

where the last step of above equation is the Newtonian limitation. Equation (20) is just the formula of gravitational redshift in ordinary textbooks of gravity.

3 Application in Cosmos with Perturbed R–W Metric

First we recall the unperturbed Robertson–Walker metric

$$ds^2 = -c^2 dt^2 + \frac{R(t)^2 \delta_{ij} dx^i dx^j}{\left(1 + \frac{k}{4} r^2\right)^2} , \tag{21}$$

where $R(t)$ is the cosmic scalar factor, $k = -1,\, 0,\, +1$ is corresponding to the open, flat, and closed universe, respectively. $R(t)$ has the dimension of length and dx^i is dimensionless. As we already know, $R(t)$ is model dependent. Since usually we do not consider the local gravitational redshift and the Doppler effect in the problem of cosmological expansion, we then have $\Delta t_A = \Delta\tau_A$ and $\Delta t_B = \Delta\tau_B$, and thus the formula for the Hubble redshift is

$$\frac{\Delta\tau_A}{\Delta\tau_B} = \frac{\Delta t_A}{\Delta t_B} = \frac{R(t_A)}{R(t_B)}\,. \tag{22}$$

The results of (22) can easily be deduced directly from the unified formula (17), if we take $v_A = v_B = w_A = w_B = 0$.

Next, we consider a linearly simplest perturbed Robertson–Walker metric of the form

$$ds^2 = -c^2\left(1 - \frac{2w}{c^2}\right)dt^2 + \left(1 + \frac{2w}{c^2}\right)\frac{R^2\delta_{ij}dx^i dx^j}{\left(1 + \frac{k}{4}r^2\right)^2}\,, \tag{23}$$

where the gravitational potential $w = w(t, x^i)$ is assumed to be a small quantity. Later we only consider the Doppler effect caused by the motion of the source, then $\mathbf{v}_B = 0$ (also possible $\mathbf{v}_A = 0$, then $\mathbf{v}_B \neq 0$). The velocity of the source A is

$$v_A^i = R(t_A)\frac{dx_A^i}{dt}\,. \tag{24}$$

$F(A)$, G_A, and G_B can be calculated as follows

$$F(A) = \frac{\left(1 + \frac{2w}{c^2}\right)\sqrt{R^2\delta_{ij}\frac{dx^i}{dx}\frac{dx^j}{dx}}}{c\left(1 + \frac{kr^2}{4}\right)} = \frac{R(t_A)}{c} + O(3)\,, \tag{25}$$

where we have neglected all of higher-order terms and consider $\delta_{ij}n^i n^j = 1$ and kr^2 as higher-order term also. Then

$$\sqrt{\frac{G_B}{G_A}} = 1 + \frac{1}{c^2}w(t_A, x_A^i) - \frac{1}{c^2}w(t_B, x_B^i) + \frac{v_A^2}{c^2}\,. \tag{26}$$

Now we calculate the last term of (17). Consider the integral in (17)

$$I \equiv \int_{x(A)}^{x(B)} \frac{\partial F}{\partial t}\Delta t dx\,, \tag{27}$$

where we have omitted the term of $\int_{x(A)}^{x(B)} \frac{\partial F}{\partial x^i}\Delta x^i dx$ because of (16). Therefore we have

$$I = \int_{x(A)}^{x(B)} \frac{\partial}{\partial t}\left[R(t)\left(1 + \frac{2w}{c^2}\right)\right]\frac{\Delta t\sqrt{\delta_{ij}dx^i dx^j}}{c\left(1 + \frac{k}{4}r^2\right)}\,. \tag{28}$$

From (22) we have $\Delta t = \Delta t_A R(t)/R(t_A)$, then (28) becomes

$$I = \int_{x(A)}^{x(B)} \frac{R(t)\Delta t_A}{R(t_A)} \left(\dot{R}(t) + \dot{R}(t)\frac{2w}{c^2} + R(t)\frac{\partial (2w/c^2)}{\partial t} \right) \frac{\sqrt{\delta_{ij}dx^i dx^j}}{c\left(1+\frac{k}{4}r^2\right)} . \quad (29)$$

Considering null geodetic line, (23) yields

$$\left(1 - \frac{w}{c^2}\right) dt = \pm \left(1 + \frac{w}{c^2}\right) \frac{R(t)\sqrt{\delta_{ij}dx^i dx^j}}{c\left(1+\frac{k}{4}r^2\right)} , \quad (30)$$

which we use to evaluate the integral I and get

$$\begin{aligned} I &= \frac{\Delta t_A}{R(t_A)} \int_{x(A)}^{x(B)} \left(\dot{R}(t) + 2\frac{R(t)}{c^2}\frac{\partial w}{\partial t} \right) dt \\ &= \Delta t_A \left[\left(\frac{R(t_B) - R(t_A)}{R(t_A)} \right) + \frac{2}{c^2 R(t_A)} \int_{x(A)}^{x(B)} R(t)\frac{\partial w}{\partial t} dt \right] . \end{aligned} \quad (31)$$

The second term of (17) then becomes

$$\sqrt{\frac{G_B}{G_A}} \left(\frac{R(t_B) - R(t_A)}{R(t_A)} + \frac{2}{c^2 R(t_A)} \int_{x(A)}^{x(B)} R(t)\frac{\partial w}{\partial t} dt \right) . \quad (32)$$

Substituting (25), (26), and (32) into (17) (pay attention to that, in cosmology the term $\frac{\mathbf{k}}{|k|} \cdot \mathbf{v}$ should be replaced by $\frac{\mathbf{k}}{|k|} \cdot \frac{\mathbf{v}}{R}$ in (11), (12), and (17)), we finally obtain

$$\begin{aligned} \frac{\Delta \tau_B}{\Delta \tau_A} &= \left[1 + \frac{1}{c^2}\left(w(t_A, x_A^i) - w(t_B, x_B^i) \right) + \frac{v_A^2}{c^2} \right] \\ &\quad \times \left\{ \frac{R(t_B)}{R(t_A)} - \frac{\mathbf{k}_A \cdot \mathbf{v}_A}{c|k_A|} + \frac{2}{c^2 R(t_A)} \int_{x(A)}^{x(B)} R(t)\frac{\partial w}{\partial t} dt \right\} , \end{aligned} \quad (33)$$

where $\frac{1}{c^2}\left(w(t_A, x_A^i) - w(t_B, x_B^i) \right)$ is the contribution from the normal gravitational redshift; $\frac{\mathbf{k}_A \cdot \mathbf{v}_A}{c|k_A|}$ and $\frac{v_A^2}{c^2}$ are the Doppler effect and transverse Doppler effect (or relativistic Doppler effect), respectively; $R(t_B)/R(t_A)$ just contributes to Hubble redshift; and the last term is related to Rees–Sciama effect [13–16]. If we put $\mathbf{v}_A = 0$ and $R(t) = R(T_A) = R(t_B)$, then

$$\frac{\Delta \tau_B}{\Delta \tau_A} = 1 + \frac{1}{c^2}\left(w(t_A, x_A^i) - w(t_b, x_B^i) \right) + \frac{2}{c^2} \int_{x(A)}^{x(B)} \frac{\partial w}{\partial t} dt , \quad (34)$$

where w is the same as u in [17] where $c = 1$ units is taken, and their results totally agree with ours. We thought that our unified formula allows to derive the Birkinshaw–Gull effect [18, 19] too, if we use suitable perturbation functions for $w(t, x^i)$ and $w_j(t, x^i)$. This will be discussed in another chapter.

4 Application in Solar System with DSX Metric

In near future, high-precision measurement will be done up to 2PN level as we mentioned before, thus allowing the coupling term (i.e., the term connecting the gravitational redshift, the Doppler redshift, and so on) to be measured. Our scheme (the unified form (17)) offers the possibility for this if an appropriate assumptions about the metric are used. Accordingly, in this section we start from DSX formalism [12,20,21] and its extension [11] and evaluate formula (17) for this metric. This extended DSX metric is described by

$$g_{00} = -\exp\left(-\frac{2w}{c^2}\right) + O(6)\,, \tag{35}$$

$$g_{0i} = -\frac{4w_i}{c^3} + O(5)\,, \tag{36}$$

$$g_{ij} = \delta_{ij}\exp\left(\frac{2w}{c^2}\right) + \frac{q_{ij}}{c^4} + O(6)\,, \tag{37}$$

$$g_{ij}g_{00} = -\delta_{ij} - \frac{q_{ij}}{c^4} + O(6)\,. \tag{38}$$

In fact, in the following calculation, q_{ij} appears only in the function $F(x^i, t)$, but in the final 2PN formula of clock rates (see (53)), q_{ij} does not exist which agree with the result in [7]. Substituting (35–38) into (17), calculating all of components, we can get a unified formula for the comparison of clock rates at 2PN level.

We begin by evaluating the terms G_A and G_B for the extended DSX metric:

$$\begin{aligned} G_A &= -g_{00}(A) - 2g_{0i}(A)\frac{v_A^i}{c} - g_{ij}(A)\frac{v_A^i v_A^j}{c^2} \\ &= 1 - \frac{2w(A)}{c^2} + \frac{2w^2(A)}{c^4} + \frac{8w_i(A)v_A^i}{c^4} - \frac{v_A^2}{c^2} - \frac{2w(A)v_A^2}{c^4} + O(6)\,, \end{aligned} \tag{39}$$

$$G_B = 1 - \frac{2w(B)}{c^2} + \frac{2w^2(B)}{c^4} + \frac{8w_i(B)v_B^i}{c^4} - \frac{v_B^2}{c^2} - \frac{2w(B)v_B^2}{c^4} + O(6)\,. \tag{40}$$

Since $\mathbf{k}_A \cdot \mathbf{v}_A/|k_A|$ and $\mathbf{k}_B \cdot \mathbf{v}_B/|k_B|$ are first order already, so $F(A)$ and $F(B)$ need to be calculated up to c^{-5} level. From (8), by using (35–38), we get

$$\begin{aligned} F(A) = \frac{1}{c}\Bigg\{1 + \frac{2w(A)}{c^2} - \frac{4w_i(A)}{c^3}\frac{dx^i}{dx}\bigg|_A + \frac{2w^2(A)}{c^4} \\ + \frac{1}{2c^4}q_{ij}(A)\left(\frac{dx^i}{dx}\frac{dx^j}{dx}\right)\bigg|_A\Bigg\} + O(6)\,, \end{aligned} \tag{41}$$

where we have neglected $g_{0i}g_{0j}(\sim O(6))$. Similarly

$$\begin{aligned} F(B) = \frac{1}{c}\Bigg\{1 + \frac{2w(B)}{c^2} - \frac{4w_i(B)}{c^3}\frac{dx^i}{dx}\bigg|_B + \frac{2w^2(B)}{c^4} \\ + \frac{1}{2c^4}q_{ij}(B)\left(\frac{dx^i}{dx}\frac{dx^j}{dx}\right)\bigg|_B\Bigg\} + O(6)\,. \end{aligned} \tag{42}$$

At last, we consider the integral (the second term) in (17)

$$\int_A^B \left(\frac{\partial F}{\partial t} \Delta t + \frac{\partial F}{\partial x^i} \Delta x^i \right) dx \,. \tag{43}$$

Because of (16) we will omit the second term in it. We only consider $\int_A^B \frac{\partial F}{\partial t} \Delta t dx$. As we know Δt_A at A and Δt_B at B, we could calculate Δt at an arbitrary point between A and B approximately. Then the definite integral can be evaluated by the median method, i.e.,

$$\int_A^B \frac{\partial F}{\partial t} \Delta t dx = \Delta \bar{t} \int_A^B \frac{\partial F}{\partial t} dx \,, \tag{44}$$

where $\Delta \bar{t}$ is the median value, for which we introduce a parameter η:

$$\Delta \bar{t} = \eta \Delta t_A \,. \tag{45}$$

η is a value closed to 1. For the term $\frac{\partial F}{\partial t}$ one finds

$$\frac{\partial F}{\partial t} = \frac{1}{c} \left(\frac{2}{c^2} \frac{\partial w}{\partial t} - \frac{4}{c^3} \frac{\partial w_i}{\partial t} \frac{dx^i}{dx} + \frac{4w}{c^4} \frac{\partial w}{\partial t} + \frac{1}{2c^4} \frac{\partial q_{ij}}{\partial t} \frac{dx^i}{dx} \frac{dx^j}{dx} \right) + O(6) \,. \tag{46}$$

In the solar system the change of the potential (the other metric much smaller) with time is very small (less then $O(2)$ level), only the leading term is considered. Therefore the term $\frac{2}{c^2} \int_A^B \frac{\partial w}{\partial t} dt$ is already on $O(4)$ level, but not $O(2)$. Furthermore, we have $dx = dl + O(2) = cdt + O(2)$. Then (46) simplifies to

$$\int_A^B \frac{\partial F}{\partial t} dx = \frac{2}{c^2} \int_A^B \frac{\partial w}{\partial t} dt + O(6) \,. \tag{47}$$

If we consider quick variable field (e.g., field in pulsar) we have to take (46) to substitute into (47).

The second integral term of (17) then becomes

$$\frac{\sqrt{G_B}}{\sqrt{G_A}} \frac{2\eta}{\left(1 - F(B)\frac{\mathbf{k}_B \cdot \mathbf{v}_B}{|k_B|}\right) c^2} \int_A^B \frac{\partial w}{\partial t} dt \,. \tag{48}$$

Gathering all evaluations done thus far in this section, we arrive at following general formula for the solar system:

$$\frac{\Delta \tau_B}{\Delta \tau_A} = \sqrt{\frac{G_B}{G_A}} \frac{1}{\left(1 - F(B)\frac{\mathbf{k}_B \cdot \mathbf{v}_B}{|k_B|}\right)} \left(1 - F(A) \frac{\mathbf{k}_A \cdot \mathbf{v}_A}{|k_A|} + \frac{2\eta}{c^2} \int_A^B \frac{\partial w}{\partial t} dt \right) . \tag{49}$$

In (49) we only consider the leading term, namely the scalar potential changing with the time, but in our scheme all of the second post-Newtonian terms (see

(46)) can be included in. Maybe in a system of binary pulsars, the higher-order terms in (46) are important. Formula (49) in static metric and in 1PN level agrees with the known formula [22].

We proceed by evaluating the remaining terms in (49). From (39) and (40), we find

$$\sqrt{G_B} = 1 - \frac{w(B)}{c^2} - \frac{v_B^2}{2c^2} + \frac{w^2(B)}{2c^4} + \frac{4w_i(B)v_B^i}{c^4} - \frac{3w(B)v_B^2}{2c^4} - \frac{v_B^4}{8c^4}, \quad (50)$$

$$\frac{1}{\sqrt{G_A}} = 1 + \frac{w(A)}{c^2} + \frac{v_A^2}{2c^2} + \frac{w^2(A)}{2c^4} - \frac{4w_i(A)v_A^i}{c^4} + \frac{5w(A)v_A^2}{2c^4} + \frac{3v_A^4}{8c^4}. \quad (51)$$

We also have

$$\begin{aligned}\left(1 - F(B)\frac{\mathbf{k}_B \cdot \mathbf{v}_B}{|k_B|}\right)^{-1} &= 1 + \frac{\mathbf{k}_B \cdot \mathbf{v}_B}{c|k_B|} + \frac{1}{c^2}\left(\frac{\mathbf{k}_B \cdot \mathbf{v}_B}{|k_B|}\right)^2 + \frac{2w(B)}{c^3}\frac{\mathbf{k}_B \cdot \mathbf{v}_B}{|k_B|} \\ &+ \frac{1}{c^3}\left(\frac{\mathbf{k}_B \cdot \mathbf{v}_B}{|k_B|}\right)^3 - \frac{4w_i(B)}{c^4}\left.\frac{dx^i}{dx}\right|_B \frac{\mathbf{k}_B \cdot \mathbf{v}_B}{|k_B|} \\ &+ \frac{4w(B)}{c^4}\left(\frac{\mathbf{k}_B \cdot \mathbf{v}_B}{|k_B|}\right)^2 + \frac{1}{c^4}\left(\frac{\mathbf{k}_B \cdot \mathbf{v}_B}{|k_B|}\right)^4. \end{aligned} \quad (52)$$

Substituting (41), (42), and (50–52) into (49), we finally have a unified formula for the comparison of clock rates in the solar system, on the 2PN level of precision

$$\begin{aligned}\frac{\Delta\tau_B}{\Delta\tau_A} = 1 &+ \left\{\frac{1}{c^2}\Big(w(A) - w(B)\Big) + \frac{1}{2c^2}\left(v_A^2 - v_B^2\right) - \left(\frac{\mathbf{k}_A \cdot \mathbf{v}_A}{c|k_A|} - \frac{\mathbf{k}_B \cdot \mathbf{v}_B}{c|k_B|}\right)\right. \\ &\left. - \frac{1}{c^2}\left(\frac{\mathbf{k}_B \cdot \mathbf{v}_B}{|k_B|}\right)\left(\frac{\mathbf{k}_A \cdot \mathbf{v}_A}{|k_A|} - \frac{\mathbf{k}_B \cdot \mathbf{v}_B}{|k_B|}\right)\right\} \\ &+ \frac{1}{c^3}\left\{\Big(w(B) - w(A)\Big)\left(\frac{\mathbf{k}_B \cdot \mathbf{v}_B}{c|k_B|} + \frac{\mathbf{k}_A \cdot \mathbf{v}_A}{c|k_A|}\right)\right. \\ &+ 2w(A)\left(\frac{\mathbf{k}_B \cdot \mathbf{v}_B}{c|k_B|} - \frac{\mathbf{k}_A \cdot \mathbf{v}_A}{c|k_A|}\right) \\ &- \frac{1}{2}\Big(v^2(B) - v^2(A)\Big)\left(\frac{\mathbf{k}_B \cdot \mathbf{v}_B}{c|k_B|} - \frac{\mathbf{k}_A \cdot \mathbf{v}_A}{c|k_A|}\right) \\ &\left. + \left(\frac{\mathbf{k}_B \cdot \mathbf{v}_B}{c|k_B|}\right)^2\left(\frac{\mathbf{k}_B \cdot \mathbf{v}_B}{c|k_B|} - \frac{\mathbf{k}_A \cdot \mathbf{v}_A}{c|k_A|}\right)\right\} \\ &+ \frac{1}{c^4}\left\{\frac{1}{2}\left(w(B) - w(A)\right)^2 + \frac{1}{2}\left(10w(A) - w(B)\right)v_A^2\right. \\ &- \frac{1}{2}\left(w(A) + 6w(B)\right)v_B^2 + \frac{1}{8}\left(3v_A^4 - 2v_A^2v_B^2 - v_B^4\right) \\ &+ 4\left(w_i(B)v_B^i - w_i(A)v_B^i - w_i(B)\frac{k_B^i}{|k_B|}\frac{\mathbf{k}_B \cdot \mathbf{v}_B}{c|k_B|} + w_i(A)\frac{k_A^i}{|k_A|}\frac{\mathbf{k}_A \cdot \mathbf{v}_A}{c|k_A|}\right)\end{aligned}$$

$$
\begin{aligned}
&+\left(\frac{\mathbf{k}_B\cdot\mathbf{v}_B}{c|k_B|}\right)^2\left(3w(B)+w(A)+\frac{1}{2}\left(v_A^2-v_B^2\right)\right)\\
&-2\left(\frac{\mathbf{k}_B\cdot\mathbf{v}_B}{c|k_B|}\right)\left(\frac{\mathbf{k}_A\cdot\mathbf{v}_A}{c|k_A|}\right)\left(w(A)+w(B)\right)\\
&+\left(\frac{\mathbf{k}_B\cdot\mathbf{v}_B}{c|k_B|}\right)^3\left(\frac{\mathbf{k}_B\cdot\mathbf{v}_B}{c|k_B|}-\frac{\mathbf{k}_A\cdot\mathbf{v}_A}{c|k_A|}\right)\Bigg\}\\
&+\frac{2\eta}{c^2}\int_A^B\frac{\partial w}{\partial t}dt+O(5)\,. \qquad (53)
\end{aligned}
$$

The Formula (53) "contains" the Doppler effect, transverse Doppler effect (relativistic Doppler effect), gravitational redshift, and their complete coupling effects to 2PN level in the solar system. In addition there is a term which is the integral of the rates of change of the scalar potential along the null geodetic line from source A to receiver B. This is probably the most interesting result in our chapter. Hopefully this integral term and the coupling effects can be tested in the future with a deep-space explorer and are confirmed.

5 Conclusion Remarks

We have synthesized all known effects for the comparison of clock rates in one formula (17). The synthesized formula contains additional coupling terms and a new integral terms and thus gives essential new but untested information. Therefore to get this synthesized formula is not an end in itself, but a starting point for the further test work. We hope that this work could contribute to the further comparison of clock rates, such as ACES mission planned in 2006. The general form may be taken as the basis for a starting point to compare clock rates at any two different space–time points. For example, the frequency shift caused by gravitomagnetic effect (or Lens–Thirring effect) can also be considered in our scheme.

The general form is valid not only for any metric gravitational theory, but also for general relativity. If we substitute the parametrized 2PN metric into the formula, (53) could include parameters. In Sect. 4, we have discussed the comparison of two clock rates both on the earth and a space station with 2PN precision. In fact, the calculation of the higher precision might be done in a similar way, if we know the metric to higher order.

Equation (17) or (49) says that the clock rates depend on the trajectory of the transmitted signal and on the metric (especially the scalar potential) varying with time. While (17) is a unified formula, (49) is valid only in the solar system; however if replace (47) by the integral of (46), a general 2PN formula results.

In the case of cosmology, the general form can be used for any linearly perturbed metric, in particular it allows to include the general Sachs–Wolfe effects.

Acknowledgments

This work was supported by the National Natural Science Foundation of China (Grant No. 10273008). In the workshop "Relativistic Astrophysics and High-Precision Astrodynamics" led by Prof. Wei-tou Ni, we had several very fruitful discussions, and especially we would like to thank Prof. Tianyi Huang for his useful discussion.

References

1. R. Holzwarth et al., Phys. Rev. Lett. **85**, 2264 (2000).
2. Th. Udem et al., Phys. Rev. Lett. **86**, 4996 (2001).
3. C. Salomon and C. Veillet, in Proceeding of the Symposium on Space Station Utilisation, ESA–SP, **385**, 295 (1996).
4. A. Spallicci et al., Class. Quantum Grav. **14**, 2971 (1997).
5. W.-T. Ni, Int. J. Mod. Phys. D **11**(7) 947 (2002).
6. W.-T. Ni, S. Shiomi and A.-C., Liao, Class. Quantum Grav. **21**, S641 (2004).
7. B. Linet and P. Teyssandier, Phys. Rev. D **66**, 024045 (2002).
8. C. Poncin-Lafitte, B. Linet and P. Teyssandier, Class. Quantum Grav. **21**, 4463 (2004).
9. E. Martinez-Gonzalez, J.L. Sanz and J. Silk, Astrophys. J. **355**, L5–L9 (1990).
10. S. Weinberg, Gravitation and Cosmology (Wiley) (1972).
11. C. Xu and X. Wu, Chin. Phys. Lett. **20**, 195 (2003).
12. T. Damour, M. Soffel and C. Xu, Phys. Rev. D **43**, 3273 (1991).
13. M. Rees and D.W. Sciama, Nature **217**, 511 (1968).
14. G.R. Blumenthal et al., Astrophys. J. **388**, 234 (1992).
15. R.B. Tully et al., Astrophys. J. **388**, 9 (1992).
16. A. Meszaros, Astrophys. J. **423**, 19 (1994).
17. R. Tuluie, P. Laguna and P. Anninos, Astrophys. J. **463**, 15 (1996).
18. M. Birkinshaw and S.F. Gull, Nature **302**, 315 (1983).
19. M. Birkinshaw, in Lecture Notes in Physics **330**, Gravitational Lenses (eds Moran JM, Hewitt JN and Lo KL, Springer, Berlin Heidelberg New York), p. 59 (1989).
20. T. Damour, M. Soffel and C. Xu, Phys. Rev. D **45**, 1017 (1992).
21. T. Damour, M. Soffel and C. Xu, Phys. Rev. D **47**, 3124 (1993).
22. R. Adler, M. Bazin and M. Schiffer, Introduction to General Relativity (McGraw-Hill Book) (1975).

Gravity Tests and the Pioneer Anomaly

Marc-Thierry Jaekel[1] and Serge Reynaud[2]

[1] Laboratoire de Physique Théorique de l'Ecole Normale Supérieure, CNRS, UPMC, 24 rue Lhomond, F75231 Paris Cedex 05, France
[2] Laboratoire Kastler Brossel, Université Pierre et Marie Curie, case 74, CNRS, ENS, Campus Jussieu, F75252 Paris Cedex 05, France

Summary. Experimental tests of gravity performed in the solar system show a good agreement with general relativity. The latter is, however, challenged by the Pioneer anomaly which might be pointing at some modification of gravity law at ranges of the order of the size of the solar system. We introduce a metric extension of general relativity which, while preserving the equivalence principle, modifies the coupling between curvature and stress tensors and, therefore, the metric solution in the solar system. The "post-Einsteinian extension" replaces Newton gravitation constant by two running coupling constants, which depend on the scale and differ in the sectors of traceless and traced tensors, so that the metric solution is characterized by two gravitation potentials. The extended theory has the capability to preserve compatibility with gravity tests while accounting for the Pioneer anomaly. It can also be tested by new experiments or, maybe, by having a new look at data of already performed experiments.

1 Introduction

Most gravitation tests performed in the solar system show a good agreement with general relativity (GR). In particular, the equivalence principle (EP), lying at the basis of GR, is one of the most accurately verified properties of nature [1]. This entails that the gravitational field has to be identified with the metric tensor $g_{\mu\nu}$ in a Riemannian space–time. Then, the parametrized post-Newtonian (PPN) formalism allows one to give a quantitative form to the agreement of observations with the metric tensor predicted by GR, through its confrontation with a family of more general metric solutions. Alternatively, GR can be tested by looking for hypothetical deviations of gravity force law from its standard form [2], as predicted by unification models although not observed up to now.

Besides these successes, GR is challenged by observations performed at galactic and cosmological scales. Anomalies have been known for some time to affect the rotation curves of galaxies. They are commonly accounted for by keeping GR as the theory of gravity at galactic scales but introducing unseen "dark matter" to reproduce the rotation curves [3, 4]. Anomalies have been seen more recently in the relation between redshifts and luminosities for type II supernovae. They are usually interpreted as an unexpected acceleration of cosmic expansion due to the presence of some "dark energy" of completely unknown origin [5]. As long as the "dark side" of the universe is not observed through other means, these galactic and cosmic anomalies may also be interpreted as deviations from GR occurring at large scales [6–8].

The Pioneer anomaly constitutes a new piece of information in this puzzling context, which may already reveal an anomalous behavior of gravity at scales of the order of the size of the solar system [9]. The anomaly was discovered when Doppler tracking data from the Pioneer 10/11 probes were analyzed during their travel to the outer parts of the solar system. After the probes had reached a quieter environment, after flying by Jupiter and Saturn, a precise comparison of tracking data with predictions of GR confirmed that the Doppler velocity was showing an anomaly varying linearly with elapsed time (see Fig. 8 of [10]). The deviation may be represented as an anomalous acceleration directed toward the Sun with an approximately constant amplitude over a large range of heliocentric distances (AU ≡ astronomical unit)

$$a_P = (0.87 \pm 0.13)\ \mathrm{nm\,s^{-2}}, \quad 20\ \mathrm{AU} \lesssim r \lesssim 70\ \mathrm{AU} \tag{1}$$

Though a number of mechanisms have been considered to this aim [11–14], the anomaly has escaped up to now all attempts of explanation as a systematic effect generated by the spacecraft itself or its environment. In particular, present knowledge of the outer part of the solar system does apparently preclude interpretations in terms of gravity [15] or drag effects [16] of ordinary matter. The inability of explaining the anomaly with conventional physics has given rise to a growing number of new theoretical propositions. It has also motivated proposals for new missions designed to study the anomaly and try to understand its origin [17]. The importance of the Pioneer anomaly for space navigation already justifies it to be submitted to further scrutiny while its potential impact on fundamental physics, especially on gravitation theory, can no more be neglected. The possibility that the Pioneer anomaly be the first hint of a modification of gravity law at large scales cannot be let aside investigations [18]. In this context, the compatibility of the Pioneer anomaly with other gravity tests appears to be a key question.

To discuss this point, we first recall that, though the interpretation of gravitation as the metric of space–time constitutes an extremely well-tested basis, the precise form of the coupling between space–time curvature and gravity sources can still be discussed [19]. Like the other fundamental interactions, gravitation may also be treated within the framework of field theory [20–22]. Radiative corrections, due to its coupling to other fields, then naturally lead to embed GR within the larger class of fourth-order theories [23–25]. Modifications are thus expected to appear [26–28] and they may affect large length scales [29–32]. This suggests to consider GR as an effective theory of gravity valid at the length scales for which it has been accurately tested but not necessarily at smaller or larger scales. Note that, in contrast to GR [33], fourth-order theories show renormalizability as well as asymptotic freedom at high energies [34]. Hence, they constitute a strong basis for extending the gravitation theory at scales not already constrained by experiments, for instance using renormalization group trajectories [35]. Renormalizability of these theories, however, comes with a counterpart, i.e., the problem of ghosts. It has, however, been convincingly argued that this problem does not constitute a definitive dead end for an effective field theory valid in a limited scale domain [36]. In particular, the departure from unitarity is expected to be negligible at ordinary scales tested in present-day universe [37].

In this chapter, we will review the main features of a phenomenological framework which has been recently developed [38–40]. It relies upon a theory of gravitation lying in the vicinity of GR, the deviation representing, for example, the radiative corrections due to the coupling of gravity with other fields [41]. It is presented below in its linearized form, where its significance is more easily given, and then in its full nonlinear version. It is also interpreted as an extension of the PPN Ansatz with the Eddington parameters β and γ being functions of heliospheric distances rather than mere constants. The extended framework is shown to have the ability to account for Pioneer anomaly while remaining compatible with other gravity tests. It also leads to the prediction of other anomalies related to Pioneer anomaly, which can be tested by new experiments or, in some cases, by having a new look at data of already performed experiments.

2 Gravity Tests in the Solar System

GR provides us with an excellent theoretical description of gravitational phenomena in the solar system. To discuss the experimental evidences in favor of this statement, we first recall a few basic features of this description.

To apply the principle of relativity to accelerated motions, Einstein [42,43] introduced what is now called the *equivalence principle*. A weak form of this principle is expressed by the universality of free fall, a property reflecting the universal coupling of all bodies to gravitation. With Einstein, this property acquires a geometrical significance, gravitation fields being identified with the metric tensor $g_{\mu\nu}$ while freely falling motions are geodesics of the associated space–time. Universality of free fall is then a consequence of the metric nature of gravitation theory.

The equivalence principle is one of the best ever tested properties of nature. Potential violations are usually parametrized by a relative difference η in the accelerations a_1 and a_2 undergone by two test bodies of different compositions in free fall at the same location. Modern experiments constrain the parameter η to stay below the 10^{-12} level. These experiments test the principle at distances ranging from the millimeter in laboratory experiments ([44] and references in) to the sizes of Earth–Moon [45] or Sun–Mars orbit [46,47].

To obtain GR, it remains to write the equations determining the metric tensor from the distribution of energy and momentum in space–time. GR corresponds to a particular choice of the form of the coupling between curvature tensor and stress tensor [48–50]: the Einstein curvature tensor $E_{\mu\nu}$ is simply proportional to the stress tensor $T_{\mu\nu}$, the proportionality constant being related to the Newton gravitation constant G_N inherited from classical physics

$$E_{\mu\nu} \equiv R_{\mu\nu} - \frac{1}{2} g_{\mu\nu} R = \frac{8\pi G_N}{c^4} T_{\mu\nu}. \tag{2}$$

Note that this relation accounts in a simple manner for the fact that $E_{\mu\nu}$ and $T_{\mu\nu}$ both have a null covariant divergence: the first property comes with Riemannian geometry (Bianchi identities) while the second one expresses conservation of energy and momentum and is a necessary and sufficient condition for motions of test masses to follow geodesics.

The metric tensor in the solar system is then deduced by solving the Einstein–Hilbert equation (2). Here we consider the simple case where the gravity source, i.e., the Sun, is described as a point-like motionless mass M so that the metric is simply written in terms of the Newton potential ϕ. The solution is conveniently written in terms of spherical coordinates (c denotes light velocity, t and r the time and radius, and θ and φ the colatitude and azimuth angles) with the gauge convention of isotropic spatial coordinates

$$\begin{aligned} ds^2 &= g_{00} c^2 dt^2 + g_{rr} \left(dr^2 + r^2 (d\theta^2 + \sin^2\theta d\varphi^2) \right) \\ g_{00} &= 1 + 2\phi + 2\phi^2 + \ldots, \qquad g_{rr} = -1 + 2\phi + \ldots \\ \phi &\equiv -\frac{\kappa}{r}, \qquad \kappa \equiv \frac{G_N M}{c^2}, \qquad |\phi| \ll 1 \end{aligned} \tag{3}$$

GR is usually tested through its confrontation with the family of PPN metric tensors introduced by Eddington [51] and then developed by several physicists [52–55]

$$g_{00} = 1 + 2\alpha\phi + 2\beta\phi^2 + \ldots, \qquad g_{rr} = -1 + 2\gamma\phi + \ldots \tag{4}$$

The three parameters α, β, and γ are constants, the first of which can be set to unity by redefining Newton constant G_N. Within the PPN family, GR corresponds to $\gamma = \beta = 1$. Anomalous values of γ or β differing from unity affect geodesic motions and can therefore be evaluated from a comparison of observations with predictions deduced from (4).

Experiments which have now been performed for more than four decades have led to more and more strict bounds on the anomalies $\gamma - 1$ and $\beta - 1$. For example, Doppler ranging on Viking probes in the vicinity of Mars [46] and deflection measurements using VLBI astrometry [56] or radar ranging on the Cassini probe [57] give smaller and smaller values of $\gamma - 1$, with presently a bound of a few 10^{-5}. Analysis of the precession of planet perihelions [58] and of the polarization by the Sun of the Moon orbit around the Earth [59] allows for the determination of linear superpositions of β and γ, resulting now to $\beta - 1$

smaller than a few 10^{-4}. These tests are compatible with GR with, however, a few exceptions, among which notably the anomalous observations recorded on Pioneer probes. We will see below that this contradiction between Pioneer observations and other gravity tests may be cured in an extended framework, thanks to the fact that the anomaly $\gamma - 1$ is no longer a constant but a function in this more general framework.

An alternative manner to test GR has been to check the $1/r$ dependence of the Newton potential ϕ, i.e., also of the component g_{00} in (3). Hypothetical modifications of its standard expression are usually parametrized in terms of an additional Yukawa potential depending on two parameters: the range λ and the amplitude α measured with respect to Newton potential ϕ

$$\Phi_N(r) = \phi(r)\left(1 + \alpha e^{-\frac{r}{\lambda}}\right). \tag{5}$$

The presence of such a Yukawa correction has been looked for at various distances ranging from the millimeter in laboratory experiments [44] to the size of planetary orbits [60]. The accuracy of short-range tests has been recently improved, as gravity experiments were pushed to smaller distances [61–63] and as Casimir forces, which become dominant at submillimeter range, were more satisfactorily taken into account [64–67]. On the other side of the distance range, long-range tests of the Newton law are performed by monitoring the motions of planets or probes in the solar system. They also show an agreement with GR with a good accuracy for ranges of the order of the Earth–Moon [45] or Sun–Mars distances [46,47,68,69]. When the whole set of results is reported on a global figure (see Fig. 1 in [70] reproduced, thanks to a courtesy of the authors of [60]), it appears that windows, however, remain open for violations of the standard form of Newton force law at short ranges, below the millimeter, as well as long ones, of the order of or larger than the size of the solar system.

One merit of the latter tests is to shed light on a potential scale dependence of violations of GR. As a specific experiment is only sensitive to a given range of distances, this has to be accounted for, especially in the context, recalled in Sect. 1, where doubts arise about the validity of GR at galactic or cosmic scales. To discuss this scale dependence, it is worth rewriting the Yukawa perturbation (5) in terms of a running constant replacing Newton gravitation constant. To this aim, we introduce the expression of the potential $\Phi_N[\mathbf{k}]$ in Fourier space, with $\mathbf{k}$ the spatial wave vector, and relate it to a coupling constant $\widetilde{G}_N[\mathbf{k}]$

$$-\mathbf{k}^2 \Phi_N[\mathbf{k}] \equiv 4\pi \frac{\widetilde{G}_N[\mathbf{k}]\, M}{c^2}, \qquad \widetilde{G}_N[\mathbf{k}] = G_N\left(1 + \alpha \frac{\mathbf{k}^2}{\mathbf{k}^2 + \frac{1}{\lambda^2}}\right) \tag{6}$$

The left-hand equation extends the standard Poisson equation ($-\mathbf{k}^2$ is the Laplacian operator), with the constant G_N replaced by a "running coupling constant" $\widetilde{G}_N[\mathbf{k}]$ which depends on spatial wave vector $\mathbf{k}$.

Note that the Yukawa correction (5) gives rise in the domain $r \ll \lambda$ to a perturbation $\delta\Phi_N$ which is linear in the distance $\sim -(\alpha/2\lambda^2)(\kappa r)$.

In (6) equivalently, the correction of the running constant scales as $\delta \widetilde{G}_N \sim (\alpha/\lambda^2)(G_N/\mathbf{k}^2)$ in the domain $|\mathbf{k}\lambda| \gg 1$. In fact, experimental constraints obtained at scales of the order of the size of the solar system can be written as bounds on the combination α/λ^2, so that they can be translated into bounds on the anomalous acceleration $\partial_r \delta \Phi_N$ or, equivalently, on $\mathbf{k}^2 \delta \widetilde{G}_N$. It is worth emphasizing that these bounds result in allowed anomalous accelerations which remain 2,000 times too small to account for the Pioneer anomaly [70]. In other words, the Pioneer anomaly cannot be due to a modification of Newton law, as such a modification would be much too large to remain unnoticed by planetary tests. Anew, this contradiction between Pioneer observations and other gravity tests may be cured by the extended framework studied below, now thanks to the fact that the existence of an anomalous space-dependent potential will not only be considered for the metric component g_{00}, but also for g_{rr}.

To summarize this section, tests performed on gravity in the solar system confirm its metric character and provide strong evidence in favor of gravitation theory being very close to GR. They, however, still leave room for alternative metric theories, which deviate from GR in a specific way. Anomalies in the metric components remain allowed, as long as they modify spatial dependencies without strongly affecting the time component g_{00}. It is shown in Sect. 3 that such extensions of GR may in fact arise naturally, in particular when effects of radiative corrections are taken into account.

3 Linearized Gravitation Theory

We come now to the description of the "post-Einsteinian" extension of GR. We first repeat that tests performed at various length scales have showed that the equivalence principle was preserved at a higher accuracy level (10^{-12}) than the EP violation which would be needed to account for the Pioneer anomaly. As a matter of fact, the standard Newton acceleration at 70 UA is of the order of 1 μm s^{-2} while the Pioneer anomaly is of the order of 1 nm s^{-2}, which would correspond to a violation level of the order of 10^{-3}. This does not mean that EP violations are excluded, and they are indeed predicted by unification models [71, 72]. However, any such violations are bound to occur at a lower level than needed to affect the Pioneer anomaly. Hence, EP violations will be ignored in the following and we shall restrict our discussion to a confrontation of GR with alternative metric theories.

Furthermore, neither PPN extensions of GR nor mere modifications of Newton force laws have the ability to account for the Pioneer anomaly (see Sect. 2). However, such extensions do not cover the totality of possible extensions of GR. In particular, there exist extended metric theories characterized by the existence of two gravitation potentials instead of a single one [38–40]. The first one merely represents a modified Newton potential while the second one can be understood in terms of a space-dependent PPN parameter γ. In this larger family of extensions, there is enough room available for accommodating the Pioneer anomaly while preserving compatibility with other gravity tests. Let us stress that this larger family is not introduced as an ad hoc solution to the Pioneer anomaly. It emerges as the natural extension of GR induced by radiative corrections due to the coupling of gravity with other fields, and some phenomenological consequences were explored [41] before noticing that they included Pioneer-like anomalies [38–40]. To present these ideas in a simple manner, we will start with the linearized version of gravitation theory, which is approximately valid for describing Pioneer-like probes having escape motions in the outer solar system [38, 39]. We will then present some salient features of the nonlinear theory [40].

In the linearized treatment, the metric field may be represented as a small perturbation $h_{\mu\nu}$ of Minkowski metric $\eta_{\mu\nu}$

$$\begin{aligned} g_{\mu\nu} &= \eta_{\mu\nu} + h_{\mu\nu} \\ \eta_{\mu\nu} &= \mathrm{diag}(1,-1,-1,-1), \quad |h_{\mu\nu}| \ll 1. \end{aligned} \tag{7}$$

The field $h_{\mu\nu}$ is a function of position x in space–time or, equivalently in Fourier space, of wave vector k

$$h_{\mu\nu}(x) \equiv \int \frac{d^4k}{(2\pi)^4} e^{-ikx} h_{\mu\nu}[k]. \tag{8}$$

Gauge-invariant observables of the metric theory, i.e., quantities which do not depend on a choice of coordinates, are given by curvature tensors. In the linearized theory, i.e., at first order in $h_{\mu\nu}$, Riemann, Ricci, scalar, and Einstein curvatures are written in momentum representation as

$$\begin{aligned} R_{\lambda\mu\nu\rho} &= \frac{1}{2}\left(k_\lambda k_\nu h_{\mu\rho} - k_\lambda k_\rho h_{\mu\nu} - k_\mu k_\nu h_{\lambda\rho} + k_\mu k_\rho h_{\lambda\nu}\right) \\ R_{\mu\nu} &= R^\lambda{}_{\mu\lambda\nu}, \quad R = R^\mu{}_\mu, \quad E_{\mu\nu} = R_{\mu\nu} - \tfrac{1}{2}\eta_{\mu\nu} R \end{aligned} \tag{9}$$

We use the sign conventions of [73], indices being raised or lowered using Minkowski metric.

These curvature fields are similar to the gauge-invariant electromagnetic fields of electrodynamics so that, while being supported by its geometrical interpretation, GR shows essential similarities with other field theories [21, 23]. This suggests that GR may be considered as the low energy effective limit of a more complete unified theory [27, 28] which should describe the coupling of gravity with other fields. In any case, this theory should contain radiative corrections to the graviton propagator, leading to a modification of gravitation equations (2) and to a momentum dependence of the coupling between curvature and stress tensors. In the weak-field approximation, it is easily seen that Einstein tensor, which is divergenceless, has a natural decomposition on the two sectors corresponding to different conformal weights [41], i.e., also on traceless (conformal weight 0) and traced components (conformal weight 1).

The general coupling between curvature and stress tensors can be written in terms of a linear response function constrained by the transversality condition

$$E_{\mu\nu}[k] = \chi_{\mu\nu\lambda\rho}[k]\; T^{\lambda\rho}[k], \quad k^{\mu}\chi_{\mu\nu\lambda\rho}[k] = 0. \tag{10}$$

We consider as above the isotropic and stationary situation with a point-like and motionless Sun of mass M. We then deduce that the general coupling (10) is described by two running constants $\tilde{G}^0$ and $\tilde{G}^1$, depending on the spatial wave vector $\mathbf{k}$ and living in the two sectors (0) and (1), so that gravitation equations (10) become [38,39]

$$\begin{aligned} E_{\mu\nu}[k] &\equiv 2\pi\delta(k_0)E_{\mu\nu}[\mathbf{k}], \qquad \pi_{\mu\nu} \equiv \eta_{\mu\nu} - \frac{k_\mu k_\nu}{k^2} \\ E_{\mu\nu}[\mathbf{k}] &= \pi^0_\mu \pi^0_\nu \tilde{G}^0[\mathbf{k}]\frac{8\pi M}{c^2} + \pi_{\mu\nu}\pi^{00}\,\frac{\tilde{G}^1[\mathbf{k}] - \tilde{G}^0[\mathbf{k}]}{3}\,\frac{8\pi M}{c^2}. \end{aligned} \tag{11}$$

The Newton gravitation constant G_N in (2) has been replaced in (11) by two running coupling constants $\tilde{G}^0$ and $\tilde{G}^1$ which are related through Poisson-like equations to two potentials Φ^0 and Φ^1 (compare with (6))

$$-\mathbf{k}^2\Phi^a[\mathbf{k}] = \tilde{G}_a[\mathbf{k}]\frac{4\pi M}{c^2}, \quad a = 0, 1. \tag{12}$$

These two potentials determine the metric, i.e., the solution of the modified equations (11), written here with spatial isotropic coordinates

$$\begin{aligned} g_{00} &= 1 + 2\Phi_N\,, & \Phi_N &\equiv \frac{4\Phi^0 - \Phi^1}{3} \\ g_{rr} &= -(1 - 2\Phi_N + 2\Phi_P)\,, & \Phi_P &\equiv \frac{2(\Phi^0 - \Phi^1)}{3}, \end{aligned} \tag{13}$$

where Φ_N is defined from the difference $(g_{00} - 1)$ and identified as an extended Newton potential while Φ_P is defined from $(-g_{00}g_{rr} - 1)$ and interpreted as measuring the difference between the potentials Φ^0 and Φ^1 in the two sectors of traceless and traced curvatures. As Φ^0 and Φ^1, Φ_N and Φ_P obey Poisson equations with running constants $\tilde{G}_N$ and $\tilde{G}_P$ written as linear combinations of $\tilde{G}^0$ and $\tilde{G}^1$

$$\begin{aligned} -\mathbf{k}^2\Phi_a[\mathbf{k}] &= \tilde{G}^a[\mathbf{k}]\frac{4\pi M}{c^2}, \quad a = N, P \\ \tilde{G}_N &\equiv \frac{4\tilde{G}^0 - \tilde{G}^1}{3}, \quad \tilde{G}_P \equiv \frac{2(\tilde{G}^0 - \tilde{G}^1)}{3}. \end{aligned} \tag{14}$$

Standard Einstein equation is recovered when the running constants $\tilde{G}^0$ and $\tilde{G}^1$ are momentum independent and equal to each other, i.e., also when

$$\begin{aligned} [\tilde{G}_N]_{\mathrm{st}} &\equiv G_N & [\tilde{G}_P]_{\mathrm{st}} &= 0 \\ [\Phi_N(r)]_{\mathrm{st}} &\equiv \phi(r) & [\Phi_P(r)]_{\mathrm{st}} &= 0. \end{aligned} \tag{15}$$

The two potentials Φ_a will be written as sums of these standard expressions and anomalies which, according to the discussions of Sect. 3, will remain small

$$\Phi_a(r) \equiv [\Phi_a(r)]_{\rm st} + \delta\Phi_a(r), \quad |\delta\Phi_a(r)| \ll 1. \tag{16}$$

4 Nonlinear Gravitation Theory

Before embarking in the discussion of phenomenological consequences of these anomalous potentials, let us recall briefly that the linearized theory presented in the Sect. 3 can be transformed into a full nonlinear theory. The linearized theory will indeed be sufficient to discuss the anomalous acceleration of Pioneer probes as well as potential effects on light-like waves [38, 39] but the nonlinear theory will be needed to address the case of planetary tests [40]. The relation between metric and curvature tensors now takes a nonlinear form going beyond (9) which was valid only at first order around Minkowski space–time. We may, nevertheless, simplify this relation by working at first order in deviations from standard Einstein theory [40].

To this aim, we write the metric, now in terms of Schwarzschild coordinates [74]

$$\begin{aligned} ds^2 &= \bar{g}_{00}(\bar{r})c^2dt^2 + \bar{g}_{rr}(\bar{r})d\bar{r}^2 - \bar{r}^2\left(d\theta^2 + \sin^2\theta d\varphi^2\right) \\ \bar{g}_{\mu\nu}(r) &\equiv [\bar{g}_{\mu\nu}(r)]_{\rm st} + \delta\bar{g}_{\mu\nu}(r), \quad |\delta\bar{g}_{\mu\nu}(r)| \ll 1. \end{aligned} \tag{17}$$

The standard GR solution is then treated exactly

$$[\bar{g}_{00}]_{\rm st} = 1 - 2\kappa\bar{u} = -\frac{1}{[\bar{g}_{rr}]_{\rm st}}, \quad \bar{u} \equiv \frac{1}{\bar{r}}. \tag{18}$$

while the anomalous metric components are taken into account at first order. Proceeding in this manner, it is possible to define in the nonlinear theory two potentials $\delta\bar{\Phi}_N$ and $\delta\bar{\Phi}_P$ which generalize (16)

$$\begin{aligned} \delta\bar{g}_{rr} &= \frac{2\bar{u}}{(1-2\kappa\bar{u})^2}(\delta\bar{\Phi}_N - \delta\bar{\Phi}_P)', \quad f' \equiv \partial_{\bar{u}} f \\ \delta\bar{g}_{00} &= 2(1-2\kappa\bar{u})\int \frac{\delta\bar{\Phi}_N{}' - 2\kappa\bar{u}\delta\bar{\Phi}_P{}'}{(1-2\kappa\bar{u})^2} d\bar{u}. \end{aligned} \tag{19}$$

In the linearized approximation, corrections in $\kappa\bar{u}$ are disregarded and the simple relations of Sect. 3 are recovered. In the general case, (19) fully describe nonlinear effects of the Newton potential $\kappa\bar{u}$. The precise form of the nonlinear version (19) has been chosen so that potentials are related in a simple way to the corresponding anomalous Einstein curvatures

$$\begin{aligned} \delta\bar{E}_0^0 &\equiv 2\bar{u}^4(\delta\bar{\Phi}_N - \delta\bar{\Phi}_P)'' \\ \delta\bar{E}_r^r &\equiv 2\bar{u}^3\delta\bar{\Phi}_P'. \end{aligned} \tag{20}$$

At this stage, it is worth noticing that the PPN metric [1] may be recovered as a particular case of the more general extension (19). This particular case corresponds to the following expressions of anomalous potentials and anomalous Einstein curvatures [40]

$$\begin{aligned} \delta\bar{\Phi}_N &= (\beta-1)\kappa^2\bar{u}^2 + O(\kappa^3\bar{u}^3), \quad \text{[PPN]} \\ \delta\bar{\Phi}_P &= (\gamma-1)\kappa\bar{u} + O(\kappa^2\bar{u}^2) \end{aligned} \tag{21}$$

$$\begin{aligned} \delta E_0^0 &= \bar{u}^2 O(\kappa^2\bar{u}^2), \quad \text{[PPN]} \\ \delta E_r^r &= \bar{u}^2\left(2(\gamma-1)\kappa\bar{u} + O(\kappa^2\bar{u}^2)\right). \end{aligned} \tag{22}$$

Note that the PPN metric already shows an anomalous behavior of Einstein curvatures which have non-null values apart from the gravity source. This is the case for δE_r^r at first order in κ, and for δE_0^0 at higher orders. Relations (20) thus extend this anomalous behavior to more general dependences of the curvatures δE_0^0 and δE_r^r. Similar statements apply as well for anomalous potentials $\delta\bar{\Phi}_N$ and $\delta\bar{\Phi}_P$, which generalize the specific dependence of PPN potentials (21) where $\beta-1$ and $\gamma-1$ are constants. In other words, the post-Einsteinian metric (19) can be thought of as an extension of PPN metric where $\beta-1$ and $\gamma-1$ are no longer constants but rather functions of space.

5 Phenomenological Consequences

As already discussed, the new phenomenological framework is characterized by two anomalous potentials: the first one $\delta\bar{\Phi}_N$ is a modification of Newton potential while the second one $\delta\bar{\Phi}_P$ represents the difference of gravitational couplings in the two sectors of traceless and traced curvatures. The first potential is not able by itself to explain the Pioneer anomaly: its anomalous part is indeed bound by planetary tests to be much smaller than would be needed to account for the Pioneer anomaly [70]. This is why we will focus the attention in the following on the second potential which can produce a Pioneer-like anomaly for probes on escape trajectories in the outer solar system [38, 39]. This second potential can also be understood as promoting the PPN parameter γ to the status of a space-dependent function and it has therefore other consequences which have to be evaluated with great care. It is clear that the modification of GR needed to produce the Pioneer anomaly should not spoil its agreement with other gravity tests.

We first discuss the effect of the second potential on Doppler tracking of Pioneer-like probes. To this aim, we calculate the Doppler velocity taking into account the perturbations on probe motions as well as on light propagation between stations on Earth and probes. We then write the time derivative of this velocity as an acceleration a and finally subtract the expression obtained in standard theory from that obtained in extended one. We thus obtain the prediction of the post-Einsteinian extension [38, 39] for the Pioneer anomalous acceleration $\delta a \equiv a - [a]_{\rm st}$. In a configuration similar to that of Pioneer 10/11 probes, which follow nearly radial trajectories with a kinetic energy much larger than their potential energy, this anomalous acceleration takes the simplified form

$$\delta a \simeq 2 \frac{d\delta\Phi_P}{dr} v_P^2. \tag{23}$$

Thus, an anomaly in Doppler tracking of Pioneer-like probes is a direct consequence of the presence of the second potential $\delta\Phi_P$. Note that the anomalous acceleration comes out as proportional to the kinetic energy, which is a remarkable prediction of the new framework. Data on probes with very different kinetic energies (unfortunately not available at the moment) could thus be used to confirm or infirm this prediction.

Using the known velocity of the Pioneer probes ($v_P \sim 12\,{\rm km\,s^{-1}}$), and identifying the acceleration (23) with the recorded Pioneer anomaly (1), we deduce the value of the derivative $d\delta\Phi_P/dr$ in the outer solar system. The constancy of recorded anomaly over a large range of distances agrees with a simple parametrization of the second potential [38]

$$\delta\Phi_P(r) \equiv -\frac{G_P M}{rc^2} + \zeta_P M \frac{r}{c^2}$$
$$\left|\frac{G_P}{G_N}\right| \ll 1, \quad \zeta_P M \sim 0.25 \text{ m s}^{-2}. \tag{24}$$

This value of the parameter $\zeta_P M$ is much larger than that allowed for the parameter $\zeta_N M$ which could be defined on the first potential [38]. This shows in a clear manner how the second potential $\delta\Phi_P$ opens the possibility to account for the Pioneer anomaly. It has to be kept in mind for the forthcoming discussions that the simple model (24) does not need to be exact in the whole solar system. However, the expression of $\delta\Phi_P$ can generally be given the form (24), provided ζ_P denotes a function of the heliocentric distance.

We come now to the discussion of the effects of the second potential $\delta\Phi_P$ on the propagation of light rays, which can be done in the linearized theory. Considering in particular deflection experiments usually devoted to the determination of Eddington parameter γ, we obtain the following expression for the anomaly $\delta\psi$ (with respect to GR) of the deflection angle of rays grazing the surface of the Sun [39] (the same decomposition as in (24) is used for $\delta\Phi_P$)

$$\delta\psi \simeq -\kappa \frac{\partial}{\partial\rho}\left(\delta\gamma(\rho)\ \ln\frac{4r_1 r_2}{\rho^2}\right)$$
$$\delta\gamma(\rho) = -\frac{G_P}{G_N} + \frac{\zeta_P(\rho)\rho^2}{2G_N}. \tag{25}$$

Terms which are not amplified near occultation have been neglected; r_1 and r_2 correspond to the heliocentric distances of the receiver and emitter, and ρ is the distance of closest approach of the light ray to the Sun; and $\delta\gamma(\rho)$ is a range-dependent anomalous part in Eddington parameter γ. These expressions are reduced to PPN ones when the function ζ_P vanishes. Otherwise, they show that Eddington deflection tests could reveal the presence of $\delta\Phi_P$ through a space dependence of the parameter γ.

We conclude this survey of phenomenological consequences of the new framework by discussing planetary tests and, in particular, those involving the perihelion precession of planets. As the latter are known to depend on the two PPN parameters β and γ, this discussion has to be presented in the context of the nonlinear theory. Still focusing the attention on the effects of the second potential, one thus obtains the following expression for the anomaly $\delta\Delta\varphi$ (with respect to GR) of the perihelion precession [40]

$$\frac{\delta\Delta\varphi}{2\pi} \simeq \bar{u}\left(\bar{u}\delta\bar{\Phi}_P\right)'' + \frac{e^2\bar{u}^2}{8}\left(\bar{u}^2\delta\bar{\Phi}_P{}'' + \bar{u}\delta\bar{\Phi}_P{}'\right)'' . \tag{26}$$

This expression has been truncated after leading and subleading orders in the eccentricity e; the function $\delta\bar{\Phi}_P$ and its derivatives have thus to be evaluated at the inverse radius $\bar{u}$ of the nearly circular ($e \ll 1$) planetary orbit. As the leading order vanishes for a contribution $\zeta_P Mr/c^2$ with ζ_P constant, the main result is thus proportional to e^2 in this case. This means that perihelion precession of planets could be used as a sensitive probe of the value and variation of ζ_P for distances corresponding to the radii of planetary orbits [40].

6 Discussion

Gravity tests which have been performed up to now in the solar system firmly support the validity of the equivalence principle, i.e., also the metric nature of gravitation. They also strongly indicate that the actual gravitation theory should be very close to GR. Nonetheless, these tests still leave room for alternative metric theories of gravitation, and the anomaly observed on the trajectories of the Pioneer 10/11 probes may well be a first indication of a modification of gravity law in the outer part of the solar system. This possibility would have such a large impact on fundamental physics, astrophysics, and maybe cosmology that it certainly deserves further investigations. We have discussed in the present chapter a new extension of GR which allows one to address these questions in a well-defined theoretical framework [39, 40].

When its radiative corrections are taken into account, GR appears as imbedded in a family of metric theories characterized, at the linearized level, by two running coupling constants which replace the single Newton gravitation constant or, equivalently, by two potentials which replace the standard

Newton potential. When applied to the solar system, this post-Einsteinian extension of GR leads to a phenomenological framework which has the ability to make the Pioneer anomaly compatible with other gravity tests. Precisely, the first potential Φ_N remains close to its standard Newtonian form to fit planetary tests but the second potential Φ_P opens a phenomenological freedom which can be understood as an Eddington parameter γ differing from unity, as in PPN metric, with now a possible space dependence.

To confirm, or infirm, the pertinence of this framework with respect to gravity tests in the solar system, it is now necessary to reanalyze the motions of massive or massless probes in this new context. Contrarily to what has been done here, it is particularly important to take into account the effects of anomalous potentials $\delta\Phi_N$ and $\delta\Phi_P$ simultaneously. Let us scan in the last paragraphs of this paper some ideas which look particularly promising.

The main novelty induced by the second potential $\delta\Phi_P$ is to produce an anomaly on Doppler tracking of Pioneer-like probes having highly eccentric motions in the outer solar system. As already discussed, if the recorded anomaly is identified with this effect, one deduces the value of the derivative $d\delta\Phi_P/dr$ of the second potential at the large distances explored by Pioneer probes. A natural idea is therefore to confront the more detailed prediction deduced from the new theory [39] against the larger set of data which will soon be available [75, 76]. It is particularly clear that the eccentricity of the orbits plays a key role in the evaluation of the Pioneer anomaly: it takes large values for Pioneer-like motions which sense $\delta\Phi_P$ whereas it is zero for circular orbits which do not. This suggests to devote a dedicated analysis to the intermediate situation, not only for the two categories of bound and unbound orbits, but also for the flybies used to bring Pioneer-like probes from the former category to the latter one. While Pioneer probes may sense the second potential at the large heliocentric distances they are exploring, planets or planetary probes may feel its presence at distances of the order of the astronomical unit. Then, it would be worth studying planetary probes on elliptical orbits, for example on transfer orbits from Earth to Mars or Jupiter. Another natural target for such a study is LISA with its three crafts on elliptical orbits [77].

The second potential $\delta\Phi_P$ also affects the propagation of light waves, and it could thus be detected as a range dependence of the anomalous Eddington parameter $(\gamma - 1)$ to be seen, for example, in deflection experiments. This might already be attainable through a reanalysis of existing data, given by the Cassini experiment [57], VLBI measurements [56], or HIPPARCOS [78]. It may also be reached in the future by higher accuracy Eddington tests, as for example the LATOR project [79], or global mapping of deflection over the sky (GAIA project [80]). Reconstruction of the dependence of γ vs. the impact parameter ρ would directly provide the space dependence of the second potential $\delta\Phi_P$. This would then either produce a clear signature of the new framework presented in this chapter or put constraints on the presence of the second potential at small heliocentric distances.

References

1. Will C.M., *Theory and experiment in gravitational physics* (Cambridge U. P., Cambridge, 1993); *Living Rev. Rel.* **4** (2001) 4.
2. Fischbach E. and Talmadge C., *The Search for Non Newtonian Gravity* (Springer, Berlin, 1998).
3. Aguirre A., Burgess C.P., Friedland A. and Nolte D., *Class. Quantum Grav.* **18** (2001) R223.
4. Riess A.G., Filippenko A.V., Challis P. *et al*, *Astron. J.* **116** (1998) 1009.
5. Perlmutter S., Aldering G., Goldhaber G. *et al*, *Astrophys. J.* **517** (1999) 565; Perlmutter S., Turner M.S. and White M., *Phys. Rev. Lett.* **83** (1999) 670.
6. Sanders R.H. and McGaugh S.S., *Annu. Rev. Astron. Astrophys.* **40** (2002) 263.
7. Lue A., Scoccimarro R. and Starkman G., *Phys. Rev.* **D 69** (2004) 044005.
8. Carroll S.M., Duvvuri V., Trodden M. and Turner M.S., *Phys. Rev.* **D 70** (2004) 043528.
9. Anderson J.D., Laing P.A., Lau E.L. *et al*, *Phys. Rev. Lett.* **81** (1998) 2858.
10. Anderson J.D., Laing P.A., Lau E.L. *et al*, *Phys. Rev.* **D 65** (2002) 082004.
11. Anderson J.D., Nieto M.M. and Turyshev S.G., *Int. J. Mod. Phys.* **D11** (2002) 1545.
12. Anderson J.D., Lau E.L., Turyshev S.G. *et al*, *Mod. Phys. Lett.* **A17** (2003) 875.
13. Nieto M.M. and Turyshev S.G., *Class. Quantum Grav.* **21** (2004) 4005.
14. Turyshev S.G., Nieto M.M. and Anderson J.D., *35th COSPAR Scientific Assembly* (2004) [gr-qc/0409117].
15. Nieto M.M., *Phys. Rev.* **D72** (2005) 083004.
16. Nieto M.M., Turyshev S.G. and Anderson J.D., *Phys. Lett.B* **613** (2005) 11.
17. The Pioneer Explorer Collaboration: H. Dittus, S.G. Turyshev *et al*, *ESA Spec. Publ.* **588** (2005) 3.
18. Bertolami O. and Paramos J., *Class. Quantum Grav.* **21** (2004) 3309; see also [astro-ph/0408216] and [gr-qc/0411020].
19. Weinberg S., *Gravitation and Cosmology* (Wiley, New York, 1972).
20. Thirring W.E., *Ann. Phys.* **16** (1961) 96.
21. Feynman R.P., *Acta Phys. Polonica* **24** 711 (1963).
22. Weinberg S., *Phys. Rev.* **138** B988 (1965).
23. Utiyama R. and De Witt B., *J. Math. Phys.* **3** (1962) 608.
24. Deser S. and van Nieuwenhuizen P., *Phys. Rev.* **D 10** (1974) 401.
25. Capper D.M., Duff M.J. and Halpern L., *Phys. Rev.* **D 10** (1974) 461.
26. Stelle K.S., *Phys. Rev.* **D 16** (1977) 953; *Gen. Rel. Grav.* **9** (1978) 353.
27. Sakharov A.D., *Doklady Akad. Nauk SSSR* **177** (1967) 70 [*Sov. Phys. Doklady* **12** 1040].
28. Adler R.J., *Rev. Modern Phys.* **54** (1982) 729.
29. Goldman T., Pérez-Mercader J., Cooper F. and Nieto M.M., *Phys. Lett.* **B 281** 219.
30. Deffayet C., Dvali G., Gabadadze G. and Vainshtein A., *Phys. Rev.* **D 65** (2002) 044026.
31. Dvali G., Gruzinov A. and Zaldarriaga M., *Phys. Rev.* **D 68** (2003) 024012.
32. Gabadadze G. and Shifman M., *Phys. Rev.* **D 69** (2004) 124032.
33. t'Hooft G. and Veltman M., *Ann. Inst. H. Poincaré* **A 20** (1974) 69.
34. Fradkin E.S. and Tseytlin A.A., *Nucl. Phys.* **B 201** (1982) 469.

35. Lauscher O. and Reuter M., *Class. Quantum Grav.* **19** (2002) 483.
36. Simon J.Z., *Phys. Rev.* **D 41** (1990) 3720.
37. Hawking S.W. and Hertog T., *Phys. Rev.* **D 65** (2002) 103515.
38. Jaekel M.-T. and Reynaud S., *Mod. Phys. Lett.* **A20** (2005) 1047.
39. Jaekel M.-T. and Reynaud S., *Class. Quantum Grav.* **22** (2005) 2135.
40. Jaekel M.-T. and Reynaud S., *Class. Quantum Grav.* **23** (2006), 777.
41. Jaekel M.-T. and Reynaud S., *Annalen der Physik* **4** (1995) 68.
42. Einstein A., *Jahrbuch der Radioaktivität und Elektronik* **4** (1907) 411.
43. Einstein A., *Annalen der Physik* **35** (1911) 898.
44. Adelberger E.G., Heckel B.R. and Nelson A.E., *Ann. Rev. Nucl. Part. Sci.* **53** (2003) 77.
45. Williams J.G., Newhall X.X. and Dickey J.O., *Phys. Rev.* **D53** (1996) 6730.
46. Hellings R.W., Adams P.J., Anderson J.D. *et al*, *Phys. Rev. Lett.* **51** (1983) 1609.
47. Anderson J.D., Gross M., Nordtvedt K.L. and Turyshev S.G., *Astrophys. J.* **459** (1996) 365.
48. Einstein A., *Sitzungsberichte der Preussischen Akademie der Wissenschaften zu Berlin* (1915) 844.
49. Hilbert D., *Nachrichten von der Gesellshaft der Wissenshaften zu Göttingen* (1915) 395.
50. Einstein A., *Annalen der Physik* **49** (1916) 769.
51. Eddington A.S., *The mathematical theory of relativity* (Cambridge U.P., Cambridge, 1957).
52. Robertson H.P., in *Space age astronomy* (Academic Press, 1962).
53. Ross D.H. and Schiff L.I., *Phys. Rev.* **141** (1966) 1215.
54. Nordtvedt K. Jr., *Phys. Rev.* **169** (1968) 1014 & 1017.
55. Will C.M. and Nordtvedt K., *Astrophys. J.* **177** (1972) 757; Nordtvedt K. and Will C.M., *ibidem* 775.
56. Shapiro S.S., Davis J.L., Lebach D.E. and Gregory J.S., *Phys. Rev. Lett.* **92** (2004) 121101.
57. Bertotti B., Iess L. and Tortora P., *Nature* **425** (2003) 374.
58. Talmadge C. *et al*, *Phys. Rev. Lett.* **61** (1988) 1159.
59. Nordtvedt K. Jr., [gr-qc/0301024].
60. Coy J., Fischbach E., Hellings R., Talmadge C. and Standish E.M., private communication (2003).
61. Hoyle C.D. *et al*, *Phys. Rev. Lett.* **86** (2001) 1418; *Phys. Rev.* **D70** (2004) 042004.
62. Long J. *et al*, *Nature* **421** (2003) 922.
63. Chiaverini J. *et al*, *Phys. Rev. Lett.* **90** (2003) 151101.
64. Bordag M., Mohideen U. and Mostepanenko V.M., *Phys. Rep.* **353** (2001) 1.
65. Lambrecht A. and Reynaud S., in *Poincaré Seminar 2002*, ed. B. Duplantier and V. Rivasseau (Birkhaüser Verlag, Basel, 2003), p. 109.
66. Decca R.S. *et al*, *Phys. Rev.* **D68** (2003) 116003.
67. Chen F., *et al*, *Phys. Rev.* **A69** (2004) 022117.
68. Reasenberg R.D. *et al*, *Astrophys. J.* **234** (1979) L219.
69. Kolosnitsyn N.I. and Melnikov V.N., *Gen. Rel. Grav.* **36** (2004) 1619.
70. Jaekel M.-T. and Reynaud S., *Int. J. Mod. Phys.* **A20** (2005) 2294.
71. Damour T., *Class. Quantum Grav.* **13** (1996) A33; Damour T., Piazza F. and Veneziano G., *Phys. Rev.* **D 66** (2002) 046007.

72. Overduin J.M., *Phys. Rev.* **D62** (2000) 102001.
73. Landau L.D. and Lifschitz E.M., *The classical theory of fields* (Butterworths, London, 1975).
74. Misner C.W., Thorne K.S. and Wheeler J.A., Gravitation (Freeman, New York, 1972).
75. Nieto M.M. and Anderson J.D., *Class. Quantum Grav.* **22** (2005) 5343.
76. Turyshev S.G., *Recent progress in the study of the Pioneer anomaly* (GREX meeting, 2005) `http://www.spectro.jussieu.fr/GREX/Paris05/Talks/Turyshev.pdf`.
77. LISA Site @ ESA `http://www.rssd.esa.int/Lisa`.
78. HIPPARCOS Site @ ESA `http://www.rssd.esa.int/Hipparcos`.
79. LATOR Collaboration: S.G. Turyshev, H. Dittus *et al*, *ESA Spec. Publ.* **588** (2005) 11.
80. GAIA Site @ ESA `http://www.rssd.esa.int/Gaia`.

Laser Ranging Delay in the Bimetric Theory of Gravity

Sergei M. Kopeikin[1] and Wei-Tou Ni[2]

[1] Department of Physics & Astronomy, University of Missouri-Columbia, Columbia, MO 65211, USA
[2] Purple Mountain Observatory, Chinese Academy of Sciences, Nanjing, 210008, China

Summary. We introduce a linearized bimetric theory of gravity with two metrics. The metric $g_{\alpha\beta}$ describes null hypersurfaces of the gravitational field while light moves on null hypersurfaces of the *optical* metric $\bar{g}_{\alpha\beta}$. Bimetrism naturally arises in vector–tensor theories with matter being nonminimally coupled to gravity via long-range vector field. We derive explicit Lorentz-invariant solution for a light ray propagating in space–time of the bimetric theory and disentangle relativistic effects associated with the existence of the two metrics. This analysis may be valuable for future spaceborne laser missions ASTROD and LATOR dedicated to map various relativistic gravity parameters in the solar system to unparalleled degree of accuracy.

Recently Carlip [1] has introduced a bimetric theory of gravity with two metrics, $g_{\alpha\beta}$ and $\bar{g}_{\alpha\beta}$, and a unit vector field w^α coupled to matter via constant parameter ϵ. This theory is a variant of a vector–tensor theory [2] where the vector field w^α obeys the source-free field equations and is responsible for the spontaneous violation of the Lorentz invariance of gravity [3, 4] in the sense that it introduces a preferred frame which effects can be observed only in gravitational experiments conducted in nonnegligible gravitational field. One can show that in the case of a bimetric theory of gravity the Huges–Drever experiments and other precision experiments constrain the asymptotic (vanishing gravity) difference between these two metrics severely [5–7]. Here we consider more complicated (nonvanishing gravity) case of geometric optics of light rays (laser beams) in the bimetric theory. Carlip's bimetric theory [1] adopts specific values of parameters in Jackobson's theory [2] and adds one more parameter ϵ which is a coupling constant between the vector field w^α and the stress–energy tensor of matter. It is useful to understand what kind of relativistic effects one can expect in the bimetric theory in application to the spaceborne laser ranging experiments like ASTROD [8,9] and LATOR [10–12]. We shall investigate this problem by calculating the time delay of light propagating in the time-dependent gravitational field of an arbitrary moving body.

In what follows, we shall consider an isolated N-body system (a solar system) resided in an asymptotically flat space–time of the gravity metric $g_{\alpha\beta}$. According to Carlip [1] the optical metric

$$\bar{g}_{\alpha\beta} = g_{\alpha\beta} + \left(1 - \frac{1}{\epsilon^2}\right) w_\alpha w_\beta \,, \tag{1}$$

$$\bar{g}^{\alpha\beta} = g^{\alpha\beta} - \left(\epsilon^2 - 1\right) w^\alpha w^\beta \,, \tag{2}$$

where ϵ is a constant parameter defining the coupling of the vector field w^α with matter and describing the degree of violation of the Lorentz invariance for electromagnetic field (and other material fields). Following Carlip [1] we assume that the Greek indices are raised and lowered with the metric $g_{\alpha\beta}$.

Let us work in a global coordinate system $x^\alpha = (ct, x^i)$, where c in Carlip's theory [1] is the speed of gravity and x^i are spatial coordinates. In the global frame the linearized expansions for the metric and the vector field are

$$g_{\alpha\beta} = \eta_{\alpha\beta} + h_{\alpha\beta} \,, \tag{3}$$

$$g^{\alpha\beta} = \eta^{\alpha\beta} - h^{\alpha\beta} \,, \tag{4}$$

$$w^\alpha = V^\alpha + \zeta^\alpha \,, \tag{5}$$

$$w_\alpha = V_\alpha + \zeta_\alpha - h_{\alpha\beta} V^\beta \,, \tag{6}$$

where $\eta_{\alpha\beta} = \mathrm{diag}(-1, +1, +1, +1)$ is the Minkowski metric, $h_{\alpha\beta}$ is the perturbation of the gravity metric, and ζ^α is the perturbation of the vector field which unperturbed value in the global frame is V^α. We emphasize that V^α remains arbitrary and our analysis is not limited to the case of the preferred frame where $V^\alpha = (1, 0, 0, 0)$.

The optical metric $\bar{g}_{\alpha\beta}$ is decomposed as follows

$$\bar{g}_{\alpha\beta} = \bar{\eta}_{\alpha\beta} + \bar{h}_{\alpha\beta} \,, \tag{7}$$

$$\bar{g}^{\alpha\beta} = \bar{\eta}^{\alpha\beta} - \bar{h}^{\alpha\beta} \,, \tag{8}$$

where the unperturbed part of the optical metric is defined by

$$\bar{\eta}^{\alpha\beta} = \eta^{\alpha\beta} - \left(\epsilon^2 - 1\right) V^\alpha V^\beta \,, \tag{9}$$

$$\bar{\eta}_{\alpha\beta} = \eta_{\alpha\beta} + \left(1 - \frac{1}{\epsilon^2}\right) V_\alpha V_\beta \,, \tag{10}$$

and the perturbation

$$\bar{h}^{\alpha\beta} = h^{\alpha\beta} + \left(\epsilon^2 - 1\right)\left(V^\alpha \zeta^\beta + V^\beta \zeta^\alpha\right) . \tag{11}$$

According to Carlip [1] light propagates in the bimetric theory along light geodesics of the optical metric $\bar{g}_{\alpha\beta}$. In geometric optics limit, the light rays are defined by a covariant equation for electromagnetic phase (eikonal) φ which reads [1,13]

$$\bar{g}^{\mu\nu}\partial_\mu\varphi\partial_\nu\varphi = 0\,. \tag{12}$$

This equation is formally equivalent to the equation of light propagating through dispersive medium with refraction index ϵ moving with respect to a preferred frame with velocity w^α. Theory of light propagation through the dispersive medium has been worked out by Synge [13] and we shall use his theory to integrate (12) and to interpret its solution.

As follows from (12), we do not need to know solutions for the metric perturbation $h^{\alpha\beta}$ and that of the vector field ζ^α separately. What we need to perform the integration of (12) is solution for the perturbation $\bar{h}^{\alpha\beta}$ of the optical metric (8). In the linearized approximation the metric perturbation $\bar{h}^{\alpha\beta}$ obeys the following gravity field equations [1]

$$\Box\bar{h}^{\alpha\beta} = -1/2\frac{16\pi G}{(1-4\varpi)\epsilon c^2}\left(S^{(\alpha}{}_{\mu}T^{\beta)\mu} - \eta^{\alpha\beta}T^\lambda_{\ \lambda}\right)\,, \tag{13}$$

where $\Box \equiv -c^{-2}\partial^2/\partial t^2 + \nabla^2$ is the wave operator in flat space–time defining null characteristics of the gravitational field, the constant tensor $S^{\alpha\beta} = \eta^{\alpha\beta} + 2\varpi V^\alpha V^\beta$, ϖ is another constant parameter of the bimetric theory ($\varpi = 0$ in general relativity), and $T^{\mu\nu}$ is the stress–energy tensor of the point-like bodies composing of the N-body system. Equation (13) is valid under imposing the following gauge conditions

$$\partial_\beta\left(h^{\alpha\beta} - \frac{1}{2}\eta^{\alpha\beta}h\right) + \left(1 - \frac{1}{\epsilon^2}\right)V_\alpha\partial_\beta\zeta^\beta = 0\,. \tag{14}$$

In the linearized approximation the stress–energy tensor reads [1]

$$T^{\alpha\beta}(t,\boldsymbol{x}) = \sum_{a=1}^{N}\frac{M_a u_a^\alpha\, u_a^\beta\, \delta^{(3)}(\boldsymbol{x} - \boldsymbol{x}_a(t))}{\gamma_a\sqrt{1-(1-\epsilon^{-2})\,(u^\mu V_\mu)^2}}\,, \tag{15}$$

where the index $a = 1, 2, \ldots, N$ enumerates gravitating bodies of the solar system, M_a is the (constant) rest mass of the ath body, $\boldsymbol{x}_a(t)$ is time-dependent spatial coordinate of the ath body, $\boldsymbol{v}_a(t) = d\boldsymbol{x}_a(t)/dt$ is velocity of the ath body, $u_a^\alpha = \gamma_a(1,\, \boldsymbol{v}_a/c)$ is the 4-velocity of the ath body, $\gamma_a = \left(1 - v_a^2/c^2\right)^{-1/2}$ is the Lorentz factor, and $\delta^{(3)}(\boldsymbol{x})$ is the three-dimensional Dirac's delta function.

Because the field equations (13) are linear, we can consider their solution as a linear superposition of the solutions for each body. It allows us to focus on the relativistic effects caused by one body (Sun, planet, etc.) only. Solving (13) by making use of the retarded Liénard–Wiechert tensor potentials [14], one obtains the metric tensor perturbation

$$\bar{h}^{\alpha\beta}(t,\boldsymbol{x}) = \frac{2GM}{(1-4\varpi)c^2}\frac{2u^\alpha u^\beta + \eta^{\alpha\beta} + 2\varpi\left(u^\alpha V^\beta + u^\beta V^\alpha\right)(u^\mu V_\mu)}{\sqrt{\epsilon^2 - (\epsilon^2-1)\,(u^\mu V_\mu)^2}}\frac{1}{r_R}\,, \tag{16}$$

where $r_R \equiv -u_\alpha r^\alpha$, $r^\alpha = x^\alpha - z^\alpha(s)$, $z^\alpha(t) = (ct, \boldsymbol{z}(t))$ is the world line of the light ray deflecting body parametrized by the coordinate time t, $u^\alpha(t) = c^{-1}dz^\alpha(t)/dt$.

Because we solved the field equations (13) in terms of the retarded Liénard–Wiechert potentials, the distance $r^\alpha = x^\alpha - z^\alpha(s)$, the body's world line $z^\alpha(s) = (cs, \boldsymbol{z}(s))$, and the body's 4-velocity $u^\alpha(s)$ in (16) are functions of the *retarded* time s. The retarded time s is found in the linearized approximation of the bimetric theory as a solution of the gravity null cone equation

$$\eta_{\mu\nu} r^\mu r^\nu \equiv \eta_{\mu\nu}\left(x^\mu - z^\mu(s)\right)\left(x^\nu - z^\nu(s)\right) = 0\,, \tag{17}$$

i.e.,

$$s = t - \frac{1}{c}|\boldsymbol{x} - \boldsymbol{z}(s)|\,, \tag{18}$$

where the fundamental constant c in (18) is the fundamental speed of propagation of gravity.

Light rays are defined by a covariant equation (12) for electromagnetic phase (eikonal) φ. Assuming that unperturbed solution of (12) is a plane wave, we can write a general solution of this equations as follows

$$\varphi(x^\alpha) = \varphi_0 + k_\alpha x^\alpha + \psi(x^\alpha)\,, \tag{19}$$

where k_α is an unperturbed (constant) wave covector of the electromagnetic wave, and $\psi(x)$ is a relativistic perturbation of the eikonal generated by the metric tensor perturbation $\bar{h}_{\alpha\beta}$ defined in (16). Substitution of (19) to (12) yields

$$\bar{\eta}^{\alpha\beta} k_\alpha k_\beta = 0\,, \tag{20}$$

$$\bar{\eta}^{\alpha\beta} k_\alpha \frac{\partial \psi}{\partial x^\beta} = \frac{1}{2} h^{\alpha\beta} k_\alpha k_\beta\,. \tag{21}$$

Let us define a vector (see Fig. 1)

$$\sigma^\alpha = \bar{\eta}^{\alpha\beta} k_\beta = k^\alpha - \left(\epsilon^2 - 1\right)\left(V^\beta k_\beta\right) V^\alpha\,, \tag{22}$$

such that

$$\bar{\eta}_{\alpha\beta}\sigma^\alpha\sigma^\beta = 0\,, \qquad \text{and} \qquad k_\alpha \sigma^\alpha = 0\,. \tag{23}$$

Vector σ^α defines the direction of propagation of light ray from a source of light (laser, star) to observer (see Fig. 1). Making use of vector σ^α simplifies (21) and reduces it to the following form

$$\sigma^\alpha \frac{\partial \psi}{\partial x^\alpha} = \frac{1}{2} h_{\alpha\beta}\sigma^\alpha\sigma^\beta\,. \tag{24}$$

Unperturbed characteristics of the eikonal equation (24) are straight lines (light rays) parametrized by the affine parameter λ in such a way that

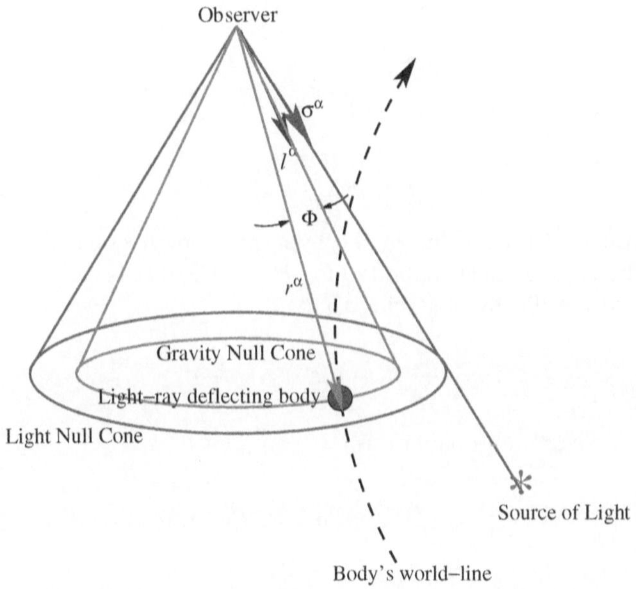

Fig. 1. The light and gravity null cones of the bimetric theory are shown. Gravity propagates from the source of light (laser) to observer along the gravity null cone defined by the metric $g_{\alpha\beta}$. Gravitationally unperturbed direction to the source of light is defined by vector σ^α lying on the null hypersurface of the optical metric $\bar{g}_{\alpha\beta}$. Gravitationally perturbed direction of the light propagation is l^α and this vector resides on the gravity null cone. Each light ray deflecting body (Sun, planet) deflects light from its retarded position defined with respect to observer by a null vector $r^\alpha = x^\alpha - x^\alpha_J(s)$ which also resides on the gravity null cone. The eikonal gravitational perturbation is $\psi = -(2GM/c^2)\chi \ln \Phi$, where $\Phi = -l_\alpha r^\alpha$. Gravitational light ray deflection experiments measure the range χ and shape Φ of the relativistic time delay of light. The range measurement allows us to pin down the parameter χ while measuring the shape Φ of the time delay gives us the components of the vector l^α under assumption that vector r^α is known. Deviation of l^α from the gravity-unperturbed direction σ^α of the light ray measures the degree of violation of the Lorentz invariance of the gravitational field.

$$\frac{d}{d\lambda} = \sigma^\alpha \frac{\partial}{\partial x^\alpha} \,. \tag{25}$$

Integration of (25) by making use of the unperturbed characteristics is straightforward (see, for example, [15]) and can be written as follows

$$\psi(x^\alpha) = -\frac{2GM}{c^2}\chi \ln\left(-l_\alpha r^\alpha\right) \,, \tag{26}$$

where

$$\chi = \frac{\left(\sigma_\alpha u^\alpha\right)^2 + (1/2)\left(\sigma_\alpha \sigma^\alpha\right) + 2\varpi \left(\sigma_\alpha u^\alpha\right)\left(\sigma_\alpha V^\alpha\right)\left(u_\alpha V^\alpha\right)}{(1-4\varpi)(\epsilon^2 - (\epsilon^{-2}-1)\left(u^\mu \zeta_\mu\right)^2)^{1/2}\left(P_{\alpha\beta}\sigma^\alpha\sigma^\beta\right)^{1/2}} \,, \tag{27}$$

$$l^\alpha = \sigma_\perp^\alpha + \sigma_\perp u^\alpha \,, \tag{28}$$

$$\sigma_\perp^\alpha = P^\alpha_\beta \sigma^\beta \,, \tag{29}$$

$$\sigma_\perp = (\sigma_{\perp\alpha}\sigma_\perp^\alpha)^{1/2} = \left(P_{\alpha\beta}\sigma^\alpha\sigma^\beta\right)^{1/2} \,, \tag{30}$$

and

$$P_{\alpha\beta} = \eta_{\alpha\beta} + u_\alpha u_\beta \,, \tag{31}$$

is the operator of projection on the hyperplane orthogonal to the 4-velocity u^α of the light ray deflecting body ($P_{\alpha\beta}P^\beta_\gamma = P_{\alpha\gamma}$). It is easy to confirm that solution (26) is valid by observing that

$$\frac{d}{d\lambda}\ln\left(-l_\alpha r^\alpha\right) = -\frac{\sigma_\perp}{r_R} \,, \tag{32}$$

where $r_R = -u_\alpha r^\alpha$, and equations

$$\partial_\alpha r^\mu = \delta^\mu_\alpha - \frac{u^\mu}{\gamma}\partial_\alpha s \,, \tag{33}$$

$$\partial_\alpha s = -\gamma\,\frac{r_\alpha}{r_R} \,, \tag{34}$$

where $\gamma = (1-\beta^2)^{-1/2}$, have been used. Equation (28) allows us to recast the argument of the logarithm in (26) as

$$l_\alpha r^\alpha = \sigma_\perp^\alpha r_\alpha - \sigma_\perp r_R \,. \tag{35}$$

It is remarkable that both vectors l^α and r^α are null vectors of the gravity metric $g_{\alpha\beta}$ (see Fig. 1). Indeed, in the linearized approximation $g_{\alpha\beta} = \eta_{\alpha\beta}$ and one can easily prove by inspection that

$$\eta_{\alpha\beta}l^\alpha l^\beta = 0 \,, \tag{36}$$

$$\eta_{\alpha\beta}r^\alpha r^\beta = 0 \,, \tag{37}$$

which are consequences of the definitions given by (17) and (28). Thus, neither l^α nor r^α belong to the null cone of the optical metric $\bar{g}_{\alpha\beta}$ but characterize the null hypersurfaces of the gravity metric $g_{\alpha\beta}$.

Solutions (19) and (26) for the electromagnetic eikonal in the bimetric theory should be compared with a similar solution for the case of propagation of light in general relativity where the gravity and light null cones coincide [16]. The reader can see that the null characteristics of the gravity metric $g_{\alpha\beta}$ enter the gravitationally perturbed part of the eikonal (26) in the bimetric theory in the form of the dot product $l_\alpha r^\alpha$ which is the argument of the logarithm, where r^α is the null distance of the metric $g_{\alpha\beta}$ between the observer and the light ray deflecting body. A remarkable fact is that both l^α and r^α are null vectors of the metric $g_{\alpha\beta}$ describing null hypersurfaces of the gravitational field. Consequently, gravitational light ray deflection experiments in the field of moving bodies are sensitive to, and can measure, the divergence between the null characteristics of the gravity metric $g_{\alpha\beta}$ and the optical metric $\bar{g}_{\alpha\beta}$ in the case of a nonstationary gravity field in contrast to other relativistic experiments limiting the PPN-preferred frame parameters $\alpha_1, \alpha_2, \alpha_3$ [7]. This allows us to measure the spontaneous violation of the Lorentz invariance of the gravitational field predicted by the vector–tensor theories of gravity that admit existence of a vector field w^α coupled to matter via parameter ϵ parametrizing the difference between the gravity and optical metrics. In conventional type of the gravitational light ray deflection experiments conducted with VLBI [16,17], one needs the angle Φ (see Fig. 1) to be as small as possible to magnify the Lorentz-invariance violation effects driven by gravity field. Currently, VLBI can measure gravitomagnetic effects of order v/c beyond the static Shapiro effect [18,19]. Further progress in measuring more subtle effects of the bimetric theory of gravity of order v^2/c^2 and higher beyond the static Shapiro time delay can be achieved with laser ranging technique in the experiments like LATOR [10–12] and/or ASTROD [8,9]. In case of laser ranging between spacecrafts with a gravitating body (Sun) located near the direction of the laser beam, the angle Φ (see Fig. 1) can vary in a large dynamical range so that relativistic effects of the bimetric theory of gravity could be explored with much better precision than in VLBI experiments. We will study this situation for laser ranging experiment in various theories, e.g., in the more general vector–metric theories [20] and in the axion electrodynamics [21].

References

1. Carlip, S. *Clas. Quant. Grav.* **21**, 3803 (2004).
2. Jacobson, T. & Mattingly D. *Phys. Rev.* **D 64**, 024028 (2001).
3. Kostelecky, A. *Phys. Rev.* **D 69**, 105009 (2004).
4. Lämmerzahl, C. *Ann. Phys.* (Leipzig) **14**, 71 (2005).
5. Ni, W.-T. “Implications of Huges-Drever Experiments”, In: *Proceedings of the 1983 International School and Symposium on Precision Measurement and Gravity Experiment*, pp. 519–529, ed. W.-T. Ni (National Tsing Hua University, 1983).
6. Ni, W.-T. *Int. J. Mod. Phys.* **D 14**, 901 (2005).
7. Mattingly, D. *Living Rev Relativity* **8**, 5 (2005).
8. Ni, W.-T. *Int. J. Mod. Phys.* **D 14**, 901 (2005).
9. Ni, W.-T., Shiomi, S. & Liao, A.-C. *Clas. Quant. Grav.* **22**, S269 (2004); and references therein.
10. Yu, J.W., Shao, M., Gursel, Y. & Hellings, R. “LATOR: laser astrometric test of relativity”, In: *Amplitude and Intensity Spatial Interferometry II*, Proc. SPIE, Vol. 2200, pp. 325–334, ed. J.B. Breckinridge (1994).
11. Turyshev, S.G., Shao, M. & Nordtvedt, K. *Class. Quant. Grav.* **21**, 2773 (2004).
12. Turyshev, S.G., Shao, M. & Nordtvedt, K. “Science, Technology and Mission Design for the Laser Astrometric Test Of Relativity”, this volume, page 479.
13. Synge J.L. *Relativity: The General Theory* (North-Holland: Amsterdam 1960).
14. Landau, L.D. & Lifshitz, E.M. *The Classical Theory of Fields* (Oxford: Pergamon) (1971).

15. Zeldovich, Ya.B. & Myskis, A.D. *Elements of Mathematical Phsycics* (Nauka: Moscow) (1973) (in Russian).
16. Kopeikin, S.M. & Schäfer, G. *Phys. Rev.* **D 60** 124002 (1999).
17. Kopeikin, S.M. & Mashhoon, B. *Phys. Rev.* **D 65** 064025 (2002).
18. Fomalont, E.B. & Kopeikin, S. M. *Astrophys. J.* **598** 704 (2003).
19. Kopeikin, S.M. "Gravitomagnetism and the Speed of Gravity", *Int. J. Mod. Phys.* **D 15**, 3005 (2006). [arXiv:gr-qc/0507001]
20. Luo, X., Xie, Y. and Huang, T. "A Metric–Vector Gravitational Theory and the Constraints on Its Parameters" *General Relativity and Gravitation* **37** in press (2006).
21. Cheng, S.L., Geng, C.Q. & Ni, W.-T. *Phys. Rev.* **D 52**, 3132 (1995).

Part III

Technologies

Measurement of the Shapiro Time Delay Between Drag-Free Spacecraft

Neil Ashby[1] and Peter L. Bender[2]

[1] Department of Physics, UCB 390, University of Colorado, Boulder, CO 80309-0390, USA
[2] JILA, UCB 440, University of Colorado, Boulder, CO 80309-0440, USA

Summary. In 1964 Shapiro pointed out that γ can be determined from measurements of the relativistic time delay for electromagnetic waves passing near a massive body such as the Sun. The delay for two-way measurements from Earth to a spacecraft passing behind the Sun can be more than 200 µs. Microwave range and range-rate measurements of this kind from Earth to several spacecraft have provided our best information so far on γ. Laser time-delay measurements and determinations of the deflection of laser beams near the Sun also have been proposed. A mission of this kind called Laser Astrometric Test of Relativity (LATOR) currently is being considered. Here we discuss a considerably different mission which would use drag-free spacecraft, whose orbits can be accurately determined, to measure the Shapiro time delay for laser beams passing near the Sun. One spacecraft would be located near the L1 Lagrange point, between the Earth and the Sun. The other would be launched into an orbit similar to the ones used in LATOR, with 1.5 year period and eccentricity such that three occultations by the Sun would occur within 2 years after launch. We also consider higher-order time-delay effects. In the present experiment laser signals are sent from a drag-free spacecraft at the L1 point, and transponded back by a drag-free spacecraft passing behind the Sun. A high-stability frequency standard located on the L1 spacecraft permits accurate measurement of the time delay. Both spacecraft are designed for extremely low spurious accelerations at periods out to roughly 20 days.

1 Introduction

Historically, the first accurate measurements of the Shapiro time delay [1] were made using microwave range measurements to the Mariner 9 spacecraft orbiting around Mars [2, 3] and to the Viking Orbiters and Landers [3–5]. Recently, a measurement of the spatial curvature parameter γ with $2.3 \cdot 10^{-5}$ accuracy was made during the Cassini mission [6]. The time derivative of the time delay was measured from Doppler shifts in microwave signals sent from Earth to a transponder on the spacecraft and back. Great care was taken to minimize spurious effects due to the Earth's atmosphere and to the interplanetary electron density. An additional improvement in the accuracy for γ to roughly $1 \cdot 10^{-6}$ is expected from measurement of the gravitational deflection of light rays during the GAIA astrometric mission of the European Space Agency.

The LATOR experiment [7–9] is intended to give further improvements in the measurement of γ. It involves placing two spacecraft in very similar solar orbits with periods of 1.5 years. The orbits can be chosen so that the spacecraft make three passes behind the Sun during a 7-month period centered on 18 months after launch, and the angular separation between them as seen from Earth is roughly 1°. An optical interferometer on the International Space Station observes laser beams from the two distant spacecraft and would measure the angular separation between the spacecraft with high accuracy. The lengths of the three sides of the triangle would be measured with lasers. From the non-Euclidian geometry of the triangle when one arm passes near the Sun, γ can be determined.

As an alternate approach, we consider making Shapiro time-delay measurements from a satellite near the L1 point of the Earth–Sun system to a single transponder spacecraft in a LATOR-type orbit when the line of sight passes near the Sun. The L1 spacecraft would have a high-stability atomic frequency standard with performance similar to that expected for the cooled Cs clocks that have been developed for the ACES [10] and the PARCS experiments [11] on the International Space Station. Both the distant spacecraft and the L1 spacecraft would be designed to have very low levels of nongravitational orbit disturbances. The atomic frequency standard would need to be quite small, but the environmental disturbances near the L1 point would be considerably lower than those on the Space Station.

The size of the Shapiro time delay and its variation with time for the LATOR-type orbit of the distant spacecraft will be discussed in Sect. 2. In Sect. 3, the expected signal-to-noise ratio will be derived for determining γ from an idealized Gravitational Time Delay (GTD) mission. For this ideal case, only white frequency noise in the clock on the L1 spacecraft is allowed for. In Sect. 4, the requirements on the drag-free systems to minimize nongravitational accelerations of the spacecraft will be considered, along with possible excess clock noise at very low frequencies. Then, the limitations from the actual time-delay measurement method will be discussed in Sect. 5. The overall results will be summarized in Sect. 6.

2 Shapiro Time Delay

Because of the sensitivity of the measurements discussed in this chapter, we have performed a more accurate calculation of the Shapiro time delay than is quoted in many textbooks. Usually a "straight line" approximation is used, in which the delay is calculated assuming the path of the photon is straight [1]. A more careful calculation including the bending of the path shows that there is an additional contribution of first order in the quantity $GM_\odot/c^3$. For example, Weinberg gives the following expression for the time delay required for a photon to pass from the point of closest approach to the Sun, at $r = b$, to the radius r [12]:

$$\Delta t^{(1)}_{\mathrm{delay}} = (1+\gamma)\frac{GM_\odot}{c^3}\ln\left(\frac{r+\sqrt{r^2-b^2}}{b}\right) + (1+\gamma)\frac{GM_\odot}{c^3}\sqrt{\frac{r-b}{r+b}}. \quad (1)$$

The last term in this result can amount to tens of microseconds. The result is expressed in the isotropic coordinates that are customarily used when discussing time-delay observations.

Because of its possible importance in solar system time-delay observations, we have extended the delay calculation to higher order. We briefly describe the method here. If the Sun is taken to be a spherically symmetric mass source, then the motion of a photon can be assumed to lie in the $\theta = \pi/2$, equatorial plane. Also, in isotropic coordinates the metric does not depend explicitly on coordinate time or on azimuthal angle. The cyclic nature of these two coordinates leads immediately to two constants of the motion, corresponding roughly to energy and angular momentum. These two constants allow the azimuthal angle and the scalar orbit parameter to be eliminated in favor of the coordinate time. Since the path followed by a photon is a null geodesic, there results a first-order differential equation for dt/dr, where t is the coordinate time ($x^0 = ct$). Contributions to this equation can be expanded in powers of $\mu = GM_\odot/c^2$ and the equation can be integrated. The result of these calculations gives for the next-order contribution to time delay [13]

$$\Delta t^{(2)}_{\mathrm{delay}} = \frac{(3\delta/2 + 4(\gamma+1) - 2\beta)(GM_\odot)^2}{bc^5}\mathrm{ArcTan}\left(\frac{\sqrt{r^2-b^2}}{b}\right) - \frac{(\gamma+1)^2(GM_\odot)^2}{2bc^5}\left(2+\frac{b}{r+b}\right), \quad (2)$$

where β is a PPN parameter measuring the strength of the nonlinear term in g_{00}, and δ is defined by expanding the isotropic metric component g_{11} to higher order:

$$g_{11}(r) = 1 + \frac{2\gamma\mu}{r} + \frac{3\delta\mu^2}{r^2}. \quad (3)$$

In general relativity, $\delta = 1$.

Figure 1 shows the second-order time-delay contributions of (2) for several different values of the distance of closest approach, b. The three curves, upper to lower, correspond, respectively, to $b = 1.1, 1.2, 1.3$ solar radii, and the quantity plotted is (2) in picoseconds, for the delay during one-way travel from the point of closest approach to the radial distance r. The horizontal axis is the final radial distance expressed in units of the solar radius. In the experiment proposed here, this delay will contribute four times, so the total second-order

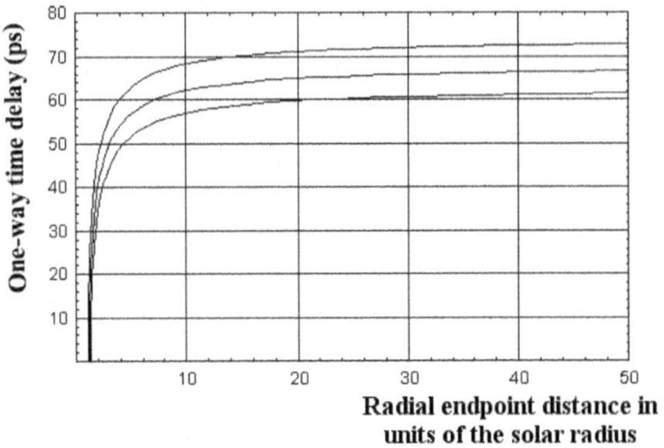

Fig. 1. Second-order contributions to time delay for one-way passage of a photon from the point of closest approach to a radial distance r from the Sun.

delay can amount to about a quarter of a microsecond, and certainly has to be taken into account. The effect of a deviation $\gamma - 1$ on these contributions is negligible.

We have also estimated the time delay due to the solar quadrupole moment. Such contributions are controlled by the parameter

$$\frac{GM_\odot J_2 a_1^2}{b^2 c^3} < 10^{-12}\ \text{s}, \tag{4}$$

where $J_2 \approx 2 \cdot 10^{-7}$ is the Sun's quadrupole moment coefficient and a_1 is the Sun's equatorial radius. There is a complicated dependence of this delay on the orientation of the Sun's rotation axis with respect to the photon's path, but the net effect is only 2 or 3 ps and we shall not consider it further.

In the present modification of the LATOR mission, the line of sight from the spacecraft at L1 to the distant spacecraft passes across the Sun three times. Figure 2 plots the total Shapiro time delay in microseconds, and the angle (in milliradians) between the line to the Sun's center and the line to the occulted spacecraft. Both first-order contributions of (1) are included. The second-order effects are too small to see on the graph.

This figure shows that during conjunctions the time delay is, to a very good approximation, symmetric about the time t_0 when the distant spacecraft is exactly behind the center of the Sun. The logarithm clearly dominates the time dependence. For purposes of analysis in the following sections, we have found that the time-delay function, within a span of ± 20 days on either side of t_0, can be fit very well by a function of the form

$$0.97 \cdot \frac{8GM}{c^3} \ln(R|t - t_0|) + \ \text{const.}, \tag{5}$$

where R is the rate of motion of the line of sight with respect to the Sun's center in solar radii per day, and t is in days. The constant is not important

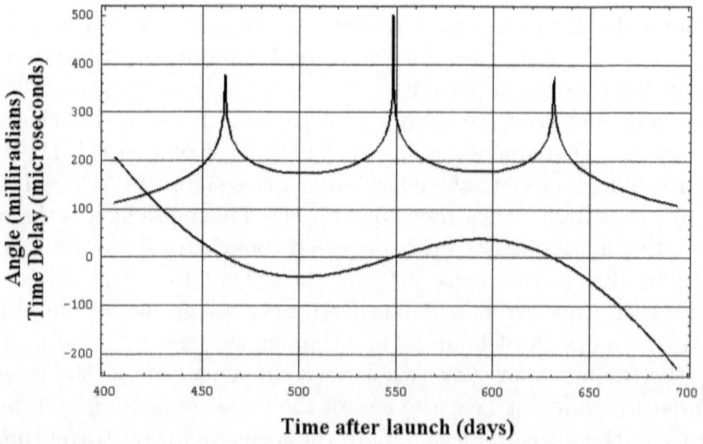

Fig. 2. Plots of the first-order time delay (in microseconds) as the occulted spacecraft passes behind the Sun, and the angle between the line from L1 to the Sun and the line from L1 to the distant spacecraft (in milliradians).

when discussing uncertainties, since it would represent a constant bias in the time delay. The logarithmic dependence is the distinctive time signature for the Shapiro effect, and we shall henceforth drop the constant.

3 Idealized Gravitational Time-Delay Mission

Most of the accuracy for determining the relativistic time delay is expected to come from a period of roughly 20 days around the time of passage of the line of sight behind the Sun, as will be discussed later. Thus the noise in the frequency standard and in displacements of the spacecraft due to nongravitational accelerations will be of interest at frequencies down to about $5 \cdot 10^{-7}$ Hz. It is assumed that the measurements will be made continuously from time $-t_2$ to $-t_1$ and from time t_1 to t_2, where $t = 0$ when the line of sight between the spacecraft passes through the center of the Sun.

For simplicity, the distance between the spacecraft is assumed to be constant except for the relativistic time delay. The time signature of $\gamma^* = (\gamma+1)/2$ is taken to be

$$g(t) = -B(\ln|Rt| - M), \tag{6}$$

where $M =< \ln|Rt| >$ is the mean value of $\ln|Rt|$ over the periods $-t_2$ to $-t_1$ and t_1 to t_2, and $B = 0.97 \cdot 8GM_\odot/c^3 = 3.82 \cdot 10^{-5}$ s.

The rate at which the line of sight to the distant spacecraft passes across the Sun varies substantially between the three conjunctions with the Sun, with the second one having a considerably lower rate than the other two. In solar radii per day, the rate for the first and third conjunctions is about 0.7, and we will assume this value for our reference case. To avoid measurements closer than 0.4 solar radii from the limb, we chose to make t_1 be 2 days after t_0 and t_2 be 10 days after t_0. For the second conjunction, the rate would be only about 0.2 solar radii per day.

In the reference case, we assume that the only noise in the measurements is white fractional frequency noise $n(f)$ at a level of $1 \cdot 10^{-13}\,\mathrm{Hz}^{-1/2}$ at all frequencies. This is close to the noise level expected for the ACES and PARCS clocks [10,11] at frequencies above $5 \cdot 10^{-7}$ Hz. This noise affects the results in two ways. One is the jitter in the measured round-trip travel time of roughly 2,000–2,200 s due to the phase jitter in the output from the frequency standard. However, this error is reduced strongly when the measurements are averaged over periods of hours. The other, more serious, noise is the variations in the frequency over the whole measurement time. If the frequency of the standard is different near the end of the measurement time from what it was earlier in the measurement period, the measured total travel time will be affected proportionately.

The usual method of optimal filtering (see, e.g., [14] and references therein) would be used to determine the value of γ^*. Let $g(f)$ be the Fourier transform of $g(t)$ over the time of the measurements. Then the square of the signal-to-noise ratio is given by

$$\left(\frac{S}{N}\right)^2 = \int_0^\infty \frac{2|g(f)|^2}{n(f)^2} df \,. \tag{7}$$

We also assume for the reference case that $t_1 = 2$ days and $t_2 = 10$ days. Because of the symmetry of the signal before and after $t = 0$, only the cosine terms of $g(f)$ are nonzero

$$g(f) = 2\int_{t_1}^{t_2} g(t)\cos(\omega t)\,dt, \tag{8}$$

where $\omega = 2\pi f$. The factor 2 comes from time symmetry of the time-delay signal.

Since $n(f)$ is assumed to be independent of frequency in the reference case, it can be taken outside the integral in (7):

$$\left(\frac{S}{N}\right)^2 = \frac{2}{n(f)^2}\int_0^\infty g(f)^2 df \,, \tag{9}$$

and Parseval's theorem implies:

$$\int_0^\infty g(f)^2\,df \approx 2\pi \int_{t_1}^{t_2} g(t)^2\,dt \,. \tag{10}$$

After some algebra,

$$\begin{aligned}\int_{t_1}^{t_2} g(t)^2\,dt = B^2 R\big(&t_2(\ln(Rt_2))^2 - t_1(\ln(Rt_1))^2 - 2\left(t_2\ln(Rt_2) - t_1\ln(Rt_1)\right)\\ &+2(t_2 - t_1) - M^2(t_2 - t_1)\big)\end{aligned} \tag{11}$$

where M is the mean value

$$M = \left(t_2 \ln(Rt_2) - t_1 \ln(Rt_1)\right)/(t_2 - t_1) - 1\,. \tag{12}$$

From the above,

$$\int_{t_1}^{t_2} g(t)^2\,dt = B^2 R\Big((t_2 - t_1) - \big(t_2 t_1/(t_2 - t_1)\big)\big(\ln(Rt_2) - \ln(Rt_1)\big)^2\Big). \tag{13}$$

For our reference case, $t_2 = 8.64 \cdot 10^5$ s, $t_1 = 1.728 \cdot 10^5$ s, and

$$\int_0^\infty g(f)^2 df = 2\pi \cdot 1.317 \cdot 10^5 B^2 (s^3). \tag{14}$$

Because only the cosine terms in the noise contribute, the effective noise level is reduced from $1 \cdot 10^{-13}$ to $0.71 \cdot 10^{-13}\,\mathrm{Hz}^{-1/2}$. Since the total round-trip time is about 2,200 s for the first and third conjunctions, the noise in measuring it is

$$n(f) = 1.56 \cdot 10^{-10}\,\mathrm{s\,Hz}^{-1/2}. \tag{15}$$

Thus, from (9), (10), and (13),

$$(S/N)^2 = 1.655 \cdot 10^6 B^2/[2.42 \cdot 10^{-20}] = 9.3 \cdot 10^{16}. \tag{16}$$

This corresponds to an idealized precision of $0.33 \cdot 10^{-8}$ for determining γ^*, or $0.66 \cdot 10^{-8}$ for γ.

4 Effects of Nongravitational Accelerations

For the GTD mission, the level of nongravitational accelerations of the spacecraft has to be kept very low out to long periods. For comparison, a joint mission of the European Space Agency and NASA called the Laser Interferometer Space Antenna (LISA) [15] has a requirement of less than $3 \cdot 10^{-15}\,\mathrm{m\,s}^{-2}\,\mathrm{Hz}^{-1/2}$ for the spurious accelerations of proof masses aboard each spacecraft at frequencies down to 0.1 mHz [16]. Each spacecraft is servocontrolled to follow the average position of two proof masses inside it to roughly $3 \cdot 10^{-9}\,\mathrm{m\,Hz}^{-1/2}$. This is done by a disturbance reduction system (DRS) ("drag-free" system) [17] that uses micronewton thrusters to cancel out the solar radiation pressure force and other nongravitational forces on the spacecraft. The relative displacements of the spacecraft with respect to the proof masses are determined by two gravitational reference sensors (GRSs) [18, 19] containing the proof masses. For the GTD mission, only a single GRS would be needed on each spacecraft.

Below 0.1 mHz, it has been suggested [20] that a spurious acceleration level increasing only as $((0.1\,\mathrm{mHz})/f)^{0.5}$ between 0.1 and 0.003 mHz could be achieved with only moderate additional experimental constraints. At still lower frequencies, we assume that the acceleration level will increase as $(0.003\,\text{mHz})/f$. If $a(f)$ is the spurious acceleration level at frequency f for the GRS on each spacecraft, the resulting noise level in the round-trip distance between the spacecraft will be

$$x(f) = 2\sqrt{2}\frac{a(f)}{\omega^2}. \tag{17}$$

The equivalent time-delay noise $p(f) = x(f)/c$ will cross the value of $n(f) = 1 \cdot 10^{-13}\,\text{Hz}^{-1/2}$ that we have adopted at a frequency of $4.0 \cdot 10^{-7}\,\text{Hz}$, and near that frequency it is given by

$$p(f) = 2 \cdot 10^{-10}\left(4.0 \cdot 10^{-7}\ \text{Hz}/f\right)^3 \text{s}\,\text{Hz}^{-1/2}. \tag{18}$$

In view of the rapid increase of $p(f)$ with decreasing f, we can approximate its effect by cutting off the integral in (2) on the lower end at $4.0 \cdot 10^{-7}\,\text{Hz}$. To see what the effect of the assumed level of spurious accelerations is, we have calculated $g(f)$ from (3) and then numerically integrated the function $(g(f)/B)^2$. The dependence of this function on f is shown in Fig. 3. The total integral is $8.1 \cdot 10^5\,\text{s}$, in good agreement with (14), and the fraction of the integral from frequencies below $4 \cdot 10^{-7}\,\text{Hz}$ is only about 5%. Thus the effect of limitations from the spurious accelerations of the spacecraft appears to be small, if the assumed performance level for the DRSs can be achieved. At the very low acceleration levels and frequencies involved, verification of the necessary performance will have to be achieved from modeling rather than direct testing, but this approach seems quite feasible for the types of forces involved.

The most serious challenge for the DRS at the lowest frequencies is likely to be the rate of change of the solar intensity at the distant spacecraft. During the 20 day periods around the first and third solar conjunctions, the equilibrium temperature of the spacecraft would change by roughly 8 K if special measures were not taken. To minimize the effect on the GRS, both active temperature

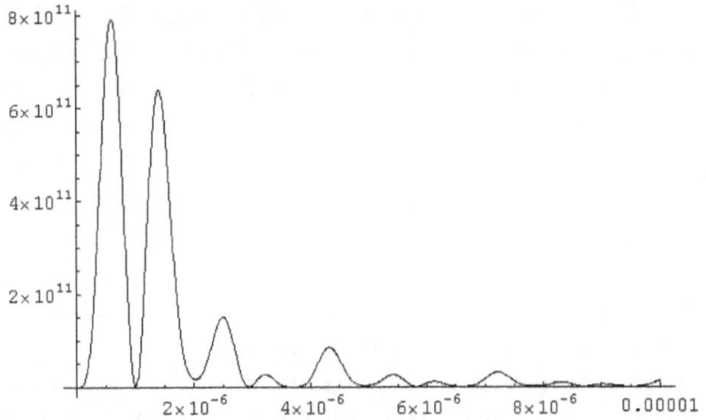

Fig. 3. Plot of the function $g(f)^2/B^2$ for frequencies up to 0.01 mHz.

control of the spacecraft and careful design to minimize temperature gradients across the GRS will be needed.

5 Other Time-Delay Measurement Errors

One method for measuring the time delay for signals sent between spacecraft that are far apart is the use of pulsed lasers. Lasers giving 100 ps pulses in the green with 0.1 J per pulse energy and 20 Hz repetition rate are available commercially. We assume that a laser giving a train of such pulses is located on each spacecraft, and that a fast detector measures the time delay between the receipt of a pulse and the next pulse generated onboard. The expected accuracy is 5 ps or better. For the round-trip travel time, the resulting contribution of the timing to the error will be 10 ps or less.

It is difficult to know what should be assumed concerning the time variation of pulse timing errors, since they may well be systematic in nature. During the time of perhaps 8 days for the line of sight between the two spacecraft to go from 0.4 to 6.0 times the solar radius from the limb, the change in the round-trip time delay will be about $6.4 \cdot 10^{-5}$ s. Thus a drift of 10 ps in the timing error over this time, and an opposite sign drift during the corresponding time before conjunction, could give as much as a roughly $3 \cdot 10^{-7}$ error in γ. This is what would result from treating the timing error as a worst case error, in terms of its time dependence. An error this large is unlikely, and some additional reduction in the timing system error probably is possible by improvements in the system design. However, to achieve an accuracy of better than $1 \cdot 10^{-7}$ for γ for the reference case would require a substantial improvement in the travel time measurement approach.

In view of this situation, it is also desirable to investigate what could be done with a cw laser travel time measurement system. The system we will consider consists of taking perhaps 0.5 W of cw output at 1,030 or 1,064 nm wavelength from a frequency-stabilized YbYAG or NdYAG laser on each spacecraft and putting it through an electrooptic phase modulator. Such modulators at frequencies up to 40 GHz are now commercially available, and probably will be space qualified in the next few years. This is because of strong interest in laser communications at high data rates between spacecraft. A short Fabry–Perot interferometer would be included after the modulator to pass the two sidebands and strongly suppress the carrier. The beam would then be sent to the distant spacecraft through a roughly 100 mm diameter transmitting telescope. A separate 100 mm diameter receiving telescope, with careful attention to reducing the effect of scattered sunlight hitting the entrance aperture, also would be provided on each spacecraft, to minimize the problem of making measurements near the limb of the Sun.

To detect changes in the round-trip travel time, it would be necessary to compensate for the Doppler shifts in the received signals. The one-way Doppler shifts near the times of the first and third solar conjunctions will be up to about 5 GHz. This can be compensated for by adding additional sidebands to the local oscillator laser beams. With corrections, the phase difference of the two beat notes measured at each end of the path would give a measure of the time delay over the path. For a modulation frequency of 40 GHz and sidebands 80 GHz apart, the sum of the phase differences would change by 1 cycle each time the round-trip time changes by 12.5 ps. Measurement of the phase differences to 5° accuracy on each spacecraft would correspond to less than a 0.35 ps measurement error for the round-trip time delay. If a drift in the error of this magnitude occurred over the 8 days of the measurements before conjunction and an opposite drift occurred during the measurements after conjunction, this would give a worst case error of $1.1 \cdot 10^{-8}$ for γ. However, a one-sigma error estimate of $0.4 \cdot 10^{-8}$ for γ from this error source seems reasonable.

Another benefit of an improved travel time measurement method is related to the stability requirements for the L1 clock. We assumed earlier that a power spectral amplitude of $1 \cdot 10^{-13}\,\mathrm{Hz}^{-1/2}$ for the performance of the L1 clock can be maintained at frequencies down to about $4 \cdot 10^{-7}$ Hz. Laboratory data on this seem quite encouraging, but as for spurious acceleration sources, considerable reliance on modeling of the disturbing effects at low frequencies may be needed. On the other hand, checks on the clock stability on orbit may be possible by sending optical timing signals from the Earth to the L1 spacecraft and back. If measurements of roughly 2 ps or better accuracy from day to day can be achieved, they can be used to compare the frequency of the L1 clock with the best available frequency standards on the ground. In principle, the average frequency difference over 1 day could be measured to roughly $5 \cdot 10^{-17}$. Since this would be a clock time comparison using two-way measurements, most of the effects of the atmospheric time-delay uncertainty and of spacecraft motion uncertainty would be avoided.

6 Summary

The earliest studies of a dedicated mission aimed mainly at determining the GTD were carried out by the European Space Research Organization during 1969–1973. The mission was called A Space Experiment on Gravitational Theories (SOREL) [21]. A drag-free spacecraft was proposed, and time-delay measurements were to be made when the spacecraft passed behind the Sun. Both microwave tracking and pulsed laser time-delay measurements were assumed. One approach studied was to measure the arrival times of laser pulses against an atomic frequency standard on the spacecraft. The other was to transmit laser pulses back to the ground, and rely less on having a high-accuracy frequency standard onboard.

The type of mission described in the present chapter has much higher accuracy goals than SOREL, and makes use of many technology developments that have occurred in the last three decades or so. One important difference is the proposal to make use of DRSs similar to those proposed for the LISA mission. Such DRSs are scheduled for flight in 2009 on the LISA Pathfinder mission of the European Space Agency [18,19,22]. However, major improvements in the expected performance at frequencies down to roughly $4 \cdot 10^{-7}$ Hz would be required. Another difference is the proposal to make the measurements from a spacecraft near the L1 point of the Earth–Sun system, to avoid the problem of going through the Earth's atmosphere.

In the preceding sections, three of the four main error sources for the suggested GTD mission have been discussed. The remaining one, which we have not investigated, is the orbit determination part of the problem. For both spacecraft, the main limitation is likely to be the performance of the disturbance compensation system over periods perhaps twice as long as the 20 days assumed for the main part of the time-delay measurement process. However, the effect of uncertainty in the motion perpendicular to the ecliptic also needs to be considered.

If the round-trip delay can indeed be determined to 0.4 ps in terms of the instantaneous clock frequency, then a measurement of γ from the GTD with an accuracy of 1 or 2 $\cdot 10^{-8}$ seems possible. However, the limitations from the disturbance compensation system and the orbit determination problem clearly need to be investigated further.

Acknowledgments

It is a pleasure to thank the following: Stefano Vitale, Bill Weber, Pierre Touboul, Tim Sumner, Diana Shaul, Stephen Merkowitz, and many others for information on the gravitational reference sensors for the LISA and LISA Pathfinder missions; Steve Jefferts, Leo Hollberg, Bill Klipstein, and Christoph Salomon for information on high-stability cesium frequency standards designed for use in space; Jun Ye, Jan Hall, and Judah Levine for discussions of the use of laser pulses, laser comb generators, and laser sideband systems for precise time transfer; Jennifer McGee for assistance with the time-delay calculations; Slava Turyshev and Mike Shao for detailed information on the LATOR mission; Bruno Bertotti, Luciano Iess, and John Armstrong for results concerning the Cassini time-delay measurements; and Irwin Shapiro, Bob Reasenberg, Ken Nordvedt, Jim Williams, Jim Faller, and Bob King for their early enthusiasm concerning testing relativity.

References

1. I.I. Shapiro. *Phys. Rev. Lett.*, 13:789, 1964.
2. J.D. Anderson et al. *Acta Astronautica*, 5:43, 1978.
3. R.D. Reasenberg, I.I. Shapiro, and P.E. MacNeil. *Ap. J.*, 234:L219, 1979.
4. I.I. Shapiro, R.E. Reasenberg, and P.E. MacNeil. *J. Geophys. Res.*, 82:4329, 1977.

5. D.L. Cain, J.D. Anderson, M.S.W. Keesey, et al. *Bull. Am. Astron. Soc.*, 10:396, 1978.
6. B. Bertotti, L. Iess, and P. Tortora. *Nature*, 425:374, 2003.
7. S.G. Turyshev, M. Shao, and K. Nordtvedt. *Class. Quantum Grav.*, 21:2773, 2004.
8. S.G. Turyshev, M. Shao, and K. Nordtvedt. *Int. J. Mod. Phys.*, D13, 2004.
9. S.G. Turyshev, M. Shao, and K. Nordvedt. *Proc. XXII Texas Symp. on Relativistic Astrophysics (in press)*. 2005.
10. S. Schiller, A. Goerlitz, A. Nevsky, A. Wicht, C. Lämmerzahl, H.-J. Dittus, S. Theil, P. Touboul, C. Salomon, P. Lemonde, U. Sterr, F. Riehle, E. Peik, G. M. Tino, L. Iorio, I. Ciufolini, E. Samain, A. Peters, W. Ertmer, E. Rasel, L. Maleki, and S. Karshenboim. *Proc. 39th ESLAB Symposium, Noordwijk, 19–21 April 2005, F. Favata, J. Sanz-Forcaca, A Giménez eds.*, pages 39–42, 2005.
11. D.B. Sullivan, N. Ashby, E.A. Donley, T.P. Heavner, L. Hollberg, S.R. Jefferts, W.M. Klipstein, W.D. Phillips, and D.J. Seidel. *Adv. Space Res.*, 36:107–113, 2005.
12. Stephen M. Weinberg. *Gravitation and Cosmology.* John Wiley & Sons, New York, 1972.
13. N. Ashby and J. McGee. *to be published*, 2006.
14. K.S. Thorne. *300 Years of Gravitation.* University of Chicago Press, 1987.
15. K. Danzmann and A. Ruediger. *Class. Quantum Grav.*, 20:S1, 2003.
16. *Proc. 5th Int. LISA Symp. & 38th ESLAB Symp., (Noordwijk, The Netherlands, 2004); Class. Quantum Grav.*, 22:S125 through S542, 2005.
17. P.G. Maghami and T.T. Hyde. *Class. Quantum Grav.*, 20:S273, 2003.
18. D. Bortoluzzi et al. *Class. Quantum Grav.*, 20:S89, 2003.
19. S. Anza et al. *Class. Quantum Grav.*, 22:S125, 2005.
20. P.L. Bender. *Class. Quantum Grav.*, 20:S301, 2003.
21. G.M. Israel et al. *SOREL, A Space Experiment on Gravitational Theories: A Report of the Mission Definition Group*, volume MS (73)4. European Space Research Organization, 1973.
22. W. Fichter et al. *Class. Quantum Grav.*, 22:S139, 2005.

Laser Transponders for High-Accuracy Interplanetary Laser Ranging and Time Transfer

John J. Degnan

Sigma Space Corporation, 4801 Forbes Blvd., Lanham, MD 20706 USA

Summary. Satellite laser ranging (SLR) and lunar laser ranging (LLR) to passive reflectors have been carried out successfully since 1964 and 1969, respectively. The single-ended SLR ranging technique, although capable of providing millimeter precision range data to satellites, is not practical over interplanetary ranges. Double-ended laser transponders for decimeter or better accuracy interplanetary ranging and subnanosecond time transfer are well within the state-of-the-art, however, as was recently demonstrated in two successful transponder experiments carried out by the NASA Goddard Space Flight Center to laser altimeters onboard the Messenger spacecraft (currently enroute to Mercury) and the Mars Global Surveyor spacecraft (presently in Mars Orbit). A high-accuracy interplanetary ranging capability would support a number of new scientific investigations (e.g., solar system and planetary physics, general relativity, etc.) and enhance deep-space mission operations and reliability through vastly improved navigation accuracy and time synchronization with Earth mission control centers. The performance of future lunar or interplanetary laser transponder and laser communications instruments can be simulated and tested at distances to Pluto and beyond using existing passive SLR and LLR targets already in space.

1 Satellite and Lunar Laser Ranging

Laser ranging to passive retroreflectors on Earth orbiting satellites was first demonstrated at the NASA Goddard Space Flight Center on 31 October 1964 [8]. The basic measurement of this single-ended instrument is both simple and unambiguous. The outgoing laser pulse starts a highly precise timer, is reflected by the satellite, and the return signal stops the timer. One then multiplies the time interval by the speed of light, correcting for satellite signature (impulse response) and atmospheric propagation delay effects, to compute a range to the satellite center of mass. Today, an international network of approximately 40 satellite laser ranging (SLR) stations routinely track two dozen space missions in Earth orbit. Over the past four decades, the ranging precision has improved from a few meters to 1 or 2 mm, and the subcentimeter

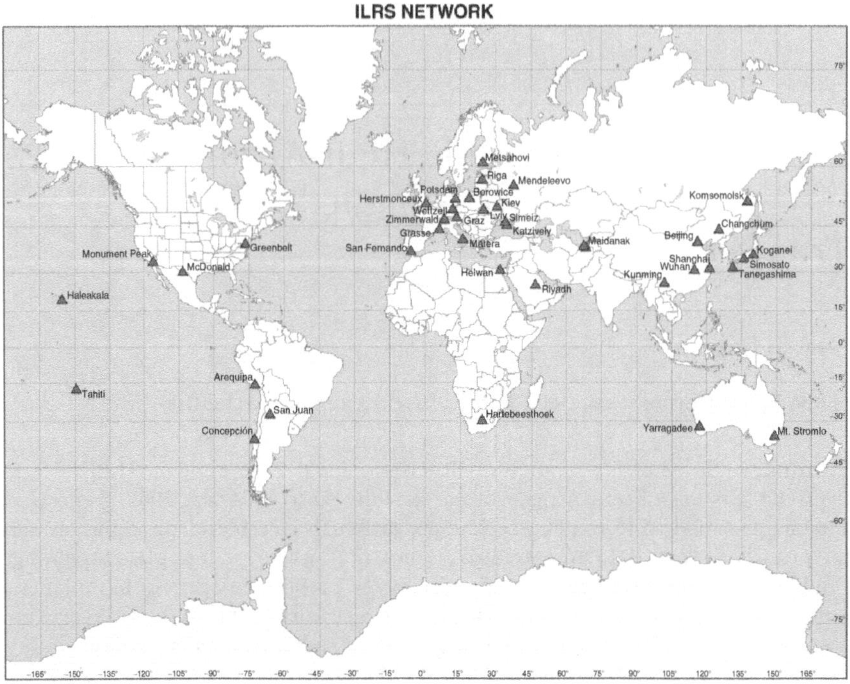

Fig. 1. Global distribution of the ILRS satellite laser ranging network.

absolute accuracy is presently limited, not by the instrumentation, but by uncertainties in the atmospheric propagation model and pulse spreading by the satellite target arrays. For more information on SLR, the reader is referred to a series of review articles devoted to SLR history [3], hardware [1], and mathematical models [2].

Since its inception in 1998, the International Laser Ranging Service (ILRS), an Official service of the International Association for Geodesy (IAG), has set mission tracking policy and managed the daily operations of the international SLR network. The global distribution of ILRS stations is shown in Fig. 1, and, as will be demonstrated later, most of these stations are potentially capable of supporting future centimeter ranging and subnanosecond time transfer to the other planets within our solar system.

A select few of the ILRS stations have successfully tracked one or more of the five retroreflectors placed on the Moon by the manned US Apollo 11, 14, and 15 and two unmanned Soviet Lunakhod missions to the Moon. Most of the operational lunar laser ranging (LLR) data over the past 36 years has come from three sites – the NASA/University of Texas station at McDonald Observatory, the French CERGA station in the coastal Mediterranean town of Grasse, and the NASA/University of Hawaii station at the top of Mt. Haleakala in Maui (which was decommissioned in 1992 due to NASA funding cuts). It is important to note that, even with meter class telescopes located at mountaintop sites with excellent atmospheric "seeing" and with moderately high subnanosecond pulse energies on the order of 100–200 mJ, LLR systems typically detect one single photon return from the lunar arrays out of every 10–20 laser fires, or roughly one photon per second at typical 10–20 Hz fire rates. This low signal photon return rate makes the extraction of the signal from background noise difficult except when the sunlit lunar surface is outside the receiver field of view (FOV). On the other hand, LLR observers have also found it necessary to offset their pointing from prominent lunar features to guide their narrow laser beam successfully to the target. The net consequence of these two constraints is to limit lunar tracking to temporal periods which are far from both "Full Moon" and "New Moon." In spite of these limitations, LLR has proved invaluable to a number of important scientific endeavors in the fields of lunar physics and general relativity [6]. These include:

Lunar Physics (LLR)

- Centimeter accuracy lunar ephemerides
- Lunar librations (variations from uniform rotation)
- Lunar tidal displacements
- Lunar mass distribution
- Secular deceleration due to tidal dissipation in Earth's oceans
- Measurement of $G(M_E + M_M)$

General Relativity

- Test/evaluation of competing theories
- Support atomic clock experiments in aircraft and spacecraft
- Verify equivalence principle
- Constrain β-parameter in the Robertson–Walker metric
- Constrain time rate of change in G

Under the Apache Point Observatory Lunar Laser-ranging Operation (APOLLO) program in New Mexico, activities have been underway to produce multiphoton lunar ranging returns through the use of larger 3.5 m diameter telescopes and more powerful lasers [7], and the first lunar returns were reported in October 2005.[1] Returns from both the strongest (Apollo 15 with 300 retroreflectors) and weakest (Apollo 11 with 100 retroreflectors) lunar targets were obtained including some successful experimental sessions near Full Moon. During the best run reported to date, 420 returns were detected out of 5,000 attempts for an 8.4% return rate. Nevertheless, the conventional SLR technique of ranging to passive retroreflectors is unlikely to be useful for targets much beyond the Earth–lunar distance (384,000 km or 0.0026 AU). This is due to the R^{-4} dependence of the received signal strength, where R is the target range.

[1] http://physics.ucsd.edu/~tmurphy/apollo/first_range.html

2 Science and Mission Benefits of Interplanetary Ranging

Transponders can overcome the distance limitations of conventional SLR and LLR systems. Transponders consist of two terminals – each with its own laser, telescope, and timing receiver [5]. Their principal advantage is that the signal strength falls off only as R^{-2}, and this greatly extends the range over which they can be used. The possibility of interplanetary ranging at the centimeter level provides new measurement opportunities in the fields of solar system and planetary science and general relativity. It also provides new operational capabilities, which can reduce the risk and cost of navigating and monitoring future spacecraft missions. Some examples follow:

Solar System and Planetary Science

- Solar physics: gravity field, internal mass distribution, and rotation
- Few millimeter accuracy lunar ephemerides and librations
- Improves ranging accuracy and temporal sampling over current LLR operations to Apollo retroreflectors on the Moon with small, low energy, ground stations
- Decimeter or better accuracy planetary ephemerides
- Mass distribution within the asteroid belt

General Relativity

It provides more accurate (2–3 orders of magnitude) tests of relativity and constraints on its metrics than LLR or microwave radar ranging to the planets, e.g.:

- Precession of Mercury's perihelion
- Constraints on the magnitude of $\dot{G}$ (1×10^{-12} from LLR)
- Gravitational and velocity effects on spacecraft clocks
- Shapiro time delay

Lunar and Planetary Mission Operations

- Decimeter or better accuracy spacecraft ranging
- Calibration/validation/backup for Deep Space Network (DSN) microwave tracking
- Subnanosecond transfer of GPS time to interplanetary spacecraft; for improved synchronization of Earth/spacecraft operations
- Transponder can serve as independent self-locking beacon for collocated laser communications systems

3 Echo vs. Asynchronous Transponders

There are two types of transponder: *echo* and *asynchronous*. The timing diagrams for echo and asynchronous transponders are shown in Fig. 2a,b, respectively. In an Earth–Moon echo transponder, for example, a pulse emitted from the Earth terminal at time t_{E1} is detected by the lunar terminal at time t_{M1} which then generates a response pulse at time t_{M2} subsequently detected by A at time t_{E2}. The delay between the received and transmitted pulse at the lunar terminal, t_d, would be either known a priori through careful calibration or controlled via active electronics and would be subtracted from the observed round-trip time before computing the target range. Alternatively, the delay can be measured locally by a timer at the lunar terminal and transmitted to the Earth terminal via a communications link. The signal return rate at the primary station is then equal to the fire rate of the laser multiplied by the joint probability that pulses are detected at both ends of the link. Thus, the simple echo approach works very well when the round-trip time-of-flight is relatively short and there is a high probability of detection at both ends of the link, i.e., when both the uplink and downlink signal is reasonably strong and pointing uncertainties are small relative to the transmitter divergence. This approach should work very well over Earth–Moon or shorter links. However, in interplanetary links where the light transit time is relatively long (several

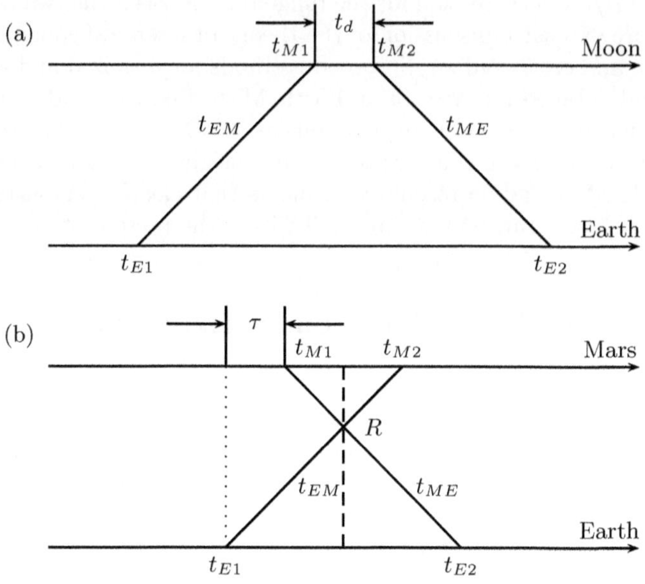

Fig. 2. Timing diagrams for (**a**) echo and (**b**) asynchronous transponder.

minutes to hours) and the probability of detection is small at one or both ends of the link, it is worthwhile considering the asynchronous laser transponder.

In an asynchronous transponder, the two terminals independently fire pulses at each other at a known laser fire rate, as illustrated by the timing diagram in Fig. 2b. For an Earth–Mars link, for example, the Earth terminal records the times of departure of its own transmitted pulses (t_{E1}) as well as the times of arrival of pulses from Mars (t_{E2}) and vice versa. In a high SNR system with good pointing, the pulses arrive at roughly the laser fire rate whereas, in low Signal-to-Noise Ratio (SNR) or photon-counting systems [5], the pulses may arrive intermittently. The departure and arrival times measured at each terminal are then communicated to, and properly paired at, an Earth-based processor which then calculates a range and clock offset between the two terminals for each set of two-way measurements occurring within a reasonably short time interval. The relevant equations are

$$R = \frac{c}{2}\left[t_{ME} + t_{EM}\right] = \frac{c}{2}\left[(t_{E2} - t_{E1}) + (t_{M2} - t_{M1})\right] \tag{1}$$

for the inter-terminal range at the time when the two photon world lines cross in Fig. 2b and

$$\tau = \frac{(t_{E2} - t_{E1}) - (t_{M1} - t_{M2})}{2(1 + \dot{R}/c)} \tag{2}$$

for the corresponding time offset between the pulses departing from each terminal, where $\dot{R}/c$ is a correction for the range rate between the two terminals.

For a more extensive discussion of the theory of laser transponders, background noise and error sources, proposed methods for terminal and signal acquisition, and detailed analyses of an Earth–Mars link, the reader is referred to a comprehensive article previously published by Degnan [5]. The remainder of the present chapter will concentrate on new insights gained by comparisons with the SLR effort and on recent experiments that clearly demonstrate that interplanetary laser transponders are well within the present state-of-the-art.

4 Recent Deep-Space Transponder Experiments

In late May 2005, NASA Goddard Space Flight Center (GSFC) was conducting the first successful two-way transponder experiments at a wavelength of 1,064 nm with a laser altimeter onboard the Messenger spacecraft, which is currently enroute to Mercury. From a distance of about 24 million km (0.17 AU), the Messenger spacecraft performed a raster scan of the Earth while firing its Q-switched Nd:YAG laser at an 8 Hz rate. Simultaneously, a ground based Q-switched Nd:YAG laser at GSFC's 1.2 m telescope was aimed at the Messenger spacecraft. During the few second periods when the Messenger raster scan passed over the Earth station, pulses were successfully exchanged between the two terminals [10]. The pulse time of departure and

Table 1. Summary of key instrument parameters for recent deep-space transponder experiments at 1,064 nm.

Experiment	MLA (cruise)		MOLA (Mars)
Range (10^6 km)	24.3		~80.0
Wavelength (nm)	1,064		1,064
	Uplink	Downlink	Uplink
Pulse width (ns)	10	6	5
Pulse energy (mJ)	16	20	150
Repetition rate (Hz)	240	8	56
Laser power (W)	3.84	0.16	8.4
Full divergence (μrad)	60	100	50
Receive area (m^2)	0.042	1.003	0.196
EA-product ($J\,m^{-2}$)	0.00067	0.020	0.0294
PA-product ($W\,m^{-2}$)	0.161	0.160	1.64

arrival data collected by the two terminals was used to estimate the Earth–spacecraft range with decimeter precision [9], a precision orders of magnitude better than could be achieved with the spacecraft microwave Doppler data.

In late September 2005, a similar experiment was conducted by the same GSFC team to the Mars Orbiter Laser Altimeter (MOLA), an instrument on the Mars Global Surveyor (MGS) spacecraft in orbit about Mars. Because the MOLA laser was no longer operable following a successful topographic mapping mission at Mars, this was necessarily a one-way experiment in which the MOLA detector saw hundreds of pulses from a modestly powered Q-switched Nd:YAG laser at GSFC. The instrument parameters for these two experiments are summarized in Table 1.

It must be stressed that the latter were experiments of opportunity, not design. The near-infrared (NIR) detectors used in these experiments are far less sensitive than the photon-counting visible detectors typically used in SLR or LLR. As a result, the energy–aperture (EA) product needed to observe a return in these preliminary experiments, although modest, was significantly higher than would be necessary for a dedicated deep-space transponder mission. Furthermore, neither spacecraft had a capability of independently acquiring and locking onto the opposite terminal and instead relied on temporary illumination during the raster scan.

5 Testing Future Transponders/Lasercom Systems in Space

Interest at NASA in laser communications has been intermittently high since the 1960s and, with the recent successful transponder experiments, interest in laser transponders is on the rise as well. Past initiatives for interplanetary transponders or laser communications often were bogged down in esoteric discussions on difficult topics such as the effects of atmospheric turbulence on beam propagation. It is well known that turbulence has several effects on laser beam propagation including beam spreading, short-term beam wander, and scintillation (fading) [2]. End-to-end ground-based experiments which can convincingly simulate all aspects of these complex systems are both difficult to envision and expensive to implement. Fortunately, atmospheric transmission and turbulence effects on the uplink and downlink beams are the same, whether the uplink beam is being reflected from a passive high altitude satellite in Earth orbit as in SLR/LLR or transmitted from a distant transponder or lasercom terminal in deep space. It may be relevant, therefore, to consider an experiment in which two closely spaced ground terminals range at different wavelengths to the same passive Earth-orbiting satellite as in Fig. 3. Each station must be located within the reflected return spot of the other station. The larger terminal, simulating the Earth station, would exchange reflected pulses from the satellite with a smaller station, simulating the remote transponder or lasercom terminal. In Fig. 3, we show SLR Station A ranging to a passive satellite (e.g., LAGEOS) in the infrared (1,064 nm) while Station B ranges to the same satellite in the green (532 nm). Each station is equipped with an additional receiver channel at the opposite wavelength to detect reflected pulses from the other station to simulate a dual wavelength transponder or lasercom experiment. The experiment is self-calibrating since the transponder measures the dogleg defined by Station A-satellite and Station B-satellite while the individual ranging systems measure the Station A-satellite and Station B-satellite

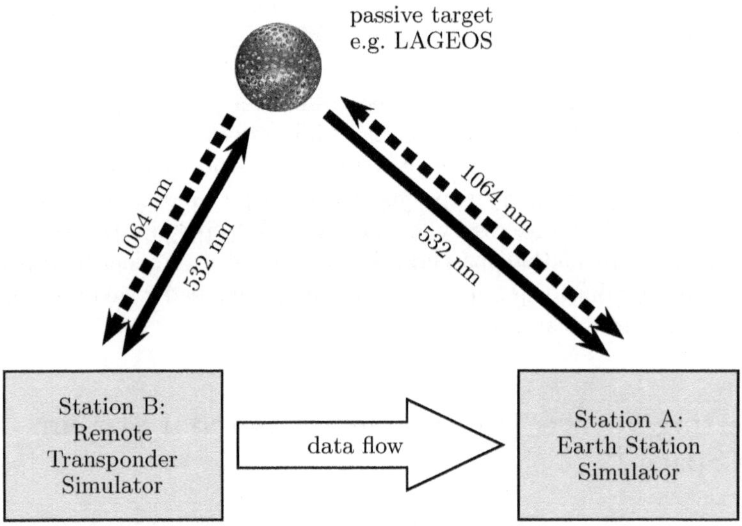

Fig. 3. Dual station laser ranging to LAGEOS with Station A simulating the Earth station and Station B simulating the remote transponder or lasercom terminal. Both stations must lie within each other's reflected spot.

distances and ground surveys typically define the interstation vector, or third leg of the triangle, to better than 2 mm. This provides an accurate way to test the ranging and time transfer algorithms. Automated acquisition of the Earth station by the remote terminal can be demonstrated by either turning off or ignoring the closed ranging loop at 532 nm while it searches for the reflected light at 1,064 nm. The ability to lock Station A onto the satellite via a closed single-ended ranging loop at 1,064 nm ensures a steady source of photons from the Earth station for the remote terminal to find and lock onto.

The link equations define the received signal strength at either station. For the infrared link from the Earth station A to the remote terminal B via a passive satellite, the link equation is given by [4]

$$\eta_R^{AB} = \frac{4\eta_q^B \eta_t^A \sigma_s \eta_r^B T_A^{2\sec\theta_A}}{h\nu_A(\theta_t^A)^2(4\pi)^2} \frac{E_t^A A_r^B}{R_R^4} \tag{3}$$

which depends on the transmitted energy E_t, the receive aperture A_r, the detector quantum efficiency η_q, the laser photon energy $h\nu$, the one-way zenith atmospheric transmission T_A, the satellite zenith angle θ_A ,the divergence half-angle of the laser beam θ_t, the target optical cross-section σ_t (measured in square meters), and the optical throughput efficiencies of the transmitter (η_t) and receiver (η_r) optics, respectively. The A and B superscripts and subscripts indicate the terminal for which the value applies, and are reversed for the opposite link from terminal B to A. The quantity R_R is the slant range to the target satellite. For the nominally circular orbits of typical SLR targets, R_R can be expressed as a function of the satellite height above sea level h, and the satellite zenith angle

$$R_R(h,\theta_A) = -R_E\cos\theta_A + \sqrt{(R_E\cos\theta_A)^2 + h(h+2R_E)}, \tag{4}$$

where $R_E = 6{,}378$ km is the mean volumetric radius of the Earth and (4) reduces to h when $\theta_A = 0$.

For interplanetary transponder (or lasercom) links, the link equation is given by [4]

$$\eta_T^{AB} = \frac{4\eta_q^B \eta_t^A \eta_r^B T_A^{\sec\theta_A} T_B^{\sec\theta_B}}{h\nu_A(\theta_t^A)^2(4\pi)} \frac{E_t^A A_r^B}{R_T^2}. \tag{5}$$

Setting the two mean signal counts equal in (3) and (5), we can derive an expression for the equivalent transponder distance, R_T, in terms of the actual slant range to the satellite, R_R, i.e.,

$$R_T(h,\theta_A,\sigma_s) = R_R^2(h,\cos\theta_A)\sqrt{\frac{4\pi}{\sigma_s}\frac{T_B^{\sec\theta_B}}{T_A^{\sec\theta_A}}} \simeq R_R^2(h,\cos\theta_A)\sqrt{\frac{4\pi}{\sigma_s}\frac{1}{T_A^{\sec\theta_A}}}, \tag{6}$$

where the approximation holds if the remote terminal is in interplanetary cruise phase, in orbit, or sitting on the surface of a planet or moon with little

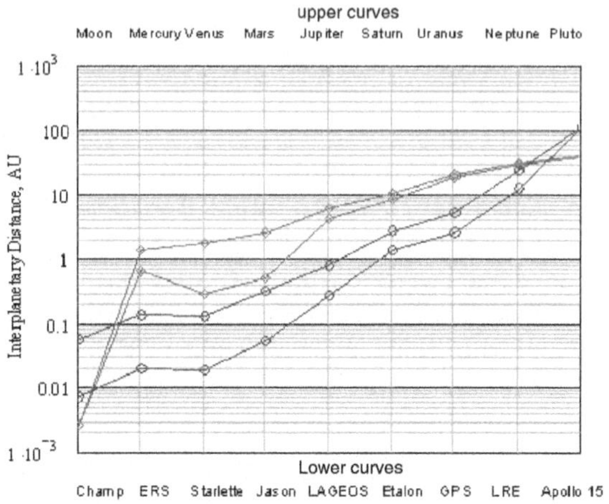

Fig. 4. The minimum and maximum distances from the Earth to the Moon and the eight planets listed at the top of the graph are illustrated by the *two upper curves* in the figure. The minimum and maximum transponder ranges simulated by the various SLR satellites listed at the bottom of the figure are indicated by the *two lower curves.*

or no atmosphere ($T_B \sim 1$). Since the SLR satellites are normally tracked over the range $0° \leq \theta_A \leq 70°$, (6) defines a maximum and minimum simulated transponder range for each satellite. These are indicated by the blue curves in Fig. 4 for our selected satellites. In the plots, we have assumed a value $T_A = 0.7$ corresponding to the one-way zenith transmission for a standard clear atmosphere at 532 nm. The red curves are plots of the minimum and maximum interplanetary distances of the Moon and other planets from Earth.

It is worthwhile to note that atmospheric turbulence can influence the effective transmitter beam divergence on the uplink, but this cancels out in our derivation of (6). Furthermore, the fading statistics for the dual station ranging experiment to the passive satellite should be comparable to that of an interplanetary transponder or lasercom experiment, at least to the extent that the satellite mimics a coherent point source of radiation.

Figure 4 shows that a dual station ranging experiment to the lowest of the SLR satellites, Champ, provides a weaker return than a two-way lunar transponder. Low elevation angle experiments to Jason are comparable to a Venus or Mars link when they are closest to Earth. Experiments to the LAGEOS and Etalon satellites would simulate ranging to Mercury, Venus, and Mars throughout their synodic cycles while experiments to GPS and LRE (at 25,000 km) would simulate links up to and beyond Jupiter and Saturn. Dual station experiments to the Apollo 15 reflector on the lunar surface would simulate transponder links to over 100 AU, well beyond the orbit of Pluto (<40 AU).

The nine SLR satellites in Fig. 4 were chosen based on their ability to simulate different transponder ranges and because the effects of target signa-

Table 2. Characteristics of selected SLR satellites which can be used to simulate deep-space transponder or lasercom links (from ILRS website).

Satellite	Altitude (km)	Mean target cross-section (10^6 m^2)	Minimum transponder range (AU)	Maximum transponder range (AU)
Champ	500	1.0	0.007	0.057
ERS 1 and 2	800	0.85	0.02	0.135
Starlette–Stella	950	1.8	0.019	0.123
Jason	1,300	0.8	0.054	0.306
LAGEOS	6,000	15	0.263	0.771
GLOHASS	19,000	55	1.38	2.72
GPS	20,000	19	2.60	5.06
LRE (elliptical)	25,000	2	12.52	23.12
Apollo 15	384,000	1, 400	111.6	

ture are minimized. The reduced pulse spreading by the target significantly improves the precision of the measured transponder range and also provides a reasonably high fidelity facsimile of the outgoing optical pulse train from a ground-based lasercom transmitter. The primary characteristics of these satellites, used in the computation of equivalent transponder ranges, are summarized in Table 2.

Another way to interpret Fig. 4 is to say that any *single* SLR station that can track the aforementioned satellites has demonstrated an adequate EA-product for the corresponding transponder link under the same noise background and atmospheric conditions. Since all of the ILRS stations are required to track LAGEOS for membership, they all have adequate EA-product to track out to about 1 AU. About a third of ILRS stations regularly track GPS, which from Fig. 4 or Table 2 implies an equivalent transponder range out to 5 AU. The workhorse NASA MOBLAS system, with an EA-product of $0.045\,\mathrm{J\,m^{-2}}$ and a power–aperture (PA) product of $0.23\,\mathrm{W\,m^{-2}}$, falls into this category as does the photon-counting Graz station in Austria with EA and PA products of $0.79 \times 10^{-5}\,\mathrm{J\,m^{-2}}$ and $0.157\,\mathrm{W\,m^{-2}}$, respectively. As mentioned previously, three stations have routinely tracked the Apollo reflectors but only at night with low noise background and single photon returns. Nevertheless, the same EA-product, which is only about 70% larger than a MOBLAS, should permit transponder links beyond 100 AU under equivalent operating conditions.

6 Concluding Remarks

It is clear from the recent successes that decimeter or better interplanetary ranging and subcentimeter time transfer is within the current state-of-the-art and can be achieved with very modest laser powers and telescope apertures. These experiments of opportunity have bolstered interest at NASA in laser transponders and laser communications. In a recent development, NASA's Lunar Reconnaissance Orbiter (LRO), tentatively scheduled for launch in 2008, will carry a small (21 mm diameter) telescope, with a relatively wide (1.15°) FOV, on its S-band microwave communications antenna. The latter will be used to view Earth-based SLR systems from lunar orbit. The incoming optical pulses at 532 nm will be transmitted from the focal plane of the telescope via fiber to one of the ranging detectors in the Lunar Orbiter Laser Altimeter (LOLA) instrument. The LOLA detectors, although designed primarily for the few nanosecond resolution altimetry channel at 1,064 nm, have sensitivity at 532 nm and will provide one-way differential range data to the altimetric mission, which requires highly accurate orbits for mapping the lunar topography and gravity field. Due to schedule constraints, there are no plans to put a transmitter on the LRO mission for full two-way transponding.

References

1. J.J. Degnan: Satellite Laser Ranging: Current Status and Future Prospects, *IEEE Transactions on Geoscience and Remote Sensing*, **GE-23**, 398 (1985).
2. J.J. Degnan: Millimeter Accuracy Satellite Laser Ranging: A Review, **in** Contributions of Space Geodesy to Geodynamics: Technology, D.E. Smith and D.L. Turcotte (Eds.), *AGU Geodynamics Series* **25**, 133 (1993).
3. J.J. Degnan: Thirty Years of Satellite Laser Ranging, Proc. Ninth International Workshop on Laser Ranging Instrumentation, pp. 1–20, Canberra, Australia, November 7–11, 1994.
4. J.J. Degnan: Photon Counting Microlaser Rangers, Transponders, and Altimeters, Surveys in Geophysics, Special Issue on *Evolving Geodesy* **22**, 431 (2001).
5. J.J. Degnan: Asynchronous Laser Transponders for Precise Interplanetary Ranging and Time Transfer, Journal of Geodynamics (Special Issue on Laser Altimetry), pp. 551–594, November, 2002.
6. J.O. Dickey, P.L. Bender, J.E. Faller, X.X. Newhall, R.L. Ricklefs, J.G. Ries, P.J. Shelus, C. Veillet, A.L. Whipple, J.R. Wiant, J.G. Williams, and C.F. Yoder: Lunar laser ranging: a continuing legacy of the Apollo Program, *Science* **265**, 482 (1994).
7. T.W. Murphy Jr., J.D. Strasburg, C.W. Stubbs, E.G. Adelberger, L. Tom, A.E. Orin, E.L. Michelson, J. Battat, C.D. Hoyle, E. Swanson, and E. Williams: APOLLO: Meeting the Millimeter Goal, 14th International Workshop on Laser Ranging, San Fernando, Spain, June 7–11 (2004).
8. H.H. Plotkin, T.S. Johnson, P.L. Spadin, and J. Moye: Reflection of Ruby Laser Radiation from Explorer XXII, *Proc. IEEE* **53** 301 (1965).
9. D.E. Smith, M.T. Zuber, X. Sun, G.A. Neumann, J.F. Cavanaugh, J.F. McGarry, and T.W. Zagwodzki: Two Way Laser Link over Interplanetary Distance, *Science* **311**, 53 (2006).
10. X. Sun, G.A. Neumann, J.F. McGarry, T.W. Zagwodzki, J.F. Cavanaugh, J.J. Degnan, D.B. Coyle, D.R. Skillman, M.T. Zuber, and D.E. Smith: Laser Ranging between the Mercury Laser Altimeter and an Earth-based Laser Satellite Tracking Stationover a 24 Million Kilometer Distance, presented at OSA Annual Meeting, Tucson, AZ, October 16–20, 2005.

Unequal-Arm Interferometry and Ranging in Space

Massimo Tinto

Jet Propulsion Laboratory, California Institute of Technology, Pasadena, CA 91109
also at: LIGO Laboratories, California Institute of Technology, Pasadena, CA 91125

Summary. Spaceborne interferometric gravitational wave detectors, sensitive in the low-frequency (millihertz) band, will fly in the next decade. In these detectors the spacecraft-to-spacecraft light travel times will necessarily be unequal, time-varying, and (due to aberration) have different time delays on up- and downlinks. By using knowledge of the interspacecraft light travel times and their time evolution, it is possible to cancel in postprocessing the otherwise dominant laser phase noise and obtain a variety of interferometric data combinations sensitive to gravitational radiation. This technique, which has been named time-delay interferometry (TDI), can be implemented with constellations of three or more formation-flying spacecraft that coherently track each other. As an example application we consider the Laser Interferometer Space Antenna (LISA) mission and show that TDI combinations can be synthesized by properly time shifting and linearly combining the phase measurements performed onboard the three spacecraft. Since TDI exactly suppresses the laser noises when the delays coincide with the light travel times, we then show that TDI can also be used for estimating the time delays needed for its implementation. This is done by performing a postprocessing nonlinear minimization procedure, which provides an effective, powerful, and simple way for making measurements of the interspacecraft light travel times. This processing technique, named time-delay interferometric ranging (TDIR), is highly accurate in estimating the time delays and allows TDI to be successfully implemented without the need of a dedicated ranging subsystem.

1 Introduction

Interferometric detectors of gravitational radiation (with frequency content $0 < f < f_u$) use a coherent train of electromagnetic waves (of nominal frequency $\nu_0 \gg f_u$) folded into several beams, and at one or more points where these intersect, monitor relative fluctuations of frequency or phase (homodyne detection). The observed low-frequency fluctuations are due to several causes (a) frequency variations of the source of the electromagnetic signal about ν_0, (b) relative motions of the electromagnetic source and the mirrors (or amplifying transponders) that do the folding, (c) temporal variations of the index of refraction along the beams, and (d) according to general relativity, to any time-variable gravitational fields present, such as the transverse–traceless metric curvature of a passing plane gravitational wave (GW). To observe gravitational waves in this way, it is thus necessary to control, or monitor, the other sources of relative frequency fluctuations, and, in the data analysis, to use optimal algorithms based on the different characteristic interferometer responses to gravitational waves (the signal) and to the other sources (the noise) [1]. By comparing phases of electromagnetic beams referenced to the same frequency generator and propagated along nonparallel equal-length arms, frequency fluctuations of the frequency reference can be removed and gravitational wave signals at levels many orders of magnitude lower can be detected.

In the present single spacecraft Doppler tracking observations, for instance, many of the noise sources can be either reduced or calibrated by implementing appropriate microwave frequency links and by using specialized electronics [2], so the fundamental limitation is imposed by the frequency (time-keeping) fluctuations inherent to the reference clock that controls the microwave system. Hydrogen maser clocks, currently used in Doppler tracking experiments, achieve their best performance at about 1,000 s integration time, with a fractional frequency stability of a few parts in 10^{-16}. This is the reason why these one-arm interferometers in space (which have one Doppler readout and a "3-pulse" response to gravitational waves [3]) are most sensitive to millihertz gravitational waves. This integration time is also comparable to the microwave propagation (or "storage") time $2L/c$ to spacecraft en route to the outer solar system (for example $L \simeq 5 - 8$ AU for the Cassini spacecraft) [2].

Next-generation low-frequency interferometric gravitational wave detectors in solar orbits, such as the Laser Interferometer Space Antenna (LISA) mission [4] and the Astrodynamical Space Test of Relativity using Optical Devices (ASTROD) mission [5], have been proposed to achieve greater sensitivity to millihertz gravitational waves. Since the arm lengths of these space-based interferometers can differ by a few percent (for LISA) to tens of percents (for ASTROD), the direct recombination of the two beams at a photodetector will not effectively remove the laser frequency noise. This is because the frequency fluctuations of the laser will be delayed by different amounts within the two unequal-length arms. To cancel the laser frequency noise, the time-varying Doppler data must be recorded and postprocessed to allow for arm-length differences [6]. The data streams will have temporal structure, which can be described as due to many-pulse responses to δ-function excitations, depending on time-of-flight delays in the response functions of the instrumental Doppler noises and in the response to incident plane-parallel, transverse, and traceless gravitational waves.

In what follows we will give an account of TDI as it will be implemented by LISA. Each of its three spacecraft orbiting the Sun will be equipped with two lasers sending beams to the other two (~0.03 AU away) while simultaneously measuring the beat frequencies between the local laser and the laser beams received from the other spacecraft. The description of TDI that will be presented in this chapter will assume a successful prior removal of any first-order Doppler beat notes due to spacecraft relative motions [7], giving six residual interspacecraft Doppler time series as the raw data of a *stationary* time-delay space interferometer. Following [8–10], we will regard LISA not as constituting one or more conventional Michelson interferometers, but rather, in a symmetrical way, a closed array of six one-arm delay lines between the test masses. This point of view is very powerful since it allows one to synthesize new data combinations that cancel laser frequency noises, and estimate achievable sensitivities of these combinations in terms of the separate and relatively simple single arm responses both to gravitational wave and instrumental noise.

In contrast to Earth-based interferometers, which operate in the long-wavelength limit (LWL) (arm lengths $\ll$ gravitational wavelength $\sim c/f_0$, where f_0 is a characteristic frequency of the GW), LISA will *not* operate in the LWL over much of its frequency band. When the physical scale of a free mass optical interferometer intended to detect gravitational waves is comparable to or larger than the GW wavelength, time delays in the response of the instrument to the waves, and travel times along beams in the instrument, cannot be ignored and must be allowed for in computing the detector response used for data interpretation.

This chapter is organized as follows. In Sect. 2 we summarize the one-arm Doppler transfer functions of an optical beam between two carefully shielded test masses inside each spacecraft resulting from (1) frequency fluctuations of the lasers used in transmission and reception, (2) fluctuations due to noninertial motions of the spacecraft, and (3) beam-pointing fluctuations and shot noise [11]. Among these, the dominant noise is from the frequency fluctuations of the lasers and is several orders (perhaps 7 or 8) above the other noises. This noise must be very precisely removed from the data to achieve the GW sensitivity at the level set by the remaining Doppler noise sources, which are at a much lower level and constitute the noise floor after the laser frequency noise is suppressed. We show that this can be accomplished by shifting and linearly combining the 12 one-way Doppler data that LISA will measure. The actual procedure can easily be understood in terms of properly defined time-delay operators that act on the one-way Doppler measurements.

As an example application, we then derive the unequal-arm Michelson interferometric combination in the simple case in which the light travel times are constant in time and independent from being up- or downlinks. The expressions for the Sagnac interferometric combinations $(\alpha, \beta, \gamma, \zeta)$, as well as all those combinations that rely only on four of the possible six interspacecraft Doppler measurements (denoted P, E, and U), are not derived in this chapter, and the reader is referred to [12,13] for details on their derivations.

In Sect. 3, we then consider the formulation of TDI when spacecraft-to-spacecraft light travel times are not constant in time, and dependent from being up- or downlinks. Reduction of data from moving interferometric laser arrays in solar orbit will in fact encounter nonsymmetric up- and downlink light time differences that are significant, and need to be accounted for to exactly cancel the laser frequency fluctuations [12, 14, 15]. In Sect. 4, we show that, by introducing a set of noncommuting time-delay operators, there exists a quite general procedure for deriving generalized TDI combinations that account for the effects of time dependence of the arms. Using this approach, it is possible to derive "flex-free" expression for the unequal-arm Michelson combinations X_1, and obtain the generalized expressions for all the TDI observables [13].

Since TDI relies on the accurate knowledge of the time delays that have to be applied to the phase measurements, in Sect. 4, we show that it is possible to estimate the delays by implementing a nonlinear least-squares minimization procedure in which a TDI combination is used for estimating the time delays. This procedure, which has been called *time-delay interferometric ranging* [16], relies on the fact that TDI nulls all the laser noises when the time delays are chosen to match the travel times experienced by the laser beams as they propagate along the sides of the array. TDIR allows the implementation of TDI without a separate interspacecraft ranging subsystem, significantly simplifying the design of the LISA instrument. At the very least, TDIR can supplement such a subsystem, allowing the synthesis of TDI combinations during ranging dropouts or glitches.

2 Time-Delay Interferometry

2.1 Statement of the Problem

Equal-arm interferometer detectors of gravitational waves can observe gravitational radiation by canceling the laser frequency fluctuations affecting the light injected into their arms. This is done by comparing phases of split beams propagated along the equal (but nonparallel) arms of the detector. The laser frequency fluctuations affecting the two beams experience the same delay within the two equal-length arms and cancel out at the photodetector where relative phases are measured. This way gravitational wave signals of dimensionless amplitude less than 10^{-20} can be observed when using lasers whose frequency stability can be as large as roughly a few parts in 10^{-13}.

If the arms of the interferometer have different lengths, however, the exact cancellation of the laser frequency fluctuations, say $C(t)$, will no longer take place at the photodetector. In fact, the larger the difference between the two arms, the larger will be the magnitude of the laser frequency fluctuations affecting the detector response. If L_1 and L_2 are the lengths of the two arms, it is easy to see that the amount of laser relative frequency fluctuations remaining in the response is equal to (units in which the speed of light $c = 1$)

$$\Delta C(t) = C(t - 2L_1) - C(t - 2L_2). \tag{1}$$

In the case of LISA, whose lasers are expected to display relative frequency fluctuations equal to about $10^{-13}\,\mathrm{Hz}^{-1/2}$ in the millihertz band, and whose arms will differ by a few percent [4], (1) implies the following expression for the amplitude of the Fourier components of the uncanceled laser frequency fluctuations (an overimposed tilde denotes the operation of Fourier transform)

$$|\widetilde{\Delta C}(f)| \simeq |\widetilde{C}(f)|\, 4\pi f |(L_1 - L_2)|. \tag{2}$$

At $f = 10^{-3}\,\mathrm{Hz}$, for instance, and assuming $|L_1 - L_2| \simeq 0.5\,\mathrm{s}$, the uncanceled fluctuations from the laser are equal to $6.3 \times 10^{-16}\,\mathrm{Hz}^{-1/2}$. Since the LISA sensitivity goal is about $10^{-20}\,\mathrm{Hz}^{-1/2}$ in this part of the frequency band, it is clear that an alternative experimental approach for canceling the laser frequency fluctuations is needed.

The solution to this problem can be understood through Fig. 1. In this idealized model the two beams exiting the two arms are not made to interfere at a common photodetector. Rather, each is made to interfere with the incoming light from the laser at a photodetector, decoupling in this way the phase fluctuations experienced by the two beams in the two arms. Now two Doppler measurements are available in digital form, and the problem becomes one of identifying an algorithm for digitally canceling the laser frequency fluctuations from a resulting new data combination.

The algorithm that performs the cancellation of the laser noise from the two Doppler measurements from the two arms, say $y_1(t)$ and $y_2(t)$, works as follows. Let us denote with $h_1(t)$ and $h_2(t)$ the gravitational wave signals entering into the Doppler data y_1 and y_2, respectively, and with n_1 and n_2 any other remaining noise affecting y_1 and y_2, respectively. The resulting

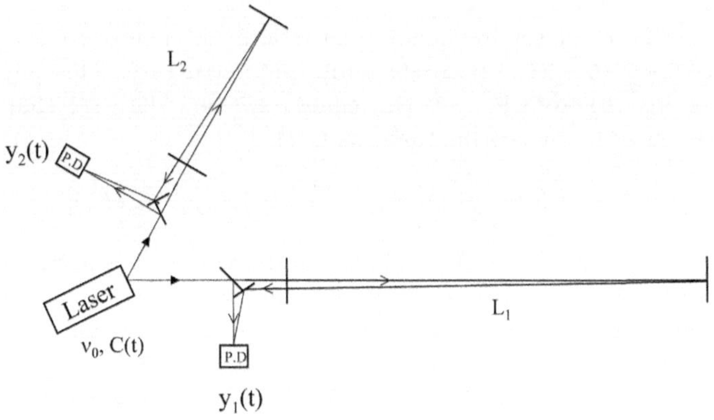

Fig. 1. Light from a laser is split into two beams, each injected into an arm formed by pairs of free-falling mirrors. Since the length of the two arms, L_1 and L_2, are different, the light beams from the two arms are not recombined at one photodetector. Instead each is separately made to interfere with the light that is injected into the arms. Two distinct photodetectors are used, and phase (or frequency) fluctuations are then monitored and recorded there.

expressions for the Doppler observables y_1 and y_2 can be written in the following form

$$y_1(t) = C(t - 2L_1) - C(t) + h_1(t) + n_1(t), \tag{3}$$

$$y_2(t) = C(t - 2L_2) - C(t) + h_2(t) + n_2(t). \tag{4}$$

From (3) and (4), it is important to note the characteristic time signature of the random process $C(t)$ in the Doppler responses y_1 and y_2. The time signature of the noise $C(t)$ in $y_1(t)$, for instance, can be understood by observing that the frequency of the signal received at time t contains laser frequency fluctuations transmitted $2L_1$ seconds earlier. By subtracting from the frequency of the received signal the frequency of the signal transmitted at time t, we also subtract the frequency fluctuations $C(t)$ with the net result shown in (3).

From (3) and (4) we may notice that, by taking the difference of the two Doppler data $y_1(t)$ and $y_2(t)$, the frequency fluctuations of the laser enter into this new data set in the following way:

$$y_1(t) - y_2(t) = C(t - 2L_1) - C(t - 2L_2) + h_1(t) - h_2(t) + n_1(t) - n_2(t). \tag{5}$$

If we now compare how the laser frequency fluctuations enter into (5) against how they appear in (3) and (4), we can further make the following observation. If we time shift the data $y_1(t)$ by the round-trip light time in arm 2, $y_1(t-2L_2)$, and subtract from it the data $y_2(t)$ after it has been time shifted by the round-trip light time in arm 1, $y_2(t - 2L_1)$, we obtain the following data set

$$\begin{aligned} y_1(t - 2L_2) - y_2(t - 2L_1) = {} & C(t - 2L_1) - C(t - 2L_2) + h_1(t - 2L_2) \\ & -h_2(t - 2L_1) + n_1(t - 2L_2) - n_2(t - 2L_1). \end{aligned} \tag{6}$$

In other words, the laser frequency fluctuations enter into $y_1(t) - y_2(t)$ and $y_1(t - 2L_2) - y_2(t - 2L_1)$ with the same time structure. This implies that, by subtracting (6) from (5), we can generate a new data set that does not contain the laser frequency fluctuations $C(t)$

$$X \equiv [y_1(t) - y_2(t)] - [y_1(t - 2L_2) - y_2(t - 2L_1)]. \tag{7}$$

The expression above of the X combination shows that it is possible to cancel the laser frequency noise in the time domain by properly time shifting and linearly combining Doppler measurements recorded by different Doppler readouts. This in essence is what TDI amounts to. In what follows we will further elaborate and generalize TDI to the realistic LISA configuration.

2.2 Time-Delay Interferometry

The description of TDI for LISA is greatly simplified if we adopt the notation shown in Fig. 2, where the overall geometry of the LISA detector is defined. There are three spacecraft, six optical benches, six lasers, six proof masses, and

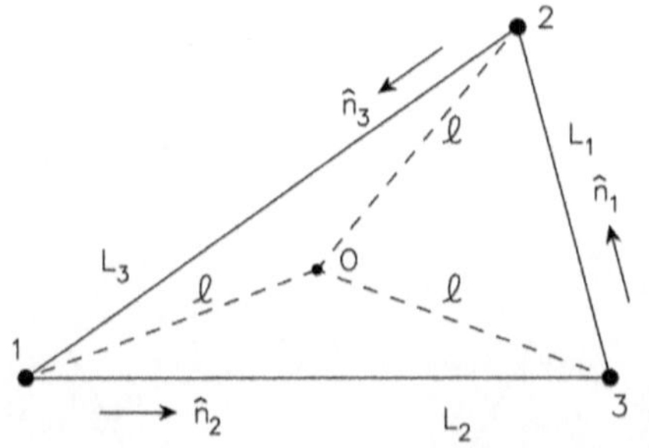

Fig. 2. Schematic LISA configuration. The spacecraft are labeled 1, 2, and 3. The optical paths are denoted by L_i, where the index i corresponds to the opposite spacecraft. The unit vectors $\hat{\mathbf{n}}_i$ point between pairs of spacecraft, with the orientation indicated.

12 photodetectors. There are also six phase difference data going clockwise and counterclockwise around the LISA triangle. For the moment we will make the simplifying assumption that the array is stationary, i.e., the back and forth optical paths between pairs of spacecraft are simply equal to their relative distances [12–15].

The spacecraft are labeled 1, 2, 3 and their separating distances are denoted L_1, L_2, L_3, with L_i being opposite spacecraft i. We orient the vertices 1, 2, 3 clockwise in Fig. 2. Unit vectors between spacecraft are $\hat{n}_i$, oriented as indicated in Fig. 2. We index the phase difference data to be analyzed as follows: the beam arriving at spacecraft i, transmitted by spacecraft j, gives rise to the phase measurement s_{ji} (along L_k). Similarly, s_{ij} is the phase difference series derived from reception at spacecraft j with transmission from spacecraft i. The other four one-way phase difference time series from signals exchanged between the spacecraft are obtained by cyclic permutation of the indices: $1 \to 2 \to 3 \to 1$. We also adopt a notation for delayed data streams, which will be convenient later for algebraic manipulations. We define the three time-delay operators $\mathcal{D}_i$, $i = 1, 2, 3$ where for any data stream $x(t)$

$$\mathcal{D}_i x(t) \equiv x(t - L_i), \tag{8}$$

where L_i, $i = 1, 2, 3$ are the light travel times along the three arms of the LISA triangle (the speed of light c is assumed to be unity in this chapter). Thus, for example, $\mathcal{D}_2 s_{13}(t) = s_{13}(t - L_2)$, $\mathcal{D}_2\mathcal{D}_3 s_{13}(t) = s_{13}(t - L_2 - L_3) = \mathcal{D}_3\mathcal{D}_2 s_{13}(t)$, etc. Note that the operators commute here. This is because the arm lengths have been assumed to be constant in time. If the L_i are functions of time, then the operators no longer commute [12, 13, 15], as will be described later.

Six more phase difference series result from laser beams exchanged between adjacent optical benches within each spacecraft; these are similarly indexed as τ_{ij}, $i, j = 1, 2, 3, i \neq j$. The proof mass plus optical bench assemblies for LISA spacecraft number 1 are shown schematically in Fig. 3. The photoreceivers that generate the data s_{21}, s_{31}, τ_{21}, and τ_{31} at spacecraft 1 are shown.

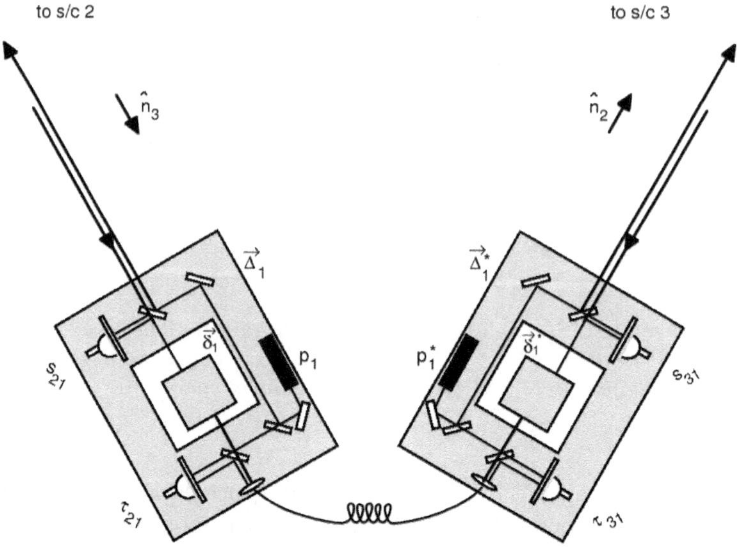

Fig. 3. Schematic diagram of proof masses plus optical benches for a LISA spacecraft. The *left-hand bench* reads out the phase signals s_{21} and τ_{21}. The *right-hand bench* analogously reads out s_{31} and τ_{31}. The random displacements of the two proof masses and two optical benches are indicated (*lower case* $\boldsymbol{\delta}_i, \boldsymbol{\delta}_i^*$ for the proof masses, *upper case* $\boldsymbol{\Delta}_i, \Delta_i^*$ for the optical benches).

The phase fluctuations from the six lasers, which need to be canceled, can be represented by six random processes p_i, p_i^*, where p_i, p_i^* are the phases of the lasers in spacecraft i on the left and right optical benches, respectively, as shown in Fig. 3. We extend the cyclic terminology so that at vertex i $(i = 1, 2, 3)$ the random displacement vectors of the two proof masses are, respectively, denoted by $\boldsymbol{\delta}_i(t), \boldsymbol{\delta}_i^*(t)$, and the random displacements (perhaps several orders of magnitude greater) of their optical benches are correspondingly denoted by $\boldsymbol{\Delta}_i(t), \boldsymbol{\Delta}_i^*(t)$ where a * added to a quantity means that it is located on the right optical bench. As pointed out in [11], the analysis does *not* assume that pairs of optical benches are rigidly connected, i.e., $\boldsymbol{\Delta}_i \neq \boldsymbol{\Delta}_i^*$, in general. The present LISA design shows optical fibers transmitting signals both ways between adjacent benches. We ignore time-delay effects for these signals and will simply denote by $\mu_i(t)$ the phase fluctuations upon transmission through the fibers of the laser beams with frequencies ν_i and ν_i^*. The $\mu_i(t)$ phase shifts within a given spacecraft might not be the same for large frequency differences $\nu_i - \nu_i^*$. For the envisioned frequency differences (a few hundred megahertz), however, the remaining fluctuations due to the optical fiber can be neglected [11]. It is also assumed that the phase noise added by the fibers is independent of the direction of light propagation through them. For ease of presentation, in what follows we will assume the center frequencies of the lasers to be the same, and denote this frequency by ν_0.

The laser phase noise in s_{23} is therefore equal to $\mathcal{D}_1 p_2(t) - p_3^*(t)$. Similarly, since s_{32} is the phase shift measured on arrival at spacecraft 2 along arm 1 of a signal transmitted from spacecraft 3, the laser phase noises enter into it with the following time signature: $\mathcal{D}_1 p_3^*(t) - p_2(t)$. Figure 3 endeavors to make the detailed light paths for these observations clear. An outgoing light beam transmitted to a distant spacecraft is routed from the laser on the local optical bench using mirrors and beam splitters; this beam does not interact with the local proof mass. Conversely, an *incoming* light beam from a distant spacecraft is bounced off the local proof mass before being reflected onto the photoreceiver where it is mixed with light from the laser on that same optical bench. The interspacecraft phase data are denoted s_{21} and s_{31} in Fig. 3.

Beams between adjacent optical benches within a single spacecraft are bounced off proof masses in the opposite way. Light to be *transmitted* from the laser on an optical bench is *first* bounced off the proof mass it encloses and then directed to the other optical bench. Upon reception it does *not* interact with the proof mass there, but is directly mixed with local laser light, and again down converted. These data are denoted τ_{21} and τ_{31} in Fig. 3.

The expressions for the s_{ji} and τ_{ji} phase measurements can now be developed from Figs. 2 and 3, and they are for the particular LISA configuration in which all the lasers have the same nominal frequency ν_0, and the spacecraft are stationary with respect to each other. Consider the $s_{31}(t)$ process (11). The photoreceiver on the right bench of spacecraft 1, which (in the spacecraft frame) experiences a time-varying displacement $\boldsymbol{\Delta}_1^*$, measures the phase difference s_{31} by first mixing the beam from the distant optical bench 3 in direction $\hat{n}_2$, and laser phase noise p_3 and optical bench motion $\boldsymbol{\Delta}_3$ that have been delayed by propagation along L_2, after one bounce off the proof mass ($\boldsymbol{\delta}_1^*$), with the local laser light (with phase noise p_1^*). Since for this simplified configuration no frequency offsets are present, there is of course no need for any heterodyne conversion [7].

In (10) the τ_{21} measurement results from light originating at the right-bench laser (p_1^*, $\boldsymbol{\Delta}_1^*$), bounced once off the right proof mass ($\boldsymbol{\delta}_1^*$), and directed through the fiber (incurring phase shift $\mu_1(t)$), to the left bench, where it is mixed with laser light (p_1). Similarly the right bench records the phase differences s_{31} and τ_{31}. The laser noises, the gravitational wave signals, the optical path noises, and proof mass and bench noises enter into the four data streams recorded at vertex 1 according to the following expressions [11]:

$$s_{21} = s_{21}^{\mathrm{gw}} + s_{21}^{\mathrm{opt.path}} + \mathcal{D}_3 p_2^* - p_1 + \nu_0 \left[-2\hat{n}_3 \cdot \boldsymbol{\delta}_1 + \hat{n}_3 \cdot \boldsymbol{\Delta}_1 + \hat{n}_3 \cdot \mathcal{D}_3 \boldsymbol{\Delta}_2^*\right], \quad (9)$$

$$\tau_{21} = p_1^* - p_1 - 2\nu_0 \hat{n}_2 \cdot (\boldsymbol{\delta}_1^* - \boldsymbol{\Delta}_1^*) + \mu_1, \quad (10)$$

$$s_{31} = s_{31}^{\mathrm{gw}} + s_{31}^{\mathrm{opt.path}} + \mathcal{D}_2 p_3 - p_1^* + \nu_0 \left[2\hat{n}_2 \cdot \boldsymbol{\delta}_1^* - \hat{n}_2 \cdot \boldsymbol{\Delta}_1^* - \hat{n}_2 \cdot \mathcal{D}_2 \boldsymbol{\Delta}_3\right], \quad (11)$$

$$\tau_{31} = p_1 - p_1^* + 2\nu_0 \hat{n}_3 \cdot (\boldsymbol{\delta}_1 - \boldsymbol{\Delta}_1) + \mu_1. \quad (12)$$

Eight other relations, for the readouts at vertices 2 and 3, are given by cyclic permutation of the indices in (9–12).

The gravitational wave phase signal components, s_{ji}^{gw}, in (9) and (11) are given by integrating with respect to time equations (1) and (2) of [9], which relate metric perturbations to optical frequency shifts. The optical path phase noise contributions, $s_{ij}^{\text{opt. path}}$, which include shot noise from the low signal-to-noise ratio (SNR) in the links between the distant spacecraft, can be derived from the corresponding terms given in [11]. The τ_{ij} measurements will be made with high SNR so that for them the shot noise is negligible.

To simplify the derivation of the expressions canceling the laser and optical bench noises, let us focus for the moment only on the laser and optical bench noises entering into the observables s_{ij} and τ_{ij}. Note that by subtracting (12) from (10), we can rewrite the resulting expression (and those obtained from it by permutation of the spacecraft indices) in the following form

$$z_1 \equiv \frac{1}{2}(\tau_{21} - \tau_{31}) = \phi_1^* - \phi_1 \,, \tag{13}$$

where ϕ_1^* and ϕ_1 are defined as,

$$\begin{aligned} \phi_1^* &\equiv p_1^* + \nu_0 \hat{\mathbf{n}}_2 \cdot \boldsymbol{\Delta}_1^*, \\ \phi_1 &\equiv p_1 - \nu_0 \hat{\mathbf{n}}_3 \cdot \boldsymbol{\Delta}_1. \end{aligned} \tag{14}$$

The importance in defining these combinations is that the expressions for the data streams s_{ij} simplify into the following form,

$$\begin{aligned} s_{21} &= \mathcal{D}_3 \phi_2^* - \phi_1, \\ s_{31} &= \mathcal{D}_2 \phi_3 - \phi_1^*. \end{aligned} \tag{15}$$

If we now combine the s_{ij} and z_i in the following way,

$$\eta_1 \equiv s_{21} - \mathcal{D}_3 z_2 = \mathcal{D}_3 \phi_2 - \phi_1 \,, \quad \eta_{1*} \equiv s_{31} + z_1 = \mathcal{D}_2 \phi_3 - \phi_1 \,, \tag{16}$$

$$\eta_2 \equiv s_{32} - \mathcal{D}_1 z_3 = \mathcal{D}_1 \phi_3 - \phi_2 \,, \quad \eta_{2*} \equiv s_{12} + z_2 = \mathcal{D}_3 \phi_1 - \phi_2 \,, \tag{17}$$

$$\eta_3 \equiv s_{13} - \mathcal{D}_2 z_1 = \mathcal{D}_2 \phi_1 - \phi_3 \,, \quad \eta_{3*} \equiv s_{23} + z_3 = \mathcal{D}_1 \phi_2 - \phi_3 \,, \tag{18}$$

we have just reduced the problem of canceling of six laser and six optical bench noises to the equivalent problem of removing the three random processes, ϕ_1, ϕ_2, and ϕ_3, from the six linear combinations, η_i and η_i^*, of the one-way measurements s_{ij} and z_i.

2.3 The Unequal-Arm Michelson Combination

To show how the delay operators can be used for deriving interferometric measurements, we will consider the simple case of the unequal-arm Michelson interferometer combination X. This TDI combination relies on the four measurements η_1, η_{1*}, η_{2*}, and η_3. Note that the two combinations $\eta_3 + \mathcal{D}_3 \eta_{2*}$

and $\eta_{1*} + \mathcal{D}_2\eta_3$, which represent the two synthesized two-way data measured onboard spacecraft 1, can be written in the following form

$$\eta_1 + \mathcal{D}_3\eta_{2*} = (D_3D_3 - I)\ \phi_1, \tag{19}$$

$$\eta_{1*} + \mathcal{D}_2\eta_3 = (D_2D_2 - I)\ \phi_1, \tag{20}$$

where I is the identity operator. Note that in the stationary case any pairs of these operators commute since up and down delays are equal. This, in general, is no longer true when the delays are functions of time.

From (19) and (20) it is easy to derive the following expression for X, by requiring the elimination of ϕ_1

$$\begin{aligned} X &= [D_2D_2 - I]\,(\eta_1 + \mathcal{D}_3\eta_{2*}) - [(D_3D_3 - I)]\,(\eta_{1*} + \mathcal{D}_2\eta_3) \\ &= [(\eta_{1*} + \mathcal{D}_2\eta_3) + \mathcal{D}_2\mathcal{D}_2(\eta_1 + \mathcal{D}_3\eta_{2*})] \\ &\quad -[(\eta_1 + \mathcal{D}_3\eta_{2*}) + \mathcal{D}_3\mathcal{D}_3(\eta_{31} + \mathcal{D}_2\eta_{13})]. \end{aligned} \tag{21}$$

After replacing (16–18) into (21), we obtain the final expression for X valid in the case of a static LISA array

$$\begin{aligned} X = {} & [(s_{31} + \mathcal{D}_2 s_{13}) + \mathcal{D}_2\mathcal{D}_2(s_{21} + \mathcal{D}_3 s_{12})] - [(s_{21} + \mathcal{D}_3 s_{12}) \\ & + \mathcal{D}_3\mathcal{D}_3(s_{31} + \mathcal{D}_2 s_{13})] + \frac{1}{2}\big[\mathcal{D}_2\mathcal{D}_2\mathcal{D}_3\mathcal{D}_3(\tau_{21} - \tau_{31}) - \mathcal{D}_3\mathcal{D}_3(\tau_{21} - \tau_{31}) \\ & - \mathcal{D}_2\mathcal{D}_2(\tau_{21} - \tau_{31}) + (\tau_{21} - \tau_{31})\big]. \end{aligned} \tag{22}$$

As pointed out in [12,17], (21) shows that X is the difference of two sums of phase measurements, each corresponding to a specific light path from a laser onboard spacecraft 1 having phase noise ϕ_1. The first square-bracket term in (21) represents a synthesized light-beam transmitted from spacecraft 1 and made to bounce once at spacecraft 3 and 2, respectively. The second square-bracket term instead corresponds to another beam also originating from the same laser, experiencing the same overall delay as the first beam, but bouncing off spacecraft 2 first and then spacecraft 3. When they are recombined they will cancel the laser phase fluctuations exactly, having both experienced the same total delays.

3 Time-Delay Interferometry with Moving Spacecraft

The rotational motion of the LISA array results in a difference of the light travel times in the two directions around a Sagnac circuit [14,15]. Two time delays along each arm must be used, say L_i' and L_i for clockwise or counter-clockwise propagation as they enter in any of the TDI combinations. Furthermore, since L_i and L_i' not only differ from one another but also can be time dependent (they “flex”), it was shown that the “first generation” TDI combinations do not completely cancel the laser phase noise (at least with present laser stability requirements), which can enter at a level above the secondary noises. For LISA, and assuming $\dot{L}_i \simeq 10\,\mathrm{m\,s^{-1}}$ [18], the estimated magnitude of the remaining frequency fluctuations from the laser can be about 30 times larger than the level set by the secondary noise sources in the center of the frequency band. To solve this potential problem, it has been shown that there exist new TDI combinations that are immune to first-order shearing (flexing, or constant rate of change of delay times). These combinations can be derived by using the time-delay operators formalism introduced in Sect. 2, although one has to keep in mind that now these operators no longer commute [13].

To derive the new "flex-free" TDI combinations, we will start by taking specific combinations of the one-way data in such a way so as to retain only one of the three noises $\phi_i, i = 1, 2, 3$ if possible. In this way we can then implement an iterative procedure based on the use of these basic combinations and of time-delay operators, to cancel the laser noises after dropping terms that are quadratic in $\dot{L}/c$ or linear in the accelerations. This iterative time-delay method, to first order in the velocity, is illustrated abstractly as follows. Given a function of time $\Psi = \Psi(t)$, time delay by L_i is now denoted either with the standard comma notation [9] or by applying the delay operator $\mathcal{D}_i$ introduced in Sect. 2

$$\mathcal{D}_i \Psi = \Psi_{,i} \equiv \Psi(t - L_i(t)) \ . \tag{23}$$

We then impose a second time delay $L_j(t)$:

$$\begin{aligned} \mathcal{D}_j \mathcal{D}_i \Psi = \Psi_{;ij} &\equiv \Psi(t - L_j(t) - L_i(t - L_j(t))) \\ &\simeq \Psi(t - L_j(t) - L_i(t) + \dot{L}_i(t) L_j) \\ &\simeq \Psi_{,ij} + \dot{\Psi}_{,ij} \dot{L}_i L_j . \end{aligned} \tag{24}$$

A third time delay $L_k(t)$ gives:

$$\begin{aligned} \mathcal{D}_k \mathcal{D}_j \mathcal{D}_i \Psi &= \Psi_{;ijk} \\ &= \Psi(t - L_k(t) - L_j(t - L_k(t)) - L_i(t - L_k(t) - L_j(t - L_k(t)))) \\ &\simeq \Psi_{,ijk} + \dot{\Psi}_{,ijk} [\dot{L}_i (L_j + L_k) + \dot{L}_j L_k] , \end{aligned} \tag{25}$$

and so on, recursively; each delay generates a first-order correction proportional to its rate of change times the sum of all delays coming after it in the subscripts. Commas have now been replaced with semicolons [12], to remind us that we consider moving arrays. When the sum of these corrections to the terms of a data combination vanishes, the combination is called *flex-free*.

Also, note that each delay operator, $\mathcal{D}_i$, has a unique inverse, D_i^{-1}, whose expression can be derived by requiring that $D_i^{-1} \mathcal{D}_i = I$, and neglecting quadratic and higher-order velocity terms. Its action on a time series $\Psi(t)$ is

$$D_i^{-1} \Psi(t) \equiv \Psi(t + L_i(t + L_i)). \tag{26}$$

Note that this is not like an advance operator one might expect, since it advances not by $L_i(t)$ but rather $L_i(t + L_i)$.

3.1 The Unequal-Arm Michelson

The unequal-arm Michelson combination relies on the four measurements η_1, η_{1*}, η_{2*}, and η_3. Note that the two combinations $\eta_1 + \eta_{2*,3}$ and $\eta_{1*} + \eta_{3,2'}$ represent the two synthesized two-way data measured onboard spacecraft 1, and can be written in the following form

$$\eta_1 + \eta_{2*,3} = (\mathcal{D}_3\mathcal{D}_{3'} - I)\ \phi_1, \tag{27}$$
$$\eta_{1*} + \eta_{3,2'} = (\mathcal{D}_{2'}\mathcal{D}_2 - I)\ \phi_1, \tag{28}$$

where I is the identity operator, and we have also used interchangeably the comma notation. Since in the stationary case any pairs of these operators commute, i.e., $\mathcal{D}_i\mathcal{D}_{j'} - \mathcal{D}_{j'}\mathcal{D}_i = 0$, from (27) and (28) it is easy to derive the following expression for the unequal-arm interferometric combination, X, which eliminates, ϕ_1

$$X = [\mathcal{D}_{2'}\mathcal{D}_2 - I]\,(\eta_1 + \eta_{2',3}) - [(\mathcal{D}_3\mathcal{D}_{3'} - I)]\,(\eta_{1*} + \eta_{3,2'}). \tag{29}$$

If, on the other hand, the time delays depend on time, the expression of the unequal-arm Michelson combination above no longer cancels ϕ_1. To derive the new expression for the unequal-arm interferometer that accounts for "flexing," let us first consider the following two combinations of the one-way measurements entering into the X observable given in (29):

$$[(\eta_{1*} + \eta_{3;2'}) + (\eta_1 + \eta_{2;3})_{;22'}] = [D_{2'}D_2D_3D_{3'} - I]\,\phi_1, \tag{30}$$
$$[(\eta_1 + \eta_{2*;3}) + (\eta_{1*} + \eta_{3;2'})_{;3'3}] = [D_3D_{3'}D_{2'}D_2 - I]\,\phi_1. \tag{31}$$

Using (30) and (31), we can use the delay technique to finally derive the following expression for the new unequal-arm Michelson combination X_1 that accounts for the flexing effect,

$$\begin{aligned} X_1 = {} & [D_2D_{2'}D_{3'}D_3 - I]\ \ [(\eta_{21} + \eta_{12;3'}) + (\eta_{31} + \eta_{13;2})_{;33'}] \\ & - [D_{3'}D_3D_2D_{2'} - I]\ \ [(\eta_{31} + \eta_{13;2}) + (\eta_{21} + \eta_{12;3'})_{;2'2}]\,. \end{aligned} \tag{32}$$

This expression is readily shown to be laser-noise-free to first order of spacecraft separation velocities $\dot{L}_i$: it is "flex-free." As usual, X_2 and X_3 are obtained by cyclic permutation of the spacecraft indices.

The reader is referred to [12, 13] for a derivation of all the other TDI combinations valid for the nonstationary LISA configuration.

4 Time-Delay Interferometric Ranging

In the case of a stationary LISA spacecraft array, it was estimated [19] that the time delays need to be known with an accuracy of about 100 ns, if the various TDI combinations are to work effectively, suppressing the residual laser phase fluctuations to a level below the secondary noises (such as the proof mass and optical path noises). For an array of spacecraft in relative motion along realistic solar orbits, the more complicated (*second-generation*) TDI combinations require an even more accurate knowledge of the time delays [20]. The most direct implementation of TDI consists in triggering the phase measurements at the correct delayed times (within the required accuracy), as suggested in [19]. This approach requires the real-time, onboard knowledge of the light travel times between pairs of spacecraft, which determine the TDI time delays. Although the triggering approach is feasible in principle, it complicates the design of the optical phase meter system, and it requires an independent onboard ranging capability. Recently, it was pointed out [21] that the phase measurements at the specific times needed by the TDI algorithm can be computed *in postprocessing* with the required accuracy, by the fractional-delay interpolation (FDI) [21, 22] of regularly sampled data (with a sampling rate of 10 Hz for a GW measurement band extending to 1 Hz). This implies that it is then possible to implement numerically a variational procedure to determine the TDI time delays from the phase difference measurements themselves, eliminating the need for an independent onboard ranging capability. Since this variational procedure relies on the TDI combinations, it has been named *time-delay interferometric ranging.*

In conventional spacecraft ranging either one-way or two-way delay times are measured. In one-way ranging, for instance, two or more tones are coherently modulated onto the transmitted carrier and their phases are measured at the receiver. By then further differencing and dividing them by the spanned bandwidth, one gets the group delay and hence the time delay (up to an ambiguity of c divided by the spanned bandwidth of the ranging tones). In two-way ranging instead a known ranging code is modulated on the transmitted carrier, transponded by a distant spacecraft back to the originator, and the received signal is then cross-correlated with the ranging code to determine the two-way time of flight.

TDIR differs from these methods in that it uses the unmodulated laser noises in a three-element array, which are canceled in TDI combinations assembled with the correct interspacecraft light travel times. This means that TDI can be used to estimate the light travel times by minimizing the laser noise power in the TDI combinations as a function of the postulated light travel times: this process defines TDIR. As an example of how TDIR works, we shall consider again the second-generation TDI combinations X_1 derived above. Note, however, that expression should now be rewritten in the following form

$$\begin{aligned} X_1 = {} & [D_{\hat{2}} D_{\hat{2}'} D_{\hat{3}'} D_{\hat{3}} - I] \left[(\eta_{21} + \eta_{12;\hat{3}'}) + (\eta_{31} + \eta_{13;\hat{2}})_{;\hat{3}\hat{3}'} \right] \\ & - [D_{\hat{3}'} D_{\hat{3}} D_{\hat{2}} D_{\hat{2}'} - I] \left[(\eta_{31} + \eta_{13;\hat{2}}) + (\eta_{21} + \eta_{12;\hat{3}'})_{;\hat{2}'\hat{2}} \right] . \end{aligned} \quad (33)$$

The time-delay indices that appear in (33) with a hat need to be provided by the data analyst (or, in the triggering approach, by the onboard ranging subsystem) with the accuracy required for effective laser noise cancellation. Thus, the X_1-based implementation of TDIR works by *minimizing the power in* X_1 *with respect to the hatted delays* $\hat{L}_k$. Since the TDI combinations constructed with the actual delays cancel laser phase noise to a level 10^8 below the secondary noises [12], it follows that if we neglect all nonlaser sources of phase noise affecting the X_1 combination, the minimum of the power integral

$$I^{(0)}(\hat{L}_k) = \frac{1}{T}\int_0^T [X_1^{(0)}(\hat{L}_k)]^2\, dt \tag{34}$$

will occur for $\hat{L}_k = L_k$ (with $k = 1, 2, 3, 1', 2', 3'$; here the superscript $^{(0)}$ denotes *laser-noise-only* quantities). The search for this minimum can be implemented in postprocessing, using FDI [21] to generate the needed s_{ij} and τ_{ij} samples at the delayed times corresponding to any choice of the $\hat{L}_k$.

In reality, the presence of nonlaser phase noises (possibly including GWs) will displace the location of the minimum from L_k. Writing $X_1 = X_1^{(0)} + X_1^{(n)}$ (with $X_1^{(n)}$ obtained by setting all ϕ_i, ϕ_i^* to zero), the power integral becomes

$$I^{(n)}(\hat{L}_k) = \frac{1}{T}\int_0^T \left[X_1(\hat{L}_k)\right]^2 dt, \tag{35}$$

or explicitly,

$$I^{(n)}(\hat{L}_k) = I^{(0)}(\hat{L}_k) + \frac{1}{T}\int_0^T \left[X_1^{(n)}\right]^2 dt + \frac{2}{T}\int_0^T X_1^{(n)} X_1^{(0)}(\hat{L}_k)\, dt. \tag{36}$$

Here we have written the nonlaser phase noise $X_1^{(n)}$ as independent of the delays $\hat{L}_k$: this holds true for a search conducted sufficiently close to the true minimum, since the ϕ_i and ϕ_i^* are much larger than the secondary noises, and so are their variations. The minimum of $I^{(n)}(\hat{L}_k)$ can be displaced from $\hat{L}_k = L_k$ because the third term of (35) (the cross-correlation integral of $X_1^{(n)}$ and $X_1^{(0)}(\hat{L}_k)$) can be negative and offset a concurrent increase in $I^{(0)}(\hat{L}_k)$. The achievable time-delay accuracies will depend on the level of the residual laser noise, the levels of the secondary noises in X_1, and the integration time T. We expect the arm-length errors to be determined by the interplay of the first and third terms in (36). By equating the variance from the imperfect cancellation of the laser with the estimation-error variance of the crossterm in (36), we can roughly estimate how well the time delays will be determined with TDIR: $\delta L_k \sim (\sigma_{X_1^{(n)}}/\sigma_{\dot{X}_1^{(0)}})\sqrt{\rho/T}$, where $\sigma_{X_1^{(n)}}$ and $\sigma_{\dot{X}_1^{(0)}}$ are the root mean squares of the secondary noises and of the time derivative of the laser noise in X_1, and ρ is the temporal width of the secondary-noise autocorrelation function. For nominal LISA noises and $T \simeq 10{,}000$ s, we thus expect δL_k of 30 ns or better to be achievable.

The TDIR concept described above was simulated, for a realistic model of the LISA orbits and instruments, with the *Synthetic LISA* software package [23]. The simulation included the generation of a number of *chunks* of contiguous data for the s_{ij} and τ_{ij} measurements, sampled at intervals of 0.25 s, and containing pseudorandom laser, proof mass, and optical path noises at the nominal level set by the LISA prephase A specification [4, 23]. The data durations considered were 8,192, 16,384, and 32,768 s.

The 18 noise processes (corresponding to the six lasers, proof masses, and optical paths) were assumed to be uncorrelated, Gaussian, and stationary, with, respectively, white, f^{-2}, and f^2 PSDs, band-limited at 1 Hz. The frequency-fluctuation measurements contained also the responses due to GWs from two circular binaries with $f_{\mathrm{GW}} \simeq 1$ and 3 mHz, located, respectively, at the vernal equinox and at ecliptic latitude 45° and longitude 120°. The strength of the two sources was adjusted to yield an optimal S/N of ~ 500 over a year (for X_1), guaranteeing that there will be times of the year when each source will be clearly visible above the noise in an observation time $\sim 10,000$ s.

By putting the three LISA spacecraft on realistic trajectories, the resulting time and direction dependence [12] of the light travel times can be written in the following form [23, 24]

$$L_k(t) = L + \frac{1}{32}(eL)\sin(3\Omega t - 3\xi_0) - \left[\frac{15}{32}(eL) \pm (\Omega RL)\right]\sin(\Omega t - \lambda_k), \quad (37)$$

where the plus (minus) refers to unprimed (primed) indices. In (37) $L/c \simeq 16.68$ s is the average light travel time, and

$$(\lambda_1, \lambda_2, \lambda_3) = (\xi_0, \xi_0 + \frac{4\pi}{3}, \xi_0 + \frac{2\pi}{3}), \quad (38)$$

with ξ_0 an arbitrary constant (set to 0 in our simulations) giving the phase of the spacecraft motion around the guiding center of the LISA array. The starting times of the chunks were spread across a year to sample the time dependence of the L_k and the directionality of the GW responses.

Separately for each chunk, we minimized $I^{(n)}[\hat{L}_k(t)]$ (35) starting from guesses for the $\hat{L}_k$ affected by errors $\gtrsim 50$ km/c, very much larger than typical accuracy of radio tracking from Earth [18]. The minimization was carried out using a Nelder–Mead simplex-based algorithm [25]. The effective cancellation of laser noise with TDI requires modeling the time dependence of the travel times *within* the chunks. In our simulations, we used two such models:

1. An orbital-dynamics model (ODM) given by (37), with $\widehat{eL}$, $\widehat{\Omega RL}$, and $\widehat{\xi_0}$ taken as the independent search parameters with respect to which $I^{(n)}$ is minimized. We excluded L and Ω from the search because the dependence of the $L_k(t)$ on such an extended parameter set is degenerate on timescales $\sim$10,000 s.
2. A linear model (LM) given by $\hat{L}_k(t) = \hat{L}_k^0 + \hat{L}_k^1(t - t_0)$ (with t_0 set to the beginning of each chunk). Because the expression for X_1 does not contain the travel times L_1 and $L_{1'}$, our independent search parameters are the constants $\hat{L}_k^0$ and $\hat{L}_k^1$ for $k = 2, 2', 3, 3'$ (eight numbers altogether).

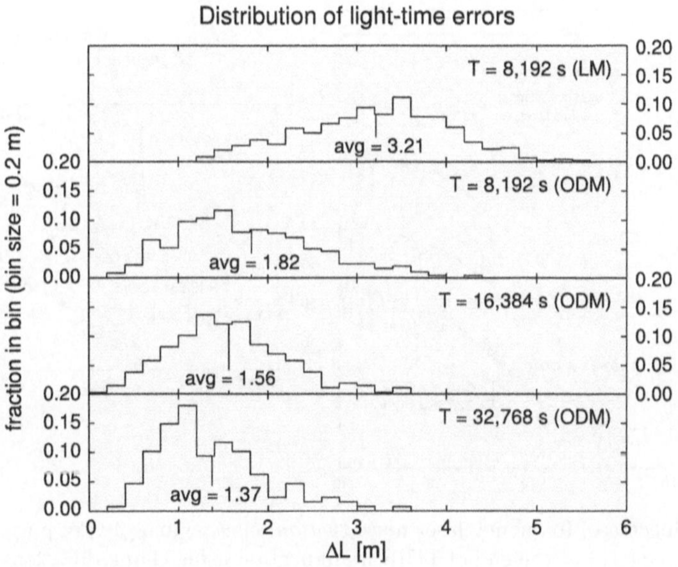

Fig. 4. Distribution of errors ΔL (see (39) and the main text above it) in the determination of light travel times, using X_1-based TDIR with chunk durations of 8,192 s (for the LM and ODM models), and 16,384 and 32,768 s (for the ODM model only). As expected, the errors are lower for longer integration times T; for the LM model, the larger errors are due to the unmodeled curvature in the time dependence of the light travel times. The distributions shown correspond to samples of 512, 256, and 128 chunks for $T = 8{,}192$, 16,384, and 32,768 s, respectively, spread across a year.

Figures 4 and 5 show the results of our simulations. The average travel time errors ΔL displayed in Fig. 4 are defined as $\Delta L = (\Delta L_2 + \Delta L_{2'} + \Delta L_3 + \Delta L_{3'})/4$, with

$$\Delta L_k = \sqrt{\frac{1}{T}\int_{t_0}^{t_0+T}\left(\hat{L}_k(t) - L_k(t)\right)^2 dt}. \tag{39}$$

Because the noises have different realizations in each chunk and because the local behavior of the $L_k(t)$ (37) changes along the year, the average error ΔL of each chunk is a random variable. Its distribution is approximated by the histograms of Fig. 4, which refer to populations of, respectively, 512 (for $T = 8{,}192$ s), 256 (for $T = 16{,}384$ s), and 128 (for $T = 32{,}768$ s) chunks (hence the roughness of the curves).

It turns out that the linear model is not quite sufficient to model the changes of the time delays during the chunk lengths considered, since the *minimum* ΔLs (computed by least-squares fitting the parameters $\hat{L}_k^0$ and $\hat{L}_k^1$ to the $L_k(t)$) are in the range 0.25–2.60 m (for $T = 8{,}192$ s), 1–10 m (for $T = 16{,}384$ s), and 4–40 m (for $T = 32{,}768$ s). Thus, in Figs. 4 and 5 we show

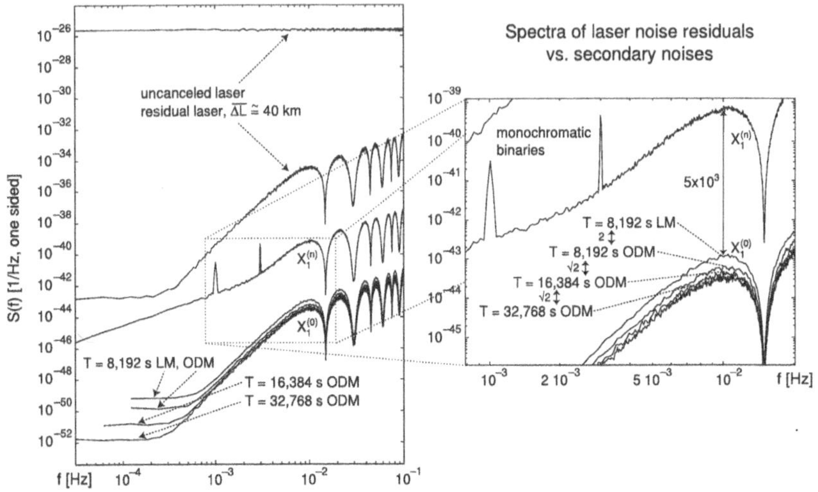

Fig. 5. Spectra of frequency laser noise (*bottom curves*) and of GW plus secondary noises (*top curve*) at the end of TDIR minimization using chunk durations of 8,192 s (for the LM and ODM models), and 16,384 and 32,768 s (for the ODM model only). We show averages of the spectra computed separately for each chunk using a triangle-windowed periodogram; the averages are taken over populations of 512, 256, and 128 chunks for $T = 8{,}192$, 16,384, and 32,768 s, respectively, spread across a year. In all cases, laser noise is suppressed to levels several orders of magnitude below the secondary noises: the cutout graph on the *right* shows that the typical laser-noise suppression factor with respect to secondary noise is $\sim 5 \times 10^3$ for the worst case considered (8,192 s LM); it improves by a factor ~ 2 for 8,192 s ODM, and by factors of $\sim\sqrt{2}$ for each successive doubling of T. The GWs from the two circular binaries stand clearly above the noise at 1 and 3 mHz.

results only for the linear model with $T = 8{,}192$ s. (The minimization of $I^{(n)}$ over the LM parameters is delicate, because for X_1 the laser-noise residuals turn out to depend strongly on ΔL_2, $\Delta L_{3'}$, and $\Delta L_{2'} - \Delta L_3$, but only weakly on $\Delta L_{2'} + \Delta L_3$. In this case, the Nelder–Mead algorithm can be made to return accurate results by using the search parameters $\hat{L}_2^0$, $\hat{L}_{3'}^0$, $\widehat{L_{2'}^0 - L_3^0}$, and $\widehat{L_{2'}^0 + L_3^0}$, plus the corresponding $\hat{L}_k^1$ parameters.)

Figure 5 shows the spectra of the *residual* laser noise (i.e., of $X_1^{(0)}$ at the minimum of $I^{(n)}(\hat{L}_k)$), as compared with spectra of GWs and secondary noises (i.e., of $X_1^{(n)}$). The spectra are computed separately for each chunk using triangle-windowed periodograms, and then averaged over the chunk populations. The two GW sources stand clearly above the secondary noises at 1 and 3 mHz. We see that the TDI cancellation of laser noise with TDIR-determined time delays is essentially complete, with the residual laser noise several orders of magnitude below the secondary noises. We conclude that for $T \sim 10{,}000$ s, with the nominal LISA noises, and even in the presence of very strong GW signals, TDIR can easily reach the time-delay accuracy required for second-generation TDI. For frequencies below 10 mHz, the residual laser-noise power decays as f^6, while the secondary noises decrease only as f^2. We attribute the flattening near 0.1 mHz (which is insignificant with respect to the LISA performance) to a combination of leakage and aliasing in the numerical estimation of the spectra and of real effects due to the first nonconstant terms in the travel time errors across the chunks.

Finally, we estimated the power in the Fourier bins containing the simulated signals using two different time series: in the first X_1 was formed using perfectly known time delays, in the second using the TDIR-determined time delays. Analyzing the 32,768 s chunks at the times along the simulated year where the signal amplitudes were maximum, we find that the signal powers in the two time series agree to the numerical precision of the calculation (about a part in 10^5).

From Figs. 4 and 5 we conclude that, for the nominal LISA noises, integration times $\sim 10,000$ s, TDIR determines the time delays with accuracies sufficient to suppress the laser phase fluctuations to a level below the LISA secondary noises, while at the same time preserving GW signals. Our simulations assumed synchronized clocks aboard the spacecraft, but we anticipate that TDIR may be extended to achieve synchronization, by minimizing noise power also with respect to clock parameters.

TDIR has the potential of simplifying the LISA design, allowing the implementation of TDI without a separate interspacecraft ranging subsystem. At the very least, TDIR can supplement such a subsystem, allowing the synthesis of TDI combinations during ranging dropouts or glitches. TDIR may be applicable in other forthcoming space science missions that rely on spacecraft formation flying and on interspacecraft ranging measurements to achieve their science objectives.

Acknowledgments

I would like to thank my JPL collaborators and friends, Frank B. Estabrook, John W. Armstrong, and Michele Vallisneri for all the stimulating conversations I had with them on the material presented in this chapter. This research was performed at the Jet Propulsion Laboratory, California Institute of Technology, under contract with the National Aeronautics and Space Administration.

References

1. M. Tinto and F.B. Estabrook, *Phys. Rev. D*, **52**, 1749 (1995).
2. M. Tinto, *Class. Quantum Grav.*, **19**, 7, 1767 (2002).
3. F.B. Estabrook and H.D. Wahlquist, *Gen. Relativ. Gravit.* **6**, 439 (1975).

4. LISA: (Laser Interferometer Space Antenna) *An international project in the field of Fundamental Physics in Space*, Pre-Phase A Report, **MPQ 233**, (Max-Planck-Institute für Quantenoptic, Garching bei München, 1998).
5. Wei-Tou Ni, This Volume.
6. M. Tinto and J.W. Armstrong *Phys. Rev. D*, **59**, 102003 (1999).
7. M. Tinto, F.B. Estabrook, and J.W. Armstrong, *Phys. Rev. D*, **65**, 082003 (2002).
8. M. Tinto, *Phys. Rev. D*, **58**, 102001 (1998).
9. J.W. Armstrong, F.B. Estabrook, and M. Tinto, *ApJ*, **527**, 814 (1999).
10. S.V. Dhurandhar, K.R. Nayak, and J.-Y. Vinet, *Phys. Rev. D*, **65**, 102002 (2002)
11. F.B. Estabrook, M. Tinto, and J.W. Armstrong, *Phys. Rev. D*, **62**, 042002 (2000).
12. D.A. Shaddock, M. Tinto, F.B. Estabrook, and J.W. Armstrong, *Phys. Rev. D*, **68**, 061303 (2003).
13. M. Tinto, F.B. Estabrook, and J.W. Armstrong, *Phys. Rev. D*, **69**, 082001 (2004).
14. D.A. Shaddock, *Phys. Rev. D*, **69**, 022001 (2004).
15. N.J. Cornish and R.W. Hellings, *Class. Quantum Grav.*, **20**, 4851 (2003).
16. M. Tinto, M. Vallisneri, and J.W. Armstrong, *Phys. Rev. D*, **71**, 041101(R) (2005).
17. D. Summers, "Algorithm tradeoffs," oral presentation, 3rd progress meeting of the ESA funded LISA PMS Project. ESTEC, NL, February 2003.
18. W.M. Folkner, F. Hechler, T.H. Sweetser, M.A. Vincent, and P.L. Bender, *Clas. Quantum Grav.*, **14**, 1543 (1997).
19. M. Tinto, D.A. Shaddock, J. Sylvestre, and J.W. Armstrong, Phys. Rev. D **67**, 122003 (2003).
20. M. Vallisneri, M. Tinto, and J.W. Armstrong, in preparation (2004).
21. D.A. Shaddock, B. Ware, R.E. Spero, and M. Vallisneri, *Phys. Rev. D*, **70**, 081101 (2004).
22. T.I. Laakso et al., IEEE Signal Processing Magazine **13**, 30 (1996).
23. M. Vallisneri, *Phys. Rev. D*, **71**, 022001 (2005).
24. N. Cornish and L. Rubbo, Phys. Rev. D **67**, 022001 (2003).
25. J.A. Nelder and R. Mead, Computer Journal **7**, 308 (1965).

Technology for Precision Gravity Measurements

Robert D. Reasenberg and James D. Phillips

Smithsonian Astrophysical Observatory, Harvard-Smithsonian Center for Astrophysics, 60 Garden Street, MS-63, Cambridge, MA 02138, USA

Summary. We discuss four technologies applicable to precision measurements in space and on the ground. The first is our tracking frequency laser distance gauge (TFG), which we developed ca. 1990 for a spaceborne astrometric optical interferometer, POINTS, and which we are using today for our principle of equivalence measurement (POEM), a laboratory test of the equivalence principle. The second is an extension of the TFG to use a semiconductor laser (SL-TFG) with the intention to make the instrument more robust and applicable to space-based experiments. In particular, we wish to apply the SL-TFG to a version of POEM that could operate in space at substantially higher accuracy. Further, some versions of the SL-TFG have reduced complexity and thus have enhanced reliability and reduced cost. The third technology is an approach to using the TFG as part of an extended space-based optical instrument. We discuss the launching of multiple beams from a single device as a means of achieving a "strong optical truss" without excess complexity or endpoint connection error. The fourth and final technology is for creating a brief period of free fall in the laboratory, and being able to repeat the free-fall rapidly. This technology is a key part of POEM.

1 Introduction

We discuss four technologies applicable to precision measurements in space and on the ground. These are all related to our principle of equivalence measurement (POEM), which requires both the accurate (ca. 1 mm) control of the motion of a mass (ca. 50 kg) over an extended range (ca. 1.5 m and 5 m s^{-1}) and advanced laser gauging.

In Sect. 2, we provide a brief introduction to POEM, a laboratory equivalence principle (EP) test with an accuracy goal of $\sigma(\Delta g)/g = 5 \times 10^{-14}$. POEM is the prime scientific driver for the technologies described in the subsequent sections. In Sect. 3, we discuss the tracking frequency laser distance gauge (TFG), starting with a brief review of the art, our reasons for developing the TFG, and its known advantages. We describe recent advances in the development of the TFG, and also its precision in measuring both incremental and absolute distance. We speculate on an improved means of measuring absolute distance. In Sect. 4, we discuss space-based applications for the TFG. We introduce a novel multibeam launcher for ultra-stable optical trusses as a component for an architecture for pointing and stabilizing a large imaging interferometer. An extension of that architecture is shown for Stellar Imager. Finally, in Sect. 5, we discuss achieving free-fall conditions in the laboratory, with emphasis on the POEM approach of moving a vacuum chamber along a vertical path under computer control.

2 Principle of Equivalence Measurement

To test the EP to an accuracy of at least $\sigma(\Delta g)/g = 5 \times 10^{-14}$, we are developing POEM in which we directly examine the relative motion of two test masses that are freely falling. The test mass assemblies (TMA) will be in free fall in a comoving vacuum chamber for about 0.8 s per "toss," i.e., motion both up and down along a vertical path of about 90 cm. Such an experiment tests both for a possible violation of the weak equivalence principle (WEP) and for new forces that might mimic a WEP violation.

Figure 1 shows the principal components of the measurement system. Inside the vacuum chamber is a single pair of TMA resting on shelves separated by 0.5 m. Each TMA contains a sample of a test substance (A or B)

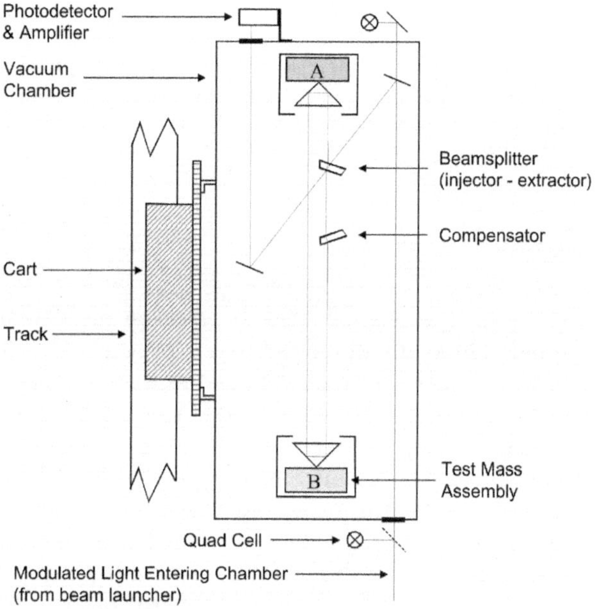

Fig. 1. Principle components of the measurement system. Both key technologies are shown: motion system and laser gauge.

and a corner-cube retroreflector. Conditioned laser light entering at the lower right reaches the beamsplitter, illuminating the optical cavity formed by the two retroreflectors, and is then passed to the detector (upper left). The compensator plate makes possible the alignment of the cavity in the presence of imperfect retroreflectors. Considerable work has been done on selecting the test substances. One approach (e.g., [4]) is to select one element from each of three groups: {Be}, {W, Pt, Au}, and {C, Mg, Si, Al}. This allows the experiment to efficiently search for EP violations due to three types of energy characterized by $\xi_1 = (N+Z)/\mu$, $\xi_2 = (N-Z)/\mu$, and $\xi_3 = E/\mu \approx Z(Z-1)/((N+Z)^{1/3}\mu)$, where N = neutron number, Z = proton number = lepton number, and μ = atomic mass.

The vacuum chamber is mounted to a cart that rides along a vertical track and contains a precision position sensor that reads a ruled bar mounted to the track assembly. Inside the track assembly is a linear motor (fixed magnets, moving coils mounted to the cart, and a coarse position sensor). The motor is driven by a control unit that includes the power amplifier and a pair of computers. The track, cart, motor, and its controller are a commercial system. In our approach, there is no need for mechanisms inside the chamber to drive the motion of the TMA and the TMA-observing devices during each toss. However, this approach entails moving tens of kg at speeds approaching $5\,\mathrm{m\,s^{-1}}$, which implies large forces and the potential for significant amounts of vibrational energy from which the measurement must be protected.

During the upper portion of the chamber motion, the linear motor and its control system serve to enforce a free-fall trajectory, overcoming friction. At the bottom of the free-fall portion of the motion, the chamber encounters a "bouncer" that passively applies an upward force, absorbs the energy of the falling chamber, and returns the chamber to upward motion with a minimum of force required from the motor. This reversal of the moving vacuum chamber takes place in about 0.4 s and the measurement will be repeated every 1.3 s. The bouncer will be discussed later.

In a more advanced version of POEM, there will be two pairs of TMA (Fig. 2) with a lateral separation of 7 cm, permitting a double difference observable that cancels many systematic errors. Prominent among these is gravity gradient, including the vertical component, $dg/dz = 3 \times 10^{-7}\,\mathrm{g\,m^{-1}}$.

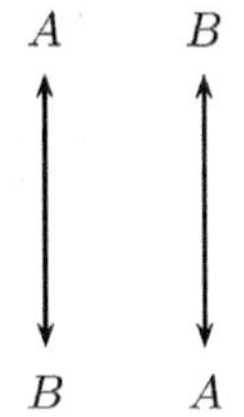

Fig. 2. Placement of test masses in a more advanced version of POEM. In the double difference observable, the EP violating signal adds and many errors cancel.

Additionally, there are small components that are time dependent, including those due to groundwater variation and parked cars on the nearby street. With a science goal of $\sigma(\Delta g)/g = 5 \times 10^{-14}$ and a TMA mass that is 30% test substance, we require a measurement accuracy of $\sigma(\Delta g)/g = 1.5 \times 10^{-14}$ for the TMA. This requirement, when combined with the vertical gravity gradient, implies a requirement for absolute distance measurement with an uncertainty under 0.05 μm. The required laser gauge will be discussed later.

The laser gauge is expected to yield a 10 pm uncertainty in a 10 ms observation. We assume white noise, making this equivalent to 1 pm in 1 s. The POEM distance measurement uses a low finesse cavity formed by two solid glass retroreflectors and fed by an intracavity beamsplitter. With the expected laser gauge precision, the derived relative acceleration from a single toss of a pair of TMA has a precision of 3×10^{-12} g. A 1-h run will have a corresponding acceleration error of 0.64×10^{-13} g (on TMA; 2.1×10^{-13} g on test substances for $R = 0.3$) due to the laser gauge. This is only a factor of 4.2 larger than the experiment goal of $\sigma(\Delta g)/g = 0.5 \times 10^{-13}$. Thus, an experiment of the intended accuracy will be quick, and we will be able to test several (compositionally) different sample pairs.

3 Tracking Frequency Laser Distance Gauge

In the late 1980s, we were working on POINTS [15], an affordable spaceborne astrometric interferometer with a nominal single-measurement accuracy of 2 μas for a pair of bright stars separated by $90 \pm 3°$. Because of the close connection among size, weight, complexity, and cost, we kept the baseline length at 2 m and thus required high precision in the metrology to achieve the nominal single-measurement astrometric accuracy. The mission requirement, shown in Fig. 3, was for a single metrology leg (laser gauge) to have an accuracy of 2 pm on timescales from 1 to 100 min. We could find no existing laser gauge that would meet the requirement. In particular, we first looked at a quadrature homodyne interferometer. These are normally described as achieving $\lambda/8$ measurement uncertainty. We could not see how to extend this approach beyond $\lambda/100$.

We next looked at the standard precision laser gauge, the heterodyne interferometer. Figure 4 shows one realization of this laser gauge. A stabilized (HeNe) laser provides a polarized beam that is split into a pair of orthogonally polarized beams by a polarizing beamsplitter oriented at 45° from the initial plane of polarization. A pair of acousto-optic modulators (AOM) shift the frequencies of these beams before they are recombined. The recombined beam is sampled and detected to provide a reference phase at the difference of the AOM drive frequencies. The recombined beam is also sent into the measurement interferometer where a polarizing beamsplitter separates the two beams, sending one along the measurement path and the other along a reference path. Finally, the beams are again combined and detected to provide a

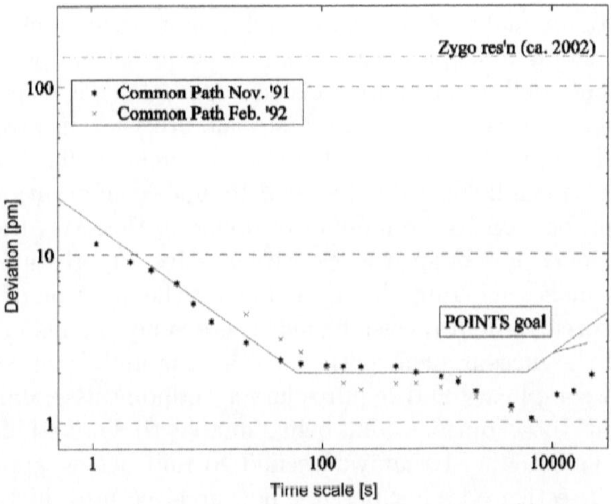

Fig. 3. Allan deviation (root variance) for the TFG.

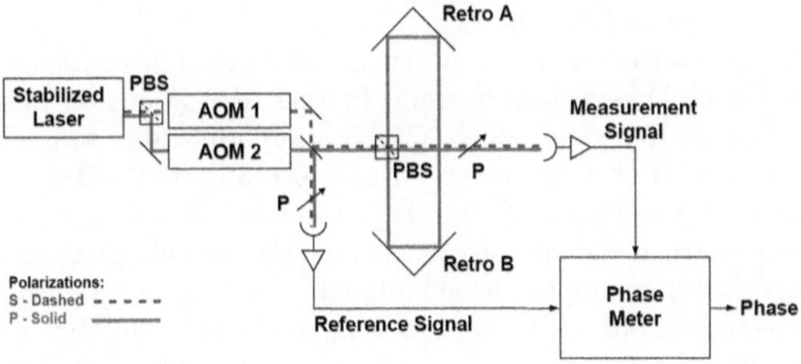

Fig. 4. Heterodyne laser gauge. This is a version of the standard commercial precision laser gauge.

measurement phase at the AOM difference frequency. The difference between the measurement and reference phases is taken as an incremental measure (in units of the laser wavelength, λ) of the distance under study.

There are three problems with this laser gauge. First, the heterodyne gauge does not lend itself easily to the measurement of absolute distance, although with enough additional complexity, this can be overcome [6]. Second, the polarizing beamsplitters are imperfect, which results in a small portion of each beam traveling the unintended path. This gives rise to a cyclic bias in the laser gauge reading with $\lambda/2$ period and a nm typical amplitude. This cyclic bias problem has been partially overcome [5] at the expense of measurement speed and instrument complexity. The bias can also be overcome by employing spatial separation of the two required beams [7,8,17,18], again at the expense of complexity. For the scheme based on the spatial separation of beams, we do not know of an analog of the space- and weight-saving technique discussed below for launching multiple beams from the same set of optics.

As we discuss below, for high-precision applications, it is useful to operate the laser gauge in a cavity. This increases the precision in proportion to the finesse. The third problem with the heterodyne gauge is that it does not operate in a cavity. It can be operated in a multibounce configuration that gains in precision in proportion to the number of bounces. However, the multibounce configuration does not enjoy the cavity's passive suppression of alignment errors. It also makes substantially more difficult the question of what distance is being measured to high precision since there is not a single reference point at the end of the measurement path. Further, the multibounce configuration has increased complexity and requires larger endpoint assemblies.

Looking at these options and being unable to find additional existing schemes left us convinced that we needed to find a new approach to laser gauging. We investigated a few possible new architectures, including one that used dichroic mirrors in much the same way as the heterodyne gauge used a polarizing beamsplitter. We selected the scheme now known as the TFG, which is the subject of the rest of this section.

Figure 5 shows the classic realization of the TFG. It is a closed-loop system based on Pound–Drever–Hall locking. In this realization, an optical signal from the variable frequency source (VFS) of an adjustable wavelength λ_{VFS} is phase modulated at a frequency f_m and introduced into the measurement interferometer whose length, L, is to be determined. When λ_{VFS} is away from λ_x, the wavelength at the intensity extremum, the optical signal emerging from the interferometer is amplitude modulated, resulting in an electrical signal at the detector output at f_m with a magnitude and sign that indicate the offset. Synchronous detection at f_m and filtering yield a signal that is used to control λ_{VFS}, driving it back to λ_x. The corresponding optical frequency shift is measured by a frequency counter.

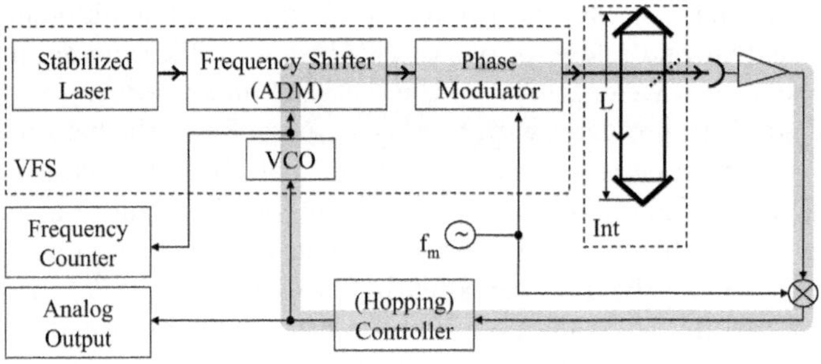

Fig. 5. Classic realization of the TFG. Pound–Drever–Hall locking is used to tie the frequency of the variable frequency source (VFS) to the changing length under measurement. (VCO: voltage controlled oscillator)

The TFG offers four advantages over the heterodyne gauge. First, it is intrinsically free of the cyclic bias that plagues the heterodyne gauges. It has only one beam and thus it cannot be subject to problems associated with separating beams of different frequencies. Second, it naturally operates in either a cavity or a nonresonant interferometer (Michelson, Mach-Zehnder, etc.). Thus, additional accuracy is accessible when needed. Third, as will be discussed below, the TFG can be used to measure absolute distance with little additional effort.

The fourth advantage is that the TFG can suppress some additional errors. Many laser gauges use corner-cube retroreflectors at their endpoints to simplify alignment. Light entering along the cube diagonal impinges on the mirror surfaces at an angle of 55° from normal incidence. At far from normal incidence, the phase shift on reflection may differ by tens of degrees between S and P polarization, and this differential phase shift will vary from mirror to mirror. Thus, after passing around a retrocavity once, the polarization state of the light may be substantially altered. Since the S–P phase difference is not identical among the mirrors, there is no option to have the polarization effect of one retroreflector canceled by the other. Finally, the light exiting the cavity will be mixed with a sample of the incoming beam, optionally passed through a polarizer, and detected.

If the polarization of the incoming light is not dictated by a requirement that it be aligned with the (polarizing) beamsplitter, one may choose it advantageously to reduce errors. In the case of a Michelson interferometer, one finds that the measured path varies smoothly as the incoming polarization is rotated. That polarization can be set to give an extreme path length, which minimizes the polarization sensitivity. In the case of a cavity, one expects (and we have found) two polarization eigenmodes, which need not be linearly polarized. If the incoming polarization is set to minimize the excitation of one of the modes, then its effect is further suppressed if the two eigenfrequencies are separated by a modest multiple of the cavity full width at half maximum (FWHM). The TFG will keep the optical signal at the center of the excited mode where the suppressed mode would have a slowly changing phase and amplitude response. Thus, even if the suppressed mode were excited and of the correct polarization to contribute to the cavity response, it would not significantly perturb the optical frequency of lock. The required mode separation becomes increasingly likely as the finesse increases.

In a similar way, operating in a cavity suppresses distance measurement error due to misalignment. If the beam entering the cavity is misaligned from the cavity axis by ϵ, there is an error of $L(1/\cos\epsilon - 1)$. In a cavity, small misalignment will decrease coupling to the principal mode and may permit coupling to other modes. However, if these modes are well separated in frequency from the principal mode, they will not be excited, and they will not corrupt the distance measurement. It is often possible to design a cavity to have good frequency separation between the principal mode and the first few higher-order modes.

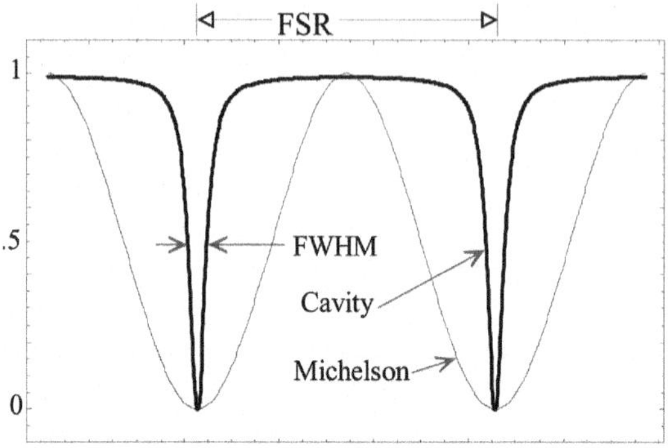

Fig. 6. Frequency response of a Michelson interferometer (*thin line*) and a cavity (*thick line*). The separation between orders is the free spectral range ($FSR = \Phi$). The finesse ($F = \Phi$/FWHM) depends on the losses.

In considering the precision of the TFG, it is natural to start with the limit set by photon counting statistics. For 1 μW of HeNe (633 nm) power detected from a Michelson interferometer, the limit is 0.06 pm after 1 s. The current TFG is limited by technical noise: σ = 10 pm at either 1 or 100 samples per second. When we completed the TFG development for POINTS, we measured the deviation as shown in Fig. 3. Note that no further improvements were made as we had met the POINTS mission requirements.

More recently, we have needed higher performance for POEM: 1 pm in 1 s of observing and the ability to sample much faster losing only by $t^{-1/2}$. For this purpose, we are operating in a low finesse cavity (Fig. 6). Note that there are limits to the use of high finesse cavities. In particular, the storage time is approximately 1/FWHM, and this limits the unit-gain bandwidth of the Pound–Drever–Hall locking.

The initial TFG was intended as a proof of concept. Recently, we have refined and improved the classic TFG. As originally implemented, the TFG frequency shift was based on passing the HeNe light twice through an acousto-optic deflector-modulator (ADM). The second pass doubles the frequency shift and cancels the deflection, yielding a beam that can be introduced into fiber. The ADM operated over a frequency range of 100 MHz which yielded a distance range of 0.2 μm over a 0.5 m path. More recently, we have introduced an ADM with a frequency range of 125 MHz and passed the light through it four times. This yields a distance range of 0.5 μm over the same path. Since the mode spacing is $\lambda/2$ = 0.316 μm, one mode is always accessible with the available frequency shift, and often there are two.

We have introduced a nonlinear aspect to the TFG loop controller. It detects that it is running out of the frequency shifter's range and hops to a mode at the far end of the range, shifting the optical frequency by the free spectral range of $c/2L = 300$ MHz. The hop is fast enough (about 1 μs) to be "unobserved" by the classical portion of the loop controller. We have demonstrated a rate of 5×10^4 hops s^{-1}, which corresponds to a linear velocity of 16 mm s^{-1}.

This speed is more than adequate for most scientific applications. However, in one commercially important application it falls far short. In the manufacture of integrated circuits, structures are photocopied onto the silicon wafer in a process that requires both the mask and silicon to move at speeds in excess of 1 m s^{-1}. This motion must be under laser-gauge control to submicron precision. In this application, the heterodyne gauge is the natural choice.

The original TFG had a unit-gain bandwidth of 100 Hz. This was adequate for the task of showing that it could meet the POINTS mission requirements. However, POEM requires a more agile laser gauge, so as to follow vibrations when the TMA are not in free fall, to reacquire lock rapidly should it be lost, and to follow accurately the TMA acceleration while in free fall. We have rebuilt the controller with a unit-gain bandwidth of 50 kHz. The limit to the unit-gain frequency with the classic TFG is the acoustic propagation delay internal to the ADM, 1.7 μs with the present unit. The semiconductor laser TFG (SL-TFG) described below can operate substantially faster.

The TFG employs a frequency measurement to determine incremental distance. Standard frequency counters have well-defined start and stop times and high time-base accuracy. However, they generally have a dead time between measurements. For many purposes, one needs to read out the laser-gauge frequency counter at 10^2–10^4 times per second, which is not possible with typical counter dead time. Further, it is often necessary to combine measurements made at high speed by fitting a model, say by the method of least squares. This treats the data locally in a manner similar to averaging. Having zero dead time results in certain of the measurement errors being correlated between adjacent samples in a way that causes them to cancel when the data are averaged. The ideal counter for the TFG does a continuous count of the incoming cycles, sampling the count at precisely defined intervals, and sending those samples to a computer for storage or analysis. Such a device was built for us at the Harvard Physics Department Electronic Instrument Design Lab. This advanced frequency counter has synchronized dual channel operation and a network (100 base-T, UDP) interface to a PC. It has a maximum counting rate of 200 MHz and internal dividers (2×, 4×, 8×) that permit counting rates up to 1,200 MHz. The counter can be read continuously, using an ordinary PC, at ∼ 10K samples per second. Timing mismatch between the effective gate times for its two channels is well under 1 ns, and variations of this mismatch are far smaller. For POEM, with 10 ms sampling, a 0.1 ns differential timing error would cause a differential distance error of 1×10^{-12} m, in 2% of the counting intervals. Due to the continuous counting, these errors would be largely compensated by an opposite error in the neighboring counting interval. The counter operates from an external commercial oven-stabilized crystal time base with a stability of $\Delta f/f < 5 \times 10^{-10}$ per 24 h.

The hopping provides an easy means of measuring absolute distance. By measuring the frequency shift before and after a hop, the TFG measures the free spectral range, Φ, of the measurement interferometer corresponding to the current length L. The estimate of L is then $c/(2\Phi)$. For the case of a dispersive medium in the path, see the Appendix of [14]. A few years ago, when we added hopping to the TFG, we demonstrated this capability to low accuracy (0.1 mm) in a preliminary test plagued by technical noise. We anticipate demonstrating a substantial improvement in this capability by using the refined TFG described below.

Recently, as we evaluated the requirements for an advanced version of POEM, we concluded that we needed a new TFG realization. There were four driving factors. First, the initial realization used a HeNe laser, which is not likely to be used in space today. There are several potential spaceborne applications for the TFG including POEM on the ISS or other platform, Stellar Imager, and MAXIM [1–3, 12]. For POEM, we would like to increase the precision to between 0.1 and 0.01 pm in 1 s to make use of an expensive platform. Second, the initial realization used free-space beams, which pose an alignment and drift problem. Third, the ADM may pose a reliability problem. Fourth, a far greater tuning range would make the instrument more flexible and increase the accuracy of the measurement of absolute distance. In addition, we were interested in the possibility of a simplified version that might find applications where the number of laser gauge links was high.

We are developing a SL-TFG on which a spaceworthy instrument could be based. In this version of the TFG, we use a distributed feedback (DFB) semiconductor diode laser operating in the 1,550 nm (190 THz) band. These lasers are available at moderate cost from several vendors, create most of their output in a single longitudinal mode, have adequate coherence length (>1 km), can be purchased with fiber-coupled output, and are rapidly tunable via their injection current. They have been used for atomic spectroscopy for more than a decade [16]. In the SL-TFG, we eliminate both the ADM and most of the free-space path in the VFS. When operating in either air or vacuum, eliminating free-space path will reduce error due to thermally driven movement of components. When testing in laboratory air, this will reduce the influence of turbulence.

In the SL-TFG, one laser (tunable laser in Fig. 7) will have its wavelength locked to the measurement interferometer. A second laser (reference laser) will be locked to a high-stability reference cavity, and the frequency difference (beat note) counted. A change of beat note indicates a change in L. The tunable laser's frequency control input is an analog signal that can be read faster, but with lower accuracy. For the SL-TFG, servobandwidth higher than 100 kHz is a secondary goal. In the SL-TFG, with a rapidly tuning DFB laser, the only intrinsic limit to bandwidth will be the light propagation time in the measurement interferometer.

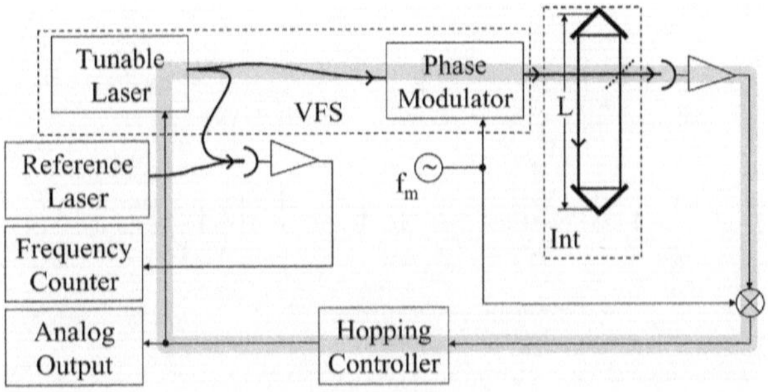

Fig. 7. TFG based on a tunable laser, e.g., a DFB laser. Pound–Drever–Hall locking is used to tie the frequency of the VFS to the changing length under measurement.

Table 1. Characteristics of the DFB lasers in use at SAO for development of the SL-TFG.

Attribute	As received at SAO	Comments
Line width	0.1 MHz	OK for $L < 1$ km, can be narrowed if needed
Bandwidth of tuning port	30 MHz (tuning range, 6 GHz)	With suitable drivers, can have 60 GHz range and $\sim$ 400 MHz BW
Range of tuning (bias port)	60 GHz (bandwidth, 1 Hz)	
Tuning sensitivity	200 MHz mA^{-1}	
Power sensitivity	$dP/dI = 0.1$ mW (optical) mA^{-1}	Maximum current, 400 mA $\Rightarrow$ 35 mW (optical); operating threshold, 40 mA

We recently purchased a pair of DFB lasers, packaged with drive electronics, from Dicos Technologies, Quebec.[1] The principal characteristics of these lasers are given in Table 1. With a TFG based on the widely tunable DFB laser, absolute distance measurements should be of intrinsically higher accuracy. It can be shown that

$$\sigma_T(L) = \frac{2}{\eta}\sqrt{\frac{\tau}{T}}(\delta L),$$

where T and τ are the integration times for absolute (L) and incremental (δL) distance, respectively. η is the fractional bandwidth, $\Delta F/F$, where ΔF is the change of laser frequency after a hop and F is the initial laser frequency. With the DFB laser, we can have $\Delta F = K\Phi$, where K is an integer and ΔF is limited by the range of the frequency counter.

How large can η be? In the initial realization of the TFG, ΔF is limited to 500 MHz by the ADM. Since $\Phi = 300$ MHz in the test interferome-

[1] This information is for technical communication only and does not constitute an endorsement of these products.

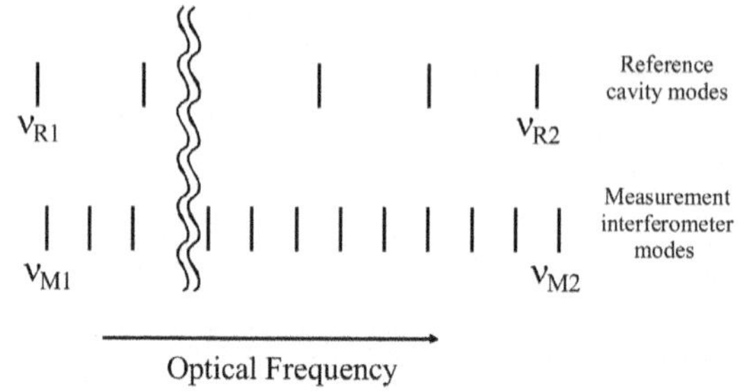

Fig. 8. Frequency map for an approach to the measurement of absolute distance using extended bandwidth.

ter and the HeNe operates at 633 nm, $\eta = 300\,\mathrm{MHz}/470\,\mathrm{THz} = 0.6 \times 10^{-6}$. In the SL-TFG, we could shift frequency both upward and downward from nominal to achieve $\Delta F = 2\,\mathrm{GHz}$. For the DFB laser operating at 1,550 nm, $\eta = 2\,\mathrm{GHz}/200\,\mathrm{THz} = 10^{-5}$. Adding additional scaling to the frequency counter would permit an increase in η, and the limit to η comes when quantization noise dominates $\sigma_\tau(\delta L)$. There are two ways around this obstacle. The first is to use a frequency counter that can count at a higher frequency without prescaling.

The second is to use the scheme shown in Fig. 8. The scheme [13] allows the measurement of absolute distance based on a frequency shift not limited by the frequency range of the counter or photodetector. There would be two reference lasers at optical frequencies ν_{R1} and ν_{R2}, both locked to modes of the reference cavity that are separated by as large a frequency span Δf_A as possible, limited by the tuning range of the lasers. The first and second measurement lasers would be locked to measurement interferometer modes at frequencies ν_{M1} and ν_{M2}, near ν_{R1} and ν_{R2}, respectively. With the lasers locked this way, the frequency differences $\nu_{Mi} - \nu_{Ri}, i = 1, 2$, would be accurately known by counting beat notes. The quantity $\nu_{M2} - \nu_{M1} = [(\nu_{M2} - \nu_{R2}) - (\nu_{M1} - \nu_{R1}) + (\nu_{R2} - \nu_{R1})]$ would be a large integer multiple, K', of Φ.

To measure $\nu_{R2} - \nu_{R1}$, the first reference laser would remain locked to a reference cavity mode at ν_{R1}, and the second reference laser would start locked to a reference cavity mode near (e.g., adjacent to) the first, at a frequency difference within the range of the photodiode and frequency counter. The free spectral range of the reference cavity would be precisely determined, taking as much time as necessary. The second reference laser would then be hopped a known number of reference cavity modes away, to lock to a mode at ν_{R2}. $\nu_{R2} - \nu_{R1}$ would now be known accurately, although at a frequency too high to count directly.

We have developed conceptually two variants of the TFG. In the first, a fast additional loop is used to narrow the linewidth of the laser. This technique has been applied to semiconductor lasers [10]. Critical aspects of the design are noted in the caption to Fig. 9. In the second, the parts count has been

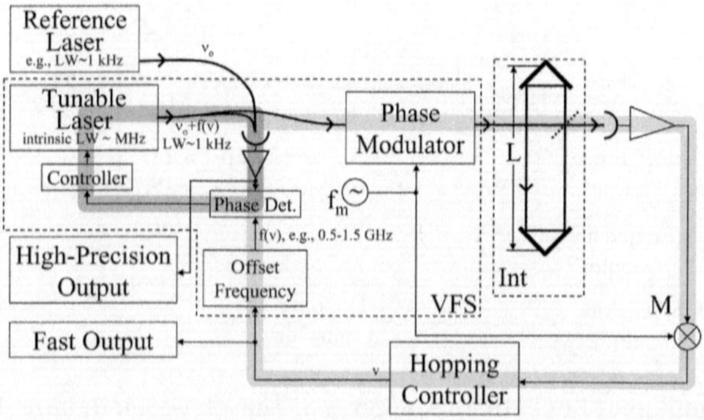

Fig. 9. TFG with narrowed laser linewidth. A tunable laser is phase locked to a reference laser, with frequency offset. This is accomplished by comparing the phase of the optical heterodyne of the lasers with that of the radio frequency (RF) signal from frequency offset, a voltage-to-frequency converter. The servoloop that does this comprises the phase detector, tunable laser, photodetector, and amplifier. The delay for a signal traveling around this loop can be conveniently kept to $\lesssim$10 ns, so the unit-gain frequency can be ~10 MHz. This is comfortably more than the typical intrinsic linewidth of the DFB laser, a few 100 kHz. Thus, the tunable laser's linewidth can be reduced to that of the reference laser, which could be ~1 kHz. The offset frequency may be tuned rapidly. The technique of narrowing the linewidth of semiconductor lasers by electrical feedback is not new [10].

reduced to yield a lower cost (and likely more reliable) laser gauge that we believe is capable of being made to work at the pm level. Critical aspects of the design are noted in the caption to Fig. 10.

4 TFG for Space Projects

In this section, we present a hypothetical metrology scheme added onto the James Webb Space Telescope (JWST; Fig. 11). (Any resemblance to actual plans of the JWST project is incidental.) We assume that the instrument is to look at targets that do not contain a bright point source suitable for serving as the pointing reference. For example, the targets could be a combination of diffuse and faint. A solution will be shown in the form of a gimbaled pointing platform that is connected by laser metrology to the instrument (since the much simpler approach of using angle encoders is unlikely to work at the mas level). In support of this approach, we introduce both a fine star tracker, showing that it is technologically achievable, and a multibeam launcher, which permits the construction of an optical truss with endpoints that are relatively simple and free of errors. Without the multibeam launcher there must be a set of endpoints attached to the structure at slightly different places, making the optical truss vulnerable to mechanical and thermal changes in that part of the structure.

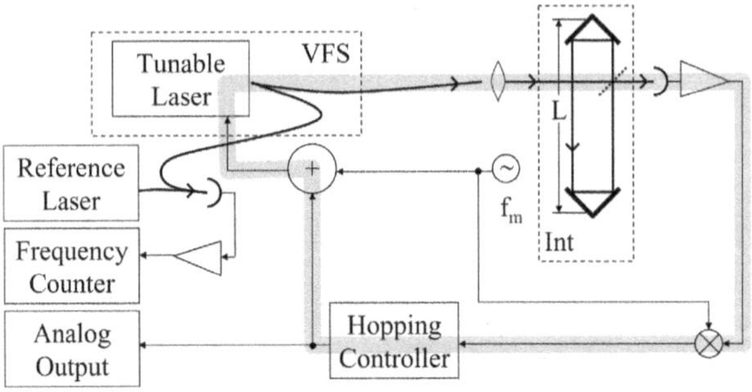

Fig. 10. Simplified TFG. In this approach, the phase modulator has been eliminated, and the modulating signal at f_m is added directly to the laser's injection current. The rest of the TFG is the same as in Fig. 7. The only optical component, aside from the measurement interferometer common to all laser gauges, is the laser diode itself. Applying the modulating signal to the laser's injection current generates a combination of amplitude and frequency modulation. In the POEM cavity, using the optimum modulating frequency, the spurious AM would generate an offset of 75 pm. This offset depends on the FM/AM ratio of the laser, which can be expected to remain largely constant with time, and depends only weakly on (DC) injection current.

Fig. 11. The front end and shield of the JWST (from STScI).

Figure 12 shows a gimbaled platform mounted in front of the JWST secondary so as not to obscure the primary. The platform contains a two axis precision star tracker with characteristics shown in Table 2. The platform is assumed to be rigid and is connected to the telescope by means of an optical truss. In the configuration shown, there are three beam launchers on the platform. Each sends multiple beams toward the telescope secondary, including two beams to retroreflectors on the back of the secondary and other beams to additional beam launchers mounted on the back of the secondary. These hang over the edge of the secondary mirror and send their beams to retroreflectors mounted on the primary mirror segments. Thus, the main optical system is tied to the platform.

Central to this architecture is the beam launcher. It must be capable of launching several beams, simultaneously or sequentially, in different directions.

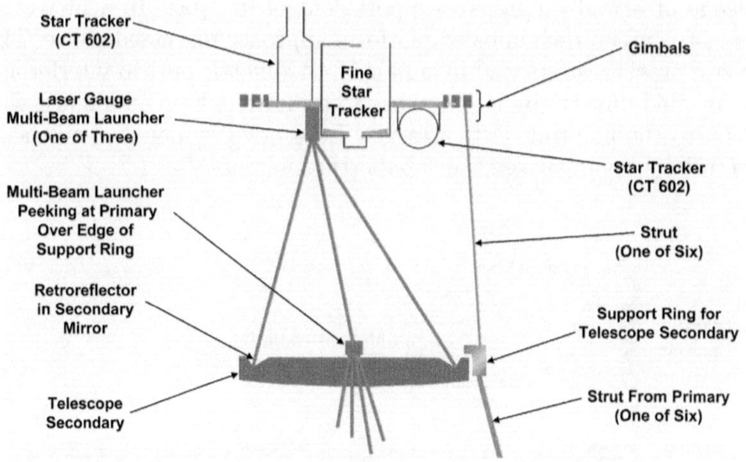

Fig. 12. Approach to pointing the JWST using a gimballed platform carrying a precision star tracker and connected to the telescope optical system by an optical truss.

Table 2. Characteristics of a precision star tracker.

Parameter	Value	Notes
Diameter	30 cm	FWHM = 0.4 arcsec at $\lambda = 0.5\,\mu$m
Target magnitude	$V = 10$	$>10^5$ photons per second detected (20% of photons entering aperture)
		$(V = 10) \Rightarrow 8$ stars per square degree (3.5 at galactic pole)
Read rate	250 Hz	400 photoelectrons per read
		$\sigma \approx 0.02$ arcsec. (Assumes cooled detectors.)

Further, all of the launched beams must be tied to a single point so as to produce a highly rigid optical truss. Finally, it is desirable that there be some degree of pointing adjustability for each beam to simplify manufacture, alignment and maintenance. Figure 13 shows the design of such a multibeam launcher. The extended caption describes the design.

Stellar Imager is a NASA Vision Study of a distributed-aperture telescope capable of resolving 30 pixels across a star's diameter. It has a maximum resolution of $\lambda/D = 0.15\,\mathrm{nrad} = 32\,\mu\mathrm{as}$ at the shortest wavelength of 155 nm. For purposes of the study, the focal length is from 1 to 10 km, the diameter is from 100 to 1,000 m, the subapertures are 1–2 m in diameter, and there will be up to 30 of them. For a telescope of this scale, pointing is a major issue.

We presented a poster with M. Karovska at AAS Meeting 204, Denver, 6/2004 on an approach to pointing the telescope [11]. We require a pointing knowledge of $5\,\mathrm{prad} = 1\,\mu\mathrm{as}$, to support control to $3\,\mu\mathrm{as}$. To achieve this, we proposed to extend the gimbaled platform approach discussed above. The precision star tracker is replaced by a pair of astrometric optical interferometers. These are tied directly by laser gauges to a metrology boom aligned with the direction to the pointing-system target. Laser gauges mounted on the end of the metrology boom survey the subaperture mirrors.

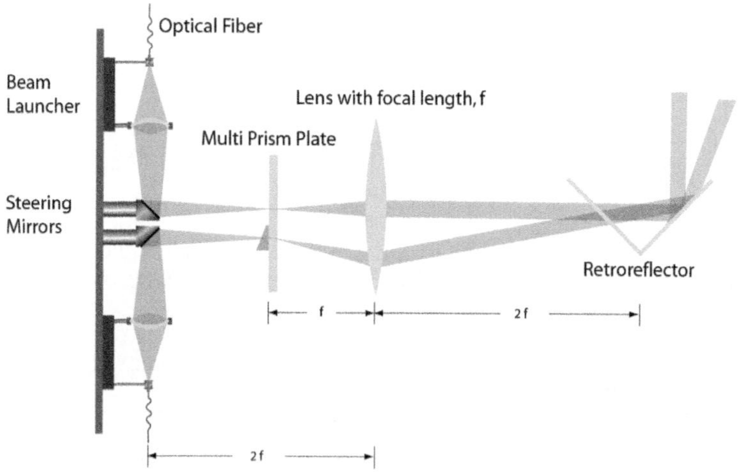

Fig. 13. Method for launching multiple beams from a single endpoint retroreflector. Each steering mirror is reimaged by the lens to a plane inside the retroreflector and thus passes through a "launch window" inside the retroreflector. Adjustment to a steering mirror causes the direction, but not the position, of the beam to change as it passes through the launch window. The prisms are necessary to make each beam appear to come from a steering mirror on the optical axis of the lens.

5 Free Fall in the Laboratory

A key technology for POEM is free fall in the laboratory. At the beginning of the project, we looked at several aspects of this issue. Two things were apparent from the start. First, if the free fall can start while the experimental apparatus is moving upward, the observing time is doubled and the signal from an EP violation is quadrupled. Second, it is important to be able to repeat the free fall quickly after the last one has ended. These ideas are embodied in figures of merit for a Galilean test, $X = Q^2 T^{1/2}/K$ and $M = X/\sigma_0$, where Q is the free-fall time, T is the total observing time, σ_0 is the measurement precision, and K is a factor that depends on the kind of motion that will be induced by an EP violation. The value of K is obtained from an analytic covariance analysis. For linear acceleration, a dropping or tossing experiment, Q is the free-fall time and $K = 12\sqrt{5} \approx 27$. (Assuming the initial position and velocity are estimated along with the acceleration, as they must be.) For a sinusoidal motion, as one expects, for example, in a rotating spacecraft or a torsion balance, the acceleration is modulated by $\sin(2\pi t/P)$. We take $Q = P/2$ and find that $K = \pi^2\sqrt{2} \approx 14$. (Alternatively, we could have taken $Q = P$, in which case we would find that $K = 4\pi^2\sqrt{2} \approx 56$. There would be no effect on X or M and the choice of $Q = P/2$ seems more natural.) In both cases, measurements must be taken several times per interval, Q.

In evaluating M or X, there may be an overall schedule factor, which favors experiments that can be run continuously (or nearly continuously), such as torsion balances and our free-fall experiment. For drop-tower experiments, there is a gap between drops, which makes the *running time* longer (in some cases, much longer) than T. In the periodic-modulation experiments, the signal has the form $\sin(u) + \cos(u)\sin(2\pi t/P)$, where u depends on the geometry. The $\cos(u)$ term may be significantly less than 1. Further, the advantage of long Q may be partially offset by a "one over f noise" in addition to the white noise assumed above. Thus, for example, in the carefully optimized STEP mission, Q is shortened by rotating the spacecraft.

There are many approaches to achieving free fall (Table 3) without going into space. Guided motion can use recirculating ball bushings or wheels with rolling bearings. These devices are generally limited to speeds of 3–5 $\mathrm{m\,s^{-1}}$. Higher speeds are possible using air bearings, which also offer a lower level of vibration. Unguided motion is limited to about 10 s of free fall near the ground and a few tens of seconds by dropping from a high altitude balloon. Longer times are possible in space at considerably higher cost. The unguided motion has the advantage of being free of the vibration generated by wheels and bushings, and the vibration passed through them from the support structure.

In the POEM approach, a vacuum chamber is in free fall for about 0.8 s. This removes the complexity of having mechanisms inside the chamber for moving the TMA and adds the complication of a moving vacuum chamber. Further, the comoving vacuum chamber provides a natural means for maintaining the required alignment of the laser gauges and the capacitance

Table 3. Characteristics of dropping facilities, approximately in increasing order of the figure of merit, X.

Quantity	N (drops)	Q	T	X	Note	
Units	(day^{-1})	(s)	(s)	note 1		
JILA EP experiment	5,760	0.2	1,152	0.2		g
JILA (toss upward)						g
GRC-short	12	2.2	26.4	3.7	2	u
GRC-tall	2	5.18	10.4	12.9	2	u
Falling inside a tube dropped from 40 km	1	2	25	1.4	3	u
POEM (drop only)	300	0.5	150	0.5	4	g
POEM (toss upward)	55,000	0.8	44,000	20	5	g
Bremen tower	3	4.8	14.4	13		u
Bremen tower (toss upward)	3	9.5	28.5	71.8		u
Parabolic flight research aircraft	40	10	400	298	6	u

In each case we consider a single day's effort.

Notes:

g Guided motion.

u Unguided motion.

1 $K = 6.71$ in all cases, except for the dropped tube (#5) for which $K = 14$.

2 NASA Glenn Research Center (GRC).

3 Balloon drop. Repeat time is nominally a month. The description of the balloon drop comes from the ambitious experiment of [9]. In this experiment, the sensor falls inside the 3 m long tube and rotates at about 0.25 Hz (Lorenzini, private communication, 5/2000). Thus, $T = 25$ s, $Q = 2$ s, $K = 14$, and $X = 1.4$. The high sensitivity of this experiment comes from the high precision (small $\sigma_0(\tau)$) of the differential detector, operating at cryogenic temperatures. The system is intended to be reused, but not likely more often than once per month.

4 POEM "drop only" used for early testing only.

5 N lowered from 65,000 to account for lateral interchange of TMA in Gen-III.

6 Figures are for NASAs "Weightless Wonder" KC-135 research aircraft, run by Johnson Space Center. This aircraft typically achieved $g/100$ for 25 s and $g/1{,}000$ for 10 s. Additional aircraft are in use. Unfortunately, the gravitational perturbations in this environment (e.g., changing orientation and fuel levels) make it unacceptable for a WEP test. In addition, short-range forces from the Earth would be lost.

gauges with the falling TMA, thus obviating the need for possible additional or enhanced mechanisms inside the chamber. This approach results in a massive moving object (ca. 50 kg) and the associated problem of large forces and large amounts of energy in vibration. Our analysis showed this to be a favorable trade.

In such a system, one must supply the energy required to launch the moving system at nearly 5 m s^{-1} and later absorb that energy at the end of the fall. Both the energy handling and the rapid recycling are facilitated by having a "bouncer" at the bottom of the chamber's path. The requirements for the bouncer are (1) it should produce only modest acceleration of the moving system, which is to contain sensitive components; (2) the acceleration should not change so rapidly that the servocontroller driving the motion system is unable to respond properly; and (3) there should be minimal loss of energy.

Our initial analysis taught us that we should avoid two things. The first is having the chamber hit a solid object, such as a leaf spring or a beam (e.g., bridging between two springs), at the bottom of the free fall because energy is lost as vibration that, at best, is dissipated and, at worst, may continue after the moving system returns to upward free fall. Further, there is an abrupt change of velocity that causes an undesirable transient response from the motion control servo. The second thing to avoid is the use of coil springs to provide an upward force on the moving system, because they have internal resonances with periods comparable with the intended turnaround time of 1/3 s. At best, the energy that goes into the internal modes is lost. At worst, the internal modes interact unfavorably with the motion control servo.

The first version of the bouncer is shown in Fig. 14. The mass on the right (10^3 kg) stores and returns the kinetic energy of the moving system via the lever and cable. The (5:1) lever shifts the energy storage to being principally gravitational potential, i.e., it reduces the kinetic energy stored in the mass. This in turn requires the mass to be larger by the same ratio. A pulley at the bottom of the chamber makes contact with the cable. The "1/4 in." cable ensures that the onset of force on the falling system is gradual and that only a very small mass needs to be accelerated as the pulley makes contact. The 1.1 m of cable has a mass of 0.18 kg, but an effective mass of 0.06 kg. In addition,

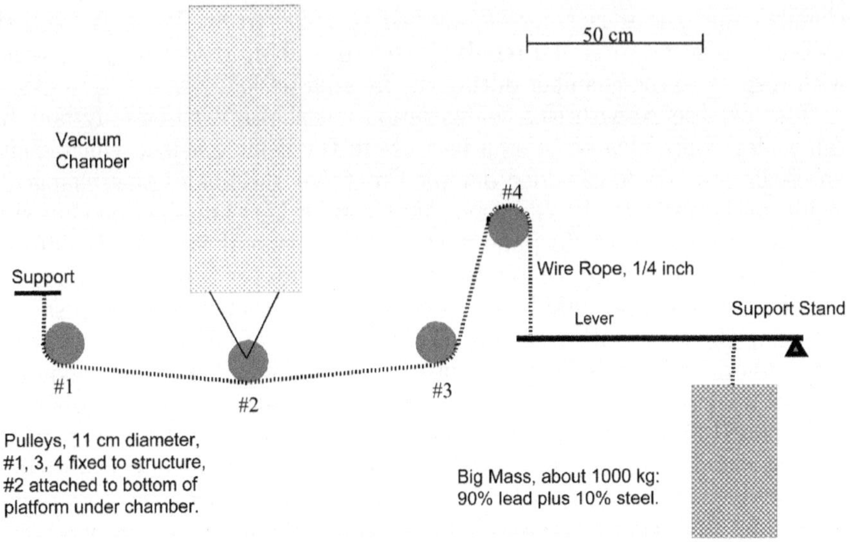

Fig. 14. Original bouncer design.

Fig. 15. Original bouncer hardware. The pulley at the bottom of the chamber (*left*) encounters a cable which lifts a ton of lead (*right*, enclosed in plastic) via the lever.

the cable probably flexes on initial pulley contact, further reducing the shock to the moving system. The hardware is shown in Fig. 15.

We have used an LVDT to investigate the ability of the motion system to put the TMA into free fall. The magnetic assembly is attached to the moving chamber and the sensor core is allowed to move freely. We have seen the LVDT core go into free fall regularly and to drift up or down by a few mm with respect to the chamber during the free-fall period.

Our experience with this configuration has shown it to be only fair. Initial pulley contact causes a step increase in the cable tension, which excites an oscillation. We have added damping material and made other changes to significantly mitigate the vibration. More serious is the dissipation caused by the cable running over the pulleys; the cable flexes and the strands shift and rub. This loss of energy causes the motion control servo to react badly and stop functioning. We could find no solution to this frictional problem short of replacing the cable with a steel band.

Instead, we have designed an entirely new bouncer based on torsion bars (made of 4340 steel, with a 75.5 inch working length, 1" working diameter, and splined ends) to apply the upward force and store the energy of the moving system. We have preserved the use of a horizontal cable to provide a gradual onset of force without requiring the sudden acceleration of significant mass. The cable ends are connected to levers on the rotating ends of a pair of torsion bars. Like coil springs, the torsion bars have internal modes. However, the lowest frequency is over 1 kHz and cannot be significantly excited because of the gentle twisting profile induced by the horizontal cable approach. Note that this design is left–right symmetric, where the previous bouncer was not.

Acknowledgments

We thank undergraduate students Brett Altschul (MIT), Giovanni DeSanti (Boston University), Dennis Feehan (Harvard), Jennifer Hoffman (Harvard), Alejandro Jenkins (Harvard), Naibi Marinas (U. Florida), and Brandon McKenna (Harvard) for skillful laboratory work. We thank colleagues Jim Faller (JILA), Robert Kimberk (SAO, Central Engineering), Tim Niebauer (Micro-g Corporation), and Doug Robertson (NGS/NOAA) for helpful discussions. We gratefully acknowledge support from the National Aeronautics and Space Administration through grant NNC04GB30G, and from the Smithsonian Institution directly, through the SAO IR&D program, and through the SI Scholarly Studies Program.

References

1. K.G. Carpenter, C.J. Schrijver, R.G. Lyon, L.G. Mundy, R.J. Allen, J.T. Armstrong, W.C. Danchi, M. Karovska, J. Marzouk, L.M. Mazzuca, D. Mozurkewich, S.G. Neff, T.A. Pauls, J. Rajagopal, G. Solyar, and X. Zhang: "The Stellar Imager (SI) Mission Concept," in Proceedings of the SPIE Conference #4854 on Future EUV/UV and Visible Space Astrophysics Missions and Instrumentation, J. Chris Blades and Oswald H. W. Siegmund, eds., 293 (2003). See also `http://hires.gsfc.nasa.gov/~si`.
2. C.K.G. Carpenter, C.J. Schrijver, R.J. Allen, A. Brown, D. Chenette, W.C. Danchi, M. Karovska, S. Kilston, R.G. Lyon, J. Marzouk, L.M. Mazzuca, R.V. Moe, F. Walter, and N. Murphy: "The Stellar Imager (SI): a revolutionary large-baseline imaging interferometer at the Sun–Earth L2 point," in Proceedings of the SPIE Conference #5491 on New Frontiers in Stellar Interferometry, Wesley Traub, ed., 243 (2004).
3. W. Cash and K.C. Gendreau: "MAXIM Science and Technology," in Proceedings of the SPIE Conference #5491 on New Frontiers in Stellar Interferometry, Wesley Traub, ed., 199 (2004).
4. T. Damour: "Testing the equivalency principle: why and how?," Class. Quantum Grav. 13, A33–A41 (1996).
5. Y. Gursel: "Metrology for spatial interferometry," in Proceedings of the SPIE Conference #2200 on Amplitude and Intensity Spatial Interferometry II, James B. Breckinridge, ed., 2200, 27–34 (1994).
6. Y. Gursel: "Metrology for spatial interferometry V," in The Proceedings of the SPIE Conference #3350 on Astronomical Interferometry, Robert D. Reasenberg, ed., 3350, 571–587 (1998), and references therein.
7. P.G. Halverson and R.E. Spero: "Signal processing and testing of displacement metrology gauges with picometre-scale cyclic nonlinearity," Journal of Optics A: Pure and Applied Optics 4(6): S304–S310 (2002).

8. J. Lawall and E. Kessler: "Michelson interferometry with 10 pm accuracy," Review of Scientific Instruments 71, 2669–2676 (2000).
9. E.C. Lorenzini, I.I Shapiro, et al.: Nuovo Cimento 109B, 1195 (1994).
10. M. Ohtsu: "Realization of ultrahigh coherence in semiconductor lasers by negative electrical feedback," Journal of Lightwave Technology 6, 245 (1988).
11. J.D. Phillips, M. Karovska, and R.D. Reasenberg: "Approaches to Metrology and Pointing for a Long Baseline UV/Optical Imaging Interferometer in Space," BAAS 36(2), #8.11 (2004a).
12. J.D. Phillips, K.G. Carpenter, K.C. Gendreau, P. Kaaret, M. Karovska, and R.D. Reasenberg: "Metrology and Pointing for Astronomical Interferometers," in the Proceedings of the SPIE Conference #5491 on New Frontiers in Stellar Interferometry, Wesley Traub, ed., 319–329 (2004b).
13. J.D. Phillips and R.D. Reasenberg: "Toward a spaceworthy picometer laser gauge," in the Proceedings of the SPIE Conference #5495 on Astronomical Telescopes and Instrumentation, 320–327 (2004).
14. J.D. Phillips and R.D. Reasenberg, "Tracking Frequency Laser Distance Gauge," Review of Scientific Instruments, 76, 064501 (2005).
15. R.D. Reasenberg, R.W. Babcock, J.F. Chandler, M.V. Gorenstein, J.P. Huchra, M.R. Pearlman, I.I. Shapiro, R.S. Taylor, P. Bender, A. Buffington, B. Carney, J.A. Hughes, K.J. Johnston, B.F. Jones, and L.E. Matson: "Microarcsecond Optical Astrometry: An Instrument and Its Astrophysical Applications," Astron. J., 32, 1731–1745 (1988).
16. C.E. Wieman and L. Hollberg: "Using diode lasers for atomic physics," Rev. Sci. Instr., 62, 1–20 (1991).
17. C.-M. Wu, J. Lawall, and R.D. Deslattes: "Heterodyne Interferometer with Subatomic Periodic Nonlinearity," Applied Optics 38, 4089–4094 (1999).
18. F. Zhao, R. Diaz, G.M. Kuan, N. Sigrist, Y. Beregovski, L.L. Ames, and K. Dutta: "SIM Internal Metrology Beam Launcher Development," in The Proceedings of the SPIE Conference #4852 on "Interferometry in Space," M. Shao, ed., 4852, 370–379 (2003), and references therein.

Clocks and Accelerometers for Space Tests of Fundamental Physics

Lute Maleki, James M. Kohel, Nathan E. Lundblad, John D. Prestage, Robert J. Thompson, and Nan Yu

Quantum Sciences and Technology Group Jet Propulsion Laboratory,
California Institute of Technology Pasadena, California, USA
lute.maleki@jpl.nasa.gov

Summary. In this chapter we discuss a technology development program at JPL to address the diminished opportunities for experimental tests of fundamental physics in space. By developing instruments that can serve multiple functions, we hope to gain flight opportunities that would otherwise be unavailable, due to recent refocusing of the space science mission in support of manned flights. We discuss the development of a liter-sized clock based on trapped mercury ions that can serve one-way navigation functions, as well as provide high stability for sensitive tests of general relativity, and possible variation of fine structure constant. We also describe progress in the development of an atom interferometer-based gravity gradiometer. This instrument is aimed at providing detailed subsurface mapping of earth and planetary bodies. It can also be used, with minor modifications, to serve as an instrument to test the equivalence principle. Finally, we report on recent progress for the development of a dual-beam atom laser based on spinor condensates, for future advanced instrumentation supporting fundamental physics studies in space.

1 Introduction

Since the early days of space exploration, the value of laboratory style experiments with clocks and accelerometers in space to test the fundamental models of physics has been well recognized. These "laboratory instruments" were identified as effective tools to test the range of validity of general relativity (GR), and soon missions such as Gravity Probe A and B were designed. It is noteworthy, however, that NASA did not select a follow-on mission to Gravity Probe A, which consisted of a hydrogen maser clock onboard a Scout rocket in a suborbital flight in 1976, until the mid-1990s. This is because tests of fundamental physics did not constitute a priority for space investigations, which mostly concentrated on observational investigations of the solar system and the space beyond.

In the 1990s two sets of unrelated developments rekindled the interest in testing physics with laboratory style experiments in space. On the theoretical front, the physics community had reached the consensus that the Standard Model fell short of providing a complete picture of the physical universe. The proper theory that would adequately address unanswered questions such as the link between gravity and quantum fields is likely to be based on radically new schemes such as M theory. This view has been reenforced by the observational astronomy that has radically changed our view of the evolution of the universe, leading to a nearly complete reformulation of cosmology. We now generally believe that our universe is flat, and is expanding at an accelerated rate. These discoveries, together with the previously known inadequacies related to, for example, the physics of black holes further amplify the need for a new physical model that can adequately describe the underlying basis for astronomical observations.

The second development pertained to the interest in conducting fundamental science on the International Space Station (ISS). The ISS was designed as a platform to conduct microgravity science, and it was apparently very well suited for laboratory style instruments as well. The use of clocks and accelerometers to test fundamental physics, then, was a natural step in the utilization of scientific potential of the ISS, especially since these instruments could indeed benefit from the microgravity environment to enhance their sensitivity. Soon, the ESA and NASA planned several clock and accelerometer experiments with instruments that were based on new technologies fueled by laser cooling of atoms, and atom wave interferometry. These planned investigations in turn spurred new interest in the physics community, and several new experiments to test GR and look for possible variation of fundamental constants were proposed. These proposals, several of which are described in this volume, are based on platforms other than the ISS, and continue to be under study today.

Despite these renewed interest, the prospects for missions based on instruments such as clocks and accelerometers are realistically rather poor. This is because of the recent plans to limit the utilization of the ISS, and focus the bulk of space missions to explore Moon and Mars. The challenge presented by sustaining the ISS for the next few years, and developing the infrastructure for manned missions to the moon within a fixed budget, has placed severe constraints on the NASA budget, with a correspondingly similar situation for ESA.

With this backdrop, the focus of fundamental physics research at the Quantum Sciences and Technology Group at JPL has been reformulated. Realizing that independent missions such as spacetime [1], and ISS investigations such as PARCS and QuITE are not realistically viable in the current budget environment of NASA, which is likely to remain constrained in the next several years, we have shifted our focus toward the development of science experiments that can be performed as adjunct to other space missions. In particular, we have focused on the development of technologies that reduce size, cost, and power of precision instruments, allowing them to become attractive for applications other than tests of fundamental physics. By designing and developing instruments that can satisfy needed functions in completely different areas of space exploration, such as navigation and subsurface mapping, we hope to create more opportunity for tests of fundamental physics.

In the following sections, we provide brief descriptions of the development of a small high performance clock with 1 l of volume and about 1 kg of mass, and an atom wave interferometer for gravity gradiometry. We will also give a short description of experiments toward more advanced accelerometers based on Bose–Einstein condensation.

2 LITE: A Liter-Sized High Performance Atomic Clock

The development of high performance atomic clocks based on ion traps is motivated by the inherent immunity of trapped ions to perturbing influences on the atomic energy structure, which in turn influence the stability of the clock. Ions in a trap do not experience collisions with the walls of a container, or with themselves. The use of heavier ions, such as mercury, allows lowering the sensitivity of the clock transition to the perturbing magnetic fields. These and other similar attributes of ion traps have led to the development of clocks that use microwave or optical transitions in a variety of ion systems.

At JPL, the work in the past two decades has been the development of a stable clock based on trapped mercury ion with performance parameters, including operational reliability, superior to that of the hydrogen maser, which has been the workhorse of navigation and communications system at NASA's deep-space network (DSN). After a successful demonstration of such an instrument dubbed the Linear Ion Trap Standard (LITS) based on mercury ions, the focus of the work has been placed on the development of a small, high performance clock suitable for direct deployment onboard planetary spacecraft [2–7]. Since this class of space platforms has limited capability for payload due to mass and power constraints, it is instrumental that any additional payload such as an advanced clock be small in volume and mass, and inherently reliable for the expected mission duration that generally average more than 5 years.

To transform the technology of LITS into a spacecraft clock with features mentioned above, several technological advances were developed. To begin with, the size and mass of the LITS as deployed in the DSN were dominated by the four layers of magnetic shielding required to keep the ions from the magnetic perturbation of the outside environment. The design of LITS required that the magnetic shields cover the trap as well as the lamp and the light detection system that is used to optically pump the ions and interrogate their clock transition. These elements include magnetic components that must be kept within the shields to ensure that their residual fields do not change by any interaction with fields external to the clock. This approach was required as the same region in the trap was used to both optically interact with the ions and apply the microwave field that induces the clock transition in the ions.

To reduce the size and weight associated with the magnetic shields that had a diameter of nearly 1 m, a new trap was designed to allow the separation of the optical interaction region from the microwave excitation region. Taking advantage of the electric charge carried by each ion, this new "shuttle trap" has two segments. In the first segment, the ions are produced through collision of electrons emitted from a hot filament with a background of mercury vapor produced by mild heating of an oxide of mercury. The ions were then subjected to the pump light from an RF-excited mercury lamp which produced the optical pumping of the ground hyperfine state ($F = 0$, $^2S_{1/2}$) via the first excited electronic state ($^2P_{1/2}$). After this step which prepared the ions in the $F = 0$ hyperfine sublevel of the ground state, the ions are "shuttled" to a second segment of the trap through the application of a small (a few volts) DC potential. In this region ions are exposed to the microwave radiation (at 40.5 GHz) that induces the $F = 0$ to $F = 1$ clock transition in the ground state before they are shuttled back to the first, optical interrogation, region. Since the ions must be kept isolated from any perturbations only during the interval that they interact with the applied microwave field, a magnetic shielding of the second segment of the trap will be all that is required. As the second segment of the trap is away from the light source and the photon detectors, the size of the shields is reduced by about an order of magnitude.

To further reduce the sensitivity of the ions to external perturbations, a second innovation was introduced to the design of the trap. The conventional design for ion traps used in atomic clocks and frequency standards is based on electrode structures that produce a quadrupole oscillating field. Such a field can be produced with the original hyperbolic electrodes, as well as with the four-rod linear trap. A feature of the quadrupole field is that it contains a node (in the case of the hyperbolic design) or a line of nodes (in the case of the linear trap) where the oscillating electric field applied to the electrodes is zero, and so is the force on the ions. When the ions move away from the node, they experience a ponderomotive force that increases quadratically as they approach any electrode. It is this ponderomotive force that keeps the ions close to the node(s) in the trap. The quadratic increase in the field away from a node, though, results in an increase in the velocity and "heating" on the ions. This effect produces a relatively large second-order Doppler shift that in turn limit the achievable accuracy with trapped ion standards.

At JPL, the microwave interrogation segment of the "shuttle" trap was designed with a multipole electrode configuration. Sixteen rods alternatively excited with the applied oscillating field produce a potential that is essentially zero at the interior of the trap and increases sharply only very close to each electrode. This flat potential allows a much larger region in which the field experienced by the ions is zero, allowing ions to move in a mostly field-free region. Thus the influence of heating and the associated second-order Doppler is greatly reduced.

These innovations have been combined with a design that utilizes small size components, and a sealed vacuum enclosure that is pumped only with getters to allow a significant reduction in the size of the clock. The current laboratory prototype is essentially smaller than 2 l in volume with a projected mass below 2 kg. The next version being developed will approach the goal of about 1 l in size and 1 kg through improvements in packaging and integration of the electronic circuitry. The projected power requirement is less than 20 W.

The prototype laboratory small clock has already demonstrated a stability performance that breaks into the 10^{-15} region at an averaging interval of about 10,000 s set by the performance of the quartz local oscillator. Similar performance is anticipated for the clock as it is engineered for a 1 kg and 1 l parameters. Such a clock can be used onboard spacecraft to allow one-way navigation. This style of spacecraft navigation, in contrast to the conventional two-way navigation that requires both an uplink and a downlink, does not require an uplink, and will significantly reduce the cost of the SDN operation, while reducing the burden of multiple spacecraft navigation with the same antennas. A modified version of this clock will operate with three different ion species with about a doubling of mass and power. A "triclock" instrument with this geometry can be placed on planetary spacecraft to perform tests of relativity similar to the scheme of the proposed spacetime mission [1].

3 The Quantum Gravity Gradiometer

The development of atom interferometers using laser light pulses to drive stimulated Raman transitions in atoms has provided a sensitive new technique for gravity gradiometry [8]. Unlike gradiometers employing mechanical accelerometers, this approach employs the individual atoms as identical drag-free test masses. The de Broglie wave associated with each atom is then utilized to perform an interferometric measurement of the local acceleration. As in conventional gradiometers, the atom interferometer-based gravity gradiometer employs two accelerometers to allow cancellation, as common-mode noise, of any vibrations of the reference platform. This dual interferometric technique holds great potential for subsurface gravity mapping and monitoring applications, including studies of planetary inner structures and dynamics, changes in ice sheets and ocean currents, changes in underground water storage, and geodesy. The accelerometers can also be used to perform tests of the equivalence principle (EP) with atomic test masses to a sensitivity rivaling proposed measurements with macroscopic test masses in free fall, such as MICROSCOPE [9] and STEP [10]. Toward this end, we are developing at JPL a compact and robust instrument by making extensive use of modular optical components and fiber optics in the laser and optics system, and a highly symmetric geometry in the atomic physics package suitable for microgravity operation.

The current laboratory-based gradiometer consists of two atom interferometers separated by 1.4 m in the vertical direction. The lower interferometer is based on an ultra-high vacuum (UHV) magneto-optic trap (MOT), which is loaded directly from a two-dimensional MOT-based cesium atom beam source. The two-dimensional MOT generates an atom flux greater than 10^9 atom s^{-1} from a volume of only $4 \times 4 \times 13\,cm^3$, including magnetic coils and optics [11]. The UHV MOT enclosure is a nonmagnetic titanium chamber with high-quality AR-coated windows for optical access. A 250-mm tall vertical tower above the trapping region allows the UHV MOT to operate as an "atom fountain" to obtain long interrogation times in ground-based operation. The same lasers used for trapping and cooling are employed for the state-normalized detection of the launched atoms upon their return. This interferometer geometry is also well suited for microgravity operation, where long-baseline interrogations can be performed simply in the central trap region and under UHV conditions.

The upper interferometer, developed as an early testbed system, is based on a vapor cell MOT. The state selection and normalized detection are performed in a separate detection region, and graphite getters are employed to minimize the background cesium vapor in this region. Due to the different geometry, atoms in the two fountains are launched at different times to operate the interferometers simultaneously.

The modular laser system consists of two frequency-stabilized external-cavity diode lasers as master oscillators and nine injection-locked slave lasers. A phase-locked loop serves the difference frequency between the master lasers to obtain a phase stability better than 1 mrad between 10 Hz and 10 kHz in the Raman lasers. The spectrum of the beatnote between the two Raman frequencies reveals that the central peak contains greater than 99% of the RF power. The master lasers, slave lasers, and frequency-control and beam-splitting modules are interconnected via fiber optics to provide a versatile yet robust laser system. The laser beams for trapping and interferometry are also delivered to the two fountains using fiber optics. The overlapping Raman beams are collimated to 24 mm diameter and aligned vertically through large-diameter (75 mm) windows at the top and bottom of the vacuum chambers for each fountain. The retro-optic for the Raman beams is mounted on a passive vibration isolation platform on the laboratory floor. A schematic of the gradiometer is shown in Fig. 1.

We launch up to 6×10^9 atoms in each atomic fountain at temperatures of 1.8 μK, as measured by stimulated Raman velocimetry. A Doppler-sensitive Raman pulse is applied to select a narrow velocity subgroup so that the remaining atoms are characterized by a one-dimensional temperature of about 100 nK. The atom interferometer is then realized by employing a $\pi/2-\pi-\pi/2$ stimulated Raman pulse sequence [12]. The difference frequency of the Raman lasers is "chirped" during this sequence to continuously track the Doppler shift of free-falling atoms. By scanning the relative phase of the final $p/2$ pulse, the relative populations of the $F = 3$ and $F = 4$ hyperfine ground states are modulated to produce the interferometer fringes that are observed via laser-induced fluorescence. Interferometer fringes are shown in Fig. 2.

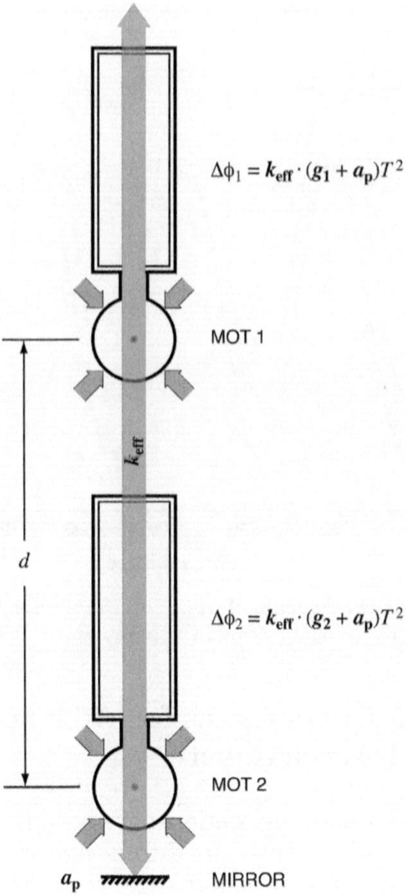

Fig. 1. Illustration of a gravity gradiometer based on dual atom interferometers. The Raman laser along the vertical axis measures the phase shift of atoms in each fountain, so that the gravity gradient along this axis is determined by $\Delta g/\Delta z = (\Delta\phi_1 - \Delta\phi_2)/(k_{\rm eff}T^2 d)$, and platform vibrations a_p are effectively canceled.

The current instrument has demonstrated a measurement sensitivity of $34\,{\rm E\,Hz}^{-1/2}$ ($1\,{\rm E} \equiv 10^{-9}\,{\rm s}^{-2}$), corresponding to a sensitivity of $5\,{\rm E\,Hz}^{-1/2}$ for a 10 m measurement baseline. A second generation instrument is currently being designed with the goal of sensitivity improved by at least an order of magnitude. This second generation instrument will be used for the design and development of a spaceborne instrument for gravity gradiometry in support of geophysical investigations. With relatively small modifications, the same instrument could also be used to test EP at a sensitivity level exceeding all ground-based experiments, and similar to those planned as independent missions.

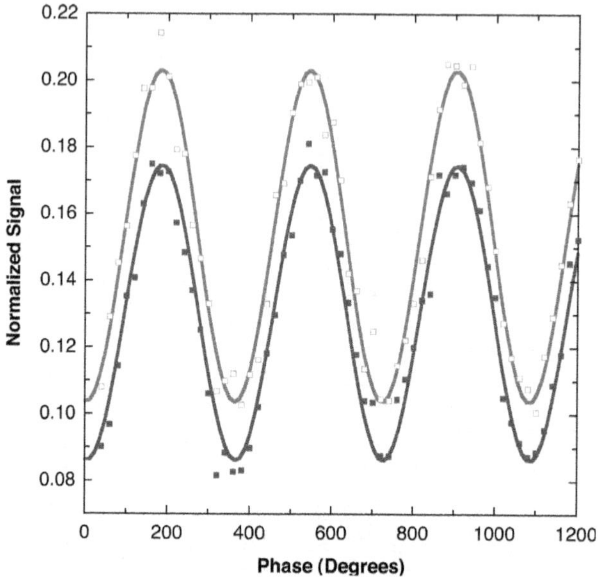

Fig. 2. Simultaneous measurements of atom interferometer fringes in two fountains using a common laser to drive the Raman transitions.

4 Bose–Einstein Condensates for Advanced Atomic-Based Accelerometers

It is widely known in atomic and optical physics community that the Bose–Einstein condensate (BEC) can be used to devise an atom laser producing coherent matter waves. Such a counterpart to the conventional photon-based lasers can be used to improve the sensitivity of matter wave interferometers that are being developed to test fundamental physics, in the same way that photon lasers improve the performance of optical interferometers [13]. Because of this potential to devise vastly improved atom wave interferometers, a program has been underway at JPL to develop a dual-beam atom laser based on spinor condensates [14]. This has led to a recent experimental scheme aimed at the generation of dual atom laser beams, oppositely propagating, with an inherent number correlation between them due to their spin-mixing origin. The novelty of the scheme lies not in the output coupler (simple field gradients that tilt the confining optical potential) but rather in the origin of the outcoupled populations in the dynamics of the coherent spin–spin interaction process. In addition, the presence of a true reservoir and the ability to control both the strength of the output coupler as well as the rate of "pumping" make this a particularly intriguing atom laser scheme.

Our apparatus is built around a single-beam running-wave dipole trap produced by a focused CO_2 laser, which provides at full power a trap depth of approximately 1.6 mK via the DC polarizability of rubidium. We load the dipole trap from a UHV magneto-optical trap (MOT) which is itself loaded by a cold atomic beam provided by an upstream two-dimensional MOT. The two-dimensional MOT exists in a rubidium vapor cell which is differentially pumped from the adjoining science chamber. All 780 nm trapping light is provided by a unique laser system based on a frequency-doubled 1,560 nm fiber amplifier, described elsewhere. The loading of the dipole trap proceeds according to established technique; we obtain initial populations in the trap of about 2×10^6 rubidium atoms at about 120 μK. The initial trap frequencies (measured via parametric resonance) are approximately 3.2 kHz transversely and 220 Hz longitudinally.

Evaporative cooling proceeds via a programmed rampdown of CO_2 laser intensity. We observed the onset of BEC at critical temperatures near 100 nK, and typically obtain condensates of 10^4 atoms with little or no discernible thermal component. Application of a magnetic field gradient along the weakly trapping axis of the trap during the first few seconds of evaporation preferentially biases out the $m_F = 1$ components, resulting in a BEC solely occupying the field-insensitive $m_F = 0$ projection. Finally, application of a small magnetic field gradient in the vertical direction provides a bias for one or the other polarized components. If this supportive gradient is only on for the first few seconds of evaporation, we obtain polarized condensates of number similar to the other options.

For the observations of the dual-beam laser, we begin with a nominally pure $m_F = 0$ condensate held in a trap whose unperturbed depth is 5 μK and is approximately a factor of 10 weaker due to gravitational tilt. We fix the background field levels at 60 mG as determined by RF spectroscopy. To coherently mix the condensate, we adiabatically compress the trapping field, typically by raising the laser power from 100 to 700 mW over 100 ms, hold the compressed condensate for a variable time, adiabatically expanding, and ballistically expanding while applying the Stern–Gerlach field. We observe that the fraction of atoms evolved into the polarized projections increases with high-density hold time and eventually reaches a static level of 50%. We also observe that the time taken to reach this steady state varies linearly in density, with an offset given by the critical density at which the quadratic Zeeman effect dominates the dynamics.

Slight changes in the offset field, beam tilt, and direction of the extraction gradient affect several variants of the dual-beam atom laser: most commonly, we observe immediate outcoupling and ballistic flight of the $m_F = -1$ component while the $m_F = +1$ component first propagates in the opposite direction (as expected), then reverses its motion, passes through the parent $m_F = 0$ condensate, and escapes along a different trajectory (see Figs. 3 and 4). We also observe the more intuitive case of both polarized components escaping into ballistic flight from opposite ends of the cigar-shaped trap.

Future work will explore the nature of correlations and entanglement in these beams, the possibility of spin-independent outcoupling, and also explore the possibilities of improving this process into the quasicontinuous regime.

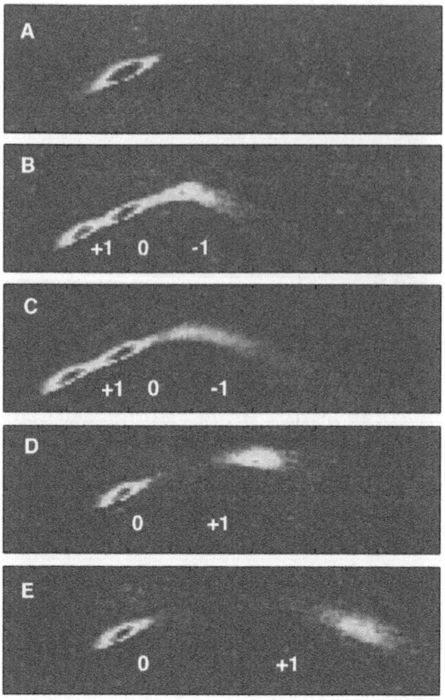

Fig. 3. A typical outcoupling run of the spinor dynamics-driven dual-beam atom laser. (**a**) 0 ms: the full condensate, in situ. (**b**) +20 ms: immediately after outcoupling. The $m_F = -1$ component immediately passes beyond the reach of the dipole trap and experiences ballistic flight and mean-field expansion. The $m_F = +1$ component remains confined in an effective guide and travels in the opposite direction. (**c**) +25 ms: the $m_F = -1$ beam continues to propagate while the $m_F = +1$ beam is turned around and returned toward the origin. (**d**) +45 ms: the $m_F = +1$ beam now falls freely and experiences mean-field expansion, like the $m_F = -1$ component before it. Note a slightly different path than $m_F = -1$. (**e**) +50 ms: continued $m_F = +1$ propagation; note the $m_F = -1$ component has traveled out of the field of view by this point. Images are 1–0.25 mm; gravity is directed toward the *lower right* and the trapping laser is directed toward the *upper right*.

Finally, it should be possible to implement a number measurement on one outcoupled component, rendering the untouched component into a known Fock state and thus ideal for Heisenberg-limited phase measurements. The scheme of the dual-beam atom laser can be tailored for applications with atom interferometry with a sensitivity beyond what is achieved with conventional laser cooled clouds.

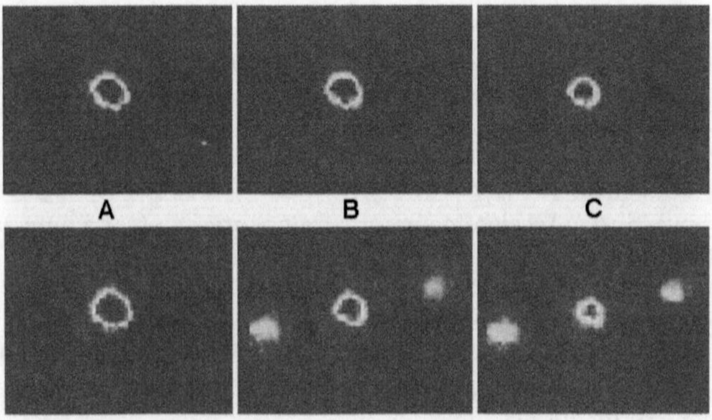

Fig. 4. The effect of adiabatic compression on the spin-mixing process; the *top row* shows condensates held for equivalent durations without compression. (**a**) 100 ms of hold time at 700 mW, (**b**) 400 ms of hold time, (**c**) 1.2 s of hold time. The slightly fewer overall number in (**c**) is due to condensate lifetime. The ballistic expansion time for all images is 17.5 ms. Images are 1–0.8 mm.

5 Summary

In this chapter we have presented a summary of technology development efforts at JPL aimed at demonstration of instruments for use in NASA missions in support of navigation and earth and planetary exploration. By designing these instruments based on approaches that support fundamental physics investigation, we hope to gain access to space platforms that are otherwise unavailable to the fundamental physics community. This strategy is aimed at keeping the prospects of laboratory style experimental tests of physics in space alive in an era when support for such investigations has been largely abandoned in favor of observational experiments, and manned flights.

The research described in this chapter was carried out at the Jet Propulsion Laboratory, California Institute of Technology, under a contract with the National Aeronautics and Space Administration.

References

1. L. Maleki and J. Prestage: Search for New Physics with Atomic Clocks Astrophysics, Clocks and Fundamental Constants Eds. S G Karshenboim and E Peik (Berlin:Springer, 2004).
2. J.D. Prestage, G.J. Dick, and L. Maleki: J. Appl. Phys. **66**, 1013 (1989).
3. J.D. Prestage, R.L. Tjoelker, L. Maleki: Phys. Rev. Lett., **74**, 3511 (1995).
4. J.D. Prestage, R.L. Tjoelker, L. Maleki: Proc. 2000 IEEE Freq. Control Symp., pp. 459–462 (2000).
5. J.D. Prestage, R.L. Tjoelker, L. Maleki: Topics Appl. Phys. **79**, 195–211, Frequency Measurement and Control, ed. A.N. Luiten (2001).

6. J.D. Prestage, S. Chung, E. Burt, L. Maleki, and R.L. Tjoelker: Proc. 2002 IEEE Freq. Control Symp., pp. 706–710 (2000).
7. R.L. Tjoelker, J.D. Prestage, P.K. Koppang, and T.B. Swanson: Proc. 2003 IEEE Freq. Control Symp., pp. 1066–1072 (2003).
8. J.M. McGuirk, G. Foster, J.B. Fixler, M.J. Snadden, and M.A. Kasevich: Phys. Rev. A **65**, 033608 (2002).
9. P. Touboul, Space Accelerometers: Present Status. In C. Lämmerzahl, C.W.F. Everitt and F.W. Hehl (eds.). *Gyros, Clocks, Interferometers ...: Testing Relativistic Gravity in Space.* (Springer, Berlin 2001).
10. N. Lockerbie, J.C. Mester, R. Torii, S. Vitale, and P. Worden. STEP: A status Report. In C. Lämmerzahl, C.W.F. Everitt and F.W. Hehl (eds.). *Gyros, Clocks, Interferometers ...: Testing Relativistic Gravity in Space.* (Springer, Berlin 2001).
11. J. Ramirez-Serrano, N. Yu, J.M. Kohel, J.R. Kellogg, and L. Maleki: Opt. Lett. **6**, 682–684 (2006).
12. M. Kasevich and S. Chu: Phys. Rev. Lett. **67**, 181 (1991).
13. N. Lundblad, R.J. Thompson, D. Aveline, and L. Maleki: Submitted to PRA (2006).
14. D.M. Stamper-Kurn, et al., Phys. Rev. Lett. **83**, 661 (1999).

Atom Interferometric Inertial Sensors for Space Applications

Philippe Bouyer[1], Franck Pereira dos Santos[2], Arnaud Landragin[2], and Christian J. Bordé[2]

[1] Laboratoire Charles Fabry de l'Institut d'Optique, CNRS, Université Paris-Sud Campus Polytechnique, RD127
91127 Palaiseau cedex, France
[2] LNE-SYRTE, UMR8630, Observatoire de Paris, 61 avenue de l'Observatoire, 75014 Paris, France
philippe.bouyer@iota.u-psud.fr http://atomoptic.fr

Summary. The techniques of atom cooling combined with atom interferometry make possible the realisation of very sensitive and accurate inertial sensors like gyroscopes or accelerometers. Besides earth-based developments, the use of these techniques in space should provide extremely high sensitivity for research in fundamental physics.

1 Introduction

Inertial sensors are useful devices in both science and industry. Higher precision sensors could find scientific applications in the areas of general relativity [1], geodesy and geology. There are also important applications of such devices in the field of navigation, surveying and analysis of earth structures. Matter-wave interferometry was envisaged for its potential to be an extremely sensitive probe for inertial forces [2]. First, neutron interferometers have been used to measure the acceleration due to gravity [3] and the rotation of the Earth [4] at the end of the seventies. In 1991, atom interference techniques [5, 31] have been used in proof-of-principle work to measure rotations [6] and accelerations [7]. In the following years, many theoretical and experimental works have been performed to investigate this new kind of inertial sensors [8]. Some of the recent works have shown very promising results leading to a sensitivity comparable to other kinds of sensors, for rotation [9,10] as well as for acceleration [11, 12].

Atom interferometry [2, 6, 8, 13, 14] is nowadays one of the most promising candidates for ultra-precise and ultra-accurate measurement of gravito-inertial forces [9–12, 15–17] or for precision measurements of fundamental constants [18]. The realisation of Bose–Einstein condensation (BEC) of a dilute gas of trapped atoms in a single quantum state [19–21] has produced the matter-wave analog of a laser in optics [22–25]. Alike the revolution brought by lasers in optical interferometry [1,26,27], it is expected that the use of Bose–Einstein condensed atoms will bring the science of atom optics, and in particular atom interferometry, to an unprecedented level of accuracy [28, 29]. In addition, BEC-based coherent atom interferometry would reach its full potential in space-based applications where micro-gravity will allow the atomic interferometers to reach their best performance [30].

2 Inertial Sensors Based on Atom Interferometry: Basic Principle

Generally, atom interferometry is performed by applying successive *coherent* phase-locked beam-splitting processes separated by a time T to an ensemble of particles (see Fig. 1) [31,32], followed by detection of the particles in each of the two output channels. The interpretation in terms of matter waves follows from the analogy with optical interferometry. The incoming matter wave is separated into two different paths by the first beam splitter. The accumulation of phases along the two paths leads to interference at the last beam splitter, producing complementary probability amplitudes in the two output channels [33–35]. The detection probability in each channel is then a sine function of the accumulated phase difference, $\Delta\phi$.

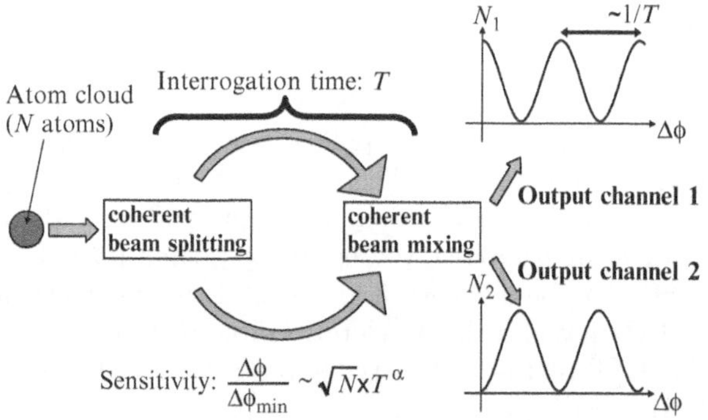

Fig. 1. Principle of an atom interferometer. An initial atomic wave packet is split into two parts by the first beam splitter. The wave packets then propagate freely along the two different paths for an "interrogation time" T, during which the two wave packets can accumulate different phases. A second pulse is then applied to the wave packets so that the number of atoms at each output is modulated with respect to this phase difference. The maximum sensitivity achievable for such an apparatus can be defined by comparing the variation of the number of atoms ΔN due to the phase difference $\Delta\phi$ at the output ($\Delta N \sim N\Delta\phi/2\pi \propto NT^{\alpha}$) with the quantum projection noise arising from atom counting $\sqrt{N}$. It scales as $\sqrt{N} \times T^{\alpha}$.

Atomic clocks [36–38] can be considered one of the most advanced application of atom interferometry [39]. In this "interferometer", the two different paths of Fig. 1 consist of the free evolution of atoms in different internal states with an energy separation $\hbar\omega_{at}$. An absolute standard of frequency is obtained by servo-locking a local oscillator to the output signal of the interferometer. The output signal of the clock then varies as $\cos(\Delta\omega \times T)$ where $\Delta\omega$ is the frequency difference between the transition frequency ω_{at} and the local oscillator frequency ω. Atom interferometers can also be used as a probe of gravito-inertial fields. In such applications, the beam splitters usually consist of pulsed near-resonance light fields which interact with the atoms to create a coherent superposition of two different *external* degrees of freedom, by coherent transfer of momentum from the light field to the atoms [2, 5, 31]. Consequently, the two interferometer paths are separated in space, and a change in the gravito-inertial field in either path will result in a modification of the accumulated phase difference. Effects of acceleration and rotation can thus be measured with very high accuracy. To date, ground-based experiments using atomic gravimeters (measuring acceleration) [11, 40], gravity gradiometers (measuring acceleration gradients) [15, 41] and gyroscopes [9, 10] have been realised and proved to be competitive with existing optical [42] or artefact-based devices [43].

3 Atom Interferometers Using Light Pulses as *Atom-Optical* Elements

The most developed atom interferometer inertial sensors are today atomic state interferometers [31, 48] which in addition use two-photon velocity selective Raman transitions [44, 45] to manipulate atoms while keeping them in long-lived ground states. With the Raman excitation, two laser beams of frequency ω_1 and ω_2 are tuned to be nearly resonant with an allowed optical transition. Their frequency difference $\omega_1 - \omega_2$ is chosen to be resonant with a microwave transition between two atomic ground-state levels. Under appropriate conditions, the atomic population Rabi flops between the ground-state levels with a rate proportional to the product of the two single-photon Rabi frequencies and inversely proportional to the optical detuning. When the beams are aligned to counter-propagate, a momentum exchange of approximately twice the single-photon momentum accompanies these transitions. This leads to a strong Doppler sensitivity of the two-photon transition frequency, and can be used to coherently divide (with a $\pi/2$ pulse) or deflect (with a π pulse) atomic wave packets. (On the other hand, when the beams are aligned to co-propagate, these transitions have a negligible effect on the atomic momentum, and the transition frequency is almost Doppler insensitive.)

Usually, an interferometer is formed using a $\pi/2 - \pi - \pi/2$ pulse sequence to coherently divide,[1] deflect and finally recombine an atomic wave packet (as in a Mach–Zehnder interferometer in optics). The resulting interference can be directly observed by measuring the atomic ground-state populations [13]. In comparison with mechanical nano-fabricated gratings [14], optical gratings can be easily vibrationally isolated from the vacuum chamber [46]. Scattering from standing waves [32, 47] can be efficient and capable of large momentum transfer. However, these beam splitters typically require a highly collimated atomic beam. In contrast, the stimulated Raman transition linewidth can be adjusted to address large transverse velocity spreads, relaxing collimation requirements and increasing interferometer count rates.

We present in this section a summary of recent work with light-pulse interferometer-based inertial sensors. We first outline the general principles of operation of light-pulse interferometers. This atomic interferometer [31, 48] uses two-photon velocity selective Raman transitions [44], to manipulate atoms while keeping them in long-lived ground states.

3.1 Principle of Light-Pulse Matter-Wave Interferometers

Light-pulse interferometers work on the principle that, when an atom absorbs or emits a photon, momentum must be conserved between the atom and the light field. Consequently, an atom which emits (absorbs) a photon of momentum $\hbar\boldsymbol{k}$ will receive a momentum impulse of $\Delta\boldsymbol{p} = -\hbar\boldsymbol{k}(+\hbar\boldsymbol{k})$. When a resonant travelling wave is used to excite the atom, the internal state of the atom becomes correlated with its momentum: an atom in its ground state $|1\rangle$ with momentum $\boldsymbol{p}$ (labelled $|1, \boldsymbol{p}\rangle$) is coupled to an excited state $|2\rangle$ of momentum $\boldsymbol{p} + \hbar\boldsymbol{k}$ (labelled $|2, \boldsymbol{p} + \hbar\boldsymbol{k}\rangle$) [31, 48]. A precise control of the light-pulse duration allows a complete transfer from one state (for example $|1, \boldsymbol{p}\rangle$) to the other ($|2, p + \hbar\boldsymbol{k}\rangle$) in the case of a π pulse and a 50/50 splitting between the two states in the case of a $\pi/2$ pulse (half the area of a π pulse). This is analogous to a polarising beam splitter (PBS) in optics, where each output port of the PBS (i.e. the photon momentum) is correlated to the laser polarisation (i.e. the photon state). In the optical case, a precise control of the input beam polarisation adjusts the balance between the output ports. In the case of atoms, a precise control of the light-pulse duration plays the role of the polarisation control.

In the $\pi/2 - \pi - \pi/2$ configuration, the first $\pi/2$ pulse excites an atom initially in the $|1, \boldsymbol{p}\rangle$ state into a coherent superposition of states $|1, \boldsymbol{p}\rangle$ and $|2, \boldsymbol{p} + \hbar\boldsymbol{k}\rangle$. If state $|2\rangle$ is stable against spontaneous decay, the two parts of the wave packet will drift apart by a distance $\hbar kT/m$ in time T. Each partial wave packet is redirected by a π pulse which induces the transitions

[1] There are other possible configurations, such as the Ramsey–Bordé $\pi/2 - \pi/2 - \pi/2 - \pi/2$ [6] which can be extended to include multiple intermediate π pulses [8] or adiabatic transfers [49] to increase the area.

$|1,\boldsymbol{p}\rangle \to |2,\boldsymbol{p}+\hbar\boldsymbol{k}\rangle$ and $|2,\boldsymbol{p}+\hbar\boldsymbol{k}\rangle \to |1,\boldsymbol{p}\rangle$. After another interval T the two partial wave packets overlap again. A final pulse causes the two wave packets to recombine and interfere. The interference is detected by measuring, for example, the number of atoms in state $|2\rangle$. We obtain a large wave packet separation by using laser-cooled atoms and velocity selective stimulated Raman transitions [44]. A very important point of these light-pulse interferometers is their intrinsic accuracy, thanks to the knowledge of the light frequency which defines the scaling factor of the interferometers.

3.2 Application to Earth-Based Inertial Sensors

Inertial forces manifest themselves by changing the relative phase of the de Broglie matter waves with respect to the phase of the driving light field, which is anchored to the local reference frame. The physical manifestation of the phase shift is a change in the probability to find the atoms, for example, in state $|2\rangle$, after the interferometer pulse sequence described above. A complete analytic treatment of wave packet phase shifts in the case of acceleration, gradient of acceleration and rotation together [34, 39, 50–52] can be realised with the ABCDξ formalism, a formalism generalising to matter waves the ABCD matrices for light optics. In these calculations, it is always important to remember that the external fields act not only on the atoms but also on other components of the experiments, such as mirrors and laser beams and that additional contributions may enter in the final expression of the phase (the final phase expression should be independent of the gauge [39, 52]). As an example, the gravitational phase shift, that can be calculated to first order using the gravitational field action integral on the atomic wave packets [31, 33, 35, 48], can be removed from the general expression of the interferometer phase shift by a simple coordinate transformation. It will then reappear in the beam-splitter phases.

The phase shift calculation obtained by an action integral along the unperturbed trajectory of the atoms works only to first order. The exact expression of the phase involves the sum of three contributions: the first one comes from the beam splitters, the second from the action integral along both paths and the third from the interferometer end points splitting under the influence of the perturbing field. When the action is calculated along the perturbed trajectories, for equal masses, one can show that it cancels for the most part with the end points splitting contribution [34, 39, 51–53], leaving the beam-splitter contribution alone with recoil correction terms. This beam-splitter contribution is a scalar product, hence invariant in coordinate transformations. If masses are unequal, the action integral produces an additional clock term which is the product of the mass difference by the mean proper time along both arms.

If the three light pulses of the pulse sequence are separated only in time, and not separated in space (usually if the velocity of the atoms is parallel to the laser beams), the interferometer is in an accelerometer (or gravimeter) configuration. In a uniformly accelerating frame with the atoms, the frequency of the driving laser changes linearly with time at the rate of $-\boldsymbol{k}\cdot\boldsymbol{a}$. The phase shift arises from the interaction between the light and the atoms [8,34,52] and can be written:

$$\Delta\phi = \phi_1(t_1) - 2\phi_2(t_2) + \phi_3(t_3) \tag{1}$$

where $\phi_i(t_i)$ is the phase of light pulse i at time t_i relative to the atoms. If the laser beams are vertical, the gravitationally induced chirp can be written to first order[2] in g:

$$\Delta\phi_{\mathrm{acc}} = -\boldsymbol{k}\cdot\boldsymbol{g}\,T^2 \tag{2}$$

It is important to note that the phase shift $\Delta\phi$ can be calculated in a more general relativistic framework [8, 54, 55], in which the atomic fields are second quantized. The starting point is the use of coupled field equations for atomic fields of a given spin in curved space–time: e.g. coupled Klein–Gordon, Dirac or Proca equations. Gravitation is described by the metric tensor $g_{\mu\nu} = \eta_{\mu\nu} + h_{\mu\nu}$ and by tetrads, which enter in these equations. By considering $h_{\mu\nu}$ as a spin-two tensor field in flat space–time [56–58] and using ordinary relativistic quantum field theory, it is possible to derive field equations that display all interesting terms coupling Dirac atomic fields, gravitational and electromagnetic fields and simple expressions of the corresponding relativistic phase shifts in atom interferometers [54]. The terms involving h_{00} lead to the gravitational shift ($h_{00} = 2\boldsymbol{g}\cdot\boldsymbol{r}/c^2$), to shifts involving higher derivatives of the gravitational potential and to the analogue of the Thomas precession (spin–orbit coupling corrected by the Thomas factor). The gravitational phase shift (2) can then be seen as the flux of a gravito-electric field $-c^2\boldsymbol{\nabla}h_{00}/2 = \boldsymbol{g}$ through the interferometer space–time area divided by a quantum of flux $\hbar/M$ in analogy with electromagnetism. It should be noted that this phase shift does not depend on the initial atomic velocity or on the mass of the particle (this is a direct consequence of the equivalence principle).

Recently, an atomic gravimeter with accuracy comparable to the best corner cube device (FG5) has been achieved [12] (Fig. 2). The main limitation of this kind of gravimeter on earth is due to spurious acceleration from the reference platform. Measuring gravity gradient may allow to overcome this problem. Indeed, using the same reference platform for two independent gravimeters enables to extract gravity fluctuations. Such an apparatus [41], using two gravimeters as described above but sharing the same light pulses, has shown a sensitivity of $3\cdot10^{-8}\,\mathrm{s}^{-2}\,\mathrm{Hz}^{-1/2}$ and has a potential on Earth as good as $10^{-9}\,\mathrm{s}^{-2}\,\mathrm{Hz}^{-1/2}$.

---

[2] A detailed calculation of the complete phase shifts can be found in [51]. Equation (1) can be simply written $\Delta\phi = -k[(z_3^{\mathrm{down}} + z_3^{\mathrm{up}})/2 - z_2^{\mathrm{down}} - z_2^{\mathrm{up}} + z_1^{\mathrm{down}}]$, where z_i^{down} and z_i^{up} represent the intersection of the wave packet classical trajectory with the ith light pulse. The notation ^{down} and ^{up} are related to the upper and lower trajectories as depicted in Fig. 1.

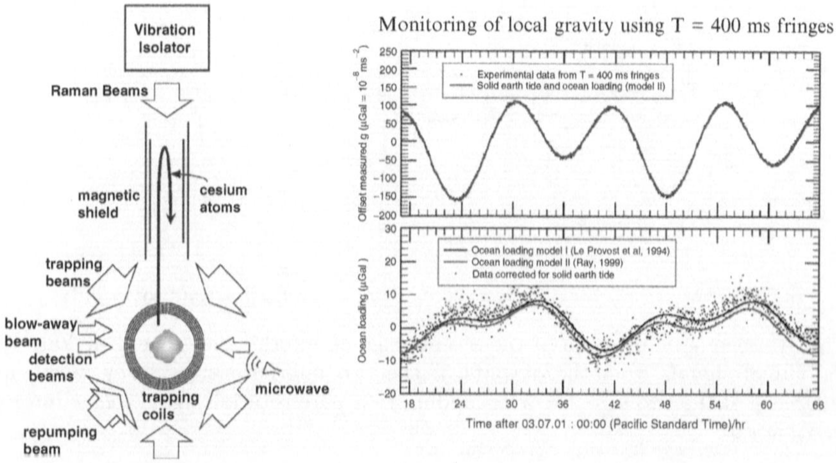

Fig. 2. Principle of the atom fountain-based gravimeter achieved in S. Chu group at Stanford. The *right figure* shows a 2 days recording of the variation of gravity. The accuracy enables to resolve ocean loading effects.

In the case of a spatial separation of the laser beams (usually if the atomic velocity is perpendicular to the direction of the laser beams), the interferometer is in a configuration similar to the optical Mach–Zehnder interferometers. Then, the interferometer is also sensitive to rotations, as in the Sagnac geometry [59] for light interferometers. For a Sagnac loop enclosing an area $\boldsymbol{A}$, a rotation $\boldsymbol{\Omega}$ produces a phase shift to first order[3] in Ω:

$$\Delta\phi_{\mathrm{rot}} = \frac{4\pi}{\lambda v_L}\,\boldsymbol{\Omega}\cdot\boldsymbol{A} \tag{3}$$

where λ is the particle wavelength and v_L its longitudinal velocity. The area A of the interferometer depends on the distance L covered between two pulses and on the recoil velocity $v_T = \hbar k/m$:

$$A = L^2 \frac{v_T}{v_L} \tag{4}$$

In the general relativistic frame, (3) corresponds to the flux of a gravitomagnetic field $c^2\boldsymbol{\nabla}\times\boldsymbol{h} = 2c\boldsymbol{\Omega}$ through an area in space $\boldsymbol{A}$ divided by a quantum of flux $\hbar c/M$. The terms that involve $\boldsymbol{h} = \{h^{0k}\}$ give the Sagnac effect in a rotating frame, the spin-rotation coupling and a relativistic correction (analogous to the Thomas precession term for h_{00}). They also describe the Lense–Thirring effects from inertial frame dragging by a massive rotating body, which is a source for $\boldsymbol{h}$.

Thanks to the use of massive particles, atomic interferometers can achieve a very high sensitivity. An atomic gyroscope [10] using thermal caesium atomic beams (where the most probable velocity is $v_L \sim 300\,\mathrm{m\,s^{-1}}$) and with an

[3] A complete calculation can be found in [39].

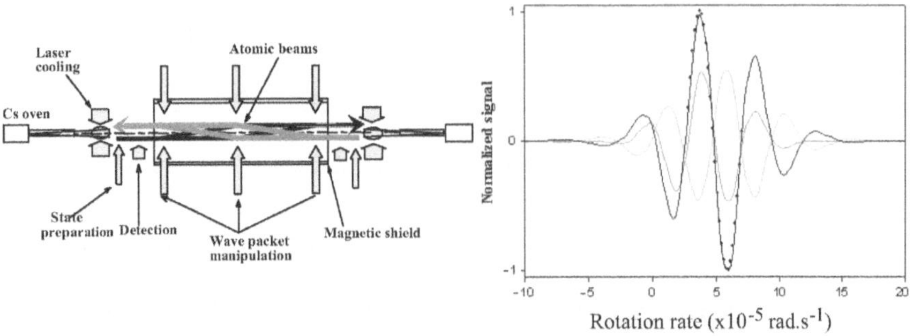

Fig. 3. Schematic diagram of the atomic Sagnac interferometer built at Yale [10]. Individual signals from the outputs of the two interferometers (*grey lines*), and difference of the two signals corresponding to a pure rotation signal (*black line*) vs. rotation rate.

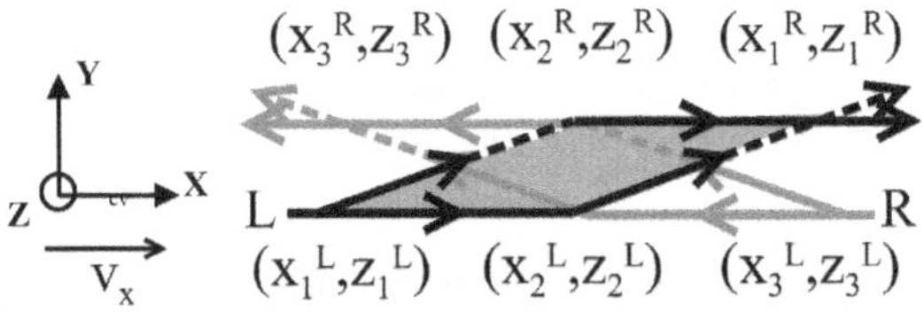

Fig. 4. General scheme of the two contra-propagating atomic interferometers. The atoms from the *left* (interferometer L) are launched with a velocity $\mathbf{v}_L = \{v_x, 0, v_z\}$ and the atoms from the *right* (interferometer R) with a velocity $\mathbf{v}_R = \{-v_x, 0, v_z\}$. They interact with the Raman beams at time t_i at position $\mathbf{r}_{i=1,2,3}^{L,R}$, respectively.

overall interferometer length of 2 m has demonstrated a sensitivity (Fig. 3) of $6 \cdot 10^{-10}\,\mathrm{rad\,s^{-1}\,Hz^{-1/2}}$. The apparatus consists of a double interferometer using two counter-propagating sources of atoms, sharing the same lasers. The use of the two signals enables to discriminate between rotation and acceleration.

Indeed, acceleration cannot be discriminated from rotation in a single atomic beam sensor, as stated above. This limitation can be circumvented by installing a second, counter-propagating, cold atomic beam (see Fig. 4). When the two atomic beams are aligned to perfectly overlap, the area vectors for the resulting interferometer loops have opposite directions, and the corresponding rotational phase shifts $\Delta\phi_{\mathrm{rot}}$ have opposite signs while the acceleration phase shift $\Delta\phi_{\mathrm{acc}}$ remains unchanged. Consequently, taking the sum of the two sensors readouts will render the sensor sensitive to acceleration only: $\Delta\phi_+ \sim 2\Delta\phi_{\mathrm{acc}}$ while taking the difference between the phase shifts of each sensor, common mode rejects uniform accelerations so that $\Delta\phi_- \sim 2\Delta\phi_{\mathrm{rot}}$. In addition, the difference $\Delta\phi_-$ common rejects the residual geometrical phase error $\delta\Phi_{\mathrm{geo}}$ if the phase fluctuations have no temporal variation on a timescale $2T$, the interferometer time. This is not the case for $\Delta\phi_+$ where a absolute phase bias $2\delta\Phi_{\mathrm{geo}}$ appears.

4 Cold Atom Sensors

4.1 Cold Atom Accelerometers

Accelerometer

Following the pioneering work of S. Chu, M. Kasevich and coworkers (see Fig. 2), new experiments have been developed to test new gravimeter configurations [60–62] or to improve previous measurements [63, 64]. We discuss in detail here the cold atom accelerometer developed at LNE-SYRTE in Paris for the watt balance experiment [65, 141] which is currently setup in the prospect of a new definition of the mass unit. This gravimeter measures the acceleration of freely falling ^{87}Rb atoms. Here, k (used in (2)) is the effective wave vector of the Raman transition, and T is the time between the interferometer Raman pulses.

This setup uses an original frequency locking system that enables to control dynamically the frequency of the two lasers, over the whole experimental sequence. First the lasers are tuned to the frequencies required to cool ^{87}Rb atoms in a magneto-optical trap (MOT). Dividing the total available laser power between a two-dimensional MOT (2D-MOT) [66] and a three-dimensional MOT (3D-MOT), loading rates of $3 \cdot 10^9$ atoms per second are obtained. Then, the magnetic field is turned off for further cooling of the atoms. Once the atoms have been released from the molasses, a frequency ramp detunes the two vertical counter-propagating beam to a detuning Δ of up to 2 GHz from the optical transition resonance. This will allow to use both the cooling and re-pumping vertical laser beam of the MOT as the Raman laser with negligible spontaneous emission (which is a decoherence process). To be used as Raman lasers, the frequency difference between these two lasers has to be subsequently phase locked with a high-bandwidth PLL. To reach an accuracy of $10^{-9} \times g$, the phase error arising from the transient evolution of their relative phase has to remain below 0.3 mrad [63]. It takes a few hundreds µs for the lock to come perfectly to the right frequency and to start phase locking (see Fig. 5), the 0.3 mrad criterion being reached in about 2 ms. (The measured spectral phase noise density in steady state [67] corresponds to a contribution of 0.56 mrad rms of phase noise in the atomic interferometer, i.e. $10^{-9} \times g$ rms.)

The Raman detuning Δ can be changed at will and other sweeps can be added in the cycle. This enables to realise first a velocity selective Raman pulse ($\sim$ 35 µs), with the detuning of 2 GHz which reduces the spontaneous emission. Then the detuning is swept back by 1 GHz for the interferometer itself, to achieve a better transfer efficiency.[4] Finally, the phased-locked Raman lasers are used to realise the interferometer. Owing to the Doppler effect, the

[4] Roughly speaking, the transfer efficiency is related to the pulse duration $\tau \propto \Delta/I$ where I is the Raman laser intensity, since for smaller τ, the Raman diffraction process will be less velocity selective [45].

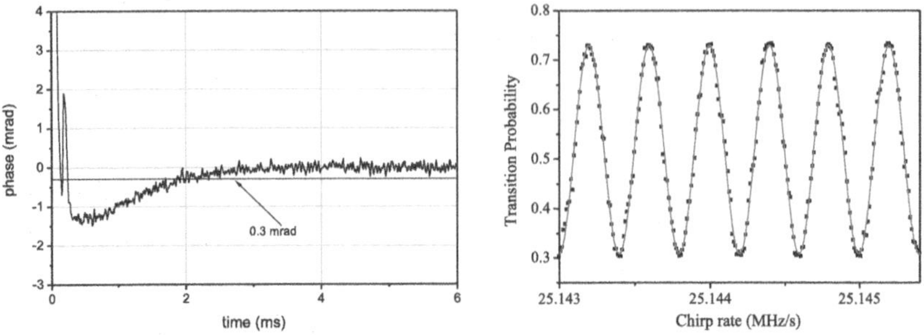

Fig. 5. *Left*: Relative Raman beams phase error. The phase lock loop (*PLL*) is closed at $t = 0$ after the 2 GHz sweep. After 0.5 ms, the phase error is exponentially decreasing with a time constant of 2 ms. *Right*: Atomic interferometer fringes obtained by scanning the Raman detuning chirp rate within the interferometer. The time between the Raman pulses is $T = 50$ ms. The *solid line* is a sinusoidal fit of the experimental points.

Raman detuning has to be chirped to compensate for the increasing vertical velocity of the atomic cloud. This chirp a, obtained by sweeping the frequency difference between the two lasers, induces an additional phase shift. The total interferometric phase is then given by: $\Delta\Phi = (kg - a)T^2$. Figure 5 displays the interferometric fringes obtained by scanning the chirp rate. In this experiment, T is 50 ms and the sensitivity is presently of $3.5 \cdot 10^{-8}$ g Hz$^{-1/2}$, limited by residual vibrations of the apparatus.

ONERA currently develops a gravimeter with cold atoms that should sustain external disturbances to make the instrument capable of being put onboard. For that, they take advantage of the abundance and the reliability of fibred components resulting from telecommunication technology. Indeed, the first limitations with the embarquability of these devices are optical: the laser sources necessarily require a good spectral quality (< 1 MHz), to be tunable near the atomic transition, operating CW with high powers (of a few tens to a few hundreds of milliwatts). The conventional techniques use lasers diode with external cavities, which makes the source sensitive to the vibrations. Moreover, the optical benches necessary to prepare the beams are generally large and hardly reducible. As shown later on in this chapter, the use of the components resulting from telecommunication technology will enable to profit from the robustness, perennity of those components and make it possible at the same time to miniaturize the optical system and to improve its reliability (Fig. 6).

Gradiometers

The measurement of the gradient of gravitational fields has important scientific and technical applications. These applications range from the measurement of G, the gravitational constant and tests of general relativity [68,69] to covert

Fig. 6. The *Girafon* gravimeter which is being constructed at ONERA, Palaiseau.

navigation, underground structure detection, oil-well logging and geodesy [70]. Initially at Stanford university in 1996, the development of a gravity gradiometer, whose operation is based on recently developed atom interference and laser manipulation techniques, has been followed by other developments for either space [71] or fundamental physics measurements [72]. A crucial aspect of every design is its intrinsic immunity to spurious accelerations.

The overall method is illustrated in Fig. 7. It uses light-pulse atom interferometer techniques [7,8,13] to measure the simultaneous acceleration of two laser-cooled ensembles of atoms. The relative acceleration of the atom clouds is measured by driving Doppler-sensitive stimulated two-photon Raman transitions [44] between atomic ground-state hyperfine levels. The geometry is chosen so that the measurement axis passes through both laser-cooled ensembles. Since the acceleration measurements are made simultaneously at both positions, many systematic measurement errors, including platform vibration, cancel as a common mode. This type of instrument is fundamentally different from current state-of-the-art instruments [73, 74]. First, the proof masses are

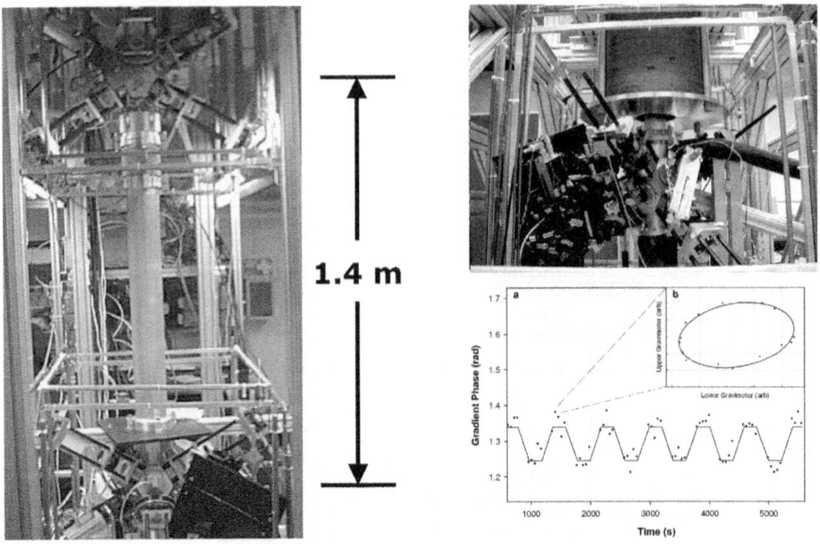

Fig. 7. Gradiometer developed in Stanford. Insert (*bottom right*), example of an application of this gradiometer to make a measurement of G [41]. A mass of 540 kg lead is alternatively brought closer to each atomic gravimeter. This preliminary measurement shows the strong potential of this system for precise measurements.

individual atoms rather than precisely machined macroscopic objects. This reduces systematic effects associated with the material properties of macroscopic objects. Second, the calibration for the two accelerometers is referenced to the wavelength of a single pair of frequency-stabilised laser beams, and is identical for both accelerometers. This provides long-term accuracy. Finally, large separations ($\gg$1 m) between accelerometers are possible. This allows for the development of high sensitivity instruments.

The relative acceleration of the two ensembles along the axis defined by the Raman beams is measured by subtracting the measured phase shifts $\Delta\phi(\boldsymbol{r}_1)$ and $\Delta\phi(\boldsymbol{r}_2)$ at each of two locations $\boldsymbol{r}_1$ and $\boldsymbol{r}_2$. The gradient is extracted by dividing the relative acceleration by the separation of the ensembles. This method determines only one component of the gravity-gradient tensor.

The Measurement of *G*

The Newtonian gravitational constant G is – together with the speed of light – the most popular physical constant. Introduced by Newton in 1686 to describe the gravitational force between two massive objects and first measured by Cavendish more than a hundred years later [75], G became more and more the subject of high-precision measurements. There are many motivations for such measurements [76], ranging from purely metrological interest for determinations of mass distributions of celestial objects to geophysical applications. In addition, many theoretical models profit from an accurate knowledge of G.

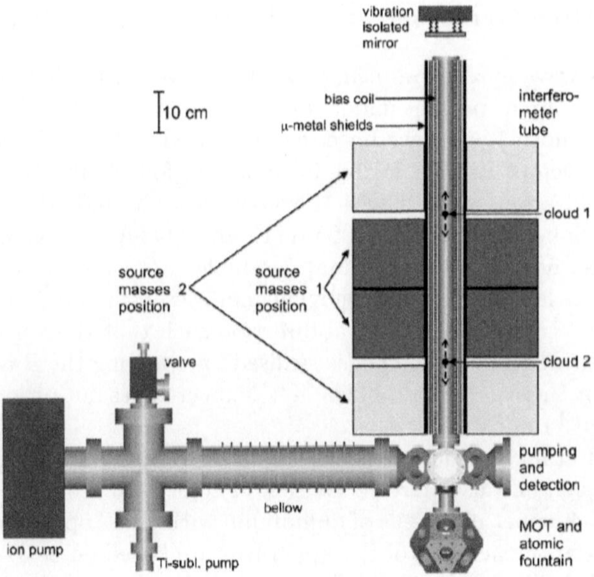

Fig. 8. Graphical illustration of the MAGIA experimental setup with the vacuum system, the atomic trajectories and the source mass positions. The laser systems, the detection units and the source mass holder are not included. The atomic trajectories during the time of the interferometer pulse sequence are sketched (*dashed arrows*).

Despite these severe motivations and some 300 measurements in the past 200 years, the 1998 CODATA [77] recommended value of $G = (6.673 \pm 0.010) \cdot 10^{11}\,\mathrm{m^3\,kg^{-1}\,s^{-2}}$ includes an uncertainty of 1,500 parts per million (ppm). Thus, G is still the least accurately known fundamental physical constant. Recently, two measurements with much smaller uncertainties of 13.7 and 41 ppm have been reported [78]. However, the given values for G still disagree on the order of 100 ppm. Therefore, it is useful to perform high resolution G measurements with different methods. This may help to identify possible systematic effects. It is worthwhile to mention that, so far, only few conceptually different methods have resulted in G measurements at the level of 1,000 ppm or better [79]. All these methods have in common that the masses, which probe the gravitational field of external source masses, are suspended (e.g. with fibres). One way to exclude this possible source of systematic effects is to perform a free-fall experiment. A high-precision measurement of G using a free-falling corner cube (FFCC) has already been performed [80] but the uncertainty remained on the order of 1,400 ppm. Experiments such as the Yale gradiometer or MAGIA developed in Italy, in which free-falling atoms are used to probe the gravitational acceleration originating from nearby source masses, are expected to surpass these results (Fig. 8).

4.2 Cold Atom Gyroscope and Cold Atom Inertial Base

Cold matter-wave gyroscopes using atomic samples with slow drift velocities of a few $\mathrm{m\,s^{-1}}$ are at present under construction at the IQ (Institut für Quantenoptik, Hannover) and have been demonstrated at LNE-SYRTE (Systèmes de Référence Temps-Espace, Paris). Both devices follow different design strategies. The cold atom sensor GOM (for Gyromètre à Onde de Matière) developed in collaboration between SYRTE and IOTA [81] is based on two caesium fountains. The two caesium ensembles are simultaneously prepared in MOTs and launched by the moving molasses technique with a speed of about $2.4\,\mathrm{m\,s^{-1}}$ and 82° in vertical direction such that they cross each other at the vertex. The interferometer is realised by applying the Raman pulses at the vertex of the atomic parabolas. The expected resolution of the setup is $4\cdot10^{-8}\,\mathrm{rad\,s^{-1}\,Hz^{-1/2}}$.

The cold atom Sagnac interferometer (CASI) at the IQ is based on a flat parabola design and uses intense sources of cold ^{87}Rb atoms. Figure 9 shows the vacuum chamber made out of aluminium with glued optical windows. The atomic sources on each end of the apparatus are based on a 3D-MOT loaded by a 2D-MOT. The 2D-MOT displays high performance with more than 10^{10} atoms per second. The typical loading rate of the 3D-MOT is a few 10^9 atoms per second such that 10^8 atoms can be loaded in the MOT in 0.1 s. Alternatively, the performance of the 3D-MOT can be further improved by Raman cooling in optical lattices. The actual interferometer will have a length of up

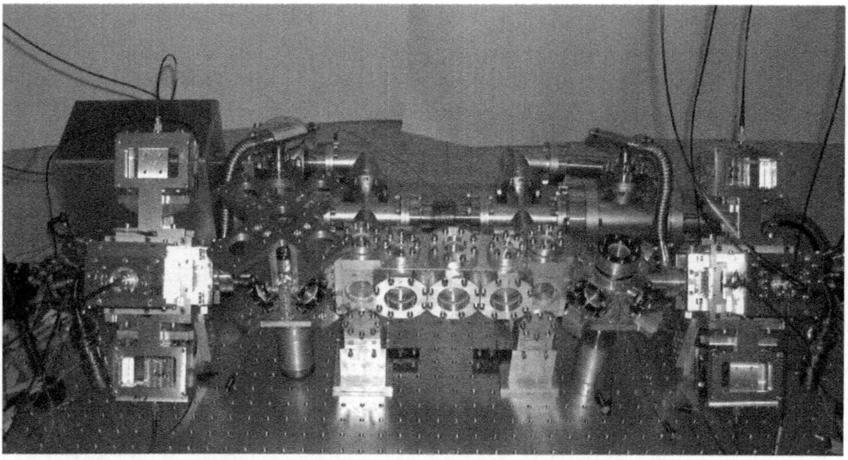

Fig. 9. The vacuum chamber of CASI. The *central part* shows the interferometry chamber with three spatially separated optical viewports for the interferometry lasers. On both sides of this chamber, a dual stage atom source is mounted which serves for the preparation of the cold atomic ensembles. The four wings on each side are the telescopes that generate elliptically shaped laser beams out of fibre-coupled lasers for cooling and trapping the Rubidium atoms.

to 15 cm. The coherent manipulation of the atoms (splitting, reflection and recombination) is performed by a temporal and/or spatially separated sequence of Raman-type interactions at the centre of the apparatus. With these parameters a shot-noise limited resolution of about $2 \cdot 10^{-9}\,\mathrm{rad\,s^{-1}\,Hz^{-1/2}}$ should be feasible using about 10^8 atoms per shot. CASI will investigate the ultimate sensitivity obtainable in cold matter-wave sensors. There is a large potential for further improvements, thanks to the expected higher stability with the use of intense cold atomic sources with a flux of more than 10^{10} atoms per second. Apart from lowering the atomic speed, the sensitivity of the apparatus can be enhanced by increasing the momentum transferred at the beam splitter as in higher-order Raman or Bragg transitions or in magneto-optical *blazed* light gratings. Their suitability for metrological applications (reproducibility, accuracy, systematic errors), however, is still to be verified. Viewing the relatively small areas achieved by present atom interferometers, an interesting alternative for such sensors may consist in waveguides (which do not deteriorate the achievable uncertainty).

The GOM is a six-axis inertial sensor. The direction of sensitivity of the setup is defined by the direction of the Raman interrogation laser with respect to the atomic trajectory. As illustrated in Fig. 10, with a classical three-pulse sequences ($\pi/2 - \pi - \pi/2$), a sensitivity to vertical rotation Ω_z and to horizontal acceleration a_y is achieved by placing the Raman lasers horizontal and perpendicular to the atomic trajectory [9] (Fig. 10a). The same sequence, using vertical lasers, leads to the measurement of horizontal rotation Ω_y and vertical acceleration a_z (Fig. 10b). Thanks to the specific setup of the GOM, it is possible to have access to the other components of acceleration and rotation which lie along the horizontal direction of propagation of the atoms (x-axis). The use of cold atoms in strongly curved trajectories allows to point the Raman lasers along the x-direction, offering a sensitivity to acceleration a_x and no sensitivity to rotation (Fig. 10c). Easy access to the horizontal rotation Ω_x is achieved by changing the pulse sequence to four pulses: $\pi/2-\pi-\pi-\pi/2$ (Fig. 10d).

The new butterfly configuration was first proposed to measure the gravity gradient [2, 82]. It can be used to measure rotations with the same Raman beams as in the previous configuration (y-axis) but in a direction (x-axis) that cannot be achieved with a standard three-pulse sequences. Four pulses, $\pi/2 - \pi - \pi - \pi/2$, are used, separated by times $T/2 - T - T/2$, respectively. The atomic paths cross each other leading to a twisted interferometer. The horizontal projection of the oriented area cancels out so that the interferometer is insensitive to rotation around the z-axis. In contrast, the vertical projection now leads to a sensitivity to rotation around the x-axis:

$$\Delta\phi = \frac{1}{2}(\boldsymbol{k} \times (\boldsymbol{g} + \boldsymbol{a})) \cdot \boldsymbol{\Omega}\, T^3\,. \tag{5}$$

This sensitivity to rotation appears from a crossed term with acceleration $(\boldsymbol{g}+\boldsymbol{a})$ and is no longer dependent on the launching velocity. This configuration

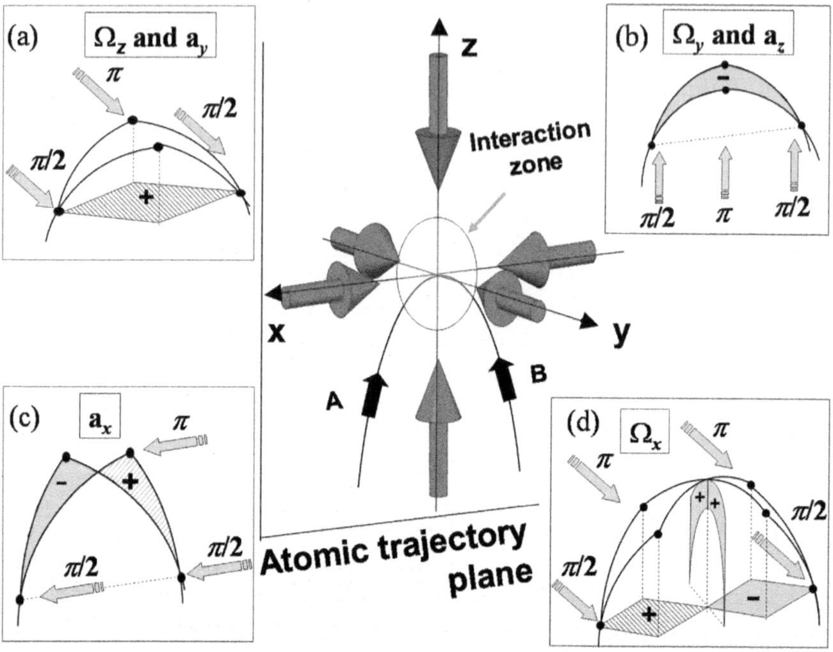

Fig. 10. Six-axis inertial sensor principle. The atomic clouds are launched on a parabolic trajectory, and interact with the Raman lasers at the top. The four configurations (**a**)–(**d**) give access to the three rotations and three accelerations. In the three pulses configuration, the Raman beams direction can be horizontal or vertical, creating the interferometer in a horizontal (**a**) or vertical (**b**, **c**) plane. With a butterfly four-pulse sequence of horizontal beams (**d**), the rotation Ω_x can be measured.

is not sensitive to DC accelerations along the direction of the Raman laser, but remains sensitive to fluctuations of horizontal and vertical accelerations. With an isolation platform, the remaining fluctuations are negligible compared to g, which does not compromise the stability of the scaling factor. The sensitivity to rotation is comparable with that of configurations (a) and (b). With $2T = 60\,\text{ms}$, this configuration leads to a interferometer area reduced by a factor 4.5, but it scales with T^3 and thus should present a higher sensitivity for longer interrogation times.

The atomic fringe patterns are presented in Fig. 11 and show contrasts of 4.9 and 4.2% for interferometers A and B, respectively. By operating the interferometer on the fringe side, as explained before, a signal-to-noise ratio from shot to shot of 18, limited by the residual vibrations, is achieved. The sensitivity to rotation is equal to $2.2\cdot10^{-5}\,\text{rad s}^{-1}$ in 1 s, decreasing to $1.8\cdot10^{-6}\,\text{rad s}^{-1}$ after 280 s of averaging time.

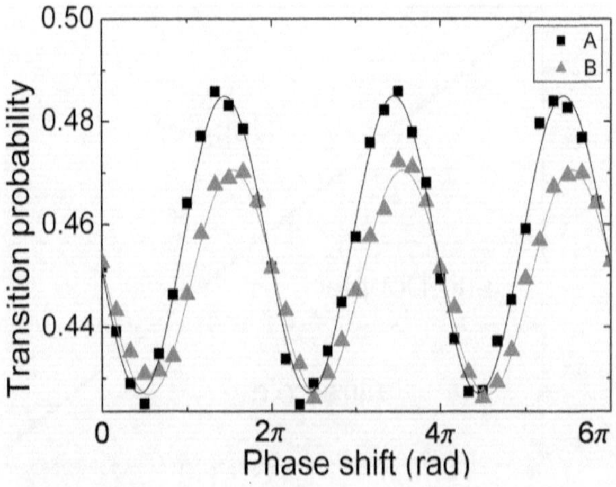

Fig. 11. Fringes obtained with both interferometers A and B in the four-pulse butterfly configuration for a total interrogation time of $2T = 60\,\mathrm{ms}$.

4.3 Ultra-Cold Sources and Applications in Space

The ultimate phase sensitivity of an atom interferometer is, aside from technical difficulties, limited by the finite number of detected particles N and scales as $\Delta\phi_{\min} = 1/\sqrt{N}$ (quantum projection noise limit [83,84]). Of course, the relation between the relative phases accumulated along the two different paths and the actual physical property to be measured is a function of the "interrogation" time T spent by the particles between the two beam splitters. Thus, the ideal sensitivity of an atom interferometer is expected to scale[5] as $\sqrt{N}\,T^{\alpha}$ with $\alpha > 0$, and it is obviously of strong interest to increase these two factors. Using cold atomic sources helps this quest for higher performances in two ways. First, a reduction of the velocity dispersion of the atomic sample (a few $\mathrm{mm\,s^{-1}}$) allows to reduce drastically the longitudinal velocity of the atoms v_L (few $\mathrm{cm\,s^{-1}}$) and enhances in the same way the enclosed area and the sensitivity for a constant length. Second, the accuracy and the knowledge of the scaling factor depend directly on the initial velocity of the atoms and can be better controlled with cold atomic sources than with thermal beams, as it has already been demonstrated with atomic clocks [85].

Nevertheless, seeking to increase the sensitivity of on-ground atom interferometers by increasing the interrogation time T, one soon reaches a limit imposed by gravity. With the stringent requirements of ultra-high vacuum and a very well-controlled environment, the current state-of-the-art in experimental realisations does not allow more than a few metres of free fall,

[5] An atomic clock or an atomic gyrometer, for example, has a sensitivity proportional to T and an on-ground gravimeter has a sensitivity proportional to T^2 due to the quadratic nature of free-fall trajectory in a constant gravitational field.

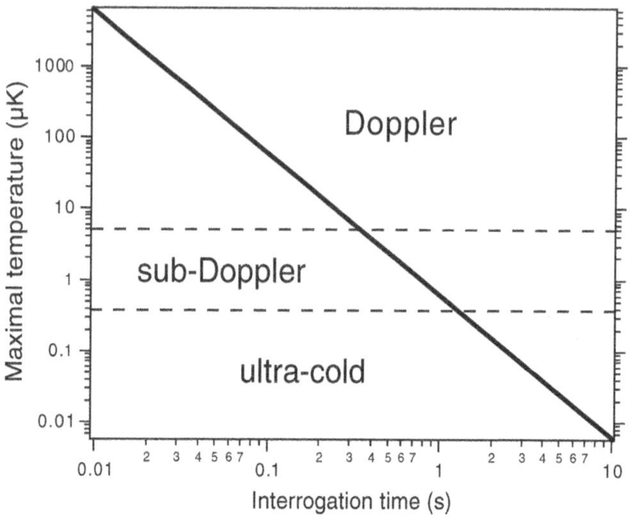

Fig. 12. Maximum temperature of atom source for a given interrogation time. The maximum interrogation time for a given initial temperature has been calculated for a detection area of $10\,\mathrm{cm}^2$ and defined as the time at which half of the atoms are no longer detected. The *dashed lines* indicate the limits of Doppler and sub-Doppler cooling. Interrogation times of several seconds are compatible only with clouds of atoms at ultra-cold temperatures, close to the quantum degenerate regime.

with corresponding interrogation times of the order of $T \sim 400\,\mathrm{ms}$. Space-based applications will allow much longer interrogation times to be used, thereby increasing dramatically the sensitivity and accuracy of atom interferometers [30].

Even in space, atom interferometry with a *classical* atomic source will not outperform the highest-precision ground-based atom interferometers that use samples of cold atoms prepared using standard techniques of Doppler and sub-Doppler laser cooling [86]. Indeed, the temperature of such sub-Doppler laser-cooled atom cloud is typically $\sim 1\,\mu\mathrm{K}$ ($v_{\mathrm{rms}} \sim 1\,\mathrm{cm\,s^{-1}}$). In the absence of gravity, the time evolution of cold samples of atoms will be dominated by the effect of finite temperature: in free space, a cloud of atoms follows a ballistic expansion until the atoms reach the walls of the apparatus where they are lost. Therefore the maximum interrogation time reasonably available for space-based atom interferometers will strongly depend on the initial temperature of the atomic source. As shown in Fig. 12, the 200 ms limit imposed by gravity for a 30 cm free fall is still compatible with typical sub-Doppler temperatures, whereas an interrogation time of several seconds is only accessible by using an "ultra-cold" source of atoms (far below the limit of laser cooling) with a temperature of the order of a few hundred nano-kelvin.

4.4 HYPER: A Proposal to Measure the Lense–Thirring Effect in Space

The HYPER project (hyper precision cold atom interferometry in space) was proposed to ESA in 2002 with the goal to benefit from the space environment, which enables very long interaction times (a few seconds) and a low spurious vibrational level. The sensitivity of the atomic interferometer is expected to reach a few $10^{-12}\,\mathrm{rad\,s^{-1}\,Hz^{-1/2}}$ for rotation and $10^{-12} \times g\,\mathrm{Hz}^{-1/2}$ for acceleration. This very sensitive and accurate apparatus offers the possibility of different tests of fundamental physics [30]. It can realise tests of general relativity by measuring the signature of the Lense–Thirring effect (magnitude and sign) or testing the equivalence principle on individual atoms. It can also be used to determine the fine structure constant by measuring the ratio of Planck's constant to an atomic mass.

The Lense–Thirring Effect

The measurement of the Lense–Thirring effect is the first scientific goal of the HYPER project and will be described in more detail in this section. The Lense–Thirring effect consists of a precession of a local inertial reference frame (realised by inertial gyroscopes) with respect to a non-local one realised by pointing the direction of fixed stars under the influence of a rotating massive body. This Lense–Thirring precession is given by

$$\boldsymbol{\Omega}_{LT} = \frac{GI}{c^2} \frac{3(\boldsymbol{\omega} \cdot \boldsymbol{r})\boldsymbol{r} - \boldsymbol{\omega} r^2}{r^5} \tag{6}$$

where G is Newton's gravitational constant, I the Earth's inertial momentum and $\boldsymbol{\omega}$ the angular velocity of the Earth. The high sensitivity of atomic Sagnac interferometers to rotation rates will enable HYPER to measure the modulation of the precession due to the Lense–Thirring effect while the satellite orbits around the Earth. In a Sun-synchronous circular orbit at 700 km altitude, HYPER will detect how the direction of the Earth's drag varies over the course of the near-polar orbit as a function of the latitudinal position θ

$$\begin{pmatrix} \Omega_x \\ \Omega_y \end{pmatrix} \propto \frac{3}{2} \begin{pmatrix} \sin(2\theta) \\ \cos(2\theta) - \frac{1}{3} \end{pmatrix}, \tag{7}$$

where e_x and e_y define the orbital plane with $\boldsymbol{e}_y$ being parallel to the Earth's inertial momentum $\boldsymbol{I}$ and $\theta \equiv \arccos(\boldsymbol{r} \cdot \boldsymbol{e}_x)$ (Fig. 13).

The HYPER Payload

HYPER carries (Fig. 14) two atomic Sagnac interferometers, each of which is sensitive to rotations around one particular axis, and a telescope used as highly sensitive star tracker (10^{-9} rad in the 0.3–3 Hz bandwidth). The two units will

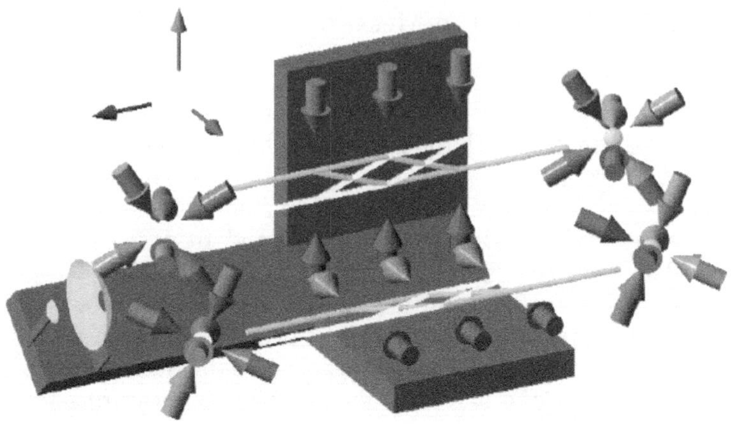

Fig. 13. Hyper gyroscopes consist of two differential atomic Sagnac interferometers in two orthogonal planes. The beam-splitting optical components are rigidly connected to the optical bench which carries the precision star tracker and serves as a non-inertial reference.

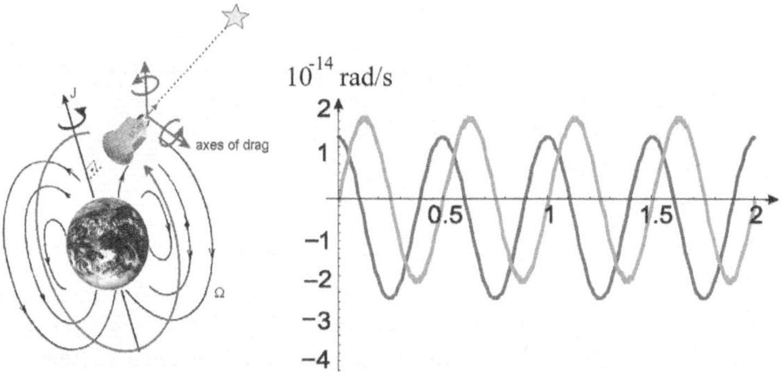

Fig. 14. Diagram of the measurement of the Lense–Thirring effect. The *black lines* visualise the vector field of the Earth's drag Ω_{LT}. The sensitive axes of the two ASUs are perpendicular to the pointing of the telescope. The direction of the Earth's drag varies over the course of the orbit showing the same structure as the field of a magnetic dipole. Due to this formal similarity, the Lense–Thirring effect is also called *gravito-magnetic effect*. The modulation of the rotation rate Ω_{LT} due to Earth's gravito-magnetism as sensed by the two orthogonal ASUs in the orbit around the Earth appears at twice the orbit frequency.

measure the vector components of the gravito-magnetic rotation rate along the two axes perpendicular to the telescope pointing direction which is directed to a guide star. The drag variation written above describes the situation for a telescope pointing in the direction perpendicular to the orbital plane of the satellite. The orbit, however, changes its orientation over the course of a year which has to be compensated by a rotation of the satellite to track continuously the guide star. Consequently the pointing of the telescope is not always directed parallel to the normal of the orbital plane. According to the equation, the rotation rate signal will oscillate at twice the frequency of the satellite revolution around the Earth. The modulated signals have the same amplitude ($3.75 \cdot 10^{-14}\,\mathrm{rad\,s^{-1}}$) on the two axes but are in quadrature. The resolution of the atomic Sagnac units (ASU) is about $3 \cdot 10^{-12}\,\mathrm{rad\,s^{-1}}$ for a drift time of about 3 s. Repeating this measurement every 3 s, each ASU will reach after one orbit of 90 min the level of $7 \cdot 10^{-14}\,\mathrm{rad\,s^{-1}}$, in the course of 1 year the level of $2 \cdot 10^{-15}\,\mathrm{rad\,s^{-1}}$, i.e. a tenth of the expected effect.

5 Coherent Atom Sensors: BEC and Atom Lasers

Dense *ultra-cold* samples of atoms are now routinely produced in laboratories all around the world. Using evaporative cooling techniques [19–21], one can cool a cloud of a few 10^6 atoms to temperatures below 100 nK [87]. At a sufficiently low temperature and high density, a cloud of atoms undergoes a phase transition to quantum degeneracy. For a cloud of bosonic (integer spin) atoms, this is known as *Bose–Einstein condensation*, in which all the atoms accumulate in the same quantum state (the atom-optical analog of the laser effect in optics). A BEC exhibits long-range correlation [24, 25, 88] and can therefore be described as a coherent "matter wave": an ideal candidate for the future of atom interferometry in space. The extremely low temperature associated with a BEC results in a very slow ballistic expansion, which in turn leads to interrogation times of the order of several tens of seconds in a space-based atom interferometer. In addition, the use of such a coherent source for atom optics could give rise to novel types of atom interferometry [28, 29, 52, 62, 89, 90].

5.1 Atom Laser: A Coherent Source for Future Space Applications

The idea for an atom laser pre-dates the demonstration of the exotic quantum phenomenon of BEC in dilute atomic gases. But it was only after the first such condensate was produced in 1995 that the pursuit to create a laser-like source of atomic de Broglie waves became intense.

In a Bose–Einstein condensate all the atoms occupy the same quantum state and can be described by the same wave function. The condensate therefore has many unusual properties not found in other states of matter. In particular, a Bose condensate can be seen as a coherent source of matter waves. Indeed, in a (photonic) laser all the photons share the same wave function. This is possible because photons have an intrinsic angular momentum, or "spin", equal to the Planck's constant $\hbar$. Particles that have a spin that is an integer multiple of $\hbar$ obey Bose–Einstein statistics. This means that more boson can occupy the same quantum state. Particles with half-integer spin – such as electrons, neutrons and protons, which all have spin $\hbar/2$ – obey Fermi–Dirac statistics. Only one fermion can occupy a given quantum state.

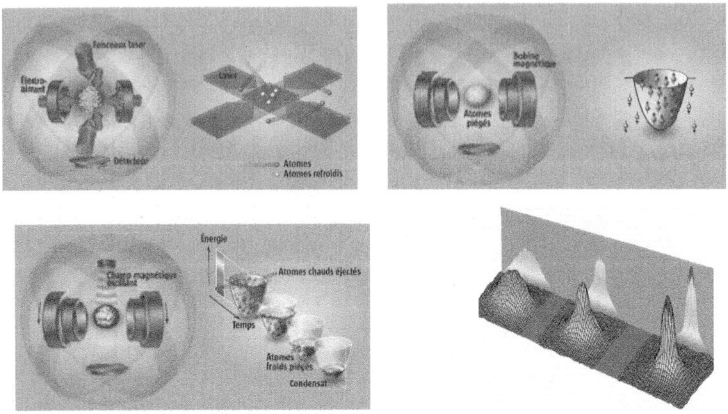

Fig. 15. Evaporative cooling towards Bose–Einstein condensation (from [91]). Initially, atoms are trapped in optical molasses using radiative forces. Then, the atoms are transferred in magnetic trap where they can stay trapped for hundred of seconds. Since no damping exists in such trap (as opposed to radiative traps), an evaporative cooling technique is used to remove the hottest atoms. In this technique, the trap is capped at a chosen height (using RF-induced spin flip) and the atoms with higher energy escape. By lowering the trap height, an ultra-cold high-density sample of atoms is obtained. The *bottom right picture* shows the BEC transition where a tiny dense peak of atoms (a coherent matter wave) appears at the centre of a Maxwell–Boltzman distribution (incoherent background).

A composite particle, such as an atom, is a boson if the sum of its protons, neutrons and electrons is an even number; the composite particle is a fermion if this sum is an odd number. Rubidium-87 or Caesium-133 atoms, for example, are bosons, so a large number of them can be forced to occupy the same quantum state and therefore have the same wave function. To achieve this, a large number of atoms must be confined within a tiny trap and cooled to sub-millikelvin temperatures using a combination of optical and magnetic techniques (see for example [92]). The Bose–Einstein condensates are produced in confining potentials such as magnetic or optical traps by exploiting either the atom's magnetic moment or an electric dipole moment induced by lasers (Fig. 15). In a magnetic trap, for instance, once the atoms have been cooled and trapped by lasers, the light is switched off and an inhomogeneous magnetic field provides a confining potential around the atoms. The trap is analogous to the optical cavity formed by the mirrors in a conventional laser. To make a laser we need to extract the coherent field from the optical cavity in a controlled way. This technique is known as "output coupling". In the case of a conventional laser the output coupler is a partially transmitting mirror. Output coupling for atoms can be achieved by transferring them from states that are confined to ones that are not, typically by changing an internal degree of freedom, such as the magnetic states of the atoms. The development of such atom laser is providing atom sources that are as different from ordinary

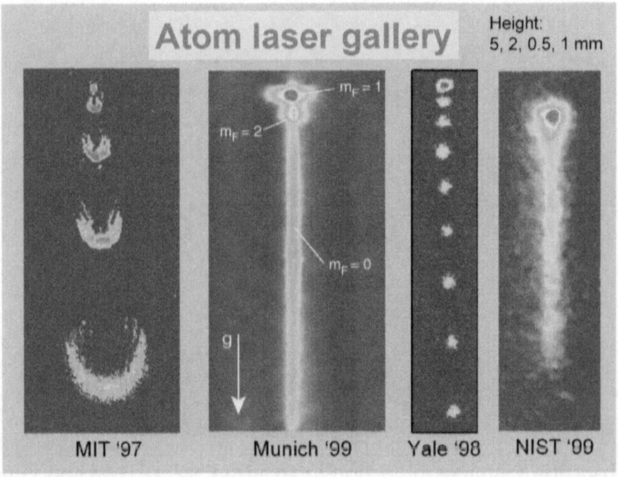

Fig. 16. Various types of atom lasers: (**a**) At MIT, intense RF pulses spin flips the atoms from a trapped state to an untrapped state. They fall under gravity. (**b**) In Yale, the condensate is loaded in an optical lattice. The combination of tunnel effect and gravity produces coherent pulses of atoms. (**c**) At NIST, Raman pulses extract atoms' pulses in a chosen direction. When the pulses overlap, a quasi-continuous atom laser is achieved. (**d**) In Munich, a weak RF coupler extracts a continuous atom wave from the condensate.

atomic beams as lasers are from classical light sources, and promises to outperform existing precision measurements in atom interferometry [28,29,90] or to study new transport properties [93–95].

The first demonstration of atomic output coupling from a Bose–Einstein condensate was performed with sodium atoms in a magnetic trap by W. Ketterle and co-workers at the Massachusetts Institute of Technology (MIT) in 1997. Only the atoms that had their magnetic moments pointing in the opposite direction to the magnetic field were trapped. The MIT researchers applied short radio-frequency pulses to "flip" the spins of some of the atoms and therefore release them from the trap (see Fig. 16a). The extracted atoms then accelerated away from the trap under the force of gravity. The output from this rudimentary atom laser was a series of pulses that expanded as they fell due to repulsive interactions between the ejected atoms and those inside the trap. Later T. Hänsch and colleagues at the Max Planck Institute for Quantum Optics in Munich extracted a continuous atom beam that lasted for 0.1 s. The Munich team used radio-frequency output coupling in an experimental setup that was similar to the one at MIT but used more stable magnetic fields (see Fig. 16b). Except for a few cases [24,96], the outcoupling methods do not allow to chose neither the direction nor the wavelength of the atom laser beam. In addition, the intrinsic repulsion between the atom laser beam and the BEC has dramatic effects [97,98] and gravity plays a significant role [99], such that the atom laser wavelength becomes rapidly small (Fig. 17).

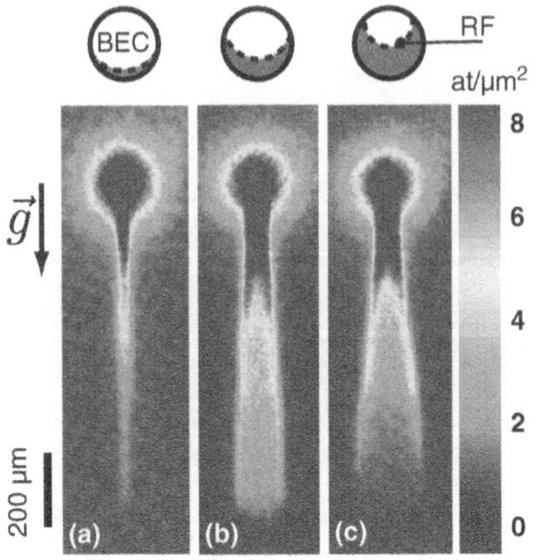

Fig. 17. Absorption images of a non-ideal atom laser, corresponding to density integration along the elongated axis x of the BEC. The figures correspond to different heights of RF-outcoupler detunings with respect to the bottom of the BEC: (**a**) $-0.37\,\mu$m, (**b**) $-2.22\,\mu$m, (**c**) $-3.55\,\mu$m. The *graph* above shows the RF-outcoupler (*dashed line*) and the BEC slice (*red*) which is crossed by the atom laser and results in the observation of caustics. The field of view is $350\times1{,}200\,\mu\mathrm{m}^2$ for each image.

The solution to overcome these limitations is either to develop coherent sources in space [90] or to suspend the atom laser during its propagation. For the latter, many atomic waveguides have been developed for cold thermal beams [100–107] or even for degenerate gases [95,108,109]. Nevertheless, as in optics, the transfer of cold atoms from magneto-optical traps into these small atom guides represents a critical step and so far, coupling attempts using either cold atomic beams [102,110] or cold atomic clouds [101,104,105,111] have led to relatively low coupling efficiency. To increase this efficiency, a solution consists in creating the atom laser directly into the guide [112], leading eventually to a continuous guided atom laser [113] analogous to the photonic fibre laser. This has been recently achieved in Orsay (LCFIO), where the BEC from which the atom laser is extracted from is *pigtailed* to the atom guide. In this setup, an atom laser is outcoupled from a hybrid opto-magnetic trap to a optical guide. The propagation direction is fixed by the propagation direction of the dipole trap laser beam and the velocity of the outcoupled atoms can be controlled by carefully adjusting the guide parameters. Using this scheme, atomic de Broglie wavelengths as high as $0.7\,\mu$m was observed (Fig. 18).

5.2 Application to $\hbar/m$ Measurement

The quantized exchange of momentum between light and atoms has opened the way to measurements of the de Broglie–Compton frequency of atomic species mc^2/h by direct frequency measurements [141]. The use of cold atom

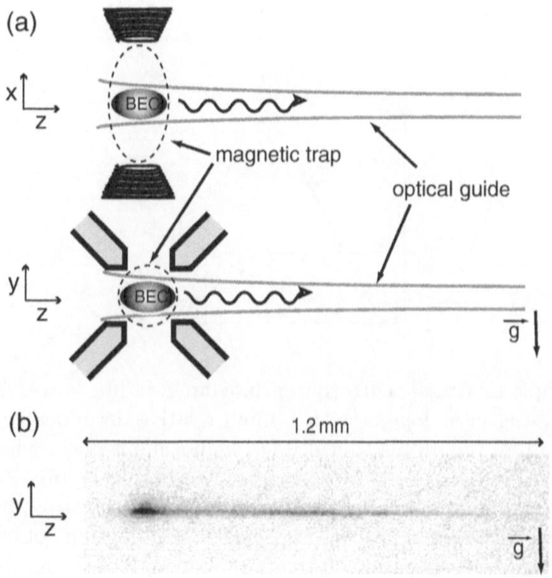

Fig. 18. (**a**) Schematic view of the setup. The BEC is obtained in a crossed hybrid magnetic and optical trap. The optical trap is horizontal. Its focus is shifted in the longitudinal direction z so as to attract the atoms. (**b**) Experimental absorption image of a guided atom laser after 50 ms of outcoupling. The imaging is along the x-axis.

interferometric techniques has subsequently led to very accurate determinations of the fine structure constant α from the ratio of the Rydberg constant to this frequency [18, 60].

Among the various new experiments aiming to improve these measurements of α via the measurement of the ratio $\hbar/m$, two experiments demonstrated a coherent matter-wave interferometer based on Bragg scattering [29, 90]. In the following, we shall review the measurement achieved in the Groupe d'Optique Atomique in Orsay (LCFIO).

Principle of Bragg Scattering

The principle of Bragg scattering is the following [114, 115]: two counter-propagating laser beams of wave vector $\pm \boldsymbol{k}_\mathrm{L}$ and frequencies ν_L and $\nu_\mathrm{L} + \delta\nu$ form a moving light grating. The common frequency ν_L is chosen to be in the optics domain but far detuned from atomic resonances to avoid spontaneous emission. A two-photon transition, involving absorption of a photon from one beam and stimulated re-emission into the other beam, results in a coherent transfer of momentum $\boldsymbol{p}_\mathrm{f} - \boldsymbol{p}_\mathrm{i} = 2\hbar \boldsymbol{k}_\mathrm{L}$ from the light field to the atoms, where $\boldsymbol{p}_\mathrm{i}$ and $\boldsymbol{p}_\mathrm{f}$ are the initial and final momenta of the atoms. Conservation of energy and momentum leads to the resonance conditions $E_\mathrm{f} = E_\mathrm{i} + h\delta\nu$,

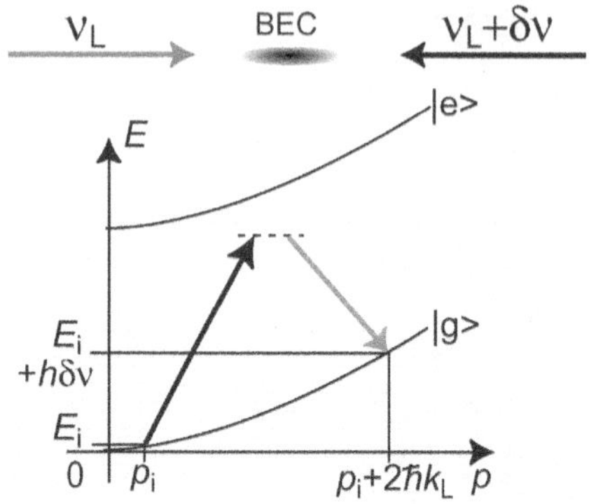

Fig. 19. Principle of Bragg scattering: a moving standing wave, formed from two counter-propagating laser beams with a small relative detuning $\delta\nu$, can coherently transfer a fraction of the atoms to a state of higher momentum when the resonance condition is fulfilled. A two-photon Bragg scattering event imparts a momentum $2\hbar k_{\mathrm{L}}$, and an energy of $h\delta\nu$ to the atoms: thus, the first-order (two photon) Bragg resonance for atoms with zero initial velocity occurs at a detuning of $h\delta\nu = 4\hbar^2 k_{\mathrm{L}}^2/2m$. This resonance condition depends on the initial velocity of the atoms relative to the optical standing wave.

where (in free space) the initial and final energies of the atoms are given by $E_{\mathrm{i}} = p_{\mathrm{i}}^2/2m$ and $E_{\mathrm{f}} = p_{\mathrm{f}}^2/2m$, respectively. Bragg scattering can be used for different types of matter-wave manipulation, depending on the pulse length τ. Using a short pulse ($\tau < 100\,\mu\mathrm{s}$), the Bragg beams are sufficiently frequency broadened that the Bragg process is insensitive to the momentum distribution within the condensate: the resonance condition is then satisfied simultaneously for the entire condensate. If the Bragg laser power and pulse duration are then selected to correspond to the $\pi/2$ condition, the probability of momentum transfer to the atoms is 50%: this is a 50/50 beam splitter for the condensate, between two different momentum states. When using longer pulses (for example $\tau = 2\,\mathrm{ms}$ in [116]), the Bragg process is velocity selective, and one can apply this technique to momentum spectroscopy [88, 116] (Fig. 19).

$\hbar/m$ Measurement

The experimental sequence proceeds as follows [116,117]: a laser-cooled sample of ^{87}Rb atoms is magnetically trapped in the $5\mathrm{S}^{1/2}|F = 1, m_F = -1\rangle$ state and then evaporatively cooled to quantum degeneracy. The magnetic trapping fields are switched off and the atoms fall for 25 ms. During this free-fall period, the coherent Bragg scattering "velocimeter" pulse is applied. In this experiment, the implementation of Bragg scattering is as follows: two orthogonally polarised, co-propagating laser beams of frequencies ν_L and $\nu_L + \delta\nu$ and wave vector $\boldsymbol{k}_{\mathrm{L}}$ are retroreflected by a highly stable mirror, with 90° polarisation rotation (see Fig. 20). With this scheme, the atoms are submitted

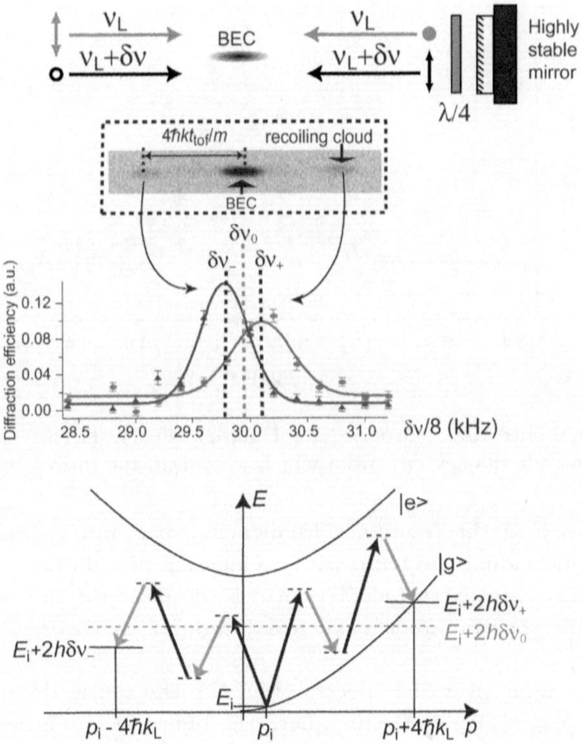

Fig. 20. Principle of our four photon, dual direction Bragg scattering scheme. *Top*: schematic of the experimental apparatus. Two retroreflected laser beams form two standing waves of orthogonal polarisations, moving in opposite directions. *Middle*: normalised number of atoms diffracted into each of the two output channels as a function of Bragg detuning $\delta\nu$. (*Inset*: typical absorption image after Bragg diffraction and free evolution during a time t_{tof}.) *Bottom*: schematic picture of the four-photon Bragg resonance condition. For zero initial momentum, the resonance condition is fulfilled by both standing waves for a detuning $\delta\nu_0$. For non-zero initial momentum p_{i}, the resonance frequency is equally and oppositely shifted for each of the two channels.

to two standing waves moving in opposite directions and with orthogonal polarisations. In addition, the relative detuning $\delta\nu$ is chosen so as to fulfill the second-order (four photon) resonance condition. This four laser Bragg scattering scheme produces a coherent transfer of momentum of $+4\hbar\boldsymbol{k}_{\text{L}}$ and $-4\hbar\boldsymbol{k}_{\text{L}}$. This scheme enables to reject the effect of a non-zero initial velocity, which can arise from imperfections in the magnetic trap switch-off. For an initial velocity p_{i}/m, the four-photon resonance conditions for the two oppositely moving standing waves are $\delta\nu_+ = \delta\nu_0(1+p_{\text{i}}/2\hbar k_{\text{L}})$ and $\delta\nu_- = \delta\nu_0(1-p_{\text{i}}/2\hbar k_{\text{L}})$, where $\delta\nu_0$ is the Doppler-free value, $\delta\nu_0 = (8/2\pi)(\hbar k_{\text{L}}^2/2m)$ (see Fig. 20). Scanning the Bragg scattering efficiency in the two directions as a function of $\delta\nu$ yields two peaks with widths corresponding to the condensate momentum width,

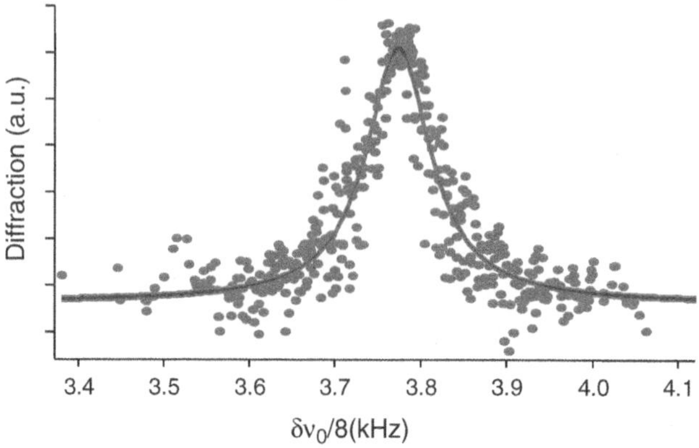

Fig. 21. Final spectrum (corrected for Doppler effect). The fit to this spectrum yields the centre frequency $\delta\nu_0$, from which we obtain the ratio h/m.

centred at each of the resonance frequencies, $\delta\nu_+$ and $\delta\nu_-$ (Fig. 20). After fitting each individual spectrum with a Gaussian distribution, the two centre frequencies $\delta\nu_\pm$ are extracted. To correct the data for the non-zero initial velocity, both spectra are then centred around the average value $\delta\nu_0 = (\delta\nu_+ + \delta\nu_-)/2$.

After averaging over 350 spectra (Fig. 21), the centre detuning was measured to be $\delta\nu_0 = 30.189(4)$ kHz where the figure in parentheses is the 68% confidence interval of the fit. This corresponds to a value $h/m \equiv \lambda^2 \times \delta\nu_0/4 = 4.5946(7) \cdot 10^{-9}\,\mathrm{m^2\,s^{-1}}$ where the wavelength $\lambda = 780.246291(2) \cdot 10^{-9}$ of the Bragg beams, slightly detuned from the $(5^2\mathrm{S}_{1/2}, \mathrm{F}=2) \to (5^2\mathrm{P}_{3/2}, \mathrm{F}=3)$ optical transition, is very accurately known from [118,119]. The offset between the measurement and the CODATA value of h/m ($4.59136 \cdot 10^{-9}\,\mathrm{m^2\,s^{-1}}$) can be explained by two major systematic effects. First, as described in [116], the frequencies ν_L and $\nu_L + \delta\nu$ of the Bragg scattering beams were obtained by using two independently driven acousto-optical modulators (AOM) of centre frequency 80 MHz. The frequency difference $\delta\nu$ was then deduced from the measurement of the frequency of each AOM driver with a high-precision frequency metre that had an accuracy of about 4×10^{-7}, giving a ± 16 Hz inaccuracy in the actual frequency difference $\delta\nu$. The resulting systematic error then gives $h/m = 4.5946(20)(7) \cdot 10^{-9}\,\mathrm{m^2\,s^{-1}}$. The second systematic effect is a collisional shift due to interactions in the high-density atomic cloud.

Effects of Interactions in a High-Density Atomic Sample

Ultra-cold ^{87}Rb atoms have repulsive interactions which modify the Bragg scattering resonance condition. The energy of an atom in the condensate is $E_\mathrm{i} = p_\mathrm{i}^2/2m + Un(\mathbf{r})$. The second term is the condensate interaction energy: $n(\boldsymbol{r})$ is the local atomic density of the condensate and $U = 5.147(5) \times 10^{-51}\,\mathrm{J\,m^3}$ is the interaction parameter. Immediately after Bragg scattering into a different momentum state, an atom experiences an effective potential $2Un(\boldsymbol{r})$ due to the surrounding condensate, and its energy is then $E_{\mathrm{f}} = p_{\mathrm{f}}^2/2m + 2Un(\boldsymbol{r})$ [88]. We can therefore replace the Bragg resonance condition (for zero initial momentum) with a *local* resonance condition which takes into account the effect of interactions:

$$2h\delta\nu_0(\boldsymbol{r}) = 16\frac{\hbar^2 k_{\mathrm{L}}^2}{2m} + Un(\boldsymbol{r}) \tag{8}$$

The parabolic density distribution of our Bose–Einstein condensate, at the time where the Bragg diffraction occurs, is

$$n(x, y, z) = n_0 \cdot \max\left[0\,;\, 1 - (x^2 + y^2)/R_\perp^2 - z^2/R_z^2\right]$$

with peak density $n_0 \simeq 3.6(4) \cdot 10^{18}\,\mathrm{m}^{-3}$ and half-lengths $R_\perp \simeq 9.8\,\mu\mathrm{m}$ and $R_z \simeq 126\,\mu\mathrm{m}$, where z is the direction of the Bragg scattering. Since the above measurement of the diffraction efficiency averages over the whole cloud, the resulting spectrum is then shifted by $U\langle n\rangle/2h \sim 4Un_0/7$ and broadened. Taking this interaction shift into account, the corrected measured value of h/m is:

$$\frac{h}{m} = \frac{\lambda^2}{4}\left(\langle\delta\nu_0\rangle - \frac{U\langle n\rangle}{2h}\right) \simeq 4.5939(21)(7) \cdot 10^{-9}\,\mathrm{m}^2\,\mathrm{s}^{-1}, \tag{9}$$

which is in agreement with the CODATA value.

5.3 The Prospect and Limits of High-Density Coherent Samples

The fact that ultra-cold bosons interact is a major drawback for precision measurements using atom interferometry. In the above experiment, interactions result in a systematic shift as well as a decrease in measurement precision. In principle, the systematic shifts can be calculated. However, the interaction parameter U is hard to measure and is generally not known to better than $\sim 10^{-4}$. The atomic density is also subject to time fluctuations and is difficult to know to better than $\sim 10^{-2}$, reducing the absolute accuracy. In addition, as shown in earlier experiments [116, 120], interactions produce a loss of coherence of the atomic samples at ultra-low, finite temperatures, limiting the maximum interrogation time of a coherent matter-wave atom interferometer. Finally, even at zero temperature, the mean-field energy due to interactions is converted into kinetic energy during free fall, giving rise to a faster ballistic expansion. This last effect will ultimately reduce interrogation times.

The Need of an Ideal Coherent Atomic Source

From the observations of both MIT and Orsay, we conclude that one should ideally use an interaction-free, ultra-cold atomic source for ultimate-precision atom interferometry in space. Using bosons, one could think of two ways of decreasing interaction effects. Close to a Feshbach resonance [121], one can control the interaction parameter U, which can be made equal to zero for a certain magnetic field [122, 123]. However, magnetic fields introduce further systematic shifts that are not controllable to within a reasonable accuracy. Alternatively, one could try to decrease the density of the sample of atoms, but the production of large atom number, ultra-low density Bose–Einstein condensate is a technical challenge not yet overcome [124].

A promising alternative solution is to use quantum-degenerate fermionic atomic sources [61]. The Pauli exclusion principle forbids symmetric two-body collision wave functions, so at zero temperature a sample of neutral atomic fermions has no interactions. An ultra-cold fermionic source may still allow very long interrogation times, even if limited by the excess energy of the Fermi pressure, and would therefore be an ideal candidate for atom interferometry in space with ultimate precision and accuracy.

6 Research and Technology: Towards a Space Atom Sensor

Intense research effort has focused on the study of atom interferometers miniaturization since their first demonstration in 1990. Atom interferometers benefit from the use of trapped ultra-cold atomic gases, gaining good signal-to-noise ratios due to the high atomic densities, and the coherence required for the visibility of interference patterns due to the low temperatures [8]. Since the recent development of atom chip-based coherent sources, efforts to incorporate interferometry on an "atom chip" [125–128] are motivated by the large physical size of a traditional apparatus and a desire to better tailor interferometer geometries. Most attempts to implement a coherent beam splitter/recombiner on a chip have used current-induced magnetic fields, typically forming double potential wells that merge and then split apart either in space, in time, or in both. Nevertheless, except in one experiment [129], various technical issues, such as noise coupled into the current and roughness or impurities of the wires, have stymied attempts to demonstrate on-chip interference. On the other hand, *traditional* light-pulse interferometer demonstrated already very high performances. Thus, efforts to reduce the size such as the CASI, GOM and Girafon scientific programs might lead to future small size, industrial atom interferometry inertial sensors. In fact, such transportable sensors are already available in the group of M. Kasevich at Stanford (Fig. 22).

The sensitivity of an interferometric measurement also depends on the interrogation time, the time during which the sample freely evolves. This time is limited by both the free fall of the atomic cloud, requiring tall vacuum chambers, and by its free expansion, demanding extra-sensitive detection systems for extremely dilute clouds. Ultra-low temperatures further reduce the expansion and should allow for more compact systems and for the full use of the

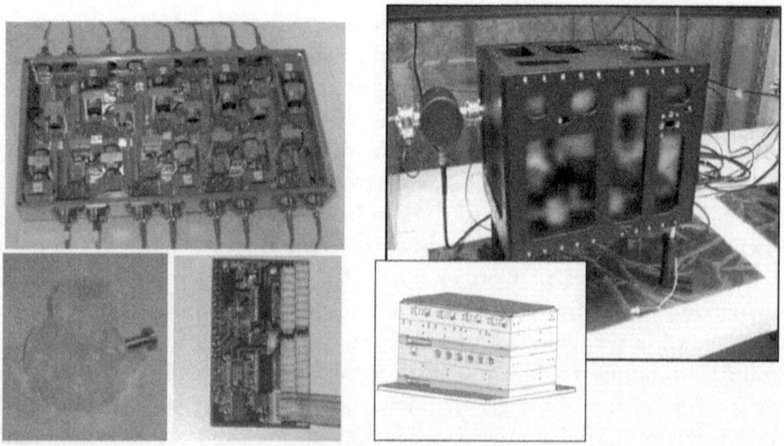

Fig. 22. The transportable interferometers developed at Stanford in M. Kasevich group (credit M. Kasevich).

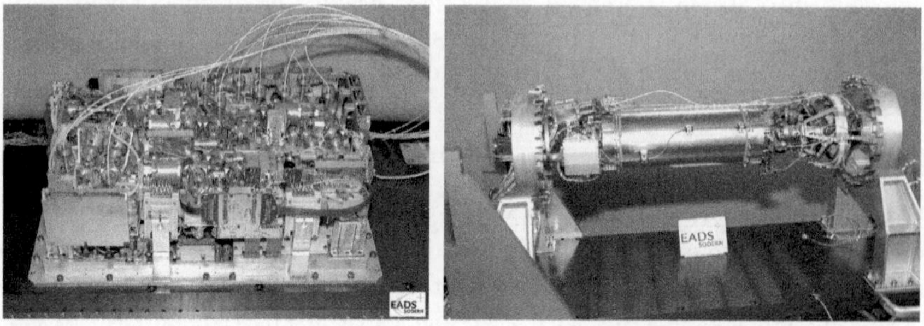

Fig. 23. The space clock PHARAO (courtesy EADS SODERN). *Left*: Photograph of the integrated laser source with cover removed. The dimensions are $530 \times 350 \times 150\,\text{mm}^3$ and the mass is 20.054 kg. The ten polarisation-maintaining optical fibres in *yellow* guide the laser beams to the caesium tube. All diode lasers (JDSU) are mounted on a Peltier cooler for temperature regulation within 2 mK. *Right*: The integrated caesium tube without the two external magnetic shields. The volume is $990 \times 336 \times 444\,\text{mm}^3$ and the total mass is 44 kg.

long free-fall time offered by a micro-gravity environment. For that purpose, the French space agency CNES is funding and acting as the prime contractor of the PHARAO clock, a micro-gravity atomic clock which was designed by SYRTE, LKB and CNES building upon several years of experience with cold atom fountain frequency standards using caesium and rubidium atoms. After the first free-fall demonstration in a zero-g Airbus, the clock industrial development began in 2002 by the realisation of an engineering model representative of the flight model in terms of interfaces, design and fully functional (Fig. 23).

As far as atom interferometry is concerned, the fact that bosons suffer from interaction shifts leading to systematic errors might prevent to achieve the ultimate limit of those sensors. As for the clock case, this problem might not be apparent in ultra-cold fermions [130]. However, degenerate fermions have an intrinsically broad momentum distribution due to Pauli blocking, limiting the visibility of interference patterns. Furthermore, to achieve quantum degeneracy, fermions must be cooled using a buffer gas, typically an ultra-cold gas of bosons, thus complicating experiments using fermions. Pairs of fermions (molecules or Cooper pairs [131]) can be created by applying a homogeneous magnetic field (Feshbach resonances [132]), offering yet more possible candidate species for atom interferometers.

A further bonus to free fall is the possibility of using weaker confining forces for the atoms, since gravity need not be compensated with additional levitation forces [124]. Temperatures achieved by evaporative cooling and adiabatic expansion are lowered as the trapping potential is reduced. Not only does the sensitivity of an interferometric measurement benefit, but also new phases of matter may be observed if the kinetic energy can be made smaller than the interatomic potential. A reduced-gravity environment will permit study of new physical phenomena, e.g. spin dynamics and magnetic ordering (see for example [133] and references therein).

6.1 ICE: Towards a Coherent Atom Sensor for Space Applications

The objective of ICE [134], a CNES-funded project that share the experience of various partners (SYRTE, ONERA and IOTA), is to produce an accelerometer for space with a coherent atomic source. It uses a mixture of Bose–Einstein condensates with two species of atoms (Rb and K) to carry out a first comparison of accelerations measured by the two different types of atomic species (with two bosons and one boson and one fermion). The central components of this project are the atomic physics vacuum system, the optics and their supports. The atomic manipulation starts with alkali-metal vapour dispensers for rubidium and potassium [135]. A slow jet of atoms is sent from the collection chamber by a dual species, 2D-MOT to the trapping chamber, for collection and cooling in a 3D-MOT. Atoms are then to be transferred to a conservative, far-off-resonance optical-dipole trap (FORT) for further cooling towards degeneracy. The sample is then ready for coherent manipulation in an atom interferometer. Raman two-photon transition will be used as atomic beam splitters and mirrors. Three-pulse sequences ($\pi/2 - \pi - \pi/2$) will be used for accelerometry.

As for the Girafon project, all light for the experiment arrives by optical fibres, making the laser sources independent of the vacuum system. Transportable fibred laser sources for laser cooling and trapping have been fabricated with the required frequency stability. The techniques for mechanically stable power distribution by free-space fibre couplers function according to specifications. The vacuum chamber is compatible with the constraints of micro-gravity in an Airbus parabolic flight. Such a flight permits total interrogation times up to 7 s, giving a potential sensitivity of better than 10^{-9} m s^{-2} per shot, limited by phase noise on the frequency reference for the Raman transitions.

6.2 Laser Systems

Continuous-Wave Fibre-Laser Source at 780 nm for Rubidium Cooling

An entirely pigtailed laser source is particularly appropriate in our case as it does not suffer from misalignments due to environmental vibrations. Moreover, telecommunication laser sources in the C-band (1,530–1,570 nm) have narrow linewidths ranging from less than 1 MHz for laser diodes, down to a few kHz for Erbium-doped fibre lasers. By second-harmonic generation (SHG) in a non-linear crystal, these 1.56 μm sources can be converted to 780 nm sources [136–138]. Such devices avoid the use of extended cavities as their linewidths are sufficiently narrow to satisfy the requirements of laser cooling.

The laser setup is sketched in Fig. 24. A 1,560 nm Erbium-doped fibre laser is amplified by a 500 mW polarisation-maintaining (PM) Erbium-doped fibre amplifier (EDFA). A 90/10 PM fibre coupler directs 10% of the pump power to a pigtailed output. Ninety per cent of light is then sent into a periodically poled Lithium–Niobate waveguide (PPLN-WG). This crystal is pigtailed on both sides with 1,560 nm single-mode fibres. The input fibre is installed in a polarisation loop system to align the electric field with principal axes of the crystal. A fibre coupler, which is monomode at 780 nm, filters pump light after the crystal and sends half of the 780 nm light into a saturated-absorption spectroscopy device for frequency servo-control. The other half is the frequency-stabilised pigtailed output. The whole device, including the frequency control electronics, was implemented in a rack for ease of transport. Typical output from the first generation device was 500 μW of 780 nm light, with more than 86 dB attenuation of 1,560 nm light after 3 m of monomode fibre. A more recent version (> 50 mW) has been used to power a MOT.

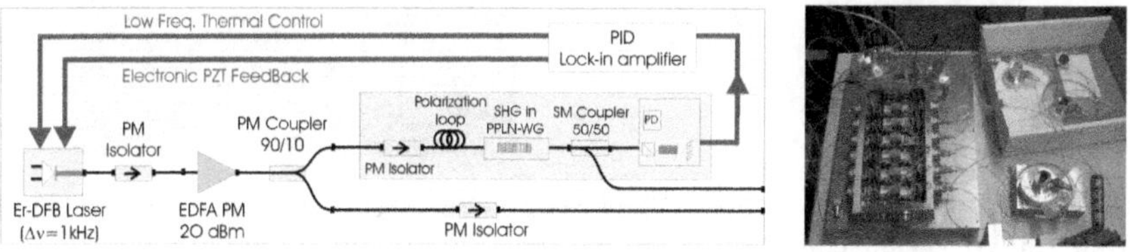

Fig. 24. *Left*: Transportable laser setup schematic. A double-loop feedback system is used for frequency control: the first returns a saturated absorption signal to the piezoelectric transducer; the second loop compensates thermal drifts of the fibre laser when the error signal of the first loop becomes large. *Right*: The fibre splitters developed at SYRTE.

Fibre Power Splitters

The optical bench and the vacuum chamber are not rigidly connected to each other, and laser light is transported to the vacuum chamber using optical fibres. Stability in trapping and coherent atom manipulation is assured by using only polarisation-maintaining fibres. Six trapping and cooling laser beams are needed for the 3D-MOT and five for the 2D-MOT, with relative power stability better than a few per cent. The fibre beam splitters are based on polarising cubes and half-wave plates with one input fibre and the relevant number of output fibres. The stability of the beam splitters has been tested by measuring the ratio of output powers between different outputs as a function of time. Fluctuations are negligible on short timescales (less than 10^{-4} relative intensity over 1 s), and very small over typical periods of experimental operation (less than 1% over a day). Even over months, drifts in power distribution are only a few per cent, which is sufficient for this experiment.

6.3 Mechanical and Vacuum Systems

The mechanical construction of the apparatus is critical to any free-fall experiment. Atomic physics experiments require heavy vacuum systems and carefully aligned optics. The ICE design is based around a cuboidal frame of foam-damped hollow bars with one face being a vibration-damped optical breadboard (see Fig. 25). The outside dimensions are $1.2 \times 0.9 \times 0.9\,\mathrm{m}^3$, and the total weight of the final system is estimated to be 400 kg (excluding power supplies, lasers, control electronics, air and water flow). The frame provides support for the vacuum system and optics, which are positioned independently of one another. The heavy parts of the vacuum system are rigged to the frame using steel chains and high-performance polymer slings under

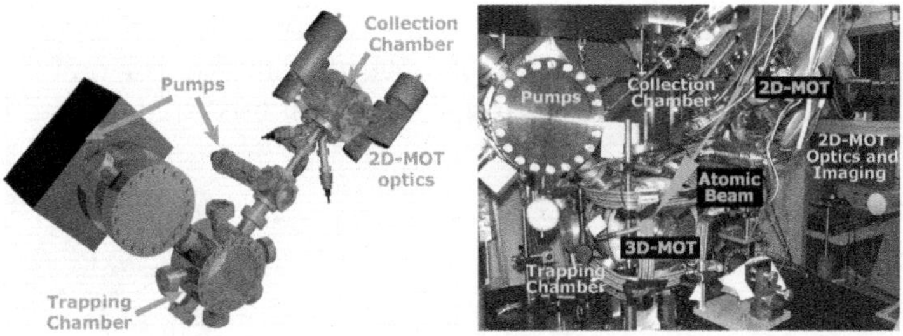

Fig. 25. *Left*: Artist's impression of the vacuum system. Atoms are transferred from the collection chamber, using a 2D-MOT, to the trapping chamber, where they are collected in a 3D-MOT. The trapping chamber has large optical accesses for the 3D-MOT, optical-dipole trap (FORT), imaging and interferometry. There is a getter pump between the two chambers to ensure a large pressure difference. The other pump is a combined ion pump–titanium sublimation pump. *Right*: The ICE mechanical structure with optics and light paths represented.

tension, adjusted using turnbuckles; most of the equipment being standard in re-creational sailing or climbing. The hollow bars have precisely positioned grooves which permit optical elements to be rigidly fixed (bolted and glued) almost anywhere in the volume within the frame. An adaptation for transportability will be to enclose the frame in a box, including acoustic and magnetic shielding, temperature control, air overpressure (dust exclusion), as well as ensuring safety in the presence of the high-power lasers.

The vacuum chamber has three main parts: the collection chamber (for the 2D-MOT), the trapping chamber (for the 3D-MOT and the FORT) and the pumps (combined ion pump and titanium sublimation pump). Between the collection and trapping chambers, there is an orifice and a getter pump, allowing for a high differential pressure, permitting rapid collection by the 2D-MOT but low trap losses in the 3D-MOT and FORT. The magnetic coils for the 2D-MOT are under vacuum, and consume just 5 W of electrical power.

To avoid heating due to vibrations in the FORT optics, or measurement uncertainties due to vibrations of the imaging system, the trapping chamber is as close as possible to the breadboard. For laboratory tests, the breadboard is at the bottom and the 2D-MOT arrives at 45° to the vertical, leaving the vertical axis available for addition of interferometry for precise measurements, e.g. a standing light wave. Around the main chamber, large electromagnet coils in Helmholtz configuration will be added to produce homogeneous stable fields up to 0.12 T (1,200 G), or gradients up to $0.6\,\mathrm{T\,m^{-1}}$ ($60\,\mathrm{G\,cm^{-1}}$).

2D MOT

The 2D-MOT is becoming a common source of cold atoms in two-chamber atomic physics experiments [66], and is particularly efficient for mixtures [139] of ^{40}K and ^{87}Rb, if isotopically enriched dispensers are used. Briefly, a 2D-MOT has four sets of beams (two mutually orthogonal, counter-propagating pairs) transversely to the axis of the output jet of atoms, and a cylindrical-quadrupole magnetic field generated by elongated electromagnet pairs (one pair, or two orthogonal pairs). Atoms are cooled transverse to the axis, as well as collimated. Implicitly, only slow atoms spend enough time in the 2D-MOT to be collimated, so the output jet is longitudinally slow. The number of atoms in the jet can be increased by the addition of the push beam, running parallel to the jet: a 2D-MOT$^+$. Typically the output jet has a mean velocity below $30\,\mathrm{m\,s^{-1}}$, with up to 10^{10} atoms per second of ^{87}Rb and 10^8 atoms per second of ^{40}K.

The ICE design uses 40 mW per species for each of the four transverse beams, each divided into two zones of about 20 mm using non-polarising beam-splitter cubes, corresponding to about three times the saturation intensity for the trapping transitions. The pushing beam uses 10 mW of power, and is about 6 mm in diameter. Each beam comes from an individual polarisation-maintaining optical fibre, with the light at 766.5 and 780 nm being superimposed on entry to the fibres. The 2D-MOT is seen as two bright lines of fluorescence in the collection chamber.

Fig. 26. *Left*: Artist's impression of the 3D-MOT (*dark, red beams*, and the electromagnets) and far-off-resonance optical-dipole trap (FORT; *pale, yellow beams*). *Right*: Photograph of the vacuum chamber, the support structure and the optics for magneto-optical traps. The main chamber has two very large viewports as well as seven side windows (and one entry for the atoms from the 2D-MOT). Thus there is plenty of optical access for the 3D-MOT, the FORT, imaging and interferometry. To preserve this optical access, the magnetic coils are outside of the chamber, although this markedly increases their weight and power consumption.

3D-MOT and Optical-Dipole Trap (Fig. 26)

The atomic jet from the 2D-MOT is captured by the 3D-MOT in the trapping chamber. At the time of writing, we have observed the transfer and capture of atoms, significantly increased by the addition of the pushing beam. The 3D-MOT uses one polarisation-maintaining fibre input per species. Beams are superimposed and split into six arms (on a small optical breadboard fixed near one face of the frame) for the three, orthogonal, counter-propagating beam pairs. Once enough number of atoms are collected in the 3D-MOT, the 2D-MOT is to be turned off, and the 3D-MOT optimised for transfer to the FORT.

The FORT consists of two nearly orthogonal (70°) beams making a crossed dipole trap using 50 W of light at 1,565 nm. Rapid control over intensity is achieved using an electro-optical modulator, and beam size using a mechanical zoom, after the design of Kinoshita et al. [140]. Optimisation of transfer from the 3D-MOT to the FORT, and the subsequent evaporative cooling, can be enhanced with strong, homogeneous, magnetic fields that will be used to control inter-species interactions via Feshbach resonances [132], to expedite sympathetic cooling of ^{40}K by ^{87}Rb. With the expected loading of the 3D-MOT during less than 5 s, then cool to degeneracy in the optical-dipole trap in around 3–10 s, ICE will be able to prepare a sample for interferometry in less than the free-fall time of a parabolic flight (around 20 s).

7 Conclusion

Previous experiments measuring the gravitational acceleration of Earth and its gradient or rotations have been demonstrated to be very promising. Sensitivities better than 1 nrad s^{-1} Hz$^{-1/2}$ for rotation measurements and

$2 \cdot 10^{-8} \times g\,\mathrm{Hz}^{-1/2}$ for a gravity measurement have already been obtained. The sensitivity of matter-wave interferometers for rotations and accelerations increases with the measurement time and can therefore be dramatically enhanced by reducing the atomic velocity. Moreover, the use of optical transitions to manipulate the atomic wave packets enables an intrinsic knowledge of the scaling factor of these inertial sensors, which is directly linked to the frequency of the transition. Therefore, combining cold atomic sources and Raman transition-based atomic interferometers results in highly sensitive and highly accurate inertial sensors.

Going to space will enhance the benefit of cold atoms by increasing the interaction time, and opens up entirely new possibilities for research in fundamental physics or for inertial navigation with unprecedented precision, and operation in space is thus strong motivation for many ongoing projects.

Several missions along this line have thus only recently been proposed by NASA as well as ESA. Therefore, quantum sensors may be used as long-term inertial references for astronomy, deep-space navigation, or in missions to precisely map and monitor Earth's gravitational field (such as GOCE, etc.). In fundamental physics these space-based cold atom sensors may be the key for ground-breaking experiments on fundamental issues, such as gravitational wave astronomy (LISA-II, etc.) or the quest for a universal theory reconciling quantum theory and gravity (e.g. tests of the equivalence principle).

Cold atom quantum sensors display an excellent sensitivity for the absolute measurement of gravity, gravity gradients, magnetic fields as well as the Earth rotation and, thus, are particularly suited for applications in Earth sciences, or more generally for future "Earth watch" facilities. The range of fascinating applications of gravity mapping extends from earthquake and volcanic eruptions prediction, earth tectonics, to the search for oil and mineral resources, to the measurement of the effect of climate changes such as variations of the ocean level. As all these topics have a large impact on society as whole, the impact of improvements generated by this new technology will be accordingly high (large "leverage factor"). On the practical side such improvements should come from alleviating the need of constantly re-calibrating gravimeters (more than 1,500 deployed) in prospecting for natural resources, as atomic quantum sensors are intrinsically free of drift – or from alleviating the need of gyroscopically stabilised inertial platforms (expensive, large and service intensive) for mounting air- or sea-borne gravity gradiometers (more than 100 complex systems deployed), as multi-axes atomic quantum sensors can be made sufficiently orientation independent.

In addition, since quantum sensors rely on well-defined quantum mechanical properties of the atomic internal structure and the precisely known interaction with light, they may be used in new definitions of base units – similarly to what has already been done for time and frequency standards (atomic clocks) or for the practical realisation of resistance (quantum Hall effect) and voltage (Josephson effect). Immediate applications would be in the re-definition of the kilogram, the only base unit in the international system of units (SI) that is defined by a material artefact of suspected stability. A very promising approach to overcome this unsatisfactory state of affairs is the use of a so-called *watt balance*, in which mechanical and electrical powers are compared. If the electrical power is measured in terms of the two quantum effects, the Josephson and the quantum Hall effect, the unit of mass can be linked to the fundamental Planck constant h through its de Broglie–Compton frequency $M_{\mathfrak{K}}c^2/h$. For proper operation, such a watt balance requires a "gravity reference" at a performance level that is difficult to achieve with classical sensors, but should be well within the range of capability of an atomic quantum gravimeter. The other way to determine the de Broglie–Compton frequency $M_{\mathfrak{K}}c^2/h$ of the kilogram is through the product of the Avogadro number by the de Broglie–Compton frequency $m_u c^2/h$ of the atomic mass unit determined by atom interferometry [141]. Unfortunately these two ways do not yet agree at the $1.3 \cdot 10^{-6}$ level and further progress is necessary. As mentioned before the determination of $m_u c^2/h$ by atom interferometry leads to a new determination of the fine structure constant α and hence to an experimental validation of the formula $R_K = h/e^2 = Z_0/2\alpha$ which is supposed to give the Von Klitzing resistance R_K compared to the vacuum impedance Z_0 in the Thompson–Lampard experiment.

Finally, handling BEC or atom lasers on ground or in space will be a leap towards the practical construction of cold coherent sources that can be used in ultra-high-precision atomic matter-wave sensors. Indeed, the long interrogation time requires a very strong collimation of the atomic source. Combining this with the high flux required for a high sensitivity leads to the need of an atom laser (like in optics, an atom laser is characterised by its high brilliance). On Earth, the best outcoupling device uses gravity to extract atoms from the magnetic cavity (except for the recent guided atom laser). Novel techniques can be explored in space, such as Raman output coupling, to extract a CW atom laser beam into a controlled propagation direction. In addition, novel types of atom interferometers using coherent sources, such as a resonant atom cavity [62] or a three-dimensional atom sensor [52], might be applied with these new sources. Ultimately, the correlation properties of the particles within the atom laser field may have a serious impact on the performance of future atom interferometer-based sensors. Hence, just as in the optical case, the sensitivity will be quantum limited by the uncertainty principle for the phase and number quadratures for single-mode operation. It is possible to go beyond this standard quantum limit with a coherent source prepared in phase–number squeezed states, i.e. Heisenberg-limited interferometry. Alternatively, entangled two-mode operation schemes, like the correlated emission laser (CEL) in laser physics, can also be used to suppress quantum noise in the relative phase.

Acknowledgements

The authors would like to express their deep thanks to the numerous colleagues who have contributed to the figures and results reported in this review and to the funding actors of the field, especially CNES, DGA and IFRAF.

References

1. W.W. Chow, J. Gea-Banacloche, L.M. Pedrotti, V.E. Sanders, W. Schleich, and M.O. Scully, *Rev. Mod. Phys.*, **72**, 61 (1985).
2. J.F. Clauser, *Physica B*, **151**, 262 (1988).
3. R. Colella, A.W. Overhauser, and S.A. Werner, *Phys. Rev. Lett.* **34**, 1472 (1975).
4. S.A. Werner, J.-L. Staudenmann, and R. Colella, *Phys. Rev. Lett.*, **42**, 1103 (1979).
5. Ch.J. Bordé, Ch. Salomon, S. Avrillier, A. Van Lerberghe, Ch. Bréant, D. Bassi and G. Scoles, *Phys. Rev. A* , **30**, 1836 (1984) and references therein.
6. F. Riehle, Th. Kisters, A. Witte, J. Helmcke and Ch.J. Bordé, *Phys. Rev. Lett.* **67**, 177 (1991).
7. M. Kasevich and S. Chu, *Appl. Phys.*, **B 54**, 321 (1992).
8. P.R. Berman (ed.), *Atom Interferometry* (Academic Press, 1997).
9. T.L. Gustavson, P. Bouyer, M.A. Kasevich, *Phys. Rev. Lett.* **78**, 2046 (1997).
10. T.L. Gustavson, *et al. Class. Quantum Grav.* **17**, 1 (2000).
11. A. Peters, K.Y. Chung, B. Young, J. Hensley and S. Chu, *Phil. Trans. R. Soc. Lond.* **A 355**, 2223 (1997).
12. A. Peters, K. Y. Chung and S. Chu, *Metrologia*, **38**, 25 (2001).
13. M. Kasevich and S. Chu, *Phys. Rev. Lett.* **67**, 181 (1991).
14. D.W. Keith, C.R. Ekstrom, Q.A. Turchette, and D.E. Pritchard, *Phys. Rev. Lett.* **66**, 2693 (1991).
15. M.J. Snadden *et al.*, *Phys. Rev. Lett.* **81**, 971 (1998).
16. Ch. J. Bordé. In P.G. Bergmann and V. de Sabbata (eds.). *Advances in the Interplay between Quantum and Gravity Physics* (Kluwer Academic Publisher, 2002) pp. 27–55.
17. M. Fattori *et al.*, *Phys. Lett.* **A 318**, 184 (2003).
18. A. Wicht *et al.*, *Proceeding of the 6th Symposium on Frequency standards and metrology* (ed. Patrick Gill, World Scientific, 2001) 193, *Physica Scripta*, **102**, 82 (2002).
19. M.H. Anderson *et al.*, *Science* **269**, 198 (1995).
20. K.B. Davis *et al.*, *Phys. Rev. Lett.* **75**, 3969 (1995).
21. C.C. Bradley, C.A. Sackett, and R.G. Hulet, *Phys. Rev. Lett.* **75**, 1687 (1995).
22. M.-O. Mewes *et al.*, *Phys. Rev. Lett.* **78**, 582 (1997).
23. B.P. Anderson and M.A. Kasevich, *Science* **282**, 1686 (1998).
24. E.W. Hagley *et al. Science* **283**, 1706 (1999).
25. I. Bloch, T.W. Hänsch, and T. Esslinger, *Phys. Rev. Lett.* **82**, 3008 (1999).
26. G. E. Stedman *et al.*, *Phys. Rev.* **A 51**, 4944 (1995).
27. G.E. Stedman, *Rep. Prog. Phys.* **60**, 615 (1997).
28. P. Bouyer and M. Kasevich, *Phys. Rev.* **A 56**, R1083 (1997).

29. S. Gupta, K. Dieckmann, Z. Hadzibabic, and D.E. Pritchard, *Phys. Rev. Lett.* **89**, 140401 (2002).
30. Hyper-Precision Cold Atom Interferometry in Space (HYPER), Assessment Study Report ESA-SCI(2000)10, European Space Agency (2000).
31. Ch.J. Bordé, *Phys. Lett.* **A 140**, 10 (1989).
32. D.M. Giltner, R.W. McGowan, and S.A. Lee, *Phys. Rev. Lett.* **75**, 2638 (1995).
33. C. Cohen-Tannoudji, *Cours au Collège de France* (1992–1993).
34. Ch. Antoine and Ch. J. Bordé, *Phys. Lett.* **A 306**, 277 (2003); *J. Opt. B: Quantum Semiclass. Opt.* **5**, S199 (2003).
35. P. Storey and C. Cohen-Tannoudji, *J. Phys. II France* **4**, 1999 (1994).
36. M. Kasevich, E. Riis, S. Chu, and R. de Voe, *Phys. Rev. Lett.* **63**, 612 (1989).
37. A. Clairon, C. Salomon, S. Guelatti, and W. Phillips, *Europhys. Lett.* **16**, 165 (1991).
38. Y. Sortais, S. Bize, and M. Abgrall, *Physica Scripta* **2001**, 50 (2001).
39. Ch.J. Bordé, *Metrologia* **39**, 435–463 (2002).
40. A. Peters, K. Chung, and S. Chu, *Metrologia* **38**, 25 (2001).
41. J.M. McGuirk *et al., Phys. Rev.* **A 65**, 033608 (2002).
42. K.U. Schreiber, A. Velikoseltsev, T. Klügel, M. Rothacher, G.E. Stedman and D.L. Wiltshire, *J. Geophys. Res.* **109** (B6): B06405 (2004).
43. T. Niebauer, G. Sasagawa, J. Faller, R. Hilt and F. Klopping, *Metrologia*, **32**, 159–180 (1995).
44. M. Kasevich, D. S. Weiss, E. Riis, K. Moler, S. Kasapi, and S. Chu, *Phys. Rev. Lett.* **66**, 2297 (1991).
45. K. Moler, D.S. Weiss, M. Kasevich, and S. Chu, *Phys. Rev.* **A 45**, 342 (1992).
46. A. Miffre *et al.*, to appear in *App. Phys.* **B**.
47. E.M. Rasel, M.K. Oberthaler, H. Batelaan, J. Schmiedmayer, and A. Zeilinger, *Phys. Rev. Lett.* **75**, 2633 (1995).
48. Ch.J. Bordé, in M. Ducloy, E. Giacobino, G. Camy (eds.), *Laser spectroscopy X* (World scientific, Singapore 1992), p. 239.
49. J.M. McGuirk, M.J. Snadden, and M.A. Kasevich, *Phys. Rev. Lett.* **85**, 4498 (2000)
50. Ch.J. Bordé, in J. Dalibard, J.-M. Raimond and J. Zinn-Justin (eds.) *Fundamental Systems in Quantum Optics*, Les Houches Lectures, Session LIII, 1990 (Elsevier Science Publishers, 1991).
51. Ch.J. Bordé, *C.R. Acad. Sci. Paris*, Série IV, **2**, 509 (2001).
52. Ch.J. Bordé, *Gen. Rel. Grav.* **36**, 475 (2004).
53. K. Bongs, R. Launais and M. Kasevich, to appear in App. Phys. B (2006).
54. Ch.J. Bordé, J.-C. Houard, and A. Karasiewicz, in C. Lämmerzahl, C.W.F. Everitt and F.W. Hehl (eds.): *Gyros, Clocks and Interferometers: Testing Relativistic Gravity in Space* (Springer, Berlin 2001), p. 403.
55. Ch.J. Bordé, A. Karasievicz, and Ph. Tourrenc, *Int. J. Mod. Phys.* **D 3**, 157 (1994).
56. S.N. Gupta, *Proc. Phys. Soc.* **A 65**, 161 (1952); *Proc. Phys. Soc.* **A 65**, 608 (1952).
57. R.P. Feynman, F.B. Morinigo, and W.G. Wagner, *Feynman Lectures on Gravitation*, edited by B. Hatfield (Addison-Wesley, Reading, Mass. 1995).
58. B.M. Barker, S.N. Gupta, and R.D. Haracz, *Phys. Rev.* **149**, 1027 (1966), and references therein.
59. M. Sagnac, *Compt. Rend. Acad. Sci.*, **157**, 708 (1913).

60. P. Cladé *et al.*, *Europhys. Lett.*, **71** (5), pp. 730–736 (2005).
61. G. Modugno *et al*, arXiv:physics/0411097.
62. F. Impens, P. Bouyer and C. Bordé, to appear in App. Phys. B (2006).
63. P. Cheinet *et al.*, to appear in App. Phys. B (2006).
64. `http://www.onera.fr/actualites/2006-02.php`
65. B.P. Kibble, *Atomic Masses and Fundamental Constants* **5 ed** J H Saunders and A H Wapstra (Plenum, New York) pp 541–51 (1976).
66. K. Dieckmann, R.J.C. Spreeuw, M. Weidemüller and J.T.M. Walraven. *Phys. Rev.* **A 58**, 3891 (1998).
67. P. Cheinet, B. Canuel, F. Pereira Dos Santos, A. Gauguet, F. Leduc, A. Landragin, submitted for publication to: IEEE Trans. on Instrum. Meas., physics/0510197.
68. B. Mashhoon and D. Theiss, *Phys. Rev.* **D 49**, 1542 (1982).
69. B. Mashhoon, H. Paik, C. Will, *Phys. Rev.* **D 39**, 2285 (1989).
70. N. Sneeuw, R. Rummel, and J. Müller, *Class. Quantum Grav.* **58**, A113 (1996).
71. N. You *et al.*, to appear in App. Phys. B (2006).
72. A. Bertoldi *et al.*, physics/0606126.
73. C. Jekeli, *Geophysics* **58**, 508 (1993).
74. M. Moody and H. Paik, *Phys. Rev. Lett.* **70**, 1195 (1993).
75. H. Cavendish, *Philos. Trans. R. Soc.* **88**, 467 (1798).
76. G.T. Gillies, *Rep. Prog. Phys.* **60**, 151 (1997).
77. P.J. Mohr, B.N. Taylor, *Rev. Mod. Phys.* **72**, 351 (2000). The values can also be found at `http://www.physics.nist.gov/cuu/Constants/index.html`.
78. J.H. Gundlach, S.M. Merkowitz, *Phys. Rev. Lett.* **85** (2000) 2869; T.J. Quinn, C.C. Speake, S.J. Richman, R.S. Davis, A. Picard, *Phys. Rev. Lett.* **87** (2001) 111101.
79. J. Luo and Z.-K. Hu, *Class. Quantum Grav.* **17**, 2351 (2000).
80. J.P. Schwarz, D.S. Robertson, T.M. Niebauer, J.E. Faller, *Science* **282**, 2230 (1998).
81. B. Canuel *et al.*, *Phys. Rev. Lett.* **97**, 010402 (2006).
82. T. Gustavson, PhD. Thesis, Stanford University (2000).
83. G. Santarelli, Ph. Laurent, P. Lemonde, A. Clairon, A.G. Mann, S. Chang, A.N. Luiten, and C. Salomon, *Phys. Rev. Lett.* **82**, 4619 (1999).
84. D. Wineland *et al.*, *Phys. Rev.* **A 46**, 6797 (1992).
85. H. Marion, F. Pereira Dos Santos, M. Abgrall, S. Zhang, Y. Sortais, S. Bize, I. Maksimovic, D. Calonico, J. Grünert, C. Mandache, P. Lemonde, G. Santarelli, Ph. Laurent, A. Clairon, and C. Salomon, *Phys. Rev. Lett.*, **90**, 150801 (2003).
86. H.J. Metcalf and P. Van Der Straten, *Laser Cooling and Trapping* (Springer Verlag, Berlin 1999).
87. W. Ketterle, Scientific American.com, Ask the Experts, January 19, 2004. Scientific American, May 2004, p. 120.
88. J. Stenger *et al.*, *Phys. Rev. Lett.* **82**, 4569 (1999).
89. Y. Shin, M. Saba, T. Pasquini, W. Ketterle, D.E. Pritchard, and A.E. Leanhardt, *Phys. Rev. Lett.* **92**, 050405 (2004).
90. Y. Le Coq *et al.*, to appear in App. Phys. B (2006).
91. C. Westbrook, P. Bouyer, and C. Michaut, *La recherche* **67**, 40 (2003).
92. *When atoms behave as waves: Bose-Einstein condensation and the atom laser.* in: Les Prix Nobel 2001 (The Nobel Foundation, Stockholm, 2002), pp. 118–154. reprinted in: *ChemPhysChem* **3**, 736–753 (2002); *Rev. Mod. Phys.*, **74**, 1131–1151 (2002).

93. W. Hänsel, J. Reichel, P. Hommelhoff, and T.W. Hänsch, *Phys. Rev. Lett.* **86**, 608 (2001).
94. T. Paul *et al., Phys. Rev. A* **72**, 063621 (2005).
95. D. Clément *et al., Phys. Rev. Lett.* **95**, 170409 (2005).
96. N.P. Robins, A.K. Morisson, J.J. Hope, and J.D. Close, *Phys. Rev.* **A 72**, 031606(R) (2005).
97. Y. Le Coq *et al., Phys. Rev. Lett.* **87**, 17 (2001).
98. J.-F. Riou *et al., Phys. Rev. Lett.* **96**, 070404 (2006).
99. F. Gerbier, P. Bouyer, and A. Aspect, Phys. Rev. Lett. **86**, 4729 (2001).
100. J. Fortàgh, A. Grossmann, C. Zimmermann, and T.W. Hänsch, *Phys. Rev. Lett.* **81**, 5310 (1998).
101. J. Denschlag, D. Cassetari, and J. Schmiedmayer, *Phys. Rev. Lett.* **82**, 2014 (1999).
102. D. Müller *et al., Phys. Rev. Lett.* **83**, 5194 (1999).
103. L. Pruvost, D. Marescaux, O. Houde, and H. T. Duong, *Opt. Comm.* **166**, 199 (1999).
104. N.H. Dekker *et al., Phys. Rev. Lett.* **84**, 1124 (2000).
105. M. Key *et al., Phys. Rev. Lett.* **84**, 1371 (2000).
106. B.K. Teo and G. Raithel, *Phys. Rev.* **A 63**, 031402 (2001).
107. P. Cren *et al., Eur. Phys. J.* **D 20**, 107 (2002).
108. K. Bongs *et al., Phys. Rev. A* **61**, R31602 (2000).
109. A.E. Leanhardt *et al., Phys. Rev. Lett.* **89**, 040401 (2002).
110. Z.T. Lu *et al., Phys. Rev. Lett.* **77**, 3331 (1996).
111. D. Müller, E.A. Cornell, D.Z. Anderson, and E.R.I. Abraham, *Phys. Rev.* **A 61**, 033411 (2000).
112. T. Lahaye *et al., Phys. Rev. Lett.* **93**, 093003 (2004).
113. E. Mandonnet *et al., Eur. Phys. Journ. D* **10**, 9–18 (2000).
114. P.J. Martin, B.G. Oldaker, A.H. Hiklich, and D.E. Pritchard, *Phys. Rev. Lett.* **60**, 515 (1988).
115. M. Kozuma *et al., Phys. Rev. Lett.* **82**, 871 (1999).
116. S. Richard *et al., cond-mat*/0303137, *Phys. Rev. Lett.* **91**, 010405 (2003).
117. M. Fauquembergue, J.-F. Riou, W. Guerin et al., *Review of Sci. Inst.* **76** (10), 103104 (2005).
118. J. Ye, S. Swartz, P. Jungner and J.L. Hall, *Opt. Lett.* **21**, 1280 (1996).
119. S. Bize, Y. Sortais, M.S. Santos, C. Mandache, A. Clairon and C. Salomon, *Europhys. Lett.* **45**, 558 (1999).
120. D. Hellweg *et al., Phys. Rev. Lett.* **91**, 010406 (2003).
121. W.C. Stwalley, *Phys. Rev. Lett.* **37**, 1628 (1976); E. Tiesinga *et al., Phys. Rev. A* **46**, R1167 (1992); P. Fedichev *et al., Phys. Rev. Lett.* **77**, 2913 (1996).
122. M. Theis, G. Thalhammer, K. Winkler, M. Hellwig, G. Ruff, R. Grimm, and J. Hecker Denschlag, *Phys. Rev. Lett.* **93**, 123001 (2004).
123. J.L. Roberts, N.R. Claussen, S.L. Cornish, E.A. Donley, E.A. Cornell, and C.E. Wieman, *Phys. Rev. Lett.* **86**, 4211 (2001).
124. A.E. Leanhardt, T.A. Pasquini, M. Saba, A. Schirotzek, Y. Shin, D. Kielpinski, D.E. Pritchard, and W. Ketterle, *Science* **301**, 1513 (2003).
125. S. Schneider *et al., Phys. Rev.* **A 67**, 023612 (2003).
126. H. Ott, J. Fortagh, G. Schlotterbeck, A. Grossmann, and C. Zimmermann, *Phys. Rev. Lett.* **87**, 230401 (2001).
127. W. Hänsel, P. Hommelhoff, T.W. Hänsch, and J. Reichel, *Nature (London)* **413**, 498 (2001).

128. M. Vengalattore, W. Rooijakkers, and M. Prentiss, *Phys. Rev. A* **66**, 053403 (2002).
129. Y. Wang *et al., Phys. Rev. Lett.* **94**, 090405 (2005).
130. S. Gupta, Z. Hadzibabic, M.W. Zwierlein, C.A. Stan, K. Dieckmann, C.H. Schunck, E.G.M. van Kempen, B.J. Verhaar, and W. Ketterle, *Science* **300**, 1723 (2003); M.W. Zwierlein, Z. Hadzibabic, S. Gupta, and W. Ketterle, *Phys. Rev. Lett.* **91**, 250404 (2004).
131. C.A. Regal, M. Greiner, and D.S. Jin, *Phys. Rev. Lett.* **92**, 040403 (2004).
132. G. Ferrari, M. Inguscio, W. Jastrzebski, G. Modugno and G. Roati, *Phys. Rev. Lett.*, **89** 053202 (2002). F. Ferlaino, C. D'Errico, G. Roati, M. Zaccanti, M. Inguscio, and G. Modugno, arXiv:cond-mat/0510630.
133. H. Schmaljohann *et al., Appl. Phys.* **B 79**, 1001 (2004).
134. R. Nyman *et al.*, to appear in App. Phys. B. See also http://www.ice-space.fr.
135. Natural-abundance dispensers are available from SAES Getters. We are interested in the isotopes ^{87}Rb and ^{40}K, which are, respectively, 28% and 0.012% naturally abundant. We will be investing in isotopically enriched ^{40}K (about 5%) dispensers in the near future.
136. V. Mahal, A. Arie, M.A. Arbore, M.M. Fejer, *Opt. Lett.* **21**, 1217 (1996).
137. R.J. Thompson, M. Tu, D.C. Aveline, N. Lundblad, and L. Maleki, *Opt. Exp.* **11**, 1709 (2003).
138. J. Dingjan, B. Darquié, J. Beugnon, M.P.A. Jones, S. Bergamini, G. Messin, A. Browaeys, P. Grangier, *App. Phys.* **B 82**, 47 (2006).
139. C. Ospelkaus, S. Ospelkaus, K. Sengstock and K. Bongs, *Phy. Rev. Lett.* **96**, 020401 (2006).
140. T. Kinoshita, T.R. Wenger and D.S. Weiss, *Phys. Rev.* **A 71**, 01162(R) (2005).
141. Ch.J. Bordé, *Phil. Trans. Roy. Soc.* 363, 2097 (2005) and references therein.

Drag-Free Satellite Control

Stephan Theil

Center of Applied Space Technology and Microgravity, University of Bremen, Am Fallturm, 28359 Bremen, Germany

Summary. Scientific satellite missions trying to investigate questions regarding geodesy and fundamental physics have become increasingly dependent on ultra-low disturbance environments. The precision demanded by the experiments has risen continuously as experimenters strive to deepen their understanding. Standard attitude and orbital control systems are not capable of providing such an ultra-low disturbance environment which lead to the introduction of so-called *drag-free control systems*.

Drag-free control is an enabling technology with the capability to provide these ultra-low disturbance environments. The application of drag-free control systems is of course not limited to geodesy and fundamental physics. It is a useful technology for every mission that requires a low disturbance free-fall environment.

Drag-free control has come a long way since the introduction of the original drag-free concept by Benjamin Lange in 1964. The aim of this chapter is to give an introduction and overview about the drag-free technology and its implications for scientific satellite missions. In addition to the original drag-free concept and its advancements, the chapter introduces key technologies in sensors and actuators whose development was fueled by the application of the drag-free concept in scientific satellite missions. Moreover, problems and challenges connected to drag-free satellite control and the technologies involved are discussed, and current drag-free missions like LISA and its technology demonstrator LISA Pathfinder, MICROSCOPE, STEP, or GOCE are presented.

1 Introduction

1.1 The Drag-Free Satellite Principle

The motion of satellites on orbits around Earth is mainly determined by the gravitational field of the Earth. In addition to that, forces and torques are acting on the satellite which affect attitude and orbit. In 1964 B. Lange proposed to compensate these disturbing forces and torques using a control system to get a "force- and torque-free" satellite. Since the force due to interaction with

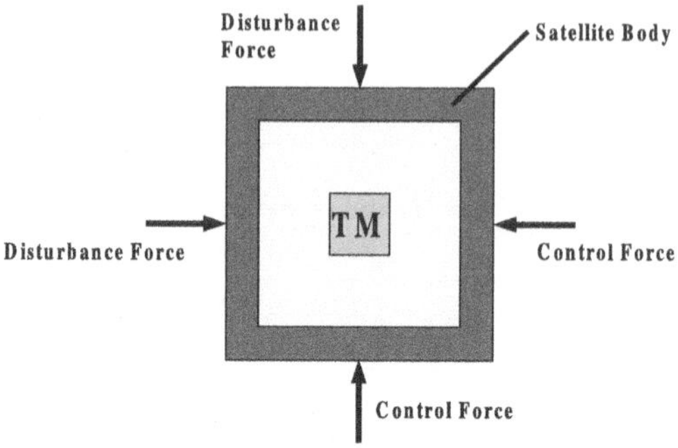

Fig. 1. Planar schematic of a drag-free satellite with its test mass (TM).

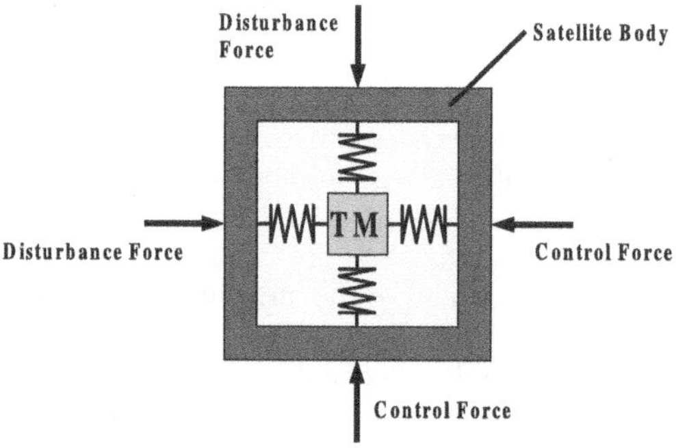

Fig. 2. Planar schematic of drag-free satellite with its test mass (TM) and coupling between satellite and test mass.

the upper atmosphere – which is denoted as air drag – is the main disturbance for satellites in low Earth orbits, the term "drag-free satellite" was introduced. The control system for compensation of the disturbance forces is called "drag-free control system."

The concept of a drag-free satellite involves centering a test mass inside a satellite. The test mass (or proof mass) is shielded by the surrounding satellite against the disturbances acting on the surface (see Fig. 1). As the test mass is free of external disturbances, it will follow a purely gravitational orbit.

To avoid a collision with the test mass, the satellite has to be controlled to follow the test mass. For that purpose the distance between satellite and test mass must be measured. This can be done by magnetic, electrostatic, or optical sensors. In all three cases the measurement cannot be obtained without applying a force on the test mass. Therefore a dynamic coupling exists between the satellite and the proof mass. This is denoted by the springs in Fig. 2. Due to this coupling the test mass inside the satellite is not free from external forces anymore. The coupling will perturb the orbit of the test mass. To minimize this effect the springs can be chosen to be very weak. But this has the disadvantage that the accuracy of the displacement measurement between test mass and satellite may become poor. As an example consider the magnetic measurement which is more accurate if a stronger field is applied. So an optimum for the equivalent stiffness of the magnetic field can be found. The second way is to reduce the displacement of the satellite w.r.t. the test mass to put the test mass at the equilibrium point of the relaxed equivalent springs. This can only be achieved by precise control. It has the advantage that the residual acceleration on the whole satellite is minimized.

1.2 Review of Drag-Free Satellite Development

Similar systems to the drag-free system are known in microgravity research where airplanes were flown on a parabolic trajectory. This was done by keeping a small object centered in free space inside the cabin which is the basic principle of parabolic flights.

The first suggestions of the drag-free control concept for a satellite were made by several investigators independently. M. Schwarzschild (1961), R.A. Ferrell, G.E. Pugh (1959), G.J.F. MacDonald, C.W. Sherwin (1962), and B.O. Lange (1961) have proposed the drag-free satellite in various forms. Lange derived in his thesis [8] the nine-degree-of-freedom equations of motion. He evaluated and discussed special cases of the equations of motions and gave a comprehensive list of applications for the drag-free satellite. In 1970, J.D. Powell developed the first analog estimator for estimating the center of mass of a spinning drag-free satellite in two-dimensional space.

The first successfully flown drag-free satellite was TRIAD I [3]. The disturbance compensation system DISCOS was developed at Stanford University under responsibility of D.B. DeBra. The drag-free control system compensated the disturbances on the satellite in three degrees of freedom. It reached a residual acceleration on the satellite of about $5 \cdot 10^{-11}\,\mathrm{m\,s^{-2}}$ when averaged over 3 days. The second drag-free satellite application was the TIP II satellite which was partially drag-free in one axis [11]. This first generation of drag-free satellites was designed to improve the ephemeris prediction of the U.S. Navy's navigation satellite system TRANSIT.

The next generation of drag-free satellites is used for scientific missions like the Gravity Field and Steady-State Ocean Circulation Explorer (GOCE) mission, Gravity Probe B (GP-B), the Satellite Test of the Equivalence Principle (STEP), and the Laser Interferometer Space Antenna (LISA) mission. The difference with respect to the first generation is, on one hand, that in some missions more than one test mass is used which can improve the overall performance of the drag-free control system; on the other hand, the quality of the "zero-g" environment is orders of magnitudes better than for the first generation. Especially this improvement made drag-free control to one of the enabling technologies for current and future fundamental physics space missions.

2 Dynamic Model

To design a control system for the dynamics of a drag-free satellite, the equations of motion are needed. This chapter will show the significant differences to conventional satellites.

2.1 Equations of Motion

Satellite Equations of Motion

The equations of motion for the satellite are similar to the dynamics of conventional satellites. One term is added to the equation for the translational motion as well as to the equation for the rotational motion. They become

$$\ddot{\boldsymbol{r}}_{i,b}^{i} = \boldsymbol{g}_{i,b}^{i}\left(\boldsymbol{r}_{i,b}^{i}\right) + \boldsymbol{f}_{\text{control}}^{i} + \boldsymbol{f}_{\text{dist}}^{i} + \boldsymbol{f}_{\text{coupl,sat}}^{i} \tag{1}$$

where $\ddot{\boldsymbol{r}}_{i,b}^{i}$ is the acceleration of the satellite relative to the inertial frame expressed in the inertial frame; $\boldsymbol{g}_{i,b}^{i}$ the gravitational acceleration as a function of the satellite's position; $\boldsymbol{f}_{\text{control}}^{i}$ the specific control force; $\boldsymbol{f}_{\text{dist}}^{i}$ the sum of all disturbance-specific forces acting on the satellite; and $\boldsymbol{f}_{\text{coupl,sat}}^{i}$ the specific force on the satellite due to the coupling between satellite and all test masses, and sum of coupling forces from each single test mass.
And

$$\dot{\boldsymbol{\omega}}_{i,b}^{b} = \left(\mathcal{I}_{b}^{b}\right)^{-1}\left[\boldsymbol{T}_{\text{control}}^{b} + \boldsymbol{T}_{\text{dist}}^{b} + \boldsymbol{T}_{\text{coupl,sat}}^{b} - \boldsymbol{\omega}_{i,b}^{b} \times \left(\mathcal{I}_{b}\boldsymbol{\omega}_{i,b}^{b}\right)\right] \tag{2}$$

$$\dot{\mathbf{q}}_{i}^{b} = \frac{1}{2}\hat{\boldsymbol{\omega}}_{i,b}^{b} \odot \mathbf{q}_{i}^{b} \tag{3}$$

where $\boldsymbol{\omega}_{i,b}^{b}$ is the angular velocity of the satellite w.r.t. inertial frame expressed in body-fixed coordinate frame; $\mathcal{I}_{b}^{b}$ the moments of inertia tensor of the satellite; $\boldsymbol{T}_{\text{control}}^{b}$ the control torques applied for attitude control expressed in the body-fixed frame; $\boldsymbol{T}_{\text{dist}}^{b}$ the disturbance torques acting on the satellite expressed in the body-fixed frame; $\boldsymbol{T}_{\text{coupl,sat}}^{b}$ the torques generated from satellite–test mass coupling expressed in the body-fixed frame, and sum of coupling torques generated by each single test mass; and $\mathbf{q}_{i}^{b}$ the attitude quaternion describing the orientation of the satellite body-fixed frame w.r.t. the inertial frame.

The term $\hat{\boldsymbol{\omega}}_{i,b}^{b}$ is the quaternion representation of the angular velocity. The operator $\odot$ is the quaternion multiplication.

Test Mass Equations of Motion

The equations of motion of a test mass relative to the satellite can be derived from the equations of motion in the inertial frame. This relative motion of the test mass is conveniently expressed in the rotating sensor frame which is fixed to the satellite body. For that reasons terms due to the rotation and acceleration of the satellite will be included in the equation of motion for the test mass. Then the translational motion of the test mass w.r.t. the satellite-fixed sensor frame can be described as

$$\begin{aligned}\ddot{\boldsymbol{r}}^b_{b,tm} =& \Delta \boldsymbol{g}^b_{b,tm} - \boldsymbol{f}^b_{\text{control}} - \boldsymbol{f}^b_{\text{dist}} - \boldsymbol{f}^b_{\text{coupl,sat}} + \boldsymbol{f}^b_{\text{coupl,tm}} \\ & - 2\,\boldsymbol{\omega}^b_{i,b} \times \dot{\boldsymbol{r}}^b_{b,tm} - \dot{\boldsymbol{\omega}}^b_{i,b} \times \boldsymbol{r}^b_{b,tm} - \boldsymbol{\omega}^b_{i,b} \times \left(\boldsymbol{\omega}^b_{i,b} \times \boldsymbol{r}^b_{b,tm}\right) ,\end{aligned} \tag{4}$$

where $\boldsymbol{r}^b_{b,tm}$ is the position of the test mass relative to the satellite body frame; $\boldsymbol{g}^b_{b,tm}$ the gravitational acceleration as a function of the test mass position; $\boldsymbol{f}^b_{\text{coupl,tm}}$ the specific force acting on the test mass due to satellite–test mass coupling; and $\boldsymbol{\omega}^b_{i,b}$ the rotation of the satellite w.r.t. the inertial frame.

The formulation of the equations of motion for the test mass attitude w.r.t. the sensor is based on the conservation of the angular momentum. The test mass inside the satellite is shielded from all external nongravitational forces and torques. So the equation for the rotational motion can be written as

$$\begin{aligned}\dot{\boldsymbol{\omega}}^{tm}_{b,tm} =& \left(\mathcal{I}^{tm}_{tm}\right)^{-1} \left[\boldsymbol{T}^{tm}_{coup,tm} + \boldsymbol{T}^{tm}_{\text{gg,tm}} - \left(\boldsymbol{\omega}^{tm}_{i,b} + \boldsymbol{\omega}^{tm}_{b,tm}\right) \left(\mathcal{I}^{tm}_{tm} \left(\boldsymbol{\omega}^{tm}_{i,b} + \boldsymbol{\omega}^{tm}_{b,tm}\right)\right)\right] \\ & - \dot{\boldsymbol{\omega}}^{tm}_{i,b} ,\end{aligned} \tag{5}$$

where $\boldsymbol{\omega}^{tm}_{b,tm}$ is the angular velocity of the test mass relative to the satellite body-fixed frame expressed in the test mass body-fixed frame; $\mathcal{I}^{tm}_{tm}$ the moments of inertia matrix of the test mass; $\boldsymbol{T}^{tm}_{\text{gg,tm}}$ the gravity-gradient torque for the test mass from Earth gravity field as well as from gravity gradient inside the satellite; and $\boldsymbol{T}^{tm}_{coup,tm}$ the torque on the test mass due to satellite–test mass coupling.

The attitude of the test mass w.r.t. the sensor frame can be expressed by quaternions. The differential equation describing the kinematics of the test mass w.r.t. the satellite is written as

$$\dot{\mathbf{q}}^{tm}_b = \frac{1}{2}\, \hat{\boldsymbol{\omega}}^{tm}_{b,tm} \odot \mathbf{q}^{tm}_b . \tag{6}$$

2.2 Forces and Torques

To model the dynamic behavior of satellite and test masses, the forces and torques acting on both have to be modeled too. There exist forces and torques acting on both – satellite and test masses – as well as forces and torques acting on the satellite only.

The first group includes the effect of gravitation between celestial body, satellite, and test masses as well as interaction of satellite and test masses via the coupling which is inherent in a measurement and/or positioning system for the test masses.

Forces and torques acting on the satellite only are:

- Controlled actuation forces and torques for satellite attitude and translation control (ATC)
- Forces and torques due to interaction with the upper atmosphere (for low Earth orbits)
- Electromagnetic radiation-induced forces and torques on the satellite surface
- Torques due to interaction of satellite components with the external magnetic field
- Force and torque impulses from space debris and meteoroid hits

The following subsections will focus on the forces acting directly on the test masses. These forces are due to the coupling between satellite and test masses as to due to gravitational attraction.

Test Mass–Satellite Coupling

If the drag-free satellite houses only one test mass, the equations of motion become relatively simple as denoted in Sect. 2.1. If more than one test mass is onboard the satellite the dynamics of all bodies (satellite and test masses) are connected via the coupling force and gravitational attraction.

We assume that the satellite is connected to a test mass via a sensor system which produces a force and a torque on the satellite. This force and torque can be described as a function of the test mass states w.r.t. the body-fixed frame. For the force and torque on the satellite produced by the coupling to test mass j, we can write:

$$\begin{aligned} \boldsymbol{F}_{sat,j} &= \boldsymbol{f}_{\mathrm{F},sat,j}\left(t, \boldsymbol{r}^b_{b,j}, \dot{\boldsymbol{r}}^b_{b,j}\right) \\ \boldsymbol{T}_{sat,j} &= \boldsymbol{r}^b_j \times \boldsymbol{f}_{\mathrm{F}}\left(t, \boldsymbol{r}^b_{b,j}, \dot{\boldsymbol{r}}^b_{b,j}\right) + \boldsymbol{f}_{\mathrm{T},sat,j}\left(t, \mathbf{q}^j_b, \boldsymbol{\omega}^j_{b,j}\right) \end{aligned} \tag{7}$$

The forces and torques for all test masses j have to be added. They render the total force and torque due to coupling which are part of the equations of motion (see (1),[1] (2), and (4)).

The force and the torque on the test mass due to the coupling with the satellite is the same but in the opposite direction. In addition the test mass might experience forces and torques from a coupling with other test masses. In case there is a coupling to a second test mass k, the force and torque become:

$$\begin{aligned} \boldsymbol{F}_j &= \boldsymbol{f}_{\mathrm{F},sat,j}\left(t, \boldsymbol{r}^b_{b,j}, \dot{\boldsymbol{r}}^b_{b,j}\right) + \boldsymbol{f}_{\mathrm{F},j,k}\left(t, \boldsymbol{r}^b_{b,j}, \dot{\boldsymbol{r}}^b_{b,j}, \boldsymbol{r}^b_{b,k}, \dot{\boldsymbol{r}}^b_{b,k}\right) \\ \boldsymbol{T}_j &= \boldsymbol{f}_{\mathrm{T},sat,j}\left(t, \mathbf{q}^j_b, \boldsymbol{\omega}^j_{b,j}\right) + \boldsymbol{f}_{\mathrm{T},j,k}\left(t, \mathbf{q}^j_b, \boldsymbol{\omega}^j_{b,j}, \mathbf{q}^k_b, \boldsymbol{\omega}^k_{b,k}\right) \end{aligned} \tag{8}$$

[1] The specific force in (1) is the force divided by the mass of the satellite, respectively, test mass.

The functions f for the coupling force and torque can be depending on the sensor type nonlinear in the states. In a first approximation a linear coupling (spring-damper system) is used.

Gravitational Attraction Between Satellite and Test Masses

For most satellite applications the gravitational field created by the satellite is usually negligible. In case of the drag-free satellite, it becomes a force which connects the dynamics of satellite and test masses. This force can be reduced or compensated by design, e.g., placing the test mass in the center of mass of the satellite. Since this configuration is not always possible – e.g., if more than one test mass is used – the gravitational force between satellite and test mass has to be taken into account.

If the structure of the satellite is stiff and has no moving parts, the gravitational attraction between satellite and test mass is constant. However, if there is a transient deformation, e.g., due to thermal expansion or sloshing of liquid (fuel), there are moving masses which make the gravitational attraction force between satellite and test mass a time-varying quantity (see [14]).

Gravity-Gradient Acceleration

In the equation of translational motion for the test mass 4, the term $\Delta \boldsymbol{g}_{b,tm}$ denotes the gravity-gradient acceleration of the test mass. It is the difference in the gravitational acceleration of satellite and test mass

$$\Delta \boldsymbol{g}_{b,tm} = \boldsymbol{g}_{tm} - \boldsymbol{g}_b. \tag{9}$$

If we neglect the gravitational attraction between satellite and test mass, the gravitational acceleration on each body can be treated independently. In contradiction to conventional satellites, the gravitational force on the satellite cannot be assumed to be independent from the attitude of the satellite. Since the center of gravity might move with the attitude of the satellite, the gravitational force resp. acceleration changes with the attitude. Though this is a very small effect it should be considered in the modeling of a drag-free satellite (see [7]).

3 Technology

The increasing interest by the scientific community in the drag-free idea and its applications has fueled the technological development in this area. In recent years a number of different concepts for sensors and actuators has been proposed and developed. This section will give a brief overview about the different technologies that are out there.

3.1 Sensors

A central part of the drag-free control system is of course the drag-free sensor. Most of the sensors available today are custom made for specific applications. Nevertheless three main categories can be identified that can be used to classify the different sensors: namely, the mode of operation the sensor is used in, the measurement principle, and the discharging mechanism.

Mode of Operation

The mode of operation is the main category used to classify different sensors as it is directly connected to the application of the sensor. In general one can distinguish between two different modes of operation, the so-called accelerometer mode (AM) and the displacement mode (DM).

The DM concept uses a free-floating test mass. The displacement of this test mass relative to its housing is measured by the sensor and this signal is used to control the satellite to follow the test mass to drive the relative displacement to zero, thus minimizing the external disturbances on the test mass. The test mass will therefore follow a purely gravitational orbit. The DM concept is used most often in satellites where the drag-free sensor is the experiment itself, because of the very high sensitivity of the sensor in this mode. However the DM concept has the drawback that it requires complicated discharging mechanisms since a permanent grounding of the test mass via a gold wire is not possible. An example for a sensor that is used in displacement mode is the LISA Pathfinder sensor (see Fig. 3).

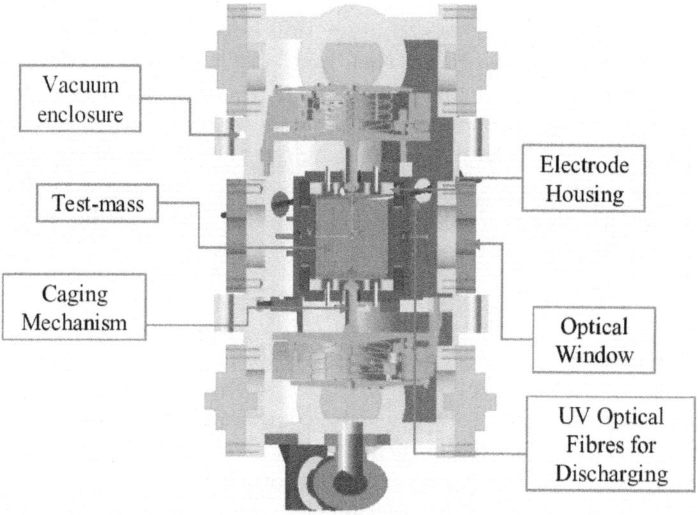

Fig. 3. LISA Pathfinder sensor.

The AM concept, on the other hand, uses the relative displacement measurements of the test mass w.r.t. the housing in an internal suspension control loop. This loop drives the displacement to zero by forcing the test mass to follow the satellite. The force that is needed to drive the displacement to zero is a measure for the acceleration on the satellite. This is the reason why the AM concept is most often used in missions that require highly sensitive accelerometers. The measurements of theses accelerometers are then fed back to the drag-free control system that uses the acceleration measurements to minimize the disturbances on the satellite thus providing a low disturbance environment for experiments onboard the spacecraft. Although less accurate the AM concept has the big advantage that permanent grounding of the test mass via a gold wire is possible.

Measurement Principle

The two most commonly used measurement principles that are used in drag-free sensors today are the electrostatic measurement principle and magnetic measurement principle.

Electrostatic Measurement Principle

In the electrostatic measurement principle, the relative displacement between the test mass and its housing is measured through a series of electrodes that are distributed around the test mass. An electrode and the opposing test mass area are forming a condenser. These condensers act as capacitive detectors. Two different methods have been proposed to measure relative displacement and attitude based on this setup, namely the gap-sensing and the slide-sensing method. Concerning the gap-sensing method, as the test mass moves relative to its housing the gap between the test mass and the electrode varies. This leads to a variation in the electric field which can be measured. Therefore the capacitive difference between the electrode and the test mass is a measure for the displacement of the test mass w.r.t. the electrode and thus the housing. If the slide-sensing method is applied, the test mass slides over the electrodes. The gap between the electrode and the test mass is constant but the overlapping area varies. The strength of the electric field depends on this overlapping area.

Magnetic Measurement Principle

A very interesting application of magnetic measurement principles is the superconducting quantum interference device (SQUID) that is used in the STEP sensor (see Fig. 4). SQUIDs combine a measurement sensitivity of as little as 10^{-15} m with the stability possible in a 2 K cryogenic environment using supercurrents.

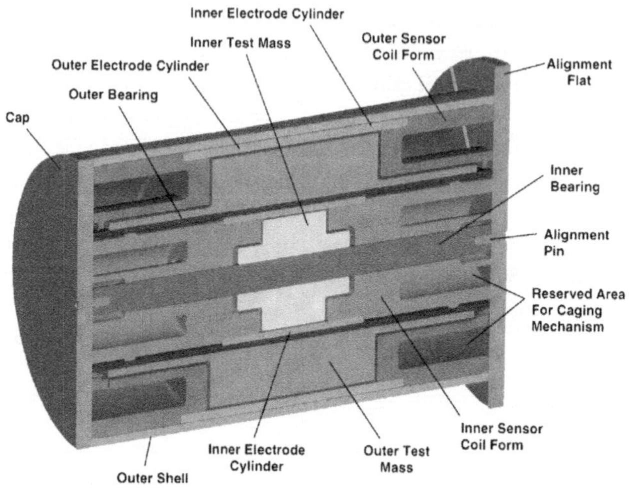

Fig. 4. STEP sensor.

Discharging Mechanism

The two most commonly used discharging methods include the ultraviolet-lamp (UV-lamp) and the gold wire. Which type of discharging mechanism has to be used is heavily influenced by the mode of operation that is used for the sensor. The most simple discharging method is the gold wire. Here the test mass is permanently grounded through the connection of the test mass with the housing via the gold wire. However the gold wire does limit the sensitivity of the sensor and cannot be used at all in case the test mass has to be free floating as in the displacement mode. This means that whenever the sensor shall be used in DM other means of discharging have to be applied. Unlike the gold wire the UV-lamp can be applied even if the test mass is free floating. The lamp emits ultraviolet photons which are used to release photoelectrons from the surface of the test mass. The freed electrons are then redistributed to neutralize the charge of the test mass. The UV-lamp does not limit the sensitivity of the sensor in the way the gold wire will but it is a very complex device that requires the charge on the test mass to be measured.

3.2 Actuators

As drag-free control systems are used to provide ultra-low disturbance environments, they usually require very low thrust levels. These low thrust demands combined with a demand for very small and accurate thrust steps gave rise to the development of new micropropulsion systems. These micropropulsion systems usually generate proportional thrust commands and operate in the thrust range from 0.1 to 100 μN. Different concepts and principles have

Fig. 5. Helium proportional thruster.

been proposed. The four most commonly used micropropulsion systems will be introduced in this section.

Helium Proportional Thrusters

The concept for Helium proportional thrusters has been developed for the Gravity Probe B mission and builds a synergy between the propulsion system and the temperature control of the cryogenic dewar. To stabilize the temperature inside the dewar, Helium is constantly vented. Instead of just venting the Helium the idea came up to use this Helium to produce the thrust for the control system. The Helium proportional thrusters produce a continuously variable thrust which is controlled by an internal control loop (Fig. 5).

Micropropulsion Cold Gas Thrusters

Micropropulsion cold gas systems are very similar to standard cold gas systems that are able to provide proportional thrust. They differ mainly in the thrust range. A very recent concept has been proposed by the Ångström Space Technology Centre at the University of Uppsala in Sweden. They have used microelectromechanical systems (MEMS) techniques to develop an all-in-one thruster that includes everything from propellant reservoir to control electronics (Fig. 6).

Field Emission Electric Propulsion

Field emission electric propulsion (FEEP) thrusters are ion thrusters that extract ions from a reservoir of liquid metal. The ions are accelerated in a strong electric field. Similar to the other propulsion concepts the FEEP thrusters are

Fig. 6. Micropropulsion cold gas thruster.

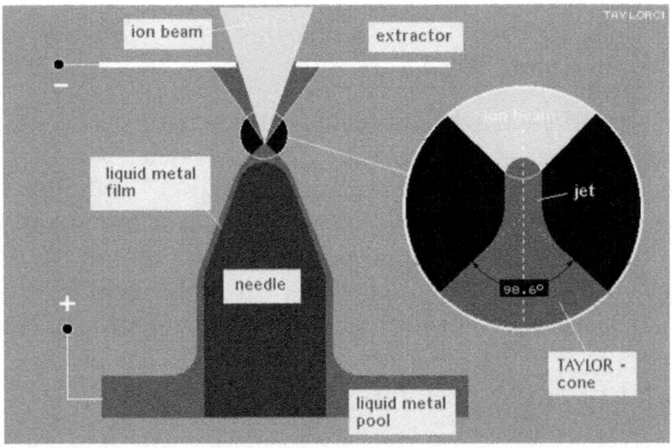

Fig. 7. FEEP concept (needle emitter).

able to produce a proportional thrust in the μN range. FEEP thrusters are capable of delivering very low thrust with very high accuracy and controllability. In addition fuel consumption is very low (Fig. 7).

Colloid Thrusters

The principle of the colloid thrusters is very similar to the FEEP concept. Colloid thrusters work by electrostatically accelerating a spray of charged, submicron diameter droplets of a conducting, nonmetallic liquid. They have a very small specific impulse and very low noise levels.

4 Missions

For the drag-free satellite several applications exist. First it can be used for geodesy. The higher harmonics of the Earth's gravitational potential perturb the orbit of a satellite. By observing the flight path of the satellite, the higher harmonics of the gravitational field can be determined. Another way for determining the gravitational potential is the measurement of the gravity gradient. The ESA mission GOCE (gravity field and steady-state ocean circulation explorer mission) has a payload consisting of two accelerometers each having a test mass which are measuring the gradient of the gravity field. The satellite is flown drag-free to reduce the disturbances and to reduce the dynamical range of the accelerometers [15].

A second application of drag-free satellites is aeronomy. Conventionally the density of the upper atmosphere is determined by observing the change in the period of a satellite's orbit. This depends on the averaging over one or more orbits. Instantaneous measurements at high frequency cannot be obtained using this method. The test mass onboard the satellite provides a signal for the drag-free control system. In operation, the control system can be used to measure instantaneously the force from the upper atmosphere as well as other forces (e.g., from radiation pressure). So every drag-free mission can obtain data for aeronomy (see [4]).

A further and most recent application is the utilization of drag-free satellites to provide a "real" zero-g or free-fall environment. This is used in a number of current and planned future missions. Especially experiments on fundamental physics are demanding a very low level of disturbances. The following missions will utilize a drag-free control system:

1. *Gravity Probe B (GP-B).* Test of relativistic effects on a gyroscope: the geodetic effect and the frame-dragging (or Lense–Thirring) effect
2. *Satellite Test of the Equivalence Principle (STEP).* Test of the weak equivalence principle
3. *MICROSCOPE.* Test of the weak equivalence principle at a lower level (precursor mission to STEP)
4. *Laser Interferometer Space Antenna (LISA).* Detection of gravitational waves
5. *HYPER.* Spatial mapping of the Lense–Thirring effect using atomic interferometers

All of them are going beyond the initial idea of one test mass shielded by the satellite. So in case of STEP, eight test masses are arranged in four differential accelerometers. Onboard the LISA spacecraft the baseline design uses two test masses. In addition to that the attitude of the test masses w.r.t. the LISA satellite is important for this experiment because the test masses are acting as mirrors for the laser beams between the satellites of the LISA constellation. It can be seen that the drag-free technology is becoming more and more important for scientific satellite missions which need a very low disturbance environment.

In the following sections, a few of the missions named above are described with a focus on the drag-free control system.

4.1 Gravity Probe B

Gravity Probe B is the relativity gyroscope experiment developed by NASA and Stanford University to test two extraordinary, unverified predictions of Albert Einstein's general theory of relativity.

The experiment will measure, very precisely, tiny changes in the direction of spin of four gyroscopes contained in an Earth satellite orbiting at 640 km altitude directly over the poles. The quality of the disturbance environment provided by the Gravity Probe B spacecraft will enable the gyroscopes to provide an almost perfect space–time reference system. They will measure how space and time are warped by the presence of the Earth, and, more profoundly, how the Earth's rotation drags space–time around with it. These effects, though small for the Earth, have far-reaching implications for the nature of matter and the structure of the Universe.

The GP-B satellite was launched on 20 April 2004 from Vandenberg Air Force Base, CA, USA. After the in-orbit checkout phase, it will start its experimental measurement phase where the ATC system maintains the residual acceleration on the satellite below a level of $10^{-12} \times g$. The ATC (= drag-free control system) uses the suspended gyroscopes (spherical test mass) as the drag-free sensor. This measurement is obtained using the magnetic measurement principle based on a SQUID. To get the superconductivity for the SQUID and to reduce thermal noise, the experiment is contained in a liquid helium filled dewar. The boil-off gas from the dewar is used for ATC of the spacecraft. The Helium proportional thrusters which were especially developed for Gravity Probe B are used as the actuators to provide a six degrees of freedom control.

4.2 LISA and LISA Pathfinder

The LISA is a joint mission with NASA. It is a three-spacecraft mission, designed to detect the gravitational waves in space given out when very massive objects undergo strong acceleration. LISA will be the first mission to try and detect them from space. To achieve that goal, the relative position of several solid blocks placed in different spacecraft, 5 million kilometers apart, will have to be constantly monitored with high accuracy using laser-based techniques. A gravitational wave passing through the spacecraft will change the separations between them, thereby revealing itself. The existence of gravitational waves follows from Einstein's theory of general relativity. When a massive body is accelerated, or its motion is disturbed, it should "radiate."

This radiation takes the form of gravitational waves, a kind of feeble emission that should affect any type of matter. In particular, a solid body should vibrate if a gravitational wave hits it. No technique yet exists to detect the resulting vibrations. However, using laser interferometry, it is possible to monitor how the distance between solid bodies varies when a gravitational wave passes by.

Each of the three LISA spacecraft will carry two telescopes with associated lasers and optical systems. Pointing in directions separated by 60°, the telescopes in each spacecraft will communicate with the other two spacecraft, located at the other two corners of an equal-sided triangle. Apart from the complexity of aiming the laser beams from one small spacecraft to another across 5 million kilometers of space, LISA has to deal with other forces besides gravitational waves which will alter the separation of the spacecraft, e.g., the solar pressure.

The spacecraft must sense the extraneous forces and counteract them. The central part of each optical system will be a cube with a side length of 4 cm, made from a gold–platinum alloy. This test mass will float freely in most of its degrees of freedom. Acting as a reflector for the laser beams, the cube will provide the benchmark for measuring the distance between spacecraft.

The forces and torques acting on the satellite have to be canceled out by the drag-free control system. The position and attitude of the two test masses onboard each spacecraft are measured using an electrostatic measurement principle. The same principle is used to suspend the masses fully or in selected degrees of freedom. The thruster system is planned to be made up of FEEP thrusters. They will provide tiny control forces and torques which are needed to control the spacecraft in the required accuracy. The residual acceleration on the test masses shall be below $10^{-16}\,\mathrm{m\,s^{-2}}$ in the bandwidth between 10^{-4} and $10^{-1}\,\mathrm{Hz}$ [5,6,12].

Currently ESA and NASA are developing and building the technology demonstrator LISA Pathfinder. Launch is scheduled in 2008. LISA Pathfinder shall demonstrate and test new technologies developed for LISA. The drag-free control system including sensors and actuators is among these new technologies to be tested.

4.3 STEP

The STEP is a joint European–U.S. space project to investigate one of the most fundamental principles in physics, the equivalence of inertial and passive gravitational mass. STEP will advance the sensitivity of the equivalence principle tests by six orders of magnitude, into regions where the principle may break down. A violation of equivalence at any level would have significant consequences for modern gravitational theory.

The STEP experiment is conceptually a modern version of Galileo Galilei's free-fall experiment, in which he is said to have dropped two weights from the

Leaning Tower of Pisa to demonstrate that they fall at the same rate. Any difference in the ratio of inertial to passive gravitational mass of the weights results in a difference in the rate of fall. In STEP, the masses are in free fall in an orbit around the Earth and if there is a violation of the equivalence principle they tend to follow slightly different orbits.

The STEP satellite will carry four of these differential accelerometers to test a range of different materials and of course for redundancy. The spacecraft will have a nearly circular orbit at an altitude of about 550 km. For thermal stability eclipses have to be avoided. Therefore a Sun-synchronous dusk-dawn orbit is chosen which will prevent the spacecraft from passing through the Earth's shadow during its 6-month lifetime.

Due to the low altitude of STEP's orbit, the interaction with the Earth's atmosphere is the main disturbance for the experiment. Unfortunately it occurs at the same frequency as the science signal. This disturbance has to be reduced to a level of $10^{-14}\,\mathrm{m\,s^{-2}}$ in the bandwidth of $10^{-6}\,\mathrm{Hz}$ around the measurement signal [13].

STEP is mainly developed at the W.W. Hansen Experimental Physics Laboratory of Stanford University. It inherits a lot of technologies from the recently launched Gravity Probe B mission. For example the test mass position is measured applying the magnetic measurement principle using SQUIDs. The experiment is also carried out in a cryogenically cooled environment to reduce the thermal noise and to enable the superconductivity needed for the magnetic suspension and the SQUIDs. The boil-off from the Helium is again used for the microthrust propulsion needed for the drag-free control system.

4.4 GOCE

The GOCE is dedicated to measuring the Earth's gravity field and modeling the geoid with extremely high accuracy and spatial resolution. It is scheduled for launch in 2006.

From its mission objective, GOCE does not need to have a free-fall environment. Nevertheless it carries a very sensitive gradiometer onboard which is sensitive to the very small gravity gradient along the spacecraft. To allow the measurement of this tiny gradient, the disturbances have to be reduced below the level of $2.5 \cdot 10^{-8}\,\mathrm{m\,s^{-2}\,Hz^{-1/2}}$ in the measurement bandwidth between 5 and 100 mHz [1, 2]. For that reason a drag-free control system is applied to cancel out the disturbances.

This control system has a second effect on the orbit of the satellite. Since it is now following a purely gravitational orbit there is no orbit decay due to the interaction with the upper atmosphere. This allows to have the GOCE satellite in a very low orbit (250 km) for a mission time of 2 years.

4.5 MICROSCOPE

MICROSCOPE (Microsatellite a trainée Compensée pour l'Observation du Principe d'Equivalence) is a CNES/ESA collaborative mission to test the equivalence principle (EP) in space to a precision of one part in 10^{15} (10^{18} for STEP, see Sect. 4.3). Even with the simplest experiment in space, the precision of the test can be improved by 2–3 orders of magnitude over the best ground-based and lunar laser-ranging tests.

The MICROSCOPE payload comprises two differential electrostatic accelerometers, one testing a pair of materials of equal composition (platinum–platinum), to provide an upper limit for systematic errors, the other testing a pair of materials of different composition (platinum–titanium) as the EP test proper. As on STEP, the test masses in the MICROSCOPE payload are concentric hollow cylinders. Unlike STEP, the problem of test mass charging is eliminated by a thin gold grounding wire.

To separate the signal frequency from error sources, the spacecraft will spin at a frequency around 10^{-3} Hz. The three-axis 120 kg MICROSCOPE satellite will be launched in 2007 by a Dnepr rocket (to be confirmed) into a Sun-synchronous, quasicircular (eccentricity 10^{-2}) orbit at 700 km altitude. The drag from the residual atmosphere and solar radiation pressure will be compensated for by a system of proportional FEEP thrusters. A total of 8–12 thrusters, each with a thrust authority of 150 μN, will be employed. Their noise level must not exceed $0.1\,\mu\mathrm{N}\,\mathrm{Hz}^{-1/2}$ to provide the required drag-free control performance of $3 \cdot 10^{-10}\,\mathrm{m\,s}^{-2}\,\mathrm{Hz}^{-1/2}$ in the measurement bandwidth [16]. The FEEP thrusters also serve as actuators for fine attitude control.

4.6 Further Applications

In the missions described in the sections before, the drag-free control system is closely connected to the experiment. In most of the missions the test masses for the drag-free control system are the central part of the experiment. Besides this configuration a more decoupled application can be imagined.

First, the drag-free control system can be used to generate a high-quality microgravity environment for experiments. These experiments are not connected to the drag-free control system. The DFC simply serves as a very accurate control system of the spacecraft bus. Thus the capabilities of a satellite are extended by adding the drag-free control system.

Secondly, the drag-free control system can be used to decouple very sensitive subsystems from external disturbances. If we consider a highly sensitive experiment in a box as the test mass, it can be shielded by the surrounding satellite. Then the experiment box is free from all external disturbances.

This concept was already studied for the Hubble Space Telescope successor the James Webb Space Telescope (or Next Generation Space Telescope – NGST; see [9, 10]). This concept offers the opportunity for high-accuracy pointing even in orbits with large disturbances.

5 Summary

The drag-free control technology has become an enabling technology for current and future space missions in the area of fundamental physics, geodesy, and also microgravity research. Although the idea of the drag-free controlled satellite is already 40 years old [8], it has taken the time to develop the needed sensor and actuator technologies for applying the idea of the drag-free satellite.

Today the drag-free technology is mainly used for fundamental physics but future space missions in other areas may utilize this technology and its variations.

References

1. E. Canuto, P. Martella, and G. Sechi. The GOCE Drag Free and Attitude Control Design Aspects and Expected Performance. In *Proceedings of the 5th International Conference on Spacecraft Guidance, Navigation and Control Systems; Frascati, Italy*, October 2002.
2. Giuseppe Catastini. The GOCE End-To-End System Simulator. In *2nd International GOCE User Workshop, ESA ESRIN*, March 2004.
3. Triad I. A Satellite Freed of all but Gravitational Forces. *Journal of Spacecraft*, 11(9), 1974.
4. Yusuf R. Jafry. *Aeronomy Coexperiments on Drag-Free Satellites with Proportional Thrusters: GP-B and STEP.* PhD thesis, Department of Aeronautics and Astronautics of Stanford University, March 1992.
5. M. Kersten and A. Schleicher. Integrated Modeling of the Laser Interferometer Space Antenna (LISA). In *Proceedings of the International Symposium on Formation Flying, Toulouse, France*, October 2002.
6. H. Klotz, H. Strauch, W. Wolfsberger, S. Marcuccio, and C. Speake. Drag-Free, Attitude and Orbit Control for LISA. In *Third International Conference on Spacecraft Guidance, Navigation and Control Systems; ESTEC, Noordwijk, Netherlands*, volume ESA SP-381. ESA, November 1996.
7. K. Kurmakaev. On the Gravitational Effects on the Motion of Extended Rigid Bodies. Master's thesis, Faculty of Applied Mathematics and Economics of the Moscow Institute of Physics and Technology, June 2003. In Russian.
8. Benjamin Lange. *The Control and Use of Drag-Free Satellites.* PhD thesis, Department of Aeronautics and Astronautics of Stanford University, June 1964.
9. N. Pedreiro. Disturbance-free Payload Concept Demonstration. In *AIAA/AAS Astrodynamics Specialist Conference, Monterey, California*, August 2002.
10. N. Pedreiro. Spacecraft Architecture for Disturbance-free Payload. In *AIAA/ AAS Astrodynamics Specialist Conference, Monterey, California*, August 2002.
11. J. Courtney Ray. *Partially Drag-Free Satellites with Application to the TIP II Satellite.* PhD thesis, Department of Aeronautics and Astronautics of Stanford University, 1976.
12. X. Sembely, L.Vaillon, and O. Vandermarq. High Accuracy Drag-Free Control for The Microscope and The Lisa Missions. In *Proceedings of the 5th International Conference on Spacecraft Guidance, Navigation and Control Systems; Frascati, Italy*, October 2002.

13. STEP. Phase A Interim Report. Technical report, NASA/ESA Joint STEP Working Group, 2001.
14. Stephan Theil. *Satellite and Test Mass Dynamics Modeling and Observation for Drag-free Satellite Control of the STEP Mission.* PhD thesis, Department of Production Engineering, University of Bremen, December 2002.
15. L. Vaillon and C. Champetier. Drag-Free & Attitude Control System for the Gravity Explorer Mission. In *Third International Conference on Spacecraft Guidance, Navigation and Control Systems; ESTEC, Noordwijk, Netherlands*, volume ESA SP-381. ESA, November 1996.
16. A. Wilson, editor. *ESA's report to the 34th COSPAR meeting*, volume 1259 of *ESA SP.* European Space Agency, Nordwijk, 2002.

Drag-Free Control Design with Cubic Test Masses

Walter Fichter[1], Alexander Schleicher[2], and Stefano Vitale[3]

[1] Universität Stuttgart, Institute of Flight Mechanics and Control, 70569 Stuttgart, Germany
[2] EADS Astrium GmbH, 88039 Friedrichshafen, Germany
[3] University of Trento, Trento, Italy

Summary. Drag-free control is an important technology required for scientific experiments in space that need free-fall conditions. This chapter describes a control design of a drag-free system that uses test masses with cubic shape (rather than spherical or cylindrical). Three interconnected control problems are considered: drag-free control of the test masses, suspension control of the test masses, and attitude control of the spacecraft. The case of two test masses is treated here rather generally, such that an application to more than two test masses or a reduction to a single test mass is straightforward. Both the derivation of the control structure and the performance optimization procedure of the feedback loops are described. It is shown that the proposed control design yields a very simple architecture of the onboard software for drag-free control and furthermore that it leads to an extreme operational flexibility for the experimentalist with respect to redefinition of control modes and performance optimization, on ground and in-flight.

1 Introduction

1.1 Free Fall and Control

Satellites or test masses in free fall open up many opportunities for missions in the areas of fundamental physics and geodesy.[1] Free fall of a test mass (TM) means that its motion is determined by gravity only, i.e., all nongravitational forces, including self-gravity from the spacecraft, is below a specified limit. Free-fall requirements are specified as acceleration noise spectral densities in $\mathrm{m\,s^{-2}\,Hz^{-1/2}}$ along a specific axis (the "sensitive axis" of the experiment), together with a frequency range, the measurement bandwidth.

The most important driver of the free-fall performance is the instrument itself. It must be designed in such a way that any nongravitational forces on the test mass are below the requirement specifications. In practice instrument design is imperfect: there will be residual forces, in the "best case" at

[1] Many examples are described in this volume.

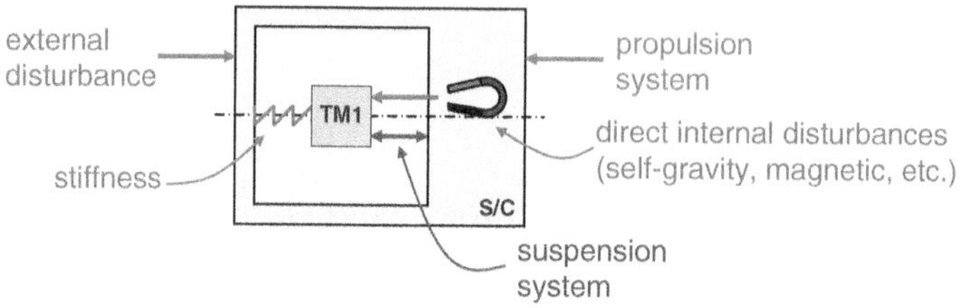

Fig. 1. Single test mass (*TM*) with disturbance forces, stiffness, and actuation systems.

zero frequency, and spatial gradients of these forces which create stiffness coupling between spacecraft and test mass [4, 5]. These effects, together with any external disturbance force on the spacecraft, require active stabilization of the test mass(es) with respect to the spacecraft, i.e., the relative motion of test mass with respect to the spacecraft must be controlled. In Fig. 1 the disturbance effects, i.e., external force, internal force, and stiffness, are shown schematically.

The relative motion between test mass and spacecraft can be controlled with two different actuation systems, according to two control principles (shown also in Fig. 1). The first principle is drag-free control, where the spacecraft follows the test mass by actuation of a propulsion system. In this case any imperfect control yields a nongravitational acceleration of the form "stiffness × displacement control error." The second principle is suspension control, where the test mass follows the spacecraft by actuation of a suspension system. Here, the residual nongravitational acceleration caused by control is of the form "stiffness × displacement + suspension control force." From an acceleration point of view, drag-free control is always the preferred option since any nongravitational forces due to suspension actuation are eliminated by definition. However, drag-free control is limited to six degrees of freedom (DoF) that represent the rigid body motion of the spacecraft which follows the test mass(es). This means that suspension control has to be applied to some DoF whenever there are two test masses or more.

In addition to the control of the test mass motion, attitude control of the spacecraft must ensure proper orientation, usually with respect to a Sun-pointing reference frame. Whenever the spacecraft's attitude changes, the test masses must follow that motion. This means there must be suspension actuation commands in at least three DoF of the test masses.

In general, the contribution to the sensitive axis acceleration of any relative coordinate between test mass and spacecraft can be expressed as

$$a_x = \sum_i \omega_{xi}^2 q_i + b_{xi} u_i \, . \tag{1}$$

Here a_x denotes the acceleration along the sensitive axis x, i.e., the axis along free fall shall be obtained. The ω_{xi}^2 are stiffness coupling coefficients from any axis i into the sensitive axis x. The b_{xi} are suspension actuation coupling coefficients from axis i into the sensitive axis x. Nominally it is $b_{xx} = 1$ and all other $b_{xi} = 0$, $i \neq x$, are zero. In practice, all of the b_{xi}, $i \neq x$, are nonzero due to imperfections.

Prior to the control design, a breakdown of the sensitive axis acceleration must be performed. This consists of a definition of the drag-free coordinates and suspension coordinates, which also determines the coefficients ω_{xi} and b_{xi}. Then for each term in (1), an acceleration contribution is allocated, which yields performance requirements (noise spectral densities) for the control errors q_i and the actuation commands u_i.

1.2 System Block Diagram

A schematic block diagram of a drag-free control system with two test masses, including measurement and actuation hardware, is shown in Fig. 2. For the measurement of the test mass motion a metrology system (electrostatic, optical, or a combination of both) provides information about all test mass displacements and attitudes. The inertial spacecraft attitude is measured with a star sensor. Therefore, 15 DoF can be measured or derived from the sensor raw data. Note that three DoF of spacecraft translation are a result of the drag-free motion and cannot be measured.

Actuation of the spacecraft motion is performed with a micropropulsion system that provides actuation authority along all six DoF. Furthermore each test mass can be actuated along six DoF with an electrostatic suspension system. In total, all 18 rigid body DoF of the spacecraft and the two test masses can be actuated.

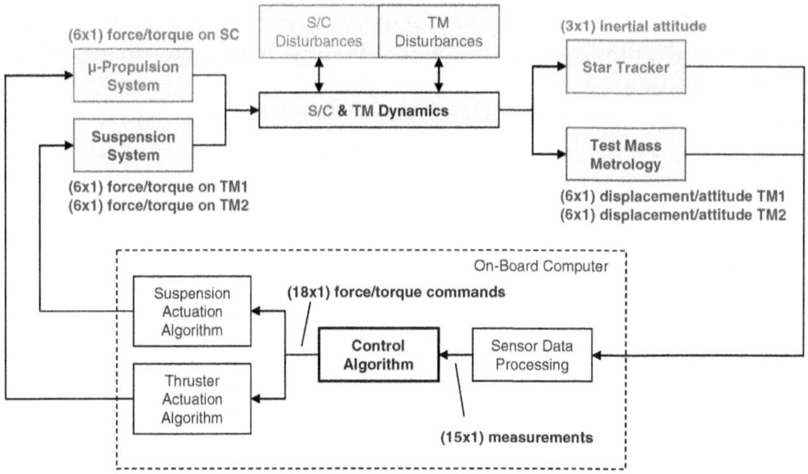

Fig. 2. System block diagram.

The onboard algorithms consist of three main units (see Fig. 2). In the sensor data processing unit displacement and rotation information of each test mass is derived from raw data. The suspension and thruster actuation algorithms convert force and torque requests from the controller into raw data that is sent to the actuator hardware, see, e.g., [1–3].[2] The remainder of this chapter is about the third set of algorithms, the control algorithms. They determine the closed-loop behavior, and thus, the acceleration contributions.

1.3 Problem Statement

The problem treated in the following is the control algorithm design for a spacecraft with two test masses.[3] Thus, in total there are 15 DoF to be controlled, which can be subdivided into three interconnected control problems:

1. Spacecraft attitude pointing (three DoF), with respect to a reference frame, typically a Sun-pointing frame. The requirement is given as a 3σ attitude control error. The 3×1 attitude is denoted by $\boldsymbol{\varphi}$.
2. Test mass drag-free control to stabilize six preselected test mass DoF (drag-free coordinates), denoted by the 6-vector $\boldsymbol{q}_{DF}$.
3. Test mass suspension control to stabilize the remaining test mass DoF, denoted by the 6-vector $\boldsymbol{q}_{SUS}$.

The control design shall take into account a predefined but "generic" set of drag-free and suspension coordinates, $\boldsymbol{q}_{DF}$ and $\boldsymbol{q}_{SUS}$. "Generic" means that any meaningful linear combination of the test mass coordinates in Cartesian coordinates shall be considered. The following performance requirements shall be considered, as a result of an acceleration breakdown procedure outlined above:

- Spacecraft attitude control error requirement given as a 3σ value for each axis. The requirements are originated mainly by thermal conditioning and communication issues.
- Requirements for the test mass control errors q_i and actuation commands u_i given as spectral densities in the measurement bandwidth. These requirements are a result of the acceleration breakdown explained above.

It shall be assumed that all 15 control variables can be measured in m and rad, and all 18 rigid body DoF can be actuated by means of appropriate algorithms for the micropropulsion and electrostatic suspension actuation system.

---

[2] Thruster actuation is a linear programming problem. For electrostatic actuation it is nonlinear, since force is proportional to voltage squared.

[3] The generalization to more than two test masses is straightforward and the single test mass case can easily be obtained by simplification.

2 Design Model

2.1 Linear Equations of Motion

Formal derivation of the linearized dynamics of a rigid body spacecraft and two test masses as it is shown schematically in Fig. 3 yields

$$\begin{pmatrix} \ddot{\boldsymbol{\varphi}} \\ \ddot{\boldsymbol{r}}_1 \\ \ddot{\boldsymbol{\varphi}}_1 \\ \ddot{\boldsymbol{r}}_2 \\ \ddot{\boldsymbol{\varphi}}_2 \end{pmatrix} = \begin{bmatrix} \mathbf{0} & \boldsymbol{E} & \mathbf{0} & \mathbf{0} & \mathbf{0} & \mathbf{0} \\ -\boldsymbol{T}_{1B} & \tilde{\boldsymbol{r}}_{01} & \boldsymbol{E} & \mathbf{0} & \mathbf{0} & \mathbf{0} \\ \mathbf{0} & -\boldsymbol{T}_{1B} & \mathbf{0} & \boldsymbol{E} & \mathbf{0} & \mathbf{0} \\ -\boldsymbol{T}_{2B} & \tilde{\boldsymbol{r}}_{02} & \mathbf{0} & \mathbf{0} & \boldsymbol{E} & \mathbf{0} \\ \mathbf{0} & -\boldsymbol{T}_{2B} & \mathbf{0} & \mathbf{0} & \mathbf{0} & \boldsymbol{E} \end{bmatrix} \cdot \begin{pmatrix} \boldsymbol{a} \\ \boldsymbol{\alpha} \\ \boldsymbol{a}_1 \\ \boldsymbol{\alpha}_1 \\ \boldsymbol{a}_2 \\ \boldsymbol{\alpha}_2 \end{pmatrix} \qquad (2)$$

This can be verified by illustration. The 3-vectors $\boldsymbol{r}_i$, $\boldsymbol{\varphi}_i$ represent the test mass displacement and rotation; the index i stands for test mass 1 and 2. The 3-vector $\boldsymbol{\varphi}$ denotes the spacecraft attitude with respect to the reference frame; $\boldsymbol{E}$ is the unit matrix with appropriate dimension; $\boldsymbol{T}_{iB}$ is the transformation matrix from the spacecraft (body) to the test mass i in nominal position; and the vector $\boldsymbol{r}_{0i}$ denotes the vector from the spacecraft CoM to the test mass i in nominal position and any vector with a $\tilde{*}$ represents a skew-symmetric crossproduct matrix, i.e.,

$$\tilde{\boldsymbol{r}}_{0i} = \begin{bmatrix} 0 & -r_{0i,z} & r_{0i,y} \\ r_{0i,z} & 0 & -r_{0i,x} \\ -r_{0i,y} & r_{0i,x} & 0 \end{bmatrix} \qquad (3)$$

The accelerations $\boldsymbol{a}$, $\boldsymbol{\alpha}$, $\boldsymbol{a}_1$, $\boldsymbol{\alpha}_1$, $\boldsymbol{a}_2$, and $\boldsymbol{\alpha}_2$ include all applied forces per unit mass and torques per unit inertia on the spacecraft and test masses, i.e., actuation signals as well as reaction forces and torque. In particular, they also include stiffness contributions. The inertial position of the spacecraft is

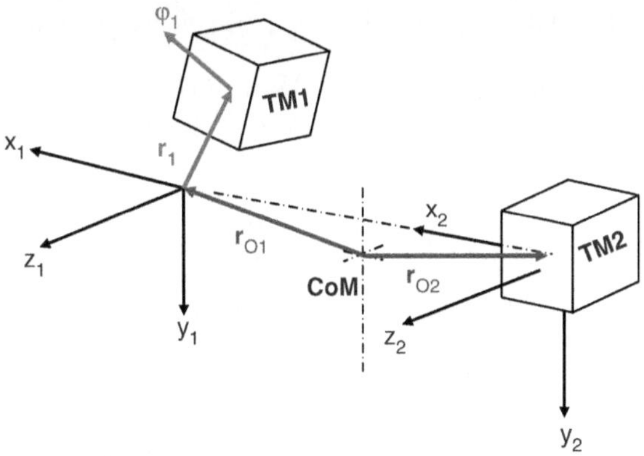

Fig. 3. Test mass nominal positions and test mass coordinates.

eliminated from the above equation since it has no coupling with all other coordinates. The equations of motion can be written in more compact notation as

$$\begin{pmatrix} \ddot{\boldsymbol{\varphi}} \\ \ddot{\boldsymbol{q}}_{TM} \end{pmatrix} = \begin{bmatrix} \boldsymbol{B}_{ATT} & \boldsymbol{0} \\ \boldsymbol{B}_1 & \boldsymbol{E} \end{bmatrix} \cdot \begin{pmatrix} \boldsymbol{a}_{SC} \\ \boldsymbol{a}_{TM} \end{pmatrix} \tag{4}$$

with the following definitions

$$\boldsymbol{B}_{ATT} = \begin{bmatrix} \boldsymbol{0} \; \boldsymbol{E} \end{bmatrix}, \qquad \boldsymbol{B}_1 = \begin{bmatrix} -\boldsymbol{T}_{1B} & \tilde{\boldsymbol{r}}_{01} \\ \boldsymbol{0} & -\boldsymbol{T}_{1B} \\ -\boldsymbol{T}_{2B} & \tilde{\boldsymbol{r}}_{02} \\ \boldsymbol{0} & -\boldsymbol{T}_{2B} \end{bmatrix} \tag{5}$$

and

$$\ddot{\boldsymbol{q}}_{TM} = \left(\ddot{\boldsymbol{r}}_1 \, \ddot{\boldsymbol{\varphi}}_1 \, \ddot{\boldsymbol{r}}_2 \, \ddot{\boldsymbol{\varphi}}_2\right)^T, \quad \boldsymbol{a}_{SC} = \left(\boldsymbol{a} \, \boldsymbol{\alpha}\right)^T, \quad \boldsymbol{a}_{TM} = \left(\boldsymbol{a}_1 \, \boldsymbol{\alpha}_1 \, \boldsymbol{a}_2 \, \boldsymbol{\alpha}_2\right)^T \tag{6}$$

2.2 Control Coordinates

The equations of motion are now transformed into control coordinates, i.e., the test mass coordinates are split up into six coordinates that shall be drag-free controlled and remaining six coordinates that shall be suspension controlled. The transformation is defined by the matrices $\boldsymbol{D}_{DF}$ and $\boldsymbol{D}_{SUS}$

$$\begin{pmatrix} \boldsymbol{q}_{DF} \\ \boldsymbol{q}_{SUS} \end{pmatrix} = \begin{bmatrix} \boldsymbol{D}_{DF} \\ \boldsymbol{D}_{SUS} \end{bmatrix} \cdot \boldsymbol{q}_{TM} \tag{7}$$

A simple example for these matrices is

$$\boldsymbol{D}_{DF} = \begin{bmatrix} 1&0&0&0&0&0&0&0&0&0&0&0 \\ 0&1&0&0&0&0&0&0&0&0&0&0 \\ 0&0&1&0&0&0&0&0&0&0&0&0 \\ 0&0&0&1&0&0&0&0&0&0&0&0 \\ 0&0&0&0&0&0&0&1&0&0&0&0 \\ 0&0&0&0&0&0&0&0&1&0&0&0 \end{bmatrix}, \quad \boldsymbol{D}_{SUS} = \begin{bmatrix} 0&0&0&0&1&0&0&0&0&0&0&0 \\ 0&0&0&0&0&1&0&0&0&0&0&0 \\ 0&0&0&0&0&0&1&0&0&0&0&0 \\ 0&0&0&0&0&0&0&0&0&1&0&0 \\ 0&0&0&0&0&0&0&0&0&0&1&0 \\ 0&0&0&0&0&0&0&0&0&0&0&1 \end{bmatrix} \tag{8}$$

Their selection must be such that the resulting drag-free controlled coordinates form a linear-independent set of coordinates. In control coordinates the dynamics equation becomes

$$\begin{pmatrix} \ddot{\boldsymbol{\varphi}} \\ \ddot{\boldsymbol{q}}_{DF} \\ \ddot{\boldsymbol{q}}_{SUS} \end{pmatrix} = \begin{bmatrix} \boldsymbol{B}_{ATT} & \boldsymbol{0} & \boldsymbol{0} \\ \boldsymbol{B}_{DF} & \boldsymbol{E} & \boldsymbol{0} \\ \boldsymbol{B}_{SUS} & \boldsymbol{0} & \boldsymbol{E} \end{bmatrix} \begin{pmatrix} a_{SC} \\ a_{TM} \end{pmatrix} \tag{9}$$

with

$$\boldsymbol{B}_{DF} = \boldsymbol{D}_{DF} \cdot \boldsymbol{B}_1, \quad \boldsymbol{B}_{SUS} = \boldsymbol{D}_{SUS} \cdot \boldsymbol{B}_1 \tag{10}$$

The applied forces $\boldsymbol{a}_{SC}$ and $\boldsymbol{a}_{TM}$ are now further subdivided into actuation, disturbance, and stiffness contributions, denoted by the variables $\boldsymbol{u}$,

$\boldsymbol{d}$, and $\boldsymbol{\Omega}^2 \cdot \boldsymbol{q}$, with appropriate indices. This yields the final model used for design purposes

$$\begin{pmatrix} \ddot{\varphi} \\ \ddot{q}_{DF} \\ \ddot{q}_{SUS} \end{pmatrix} = \begin{bmatrix} \boldsymbol{B}_{ATT} & \mathbf{0} & \mathbf{0} \\ \boldsymbol{B}_{DF} & \boldsymbol{E} & \mathbf{0} \\ \boldsymbol{B}_{SUS} & \mathbf{0} & \boldsymbol{E} \end{bmatrix} \left(\begin{pmatrix} \boldsymbol{u}_{THR} \\ \boldsymbol{u}_{ES1} \\ \boldsymbol{u}_{ES2} \end{pmatrix} + \begin{pmatrix} \boldsymbol{d}_{SC} \\ \boldsymbol{d}_{TM1} \\ \boldsymbol{d}_{TM2} \end{pmatrix} \right) + \begin{bmatrix} \mathbf{0} & \mathbf{0} & \mathbf{0} \\ \mathbf{0} & -\boldsymbol{\Omega}^2_{DF} & \mathbf{0} \\ \mathbf{0} & \mathbf{0} & -\boldsymbol{\Omega}^2_{SUS} \end{bmatrix} \begin{pmatrix} \varphi \\ q_{DF} \\ q_{SUS} \end{pmatrix} \tag{11}$$

where $\boldsymbol{u}_{THR}$ are the forces per unit mass generated by the micropropulsion system whereas $\boldsymbol{u}_{ES1}$ and $\boldsymbol{u}_{ES2}$ are forces per unit mass from the electrostatic actuation system along drag-free and suspension coordinates; the $\boldsymbol{d}_{SC}$, $\boldsymbol{d}_{TM1}$, and $\boldsymbol{d}_{TM2}$ are the corresponding external disturbance forces per unit mass; and the $\boldsymbol{\Omega}^2_i \cdot \boldsymbol{q}_i$ are stiffness forces along drag-free and suspension coordinates. For design purposes any stiffness crosscoupling is negligible, i.e., the stiffness matrices $\boldsymbol{\Omega}^2_{DF}$ and $\boldsymbol{\Omega}^2_{SUS}$ are considered to be diagonal.

3 Control Structure Design

3.1 Feedback Interconnection

First, new control signals (actuation commands) are introduced that lead to input decoupling of the drag-free coordinates:

$$\boldsymbol{u}_T = \boldsymbol{B}_{DF} \cdot \boldsymbol{u}_{THR} \iff \boldsymbol{u}_{THR} = \boldsymbol{B}_{DF}^{-1} \cdot \boldsymbol{u}_T \tag{12}$$

The model for further design purposes is then given as

$$\begin{pmatrix} \ddot{\varphi} \\ \ddot{q}_{DF} \\ \ddot{q}_{SUS} \end{pmatrix} = \begin{bmatrix} \boldsymbol{B}_{ATT} \cdot \boldsymbol{B}_{DF}^{-1} & \mathbf{0} & \mathbf{0} \\ \boldsymbol{E} & \boldsymbol{E} & \mathbf{0} \\ \boldsymbol{B}_{SUS} \cdot \boldsymbol{B}_{DF}^{-1} & \mathbf{0} & \boldsymbol{E} \end{bmatrix} \left(\begin{pmatrix} \boldsymbol{u}_{T} \\ \boldsymbol{u}_{ES1} \\ \boldsymbol{u}_{ES2} \end{pmatrix} + \begin{pmatrix} \boldsymbol{d}_{T} \\ \boldsymbol{d}_{TM1} \\ \boldsymbol{d}_{TM2} \end{pmatrix} \right) + \begin{bmatrix} \mathbf{0} & \mathbf{0} & \mathbf{0} \\ \mathbf{0} & -\boldsymbol{\Omega}^2_{DF} & \mathbf{0} \\ \mathbf{0} & \mathbf{0} & -\boldsymbol{\Omega}^2_{SUS} \end{bmatrix} \begin{pmatrix} \varphi \\ q_{DF} \\ q_{SUS} \end{pmatrix} \tag{13}$$

with a unit matrix at the block entry (2, 1) of the input matrix and where it is $\boldsymbol{d}_T = \boldsymbol{B}_{DF} \cdot \boldsymbol{d}_{SC}$. Laplace transformation yields the following system

$$\begin{pmatrix} \varphi \\ q_{DF} \\ q_{SUS} \end{pmatrix} = \begin{bmatrix} \boldsymbol{G}_{SC} \cdot \boldsymbol{B}_{ATT} \cdot \boldsymbol{B}_{DF}^{-1} & \mathbf{0} & \mathbf{0} \\ \boldsymbol{G}_{DF} & \boldsymbol{G}_{DF} & \mathbf{0} \\ \boldsymbol{G}_{SC} \cdot \boldsymbol{B}_{SUS} \cdot \boldsymbol{B}_{DF}^{-1} & \mathbf{0} & \boldsymbol{G}_{SUS} \end{bmatrix} \left(\begin{pmatrix} \boldsymbol{u}_{T} \\ \boldsymbol{u}_{ES1} \\ \boldsymbol{u}_{ES2} \end{pmatrix} + \begin{pmatrix} \boldsymbol{d}_{T} \\ \boldsymbol{d}_{TM1} \\ \boldsymbol{d}_{TM2} \end{pmatrix} \right) \tag{14}$$

with

$$\boldsymbol{G}_{SC} = \frac{1}{s^2} \cdot \boldsymbol{E}, \quad \boldsymbol{G}_{DF} = \begin{bmatrix} \frac{1}{s^2+\omega_{DFi}^2} & \\ & \ddots \end{bmatrix}, \quad \boldsymbol{G}_{SUS} = \begin{bmatrix} \frac{1}{s^2+\omega_{SUSi}^2} & \\ & \ddots \end{bmatrix} \tag{15}$$

This system has the following properties:

- First, the dynamics of the drag-free coordinates $\boldsymbol{q}_{DF}$ consists of six decoupled undamped or instable second-order systems, assuming parasitic off-diagonal stiffness to be negligible for design (not analysis) purposes. Each coordinate has two inputs, one with thrusters and one with electrostatic suspension.
- Second, there are crosscouplings from the drag-free control inputs to the attitudes due to the matrix $\boldsymbol{B}_{ATT} \cdot \boldsymbol{B}_{DF}^{-1}$. This means that more than one thruster actuation signal acts on one (scalar) attitude coordinate. This is the price to pay for decoupling the drag-free coordinates.
- Third, the attitude and drag-free coordinates (nine coordinates) are not controllable with thrusters alone (in total six inputs). This fact is physically obvious.[4] Controllability of the total system is obtained only through the suspension loops.

The feedback loops are now closed in the following way: the drag-free controller $\boldsymbol{K}_{DF}$ feeds back the drag-free coordinates via the thruster actuation, the suspension controller $\boldsymbol{K}_{SUS}$ feeds back the suspension coordinates via the suspension actuation. Both $\boldsymbol{K}_{DF}$ and $\boldsymbol{K}_{SUS}$ can be diagonal, since the associated plant transfer functions (14) and (15) are diagonal. The inertial attitude can be fed back in two different ways: either by thruster actuation $\boldsymbol{K}_{ATT}^{(1)}$, i.e., the satellite is "moved," or by suspension actuation $\boldsymbol{K}_{ATT}^{(2)}$, i.e., the test masses are "moved." Therefore a general controller for the input-decoupled system is given by

$$\begin{pmatrix} \boldsymbol{u}_T \\ \boldsymbol{u}_{ES1} \\ \boldsymbol{u}_{ES2} \end{pmatrix} = \begin{bmatrix} -\boldsymbol{K}_{ATT}^{(1)} & -\boldsymbol{K}_{DF} & \boldsymbol{0} \\ -\boldsymbol{K}_{ATT}^{(2)} & \boldsymbol{0} & \boldsymbol{0} \\ \boldsymbol{0} & \boldsymbol{0} & -\boldsymbol{K}_{SUS} \end{bmatrix} \begin{pmatrix} \boldsymbol{\varphi} \\ \boldsymbol{q}_{DF} \\ \boldsymbol{q}_{SUS} \end{pmatrix} \tag{16}$$

with $\boldsymbol{K}_{DF}$ and $\boldsymbol{K}_{SUS}$ being diagonal.

To asses the properties of the two attitude control strategies, first the drag-free and suspension loops only are closed, i.e., attitude controller $\boldsymbol{K}_{ATT}^{(1)} = \boldsymbol{K}_{ATT}^{(2)} = \boldsymbol{0}$ are set to zero. This results in the following partially closed-loop system

$$\begin{pmatrix} \boldsymbol{\varphi} \\ \boldsymbol{q}_{DF} \\ \boldsymbol{q}_{SUS} \end{pmatrix} = \begin{bmatrix} \boldsymbol{G}_{11} & \boldsymbol{G}_{12} & \boldsymbol{0} \\ \boldsymbol{G}_{21} & \boldsymbol{G}_{22} & \boldsymbol{0} \\ \boldsymbol{G}_{31} & \boldsymbol{G}_{32} & \boldsymbol{G}_{33} \end{bmatrix} \begin{pmatrix} \boldsymbol{d}_T \\ \boldsymbol{d}_{TM1} \\ \boldsymbol{d}_{TM2} \end{pmatrix} \tag{17}$$

[4] A proper mathematical check of the observability matrix yields the same result.

where

$$\begin{aligned}
\boldsymbol{G}_{11} &= \boldsymbol{G}_{SC} \cdot \boldsymbol{B}_{ATT} \cdot \boldsymbol{B}_{DF}^{-1} \cdot \boldsymbol{S}_{DF} \\
\boldsymbol{G}_{12} &= -\boldsymbol{G}_{SC} \cdot \boldsymbol{B}_{ATT} \cdot \boldsymbol{B}_{DF}^{-1} \cdot \boldsymbol{T}_{DF} \\
\boldsymbol{G}_{21} &= \boldsymbol{S}_{DF} \cdot \boldsymbol{G}_{DF} \\
\boldsymbol{G}_{22} &= \boldsymbol{S}_{DF} \cdot \boldsymbol{G}_{DF} \\
\boldsymbol{G}_{31} &= \boldsymbol{S}_{SUS} \cdot \boldsymbol{G}_{SUS} \cdot \boldsymbol{B}_{SUS} \cdot \boldsymbol{B}_{DF}^{-1} \cdot \boldsymbol{S}_{DF} \\
\boldsymbol{G}_{32} &= -\boldsymbol{S}_{SUS} \cdot \boldsymbol{G}_{SUS} \cdot \boldsymbol{B}_{SUS} \cdot \boldsymbol{B}_{DF}^{-1} \cdot \boldsymbol{T}_{DF} \\
\boldsymbol{G}_{33} &= \boldsymbol{S}_{SUS} \cdot \boldsymbol{G}_{SUS}
\end{aligned}$$

Here $\boldsymbol{S}_{SUS}$ denotes the diagonal matrix of sensitivity function of the suspension loop whereas $\boldsymbol{S}_{DF}$ and $\boldsymbol{T}_{DF}$ are the diagonal matrices of sensitivity and complementary sensitivity functions of the drag-free loop, respectively.[5] From the partially closed-loop system, the plant dynamics preferred for attitude control can be identified. In case of attitude control with thrusters the plant dynamics is given by $\boldsymbol{G}_{11}$, i.e., by

$$\boldsymbol{G}_{SC} \cdot \boldsymbol{B}_{ATT} \cdot \boldsymbol{B}_{DF}^{-1} \cdot \boldsymbol{S}_{DF} \tag{18}$$

In the case of attitude control with suspension actuation, the design plant is given by $\boldsymbol{G}_{12}$. Moreover, under the assumption that the bandwidth of the attitude controllers is much smaller than the bandwidth of the drag-free controllers,[6] $\boldsymbol{T}_{DF}$ can be approximated by $\boldsymbol{E}$. Then the plant dynamics is given by

$$-\boldsymbol{G}_{SC} \cdot \boldsymbol{B}_{ATT} \cdot \boldsymbol{B}_{DF}^{-1} \tag{19}$$

Comparing (18) and (19), it becomes clear that suspension actuation for attitude control is the preferred approach, because the design plant (19) is very simple and does not rely on the closed loop of the drag-free feedback. Physically speaking, spacecraft attitude control using suspension actuation corresponds to inertial "orientation" of the test masses (with very low bandwidth though), the spacecraft follows due to the relatively fast drag-free control loop. Subsequently, attitude control via thrusters is discarded, i.e., $\boldsymbol{K}_{ATT}^{1}$ in (16) is set to zero.[7]

To eliminate the coupling of the input matrix $\boldsymbol{B}_{ATT}\boldsymbol{B}_{DF}^{-1}$ in the design plant (19) the pseudoinverse of the input matrix has to be considered. Then the overall controller becomes

---

[5] The usual definitions $S = [E + GK]^{-1}$ and $T = E - S$ are used here.

[6] This is virtually no restriction, since it is in line with overall acceleration minimization.

[7] If attitude control is realized by thrusters, the controllers for attitude and the drag-free controller are in conflict with each other at low frequencies. The design is much more complicated.

$$\begin{pmatrix} \boldsymbol{u}_T \\ \boldsymbol{u}_{ES1} \\ \boldsymbol{u}_{ES2} \end{pmatrix} = \begin{bmatrix} 0 & -\boldsymbol{K}_{DF} & 0 \\ -\left(\boldsymbol{B}_{ATT} \cdot \boldsymbol{B}_{DF}^{-1}\right)^I \boldsymbol{K}_{ATT} & 0 & 0 \\ 0 & 0 & -\boldsymbol{K}_{SUS} \end{bmatrix} \begin{pmatrix} \boldsymbol{\varphi} \\ \boldsymbol{q}_{DF} \\ \boldsymbol{q}_{SUS} \end{pmatrix} \quad (20)$$

Note that the attitude control is performed by feeding back the attitude error to suspension commands along drag-free controlled axes. The method can further be refined by introducing an additional selection matrix $\boldsymbol{Z}$ that selects the (drag-free) coordinates that shall be used for suspension commands of the attitude control. By defining

$$\boldsymbol{u}_z = \boldsymbol{Z} \cdot \boldsymbol{u}_{ES1} \quad (21)$$

where $\boldsymbol{Z}$ is a diagonal matrix with zeros and ones in the main diagonal selecting the appropriate coordinates, the plant can be written as

$$\begin{pmatrix} \boldsymbol{\varphi} \\ \boldsymbol{q}_{DF} \\ \boldsymbol{q}_{SUS} \end{pmatrix} = \begin{bmatrix} \boldsymbol{G}_{SC} \cdot \boldsymbol{B}_{ATT} \cdot \boldsymbol{B}_{DF}^{-1} & 0 & 0 \\ \boldsymbol{G}_{DF} & \boldsymbol{G}_{DF} \cdot \boldsymbol{Z}^{-1} & 0 \\ \boldsymbol{G}_{SC} \cdot \boldsymbol{B}_{SUS} \cdot \boldsymbol{B}_{DF}^{-1} & 0 & \boldsymbol{G}_{SUS} \end{bmatrix} \left(\begin{pmatrix} \boldsymbol{u}_T \\ \boldsymbol{u}_{ES1} \\ \boldsymbol{u}_{ES2} \end{pmatrix} + \begin{pmatrix} \boldsymbol{d}_T \\ \boldsymbol{d}_{TM1} \\ \boldsymbol{d}_{TM2} \end{pmatrix} \right) \quad (22)$$

With a controller structure as shown in (20) and applying the same derivation procedure as shown above, the overall controller can be expressed as

$$\begin{pmatrix} \boldsymbol{u}_T \\ \boldsymbol{u}_{ES1} \\ \boldsymbol{u}_{ES2} \end{pmatrix} = \begin{bmatrix} 0 & -\boldsymbol{K}_{DF} & 0 \\ -\left(\boldsymbol{B}_{ATT} \cdot \boldsymbol{B}_{DF}^{-1} \cdot \boldsymbol{Z}^{-1}\right)^I \boldsymbol{K}_{ATT} & 0 & 0 \\ 0 & 0 & -\boldsymbol{K}_{SUS} \end{bmatrix} \begin{pmatrix} \boldsymbol{\varphi} \\ \boldsymbol{q}_{DF} \\ \boldsymbol{q}_{SUS} \end{pmatrix} \quad (23)$$

Now the remaining task is the design of single-input–single-output (SISO) controllers for drag-free, suspension and attitude control. For the attitude control loops, taking into account decoupling, the design plant can be chosen simply as

$$G_{ATT,i} = -\frac{1}{s^2} \,. \quad (24)$$

The expected closed-loop bandwidths of the drag-free controllers are much higher than the corresponding stiffness frequencies, thus the plant stiffness can be neglected for the design of the drag-free controllers and the resulting design plant becomes

$$G_{DF,i} = \frac{1}{s^2} \,. \quad (25)$$

The same, however, is not true for the design of the suspension controllers as the expected bandwidths will be in the same region as the stiffness frequencies. Thus the stiffness cannot be neglected and the design plant becomes

$$G_{SUS,i} = \frac{1}{s^2 + \omega_{SUS,i}^2} \,. \quad (26)$$

3.2 Analytical Closed-Loop Description

Taking into account the feedback interconnection presented in Sect. 3.1, an analytical expression of the closed-loop system can be derived. This closed-loop expression shall include the transfer functions from noise inputs to control coordinates as well as actuation commands, thus providing the means to analyze the effect of any input variable on the system states and the actuation signals. To derive the closed-loop expression the plant model including disturbance noise and the controller including sensor noise have to be considered

$$\begin{pmatrix} \varphi \\ q_{DF} \\ q_{SUS} \end{pmatrix} = \begin{bmatrix} G_{SC} \cdot \alpha & 0 & 0 \\ G_{DF} & G_{DF} & 0 \\ G_{SUS} \cdot \beta & 0 & G_{SUS} \end{bmatrix} \begin{pmatrix} u_T \\ u_{ES1} \\ u_{ES2} \end{pmatrix} + \begin{bmatrix} G_{SC} \cdot \alpha & 0 & 0 \\ G_{DF} & G_{DF} & 0 \\ G_{SUS} \cdot \beta & 0 & G_{SUS} \end{bmatrix} \begin{pmatrix} \eta_T \\ \eta_{ES1} \\ \eta_{ES2} \end{pmatrix} \tag{27}$$

$$\begin{pmatrix} u_T \\ u_{ES1} \\ u_{ES2} \end{pmatrix} = \begin{bmatrix} 0 & -K_{DF} & 0 \\ \alpha^I \cdot K_{ATT} & 0 & 0 \\ 0 & 0 & -K_{SUS} \end{bmatrix} \begin{pmatrix} \varphi \\ q_{DF} \\ q_{SUS} \end{pmatrix} + \begin{bmatrix} 0 & -K_{DF} & 0 \\ \alpha^I \cdot K_{ATT} & 0 & 0 \\ 0 & 0 & -K_{SUS} \end{bmatrix} \begin{pmatrix} \eta_{STR} \\ \eta_{DF} \\ \eta_{SUS} \end{pmatrix}, \tag{28}$$

where we used the following definitions

$$\alpha = B_{ATT} \cdot B_{DF}^{-1}, \quad \beta = B_{SUS} \cdot B_{DF}^{-1} \tag{29}$$

Taking into account that

$$\begin{aligned} \eta_{DF} &= D_{DF} \cdot \eta_{IS}, & \eta_{SUS} &= D_{SUS} \cdot \eta_{IS} \\ \eta_{ES1} &= D_{DF} \cdot \eta_{ES}, & \eta_{ES2} &= D_{SUS} \cdot \eta_{ES} \\ \eta_T &= B_{DF} \cdot \eta_{THR} \end{aligned} \tag{30}$$

and defining the noise input vector η as

$$\eta = \left(\eta_{STR}\ \eta_{IS}\ \eta_{THR}\ \eta_{ES}\right)^T \tag{31}$$

yields the following modified equations

$$\begin{pmatrix} \varphi \\ q_{DF} \\ q_{SUS} \end{pmatrix} = \begin{bmatrix} G_{SC} \cdot \alpha & 0 & 0 \\ G_{DF} & G_{DF} & 0 \\ G_{SUS} \cdot \beta & 0 & G_{SUS} \end{bmatrix} \begin{pmatrix} u_T \\ u_{ES1} \\ u_{ES2} \end{pmatrix} + \begin{bmatrix} 0 & 0 & G_{SC} \cdot B_{ATT} & 0 \\ 0 & 0 & G_{DF} \cdot B_{DF} & G_{DF} \cdot D_{DF} \\ 0 & 0 & G_{SUS} \cdot B_{SUS} & G_{SUS} \cdot D_{SUS} \end{bmatrix} \begin{pmatrix} \eta_{STR} \\ \eta_{IS} \\ \eta_{THR} \\ \eta_{ES} \end{pmatrix} \tag{32}$$

$$\begin{pmatrix} \boldsymbol{u}_T \\ \boldsymbol{u}_{ES1} \\ \boldsymbol{u}_{ES2} \end{pmatrix} = \begin{bmatrix} \mathbf{0} & -\boldsymbol{K}_{DF} & \mathbf{0} \\ \boldsymbol{\alpha}^I \cdot \boldsymbol{K}_{ATT} & \mathbf{0} & \mathbf{0} \\ \mathbf{0} & \mathbf{0} & -\boldsymbol{K}_{SUS} \end{bmatrix} \begin{pmatrix} \boldsymbol{\varphi} \\ \boldsymbol{q}_{DF} \\ \boldsymbol{q}_{SUS} \end{pmatrix}$$
$$+ \begin{bmatrix} \mathbf{0} & -\boldsymbol{K}_{DF} \cdot \boldsymbol{D}_{DF} & \mathbf{0}\,\mathbf{0} \\ \boldsymbol{\alpha}^I \cdot \boldsymbol{K}_{ATT} & \mathbf{0} & \mathbf{0}\,\mathbf{0} \\ \mathbf{0} & -\boldsymbol{K}_{SUS} \cdot \boldsymbol{D}_{SUS} & \mathbf{0}\,\mathbf{0} \end{bmatrix} \begin{pmatrix} \boldsymbol{\eta}_{STR} \\ \boldsymbol{\eta}_{IS} \\ \boldsymbol{\eta}_{THR} \\ \boldsymbol{\eta}_{ES} \end{pmatrix} . \tag{33}$$

Substituting these equations into each other and solving first for the control coordinates and then for control signals finally yield the desired closed-loop expression

$$\begin{pmatrix} \boldsymbol{\varphi} \\ \boldsymbol{q}_{DF} \\ \boldsymbol{q}_{SUS} \\ \boldsymbol{u}_T \\ \boldsymbol{u}_{ES1} \\ \boldsymbol{u}_{ES2} \end{pmatrix} = \begin{bmatrix} M_{11} & M_{12} & M_{13} & M_{14} \\ M_{21} & M_{22} & M_{23} & M_{24} \\ M_{31} & M_{32} & M_{33} & M_{34} \\ M_{41} & M_{42} & M_{43} & M_{44} \\ M_{51} & M_{52} & M_{53} & M_{54} \\ M_{61} & M_{62} & M_{63} & M_{64} \end{bmatrix} \begin{pmatrix} \boldsymbol{\eta}_{STR} \\ \boldsymbol{\eta}_{IS} \\ \boldsymbol{\eta}_{THR} \\ \boldsymbol{\eta}_{ES} \end{pmatrix} , \tag{34}$$

where the elements M_{ij} are lengthy expressions of transfer matrices that will not be repeated here. A schematic representation of this closed-loop system is shown in Fig. 4.

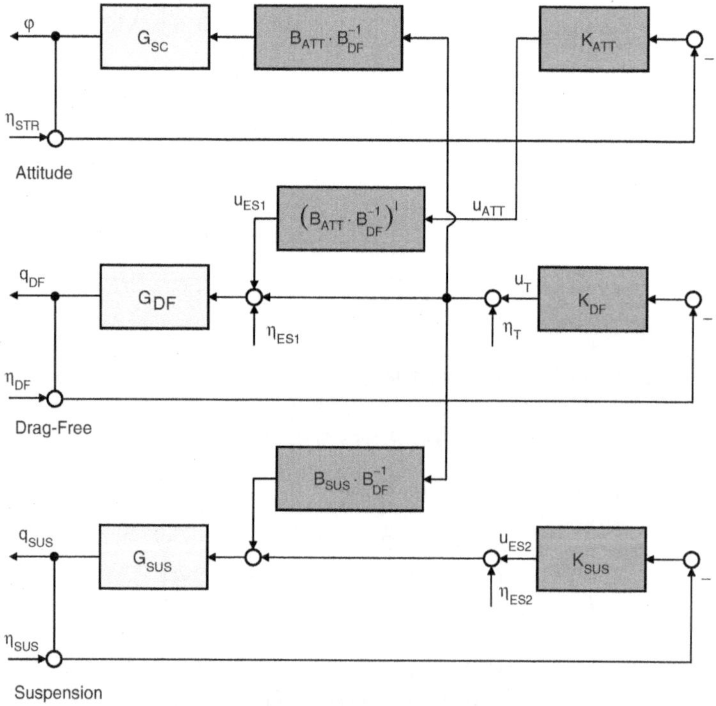

Fig. 4. Closed-loop system.

4 Controller Optimization Procedure

4.1 Control Loop Design Methodology

Due to the decoupling scheme presented in Sect. 3.1, the controller design simplifies to closing a number of simple SISO loops. Many techniques are available for this design problem. For the problem treated here, H_∞ design techniques are considered to be best suited, because closed-loop responses can be optimized and performance requirements given in the form of spectral densities valid in a certain measurement bandwidth can be directly taken into account in the design process [6]. Another advantage of this scheme is that it can easily be extended by considering model uncertainties in the design to increase robustness. However, this chapter will focus on the nominal design without uncertainty model.

H_∞ optimization techniques rely on extending the nominal design plant with specific weighting functions that act as bounds for the respective closed-loop sensitivity function. Especially the S/T weighting scheme has been widely used throughout the literature, where the sensitivity S which is a measure for the disturbance rejection in the closed-loop system and the complementary sensitivity function T that specifies the noise rejection are bounded through the use of (frequency-dependent) weighting functions. In bounding S and T the weighting functions will define cut-off frequency and roll-off of the sensitivity functions and will thus also define the bandwidth ω_{bw} of the controller. However, the sensitivity functions always have to meet the following additional constraint

$$S(s) + T(s) = 1 \tag{35}$$

Typical plots of the sensitivity function S and the complementary sensitivity function T are shown in Fig. 5.

Nevertheless, the slightly modified GS/T weighting scheme shall be used here as the standard S/T weighting scheme may result in a controller that inverts the plant. Instead of shaping S directly this scheme shapes S via shaping

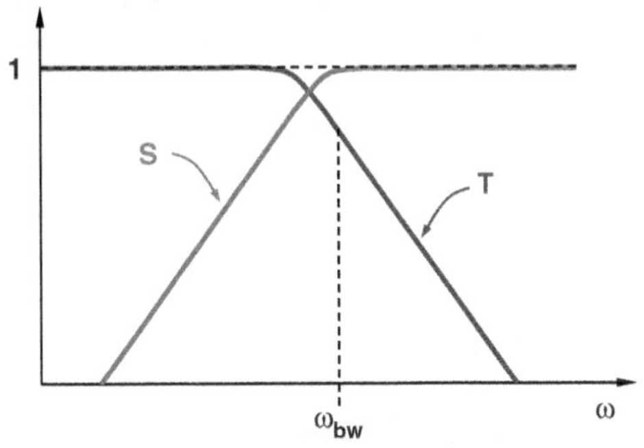

Fig. 5. Typical sensitivity functions.

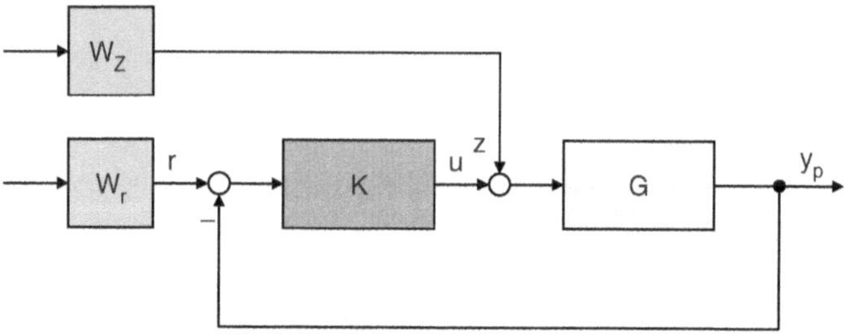

Fig. 6. *GS/T* weighting scheme.

the combination of the plant G and T. A block diagram of the GS/T weighting scheme is shown in Fig. 6. Here W_z and W_r are the weighting functions that bound GS and T, respectively.

The control structure design described above provides not perfect decoupling. There are still feedforward coupling in the overall system (drag-free to suspension) and frequency decoupling (attitude and drag-free) will not be perfect. Therefore, it is important to define the proper design sequence. As the drag-free loops are the loops most important for the overall performance of the drag-free system, they are closed first. The suspension loops are closed next taking into account the influences of the closed drag-free loops. This is very important as the suspension loops are very sensitive to crosscoupling. The attitude loops are closed last because the requirements on the spacecraft attitude are not very stringent compared to the requirements on the drag-free and suspension loops. This additional freedom in the design can be used to take into account additional requirements imposed on the attitude loop by the drag-free and the suspension loop, i.e., the influence of the attitude control loop on drag-free and suspension performance can be made negligible.

4.2 Derivation of Closed-Loop Specifications

The requirements on test mass control errors and actuation signals need to be translated into closed-loop specifications on S and T before the actual controller synthesis. Thus the analytical closed-loop expression derived in Sect. 3.2 shall be revisited.

The first step in the derivation of a drag-free system is the design of the drag-free loops. For the design of the drag-free loops a simple double integrator design plant is sufficient as has been shown earlier. However, depending on the system under consideration, it may be necessary to consider a delay in the design plant as the drag-free loops are sensitive to delays due to their higher bandwidth. The higher bandwidth of the drag-free loops in combination with the proposed feedback structure causes the star tracker noise η_{STR} and the electrostatic actuation noise η_{ES} to have only a minor influence on the drag-free loop. So, for the design of the drag-free loop the following subset of the closed-loop expression can be considered

$$\boldsymbol{q}_{DF} = \left[\boldsymbol{M}_{22}\ \boldsymbol{M}_{23}\right] \begin{pmatrix} \boldsymbol{\eta}_{IS} \\ \boldsymbol{\eta}_{THR} \end{pmatrix} \tag{36}$$

Under the assumption that the bandwidth of the drag-free controllers is above the measurement bandwidth, the closed-loop expressions M_{22} and M_{23} can be approximated as

$$\begin{aligned} \boldsymbol{M}_{22} &= \boldsymbol{T}_{DF} \cdot \boldsymbol{D}_{DF} \\ \boldsymbol{M}_{23} &= \boldsymbol{S}_{DF} \cdot \boldsymbol{G}_{DF} \cdot \boldsymbol{B}_{DF} \end{aligned} \tag{37}$$

Now the specification for S shall be derived from the respective closed-loop expression as an example to explain the process in more detail. Thus the closed-loop transfer function from thruster noise to control coordinates (M_{23}) must be considered

$$\boldsymbol{q}_{DF} = \boldsymbol{S}_{DF} \cdot \boldsymbol{G}_{DF} \cdot \boldsymbol{B}_{DF} \cdot \boldsymbol{\eta}_{THR} \,. \tag{38}$$

Here $\boldsymbol{S}_{DF}$ and $\boldsymbol{G}_{DF}$ are diagonal transfer function matrices and $\boldsymbol{B}_{DF}$ is a matrix mapping the thruster noise into the drag-free coordinates. Now substituting the test mass jitter requirements for $\boldsymbol{q}_{DF}$ and the noise shape filter for $\boldsymbol{\eta}_{THR}$ and considering $\boldsymbol{S}_{DF} \cdot \boldsymbol{B}_{DF} \cdot \boldsymbol{\eta}_{THR}$ to be a modified expression for the thruster noise, the specifications for the sensitivity of the drag-free loops are easily derived for each of the SISO axes. The complete set of S and T specifications that can be derived for the drag-free loops are summarized in Table 1 along with their origin.

The next step in the derivation of the drag-free system is the design of the suspension loops. To derive specifications for S and T the following subset of the closed-loop expressions shall be considered

$$\begin{pmatrix} \boldsymbol{q}_{SUS} \\ \boldsymbol{u}_{ES2} \end{pmatrix} = \begin{bmatrix} \boldsymbol{M}_{32}\ \boldsymbol{M}_{33}\ \boldsymbol{M}_{34} \\ \boldsymbol{M}_{62}\ \boldsymbol{M}_{63}\ \boldsymbol{M}_{64} \end{bmatrix} \begin{pmatrix} \boldsymbol{\eta}_{IS} \\ \boldsymbol{\eta}_{THR} \\ \boldsymbol{\eta}_{ES} \end{pmatrix} \tag{39}$$

Note that the influence of the star tracker noise on the suspension loop is considered in the design of the attitude control loop, i.e., the attitude loop is designed such that the influence on the suspension loop is minimized. The complete set of S and T specifications that can be derived form this subset is summarized in Table 2.

The attitude loops are closed last. They have to be designed taking into account the already closed drag-free and suspension loops with the goal to

Table 1. Drag-free loop specifications.

Specification	Origin	Description
S_{spec}	M_{23}	Input disturbance noise suppression
T_{spec}	M_{22}	Readout noise suppression

Table 2. Suspension loop specifications.

Specification	Origin	Description
S_{spec1}	M_{34}	Suspension noise suppression
S_{spec2}	M_{33}	Micropropulsion noise suppression
T_{spec1}	M_{32}	Readout noise suppression
T_{spec2}	M_{62}	Limiting the control signal jitter due to readout noise
T_{spec3}	M_{64}	Limiting the control signal jitter due to suspension noise
T_{spec4}	M_{63}	Limiting the control signal jitter due to micropropulsion noise

Table 3. Attitude loop specifications.

Specification	Origin	Description
S_{spec}	M_{14}	Input disturbance noise suppression
T_{spec1}	M_{51}	Control signal requirement derived from attitude actuation
T_{spec2}	M_{51}	Control signal requirement derived from drag-free jitter
T_{spec3}	M_{51}	Control signal requirement derived from suspension actuation
T_{spec4}	M_{51}	Control signal requirement derived from suspension jitter

minimize the influence of the attitude loops on the other loops. To derive the S and T specifications for the attitude loops the following subset of the closed-loop expression shall be considered

$$\begin{pmatrix} \varphi \\ \boldsymbol{u}_{ES1} \end{pmatrix} = \begin{bmatrix} * & \boldsymbol{M}_{14} \\ \boldsymbol{M}_{51} & * \end{bmatrix} \begin{pmatrix} \eta_{STR} \\ \eta_{ES} \end{pmatrix} \tag{40}$$

Note that more than one specification for T can be derived from the transfer function M_{51} from star tracker noise to the suspension actuation on the drag-free axes. The difference between the T specifications lies in the derivation of the control signal requirement that is used to formulate the specification. Each specification covers one of the influences of the attitude loops on drag-free and suspension loops, thus ensuring that this influence is negligible. The complete set of S and T specifications for the attitude loops is summarized in Table 3.

For all loop designs the specifications derived here are used to derive the weighting functions that bound S and T such that the specifications are fulfilled.

5 Software and Operational Design

Drag-free control systems are usually a part of experiments. This puts special requirements on the operational aspects. In particular, an exceptionally large number of different operational modes may be required to fulfill the scientific mission objectives. Traditional control systems are often designed and implemented on a mode-by-mode basis. Applying this approach to experimental drag-free systems would result in an unacceptable software complexity. A mode of a drag-free system is defined by:

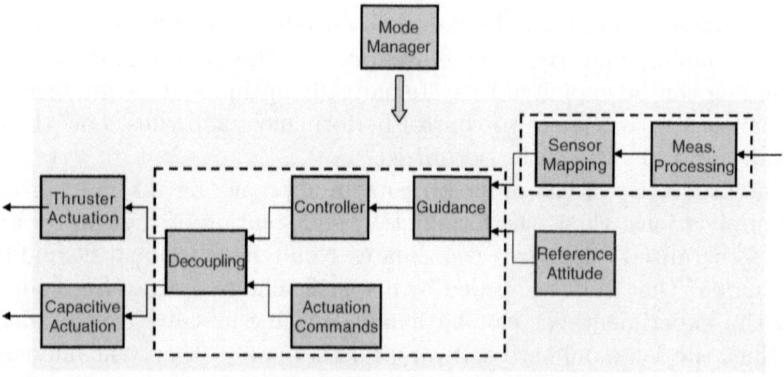

Fig. 7. Functional units in a drag-free system.

- Coordinates to be controlled, and the associated control principle to be applied
- Sensor data to be used
- Settings of the actuation system, e.g., low/high voltage range for electrostatic actuation

The first point is most important. With the proposed control design, any "new" modes introduced due to coordinate definition are covered by the single implementation of (23), with an appropriate parameterization obtained by the design procedure described in Sect. 4. Thus, the software can be divided into a number of functional units that are configured/parameterized to map any mode requirements. A mode is then represented by a set of configuration parameters for the individual functional units. This results in a large reduction of software complexity. A block diagram representation of the functional units in a drag-free system is shown in Fig. 7.

To simplify the operational interface, sets of configuration parameters representing the baseline modes can be stored onboard the S/C and can then be called through a single-mode command. By including a number of custom modes that can be freely configured and called through the mode command, the system provides operational flexibility. Even during in-orbit operations the system can be easily reconfigured to respond to any change requests from the scientists without having to change any software module.

6 Conclusion

The control design method proposed in this chapter solves the problem of drag-free, suspension, and attitude control of a spacecraft with two cubical test masses. The method can readily be generalized to more than two test masses or reduced to a single test mass case, respectively. The design procedure is independent of any specific definition/selection of drag-free coordinates (the coordinates to be drag-free controlled). It is demonstrated that the overall problem can always be decomposed into a set of 15 SISO systems by input decoupling and frequency separation. Furthermore it is shown how this decoupling can be exploited to systematically optimize the closed-loop transfer functions with respect to control performance, and thus, how the overall free-fall performance can be optimized.

The properties of the proposed control approach have a number of practical implications. First, the formulation and solution of the control problem with "generalized" drag-free coordinates result in functional flexibility: any "new mode" that may be needed by a specific choice of drag-free coordinates from the experimentalist can be handled with the same control algorithm and thus, the same onboard software. Furthermore it keeps the onboard software architecture simple which leads to efficient software coding and testing. Second, the optimization with respect to control performance becomes very simple and efficient, i.e., any further optimization and/or redesign can be accomplished very quickly, even in-flight. Furthermore it allows a traceable ("axis by axis") assessment of the overall free-fall performance.

All of these implications support a large degree of flexibility for the experimentalist and a high reuse potential for future missions.

References

1. D. Bortoluzzi, M. DaLio, W. Weber, S. Vitale: Test-mass electrostatic suspension force to voltage conversion laws and actuation strategy. Technical Report LTPA-ACT-UTN Issue 1, Revision 0, University of Trento, 2004
2. D. Bindel, A. Schleicher: Thruster Actuation Algorithm Design and Analysis. Technical Note S2-ZAR-TN-2001 Issue 2.0, ZARM, 2005
3. P. Wiktor: The Design of a Propulsion System Using Vent Gas from a Liquid Helium Cryogenic System, PhD thesis, Stanford University (1992)
4. M. Armano, D. Bortoluzzi, C.D. Hoyle, S. Vitale: Gravitational Compensation for the LISA Pathfinder. *Class. Quantum Grav.* **22**, S501 (2005)
5. S. Vitale: The LISA Technology Package On Board SMART-2. Technical Report Unitn-Int-10-2002, Release 1.3, University of Trento, 2002
6. S. Skogestad, I. Postlethwaite: *Multivariable Feedback Control* (John Wiley & Sons), 1996
7. W. Fichter, A. Schleicher, M. Schlotterer: DFACS General Design, Technical Note LISA Pathfinder, S2-ASD-TN-2001, Issue 2.1, EADS Astrium, 2005

Solar Sail Propulsion: An Enabling Technology for Fundamental Physics Missions

Bernd Dachwald[1], Wolfgang Seboldt[2], and Claus Lämmerzahl[3]

[1] DLR, Mission Operations Section, Oberpfaffenhofen, 82234 Wessling, Germany
[2] DLR, Institute of Space Simulation, Linder Höhe, 51147 Köln, Germany
[3] ZARM, University of Bremen, Am Fallturm, 28359 Bremen, Germany

Summary. Solar sails enable a wide range of high-energy missions, many of which are difficult or even impossible to accomplish with any other type of conventional propulsion system. They are also an enabling propulsion technology for two types of deep-space missions that are very favorable for testing current gravitational theories and the large-scale gravitational field of the solar system: the first type comprises missions that go very close to the Sun (<8 solar radii) and the second one comprises missions that go fast very far away from the Sun (~200 AU). Being propelled solely by the freely available solar radiation pressure, solar sails do not consume any propellant. Therefore, their capability to gain (or reduce) orbital energy is theoretically unlimited and practically only limited by their lifetime in the space environment and their distance from the Sun (because the solar radiation pressure decreases with the square of solar distance). Nevertheless, solar sails make also missions that go far away from the Sun feasible because they can gain a large amount of orbital energy by first making one or more close solar approaches that turn the trajectory hyperbolic. For both mission types, the temperature limit of the sail film is a critical issue. In this chapter, we briefly review the physics and the current technological status of solar sails, and then present mission outlines and trade-offs for both mission types. Thereby, we will show that even near- or medium-term solar sails with a relatively moderate performance enable these kinds of missions.

1 Introduction

Innovative deep-space missions require ever larger orbital energy changes (typically expressed as velocity increments, ΔVs[1]) and thus ever more demanding propulsion capabilities. The so-called *rocket equation* gives the total ΔV that spacecraft can gain as

$$\Delta V = V_e \ln \frac{m_0}{m_f} \tag{1}$$

[1] The ΔVs are always positive, no matter whether they are used for gaining or reducing the spacecraft's orbital energy.

where V_e is the exhaust velocity of the propellant, m_0 is the initial spacecraft mass, and m_f is the final spacecraft mass. Due to the energy barrier inherent in chemical combustion, chemical high-thrust propulsion systems (rocket engines) have a limited V_e and thus a limited ΔV-capability (the exhaust velocity of chemical rocket engines is less than $5\,\mathrm{km\,s^{-1}}$, typically about $4\,\mathrm{km\,s^{-1}}$).

Utilizing solely the freely available solar radiation pressure (SRP) for propulsion, solar sails do not consume any propellant. Therefore, their ΔV-capability is theoretically unlimited and practically only limited by their lifetime in the space environment and their distance from the Sun (because the SRP decreases with $1/r^2$, where r is the solar distance). Solar sails enable a wide range of high-ΔV missions, many of which are difficult or even impossible to accomplish with any other type of conventional propulsion system. Solar sails are also an enabling propulsion technology for two types of deep-space missions that are very favorable for testing current gravitational theories and the large-scale gravitational field of the solar system: the first type comprises missions that go very close to the Sun (<8 solar radii) and the second one comprises missions that go fast very far away from the Sun (~200 AU). For a mission that goes into a very close solar orbit (8 solar radii[2] $= 5.568 \times 10^6\,\mathrm{km} = 0.03722\,\mathrm{AU}$, where 1 AU is 1 astronomical unit, which is Earth's mean distance from the Sun), the ΔV that is required from a low Earth orbit for a Hohmann transfer orbit is $\Delta V_1 = 22.0\,\mathrm{km\,s^{-1}}$. The ΔV that is required to circularize the transfer orbit at the Sun is $\Delta V_2 = 60.0\,\mathrm{km\,s^{-1}}$. Thus, using (1) and $V_e = 4\,\mathrm{km\,s^{-1}}$, the payload ratio of a rocket would be $m_0/m_f = \exp(\Delta V/V_e) = 1.3 \times 10^9$ (or, skipping the circularization maneuver, $m_0/m_f = 245$, which is still extremely large). Because solar sails are particularly effective within the inner solar system, they are able to attain very close solar orbits rapidly. Afterwards, to allow high-precision measurements, the solar sail would be jettisoned. For a mission that goes very fast far away from the Sun (e.g., with a solar system escape velocity of $v_{\mathrm{esc}} = 5\,\mathrm{AU\,year^{-1}}$),[3] the required ΔV at Earth is $19.2\,\mathrm{km\,s^{-1}} \Rightarrow m_0/m_f = 122$, which is also very large. Although the SRP decreases with $1/r^2$, solar sails enable not only missions that go very close to the Sun but also missions that go fast very far away from the Sun. Sauer [1] observed that the solar sail may gain a large amount of energy by making a close approach to the Sun that turns the trajectory hyperbolic, a maneuver for which Leipold [2, 3] coined the term "solar photonic assist" (SPA). Also for this mission type, after the last SPA, the solar sail can be jettisoned at about 5 AU to allow high-precision measurements.

---

[2] We will show later that this can be achieved with a near- to medium-term solar sail.

[3] We will show later that this can be achieved with a medium-term solar sail.

The chapter is organized as follows. First, we briefly review the physics of solar sails (Sect. 2) and introduce the commonly used performance parameters for solar sails (Sect. 3). Then, using the DLR solar sail ground demonstration as an example, the current technological status of solar sails is outlined (Sect. 4). Finally, after having outlined our solar sail simulation model (Sect. 5), we propose two types of deep-space missions that are very favorable for testing current gravitational theories and the large-scale gravitational field of the solar system: the first type comprises missions that go very close to the Sun (Sect. 6) and the second one comprises missions that go fast very far away from the Sun (Sect. 7).

2 Solar Sail Orbital Mechanics

2.1 Solar Sail Force Model

For the description of the SRP force exerted on a solar sail, it is convenient to introduce two unit vectors. The first one is the sail normal vector $\boldsymbol{n}$, which is perpendicular to the sail surface and always directed away from the Sun. In the orbit frame[4] $\mathcal{O}$, the direction of $\boldsymbol{n}$, which describes the sail attitude, is expressed by the pitch angle α and the clock angle δ (Fig. 1). The second unit vector is the thrust unit vector $\boldsymbol{m}$, which points along the direction of the SRP force. Its direction is described likewise by the cone angle θ and, again, by the clock angle δ.

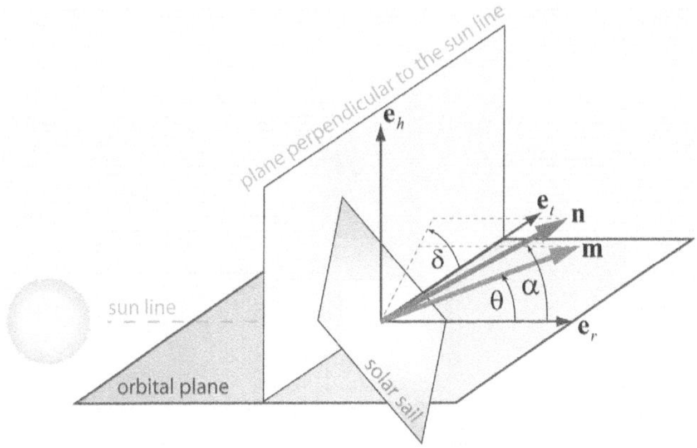

Fig. 1. Definition of the sail normal vector $\boldsymbol{n}$ and the thrust normal vector $\boldsymbol{m}$.

[4] $\mathcal{O} = \{\boldsymbol{e}_r, \boldsymbol{e}_t, \boldsymbol{e}_h\}$ is an orthogonal right-handed polar coordinate frame. $\boldsymbol{e}_r$ points always along the Sun–spacecraft line, $\boldsymbol{e}_h$ is the orbit plane normal (pointing along the spacecraft's orbital angular momentum vector), and $\boldsymbol{e}_t$ completes the right-handed coordinate system ($\boldsymbol{e}_r \times \boldsymbol{e}_t = \boldsymbol{e}_h$).

At a distance r from the Sun, the SRP is

$$P = \frac{S_0}{c}\left(\frac{r_0}{r}\right)^2 = P_0\left(\frac{r_0}{r}\right)^2 \tag{2}$$

where $S_0 = 1{,}368\,\mathrm{W\,m^{-2}}$ is the solar constant, c is the speed of light in vacuum, and $r_0 = 1\,\mathrm{AU}$, so that $P_0 = 4.563\,\mu\mathrm{N\,m^{-2}}$. Because the SRP is so small, solar sails have to be very large and lightweight.

In this chapter, the standard nonperfectly reflecting SRP force model[5] by Wright is employed, which uses the set of optical coefficients $\mathcal{P} = \{\rho, s, \varepsilon_\mathrm{f}, \varepsilon_\mathrm{b}, B_\mathrm{f}, B_\mathrm{b}\}$ to parameterize the optical characteristics of the sail film, where ρ is the reflection coefficient, s is the specular reflection factor, ε_f and ε_b are the emission coefficients of the front and back side, respectively, and B_f and B_b are the non-Lambertian coefficients of the front and back side, respectively, which describe the angular distribution of the emitted and the diffusely reflected photons. The optical coefficients for a solar sail with a highly reflective aluminum-coated front side and with a highly emissive chromium-coated back side (to keep the sail temperature at a moderate limit) are $\mathcal{P}_{\mathrm{Al|Cr}} = \{\rho = 0.88, s = 0.94, \varepsilon_\mathrm{f} = 0.05, \varepsilon_\mathrm{b} = 0.55, B_\mathrm{f} = 0.79, B_\mathrm{b} = 0.55\}$ [4]. It can be shown[6] that in a sail-fixed two-dimensional coordinate frame[7] $\mathcal{S} = \{\boldsymbol{n}, \boldsymbol{t}\}$ (see Fig. 2), the SRP force exerted on the solar sail with the area A has a normal component $F_\perp$ (along $\boldsymbol{n}$) and a tangential component $F_{||}$ (along $\boldsymbol{t}$) with

$$F_\perp = \boldsymbol{F}_{\mathrm{SRP}} \cdot \boldsymbol{n} = 2PA\cos\alpha\,(a_1\cos\alpha + a_2) \tag{3a}$$
$$F_{||} = \boldsymbol{F}_{\mathrm{SRP}} \cdot \boldsymbol{t} = -2PA\cos\alpha\, a_3 \sin\alpha \tag{3b}$$

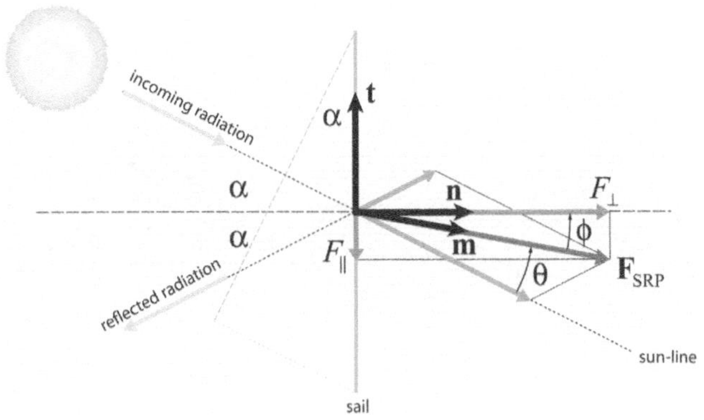

Fig. 2. SRP force on a solar sail according to the nonperfectly reflecting force model.

[5] See, e.g., [4], pp. 223–233 or [5], pp. 47–51 for a more detailed description of this model.

[6] See, e.g., [5], pp. 47–49 for derivation.

[7] Because of the symmetry, the third dimension is not relevant here.

with

$$a_1 := \frac{1}{2}(1 + s\rho) \tag{4a}$$

$$a_2 := \frac{1}{2}\left[B_\mathrm{f}(1-s)\rho + (1-\rho)\frac{\varepsilon_\mathrm{f} B_\mathrm{f} - \varepsilon_\mathrm{b} B_\mathrm{b}}{\varepsilon_\mathrm{f} + \varepsilon_\mathrm{b}}\right] \tag{4b}$$

$$a_3 := \frac{1}{2}(1 - s\rho) \tag{4c}$$

The total SRP force vector may then be written as

$$\boldsymbol{F}_\mathrm{SRP} = 2PA\cos\alpha\,\Psi\,\boldsymbol{m} \tag{5}$$

with

$$\Psi := \left[(a_1\cos\alpha + a_2)^2 + (a_3\sin\alpha)^2\right]^{1/2} \tag{6}$$

where Ψ depends only on the pitch angle α and the optical coefficients $\mathcal{P}$ of the sail film. This way, the solution for nonperfect reflection has a similar structure than the solution for ideal reflection, $\boldsymbol{F}_\mathrm{SRP}^\mathrm{ideal} = 2PA\cos\alpha\cos\alpha\,\boldsymbol{n}$. However, the SRP force is smaller and not more perpendicular to the sail (i.e., along $\boldsymbol{n}$). The angle between $\boldsymbol{m}$ and $\boldsymbol{n}$, $\phi = \arctan\left[a_3\sin\alpha/(a_1\cos\alpha + a_2)\right]$, is referred to as *centerline angle*. The cone angle, i.e., the angle between $\boldsymbol{m}$ and $\boldsymbol{e}_r$, is then obtained as $\theta = \alpha - \phi$.

$\boldsymbol{F}_\mathrm{SRP}$ is constrained to lie on the surface of a bubble that is always directed away from the Sun (Fig. 3). Nevertheless, by controlling the sail orientation relative to the Sun, a solar sail can gain orbital energy (if $\boldsymbol{F}_\mathrm{SRP}\cdot\boldsymbol{e}_t > 0$) and spiral outward, away from the Sun, or lose orbital energy (if $\boldsymbol{F}_\mathrm{SRP}\cdot\boldsymbol{e}_t < 0$) and spiral inward, toward the Sun.

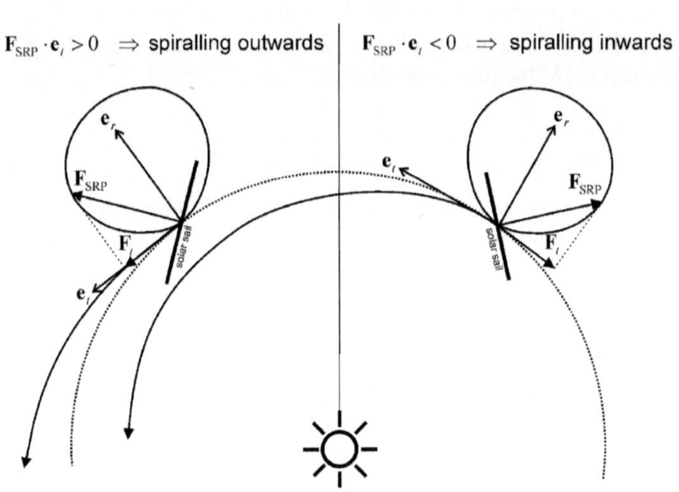

Fig. 3. Spiralling toward and away from the Sun.

3 Solar Sail Performance Parameters

The performance of solar sailcraft can be expressed by the following parameters:

- The *sail assembly loading*
$$\sigma_s = \frac{m_s}{A} \tag{7}$$
is defined as the mass of the sail assembly (the sail film and the required structure for storing, deploying, and tensioning the sail, index s) per unit area. Thus, the sail assembly loading is the key parameter for the performance of a solar sail and the efficiency of its structural design.
- The *sailcraft loading*
$$\sigma = \frac{m}{A} = \frac{m_s + m_p}{A} = \sigma_s + \frac{m_p}{A} \tag{8}$$
is defined accordingly as the specific mass of the whole sailcraft including the payload (index p). Note that the term "payload" stands for the total sailcraft except the solar sail assembly (i.e., except the propulsion system).
- The most commonly used performance parameter is the *characteristic acceleration* a_c. It is defined as the SRP acceleration acting on a sailcraft that is oriented perpendicular to the Sun line ($\boldsymbol{n} \equiv \boldsymbol{e}_r$) at r_0 (1 AU). For the nonperfectly reflecting SRP force model, it is
$$a_c = \frac{2P_0(a_1 + a_2)A}{m} = \frac{2P_0(a_1 + a_2)A}{m_s + m_p} = \frac{2P_0(a_1 + a_2)}{\sigma_s + \frac{m_p}{A}} \tag{9}$$
- The *lightness number* β, which is independent from solar distance, is defined as the ratio of the SRP acceleration experienced by a sailcraft that is oriented perpendicular to the Sun line and the solar gravitational acceleration ($5.93\,\mathrm{mm\,s^{-2}}$ at 1 AU)
$$\beta = \frac{a_c}{5.93\,\mathrm{mm\,s^{-2}}} \tag{10}$$
- Another important performance parameter is the *sail temperature limit* T_{lim}. A solar sail's equilibrium temperature at a distance r from the Sun is [5]
$$T = \left(\frac{1-\rho}{\varepsilon_{\mathrm{f}} + \varepsilon_{\mathrm{b}}} \frac{S_0 r_0^2}{\tilde{\sigma}} \frac{\cos\alpha}{r^2} \right)^{1/4} \tag{11}$$
where $\tilde{\sigma} = 5.67051 \times 10^{-8}\,\mathrm{W\,m^{-2}\,K^{-4}}$ is the Stefan–Boltzmann constant. Therefore, the minimum distance to the Sun is, for a given sail attitude α, limited by the temperature limit of the sail film,
$$r_{\mathrm{min}} = \left(\frac{1-\rho}{\varepsilon_{\mathrm{f}} + \varepsilon_{\mathrm{b}}} \frac{S_0 r_0^2}{\tilde{\sigma}} \frac{\cos\alpha}{T_{\mathrm{lim}}^4} \right)^{1/2} \tag{12}$$

The sail temperature limit T_{lim} is sometimes also expressed as the solar distance r_{lim}, where the equilibrium temperature of the Sun-facing solar sail ($\alpha = 90°$) equals the sail temperature limit, e.g., $r_{\text{lim}} = 0.264\,\text{AU}$ for an Al|Cr-coated sail with a temperature limit of $T_{\text{lim}} = 240°\text{C}$. Note, however, that the sail can fly closer to the Sun, provided

$$\alpha \geq \alpha_{\text{lim}} = \arccos\left(\frac{\varepsilon_{\text{f}} + \varepsilon_{\text{b}}}{1-\rho}\frac{\tilde{\sigma}}{S_0 r_0^2} T_{\text{lim}}^4 r^2\right) \tag{13}$$

4 Solar Sail Hardware Development

In December 1999, a ground-based demonstration of solar sail technology was performed at the German Aerospace Center (DLR) at Cologne, where a $20 \times 20\,\text{m}^2$ solar sail was successfully deployed in a simulated zero-g environment and ambient environmental conditions (Fig. 4a) [6, 7].

The solar sail consisted of four carbon fiber-reinforced plastics (CFRP) booms with a specific mass of $101\,\text{g}\,\text{m}^{-1}$ and of four triangular sail segments made of 0.1-μm aluminum-coated plastic films with a thickness between 4 and 12 μm. The booms consisted of two CFRP shells that were bonded at the edges to form a tubular shape, so that they can be pressed flat and rolled up (Fig. 4b). The booms were rolled up in a $60 \times 60 \times 65\,\text{cm}^3$-sized deployment module, from where they unfolded automatically. After deployment they returned to their tubular shape with high bending and buckling strength. Subsequently, the four sail segments were deployed by ropes. To assess the handling behavior of different sail materials, the sail segments were made of three different aluminum-coated plastic films: 12 μm polyethylene terephtalate (PET, Mylar®), 7.5 μm polyimide (PI, Kapton®), and 4 μm polyethylene naphthalate (PEN). All segments were reinforced along the three edges of the triangle to prevent rips. The specific mass of the sail film was

(a) Fully deployed 20 m × 20 m solar sail (b) Deployable CFRP boom

Fig. 4. Solar sail hardware development at DLR.

$18.9\,\mathrm{g\,m^{-2}}$ for the Mylar® segment, $12.4\,\mathrm{g\,m^{-2}}$ for the Kapton® segment, and $10.5\,\mathrm{g\,m^{-2}}$ for the PEN segment. The total mass of the solar sail assembly was 35 kg (5 kg Kapton® film,[8] 6 kg booms, and 24 kg deployment module), which yields a sail assembly loading of $87.5\,\mathrm{g\,m^{-2}}$ and a characteristic acceleration of $a_c = 0.09\,\mathrm{mm\,s^{-2}}$, however, without payload. The deployment module and the cross-section of the booms for this ground-based demonstration were dimensioned for a $40 \times 40\,\mathrm{m^2}$ solar sail (which was too large for an indoor demonstration). Two load cases were considered for the structural sizing of the booms, bending – due to the SRP force – and buckling – due to sail deployment and sail tensioning forces. According to FEM calculations, similar booms could be used also for larger sails [8].

5 Solar Sail Simulation Model

Besides the gravitational forces of all celestial bodies and the SRP force, many disturbing forces influence the motion of solar sails in space, as they are caused, e.g., by the solar wind, the finiteness of the solar disk, the reflected light from close celestial bodies, and the aberration of solar radiation (Poynting–Robertson effect). Furthermore, a real solar sail bends and wrinkles depending on the actual solar sail design. All these issues have to be considered for high-precision trajectory determination and control. For mission feasibility analysis, however, as it is done within this chapter, the following simplifications can be made:

1. The solar sail is flat.
2. The solar sail is moving under the sole influence of solar gravitation and radiation.
3. The Sun is a point mass and a point light source.
4. The solar sail attitude can be changed instantaneously.
5. The optical characteristics of the sail film do not degrade over time.

Let the reference frame $\mathcal{I} = \{\boldsymbol{e}_x, \boldsymbol{e}_y, \boldsymbol{e}_z\}$ be a heliocentric inertial right-handed Cartesian coordinate frame. The equations of motion for a solar sail in the $\mathcal{I}$-frame are:

$$\dot{\boldsymbol{r}} = \boldsymbol{v}, \qquad \dot{\boldsymbol{v}} = -\frac{\mu}{r^3}\boldsymbol{r} + \frac{\boldsymbol{F}_{\mathrm{SRP}}}{m} + \boldsymbol{a}_d \tag{14}$$

where $\boldsymbol{r} = (r_x, r_y, r_z)$ is the solar sail position, $\boldsymbol{v} = (v_x, v_y, v_z)$ is the solar sail velocity, μ is the Sun's gravitational parameter, and $\boldsymbol{a}_d$ is the disturbing acceleration, which is – according to the simplifications made above – neglected within this chapter.

[8] Assuming that all sail segments are made of Kapton®.

6 Missions to Very Close Solar Orbits

6.1 Mission Rationale

The peculiarity of being very close to the Sun is the large value of the gravitational potential,[9] being the best what we get within light years. The gravitational potential mainly influences the rate of clocks. Therefore, missions close to the Sun are a good environment for testing the behavior of clocks, i.e., to test the validity of the universality of the gravitational redshift, and, furthermore, to measure the redshift. It is a nice coincidence that one does not need a good drag-free environment for clock tests. The important thing to know is the position of the clock.

A test of the universality of the gravitational redshift includes a search for a time and position dependency of constants (e.g., the fine structure constant, the electron-to-proton mass ratio, the fine structure constant of the weak interaction, etc.). Such a time and position dependence of constants is predicted by theories describing the low-energy limit of quantum gravity theories like the string theory [9,10]. Constants also will become time dependent in quintessence scenarios [11] where the value of the constants is related to the cosmological evolution.

6.2 Solar Sail Technology

As the trajectory calculations in Sect. 6.3 will show, a quite conservative near- to medium-term solar sail technology is sufficient for this mission type: a $40 \times 40\,\mathrm{m}^2$ solar sail with a total mass of 100 kg (19.8 kg Kapton® film with $12.4\,\mathrm{g\,m}^{-2}$, 11.4 kg CFRP booms with $101\,\mathrm{g\,m}^{-1}$, and 68.8 kg deployment module + scientific spacecraft). According to (8), the sailcraft loading is $\sigma = 62.5\,\mathrm{g\,m}^{-2}$ and according to (9), the characteristic acceleration is $a_c = 0.1326\,\mathrm{mm\,s}^{-2}$.

6.3 Mission Design

A local steering law (LSL) was used for the trajectory design of this mission. LSLs give the locally optimal thrust direction to change some specific osculating orbital element of the spacecraft with a locally maximum rate. To obtain LSLs, Lagrange's planetary equations in Gauss' form may be used, which describe the rate of change of a body's osculating orbital elements due to some (propulsive and/or disturbing) acceleration. This can best be done in the orbit frame $\mathcal{O} = \{\mathbf{e}_r, \mathbf{e}_t, \mathbf{e}_h\}$. According to [12], the equation for the semimajor axis a can be written as

$$\frac{da}{dt} = \frac{2a^2}{h}\left(e \sin f a_r + \frac{p}{r} a_t\right) \tag{15}$$

[9] Note that it is the large value of the gravitational potential and not the strength of the gravitational acceleration what is of use.

where a_r and a_t are the acceleration components along $\boldsymbol{e}_r$ and $\boldsymbol{e}_t$, respectively, $h = |\boldsymbol{h}|$ is the orbital angular momentum per spacecraft unit mass, e is the eccentricity, f is the true anomaly, and p is the semilatus rectum of the orbit. Because (15) can be written as

$$\frac{da}{dt} = \begin{pmatrix} \frac{2a^2}{h} e \sin f \\ \frac{2a^2}{h} \frac{p}{r} \\ 0 \end{pmatrix} \cdot \begin{pmatrix} a_r \\ a_t \\ a_h \end{pmatrix} = \boldsymbol{k}_a \cdot \boldsymbol{a} \tag{16}$$

it is clear that to decrease the semimajor axis with a maximum rate, the thrust vector has to be along the direction $-\boldsymbol{k}_a$, a vector that is in the orbital plane ($\Rightarrow a_h = 0$).

For all trajectory calculations within this chapter, a direct interplanetary insertion of the solar sail by the launch vehicle with zero hyperbolic excess energy ($C_3 = 0\,\mathrm{km^2\,s^{-2}}$) is assumed. By applying the LSL for decreasing the semimajor axis with a maximum rate to a solar sail with $a_c = 0.1326\,\mathrm{mm\,s^{-2}}$, the trajectory shown in Fig. 5 is obtained. The solar sail would allow to deliver the scientific spacecraft into a circular orbit at ~5.4 solar radii[10] within 6.9 years, but the solar sail film would not sustain the high temperatures that are associated with this steering profile.

Therefore, we have used $T_{\mathrm{lim}} = 300°\mathrm{C}$, the temperature limit of Kapton® [13], as a path constraint for the following calculations. This was realized by constraining the sail attitude so that the light incidence angle cannot become smaller than the critical one, $\alpha \geq \alpha_{\mathrm{lim}}$, where the sail temperature limit of

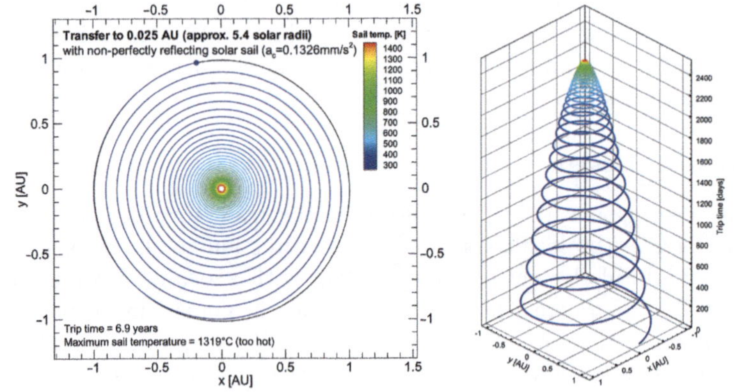

(a) Trajectory

(b) Trajectory over trip time

Fig. 5. Trajectory to a close solar orbit without temperature constraint.

[10] For closer solar distances the numerical integrator, a Runge–Kutta–Fehlberg method of order 4(5), failed to meet the required numerical accuracies (10^{-12} relative error, 10^{-12} absolute error).

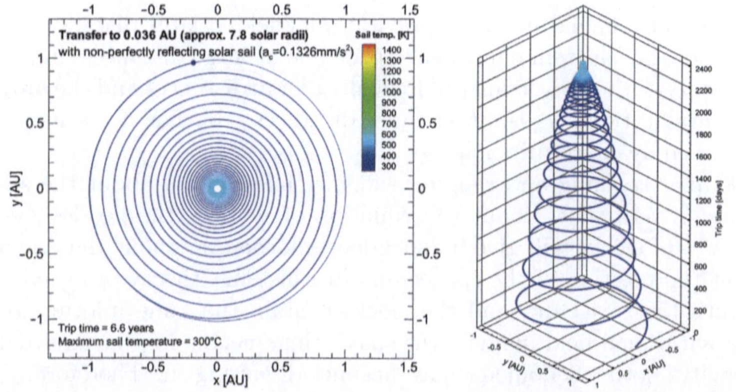

(a) Trajectory (b) Trajectory over trip time

Fig. 6. Trajectory to a close solar orbit with temperature constraint.

300°C would be exceeded. This way, the trajectory shown in Fig. 6 is obtained. The solar sail allows to deliver the scientific spacecraft into a circular orbit at ~7.8 solar radii[11] within 6.6 years. At this distance, the solar sail would be jettisoned to prevent disturbing accelerations on the scientific payload during the measurements.

One can see from Fig. 6b that the orbits close to the Sun take longer because of the larger pitch angles, which keep the sail colder but also reduce the SRP force.

7 Fast Solar System Escape Missions

7.1 Mission Rationale

Recently, the physics in weak gravitational fields attracted much interest because of some unexplained phenomena that occur under these conditions. These weak gravity phenomena are the galactic rotation curves and the Pioneer anomaly (see the review by Lämmerzahl, Preuss, and Dittus on page 75 in this volume, and the article on the Pioneer anomaly by Johann, Dittus, and Lämmerzahl on page 577). Because these phenomena may have a common origin, there are some attempts to describe both phenomena with one theory. Such theories are a Yukawa modification of the gravitational field [14] or the MOND theory [15]. The knowledge of the nature of weak gravitational fields is very important for the understanding the dynamics of galaxies and the universe as a whole.

---

[11] Again, for closer solar distances the numerical integrator failed to meet the required numerical accuracies.

Related to the nature of the gravitational field at large distances is also the validity of the Einstein equivalence principle at large distances, in particular, the validity of the universality of free fall at large distances and the universality of the gravitational redshift at large distances [16]. For a discussion of the Einstein equivalence principle, see [17].

Doppler tracking and ranging will give information about the spacecraft trajectory and, thus, about the equation of motion, the geodesic equation. The rates of clocks will give independent information about the gravitational field at the position of the spacecraft. In Einstein's theory of general relativity, both the trajectory and the clocks contain the same information about the gravitational field, namely the space–time metric. In a generalized theory of gravity, both phenomena may become disentangled. Therefore, either to have an independent tool to explore the gravitational field or to explore new components of gravity, it is important to have a clock onboard of the spacecraft. An additional nice feature about clocks is that – according to standard theory – clocks are sensitive to the gravitational field independently of the state of motion of the clock. Such a scenario yields, in particular, a confirmation or falsification of the Pioneer anomaly and, in the case of confirmation, the resolution of the direction of the anomalous acceleration.

To sum up, solar system escape missions are important for:

- Testing the universality of free fall at large distances
- Testing the universality of the gravitational redshift at large distances
- Measuring the gravitational field via Doppler tracking and ranging (in particular to confirm the Pioneer anomaly and to resolve the direction of the anomalous acceleration)
- Measuring independently the gravitational fields via clocks

Of course, beyond these fundamental quests, there is also much interest in the exploration of the outer planets and the nature of near-interstellar space (e.g., particle content), the heliopause, and trans-Neptunian objects.

7.2 Solar Sail Technology

As the trajectory calculations in Sect. 7.3 will show, a more advanced solar sail technology with $a_c \geq 0.4\,\mathrm{mm\,s^{-2}}$ is required for this mission type. This could be achieved by using larger solar sails, thinner sail films, more advanced materials for the films and the booms, e.g., carbon nanotube sheets [18], and by reducing the mass of the deployment mechanism.

7.3 Mission Design

Evolutionary neurocontrol (ENC) was used for the calculation of near-globally optimal trajectories. This method is based on artificial neural networks (ANNs) and evolutionary algorithms (EAs) and attacks low-thrust trajectory optimization problems from the perspective of artificial intelligence and

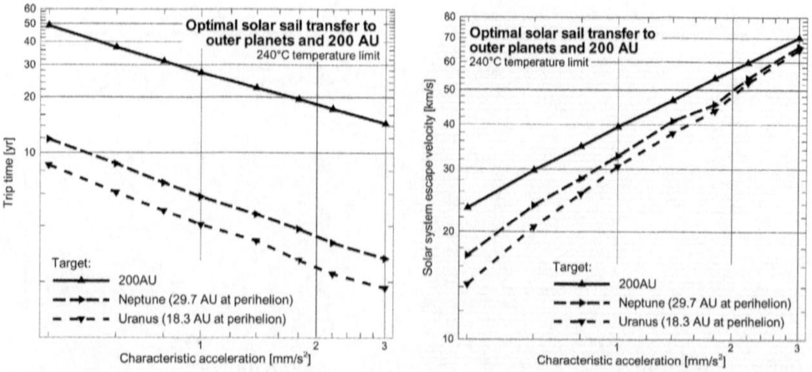

(a) Trip time (b) Solar system escape velocity

Fig. 7. Optimal transfer to the outer planets and to 200 AU.

machine learning. ENC was implemented by one of the authors (Dachwald) within a low-thrust trajectory optimization program called *InTrance*, which stands for "*In*telligent *Tra*jectory optimization using *n*euro*c*ontroller *e*volution" [19–21].

Figure 7 shows the minimum trip times τ and the achieved solar system escape velocities v_{esc} for optimal trajectories to Uranus distance, Neptune distance, and 200 AU using a sail film temperature limit of $T_{lim} = 240°C$.

Figure 7b shows that v_{esc} increases for more distant targets because it is beneficial, in this case, to spend more time in the inner solar system to gain more energy. Figure 7a shows that even medium-term solar sails ($a_c \approx 0.4\,mm\,s^{-2}$) are able to reach Uranus within less than 10 years, and that a little more advanced solar sails ($a_c \approx 0.6\,mm\,s^{-2}$) are able to reach Neptune and the inner Edgeworth–Kuiper belt within less than 10 years. A very advanced solar sail, $a_c = 3.0\,mm\,s^{-2}$, can reach 200 AU within less than 15 years.

Figure 8 shows the trajectories for optimal transfers to 200 AU for four different characteristic accelerations, the solar sail film temperature being limited to $T_{lim} = 240°C$ in all four cases. One can see that more and more SPAs are required as the characteristic acceleration of the sail decreases. The optimal trajectory for the lightest solar sail ($a_c = 3.0\,mm\,s^{-2}$) makes only a single SPA, whereas the optimal trajectory for the heaviest solar sail ($a_c = 0.5\,mm\,s^{-2}$) first spirals toward the Sun and then requires four SPAs to reach 200 AU in minimum time. The lower the characteristic acceleration of the solar sail, the larger also the fraction of flight time that must be spent in the inner solar system for gaining orbital energy [22]. For solar sailcraft with a low characteristic acceleration, the larger flight time fraction in the inner solar system may – additionally to the longer total flight time – render sail film degradation a serious problem [22, 23].

Figure 9 shows for three different sail temperature limits (200, 240, and 280°C) the minimum trip times and the achieved solar system escape velocities

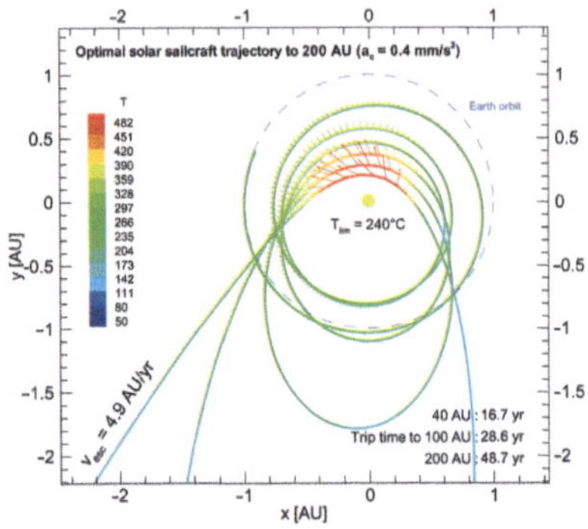

(a) $a_c = 0.4\,\mathrm{mm/s^2}$

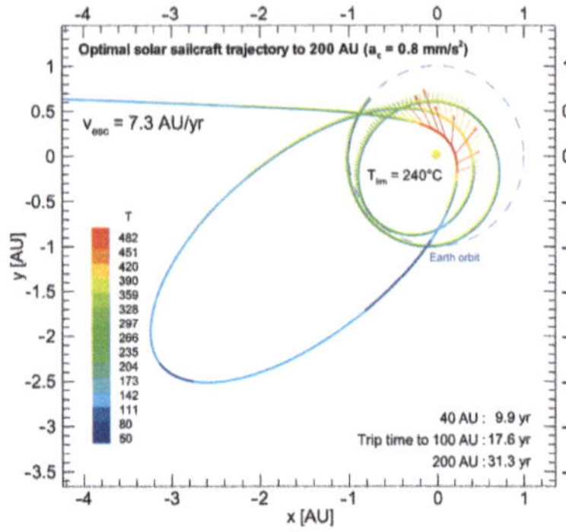

(b) $a_c = 0.8\,\mathrm{mm/s^2}$

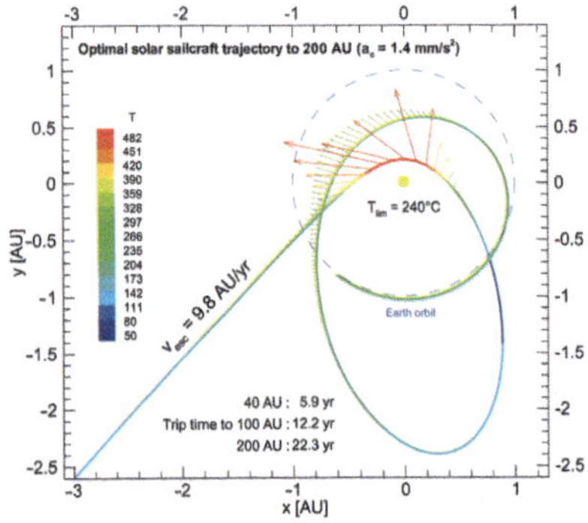

(c) $a_c = 1.4\,\mathrm{mm/s^2}$

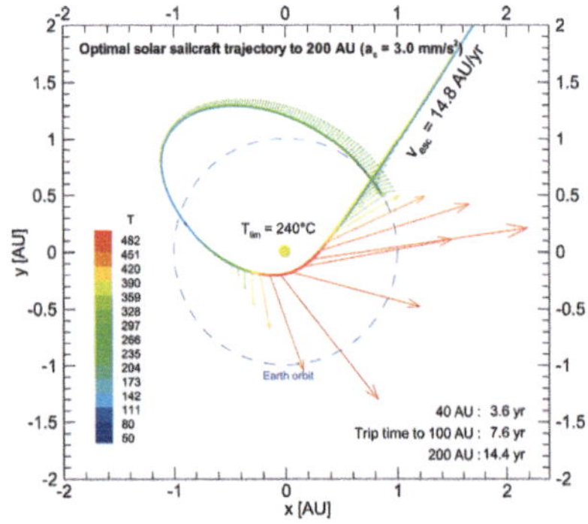

(d) $a_c = 3.0\,\mathrm{mm/s^2}$

Fig. 8. Geometry of optimal transfer trajectories to 200 AU for different characteristic accelerations.

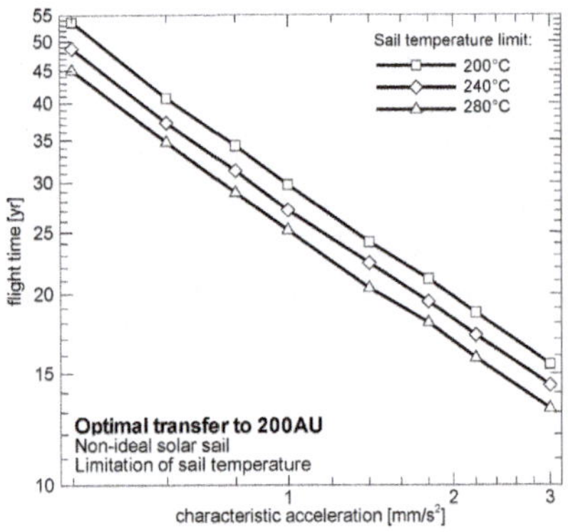

(a) Trip time

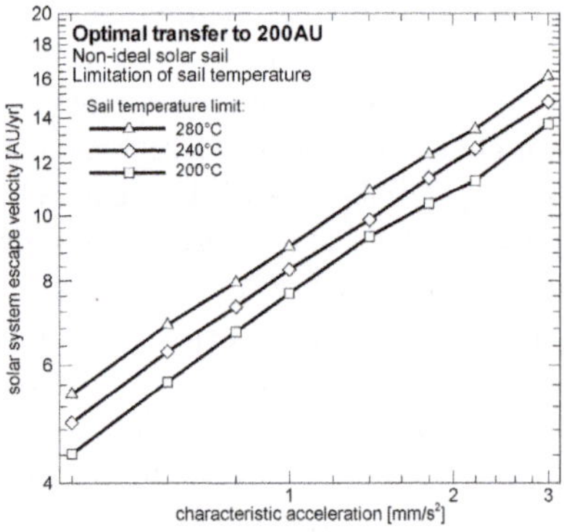

(b) Solar system escape velocity

Fig. 9. Optimal transfer to 200 AU for different solar sail temperature limits.

for optimal transfers to 200 AU. A translation of the sail temperature limits into sail film materials is not within the scope of this chapter because the allowed maximum sail film temperature depends not only on the film material, but also on the sail design (stresses, wrinkles, etc.).

Figure 9 shows that the minimum trip times and the achieved solar system escape velocities obey a potential law for all sail temperature limits,

$$\tau(a_c, T_{\text{lim}}) = c_1(T_{\text{lim}})a_c^{c_2}$$

with $c_2 = 0.543^{\pm 0.011}$ and

$$v_{\text{esc}}(a_c, T_{\text{lim}}) = c_3(T_{\text{lim}})a_c^{c_4}.$$

with $c_4 = -0.605^{\pm 0.006}$. Note that both c_2 and c_4 do not depend on T_{lim}. Figure 9 shows that the minimum trip times to 200 AU depend considerably on the sail temperature limit (r_{min} varies also considerably with T_{lim}, but little with a_c: $0.222\,\text{AU} < r_{\text{min}} < 0.238\,\text{AU}$ for $T_{\text{lim}} = 200°\text{C}$, $0.195\,\text{AU} < r_{\text{min}} < 0.204\,\text{AU}$ for $T_{\text{lim}} = 240°\text{C}$, and $0.165\,\text{AU} < r_{\text{min}} < 0.181\,\text{AU}$ for $T_{\text{lim}} = 280°\text{C}$). For $T_{\text{lim}} = 280°\text{C}$, a characteristic acceleration of about $1.0\,\text{mm}\,\text{s}^{-2}$ is required to reach 200 AU within 25 years from launch, whereas a characteristic acceleration of about $1.4\,\text{mm}\,\text{s}^{-2}$ is required for $T_{\text{lim}} = 200°\text{C}$. Note that the solar sail design parameters are typically very sensitive with respect to the characteristic acceleration, as the following example may show: if a solar sail with a sail assembly loading of $\sigma_s = 5\,\text{g}\,\text{m}^{-2}$ should be used to transport a payload (including spacecraft bus) of $m_p = 50\,\text{kg}$ to 200 AU, $a_c = 1.0\,\text{mm}\,\text{s}^{-2}$ yields according to (9) a sail area of $A = 123 \times 123\,\text{m}^2$, whereas $a_c = 1.4\,\text{mm}\,\text{s}^{-2}$ yields a sail area of $A = 233 \times 233\,\text{m}^2$. If the sail size is held fixed at $123 \times 123\,\text{m}^2$, the payload reduces to $m_p = 14\,\text{kg}$. Another possibility is to decrease the sail assembly loading to $\sigma_s = 2.6\,\text{g}\,\text{m}^{-2}$, which can only be done with some much more advanced sail fabrication/deployment technology.

8 Conclusions

Solar sails allow high-energy trajectories that are difficult or even impossible to accomplish with any other type of conventional propulsion system. We have shown that solar sails are also an enabling propulsion technology for missions that go very close to the Sun and for missions that go fast very far away from the Sun. Those two types of deep-space missions are very favorable for doing fundamental physics in unique gravitational environments. Even near- to medium-term solar sails ($a_c \approx 0.13\,\text{mm}\,\text{s}^{-2}$) are able to deliver spacecraft into very close (circular) solar orbits (≈ 8 solar radii in ≈ 7 years). Medium-term solar sails ($a_c \approx 0.4\,\text{mm}\,\text{s}^{-2}$) are required to deliver spacecraft fast to the outer solar system (≈ 40 AU in ≈ 16 years), but very advanced solar sails ($a_c \approx 3.0\,\text{mm}\,\text{s}^{-2}$) are necessary to deliver spacecraft fast into near-interstellar space (200 AU in ≈ 15 years).

References

1. C.G. Sauer. Optimum solar-sail interplanetary trajectories. San Diego, USA, August 1976. AIAA/AAS Astrodynamics Conference. AIAA Paper 76-792.
2. M. Leipold and O. Wagner. 'Solar photonic assist' trajectory design fo solar sail missions to the outer solar system and beyond. In T. H. Stengle, editor, *Spaceflight Dynamics 1998*, volume 100 Part 2 of *Advances in the Astronautical Sciences*, pages 1035–1045. Univelt, Inc., 1998.
3. M. Leipold. To the sun and Pluto with solar sails and micro-sciencecraft. *Acta Astronautica*, 45(4-9):549, 1999.
4. J. L. Wright. *Space Sailing.* Gordon and Breach Science Publishers, Philadelphia, 1992.
5. C. R. McInnes. *Solar Sailing. Technology, Dynamics and Mission Applications.* Springer–Praxis Series in Space Science and Technology. Springer–Praxis, Berlin, Heidelberg, New York, Chicester, 1999.
6. M. Leipold, M. Eiden, C.E. Garner, L. Herbeck, D. Kassing, T. Niederstadt, T. Krüger, G. Pagel, M. Rezazad, H. Rozemeijer, W. Seboldt, C. Schöppinger, C. Sickinger, and W. Unkenbold. Solar sail technology development and demonstration. Laurel, USA, 2000. 4th IAA International Conference on Low-Cost Planetary Missions. IAA-L-0707.
7. W. Seboldt, M. Leipold, M. Rezazad, L. Herbeck, W. Unkenbold, D. Kassing, and M. Eiden. Ground-based demonstration of solar sail technology. Rio de Janeiro, Brazil, 2000. 51st International Astronautical Congress. IAF-00-S.6.11.
8. M. Leipold, C.E. Garner, R. Freeland, A. Herrmann, M. Noca, G. Pagel, W. Seboldt, G. Sprague, and W. Unckenbold. ODISSEE – A proposal for demonstration of a solar sail in Earth orbit. *Acta Astronautica*, 45(4-9):557, 1999.
9. T. Damour. Equivalence principle and clocks. In J. Trân Tanh Vân, J. Dumarchez, J. Reynaud, C. Salomon, S. Thorsett, and J.Y. Vinet, editors, *Gravitational Waves and Experimental Gravity*, page 357. World Publishers, Hanoi, 2000.
10. T. Damour, F. Piazza, and G. Veneziano. Violations of the equivalence principle in a dilaton–runaway scenario. *Phys. Rev.*, D 66:046007, 2002.
11. C. Wetterich. Probing quintessence with time variation of couplings. *Astropart. Phys.*, 10:2, 2003.
12. R. H. Battin. *An Introduction to the Mathematics and Methods of Astrodynamics.* AIAA Education Series. American Institute of Aeronautics and Astronautics, Reston, revised edition, 1999.
13. M. Leipold. *Solar Sail Mission Design.* Doctoral thesis, Lehrstuhl für Flugmechanik und Flugregelung; Technische Universität München, 1999. DLR-FB-2000-22.
14. R.H. Sanders. Anti–gravity and galaxy rotation curves. *Astron. Astroph.*, 136:L21, 1984.
15. M. Milgrom. MOND - theoretical aspects. *New Astr. Rev.*, 46:741, 2002.
16. J.D. Anderson and J.G. Williams. Long-range tests of the equivalence principle. *Class. Quantum Grav.*, 18:2447, 2001.
17. C. Lämmerzahl. The Einstein Equivalence Principle and the search for new physics. In D. Giulini, C. Kiefer, and C. Lämmerzahl, editors, *Quantum Gravity – From Theory to Experimental Search*, page 367. Springer Verlag, Berlin, 2003.

18. M. Zhang, S. Fang, A.A. Zakhidov, S.B. Lee, A.E. Aliev, C.D. Williams, K.R. Atkinson, and R.H. Baughman. Strong, transparent, multifunctional, carbon nanotube sheets. *Science*, 309:1215, 2005.
19. B. Dachwald. *Low-Thrust Trajectory Optimization and Interplanetary Mission Analysis Using Evolutionary Neurocontrol.* Doctoral thesis, Universität der Bundeswehr München; Fakultät für Luft- und Raumfahrttechnik, 2004.
20. B. Dachwald. Optimization of interplanetary solar sailcraft trajectories using evolutionary neurocontrol. *Journal of Guidance, Control, and Dynamics*, 27(1):66, 2004.
21. B. Dachwald. Optimization of very-low-thrust trajectories using evolutionary neurocontrol. *Acta Astronautica*, 57(2-8):175, 2005.
22. B. Dachwald. Optimal solar sail trajectories for missions to the outer solar system. *Journal of Guidance, Control, and Dynamics*, 28(6):1187, 2005.
23. B. Dachwald, G. Mengali, A.A. Quarta, and M. Macdonald. Parametric model and optimal control of solar sails with optical degradation. *Journal of Guidance, Control, and Dynamics.* 29(5):1170, 2006.

Part IV

Missions and Projects

Testing Relativity with Space Astrometry Missions

Sergei A. Klioner

Lohrmann Observatory, Dresden Technical University, Mommsenstr. 13, 01062 Dresden, Germany

Summary. The relativistic aspects of space astrometry missions like Gaia and SIM are summarized. After a short overview of the relativistic modelling of high-accuracy positional observations, various relativistic tests with astrometric data are discussed.

1 Introduction

The quick progress in accuracy of positional observations that we witness in the last decades (the accuracy of astrometric VLBI in radio band and HIPPARCOS in optical band is 50–100 times better than astrometric accuracy 20 years ago) is expected to continue even faster due to the space astrometry projects [3,12,47], SIM [54] and JASMINE [16] to be launched within a decade. The positional accuracy should attain the level of 1 μas, which is the angle at which an observer in Europe would see the thickness of a sheet of paper on which this book in printed (about 30 μm) if the book is in New Zealand. This accuracy, which we could not dream of 20 years ago, makes it possible to boost our knowledge in many fields of astronomy and also in gravitational physics. Especially, promising for gravitational physics is Gaia because of its 10^9 objects observed as close as 45° from the Sun. In this paper, we deal mostly with relativistic experiments with Gaia, although the relativistic modelling and the relativistic experiments for other missions can be considered along the same lines.

The ESA project Gaia [12, 47] aimed at measuring positions, proper motions, parallaxes, radial velocities and photometric parameters of about 10^9 celestial objects can be used not only for the declared principal scientific goals of the mission (reference frame, stellar physics and evolution, Galaxy dynamics), but also as a tool to test relativity in a variety of ways. Considering the experience with similar but less ambitious ESA project HIPPARCOS [14], the merit of Gaia for testing relativity has been recognized from the very beginning of the mission planning [12]. The main expected relativistic experiments with Gaia are the measurements of gravitational light deflection with a precision of about 10^{-6} and of relativistic perihelion advances of asteroids with a precision of about 10^{-4}. These relativistic experiments were discussed in detail in a number of publications [12, 18, 19, 40, 61]. However, a number of additional experiments can be performed provided that the whole Gaia data processing is made compatible within a single consistent relativistic framework. Several such experiments are suggested and discussed below.

Needless to say how important it is for fundamental physics and astronomy to measure the predictions of general relativity as reliably and as precisely as possible. Recently, theoretical arguments were put forward suggesting that the present agreement between general relativity and experiment may be naturally compatible with the existence of a scalar contribution to gravity [7, 43]. Later similar mechanisms were formulated in a non-metric framework by Damour and Polyakov [8] and in the framework of string theory and inflatory cosmology (the so-called *dilaton-runaway scenario*) by Damour et al. [9]. In particular, Damour and Nordtvedt [6, 7] have found that a scalar–tensor theory of gravity may contain a "built-in" cosmological attractor mechanism towards general relativity. A possible scenario for cosmological evolution of the scalar field was given by Nordtvedt [43] and Damour and Nordtvedt [7]. All these arguments assume that the parameter $1-\gamma$ was of the order of 1 in the early universe, at the time of inflation, and has evolved to be close to zero at the present time. The analyses discussed above predict that small deviations from general relativity are currently present in the range from 4×10^{-5} to $\sim 5\times10^{-8}$ of the post-Newtonian effects. These predictions strongly motivate new searches for very small deviations of relativistic gravity in the solar system. Although Gaia will not be able to test the most stringent limit of that prediction, it is expected to produce a determination of γ with an accuracy of 10^{-6} – 5×10^{-7}, which is the most precise determination of that PPN parameter expected by 2015.

Let us first consider the relativistic modelling of positional observations with an accuracy of 1 μas and then discuss the possibilities to test relativity with these observations.

2 Modelling of Positional Observations in Newtonian Physics

Reduction scheme of positional observations in Newtonian physics is rather simple. Absolute Euclidean space and absolute time of Newtonian physics lead to the existence of global preferred coordinates: inertial coordinates that are unique up to a constant shift of the origin of the time coordinate, time-independent rotation of spatial axes and a shift of the origin of spatial coordinates, which is at most linear in time. Although already in Newtonian physics one can introduce arbitrary coordinates (e.g. some curvilinear coordinates), inertial coordinates are certainly preferred, since the laws of physics look especially simple when expressed in an inertial reference system. Moreover, observed quantities (distances, directions, etc.) are directly related to those global inertial coordinates.

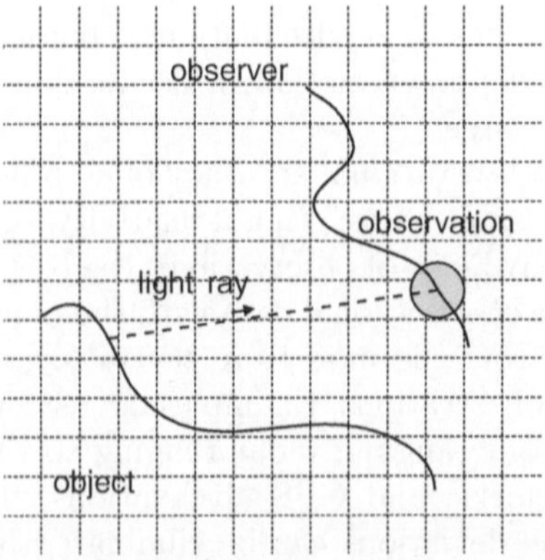

Fig. 1. Four parts of an astronomical event from the point of view of Newtonian physics: (1) motion of the observed object; (2) motion of the observer; (3) trajectory of an electromagnetic signal from the observed object to the observer, which is tacitly assumed to be a straight line in Newtonian astronomy and (4) the process of observation responsible for Newtonian aberration. The coordinate grid in the background symbolizes a global inertial reference system.

Let us briefly consider the Newtonian scheme of reduction of astronomical observations. Figure 1 sketches the four constituents of an astronomical observation from the point of view of Newtonian physics (1) motion of the observed object, (2) motion of the observer, (3) propagation of an electromagnetic signal from the object to the observer and (4) the process of observation. The last two parts can be formulated in a quite simple way in Newtonian physics. It is normally tacitly assumed here that the light rays are straight lines in some inertial coordinates. As for "the process of observation", it is responsible for the appearance of Newtonian aberration that reflects the difference in observed directions to the source as seen by a moving observer and by an observer at rest relative to the chosen coordinates.

The goal of Newtonian reduction of astronomical observations is to model (to predict) the results of observations performed by a fictitious observer (normally situated at the origin of the chosen reference system, e.g. at the barycenter of the solar system), at some given moment of time. One attempts here to correct for all the effects in observations that are produced by the motion and the barycentric position of the real observer (aberration and, e.g. parallax, respectively) and by the motion of the object (proper motion and, possibly, light travel time effects). The structure of a Newtonian reduction scheme does not depend on the goal accuracy of reduction and can be described as follows (1) aberration, (2) parallax, (3) proper motion and/or (4) light travel time effects. For lower accuracies when only linear effects from aberration, parallax and proper motion are of interest, one could apply the corresponding corrections in arbitrary order. On the contrary, for higher accuracies the order of these reductions is important. All parameters of the model, i.e. the coordinates of the observer and the object as functions of time, are defined in the chosen inertial reference system. That is, the five standard astrometric parameters of the object (right ascension α, declination δ, parallax π, proper motion in right ascension μ_α and proper motion in declination μ_δ) are also defined in the chosen reference system.

Rapid increase of observational accuracy of astronomical observations has already made indispensable to use general relativity for modelling of the observational data. For many kinds of observations, the Newtonian scheme sketched above fails to describe observational data with the required accuracy. In many cases the deviations from the model are several orders of magnitude larger than the accuracy of observations. Examples are astrometric (geodetic) VLBI observations, lunar laser ranging, radar ranging to the planets, experiments with high-accuracy clocks, and GPS observations. It is also widely known and accepted that the deviations can be eliminated by using Einstein's general theory of relativity (instead of Newtonian physics) for the modelling of observations.

The accuracy of positional observations to be produced by Gaia is expected to attain 2–3 μas for the stars with magnitude $V < 10$ mag and 15 μas for the stars of $V = 15$ mag. It is clear that not only the largest relativistic effects but also many additional subtle effects should be taken into account to attain that accuracy. It is also quite clear that relativistic effects cannot be considered as small corrections to a Newtonian model as has been often done earlier when the accuracy was not so high. The whole model should be formulated in a language compatible with general relativity. In such a relativistic framework, many Newtonian concepts must be abandoned and the meaning of astrometric parameters such as position, parallax and proper motion of a star should be redefined.

3 Relativistic Modelling of Astronomical Observations

Let us now outline general principles of relativistic modelling of astronomical observations. It is interesting that in spite of a deep conceptual difference between Newtonian physics and general relativity, the structure of the reduction scheme changes, in principle, only in one point: light rays are no longer straight lines and should be carefully modelled. Figure 2 shows the four constituents of an astronomical observation in the relativistic framework. In curved space–time, there is no preferred coordinates where the laws of physics would have substantial simpler form than in other coordinates. Therefore, any reference system covering the space–time region under study can be used. Instead of Newtonian inertial coordinates, one has to choose some reference system in curved space–time, which is sketched symbolically on Fig. 2 as a grid of curved coordinates.

3.1 General Scheme of Relativistic Modelling

General scheme of relativistic modelling is presented on Fig. 3. Starting from general theory of relativity, any other metric theory of gravity or the PPN

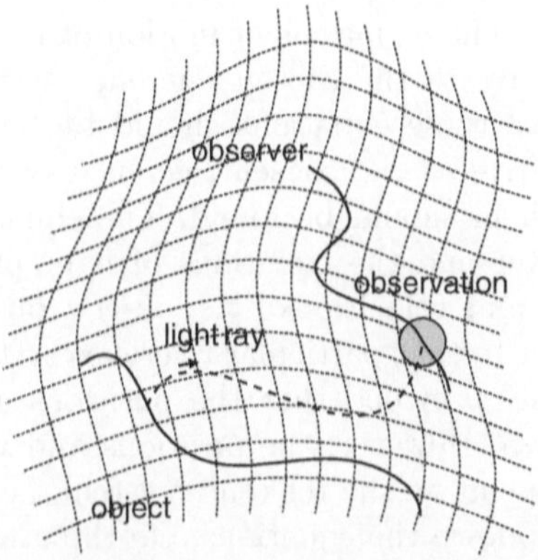

Fig. 2. Four parts of an astronomical event from the point of view of relativistic physics: (1) motion of the observed object; (2) motion of the observer; (3) trajectory of an electromagnetic signal from the observed object to the observer, which represents a geodetic line (i.e. a *"curved" line*) in the chosen reference system and (4) the process of observation. The grid of curved coordinates in the background symbolizes the chosen relativistic reference system.

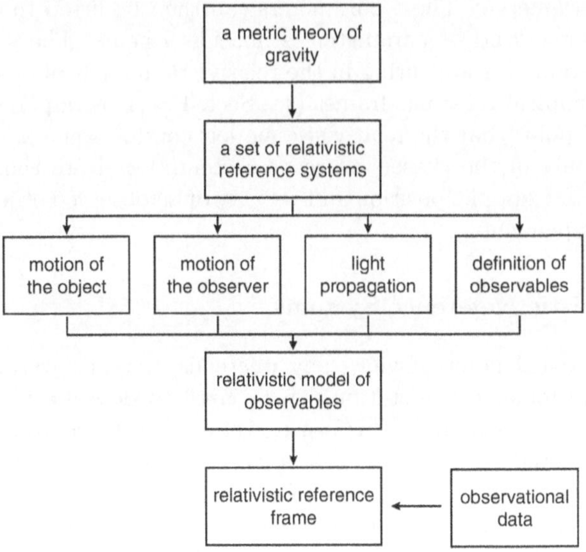

Fig. 3. General principles of relativistic modelling of astronomical observations (see text for further explanations).

formalism one should define at least one relativistic four-dimensional reference system covering the region of space–time where all the processes constituting particular kind of astronomical observations are located. Each of four constituents of an astronomical observation should be modelled in the relativistic framework. The equations of motion of both the observed object and the observer relative to the chosen reference system should be derived and a method to solve these equations should be found. The equations of light propagation relative to the chosen reference system should be derived and a way to solve them should be found. The equations of motion of the object and the observer and the equations of light propagation enable one to compute positions and velocities of the object, observer and the photon (light ray) with respect to that particular reference system at a given moment of the coordinate time, provided that the positions and velocities at some initial epoch are known. However, the positions and velocities calculated in this way obviously depend on the reference system, i.e. on the preferences of the person who writes down the equations. On the other hand, the results of observations cannot depend on the choice of the reference system. Therefore, it is clear that one more step of the modelling is needed: a relativistic description of the process of observation. This part of the model allows one to compute a coordinate-independent theoretical prediction of observables starting from the coordinate-dependent quantities mentioned earlier.

These four components can now be combined into relativistic models of observables. The models give expressions for relevant observables as functions of a set of parameters. These parameters can then be fitted to observational data using some kind of parameter estimation scheme. The sets of certain estimated parameters appearing in the relativistic models of observables represent astronomical reference frames (see Sect. 4.2). It is important to understand at this point that the relativistic models contain some parameters that are defined only in the chosen reference system(s) and are thus coordinate-dependent. For example, position and velocity of an observed object are clearly coordinate-dependent.

3.2 Relativistic Reference Systems

From the physical point of view, any reference system covering the region of space–time under consideration can be used to describe physical phenomena within that region. In this sense, we are free to choose the reference system to be used to model the observations. However, reference systems, in which mathematical description of physical laws is in one sense or another simpler than in some other reference systems, are more convenient for practical calculations. Therefore, one can use the freedom to choose the reference system to make the parametrization as convenient and reasonable as possible.

Three relativistic reference systems play a role for modelling of high-accuracy positional observations: Barycentric Celestial Reference System (BCRS), Geocentric Celestial Reference System (GCRS) and Centre-of-Mass Reference System (CoMRS) of the satellite. The latter represents a local proper reference system of the satellite considered as a massless observer.

The Barycentric Celestial Reference System

Two working groups on relativity in astrometry, celestial mechanics and metrology established in 1997 by the International Astronomical Union (IAU) and Bureau International des Poids et Mesures (BIPM) have come to the conclusion that the most convenient relativistic reference system for the applications in astrometry, solar system dynamics, and time keeping and dissemination is defined by the following metric tensor [59]:

$$\begin{aligned} g_{00} &= -1 + \frac{2w}{c^2} - \frac{2w^2}{c^4} + \mathcal{O}(c^{-5}), \\ g_{0i} &= -\frac{4}{c^3} w^i + \mathcal{O}(c^{-5}), \\ g_{ij} &= \delta_{ij} \left(1 + \frac{2}{c^2} w\right) + \mathcal{O}(c^{-4}) \end{aligned} \tag{1}$$

with the post-Newtonian potentials w and w^i defined by

$$\begin{aligned} w(t, \boldsymbol{x}) = G \int d^3x' \, \frac{\sigma(t, \boldsymbol{x}')}{|\boldsymbol{x} - \boldsymbol{x}'|} \\ + \frac{1}{2c^2} G \frac{\partial^2}{\partial t^2} \int d^3x' \, \sigma(t, \boldsymbol{x}') |\boldsymbol{x} - \boldsymbol{x}'| \, , \end{aligned} \tag{2}$$

$$w^i(t, \boldsymbol{x}) = G \int d^3x' \frac{\sigma^i(t, \boldsymbol{x}')}{|\boldsymbol{x} - \boldsymbol{x}'|} \, , \tag{3}$$

σ and σ^i being related to the components of the energy–momentum tensor $T^{\alpha\beta}$ as

$$\sigma = \frac{1}{c^2} \left(T^{00} + T^{ss}\right), \quad \sigma^i = \frac{1}{c} \, T^{0i} \, . \tag{4}$$

The origin of spatial coordinates of this reference system is chosen to coincide with the barycenter of the solar system. The reference system defined in this way is called Barycentric Celestial Reference System (BCRS). The BCRS has been explicitly recommended by the IAU for the modelling of high-accuracy astronomical observations [20,52,59]. With its current definition, the BCRS is a post-Newtonian reference system with higher-order terms neglected in the metric tensor (1). This is satisfactory since the post-Newtonian approximation is sufficient to model any observations in the foreseeable future (including micro-arcsecond astrometry as long as the observations are made further than about one degree from the Sun). Post-post-Newtonian terms can be added to the metric tensor as soon as they are necessary for some applications (for example for LATOR [60]). The word "celestial" in the name of BCRS is used to stress that the BCRS does not rotate with the Earth and that remote sources do not move relative to the BCRS in some averaged sense. The PPN version of the BCRS valid for certain class of metric theories of gravity can be found in Klioner and Soffel [27] and Will [63]. The BCRS will be used to model Gaia observations. This is a reference system underlying the resulting Gaia catalogue (see Sect. 4.2). The coordinate time of the BCRS is called Barycentric Coordinate Time (TCB). The TCB will be used to parametrize the Gaia catalogue.

The Geocentric Celestial Reference System

The second reference system defined by the same IAU resolutions [52] is the GCRS. This reference system is only marginally important for Gaia (mostly for modelling of orbit tracking data and relating the Gaia onboard clock to TAI [21]). Nevertheless, it is important to understand the principles of GCRS, since they are important for the CoMRS of the satellite.

The GCRS is constructed for any of the bodies belonging to an N-body system. A body is just a region of space–time (a "world tube") where the energy–momentum tensor is not zero (i.e. a body is a blob of matter), the energy–momentum tensor being assumed to be zero between the bodies. The GCRS possesses two remarkable properties:

(A) The gravitational field of external bodies is represented only in the form of a relativistic tidal potential, which is at least of second order in the local spatial coordinates and coincides with the usual Newtonian tidal potential in the Newtonian limit.

(B) The internal gravitational field of the subsystem coincides with the gravitational field of the corresponding isolated source provided that the tidal influence of the external matter is neglected.

These two requirements can simultaneously be satisfied in general relativity as a consequence of the validity of the Strong Equivalence Principle [27,59]. The GCRS is the reference system where the influence of external matter (all other bodies of the N-body system except for the arbitrarily selected central body) is effaced as much as it is allowed by general relativity. The GCRS can be constructed for any body of the N-body system, but is, currently, especially important for the Earth (hence the word "geocentric"). The metric tensor of in the GCRS coordinates $(T, \boldsymbol{X})$ has the same form as (1), but with geocentric gravitational potentials W and W^a instead of w and w^i, respectively. Both these geocentric potentials consists of three parts:

$$W(T,\boldsymbol{X}) = W_E + Q_a\,X^a + W_T, \tag{5}$$
$$W^a(T,\boldsymbol{X}) = W_E^a + \tfrac{1}{2}\,\varepsilon_{abc}\,C_b\,X^c + W_T^a\,, \tag{6}$$

where W_E and W_E^a is the gravitational potential of the Earth, W_T and W_T^a are tidal gravitational potentials due to external matter and the two other terms $Q_a(T)\,X^a$ and $\frac{1}{2}\,\varepsilon_{abc}\,C_b(T)\,X^c$, Q_a and C_a being arbitrary functions of time T, describe inertial forces. The Q_a defines the translational motion of the GCRS origin relative to the geocentric momentarily co-moving locally inertial reference system. In other words, an accelerometer attached at the GCRS origin measures Q_a [27, Sect. VIII]. The C_a defines the rotational motion of the spatial axes of the GCRS relative to the momentarily co-moving locally inertial reference system. Clearly, the equations of test particles relative to the GCRS with $C_a \neq 0$ contain Coriolis forces. Possible choices of C_a and its relation to the rotational matrix in the coordinate transformations between the GCRS and BCRS are discussed, e.g. in [59].

Local Reference System of the Observer

The local reference system of a test (massless) observer is a version of the GCRS for a massless body. As discussed by Klioner [24], this reference system is called CoMRS in the Gaia nomenclature and can be constructed from the BCRS in totally the same way as the GCRS except that all the internal potentials should be set to zero. The CoMRS possesses property (B) formulated above and, therefore, is adequate for modelling physical processes in the immediate vicinity of the satellite (to define observables, and kinematical and/or dynamical description of the satellite's attitude).

BCRS as a Local Reference System of the Solar System

Normally, one believes that the BCRS is constructed for the solar system considered to be isolated. One neglects herewith two kinds of effects (1) tidal forces due to any particular external body (e.g. nearby stars or the Galaxy) and (2) effects of cosmological background. The tidal forces can be easily taken into account using the same theoretical framework as that used to construct the GCRS (e.g. [59]): the BCRS for the solar system is then the same as the GCRS for the Earth. The BCRS metric tensor would contain some tidal potentials from other stars or the Galaxy as a whole, but those potentials can be demonstrated to be numerically negligible.

To include effects from the cosmological background a new approach is necessary. Klioner and Soffel [28] have argued that neglecting the interaction of the cosmological fluid (including the cosmological constant) with the solar system matter, one gets the following simple version of the BCRS metric with cosmological terms

$$g_{00} \approx -1 + \frac{2}{c^2}\, w - \frac{2}{c^4}\, w^2 + \frac{1}{c^2}\, A_1(t)\, |\boldsymbol{x}|^2,$$

$$g_{0i} \approx -\frac{4}{c^3}\, w^i,$$

$$g_{ij} \approx \delta_{ij} \left(1 + \frac{2}{c^2}\, w + \frac{1}{c^2}\, B_1(t)\, |\boldsymbol{x}|^2\right), \tag{7}$$

$$A_1 = \frac{\ddot{a}}{a}, \tag{8}$$

$$B_1 = -\frac{1}{2} \left(\left(\frac{\dot{a}}{a}\right)^2 + \frac{k\, c^2}{a^2}\right), \tag{9}$$

$a(t)$ and $k = -1, 0, +1$ being the usual parameters of the Robertson–Walker metric. The details of the derivation and a discussion of the neglected terms can be found in [28].

3.3 Motion of the Objects and the Observer

Typically, for objects situated in the solar system (asteroids, planets, space vehicles) the equations of motion are ordinary differential equations of second order and numerical integration with suitable initial or boundary conditions can be used to solve them. For objects outside of the solar system, one use often simple models like uniform and rectilinear motion in space or more complicated ones, e.g. for binary stars. In any case one should understand that in the relativistic framework, all these ad hoc models give positions and velocities of observed objects in the chosen relativistic reference system.

The principal relativistic effects in the translational motion of solar system bodies (including Gaia satellite, asteroids, etc.) in the BCRS are contained in the so-called Einstein–Infeld–Hoffmann (EIH) equations of motion of N gravitating bodies, whose gravitational fields can be described by their masses M_A only:

$$\ddot{\boldsymbol{x}}_A = -\sum_{B\neq A} G M_B \frac{\boldsymbol{x}_A - \boldsymbol{x}_B}{|\boldsymbol{x}_A - \boldsymbol{x}_B|^3} + \frac{1}{c^2}\mathbf{F}_{pN}(M_B, \boldsymbol{x}_B, \dot{\boldsymbol{x}}_B) + \mathcal{O}(c^{-4}) \,. \quad (10)$$

The Newtonian part of these equations (shown explicitly above) follows from the term of order c^{-2} in g_{00}. The relativistic terms require all other terms in the BCRS metric tensor specified above. Various parts of these equations represent (1) relativistic perihelion advance ($\sim 43''$ per century for Mercury, $\sim 10''$ per century for Icarus, etc.), (2) geodetic precession ($\sim 2''$ per century for Lunar orbit) and (3) various periodic relativistic effects (important mostly for LLR and binary pulsar timing observations). Further effects not contained in the EIH equations are the effects due to rotation of the bodies (Lense–Thirring or gravitomagnetic effects) and those due to non-sphericity of the gravitating bodies. These additional effects are marginal for the current accuracy of LLR and SLR, but negligible for Gaia.

In case of Gaia satellite, one should use a slightly simplified version of the EIH equations, since the influence of the mass of the satellite on the motion of other gravitating bodies can be neglected. The influence of the post-Newtonian force $\mathbf{F}_{pN}$ on the motion of Gaia has been investigated in detail by Klioner [25]. Klioner [25] has considered six different dynamical models for the motion of Gaia: the purely Newtonian mode (N), the full post-Newtonian model (pN) and four restricted post-Newtonian models with the following bodies in the post-Newtonian force $\mathbf{F}_{pN}$: the Sun only (S), the Sun and the Earth (S+E), the Sun, Earth and Jupiter (S+E+J) and the Sun, Earth and the Moon (S+E+M). The errors of the Newtonian and restricted

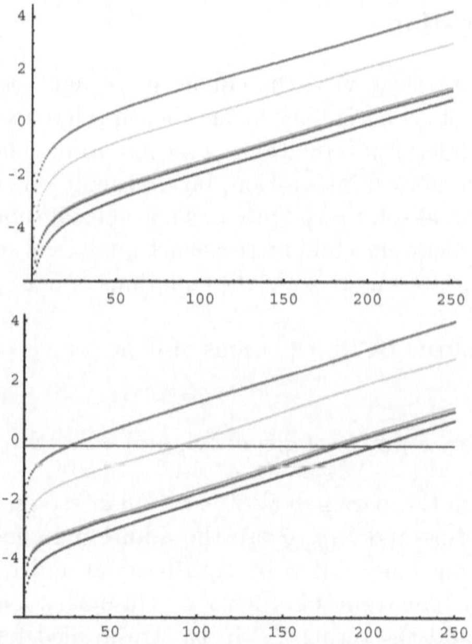

Fig. 4. The maximal errors in positions (in km, logarithmic scale (e.g. −2 corresponds to 10^{-2} km), *upper panel*) and velocity (in mm s^{-1}, logarithmic scale (e.g. 0 corresponds to $10^0 = 1$ mm s^{-1}), *lower panel*) of models N (*upper line*), S (*second upper line*), S+E (*third upper line*), S+E+J (*fourth upper line*), and S+E+M (*lower line*) grow in time (abscissa on both plots is the time in days from the start of integrations). The plots show maximal errors for the given time span among the 12,000 integrations performed by Klioner [26]. Logarithmic growth for all the models is easy to see. The black horizontal line on the lower plot corresponds to the goal velocity accuracy of 0.6 mm s^{-1} adopted for Gaia. The intersections of that line and the curves give the maximal time spans during which the corresponding errors remains below 0.6 mm s^{-1}.

post-Newtonian models are then found by comparing the results of numerical integrations to the results obtained with the full post-Newtonian model using the same initial conditions on a Lissajous orbit around the Lagrange point L_2 of the Sun–Earth–Gaia problem. The main results of this study, which will help to optimize the Gaia orbit modelling and determination, are represented on Fig. 4.

The GCRS metric tensor allows one also to derive the equations of rotational motion of an extended body (i.e. of the Earth). In the same way, the CoMRS allows one to model the rotational motion of the satellite [24]. These equations will not be discussed here, since they are only marginally important for Gaia.

3.4 Light Propagation

In any metric theory of gravity, the equations of light propagation coincide with the equations of geodetic lines in the chosen reference system. The equations are ordinary differential equations of second order. These equations could also be solved by numerical integration, but normally one prefers to use some approximate analytical solutions. Only in some special (normally, highly symmetrical) cases like Schwarzschild metric exact analytical solutions are known. Anyway, an appropriate way to solve the equations of light propagation should be found.

The structure of the BCRS equations of light propagation can be written as follows:

$$\boldsymbol{x}(t) = \boldsymbol{x}_0 + c\,\boldsymbol{\sigma}\,(t - t_0) + c^{-2}\,\mathbf{S}_{pN}(t) + c^{-3}\,\mathbf{S}_{1.5pN}(t)\,, \tag{11}$$

where $\boldsymbol{x}_0$ and $\boldsymbol{\sigma}$ are the parameters of Newtonian straight line, $\mathbf{S}_{pN}$ are the post-Newtonian terms and $\mathbf{S}_{1.5pN}$ are the additional effects induced by the motion of gravitating matter (i.e. by translational and rotational motion of gravitating bodies). The terms of order of c^{-2} in both g_{00} and g_{ij} are required to derive $\mathbf{S}_{pN}(t)$, and the terms c^{-3} in g_{0i} are needed for $\mathbf{S}_{1.5pN}$. The next order effects, the so-called post-post-Newtonian effects, would require terms of order of c^{-4} in both g_{00} and g_{ij} (the c^{-4} terms in g_{ij} are not in the current definition of the BCRS metric tensor). The principle observable effects in the light propagation are (1) the gravitational light deflection (amounting to $1.75''$ for a light ray grazing the Sun) and (2) the gravitational signal retardation (the Shapiro effect; this effect amounts to $\sim$240 μs for the radar ranging of Venus in upper conjunction).

3.5 Conversion to Observables: Proper Direction

As mentioned above, the conversion of the coordinate-dependent quantities into coordinate-independent observables is an important part of relativistic modelling. From the mathematical point of view the coordinate-independent quantities are scalars. Special mathematical techniques are known to perform the suitable conversion in each particular case. One of the most important application of this conversion procedure is a conversion of the coordinate direction $\boldsymbol{n}$ into the source into the corresponding observable direction $\boldsymbol{s}$. The observable direction is often called "proper direction" in gravitational physics. Proper direction is a direction relative to the proper reference frame of the observer (see Sect. 4.2 about the difference of the concept of "reference frame" in astronomy and gravitational physics). A proper reference frame is a mathematical model of an ideal clock and three orthogonal rigid rods, which the observer uses to measure time intervals, distances and directions in his vicinity. Such a proper reference frame represents the coordinate basis of the CoMRS at its origin [24]. In special theory of relativity, the proper reference frame of an observer is related to some inertial reference system by a Lorentz transformation. It is therefore, sufficient to use Lorentz transformations to convert $\boldsymbol{n}$ into $\boldsymbol{s}$. The parameter of the Lorentz transformation in this case coincides with the velocity of the observer relative to the chosen reference system. In general relativity, it is also sufficient to use Lorentz transformations, but the parameter $\boldsymbol{\nu}$ of the transformations should be related to the BCRS velocity of the observer as

$$\boldsymbol{\nu} = \dot{\boldsymbol{x}}_o \left(1 + \frac{2}{c^2} w(t, \boldsymbol{x}_o)\right) + \mathcal{O}(c^{-4}), \tag{12}$$

where $\boldsymbol{x}_o$ and $\dot{\boldsymbol{x}}_o$ are the BCRS positions and velocity of the observer, respectively. A detailed discussion of this conversion and a comparison of different approaches can be found in [24]. The relativistic terms in (12) are derived from the c^{-2} terms in g_{00} and g_{ij} of the BCRS metric tensor. The difference between $\boldsymbol{n}$ and $\boldsymbol{s}$ can be called *relativistic aberration*. The difference between the Newtonian aberration and the relativistic one may amount to several milliarcsecond for Gaia observations.

3.6 Conversion to Observables: Proper Time

Another important case is the conversion of intervals of the coordinate time t into the corresponding intervals of the proper time τ of the observer. The general form of this conversion reads

$$\frac{d\tau}{dt} = 1 + c^{-2} A_{pN} + c^{-4} A_{ppN} + \mathcal{O}(c^{-5}), \tag{13}$$

where A_{pN} and A_{ppN} are the post-Newtonian and post-post-Newtonian terms, respectively. Explicit form of these two functions depends on the metric tensor: to compute for A_{pN} the c^{-2} terms in g_{00} are needed, while the c^{-4} terms in g_{00}, the c^{-3} terms in g_{0i}, and the c^{-2} ones in g_{ij} are required to compute A_{ppN}. Typically in the solar system and in particular for Gaia onboard clocks $|c^{-2} A_{pN}| \sim 10^{-8}$ and $|c^{-4} A_{ppN}| \sim 10^{-16}$.

3.7 Additional Effects: Relativistic Effects on the Imaging by a Rotating Telescope

For high-accuracy astrometry, special care should be taken to define the observables in accordance with the technical design of the instrumentation. One of the problems is related to the non-uniform motion of the instrumentation in space. In the framework of relativity, one usually considers point-like observers. The methods to calculate observed quantities for such observers were outlined earlier. It is assumed herewith that the instrumentation of the observer is so small that each part of the instrument has the same position and velocity. However, it is clear that the velocities of different parts of the primary mirror in "inertial" coordinates (e.g. not rotating with the Earth) are slightly different. However, normally the observational accuracy is considered to be "too low" and the telescope size "too small" for those differences to be of practical relevance. In case of Gaia, we deal with a scanning satellite that permanently rotates in space with a period of 6 h. The size of the primary mirror of Gaia is comparable with the size of the spacecraft itself and amounts to a few metres. Therefore, at the Gaia's goal accuracy of about 1 μas, one cannot neglect the difference of the velocities of various parts of the instruments. Considering the situation in the CoMRS, one can argue that for the purposes of imaging one can consider that we deal with an inertial reference system of special relativity. Then, considering the special-relativistic light reflection law and finite light velocity, one can calculate the deviation of the images produced by a rotating telescope from the images from the same telescope at rest. In some cases, these effects should be taken into account. These effects are discussed in details in [1].

4 Relativity for Gaia

Now, having all these theoretical tools one can formulate the relativistic model for Gaia. The relativistic model for Gaia is well documented [21], so that we just outline the overall structure of the model here.

4.1 Structure of the Standard Relativistic Model

The model consists essentially in subsequent transformations between five following vectors (Fig. 5):

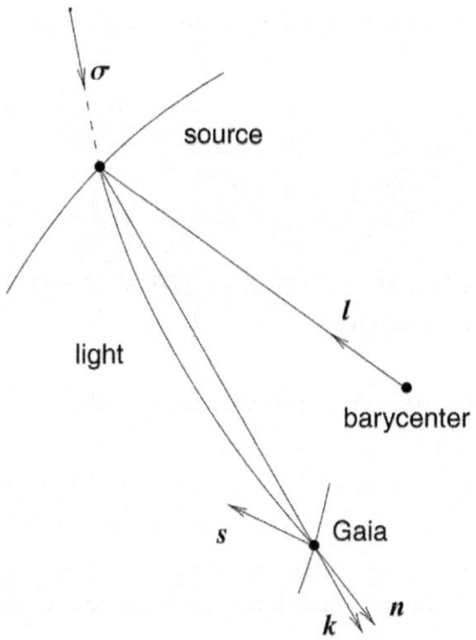

Fig. 5. Five principal vectors used in the model (see text for explanations).

remote sources: $\boldsymbol{s} \overset{(1)}{\longleftrightarrow} \boldsymbol{n} \overset{(2)}{\longleftrightarrow} \boldsymbol{\sigma} \overset{(3)}{\longleftrightarrow} \boldsymbol{k} \overset{(4)}{\longleftrightarrow} \boldsymbol{l}, \pi \overset{(5)}{\longleftrightarrow} \boldsymbol{l}(t_0), \pi(t_0), \boldsymbol{\mu}(t_0), \ldots$

solar system objects: $\boldsymbol{s} \overset{(1)}{\longleftrightarrow} \boldsymbol{n} \overset{(2,3)}{\longleftrightarrow} \boldsymbol{k} \overset{(6)}{\longleftrightarrow} \text{orbit}$

Fig. 6. Transformation sequences (see text for explanations).

(a) $\boldsymbol{s}$ is the unit observed direction (the word "unit" means here and below that the formally Euclidean scalar product $\boldsymbol{s} \cdot \boldsymbol{s} = s^i\, s^i$ is equal to unity).
(b) $\boldsymbol{n}$ is the unit vector tangential to the light ray at the moment of observation.
(c) $\boldsymbol{\sigma}$ is the unit vector tangential to the light ray at $t = -\infty$.
(d) $\boldsymbol{k}$ is the unit coordinate vector from the source to the observer.
(e) $\boldsymbol{l}$ is the unit vector from the barycenter of the solar system to the source.

Note that the last four vectors should be interpreted as sets of three numbers characterizing the position of the source with respect to the BCRS. All these vectors would change their numerical values if some other relativistic reference system is used instead of the BCRS. Vector $\boldsymbol{s}$ represents components of the observed direction relative to the CoMRS. The model consists then in a sequence of transformations between these vectors as shown on Fig. 6. The physical meaning of each transformation can be summarized as follows (the numbering here coincides with the numbering on Fig. 6):

(1) Aberration (effects vanishing together with the barycentric velocity of the observer): this step converts the observed direction to the source $\boldsymbol{s}$ into the unit BCRS coordinate velocity of the light ray $\boldsymbol{n}$ at the point of observation.
(2) Gravitational light deflection for the source at infinity: this step converts $\boldsymbol{n}$ into the unit direction of propagation $\boldsymbol{\sigma}$ of the light ray infinitely far from the solar system at $t \to -\infty$.
(3) Coupling of finite distance to the source and the gravitational light deflection in the gravitational field of the solar system: this step converts $\boldsymbol{\sigma}$ into a unit BCRS coordinate direction $\boldsymbol{k}$ going from the source to the observer.
(4) Parallax: this step converts $\boldsymbol{k}$ into a unit BCRS direction $\boldsymbol{l}$ going from the barycenter of the solar system to the source.
(5) Proper motion, etc.: this step provides a reasonable parametrization of the time dependence of $\boldsymbol{l}$ (and, possibly, of the parallax π) caused by the motion of the source relative to the barycenter of the solar system.
(6) Orbit determination process for solar system objects.

These transformations have already been discussed in full detail [21, 22, 24, 29]. The most complicated part of the model is the light deflection model where the effects of (1) monopole fields of all major solar system bodies, (2) quadrupole fields of the giant planets and (3) gravitomagnetic fields due to translational motion of all major bodies should be taken into account to attain the accuracy of 1 µas [29]. Moreover, each body with a mean density ρ and radius $R \geq (\rho/1\,\mathrm{g\,cm^{-3}})^{-1/2} \times 650\,\mathrm{km}$ produces a light deflection of

Table 1. Various gravitational effects in the light propagation in µas. δ_{pN} and δ_{ppN} are the post-Newtonian and post-post-Newtonian effects due to the spherically symmetric field of each body, δ_Q are the effects due to the quadrupole gravitational fields, respectively. Symbol "–" means that the corresponding effect is smaller than 0.1 µas. Physical parameters of the bodies are taken from [62]. The angle $\psi_{\max}$ is the maximal angular distance between the body and the source at which the corresponding effect still attains 1 µas. For these estimates the observer is supposed to be within a few million kilometres from the Earth orbit.

Body	δ_{pN}	$\psi_{\max}$	δ_Q	$\psi_{\max}$	δ_{ppN}	$\psi_{\max}$
Sun	$1.75 \cdot 10^6$	180°	~1		11	53′
Mercury	83	9′	–		–	
Venus	493	4.5°	–		–	
Earth	574	178°	0.6		–	
Moon	26	9°	–		–	
Mars	116	25′	0.2		–	
Jupiter	16,270	90°	240	152″	–	
Saturn	5,780	17°	95	46″	–	
Uranus	2,080	71′	8	4″	–	
Neptune	2,533	51′	10	3″	–	
Ganymede	35	32″				
Titan	32	14″				
Io	31	19″				
Callisto	28	23″				
Europe	19	11″				
Triton	10	0.7″				
Pluto	7	0.4″				
Titania	2.8	0.2″				
Oberon	2.4	0.2″				
Rhea	1.9	0.3″				
Charon	1.7	0.05″				
Iapetus	1.6	0.2″				
Ariel	1.4	0.1″				
Ceres	1.2	0.3″				
Dione	1.2	0.2″				
Umbriel	1.2	0.1″				

at least 1 µas. Therefore, a few ten of minor bodies (mainly, satellites of the giant planets) should also be taken into account in certain rare cases [21]. Table 1 summarizes the deflection effects due to various bodies.

4.2 Gaia Reference Frame

It is important to remember that all astrometric parameters of sources obtained from Gaia observations will be defined in the BCRS coordinates:

positions, proper motions, parallaxes, radial velocities, orbits of minor planets, binaries, etc. All these parameters will represent the Gaia reference frame, which is a materialization of the BCRS. The Gaia reference frame is, so to say, a model of the universe in the BCRS. Thus, the goal of astrometry in the relativistic framework is not to find "the" barycentric inertial reference frame, which is unique in the Newtonian formulation, but to find a materialization of some chosen relativistic reference system.

Let us note here that the meaning of words "reference system" and "reference frame" in relativistic astronomy is different from the meaning normally used in gravitational physics. *Reference system* is a purely mathematical construction (a chart) giving "names" to space–time events. A *reference frame* is, in contrast, some materialization (realization) of a reference system. In astronomy, the materialization is normally given in a form of a catalogue (or ephemeris) containing positions of some celestial objects relative to the selected reference system. Any astronomical reference frame (a catalogue, an ephemeris, etc.) is defined only through the reference system(s) used to construct physical models of observations.

4.3 Beyond the Standard Relativistic Model

In the model described above, any influence of gravitational fields generated outside of the solar system is ignored. For the majority of the sources, the external field can indeed be fully neglected, but there are a number of cases when the external gravitational fields produce observable effects. Several authors have discussed these additional effects in detail (see, e.g. [21,30]). Let us briefly list here the main effects of this kind: (1) gravitational light deflection caused by the masses situated outside of the solar system (a) weak micro-lensing on the stars of the Galaxy [2], (b) lensing on gravitational waves (both primordial ones and those from compact sources) and (c) lensing of the companions of edge-on binary systems; (2) cosmological effects and (3) more complicated models for the motions of observed objects in the BCRS that are necessary for the case of binary stars, etc.

Note that all these effects can be easily taken into account by a simple additive extension of the standard model, since at the required accuracy the external gravitational fields can be linearly superimposed on the solar system gravitational field. The only exception could be the effects of cosmological background, but a preliminary study by [28] shows that even here the coupling of the local solar system fields and the external ones can be neglected.

5 Gaia for Relativity

An important part of Gaia is testing relativity. Although it is quite clear how to use general relativity to model Gaia observations, the question how to use Gaia data to test relativity in the most efficient way is by no means trivial and still has no final answer. It would be too simplistic if Gaia confines itself to fitting the PPN parameter γ, i.e. just to fitting the amplitude of one particular signature in the light deflection. A much more ambitious approach is possible and desirable for Gaia. First of all, the whole Gaia data processing – from planetary ephemerides and orbit determination to the final parameter determination – should be consistent in the framework of a formalism chosen for testing relativity with Gaia (e.g. with the PPN formalism or some extension of it). To ensure the reliability of relativistic parameter estimations, special efforts should be made to monitor this relativistic consistency during the whole project.

The whole set of relativistic experiments with Gaia can be divided into two groups (1) core (or global) tests that are related with Gaia global astrometric solution and should use the whole Gaia data or at least as much data as possible and (2) shell (or local) tests that are related to some specially designed differential solutions and involve a relatively small amount of specially selected data. Below some of these experiments are discussed.

5.1 Core Tests: Gravitational Redshift from the Gaia Onboard Clock

Depending on the final design of Gaia, it may be possible to use Gaia onboard clock to test gravitational redshift. As shown by Mignard et al. [41], the total rate difference between TCG and the onboard clock is rather small ($\sim 5 \times 10^{-12}$) because of a subtle cancellation of the second-order Doppler term and the gravitational redshift for an observer on a Lissajous orbit around the Lagrange libration point L_2 of the three-body system Sun–Earth–satellite. The value of the gravitational redshift terms themselves is about 6×10^{-10}. Writing the well-known relation between the proper time τ of Gaia and $t = TCB$ with two additional parameters α_1 and α_2 as

$$\frac{d\tau}{dt} = 1 - \frac{1}{c^2}\left(\alpha_1 \frac{1}{2} |\boldsymbol{v}_o|^2 + \alpha_2 U\right) + \dots ,$$

where $\boldsymbol{v}_o$ is the velocity of Gaia and U is the gravitational potential at the Gaia location, one can argue that with the Gaia nominal orbit the test of the gravitational redshift is most sensitive to $\alpha_1 - \alpha_2$.

5.2 Core Tests: Robertson–Mansouri–Sexl Parameters from Aberration

The Robertson–Mansouri–Sexl parameters [39, 51] are numerical parameters of the simplest version of a test theory for special relativity (just as the PPN formalism is a test theory for general relativity in the post-Newtonian approximation). These parameters reflect possible violations of special relativity (or local invariance of physical laws under Lorentz transformation). These parameters were constrained using many different experiments (Michelson–Morley, Kennedy–Thorndike, and many newer laboratory experiments and even LLR). The idea for Gaia is to test special relativity's expression for aberration and estimate the Robertson–Mansouri–Sexl parameters. It seems that Gaia will be unable to improve the current best estimates, but since Gaia is a totally different kind of experiment it is still interesting to do this.

5.3 Core Tests: The PPN γ from Light Deflection

The most precise and, in this sense, the most important relativistic test possible with Gaia is the global test of the gravitational light deflection. As it was mentioned earlier, it is expected that Gaia will be able to measure the gravitational light deflection with a precision of $10^{-6} - 5 \times 10^{-7}$ [12, 40]. It is clear that such subtle estimates that could potentially have very deep physical consequences should be made with the highest possible quality and reliability. A number of recent estimates of the PPN parameters are based on a special post-processing of the post-fit residuals of the "standard fits" for which general relativity was assumed to be valid. Examples are the estimate of γ from the differential Shapiro delay observed with geodetic VLBI [57] and the estimate of the Lense–Thirring precession from post-fit residuals of SLR observations of geodetic satellites LAGEOS and LAGEOS II [4]. This situation is unsatisfactory since it is impossible to give the realistic accuracy of the estimates (e.g. possible correlations with other fitted parameters are not taken into account directly during the fit, but can only be verbally discussed a posteriori). Moreover, in current practice no realistic simulations with faked observational data are performed for VLBI or SLR observations. The simulations could give us the possibility to check for which kind of effects particular kind of observational data is sensitive and with which accuracy. Both these drawbacks should be addressed during the Gaia data processing, especially since the data processing in Gaia is a highly complex scientific problem.

Many calibration and instrumental parameters should be derived from the observations simultaneously with the desired astrometric parameters of sources and global (e.g. relativistic) parameters. Considering the enormous number of parameters to be fitted and observations to be processed (5×10^9 source parameters, 4×10^6 instrumental parameters and 10^8 attitude parameters, a significant number of global parameters should be determined from about 10^{12} equations), a single least-squares fit is not feasible. A solution for this problem is the so-called Global Iterative Solution (GIS) proposed in [12, Sect. 9.5.2]. The basic idea of GIS goes back to [32]. The GIS is an iterative procedure aimed at step-by-step improvement of the parameters, which are split into four groups:

(C)Calibration parameters (instrumentation parameters characterizing each CCD)

(A)Attitude parameters (characterizing the orientation of the satellite relative to a kinematically non-rotating co-moving triad [24])
(S) Source parameters (characterizing position, proper motion and parallax of each source)
(G)Global parameters (e.g. PPN parameter γ)

The process consists in cyclical repetition of the steps C, A, S and G until convergence. A detailed discussion of the GIS is given in [33–35] and [12]. Currently, this process is considered as the only possibility to process the Gaia data. A disadvantage of this process is that the full covariant matrix remains unknown and that the practical convergence behaviour is not yet fully understood. First realistic experiments with the GIS [13] show that this approach really works. It is clear that in this situation one must perform full-scale simulations of the data processing in order to understand to what kinds of deviations from general relativity Gaia observations are really sensitive. The situation is further complicated by fact that the light deflection due to the Sun is known to be correlated with the parallax zero point at a level of up to 0.9. These full-scale simulations will allow us to check the accuracy of γ determinations at least in the following versions:

1. *Standard light deflection test: one parameter γ for all bodies and all data.*

   Because of its magnitude the light deflection due to the Sun will bring most of the Gaia sensitivity to the possible violations of general-relativistic light deflection law. Other bodies will help to de-correlate γ from other parameters like parallax zero point. This approach is expected to provide the highest overall precision for γ, but should be supported and augmented by additional experiments described below.

2. *Separate deflection parameters γ for each deflecting body (at least for the Sun, Earth, Jupiter and Saturn).*

   The deflection due to other bodies (first of all due to the Earth, Jupiter and Saturn) is big enough to be tested with a significant precision. One can expect a precision of at least 10^{-3} for Jupiter and Saturn and 10^{-2} for the Earth. This is especially important since independent light deflection tests for Jupiter, Saturn and Earth provide a test of the Equivalence Principle (we effectively test that the motion of massive bodies that provide us with the value of the mass of the corresponding body is governed by the same mass as the motion of massless photons).

3. *Stability check: time dependence of γ.*

   To check the stability of the solution, it is useful to check if the Gaia observations are compatible with the hypothesis of γ changing with time. The data could be first divided into a number of pieces and γ could be determined for each of these independent data sets. A linear model for γ could then be also tested.

4. *Alternative deflection patterns: ψ^n.*

   It is also important not to restrict ourselves to testing only the general-relativistic light deflection law (angular dependence of the deflection), but also to test other possibilities. This will improve the reliability of the light deflection test. Here we suggest to test other angular dependence of the deflection. Namely, to test if the light deflection contains a signal that falls off as ψ^n, n being an integer or real number and ψ being the angular distance to the deflecting body. In general relativity, the deflection is proportional to ψ^{-1}.

5. *Alternative deflection patterns: non-radial patterns.*

   Another possibility is to look for some non-radial patterns in the light deflection. The most general approach here is to expand the deflection into vector spherical harmonics and fit the coefficients of these harmonics from observations.

6. *Higher-order effects.*

   Although Gaia is not able to observe close enough to the Sun (the minimal Sun avoidance angle is expected to be 45°) and cannot measure the expected post-post-Newtonian light deflection effect due to the Sun (see Table 1), it is useful to fit the data against a parametrized post-post-Newtonian model of light deflection. One reason is again a check of stability of the main solution of γ. Another reason is to check if the patterns predicted by higher-order terms in alternative theories of gravity are compatible with the Gaia observational data.

5.4 Core Tests: Pattern Matching in the Individual Positions and Proper Motions

Several interesting effects could be searched for by matching certain patterns in individual positions and/or proper motions of celestial objects in their distribution over the sky:

1. *Secular change of the secular aberration due to acceleration of the solar system with respect to the remote sources (or our Galaxy).*

   The observational accuracy of 1 μas together with the mission lifetime of at least 5 years allow one to see the apparent proper motions of QSOs (and perhaps other remote sources) due to the acceleration of the solar system barycenter relative to the centre of the Galaxy [12, Sect. 1.8.10]. These proper motions should be of order $\sim$4 μas year^{-1}, if one assumes the solar system to be on a circular orbit around the Galactic centre, but could be larger if one considers other options. Attempts to measure this effect from the geodetic VLBI measurements, which have been undertaken since

1992 at least, failed up to now due to a very high complexity of the noise sources in VLBI data processing. On the other hand, a determination of this parameter would be very interesting for both galactic dynamics and relativity (especially since the accuracy of the famous double pulsar test [10] is based on the *assumption* that the acceleration in question is known). Several options to measure the acceleration with Gaia should be checked: additional global parameters in the GIS and subsequent processing, and fitting the acceleration-induced pattern in the individual proper motions.

2. *Gravitational light deflection due to a hypothetical unknown massive body in the vicinity of the solar system.*

   A preliminary study [15] shows that for a hypothetical massive body moving around the Sun far enough from the barycenter of the solar system (so that the effect of its proper motion is much less that of its parallax), one can hope (if the body is massive enough) to discover its presence without seeing the body itself. The idea is to match a light deflection pattern to individual positions of the observed sources.

3. *Gravitational light deflection on primordial gravitational radiation.*

   Pyne et al. [50] have demonstrated how to estimate the flux of ultra-low-frequency primordial gravitational radiation from geodetic VLBI observations. Similar approach for high-accuracy positional observations is clearly possible. One can expect that Gaia will significantly improve the estimates of the flux given by Pyne et al. [50].

4. *Some additional parameters of cosmological interest.*

   Estimates of shear and vorticity from the apparent proper motions of QSOs. This brings a few additional global parameters to the GIS or the corresponding post-processing.

5.5 Shell Tests: Differential Light Deflection Due to the Giant Planets

The accuracy of modern and expected solar system ephemerides were assessed by Klioner [23]. It was concluded that because of relatively large uncertainties in positions of the giant planets (Jupiter, Saturn, Uranus and Neptune) the light deflection cannot be predicted at the level of 1 μas for observations performed too close to these bodies. The clear consequence of this fact is the necessity to exclude observations too close to the giant planets from the global solution and the possibility to process those excluded observations in special differential solutions. These differential solutions will allow us:

1. to measure the light deflection parameters γ for each of these planets (note that this solution is independent from the determination of parameters γ for each deflecting body as discussion in Sect. 5.3),

2. to detect the quadrupole light defection from Jupiter and, possibly, Saturn (this effect has never been observed up to now),
3. to measure the light deflection due to the gravimagnetic gravitational field induced by translational motion of the planets [29, 31],
4. to determine the positions of the planets themselves (which cannot be observed directly with competitive accuracy).

Some preliminary analysis of the quadrupole deflection experiment was done by Crosta and Mignard [5].

5.6 Shell Tests: Relativistic Effects in the Motion of Asteroids

Already 4 years after discovery of Icarus it was suggested [16] to use its motion to test general relativistic perihelion precession. The idea was then used many times [36,55,56,58,64] and leaded to an independent determination of the relativistic perihelion precession with a precision of currently 4%. Although the perihelion precession for Icarus (10.05″ per century) is significantly lower than for Mercury ($\sim 43''$ per century), it has been recognized already by Dicke [11] that asteroids with their large inclinations and their range of semi-major axes allow one to distinguish between the general-relativistic perihelion precession and the possible effects of the solar quadrupole. (It is well known that from the motion of Mercury only such a distinction is virtually impossible.) Although in the recent years the analysis of motion of the whole system of the inner planets did allow to determine separately the solar quadrupole and the relativistic precession [48], it remains unclear how reliable these estimates are. Anyway, it is clear that high-accuracy observations of about half a million of asteroids expected from Gaia will allow us not only to boost our knowledge of the solar system but also to significantly improve the tests of general relativity related to the motion of solar system bodies.

The largest relativistic effects in the motion of asteroids come from the Schwarzschild terms due to the Sun, the largest effect here being the relativistic perihelion precession. Although for all 253,000 asteroids registered by the Minor Planet Centre of the International Astronomical Union by September 2005 the relativistic perihelion precession $\Delta\omega$ is smaller than for Mercury, as it is shown in Table 2, the really observable effect proportional to the product of $\Delta\omega$ and the orbital eccentricity e is for many asteroids larger than for Mercury. Especially near-Earth objects (NEOs) have large relativistic perihelion precession. Analysis [19] shows that even with a relatively small number of asteroids (a few thousands), one can achieve the accuracies similar to the recent determinations from planetary data ($\sim 10^{-3}$ for the PPN parameter β, and $\sim 10^{-7}$ for the solar quadrupole). Even better accuracies could be expected from a global solution with all asteroids observed by Gaia.

These accuracy estimates are to be compared to the much lower accuracies expected for some specially selected asteroids from Earth-bound observations discussed by Shahid-Saless and Yeomans [53]. The reason for such a significant increase of accuracy is not only the amount of observed asteroids but

Table 2. Orbital elements (semi-major axes a, eccentricity e and inclination i) and relativistic perihelion precession $\Delta\omega$ for some planets and asteroids.

Name	Number	a (AU)	e	i (°)	$\Delta\omega$ (″ per century)	$e\Delta\omega$ (″ per century)
Major planets						
Mercury		0.39	0.21	7.00	42.98	8.84
Venus		0.72	0.01	3.39	8.62	0.06
Earth		1.00	0.02	0.00	3.84	0.06
Mars		1.52	0.09	1.85	1.35	0.12
...						
Main belt asteroids						
Phaethon	3200	1.27	0.89	22.17	10.13	9.01
Icarus	1566	1.08	0.83	22.85	10.06	8.31
Talos	5786	1.08	0.83	23.24	9.98	8.25
Hathor	2340	0.84	0.45	5.85	7.36	3.31
Ra-Shalom	2100	0.83	0.44	15.75	7.51	3.28
...						
Near-Earth objects (NEOs)						
2004 XY60		0.64	0.80	23.79	32.14	25.63
2000 BD19		0.89	0.90	25.67	26.83	24.01
1995 CR		0.91	0.87	4.04	19.97	17.36
1999 KW4	66,391	0.64	0.69	38.89	22.06	15.19
2004 UL		1.27	0.93	23.70	15.08	13.98
2001 TD45		0.80	0.78	25.42	17.12	13.31
1999 MN		0.67	0.67	2.02	18.45	12.28
2000 NL10		0.91	0.82	32.51	14.44	11.80
1998 SO		0.73	0.70	30.35	16.39	11.45
1999 FK21	85,953	0.74	0.70	12.60	16.18	11.38
2005 HC4		1.82	0.96	8.39	11.27	10.83
2004 QX2		1.29	0.90	19.07	11.04	9.97
2002 AJ129		1.37	0.91	15.52	10.70	9.79
2000 WO107		0.91	0.78	7.78	12.39	9.67
2005 EP1		0.89	0.77	16.19	12.50	9.60
...						

also the accuracy of Gaia. Up to now typical accuracy of Earth-bound positional observations is about 1″. Gaia is expected to boost that accuracy by a factor of 10^3–10^4 at least (the observational accuracy for asteroids will be lower than for stars because of the fact that asteroids are non-point objects of complicated (mostly even unknown) shape that makes it difficult to fully correct for the phase effects). This accuracy makes it indispensable to account for many subtle effects in the corresponding dynamical models. In particular, non-Schwarzschild (N-body problem) relativistic effects described by the EIH equations (possibly with PPN parameters) are expected to play an important role. The consequences of the EIH equations for orbital evolution of major planets are relatively well known. However, for asteroids having large eccentricities and inclinations and moving in various resonances with major planets, the non-Schwarzschild relativistic effects have never been investigated in details. Resonant phenomena [49] in the motion of asteroids may lead to an enhancement of the smaller relativistic perturbations (e.g. due to Jupiter) and should be very carefully taken into account. Preliminary analysis [26] shows that the non-Schwarzschild perturbations typically amount to several metres (up to a hundred metres in some cases) after a few hundred days, but in some special resonant cases can grow fast reaching a few kilometres after several years of motion.

Up to quite recently either the data processing for asteroids was purely Newtonian or the Schwarzschild terms due to the Sun were taken into account. The relativistic effects could mimic non-gravitational forces and also deteriorate the estimated parameters. To exploit the full accuracy of Gaia, one should make the asteroid data processing fully compatible with general relativity. It is clear that the full EIH equations must be used as relativistic dynamical model.

Special attention for the motion of asteroids should be paid to the relativistic effects in the framework of a restricted three-body problem Sun–Jupiter–asteroid. The relativistic effects in the positions of the stable triangular Lagrangian libration points L_4 and L_5 were historically the first example of the Nordtvedt effect given by Nordtvedt [42] (see also [46]). The Nordtvedt's idea has been used in practice by Orellana and Vucetich [44, 45] to obtain an estimate $\eta = -0.56 \pm 0.48$ for the Nordtvedt parameter from Earth-bound optical observations of the Trojan asteroids. Hestroffer [18] has pointed out that this effect could be used with Gaia data to improve the determination accuracy of the "asteroidal" Nordtvedt effect. One can expect an accuracy increase by a factor of 10^2–10^3. The Lagrangian motion within the general-relativistic restricted three-body problem was investigated by Maindl [37] and Maindl and Dvorak [38]. Similar detailed study in the framework of the PPN formalism will be published elsewhere.

References

1. G. Anglada-Escudé, S.A. Klioner, M. Soffel, and J. Torra: Relativistic effects on the imaging by a rotating optical instrument, *Astron. Astrophys.*, 462, 371 (2007).
2. V.A. Belokurov and N.W. Evans: *MNRAS* **331**, 649 (2002).
3. O. Bienaymé and C. Turon: *GAIA: A European Space Project*, EAS Publications Series, Vol. 2 (EDP Sciences, Les Ulis 2002).
4. I. Ciufolini and E.C. Pavlis: *Nature* **431**, 958 (2004).

5. M.T. Crosta and F. Mignard: Proc. of the Symposium "The Three-Dimensional Universe with Gaia", 4–7 October 2004, Observatoire de Paris-Meudon, France, ESA SP-576, 281 (2004).
6. T. Damour and K. Nordtvedt, Jr.: *Phys. Rev. Lett.* **70**, 2217 (1993).
7. Damour, T., Nordtvedt, K., Jr.: *Phys. Rev.* **D 48**, 3436 (1993).
8. T. Damour and A.M. Polyakov: *Nucl. Phys.* **B 423**, 532 (1994).
9. Damour, T., Piazza, F., Veneziano, G.: *Phys. Rev.* **D 66**, 046007 (2002).
10. Damour, T., Taylor, J.H.: *Phys. Rev.* **D 45**, 1840 (1992).
11. R.H. Dicke: *Astron. Journal* **70**, 395 (1965).
12. GAIA: Composition, Formation and Evolution of the Galaxy, Concept and Technology Study Report, ESA-SCI 4 (Noordwijk: European Space Agency, 2000).
13. F. Figueras, B. López-Marti, C. Fabricius, J. Torra, C. Jordi, P. Llimona, E. Masana, X. Luri: Proc. of the Symposium "The Three-Dimensional Universe with Gaia", 4–7 October 2004, Observatoire de Paris-Meudon, France, ESA SP-576, 369 (2004).
14. M. Frœschlé, F. Mignard, and F. Arenou: Proc. of the ESA symposium "Hipparcos – Venice 97", ESA-SP402 (1997).
15. B.S. Gaudi and J.S. Bloom: Astrometric microlensing constrains on a massive body in the outer solar system with Gaia, astro-ph/0506426
16. N. Gouda, T. Tsujimoto, Y. Kobayashi *et al.*: *Astrophys. Space Science* **280**, 89 (2002).
17. J.J. Gylvarry: *Phys. Rev.* **89**, 1046 (1953).
18. D. Hestroffer: in: O. Bienaymé, C. Turon (eds.), "GAIA: a European Space Project", EAS Publications Series, Vol. 2 (Proceeding of Les Houches Summer School "GAIA: a European Space Project", 14–17 May 2001), EDP Sciences, Les Ulis, 359 (2002).
19. D. Hestroffer and J. Berthier: Proc. of the Symposium "The Three-Dimensional Universe with Gaia", 4–7 October 2004, Observatoire de Paris-Meudon, France, ESA SP-576, 297 (2004).
20. IAU 2001, Information Bulletin, 88 (errata in IAU Information Bulletin, 89)
21. S.A. Klioner: *Astron. Journal* **125**, 1580 (2003).
22. S.A. Klioner: *Astron. Astrophys.* **404**, 783 (2003).
23. S.A. Klioner: Requirements on the accuracy of GAIA orbit determination from the relativistic light deflection effects (2003), available from the Gaia Livelink archive `http://astro.estec.esa.nl/llink/livelink`
24. S.A. Klioner: *Phys. Rev.* **D 69**, 124001 (2004).
25. S.A. Klioner: Relativistic perturbations for the Lissajous Orbits around L2, (2005), available from the Gaia Livelink archive `http://astro.estec.esa.nl/llink/livelink`
26. S.A. Klioner: Relativistic N-body effects in the motion of asteroids, in preparation.
27. S.A. Klioner and M. Soffel: *Phys. Rev.* **D 62**, ID 024019 (2000).
28. S.A. Klioner and M.H. Soffel: Proc. of the Symposium "The Three-Dimensional Universe with Gaia", 4–7 October 2004, Observatoire de Paris-Meudon, France, ESA SP-576, 305 (2004).
29. S.A. Klioner and M. Peip: *Astron. Astrophys.* **410**, 1063 (2003).
30. S.M. Kopeikin and C. Gwinn: in Towards Models and Constants for Sub-Microarcsecond Astrometry, ed. K.J. Johnston, D.D. McCarthy, B.J. Luzum, & G.H. Kaplan (Washington: US Naval Observatory), 303 (2000).

31. S.M. Kopeikin and G. Schäfer: *Phys. Rev.* **D 60**, 124002 (1999).
32. P. Lacroute: **in** C. Barbieri and P.L. Bernacca (eds.): *European Satellite Astrometry*, Proceedings of a Colloquium, held in Padova, Italy, June 5–7, 1978, Padova: Consiglio Nazionale della Ricerche, Universiota di Padova, 1979, p. 1.
33. L. Lindegren: Proposed prototype processes for the Gaia Global Iterative Solution. GAIA-LL-34(v. 2), (2001), available from the Gaia Livelink archive `http://astro.estec.esa.nl/llink/livelink`
34. L. Lindegren: Global Inerative Solution – Distributed processing of the attitude updating. SAG-LL-37, (2001), available from the Gaia Livelink archive `http://astro.estec.esa.nl/llink/livelink`
35. L. Lindegren: An alternative scheme for the GIS processing. GAIA-LL-059, (2005), available from the Gaia Livelink archive `http://astro.estec.esa.nl/llink/livelink`
36. J.H. Lieske and G.W. Null: *Astron. Journal* **74**, 297 (1969).
37. T.I. Maindl: **in** T.W. Rettig & J.M. Hahn (eds.): *Completing the Inventory of the Solar System*, APS Conference Series **107**, 147 (1996).
38. T.I. Maindl and R. Dvorak: *Astron. Astrophys.* **290**, 335 (1994).
39. R. Mansouri and R.U. Sexl: *Gen. Rel. Grav.* **8**, 497 (1977); ibid., **8**, 515 (1977); ibid. **8**, 809 (1977).
40. F. Mignard: in: O. Bienaymé, C. Turon (eds.), "GAIA: a European Space Project", EAS Publications Series, Vol. 2 EDP Sciences, Les Ulis, 107 (2002).
41. F. Mignard, M.T. Crosta, and S.A. Klioner: Relation between the Gaia proper time and TCB, GAIA_FM_020, (2004), available from the Gaia Livelink archive `http://astro.estec.esa.nl/llink/livelink`
42. K. Nordtvedt, Jr.: *Phys. Rev.* **169**, 1014 (1968).
43. K. Nordtvedt, Jr.: Proc. of Villa Mondragone International School of Gravitation and Cosmology, September 2002, [gr-qc/0301024].
44. R.B. Orellana and H. Vucetich: *Astron. Astrophys.* **200**, 248 (1988).
45. R.B. Orellana and H. Vucetich: *Astron. Astrophys.* **273**, 313 (1992).
46. J.M. Overduin: *Phys. Rev.* **D 62**, 102001 (2000).
47. M.A.C. Perryman *et al.*: *Astron. Astrophys.* **369**, 339 (2001).
48. E.V. Pitjeva: *Astron. Lett.* **31**, 340 (2005).
49. A.R. Plastino and H. Vucetich: *Astron. Astrophys.* **262**, 321 (1992).
50. T. Pyne, C. Gwinn, M. Birkinshaw, T.M. Eubanks, and D.N. Matsakis: *Astrophys. J.* **465**, 566 (1996).
51. H.P. Robertson: *Rev. Mod. Phys.* **21**, 378 (1949).
52. H. Rickman: *Reports on Astronomy, Trans. IAU*, **XXIV B** (2001).
53. B. Shahid-Saless and D.K. Yeomans: *Astron. J.* **107**, 1885 (1994).
54. M. Shao **in** R.D. Reasenberg (ed.) *Astronomical Interferometry*, *Proc. SPIE*, 3350, 536 (1998).
55. I.I. Shapiro, M.E. Ash, and W.B. Smith: *Phys. Rev. Lett.* **20**, 1517 (1968).
56. I.I. Shapiro, W.B. Smith, M.E. Ash, and S. Herrick: *Astron. J.* **76**, 588 (1971).
57. S.S. Shapiro, J.L. Davis, D.E. Lebach, and J.S. Gregory: *Phys. Rev. Lett.* **92**, 121101 (2004).
58. G. Sitarski: *Astron. J.* **104**, 1226 (1992).
59. M. Soffel, S.A. Klioner, G. Petit, P. Wolf, S.M. Kopeikin, S.M. *et al.*: *Astron. J.* **126**, 2687 (2003).
60. S.G. Turyshev, M. Shao, and K. Nordtvedt, Jr.: *Class. Quant. Grav.* **21**, 2773 (2004).

61. A. Vecchiato, M.G. Lattanzi, B. Bucciarelli, M. Crosta, F. de Felice, F., and M. Gai: *Astron. Astrophys.* **399**, 337 (2003).
62. P.R. Weissman, L.-A. McFadden, and T.V. Johnson, T.V. (eds.) *Encyclopedia of the Solar System* (Academic Press, San Diego 1999).
63. C.M. Will: *Theory and experiment in gravitational physics* (Cambridge University Press, Cambridge 1993).
64. J. Zhang: *Chinese Astron. Astrophys.* **18**, 108 (1994).

LISA, the Laser Interferometer Space Antenna, Requires the Ultimate in Lasers, Clocks, and Drag-Free Control

Albrecht Rüdiger, Gerhard Heinzel, and Michael Tröbs

Max-Planck-Institut für Gravitationsphysik, Albert-Einstein-Institut,
Callinstr. 38, D – 30176 Hannover, Germany

Summary. The existence of gravitational waves is the most prominent of Einstein's predictions that has not yet been directly verified. The space project LISA shares its goal and principle of operation with the ground-based interferometers currently being operated, the detection and measurement of gravitational waves by laser interferometry. Ground and space detection differ in their frequency ranges, and thus in the detectable sources. Toward low frequencies, ground-based detection is limited by seismic noise, and yet more fundamentally by "gravity-gradient noise," thus covering the range from a few Hz on upward to a few kHz. It is only in space that detection of signals below, say, 1 Hz is possible, opening a wide window to a different class of interesting sources of gravitational waves. The project LISA consists of three spacecraft in heliocentric orbits, forming a triangle of 5 million km sides. A technology demonstrator, the LISA Pathfinder, designed to test vital LISA technologies, is to be launched by ESA in 2009.

LISA will face great challenges in reducing measurement noise, and thus, it will very strongly depend on the technologies of lasers, clocks, and drag-free control.

1 Introduction

The talks on which this chapter is based dealt with a new window in astronomic observation presently being opened: the detection and measurement of gravitational waves (GW) and, in particular, using laser interferometry in space. This is one of the great challenges to modern physics. Although predicted by Einstein in 1916, a direct observation of these waves has yet to be accomplished.

Great hopes of such detection lie in the ground-based laser-interferometric detectors that are currently in the final phases of commissioning, approaching their design sensitivities. These ground-based detectors are sensitive in the "audio" frequencies of a few Hz up to a few kHz.

Perhaps even more promising are the spaceborne interferometers, where we will mainly have to think of the joint ESA–NASA project Laser Interferometer

Space Antenna (LISA), which would cover the frequency range from about 10^{-4} to 1 Hz. It is in that frequency range that the most violent cosmic events occur, and their detection will provide new insights into the cosmology of our universe.

Gravitational waves share their elusiveness with neutrinos: they have very little interaction with the measuring device, which is why these gravitational waves have not yet directly been detected. But that same feature also is a great advantage: because of their exceedingly low interaction with matter, gravitational waves can give us an unobstructed view into astrophysical and cosmological events that will forever be obscured in the electromagnetic window.

The price we have to pay is that, to detect and measure these minute effects of the gravitational waves, we will require the most advanced technologies in optics, lasers, and interferometry. LISA, the Laser Interferometer Space Antenna, will in unprecedented fashion depend on the advances made in the fields that gave this conference its title: *lasers, clocks, and drag-free.*

1.1 Gravitational Wave Detection

Several projects to observe gravitational waves with ground-based interferometers have gone into their final phase of commissioning and are approaching their design sensitivity, and next-generation detectors are already being extensively investigated.

Furthermore, and constituting the topic of this chapter, an international collaboration on placing a huge interferometer, LISA, into an interplanetary orbit is close to reaching final approval.

At the start, we will briefly discuss the characteristics of the large terrestrial GW detectors of the current generation.

In this way, we will learn how the detectors on ground and in space differ, in how far aims and technologies overlap, and what can scientifically be gained from the complementarity of these researches.

The main interest will then be on the laser interferometers in space, exemplified by the joint ESA–NASA project LISA, and sketched in Fig. 1.

1.2 Gravitational Waves

In two publications [1, 2], Albert Einstein has predicted the existence and estimated the strength of gravitational waves. They are a direct outcome not only of his Theory of General Relativity, but also a necessary consequence of *all* theories with finite velocity of interaction. Good introductions to the nature of gravitational waves, and on the possibilities of measuring them are given in two chapters by Kip Thorne [3, 4].

It can be shown that gravitational waves of measurable strengths are emitted only when large cosmic masses undergo strong accelerations, for instance – as shown schematically in Fig. 2 – in the orbits of a (close) binary system. The effect of such a gravitational wave is an apparent strain in space,

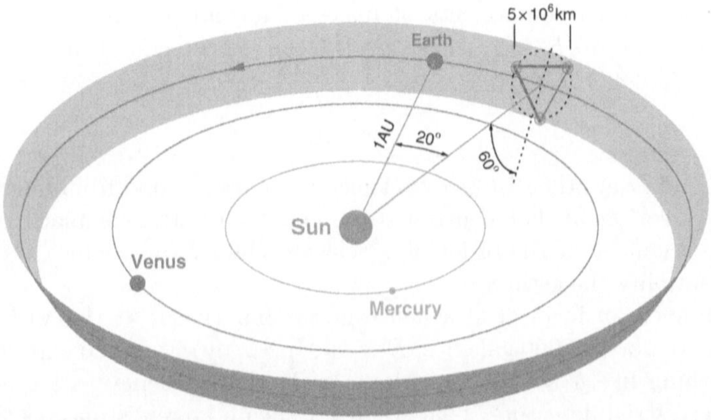

Fig. 1. Orbits of the three spacecraft of LISA, trailing the Earth by 20°. The triangle "rolls" on a cone tilted by 60° out of the ecliptic. The triangle arms (of 5 million km in length) are scaled by a factor 5.

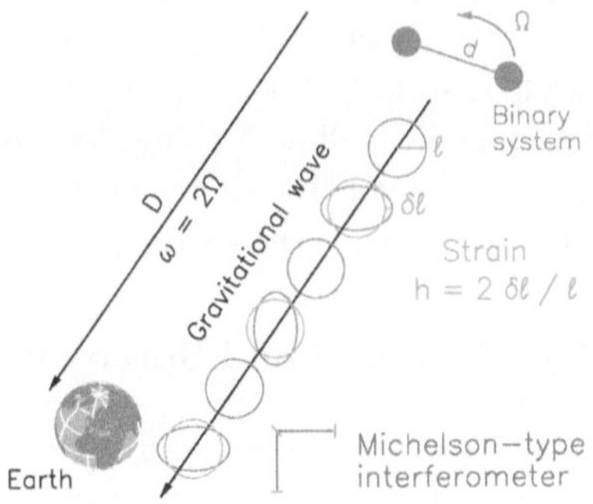

Fig. 2. Generation and propagation of a gravitational wave emitted by a binary system.

transverse to the direction of propagation, that makes distances ℓ between test bodies shrink and expand by small amounts $\delta\ell$, at twice the orbital frequency: $\omega = 2\,\Omega$. The strength of the gravitational wave, its "amplitude," is generally expressed by $h = 2\,\delta\ell/\ell$. An interferometer of the Michelson type, typically consisting of two orthogonal arms, is an ideal instrument to register such differential strains in space.

But what appears so straightforward in principle turns out to be an almost insurmountable problem. The difficulty lies in the magnitude, or rather the smallness, of the effect.

1.3 Strength of Gravitational Waves

In a linearized approximation, the so-called "quadrupole formula," the strength of the gravitational wave emitted by a mass quadrupole can be estimated.

For a binary with components of masses M_1 and M_2, or their respective Schwarzschild radii R_1, R_2, the strain h to be expected is of the order

$$h \approx \frac{R_1 R_2}{d\,D} \tag{1}$$

where d and D are the distances between the partners and from binary to the observer (see Fig. 2). For neutron stars, and even better for black holes, the distance d can be of the order of a Schwarzschild radius, which then would further simplify the estimate.

From such an inspiral of a neutron star binary out at the Virgo cluster (a cluster of about 2,000 galaxies, $D \sim 15$ Mpc away), we could expect a strain of something like $h \approx 10^{-22}$, in this case in the frequency region accessible to the terrestrial detectors. That we insert such a large distance as the Virgo cluster is to have a reasonable rate of a few events per year. Inside a single galaxy (as ours), we would count at most a few detectable events per century.

Equation (1) lends itself to an extension to more massive binary partners, such as (super)massive Black Holes. With massive black holes of, say, $10^5\,M_\odot$, the numerator would rise by a factor of 10^{10}, whereas the closest distance d in the denominator would rise only linearly, by 10^5. This would then, even though at much lower frequencies, allow "seeing" farther out into the universe by a factor 10^5, i.e., one could with the same sensitivity in h, cover the whole universe, and with a high signal-to-noise ratio.

1.4 Complementarity of Ground and Space Observation

Shown in Fig. 3 are some typical expected sources of gravitational radiation. They range in frequency over a vast spectrum, from the kHz region of supernovae and final mergers of compact binary stars down to mHz events due to formation and coalescence of supermassive black holes. Indicated are sources in two clearly separated regimes: events in the range from, say, 5 Hz to several kHz (detectable with terrestrial antennas), and a low-frequency regime,

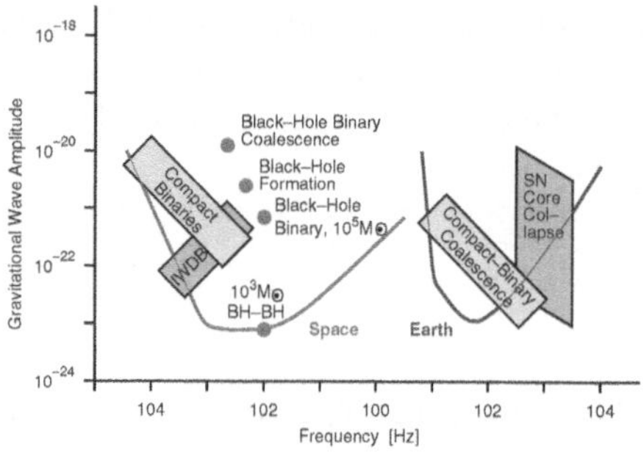

Fig. 3. Some sources of gravitational waves, with sensitivities of *Earth* and *Space* detectors.

10^{-5} to 1 Hz, accessible only with a space project such as LISA. In the following sections, we will see how the sensitivity profiles of the detectors come about. No single detector covering the whole spectrum shown could be devised.

Events Observable with Ground-Based Detectors

Clearly, one would not want to miss the information of either of these two (rather disjoint) frequency regions. The upper band ("Earth"), with supernovae and compact binary coalescence, can give us information about relativistic effects and equations of state of highly condensed matter, in highly relativistic environments. Binary inspiral is an event type than can be calculated to high post-Newtonian order, as shown, e.g., by Buonanno and Damour [5]. This will allow tracing the signal, possibly even by a single detector, until the final merger, a much less predictable phase. The ensuing phase of a ring-down of the combined core does again lend itself to an approximate calculation, and thus to an experimental verification. Chances for detection are reasonably good, particularly with the advanced detectors now being planned.

Events Observable with Space Detectors

As can be seen from Fig. 3, the events to be detected by the space project LISA, on the other hand, may have extremely high signal-to-noise ratios, and failure to find them would shatter the very foundations of our present understanding of the universe. The strongest signals will come from events involving (super)massive black holes, their formation as well when galaxies with their BH cores collide. Mergers of supermassive black hole (SMBH) binaries will produce signals so strong that they can be detected by LISA no matter where in the universe they originate. But also the (quasicontinuous) signals from neutron-star and black-hole binaries are among the events to be detected ("Compact Binaries" in Fig. 3). Interacting white dwarf binaries inside our galaxy ("IWDB" in Fig. 3) may turn out to be so numerous that they cannot all be resolved as individual events, but rather form a noise background above the instrumental noise in some frequency range around 1 mHz. While catastrophic events such as the Gamma-ray bursts are not yet well enough understood to estimate their emission of gravitational waves, there is a potential of great usefulness of GW detectors for their study, mainly at low frequencies. In addition, signals from unexpected sources are probable.

Combined observation with electromagnetic and gravitational waves could lead to a deeper understanding of the violent cosmic events in the far reaches of the universe [6].

2 Ground-Based Interferometers

The underlying concept of all ground-based laser detectors is the Michelson interferometer (see schematic in Fig. 4), in which an incoming laser beam is divided into two beams traveling along different (usually perpendicular) arms.

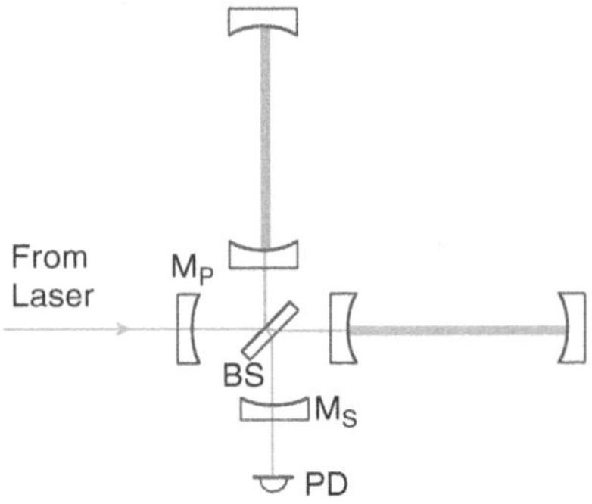

Fig. 4. Advanced Michelson interferometer with Fabry–Perots in the arms and extra mirrors M_P, M_S for power and signal recycling.

On their return, these two beams are recombined, and their interference (measured with a photodiode PD) will depend on the difference in the gravitational wave effects that the two beams have experienced. It is a very essential feature of the ground-based detectors that the beams are reflected back to the beam splitter, with practically unreduced power, to perform the interference at that beam splitter. This feature will, due to the immense distance (5 million km) between the spacecraft, no longer be possible in the space detector LISA.

A gravitational wave of frequency f propagating normal to the plane of the interferometer would give rise to a path difference δL between the two arms of

$$\delta L = h_+ \cdot L \cdot \frac{\sin(\pi f \tau)}{\pi f \tau} = h_+ \cdot L \cdot \frac{\sin(\pi L/\Lambda)}{\pi L/\Lambda} \, . \tag{2}$$

The changes δL in optical path increase with the optical paths L, until the optimum is reached at about half the wavelength $\Lambda = c/f$ of the gravitational wave, which amounts to a seemingly unrealistic 150 km for a 1 kHz signal. Schemes were devised to make the optical path L significantly longer than the geometrical arm length ℓ, which is limited on Earth to only a few km. One way is to use "optical delay lines" in the arms, with the beam bouncing back and forth in a zigzag pattern between two concave mirrors (the simplest version of this is used in GEO600 and shown in Fig. 5).

The other scheme is to use Fabry–Perot cavities (Fig. 4), again with the aim of increasing the interaction time of the light beam with the gravitational wave. For GW frequencies f beyond the inverse of the storage time τ, the response of the interferometer will, however, roll off with frequency, as $1/f\tau$, or actually with the sinc function as in Eq. (2).

2.1 The Large-Scale Projects

To give an impression of the wide international scope of the interferometer efforts, the current large-scale detectors are listed below, ordered by size. All of these projects will use low-noise Nd:YAG lasers ($\lambda = 1.064\,\mu\text{m}$), pumped with

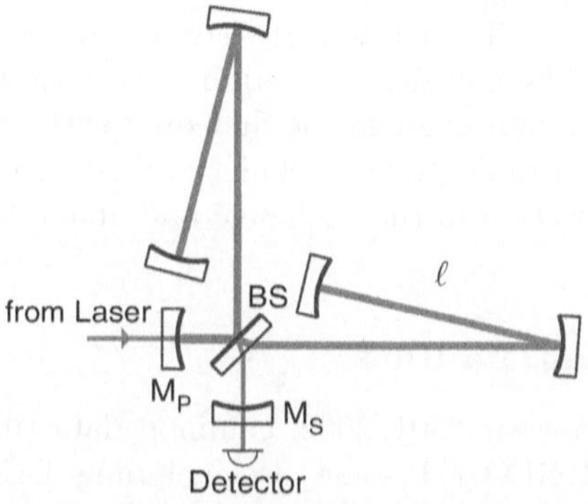

Fig. 5. The DL4 configuration with dual recycling to be used in GEO 600.

laser diodes for high overall efficiency. A wealth of experience has accumulated on highly stable and efficient lasers, from which the space missions will also profit. More details about the laser source in Sect. 4.5.

LIGO The largest is the US project named LIGO [7]. It comprises *two* facilities at two widely separated sites, in the states of Washington and Louisiana. Both house a 4 km interferometer, Hanford an additional 2 km one.

VIRGO Next in size (3 km) is the French–Italian project VIRGO [8] near Pisa, Italy. An elaborate seismic isolation system, with six-stage pendulums, will allow measurement down to GW frequencies of 10 Hz or even below, but still no overlap with the space interferometer LISA.

GEO 600 The detector of the British–German collaboration, GEO 600 [9], with an arm length of 600 m, is located near Hannover in northern Germany. It employs the advanced optical technique of "signal recycling," SR [10, 11] to make up for the shorter arms.

TAMA 300 In Japan, on a site at the National Astronomical Observatory near Tokyo, not a very quiet site, TAMA 300 has had several successful data runs and exhibited encouragingly long in-lock duty cycles [12]. TAMA is, just as LIGO and VIRGO, equipped with standard Fabry–Perot cavities in the arms. A large-scale cryogenic detector (LCGT) to be built underground, in a mine near Kamioka in central Japan, is in planning [17].

AIGO Australia also had to cut back from earlier plans of a 3 km detector, due to lack of funding. Currently a 80 m prototype detector is being built near Perth, Western Australia, with the aim of investigating new interferometry configurations [13].

2.2 International Collaboration

It is fortunate that the progress of these projects is rather well in synchronism. For the received signal to be meaningful, coincident recordings from at least two detectors at well-separated sites are essential. A minimum of three detectors (at three different sites) is required to locate the position of the source, and there is general agreement that only with at least four detectors can we speak of a veritable gravitational wave *astronomy*, based on a close international collaboration in the exchange and analysis of the experimental data.

2.3 First Common Data Runs

Since the turn of the year 2001/2002, common data runs between all three LIGO detectors and GEO 600, some also including TAMA 300, were undertaken. Since then, repeated *Science runs* have been successfully performed, with ever-improving sensitivity and duty cycle. The LIGO interferometers are now very close to their design sensitivity, and also GEO 600 and TAMA 300 are approaching theirs. The data accumulated are being analyzed for evidence of gravitational wave events of different types, and improved upper limits for the strengths of such event types have been established. Upgrades of these current detectors are envisaged and partially already firmly approved, and these will, with great certainty, be sensitive enough to observe numerous events per year.

After the first direct detection of gravitational waves will have been accomplished in hopefully no more than a few years from the time of this writing (2006), the real goal of gravitational wave detectors, starting a completely new branch of astronomy, can begin.

3 Noise and Sensitivity

The measurement of gravitational wave signals is a constant struggle against the many types of noise entering the detectors. These noise sources have presented a great technological challenge, and interesting schemes of reducing their effects have been forwarded. Two very prominent noise sources, ones that also play a decisive role in space detectors, will be discussed below.

3.1 Laser Noise

The requirements on the quality ("purity") of the laser light used for the GW interferometry are extraordinarily demanding. As it happens, the light sources for the ground-based and the spaceborne interferometers will both be Nd:YAG lasers emitting at 1064 nm in the near-infrared. High stability is achieved by unidirectional operation of nonplanar ring oscillators (NPROs) [14] (see Fig. 6). Pumped by laser diodes, they exhibit a high overall efficiency. Their good tunability allows efficient stabilization schemes.

How the very high requirements for the laser are met will be discussed in detail in Sect. 4.6. A few straightforward requirements will just briefly be listed here:

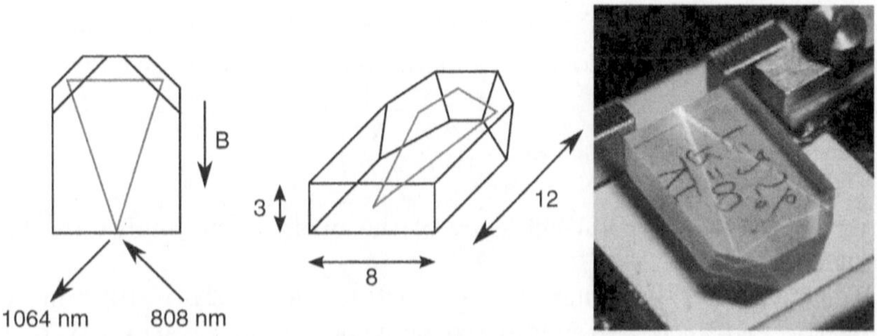

Fig. 6. NPRO laser, scheme, dimensions in mm (*left*), photo (*right*).

Frequency Stability

A perfect Michelson interferometer (with exactly matching arms) would be insensitive to frequency fluctuations of the light used. The detectors will, however, by necessity have unequal arms, the ones on the ground due to civil engineering tolerances and a particular modulation scheme chosen, the space detector due to orbital dynamics of the individual spacecraft.

Therefore, a very accurate control of the laser frequency is required, with (linear) spectral densities of the frequency fluctuations of the order $\widetilde{\delta\nu} = 10^{-4}\,\mathrm{Hz\,Hz^{-1/2}}$ for the ground-based detectors and even less for LISA. Control schemes have been devised to reach such extreme stability, albeit only in the frequency band required, and not all the way down to DC.

Beam Purity

Any geometrical asymmetry of the Michelson interferometer will make it prone to noise from geometrical fluctuations of the laser beam. Ideally the illumination of the Michelson would be a pure TEM_{00} mode. For small light powers, below 1 W as in the space project, a clean circular beam can be obtained by passing the light through a single-mode fiber. For the laser powers needed in the ground-based interferometers, however, a "mode-cleaner" is used: a non-degenerate cavity that is tuned for the TEM_{00} mode, but suppresses the (time dependent) lateral modes that represent fluctuations in position, orientation, and width of the beam [15].

3.2 Shot Noise

Particularly at higher frequencies, the sensitivity is limited by a rather fundamental source of noise, the so-called *shot noise*, a fluctuation in the measured interference coming from the "graininess" of the light.

These statistical fluctuations fake apparent fluctuations in the optical path difference ΔL that are inversely proportional to the square root of the light power P used in the interferometer. The spectral density (in the "linear" form we prefer) of the fluctuations of the path difference, $\widetilde{\Delta L}$, is given by

$$\widetilde{\Delta L} = \left(\frac{\hbar c}{2\pi} \frac{\lambda}{\eta P} \right)^{\frac{1}{2}} \tag{3}$$

where η is the conversion efficiency of the photo diode, and λ the laser's wavelength.

For measuring the minute changes of the order of $\Delta L \sim 10^{-18}$ m in our kilometric "advanced" ground-based detectors, as much as 1 MW of light power, in the visible or in the near-infrared, would be required. This is not as unrealistic as it may sound; using the concept of "power recycling" [16], such high effective powers circulating in the interferometer can be realized with modest laser powers, as already indicated in Figs. 4 and 5.

The Shot Noise Limit

Shot noise is a "white" noise, but as the response in (2) rolls off as $1/f\tau$ at frequencies above the inverse storage time τ, the apparent strain noise rises linearly with frequency, as shown in the curves "Space" and "Earth" in Fig. 3. As we will see later, this frequency-proportional rise of the sensitivity curve will limit the sensitivity in spaceborne interferometers in a similar way as in the ground-based detectors.

4 The Space Interferometer LISA

Only a space mission allows us to investigate the gravitational wave spectrum at very low frequencies. For all ground-based measurements, there is a natural, insurmountable boundary toward lower frequencies. This is given by the (unshieldable) effects due to varying gravity gradients of terrestrial origin: moving objects, meteorological phenomena, as well as motions inside the Earth. To overcome this "brick wall," the only choice is to go far enough away, either into a wide orbit around the Earth, or better yet further out into interplanetary space. Once we have left our planet behind and find ourselves in outer space, we have some great benefits for free: to get rid of terrestrial seismic and gravity-gradient noise, to have excellent vacuum along the arms, and in particular to be able to choose the arm length large enough to match the frequency of the astrophysical sources we want to observe.

4.1 The LISA Configuration

The European Space Agency (ESA) and NASA have agreed to collaborate on such a space mission called LISA, "Laser Interferometer Space Antenna" [18, 19].

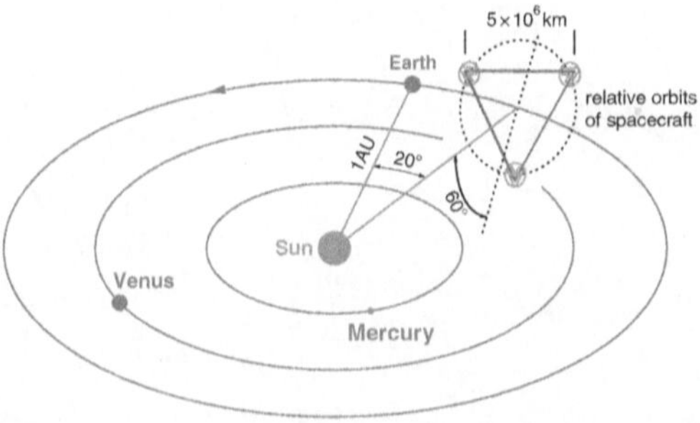

Fig. 7. Orbits of the three spacecraft of LISA, trailing the Earth by 20°. The triangle arms are scaled by factor 10.

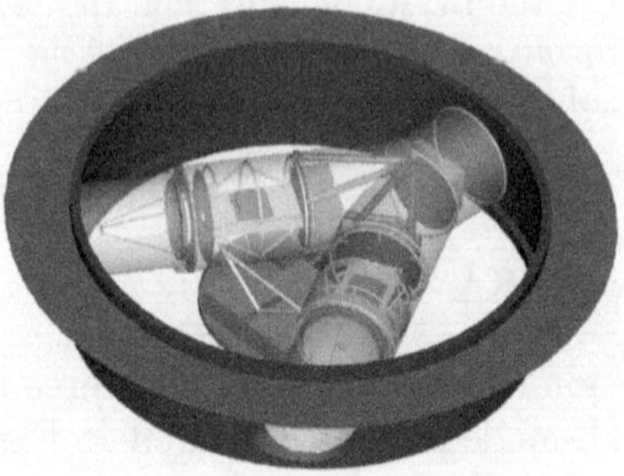

Fig. 8. View of one LISA spacecraft, housing two optical assemblies. The solar panel at top not shown, the thermal shield shown as semitransparent.

LISA consists of three identical spacecraft, placed at the corners of an equilateral triangle (Fig. 7). The sides are to be 5 million km long ($5 \cdot 10^9$ m). This triangular constellation is to revolve around the Sun in an Earth-like orbit, about 20° (i.e., roughly 50 million km) behind the Earth. The plane of this equilateral triangle needs to have an inclination of 60° with respect to the ecliptic to make the common rotation of the triangle most uniform. The three spacecraft form a total of three, but not independent, Michelson-type interferometers, here of course with 60° between the arms.

The spacecraft at each corner will have two optical assemblies that are pointed, subtending an angle of 60°, to the two other spacecraft (indicated in Fig. 8, with the Y-shaped thermal shields shown semitransparent). An optical bench, with the test mass housing in its center, can be seen in the middle of each of the two arms, and a telescope of 40 cm diameter at the outer ends. Each of the spacecraft has two separate lasers that can be phase-locked so as to represent the "beam splitter" of a Michelson interferometer. However, a different scheme of data analysis would relax that requirement of locked phase considerably [20].

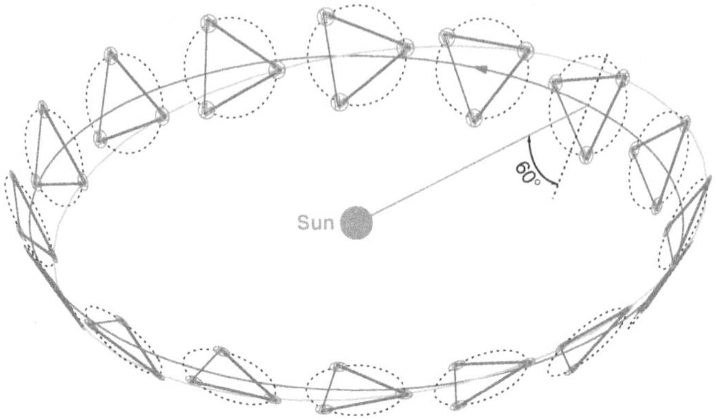

Fig. 9. Annual motion of the LISA configuration: the heavier orbit represents the Earth's orbit, and the orbit on which the center of the LISA triangle circles the Sun. The motion of one of the LISA spacecraft is indicated by the (slightly inclined) lighter orbit.

4.2 Annual Orbit of LISA

During its yearly motion around the Sun, the three spacecraft of LISA will "roll" on a cone of half-angle 60°, as indicated in Figs. 1 and 9. Each spacecraft moves on a slightly elliptic orbit around the Sun, as indicated for one spacecraft by the lighter orbit, slightly tilted with respect to the (heavier) Earth orbit.

This configuration has a number of advantages that make several of the design requirements less stringent.

Constant Angle to Sun

The spacecraft face the Sun by a constant angle of incidence of 30°, which provides a very stable thermal environment for the sensitive parts (optical assembly, the sensors) of the spacecraft. It also allows a design of the spacecraft such that no sunlight will ever enter the sensitive optical assembly.

Constant Triangle Shape

The orbits of the three spacecraft provide a stable configuration, close to an equilateral triangle. The maximum changes in the ($\approx$60°) angles subtended by the lines of sight to the other two spacecraft are in the order of 1° (Fig. 10). Thus it becomes possible to devise articulation schemes for the two "telescopes" in each spacecraft to follow these deviations.

The maximum distance variations are in the order of 100,000 km, which is also small when compared with the very large baseline of 5 million km: $\approx$2% The velocity along the line-of-sight between the spacecraft varies by about

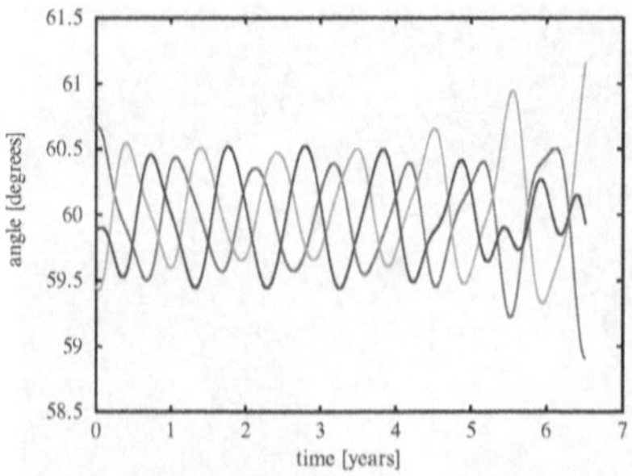

Fig. 10. Variation of the angles during the mission lifetime for one typical orbit configuration.

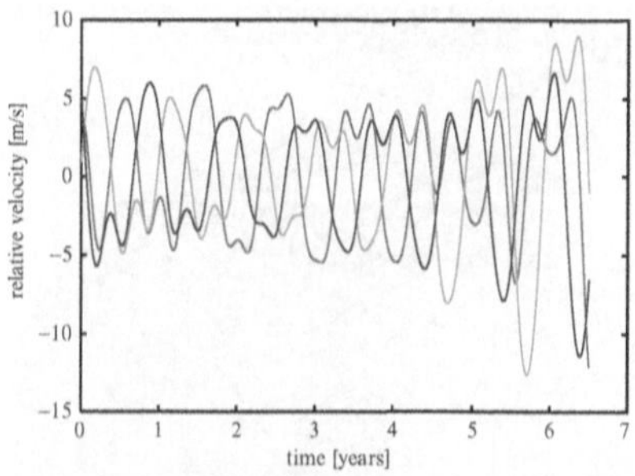

Fig. 11. Line-of-sight velocities for one typical orbit configuration.

$\pm 10\,\mathrm{m\,s^{-1}}$ during each year (Fig. 11), giving rise to Doppler shifts of about $\pm 10\,\mathrm{MHz}$ in the received beams, which must be taken into account in the interferometer design. The precise design of the orbits is subject of ongoing optimizations, with the aim to minimize these deviations from a constant triangle.

Constant Distance to Earth

The center of the LISA triangle trails the Earth in its orbit by 20°, or about 50 million km. This makes the distance to Earth, for radio communication, also quite stable, which reduces the problems of radio antenna design and radio transmission power. The radio antennas must, however, provide a rotational degree of freedom that allows them to be pointed toward the Earth for the (intermittent) data transmissions, as well as for the (probably continuous) preparedness for control signals from Earth.

An auxiliary modulation on the laser link will provide for data transfer between the spacecraft, thus allowing to swap the data to be downlinked between the spacecraft for increased redundancy and reduced downtime due to antenna repositioning.

4.3 Gravitational Reference Sensors

The distances between the different spacecraft are measured from test masses housed *drag-free* in these three spacecraft. The three LISA spacecraft each contain two test masses, one for each arm forming the link to another LISA spacecraft. The test masses, 4 cm cubes made of an Au/Pt alloy of low magnetic susceptibility, reflect the light coming from the YAG laser and define the reference mirror of the interferometer arm. These test masses are to be freely floating in space, subject only to gravity.

For this purpose, these test masses are also used as inertial references for the drag-free control of the spacecraft that constitutes a shield to external

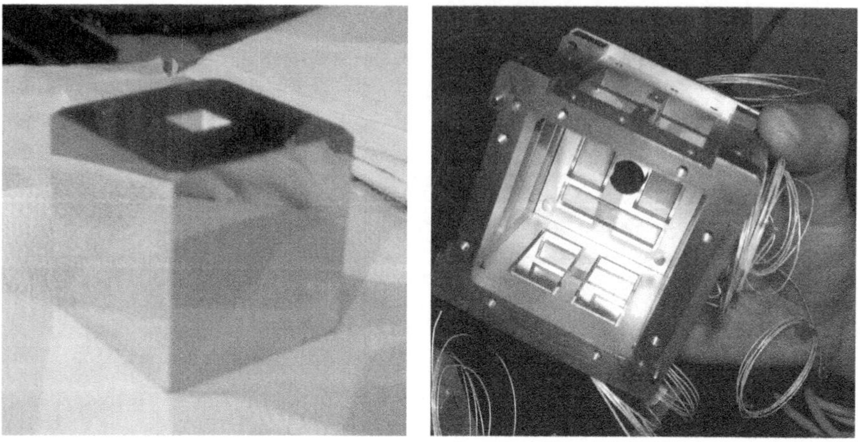

Fig. 12. Layout of gravitational sensor: (**a**) test mass and (**b**) electrode housing

forces. Development of these sensors is done at various institutions. Figure 12 shows an engineering model of a test mass and its electrode housing [22]. These sensors feature a three-axis electrostatic suspension of the test mass with capacitive position and attitude sensing.

A noise level of $10^{-9}\,\mathrm{m\,Hz^{-1/2}}$ is needed to limit the disturbances induced by relative motions of the spacecraft with respect to the test mass: for instance the disturbances due to the spacecraft self-gravity or to the test mass charge.

4.4 Micronewton Thrusters

The very weak forces required to keep up drag-free operation, less than $100\,\mu\mathrm{N}$, are to be supplied by electrical propulsion devices. In the so-called field-effect electrical propulsion (FEEP) devices, a strong electrical field forms the surface of liquid metal (Cs or In) into a cusp from which ions are accelerated to propagate into space with a velocity (of the order $60\,\mathrm{km\,s^{-1}}$) depending on the applied voltage. Another technology ("colloidal thrusters") uses small droplets of a conducting organic colloidal liquid instead. At least two of these technologies will be tested in space aboard the LISA Pathfinder (LPF) mission.

4.5 Lasers

In ground-based interferometric gravitational wave detectors, diode-pumped NPROs have proven as reliable laser sources, and they have also been identified as suitable laser candidates for LISA. Their operation has been described in detail in [23–25]. The principle is readily described: In an NPRO, the laser crystal alone forms the resonator. No additional external mirrors are necessary. This monolithic setup results in very stable operation. A typical NPRO laser crystal is shown in Fig. 6. Inside the laser crystal a ring resonator is formed. The beam is reflected at the dielectrically coated front facet of the laser crystal and by total internal reflection at three faces of the crystal. The fact that the beam does not stay in one plane results in a reciprocal polarization rotation during a round-trip. If the crystal is placed in a magnetic field as indicated in Fig. 6, then the nonvanishing Verdet constant of the crystal material Nd:YAG results in nonreciprocal polarization rotation. This causes different eigenpolarizations for the two round-trip directions. Since the dielectrically coated front facet has higher losses for one round-trip direction, the laser operates unidirectionally and hence in single-longitudinal mode as is required for most stable operation.

NPROs are longitudinally pumped by laser diodes through the front facet. The front facet is hence antireflection (AR) coated for the pump wavelength of 808 nm and partially reflecting for the laser wavelength (1064 nm).

The most promising configurations for a LISA laser are a stand-alone high-power NPRO or a fiber amplifier seeded by a low-power NPRO. For LISA, approximately 1 W of output power is required. To obtain such output powers from a single NPRO, the thermally induced lens of the NPRO has to be partially compensated. This can be achieved by a concave front facet of the laser crystal. Additionally, the region directly behind the front facet has to be made of undoped YAG to avoid spatial hole burning and to ensure single-frequency operation [25].

Both power noise and frequency noise couple into the phase measurement of LISA. Although NPROs show intrinsically low-power noise and frequency noise, both need to be further suppressed (for details see Sect. 4.6). This requires that the laser needs to have actuators for both frequency and power. In the case of the single-stage laser system for LISA, a piezoelectrical crystal (PZT) glued onto the laser crystal can be used as fast frequency actuator. A second actuator is the temperature of the laser crystal. It is slower but offers a higher tuning range than the PZT. The current of the pump diodes is commonly used as power actuator. Other requirements to the lasers include a minimum lifetime of 5 years and space qualification (vacuum, vibration, temperature cycles).

An alternative concept for a LISA laser system consists of a low-power NPRO amplified by a fiber amplifier. An advantage of such a system would be that a low-power NPRO is being built for LISA Pathfinder and thus already space qualified. The fiber amplifier in this configuration would typically use an Ytterbium-doped fiber that is seeded by the output of the low-power NPRO and pumped by laser diodes at a wavelength of 976 nm. Suitable frequency actuators for the two-stage system are the PZT of the seed laser and the laser crystal temperature. As power actuator for low frequencies, the current of the amplifier pump diodes can be used. A detailed investigation on power and frequency actuators in a laser amplifier system can be found in [26].

4.6 Power and Frequency Stabilization

As already mentioned, stabilizations for laser power and laser frequency are required by LISA (see Sect. 3.1). The stability that can be achieved is not a property of the laser itself. Instead, it depends on a number of properties of the overall control loop (reference, sensor, loop gain).

For frequency stabilization, optical resonators will be used in LISA as the first of three methods of increasing performance (Fig. 13). Their lengths define reference frequencies. Using an RF technique named after its inventors Pound, Drever, and Hall [27], the difference between laser frequency and optical resonator eigenfrequency will be measured and used as error signal for the control loop. Figure 14 shows the goal for the frequency prestabilization with measured data.

The frequencies of two identical NPROs were stabilized to two independent optical resonators made of ULE, a material with a low thermal expansion coefficient ($2{\cdot}10^{-8}\,\mathrm{K}^{-1}$). Each resonator was located in its own vacuum chamber and surrounded by thermal shields. The thermal shields consisted of four gold-coated steel cylinders separated by ceramic spacers. The cylinders acted as thermal capacitors, the ceramics spacers as thermal resistors. The thermal

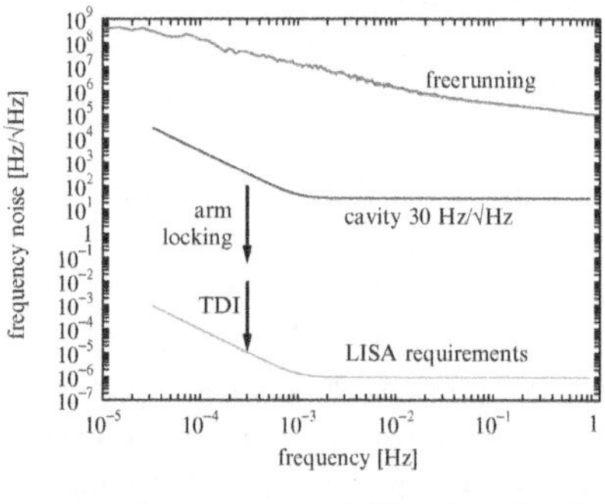

Fig. 13. Three-step frequency stabilization for the LISA laser.

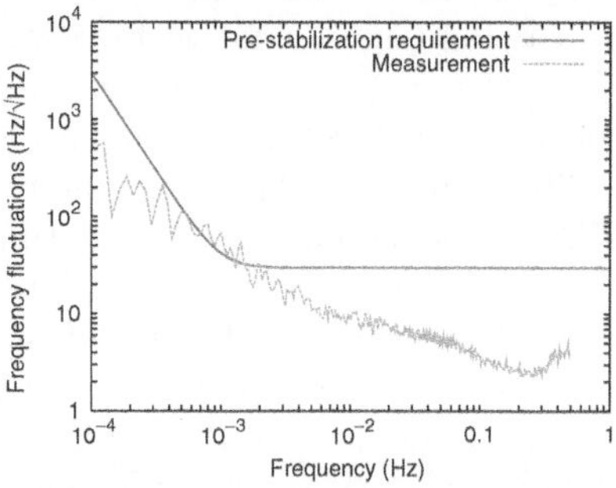

Fig. 14. Laser prestabilization to a stable cavity: requirement and laboratory results.

stability at the location of the optical resonators and their thermal expansion coefficient limit the frequency stability that can be achieved with this technique.

The remaining 7–8 orders of magnitude in laser frequency stability will be bridged by two more steps: arm locking and time-delay interferometry (TDI). The latter is a data processing technique that synthesizes a virtual equal-arm interferometer and thus reduces the effect of laser frequency fluctuations (see Sect. 6).

Arm locking, on the other hand, is a novel technique that uses the 5 million km LISA arms as reference for a real stabilization of the LISA laser frequency. The novelty lies in the fact that a unity-gain frequency of more than 10 kHz is necessary while the "sensor" has a delay of 33 s (the round-trip travel time). Meanwhile, several laboratory experiments have shown the feasability of such an unusual control loop [28–30].

For power stabilization, a fraction of the laser beam is split off using a glass wedge and detected by a photo diode. The resulting signal is compared with a stable reference voltage, and the difference is amplified and used for feedback. Special care has to be taken with temperature fluctuations and their effect on beam splitting and beam detection. In particular, the temperature of photodiodes must be stabilized to achieve the required stability [31].

4.7 Noise in LISA

This section will cover some of the most worrying noise sources in the LISA project, which then will also be relevant for other planned space projects such as Big Bang Observer [32], DECIGO [33], and the Chinese project ASTROD [34,35].

Figure 3 showed sensitivity curves for the ground-based interferometers, as well as for LISA. In both cases the shape is that of a trough, with a steeper slope at the left than on the right. The curve for LISA is again shown in Fig. 15, enlarged and in greater detail. That LISA sensitivity curve consists of three main parts, as indicated by the three differently shaded frequency regions, in which different noise mechanisms take hold.

Shot Noise

With the 40 cm optics planned, from 1 W of infrared laser power transmitted, only some 10^{-10} W will be received after 5 million km, and it would be hopeless to have that light reflected back to the central spacecraft.

Instead, also the distant spacecraft are equipped with lasers of their own, which are phase-locked with an offset to either the incoming light or the second laser on the same spacecraft. One laser in the configuration serves as master and is stabilized with both a cavity and the arm locking method [36].

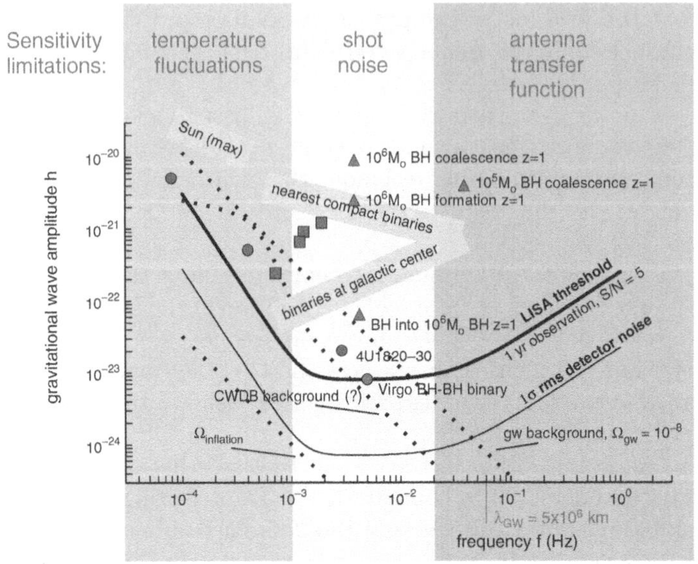

Fig. 15. Sensitivity of LISA: the heavy curve "LISA threshold" represents the signal strength that would provide a signal-to-noise ratio of 5 if averaged over 1 year, and over all possible directions and polarization angles. Major noise contributions are indicated by *different shading.*

Because of the low level of light power received, shot noise plays a decisive role in the total noise budget above 5 mHz. In the LISA noise budget above 5 mHz, $12\,\mathrm{pm\,Hz^{-1/2}}$ are allocated for the optical metrology in a single optical link, and $7.6\,\mathrm{pm\,Hz^{-1/2}}$ out of these are allocated to shot-noise alone.

The effect of shot noise is a spurious "path difference" $\widetilde{\delta L}$ inversely proportional to the square root of the power P available for interferometry. In the case of LISA, at arm lengths of 5 million km, this received power is of the order 10^{-10} W. With an increased arm length, perhaps to the order of 2 AU, i.e., 300 million km, the power would decrease by a factor of 60^2, and both the apparent spurious path differences $\widetilde{\delta L}$ and the optical path L would thus increase by an identical factor of 60. This means that the sensitivity of a space probe, other characteristics remaining the same, would have a shot noise limit for the strain $h \sim \delta L/L$ that is independent of the arm length, $L/2$. This fact will be of importance in estimating also the sensitivities of other space projects being discussed.

Antenna Transfer Function

Again, as shown in the section "The Shot Noise Limit," we have to consider that the antenna response rolls off as $1/f\tau$ at frequencies f above the inverse of the round-trip time τ. Thus at these frequencies the shot noise leads to the frequency-proportional rise at the right-hand side of the sensitivity curve in Fig. 3 and, in more detail, in Fig. 15.

Acceleration Noise

At frequencies below 5 mHz, the noise is mainly due to accelerations of the test mass that cannot be shielded even by the drag-free scheme: forces due to gravitating masses on the spacecraft when temperature changes their distances, charging of the test masses due to cosmic radiation, residual gas in the test mass housing, and nonzero "stiffness" that couples spacecraft motion to the test mass. Except for the cosmic ray charging, the acceleration noise contributions are dependent on temperature variations, and this is why in Fig. 15 they come under the heading "temperature fluctuations." These accelerations have a rather "white" spectral distribution, which thus results in position errors rolling off roughly as $1/f^2$.

Gravity-Gradient Noise

The test mass, housed in the LISA spacecraft, is subject to the gravity field of the other masses that form part of the spacecraft. These masses, though "rigidly" connected to each other, will undergo small changes in their positions, due, e.g., to the changes in temperature distribution. This thermal distortion of the spacecraft actually is one of the most prominent sources of "acceleration noise." Elaborate calculations on the temperature fluctuations to be expected (e.g., from variations in the solar radiation) and on the thermal behavior of the spacecraft's masses have resulted in a set of requirements for the LISA design [19]. Also, a drift of the spacecraft with respect to the freely falling test mass must be avoided. The technique to do that is what is termed "drag-free control."

Noise Due to Charging of the Test Mass

Cosmic radiation will cause the test mass to acquire an electrical charge, which will result in a number of noise effects. A broad discussion is given in the LISA Pre-Phase A Study (PPA2) [18]. These charges will give rise to electrostatic forces of attraction to the cage walls. The charges will also, if not perfectly shielded by the cage and the spacecraft shields, be subject to Lorentz forces due to LISA's motion in the interplanetary magnetic field. And, similarly, changes in that magnetic field will also produce forces on the test mass. As remedies, the test mass will be quite well shielded from outside fields, and in particular, the charge that has accumulated on the test mass will be monitored, and from time to time a discharge by shining ultraviolet light on the test mass and its housing, will be carried out [18].

Noise Due to Residual Gas

A very wide field of acceleration noise contributions is due to the residual gas inside the sensor. Although the vacuum will have high quality, 10^{-8} mbar = 10^{-6} Pa, the test mass will be subject to several nonnegligible accelerations. Foremost among these can be the stochastic noise due to the buffeting by the impinging residual gas molecules. This statistical noise is proportional to the square root of the residual gas pressure, p. If the casing of the sensor has a temperature gradient, due, e.g., to changes in solar radiation or in the power dissipation in the spacecraft electronics, differences in gas pressure inside the sensor will build up. Here we must mention the so-called *radiometer effect*, but perhaps even more worrisome the effect of temperature-dependent outgassing of the cage walls.

Noise Total

With a myriad of other, smaller, noise contributions the total apparent path noise amounts to something like $\widetilde{\delta L} \approx 40 \cdot 10^{-12}\,\mathrm{m\,Hz}^{-1/2}$ at the lowest part, the bottom of the trough. For signals monitored over a considerable fraction of a year, and taking into account a reasonable signal-to-noise ratio of 5, and furthermore averaging over all angles of incidence ($\sqrt{5}$), the best sensitivity is about $h \approx 10^{-23}$, indicated in Fig. 3 by the curve marked "Space," and in more detail in Fig. 15.

The LISA Prospects

Some of the gravitational wave signals are guaranteed to be much larger, resulting in signal-to-noise ratios of 1,000 or higher. Failure to observe them would cast severe doubts on our present understanding of the laws that govern the universe. Successful observation, on the other hand, would give new insight into the origin and development of galaxies, existence and nature of dark matter, and other issues of fundamental physics.

4.8 Status of LISA

In 2007, LISA has been unanimously approved by ESA science programme committee as the first large mission candidate L1 in the new Cosmic Visions programme (2005–2025) with an anticipated launch date in 2018. A System and Technology Study [19] has substantiated that improved technology, light-weighting, and collaboration with NASA will lead to a considerable reduction of cost. Since January 2005 LISA is in the mission formulation phase of the ESA mission life cycle. LISA has a nominal lifetime of 5 years, but the equipment and thruster supply are chosen to allow even 10 years of operation. A collection of papers given at the Sixths International LISA Symposium, 2006, is presented in [37].

4.9 Technology Demonstrator

Some of LISA's essential technologies (gravitational sensor, interferometry, micronewton thrusters) are to be tested in the LISA Pathfinder (LPF) mission to be launched by ESA in 2009. The core payload is the European LISA Technology package (LTP). The package will contain, on a common optical bench, two gravitational sensors, similar to the one of Sect. 4.3.

The relative motion between the two freely floating test masses will be monitored with high accuracy by interferometry [38–40]. The sensitivity in this (scaled-down) experiment will come to within a factor of ten of the proposed LISA sensitivity.

This package is to be flown in an orbit near the Sun–Earth Lagrange point L_1, relatively far away from Earth, so as to avoid the many disturbances near the Earth. The same mission LPF will also host a NASA contribution, the Disturbance Reduction System (DRS).

4.10 LISA Follow-Ons

Even as early as now concepts are being discussed for a successor to LISA, on the possible enhancements in sensitivity and/or frequency band. One scheme, the Japanese project DECIGO [33] would try to bridge the frequency gap between ground and space detectors, by reducing the arm lengths, leaving the general configuration unchanged.

The Big Bang Observer BBO [32] would have an increased number (four) of LISA-type triangles, such that independent interferometers result. These can be used to detect and measure a stochastic background of gravitational waves, similar to, but reaching much further back than the 3 K electromagnetic background radiation.

ASTROD will extend to a low-frequency range not fully covered by LISA, and thus it would be – given similar sensitivity – a further useful extension in the search for and measurement of gravitational waves from, e.g., supermassive black holes.

5 LISA Data Analysis

Because of the low-frequency band of the LISA detection, the data rate, and thus also the total amount of data, is rather low. Data will be collected onboard, and transmitted to Earth once per two to three days.

5.1 Directivity

LISA, as all interferometric GW detectors, has a preferred direction and a preferred polarization of the incoming gravitational wave. This would cause an antenna, fixed in space, to be particularly sensitive in some directions, and totally blind in others.

The annual motion of LISA will, however, average out these types of directivity, as LISA is facing different locations at the sky, and with different preferred polarization directions at different times, see Fig. 9. This is why the sensitivity curve in Fig. 15 for a signal-to-noise ratio of 5 is drawn by factor of $5\sqrt{5} = 11.2$ higher than the lower curve.

On the other hand, LISA's detection can make use of the "signature" that continuous-wave signals will have, due to the changing response sensitivity, and due to the Doppler shifts that the signal will undergo as LISA approaches and recedes from the source during its annual orbit.

A detailed analysis of the LISA sensitivity under these assumptions was made by Schilling [18, 41]. One important result was that the drastic drops in sensitivity for gravitational waves with wavelengths fitting into the arm lengths are benignly smoothed out in this averaging, as will be seen in the curve of Fig. 17.

5.2 Noise Due to Fluctuating Laser Frequency

The strength of the Michelson-interferometer scheme is that the high symmetry between the two arms makes the interferometer insensitive to a number of fluctuations of the illuminating light source. The most serious of these is the fluctuation in laser phase, $\delta\phi$, or in frequency, $\delta\nu$. Any change in laser frequency will cause spurious signals proportional to the difference in arm lengths. In the simplest case of a Michelson interferometer, the phases ϕ_i accumulated in the round-trips in the two arms are measured, $s_i(t) = \phi_i(t) - \phi_i(t - 2T_i)$, and then compared with each other:

$$\Phi(t) = s_1(t) - s_2(t) = \phi_1(t) - \phi_1(t - 2T_1) - \phi_2(t) + \phi_2(t - 2T_2) \,. \quad (4)$$

In each arm, the current laser phase is compared with the (echoed) phase of one round-trip ago, these times $2T_1$ and $2T_2$ differing by a relatively small misalignment $\Delta = 2T_1 - 2T_2$. For the sake of simplicity, let us also assume equal laser phase $\phi_1(t) = \phi_2(t) = \phi_0 + \delta\phi(t)$ in the two arms, with a phase noise component $\delta\phi(t)$. Clearly, for unequal round-trip times T_i, the error $\delta\Phi(t)$ would become

$$\delta\Phi(t) = \frac{\mathrm{d}}{\mathrm{d}t}\,\phi(t - 2T)\,(2T_1 - 2T_2) = 2\pi\delta\nu(t - 2T)\,\Delta \,. \quad (5)$$

The celestial mechanics of the LISA orbits will cause relative arm length variations in the order of 10^{-2}, and these would produce spurious signals from the natural laser frequency fluctuations well above the true gravitational wave signals.

6 Unequal-Arm length Interferometry

Even if the laser frequency is well stabilized to the best of current technology, perhaps to $30\,\mathrm{Hz}\,\mathrm{Hz}^{-1/2}$, a drastic further reduction of the effect is required. Here, a scheme first proposed by Giampieri et al. [42], and then optimized with respect to the suppression of several LISA error sources [20,43,44], promises a significant improvement. The concept of Giampieri et al. was to estimate, from the phases measured separately for each arm and each spacecraft, the underlying laser phase noise and appropriately correct for it. This scheme operates in the frequency domain. The approach to be discussed below, operating in the time domain, will offer even better compensation of the laser noise, and it is the current baseline for LISA [19].

6.1 Time-Delay Interferometry

The basic principle of the method is best demonstrated using the simplified case of a Michelson interferometer with only one master laser, and the phase measurements done in only one spacecraft. What is used is a linear combination of the readout data s_i with data additionally delayed, in each arm by the travel time in the other arm

$$X(t) = s_1(t) - s_2(t) - s_1(t - 2\tau_2) + s_2(t - 2\tau_1)\,, \tag{6}$$

where the delays τ_i are chosen to equal the true travel times T_i. It is easily verified that this algorithm can fully cancel the laser phase noise $\delta\phi(t)$. One can estimate what degree of cancelation could be achieved if there were slight deviations $\delta_i = 2T_i - 2\tau_i$ between the true round-trip times T_i and the delay times τ_i used in (6).

The laser phase noise in the measurements taken in the two arms will have the general form $\delta s_i(t) = \delta\phi(t) - \delta\phi(t - 2T_i)$, so that (6) will lead to a total phase error of

$$\begin{aligned}\delta\Phi(t) = \delta\phi(t) - \delta\phi(t - 2T_1) - \delta\phi(t - 2\tau_2) + \delta\phi(t - 2T_1 - 2\tau_2)\\ -\delta\phi(t) + \delta\phi(t - 2T_2) + \delta\phi(t - 2\tau_1) - \delta\phi(t - 2T_2 - 2\tau_1)\,.\end{aligned} \tag{7}$$

The undelayed terms $\delta\phi(t)$ cancel right away. And clearly, for $\tau_i = T_i$, this combination of noise terms cancels fully, regardless of any difference in the values for T_1, T_2. If, however, we have small deviations of the assumed values τ_i from the true round-trip times T_i, we must evaluate (7) for plausible values of T_i, τ_i.

6.2 The LISA Case of Almost Equal Arms

The typical LISA case would be a relatively small difference Δ between the two round-trip times, $\Delta = 2T_1 - 2T_2$, and also we will assume the delay time errors $\delta_i = 2\tau_i - 2T_i$ to be relatively small. Then, we can consider appropriate difference terms in (7) as derivatives at a mean time $t - 2T$:

$$\begin{aligned} -\delta\phi(t-2T_1) + \delta\phi(t-2T_2) &\approx \delta\omega(t-2T)\cdot(\Delta) \\ +\delta\phi(t-2\tau_1) - \delta\phi(t-2\tau_2) &\approx \delta\omega(t-2T)\cdot(-\Delta-\delta_1+\delta_2)\,. \end{aligned} \tag{8}$$

Thus, the terms with delays of $2T$ and $4T$ result in phase errors of

$$\begin{aligned} \delta\Phi(t)\big|_{2T} &\approx \delta\omega(t-2T)\cdot(-\delta_1+\delta_2) \\ \delta\Phi(t)\big|_{4T} &\approx \delta\omega(t-4T)\cdot(\delta_1-\delta_2)\,. \end{aligned} \tag{9}$$

In this approximation, the errors δ_1, δ_2 in guessing the round-trip times T_1, T_2 would still not result in an error $\delta\Phi$ if they happened to be identical: $\delta_1 = \delta_2$, and they would be disturbing the most if they had opposite sign.

Furthermore, at very low frequencies, the laser frequency noise $\delta\omega$ would not change drastically from delay $2T$ to $4T$. So then the terms of delays $2T$ and $4T$ would cancel to a large extent, regardless of the error difference $\delta_1 - \delta_2$. This is, however, very similar to the reduction in response to the genuine GW signals and will thus lead to neither an improvement nor a deterioration of the noise introduced by the misestimates $\delta_1 - \delta_2$. For noise frequencies f at which the argument $2\pi f\,\Delta$ becomes significant (say, ≈ 1), this low-frequency cancelation ceases.

With the allowance in optical-path noise for the laser phase noise of $\widetilde{\delta\mathcal{L}} = 10\cdot 10^{-12}\,\mathrm{m\,Hz^{-1/2}}$ (total, from four spacecraft, see PPA2 [18]), and with the LISA laser stability of $\widetilde{\delta\nu} < 100\,\mathrm{Hz\,Hz^{-1/2}}$, the allowable delay-time error $\delta_1 - \delta_2$ would be

$$|\delta_1 - \delta_2| < \frac{\widetilde{\delta\mathcal{L}}}{c}\Big/\frac{\widetilde{\delta\nu}}{\nu} = 10^{-11}\,\mathrm{m\,Hz^{-1/2}}\Big/\left(100\,\mathrm{Hz\,Hz^{-1/2}}\times 10^{-6}\,\mathrm{m}\right) \approx 10^{-7}\,\mathrm{s} \tag{10}$$

corresponding to 30 m. A more detailed analysis is given in Tinto et al. [20]

6.3 The LISA Analysis Algorithms

How powerfully the *time-delay interferometry* cancels out not only laser phase noise but also other instrumental errors is shown in various papers by Armstrong, Estabrook, and Tinto [20,43,44]. These form the baseline for the LISA procedure [19]. It is assumed that phase measurements are made in all three spacecraft, each equipped with independent lasers, with independent highly stable clocks (USOs: ultra-stable oscillators), and with an intraspacecraft link between the two lasers onboard each spacecraft. Figure 16 shows four types of such configurations. The nominal LISA configuration is an unequal-arm Michelson interferometer, as in Fig. 16, top left.

The links from one spacecraft to another are specified by two indices, of which the first one indicates the arm (via the number of the spacecraft opposite that arm), and the second one the direction (via the target spacecraft).

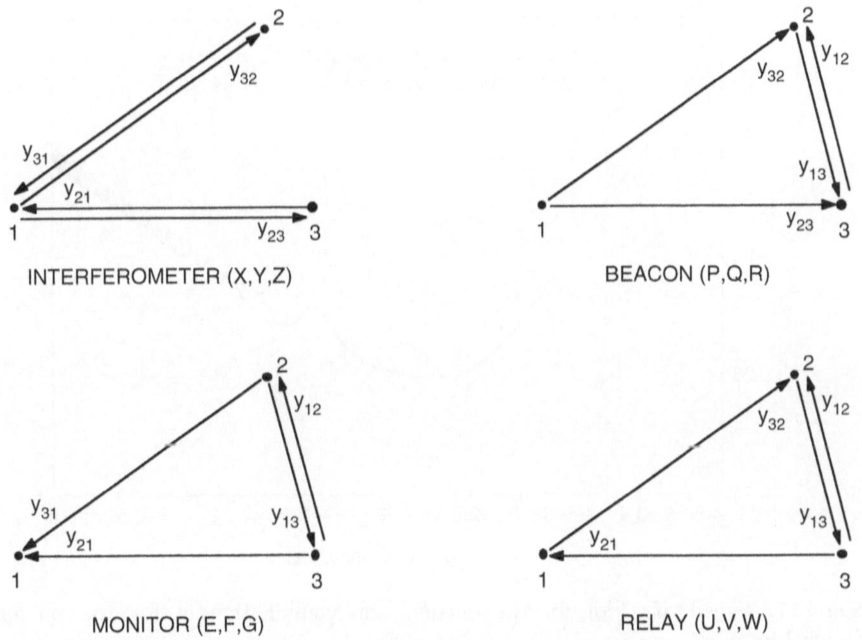

Fig. 16. The four types of four-link LISA data combining possibilities.

Indices after a comma will indicate the individual delays of the data, again by specifying the arm via the opposite spacecraft. Multiple (up to fourfold) delays are used.

One typical example (the Michelson configuration of Fig. 16, upper left) would look like this:

$$\begin{aligned} X = {} & y_{32,322} - y_{23,233} + y_{31,22} - y_{21,33} + y_{23,2} - y_{32,3} + y_{21} - y_{31} \\ & + \frac{1}{2}\left(-z_{21,2233} + z_{21,33} + z_{21,22} - z_{21}\right) \\ & + \frac{1}{2}\left(+z_{31,2233} - z_{31,33} - z_{31,22} + z_{31}\right) \end{aligned} \tag{11}$$

This set of time-domain combinations of the y_{ij} from the two arms and from the intraspacecraft signals z_{ij} cancels all noise due to laser phase fluctuations and to motions of the optical benches [43,44]. In a different approach [20], the frequency fluctuations of lasers and USOs can all be canceled.

It is assumed that shot noise and optical path noise (i.e., total optical path noise, as specified in PPA2 [18]) have the same transfer functions. The LISA sensitivity would then have the form given in Fig. 17, again averaged over one year, and over all directions of propagation and polarization, and for SNR $= 5$. This is where the shape of the (simplified) sensitivity curve of Fig. 15 comes from.

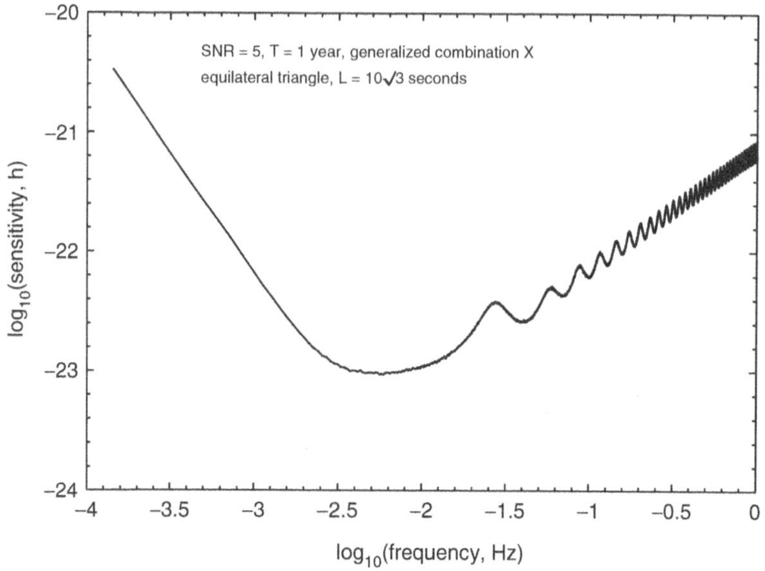

Fig. 17. Sensitivity plot for the unequal-arm combination of Fig. 16, *top left* (Michelson).

6.4 Constancy of Arm Lengths

A sufficient equality of the (three) arm lengths cannot be maintained by LISA, the arms of which will have annual changes in length of the order 100,000 km. The variation in ASTROD will be even larger, by orders of magnitude. Thus the above data analysis needs to be employed to suppress faked signals resulting from short-term fluctuations in laser frequency. These schemes require a *knowledge* of the lengths of the arms to better than 30 m to be able to apply the proper time delays to the various time series. The auxilliary modulation on the carrier light provides for ranging with sufficient accuracy in addition to clock synchronization and data transfer.

The processing techniques [20, 44] required for LISA to cancel out fluctuations in laser frequency and position of the test mass inside the sensor will also have to be applied in the case of ASTROD. In LISA, the application is relatively easy, as the arm lengths are rather well constant throughout the course of the year. For ASTROD, the arm lengths change much more rapidly, and by much larger amounts. So an increased effort has to be made to render the LISA data-analysis routines applicable also under these more challenging conditions.

7 Conclusion

The difficulties (and thus the great challenges) of gravitational wave detection stem from the fact that gravitational waves have so little interaction with matter (and space), and thus also with the measuring apparatus. Great scientific and technological efforts, large detectors, and a working international collaboration are required to detect and to measure this elusive type of radiation. And yet – just on account of their weak interaction – gravitational waves can give us knowledge about cosmic events to which the electromagnetic window will be closed forever. This goes for the processes in the (millisecond) moments of a supernova collapse, as well as of the many mergers of binaries that might be hidden by galactic dust. Such high-frequency events (a few Hz up to a few kHz) will be accessible from the detectors on Earth. For the signals to be significant, a number of ground-based detectors should be operated in coincidence, and only such joint analyses will allow to locate the source in the sky.

The perspective of detecting events with gravitational wave radiation also holds for the distant, but violent, mergers of galaxies and their central (super)massive black holes. The low frequencies (10^{-5} to 1 Hz) characteristic of such sources are accessible only from space, e.g., with LISA. The expected high signal-to-noise ratios will allow unquestionable detection with only one detector, and will even allow to locate the source in a narrow region in the sky.

A LISA follow-on mission, and also combinations of terrestrial detectors, might probe the GW background from the very beginning of our universe (10^{-14} s or even only 10^{-22} s after the big bang) [45]. In this way, gravitational wave detection can be regarded as a new window to the universe, but to open this window we must continue on our way in building and perfecting our antennas. It will only be after these large interferometers are completed (and perhaps even only after the next generation of detectors) that we can reap the fruits of this enormous effort: a sensitivity that will allow us to look far beyond our own galaxy, perhaps to the very limits of the universe.

References

1. A. Einstein, *Sitzungsber. Preuss. Akad. Wiss.* (1916) 688.
2. A. Einstein, *Sitzungsber. Preuss. Akad. Wiss.* (1918) 154.
3. K. Thorne: Gravitational Radiation, bf in: S.W. Hawking and W. Israel (eds.) *300 Years of Gravitation* (Cambridge University Press, Cambridge 1987), p.330.
4. K. Thorne: Gravitational Radiation – A New Window Onto the Universe, *Rev. Mod. Astron.* **10**, 1 (1997).
5. A. Buonanno and T. Damour: Effective one-body approach to general two-body dynamics, *Phys. Rev.* **D 59**, 084006 (1999).
6. B.F. Schutz: Lighthouses of gravitational wave astronomy – Prospects with LIGO and LISA, **in:** M. Gilfanov, R. Sunyaev, E. Churakov (Eds.) *Lighthouses of the Universe*, ESO Astrophysics Symposia, (2002), p.207.
7. D. Sigg: Status of the LIGO detectors, *Class. Quantum Grav.* **23**, S51 (2006).
8. F. Acernese, P. Amico, M. Al-Shourbagy, S. Aoudia, *et al.*: The status of VIRGO, *Class. Quantum Grav.* **23**, S63 (2006).
9. H. Lück, M. Hewitson, P. Ajith, B. Allen, *et al.*: Status of the GEO600 detector, *Class. Quantum Grav.* **23**, S71 (2006).

10. B.J. Meers: Recycling in laser-interferometric gravitational-wave detectors, *Phys. Rev.* **D 38**, 2317 (1988).
11. G. Heinzel *et al.*: Dual recycling for GEO 600, *Class. Quantum Grav.* **19**, 1547 (2002).
12. M. Ando: Current status of the TAMA300 gravitational-wave detector, *Class. Quantum Grav.* **22**, S881 (2005).
13. D.E. McClelland, S.M. Scott, M.B. Gray, A.C. Searle, *et al.*: Status of the Australian consortium for interferometric gravitational astronomy, *Class. Quantum Grav.* **23**, S41 (2006).
14. I. Zawischa *et al.*: The GEO 600 laser system, *Class. Quantum Grav.* **19**, 1775 (2002).
15. A. Rüdiger *et al.*: A mode selector to suppress fluctuations in laser beam geometry, *Opt. Acta* **28**, 641 (1981).
16. D. Schnier *et al.*: Power recycling in the Garching 30-m prototype interferometer for gravitational-wave detection, *Phys. Lett.* **A 225**, 210 (1997).
17. K. Kuroda: Large-scale cryogenic gravitational wave telescope and R&D, **in:** S. Kawamura and N. Mio (Eds.) *Gravitational Wave Detection II*, (Universal Academy Press 2000), p.45.
18. LISA Pre-Phase A Report, 2nd edition, Max-Planck-Institut für Quantenoptik, Report 233 (July 1998); often referred to as PPA2.
19. LISA: System and Technology Study Report, ESA document ESA-SCI(2000)11, July 2000, revised as `ftp://ftp.rzg.mpg.de/pub/grav/lisa/sts/sts_1.05.pdf`
20. M. Tinto *et al.*: Time-delay interferometry for LISA, *Phys. Rev.* **D 65**, 082003 (2002).
21. V. Josselin, M. Rodrigues, and P. Touboul: Inertial sensor concept for the gravity wave missions, *Acta Astronautica* **49/2**, 95 (2001).
22. A. Cavalleri *et al.*: Progress in the development of a position sensor for LISA drag-free control, *Class. Quantum Grav.* **18**, 4133 (2001).
23. T.J. Kane and R.L. Byer: Monolithic, unidirectional single-mode Nd:YAG ring laser, *Opt. Lett.* **10**, 65 (1985).
24. A.C. Nilsson, E.K. Gustafson, and R.L. Byer: Eigenpolarization Theory of Monolithic Nonplanar Ring Oscillators, *IEEE J. Quantum Electron.* **25**, 767 (1989).
25. I. Freitag, A. Tünnermann, and H. Welling: Power scaling of diode-pumped monolithic Nd:YAG lasers to output powers of several watts, *Opt. Commun.* **115**, 511 (1995).
26. M. Tröbs, P. Weßels, and C. Fallnich: Power- and frequency-noise characteristics of an Yb-doped fiber amplifier and actuators for stabilization, *Opt. Express.* **13**, 2224 (2005).
27. R.W.P. Drever, J.L. Hall, F.V. Kowalski, J. Hough, G.M. Ford, A.J. Munley, and H. Ward: Laser Phase and Frequency Stabilization Using an Optical Resonator, *Appl. Phys.* **B 31**, 97 (1983).
28. B.S. Sheard, M.B. Gray, D.E. McClelland, and D.A. Shaddock: Laser frequency stabilization by locking to a LISA arm, *Phys. Lett.* **A 320**, 9 (2003).
29. J. Sylvestre: Simulations of laser locking to a LISA arm, *Phys. Rev.* **D. 70**, 102002 (2004).
30. A.F.G. Marin, G. Heinzel, R. Schilling, A. Rüdiger, *et al.*: Phase locking to a LISA arm: first results on a hardware model, *Class. Quantum Grav.* **22**, S235 (2005),

31. M. Tröbs: *Laser development and stabilization for the spaceborne interferometric gravitational wave detector LISA*, PhD thesis, (University of Hannover, 2005).
32. V. Corbin and N.J. Cornish: Detecting the cosmic gravitational wave background with the Big Bang Observer, *Class. Quantum Grav.* **23**, 2435 (2006).
33. S. Kawamura, T. Nakamura, M. Ando, N. Seto, *et al.*: The Japanese space gravitational wave antenna - DECIGO, *Class. Quantum Grav.* **23**, S125 (2006).
34. W.-T. Ni: ASTROD – an overview, *Int. J. Mod. Phys.* **D 11**, 947 (2002).
35. Ni, W.: This volume.
36. G. Heinzel, C. Braxmaier, K. Danzmann, P. Gath, *et al.*: LISA interferometry: recent developments, *Class. Quantum Grav.* **23**, S119 (2006).
37. LASER INTERFEROMETER SPACE ANTENNA: 6th International LISA Symposium, Greenbelt, Maryland (USA), June 2006, AIP conference proceedings **873**, 3–706 (2007).
38. G. Heinzel, C. Braxmaier, R. Schilling, A. Rüdiger, *et al.*: Interferometry for the LISA technology package (LTP) aboard SMART-2, *Class. Quantum Grav.* **20**, S153 (2003).
39. G. Heinzel, V. Wand, A. Garcia, O.P. Jennrich, *et al.*: The LTP interferometer and phasemeter, *Class. Quantum Grav.* **21**, S581 (2004).
40. G. Heinzel, C. Braxmaier, M. Caldwell, K. Danzmann, *et al.*: Successful testing of the LISA Technology Package (LTP) interferometer engineering model, *Class. Quant. Grav.* **22**, S149 (2005).
41. R. Schilling: Angular and frequency response of LISA, *Class. Quantum Grav.* **14**, 1513 (1997).
42. G. Giampieri, R. Hellings, M. Tinto, and J. Faller: Algorithms for unequal-arm Michelson interferometers, *Opt. Comm.* **123**, 669 (1996).
43. M. Tinto and J.W. Armstrong: Cancellation of laser phase noise in an unequal-arm interferometer detector of gravitational radiation, *Phys. Rev.* **D 59**, 102003 (1999).
44. F.B. Estabrook, M. Tinto, and J.W. Armstrong: Time-delay analysis of LISA gravitational wave data: Elimination of Spacecraft motion effects, *Phys. Rev.* **D 62**, 042002 (2000).
45. B. Allen: The stochastic gravity-wave background: sources and detection, **in:** *Relativistic gravitation and gravitational radiation*, (Cambridge University Press, Cambridge 1997), p.373 (p. 381/382).

Lunar Laser Ranging Contributions to Relativity and Geodesy

Jürgen Müller[1], James G. Williams[2], and Slava G. Turyshev[2]

[1] Institut für Erdmessung (IfE), University of Hannover, Schneiderberg 50, 30167 Hannover, Germany
mueller@ife.uni-hannover.de

[2] Jet Propulsion Laboratory, California Institute of Technology, 4800 Oak Grove Drive, Pasadena, CA 91109, USA

Summary. Lunar laser ranging (LLR) is used to conduct high-precision measurements of ranges between an observatory on Earth and a laser retroreflector on the lunar surface. Over the years, LLR has benefited from a number of improvements both in observing technology and data modeling, which led to the current accuracy of postfit residuals of ~2 cm. Today LLR is a primary technique to study the dynamics of the Earth–Moon system and is especially important for gravitational physics, geodesy, and studies of the lunar interior. When the gravitational physics is concerned, LLR is used to perform high-accuracy tests of the equivalence principle, to search for a time variation in the gravitational constant, and to test predictions of various alternative theories of gravity. The gravitational physics parameters cause both secular and periodic effects on the lunar orbit that are detectable with the present day LLR; in addition, the accuracy of their determination benefits from the 35 years of the LLR data span. On the geodesy front, LLR contributes to the determination of Earth orientation parameters, such as nutation, precession (including relativistic precession), polar motion, and UT1, i.e., especially to the long-term variation of these effects. LLR contributes to the realization of both the terrestrial and selenocentric reference frames. The realization of a dynamically defined inertial reference frame, in contrast to the kinematically realized frame of VLBI, offers new possibilities for mutual crosschecking and confirmation. Finally, LLR also investigates the processes related to the Moon's interior dynamics. Here, we review the LLR technique focusing on its impact on relativity and give an outlook to further applications, e.g., in geodesy. We present results of our dedicated studies to investigate the sensitivity of LLR data with respect to the relativistic quantities; we also present the computed corresponding spectra indicating the typical periods related to the relativistic effects. We discuss the current observational situation and the level of LLR modeling implemented to date. We emphasize the need for modeling improvement for near future LLR opportunities. We also address improvements needed to fully utilize the scientific potential of LLR.

1 Introduction

Being one of the first space geodetic techniques, lunar laser ranging (LLR) has routinely provided observations for more than 35 years. The LLR data are collected as normal points, i.e., the combination of lunar returns obtained over a short time span of 10–20 min. Out of $\approx 10^{19}$ photons sent per pulse by the transmitter, less than 1 is statistically detected at the receiver [14]; this is because of the combination of several factors, namely energy loss (i.e., $1/R^4$ law), atmospherical extinction, and geometric reasons (rather small telescope apertures and reflector areas). Moreover, the detection of real lunar returns is rather difficult as dedicated data filtering (spatially, temporally, and spectrally) is required. These conditions are the main reason, why only a few observatories worldwide are capable of laser ranging to the Moon.

Observations began shortly after the first Apollo 11 manned mission to the Moon in 1969, which deployed a passive retroreflector on its surface. Two American and two French-built reflector arrays (transported by Soviet spacecraft) followed until 1973.[1] Most observations are taken to the largest reflector array that of the Apollo 15 mission. Over the years, more than 16,000 LLR measurements by now have been made of the distance between Earth observatories and lunar reflectors. Most LLR data have been collected by the Observatoire de la Côte d'Azur (OCA, France), the McDonald observatory (TX, USA) and – until 1990 – Haleakala (HI, USA). The new data are still coming, but today only the first two stations operate regularly. Understanding unexpected and small effects is very difficult with only one or two operating stations, because possible instrumental systematics of the ranging system cannot be separated from real scientific effects reliably. To further increase the impact of LLR in Relativity and Earth sciences more stations, with a wide geographical distribution, are needed. Therefore, the Italian colleagues have setup a new site in Matera, which has provided first LLR data quite recently. A new site with lunar capability is currently being built at the Apache Point Observatory, New Mexico, USA. This station, called APOLLO, is designed for a millimeter level accuracy ranging [13, 16]. However, to fully exploit the available LLR potential, a few more sites capable of tracking the Moon are needed, especially at diverse locations including the Southern hemisphere.

Figure 1a shows the number of LLR normal points per year since 1970. As shown there and in Fig. 1b, the range data have not been accumulated uniformly; substantial variations in data density exist as a function of synodic angle D, these phase angles are represented by 36 bins of 10° width. In Fig. 1b, data gaps are seen near new Moon (0° and 360°) and full Moon (180°) phases, and asymmetry about quarter Moon (90° and 270°) phases also is exhibited. The former properties of this data distribution are a consequence of

[1] One of the reflector arrays (of the Soviet Luna 17 mission, see also Fig. 8) has not been tracked operationally. The reason could be that the coordinates are not known well enough or that the reflectors are covered by dust or the transport cap.

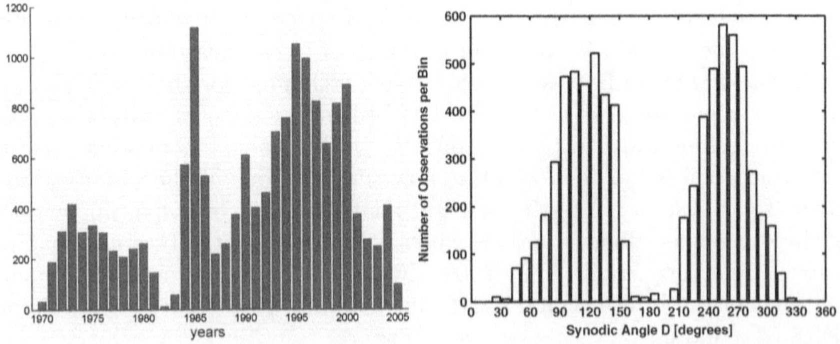

Fig. 1. (**a**) Lunar observations per year, 1970–2005. (**b**) Data distribution as a function of the synodic angle D.

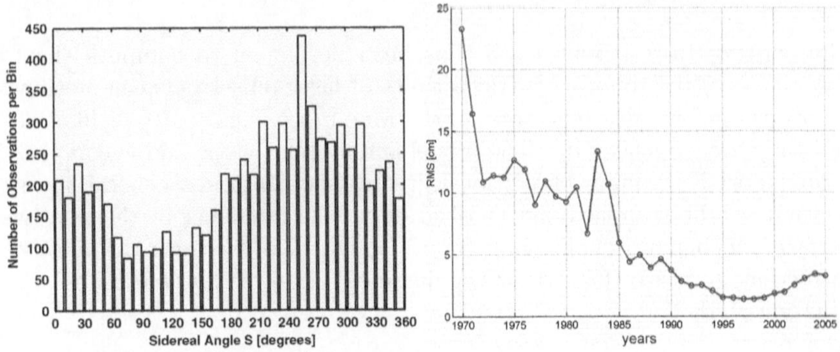

Fig. 2. (**a**) Data distribution as a function of the sidereal angle S, where the Moon is south of equator from 0° to 180°. (**b**) Weighted residuals (observed-computed Earth–Moon distance) annually averaged.

operational restrictions, such as difficulties to operate near the bright Sun in daylight (i.e., new Moon) or of high background solar illumination noise (i.e., full Moon). Also note asymmetry about quarter Moon phases. The uneven distribution with respect to the lunar sidereal angle shown in Fig. 2a represents the increased difficulty of making observations from Northern hemisphere observatories to the Moon when it is located over the Southern hemisphere. Here, the situation will only be improved if a southern observatory (e.g., in Australia) will start to track the Moon. It might be possible that new missions to the Moon could be helpful in this respect; the deployment of active laser transponders could allow satellite laser ranging systems to participate in LLR.

Although measurement precision for all model parameters benefits from the ever-increasing improvement in precision of individual range measurements (which now is at the few centimeter level, see also Fig. 2b), some parameters of scientific interest, such as time variation of Newton's coupling parameter $\dot{G}/G$ or precession rate of lunar perigee, particularly benefit from the long time period (35 years and growing) of range measurements.

In the 1970s, LLR was an early space technique for determining Earth orientation parameters (EOPs). Today, LLR still competes with other space geodetic techniques, and because of large improvements in ranging precision (30 cm in 1969 to ∼2 cm today), it now serves as one of the strongest tools in the solar system for testing general relativity. Moreover, parameters such as the station coordinates and velocities contributed to the International Terrestrial Reference Frame ITRF2000. EOP quantities were used in combined solutions of the International Earth Rotation and Reference Systems Service IERS ($\sigma = 0.5$ mas).

2 LLR Model and Relativity

The existing LLR model at IfE has been developed to compute the LLR observables – the round-trip travel times of laser pulses between stations on the Earth and passive reflectors on the Moon (see, e.g., [8, 10] or [6, 7, 11, 18] and the references therein). The model is fully relativistic and is complete up to first post-Newtonian ($1/c^2$) level; it uses the Einstein's general theory of relativity – the standard theory of gravity. The modeling of the relativistic parts is much more challenging than, e.g., in SLR, because the relativistic corrections increase the farther the distance becomes. The basic observation equation reads

$$d = c\frac{\tau}{2} = \left|\boldsymbol{r}_{\text{em}} - \boldsymbol{r}_{\text{station}} + \boldsymbol{r}_{\text{reflector}}\right| + c\,\Delta\tau \tag{1}$$

where d is the station-reflector distance, c the speed of light, τ the pulse travel time, $\boldsymbol{r}_{\text{em}}$ the vector connecting the geocenter and the selenocenter, $\boldsymbol{r}_{\text{station}}$ the geocentric position vector of the observatory, and $\boldsymbol{r}_{\text{reflector}}$ the selenocentric position vector of the reflector arrays. $\Delta\tau$ describes corrections of the travel time caused by atmospheric effects, but also (relativistic) transformations into the right time system as well as the light time equation (Shapiro effect). To apply (1), all vectors have to be transformed in one common reference frame (in our case the inertial frame), which requires consistent relativistic transformations, so-called pseudo-Lorentz transformations. The Earth–Moon vector $\boldsymbol{r}_{\text{em}}$ can only be obtained by numerical integration of the corresponding equation of motion (here in simplified version)

$$\ddot{\boldsymbol{r}}_{\text{em}} = -\frac{GM_{\text{e+m}}}{r_{\text{em}}^3}\boldsymbol{r}_{\text{em}} + GM_s\left(\frac{\boldsymbol{r}_{\text{s}} - \boldsymbol{r}_{\text{em}}}{\left|\boldsymbol{r}_{\text{s}} - \boldsymbol{r}_{\text{em}}\right|^3} - \frac{\boldsymbol{r}_{\text{s}}}{r_{\text{s}}^3}\right) + \boldsymbol{b}_{\text{newtonian}} + \boldsymbol{b}_{\text{relativistic}}, \tag{2}$$

where $\ddot{\boldsymbol{r}}_{\text{em}}$ is the relative acceleration vector between Earth and Moon, $GM_{\text{e+m}}$ is the Earth–Moon mass times the gravitational constant, $\boldsymbol{r}_{\text{s}}$ the geocentric position vector of the Sun, r_s the Earth–Sun distance, M_s the solar mass, $\boldsymbol{b}_{\mathrm{newtonian}}$ comprises all further Newtonian terms like the effect of the other planets or the gravitational fields of Earth and Moon, and $\boldsymbol{b}_{\mathrm{relativistic}}$ indicates all "relativistic" terms, i.e., those entering the Einstein–Infeld–Hofmann (EIH) equations. Corresponding relativistic equations are applied to describe the rotational motion of the Moon. The rotation angles are then applied to rotate the selenocentric reflector coordinates of (1) into the inertial frame. For the transformation of the geocentric station coordinates, the well known rotation matrices (precession, nutation, GAST, etc.) are used, see IERS 2003.

The modeling of the "Newtonian" parts has been setup according to IERS Conventions (IERS 2003) [3], but it is restricted to the 1 cm level. The weights are based upon the accuracy estimates for the normal points as provided by the observatories. On the basis of this model, two groups of parameters (170 in total) are determined by a weighted least-squares fit of the observations. The first group includes the so-called "Newtonian" parameters such as:

- Geocentric coordinates of three Earth-based LLR stations and their velocities
- A set of EOPs (luni-solar precession constant, nutation coefficients of the 18.6 years period, Earth's rotation UT0, and variation of latitude by polar motion)
- Selenocentric coordinates of four retroreflectors
- Rotation of the Moon at one initial epoch (physical librations)
- Orbit (position and velocity) of the Moon at this epoch
- Orbit of the Earth–Moon system about the Sun at one epoch
- Mass of the Earth–Moon system times the gravitational constant
- The lowest mass multipole moments of the Moon
- Lunar Love number and a rotational energy dissipation parameter
- Lag angle indicating the lunar tidal acceleration responsible for the increase of the Earth–Moon distance (about 3.8 cm year^{-1}), the increase in the lunar orbit period, and the slowdown of Earth's angular velocity

The second group of parameters is used to perform LLR tests of viable modifications of the general theory of relativity. The general theory of relativity is not expected to be perfect, because Einstein's theory and quantum mechanics are fundamentally incompatible. Therefore, it is important in physics to find out at what level of accuracy it fails. Relativistic parameters to be determined by LLR analyses are (values for general relativity are given in parentheses) mentioned below:

1. Strong equivalence principle (EP) parameter η, which for metric theories is $\eta = 4\beta - 3 - \gamma$ $(= 0)$. In LLR, a violation of the equivalence principle would show up as a displacement of the lunar orbit along the direction to the Sun. LLR is the dominant test of the strong equivalence principle, i.e., for self-gravitating bodies.

2. Space-curvature parameter γ (= 1) and nonlinearity parameter β (= 1). LLR also has the capability of determining the PPN parameters β and γ directly from the point-mass orbit perturbations (i.e., as described by the EIH equations), but, e.g., β may be derived much better by combining the EP parameter η with an independent determination of γ (see later).
3. Time variation of the gravitational coupling parameter $\dot{G}/G$ (= 0 year^{-1}). This is the second most important gravitational physics result that LLR provides. Einstein's theory does not predict a changing G, but some other theories do. So it is important to measure this as well as possible. The sensitivity improves like the square of the data span.
4. Geodetic de Sitter precession Ω_{dS} of the lunar orbit ($\simeq 1.92''$ per century). LLR has also provided the only accurate determination of the geodetic precession. The dedicated space mission GP-B will provide an improved accuracy, if that mission is successfully completed.
5. Coupling constant α (= 0) of Yukawa potential for the Earth–Moon distance that corresponds to a test of Newton's inverse-square law.
6. α_1 (= 0) and α_2 (= 0) which parametrize "preferred-frame" effects in metric gravity.
7. Combination of parameters $\zeta_1 - \zeta_0 - 1$ (= 0) derived in the Mansouri and Sexl [5] formalism indicating a violation of special relativity (there: Lorentz contraction parameter $\zeta_1 = 1/2$, time dilation parameter $\zeta_0 = -1/2$).
8. EP-violating coupling of normal matter to "dark matter" at the galactic center. It is a similar test to item 1 above, but now different periods are affected (mainly the sidereal month).
9. A further application is the detection of the Sun's J_2 ($\approx 10^{-7}$) from LLR data (independent to other methods such as solar seismology), which affects the anomalous perihelion shift of Mercury, one of the classical tests of relativity. The present uncertainty ($\approx 10^{-6}$) is larger than the expected value. The parameter J_2 is not further discussed in this chapter and will be addressed in more detail in an upcoming study.

The determination of the relativistic parameters indicated earlier can be accomplished by either modifying the equations of motion (i.e., parametrizing present terms or adding new ones) or deriving analytical expressions for their effect on the Earth–Moon distance. In the first case, the needed partial derivatives can be computed by numerical differentiation and in the second case by direct derivation of the analytical terms.

Many relativistic effects produce a sequence of periodic perturbations of the Earth–Moon range

$$\Delta r_{EM} = \sum_{i=1}^{n} \Big[A_i \cos(\omega_i \Delta t + \Phi_i) + B_i \Delta t \sin(\omega_i \Delta t + \Phi_i) \Big]. \tag{3}$$

Table 1. Typical periods of some relativistic quantities, taken from [9].

Parameter	Typical periods
η	Synodic ($29^d 12^h 44^m 2.9^s$)
$\dot{G}/G$	Secular + emerging periodic
α_1	Sidereal, annual, sidereal-2·annual, anomalistic ($27^d 13^h 18^m 33.2^s$) ± annual, synodic
α_2	2·sidereal, 2·sidereal-anomalistic, nodal ($6{,}798^d$)
$\zeta_1 - \zeta_0 - 1$	Annual (365.25^d)
$\delta g_{\text{galactic}}$	Sidereal ($27^d 7^h 43^m 11.5^s$)

A_i and B_i, ω_i, and ϕ_i are the amplitudes, frequencies, and phases, respectively, of the various perturbations. Some sample periods of perturbations important for the measurement of various parameters are given in Table 1.[2]

3 Sensitivity Study

As indicated in (1) and (2), LLR is affected in various ways and at various levels by relativity. Relativity enters the equation of motion, i.e., the orbit and the rotation of the Moon. More precisely, the Earth–Moon system behaves according to relativity. But also the light propagation and the transformations between the reference and time frames that are used have to be modeled in agreement with general relativity. The lunar measurements contain the summed signal of all effects in one, so that the separation of the individual effects is a big challenge. To better understand what the individual contributions of the different relativistic effects are, sensitivity studies have been carried out, as

$$\Delta r_{em}^{p} = \frac{\delta r_{em}}{\delta p} \Delta p, \tag{4}$$

where Δr_{em}^{p} is the perturbation of the Earth–Moon distance caused by a parameter p, which is one of the relativistic parameters described in Sect. 2. $\delta r_{em}/\delta p$ is the partial derivative of the Earth–Moon distance with respect to p; it is obtained by numerical differentiation. Δp is a small value indicating the variation in p, here we have used the present realistic error as derived from LLR analyses (see Table 2). As an example, Fig. 3a represents the sensitivity of the Earth–Moon distance with respect to a possible temporal variation of the gravitational constant in the order of $8 \cdot 10^{-13}\,\text{year}^{-1}$, the present accuracy of that parameter. It seems as if perturbations of up to 9 m are still caused,

[2] Note: the designations should not be used as formulae for the computation of the corresponding periods, e.g., the period "sidereal-2·annual" has to be calculated as $1/(1/27.32^d - 2/365.25^d) \approx 32.13^d$. In addition, "secular + emerging periodic" means the changing orbital frequencies induced by $\dot{G}/G$ are starting to become better signals than the secular rate of change of the Earth–Moon range in LLR.

Table 2. Determined values for the relativistic quantities and their realistic errors.

Parameter	Results
Equivalence principle parameter η	$(6 \pm 7) \,.\, 10^{-4}$
Metric parameter $\gamma - 1$	$(4 \pm 5) \,.\, 10^{-3}$
Metric parameter $\beta - 1$: direct measurement	$(-2 \pm 4) \,.\, 10^{-3}$
and from $\eta = 4\beta - 3 - \gamma_{\mathrm{Cassini}}$	$(1.5 \pm 1.8) \,.\, 10^{-4}$
Time-varying gravitational constant $\dot{G}/G$ (year^{-1})	$(6 \pm 8) \,.\, 10^{-13}$
Differential geodetic precession Ω_{GP} - Ω_{deSit} ($''$ per century)	$(6 \pm 10) \,.\, 10^{-3}$
Yukawa coupling constant α (for $\lambda = 4 \cdot 10^5$ km)	$(3 \pm 2) \,.\, 10^{-11}$
"Preferred-frame" parameter α_1	$(-7 \pm 9) \,.\, 10^{-5}$
"Preferred-frame" parameter α_2	$(1.8 \pm 2.5) \,.\, 10^{-5}$
Special relativistic parameters $\zeta_1 - \zeta_0 - 1$	$(-5 \pm 12) \,.\, 10^{-5}$
Influence of dark matter $\delta g_{\mathrm{galactic}}$ (cm s^{-2})	$(4 \pm 4) \,.\, 10^{-14}$

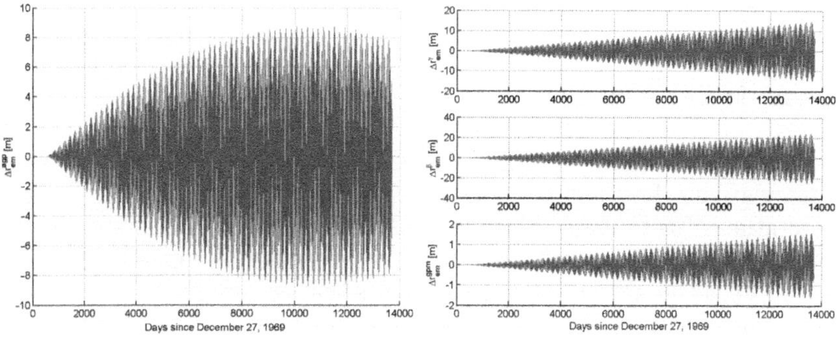

Fig. 3. (**a**) Sensitivity of LLR with respect to $\dot{G}/G$ assuming $\Delta\dot{G}/G = 8 \cdot 10^{-13}$ year^{-1}. (**b**) Sensitivity of LLR with respect to space curvature γ, nonlinearity couplings β, and geodetic precession using their present LLR accuracy (see Table 2) as perturbation value.

but this range (compared with the ranging accuracy at the centimeter level) cannot fully be exploited, because the lunar tidal acceleration perturbation is similar. Figures 3b and 4a,b show the results of corresponding investigations for all relativistic parameters that were investigated during the present study (i.e., without the parameters discussed in items 7–9 of Sect. 2). The perturbations vary between 5 cm (α_1) and 25 m (β), which indicates that the former parameter is determined quite well from LLR data, as the sensitivity values are close to the observation accuracy, whereas the latter is only poorly determined because of its high correlation with the Newtonian orbit perturbations. Nevertheless, the continuation of LLR over a longer time span will help to further decorrelate the various parameters.

To better understand those couplings, corresponding power spectra have been computed. The largest periods for the EP parameter are shown in Fig. 5a and for $\dot{G}/G$ in Fig. 5b. Obviously, many periods are affected simultaneously,

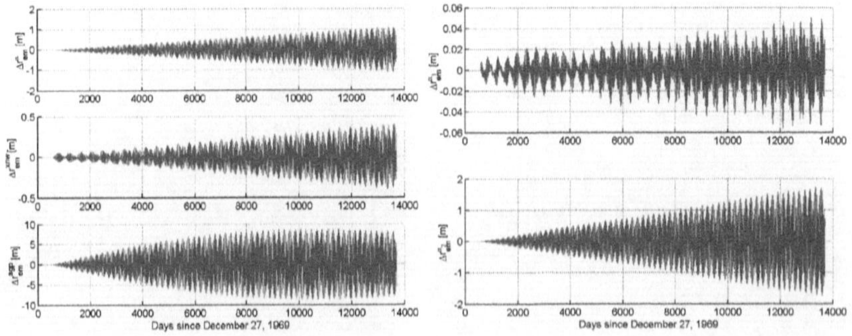

Fig. 4. (**a**) Sensitivity of LLR with respect to Yukawa interaction parameter α, equivalence principle violation η and time-variable gravitational constant $\dot{G}/G$ using their present LLR accuracy (see Table 2) as perturbation value. (**b**) Sensitivity of LLR with respect to preferred-frame effects α_1 and α_2 using their present LLR accuracy (see Table 2) as perturbation value.

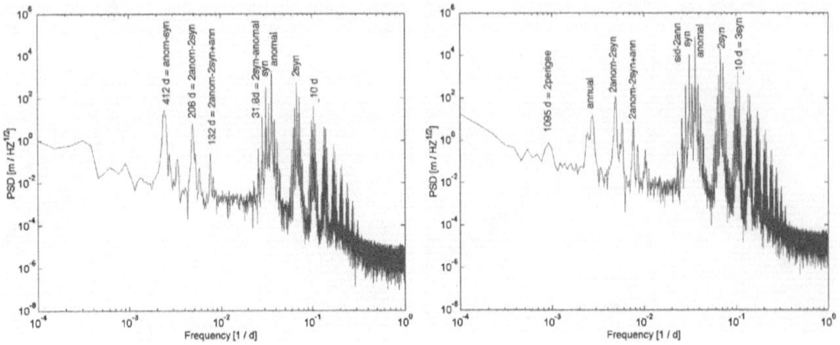

Fig. 5. (**a**) Power spectrum of a possible equivalence principle violation assuming $\Delta(m_G/m_I) \approx 10^{-13}$. (**b**) Power spectrum of the effect of $\dot{G}/G$ in the Earth–Moon distance assuming $\Delta\dot{G}/G = 8 \cdot 10^{-13}\,\text{year}^{-1}$.

because the perturbations, even if caused by a single beat period only (e.g., the synodic month for η parameter[3]), change the whole lunar orbit (and rotation) and, therefore, excite further frequencies. For comparisons also the spectra of the geodetic precession *gpm* and the Yukawa coupling parameter α are indicated (Fig. 6a,b). Again a different combination of periods is visible. As before, mainly the monthly (e.g., sidereal, synodic, anomalistic), half-monthly, etc., periods dominate, but longer periods (low frequencies) are also present, e.g., annual, 3 years, or combinations of the monthly and annual frequencies. Similar pictures are obtained when considering the preferred-frame parameters α_1 and α_2 (Fig. 7a,b). A huge variety of periods show up again, but

[3] Note, here and throughout the chapter, the relation $m_G/m_I = 1 + \eta(U/Mc^2)$ has been used equivalently, where the second term describes the self-energy of a body, cf. [18].

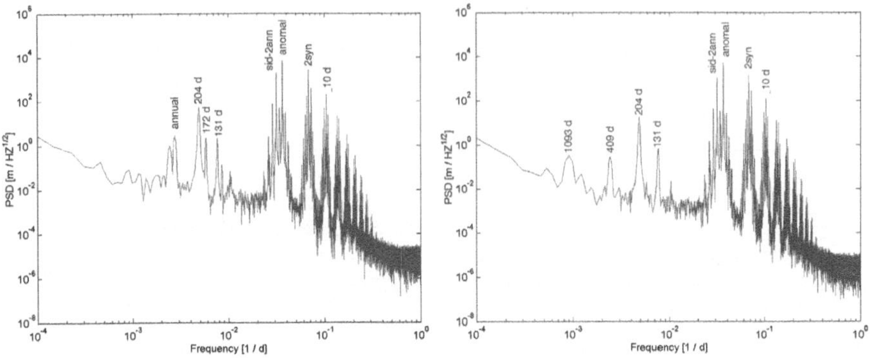

Fig. 6. (**a**) Power spectrum of an additional (deviating from Einstein's value) geodetic precession assuming $\Delta gpm = 10^{-2}$. (**b**) Power spectrum of a possible Yukawa term assuming $\Delta\alpha = 2 \cdot 10^{-11}$.

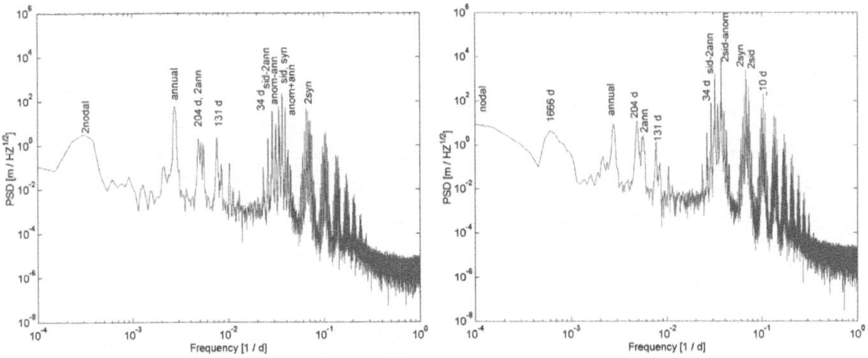

Fig. 7. (**a**) Power spectrum of a possible preferred-frame effect by α_1 assuming $\Delta\alpha_1 = 9 \cdot 10^{-5}$. (**b**) Power spectrum of a possible preferred-frame effect by α_2 assuming $\Delta\alpha_2 = 2.5 \cdot 10^{-5}$.

they differ partly from each other (note, e.g., the very low-frequency contributions). The spectra of γ and β (not shown here) are very similar to the geodetic precession spectrum. Although there are big similarities between the various spectra, the typical properties of each can be used to identify and separate the different effects and to determine corresponding parameters.

4 Results

The global adjustment of the model by least-squares fit procedures gives improved values for the estimated parameters and their formal standard errors, while consideration of parameter correlations obtained from the covariance analysis and of model limitations lead to more "realistic" errors. Incompletely modeled solid Earth tides, ocean loading or geocenter motion, and uncertainties in values of fixed model parameters have to be considered in those estimations. See Table 2 for the determined values for the relativistic quantities and their realistic errors.

The EP parameter η ($= (6 \pm 7) \cdot 10^{-4}$) benefits mostly from highest accuracy over a sufficient long time span (e.g., 1 year) and a good data coverage over the synodic month, as far as possible. Also any observations reducing the asymmetry about the quarter Moon phases (compare Fig. 1b) would improve the fit of η. The improvement of the EP parameter was not so big since 1999, as the LLR RMS residuals increased a little bit in the past years, see Fig. 2b. The reason for that increase is not completely understood and has to be investigated further.

In combination with the recent value of the space-curvature parameter γ_{Cassini} ($\gamma - 1 = (2.1 \pm 2.3) \cdot 10^{-5}$) derived from Doppler measurements to the Cassini spacecraft [1], the nonlinearity parameter β can be determined by applying the relationship $\eta = 4\beta - 3 - \gamma_{\text{Cassini}}$. One obtains $\beta - 1 = (1.5 \pm 1.8) \cdot 10^{-4}$ (note that using the EP test to determine parameters η and β assumes that there is no composition-induced EP violation). The LLR result for the space-curvature parameter γ as determined from the EIH equations is less accurate than the results derived from other measurements, because its effect is very similar to the Newtonian orbit perturbations.

For the temporal variation of the gravitational constant, $\dot{G}/G = (6 \pm 8) \cdot 10^{-13}$ has been obtained, where the formal standard deviation has been scaled by a factor 3 to yield the given value. This parameter benefits mostly from the long time span of LLR data and has experienced the biggest improvement over the past years (cf. [9]).

For the estimation of the de Sitter precession of the lunar orbit, a Coriolis-like term is added to the equation of motions, which adds the precession effect as predicted by Einstein for a second time. This term is scaled by a parameter called *gpm*, which has to give 0 if Einstein is correct. $gpm = 0$ is obtained with an accuracy of about 1%.

The Yukawa coupling parameter α has been determined by adding corresponding perturbation terms to the equations of motion, where the partials were computed by numerical differentiation. The recently determined value shows small deviations from the expected value; it will be further investigated in future.

The preferred-frame parameters α_1 and α_2 can be either determined by extending the equations of motion or by adding analytical terms to the Earth–Moon distance. In both cases, quite similar results are obtained (see [9, 10]). Recent determinations are given in Table 2.

The Mansouri–Sexl parameters ζ_0 and ζ_1 as well as the quantity indicating a possible EP-violating coupling with dark matter were not investigated during our present studies; the values given in Table 2 are taken from [9].

A further question to be considered in more detail in future is the right combination of the parameters. That means, shall all relativistic parameters be estimated together in one global adjustment, or each one alone (together with the parameters of the standard solution)? We carried out several tests considering the correlation of the relativistic quantities with each other, but also with the "classical" ones, e.g., with the orbital parameters or site velocities [4]. It is too early to make a final decision. On the one hand, "overestimation" of an effect has to be avoided, on the other hand over constraining by fixing too many parameters should also be avoided.

Final results for all relativistic parameters obtained from the IfE analysis are shown in Table 2 (see also [12]). The realistic errors are comparable with those obtained in other recent investigations, e.g., at JPL (see [14–16, 18]).

5 Further Applications

In addition to the relativistic phenomena discussed earlier, more effects related to lunar physics, geosciences, and geodesy can be investigated. The following items are of special interest (see also [12]), as they touch recent activities in the afore-mentioned disciplines:

1. *Celestial reference frame.* A dynamical realization of the International Celestial Reference System (ICRS) by the lunar orbit is obtained ($\sigma = 0.001''$) from LLR data. This can be compared and analyzed with respect to the kinematical ICRS from very long baseline interferometry (VLBI). Here, the very good long-term stability of the lunar orbit is of great advantage.
2. *Terrestrial reference frame.* The results for the station coordinates and velocities, which are estimated simultaneously in the standard solution, contribute to the realization of the international terrestrial reference frame, e.g., to the last one, the ITRF2000.
3. *Earth rotation.* LLR contributes, among others, to the determination of long-term nutation parameters, where again the stable, highly accurate orbit, and the lack of nonconservative forces from air-drag or solar radiation pressure (which affect satellite orbits substantially) is very convenient. Additionally, UT0 and VOL values are computed, which stabilize the combined EOP series, especially in the 1970s when less data from other space geodetic techniques were available. The precession rate is another example in this respect.
4. *Relativity.* As discussed in the previous sections, the dedicated investigation of Einstein's theory of relativity is of major interest. With an improved accuracy of the LLR measurements and the modeling (see Sect. 6), the investigation of further effects (e.g., the Lense–Thirring precession) or those of alternative theories might become possible.
5. *Lunar physics.* By the determination of the libration angles of the Moon, LLR gives access to underlying processes affecting lunar rotation (e.g.,

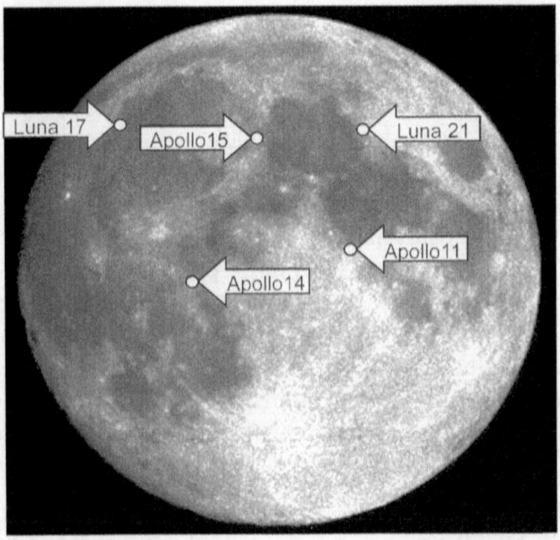

Fig. 8. Distribution of retroreflectors on the lunar surface.

Moon's core, dissipation), cf. [17]. A better distribution of the retroreflectors on the Moon (see Fig. 8) would be very helpful.

6. *Selenocentric reference frame.* The determination of a selenocentric reference frame, the combination with high-resolution images, and the establishment of a better geodetic network on the Moon is a further big item, which then allows accurate lunar mapping.
7. *Earth–Moon dynamics.* The mass of the Earth–Moon system, the lunar tidal acceleration, possible geocenter variations and related processes as well as further effects can be investigated in detail.
8. *Timescales.* The lunar orbit can also be considered as a long-term stable clock so that LLR can be used for the independent realization of timescales, which can then be compared or combined with other determinations.

Those features shall be addressed in the future, when more and better LLR data are available and the analysis models have been improved to the millimeter level, see Sect. 6.

6 Model and Observation Refinement

To exploit the full available potential of LLR, the theoretical models as well as the measurements require optimization. Using the 3.5 m telescope at the APOLLO site in New Mexico, USA, a millimeter-level ranging becomes possible. To allow an order of magnitude gain in determination of various quantities in the complete LLR solution, the current models have to be updated according to the IERS conventions 2003, and made compatible with the IAU 2000 resolutions [2]. This requires, e.g., to better model:

- Higher degrees of the gravity fields of Earth and Moon and their couplings
- The effect of the asteroids (up to 1,000)
- Relativistically consistent torques in the rotational equations of the Moon
- Relativistic spin–orbit couplings
- Torques caused by other planets like Jupiter
- The lunar tidal acceleration with more periods (diurnal and semidiurnal)
- Ocean and atmospheric loading by updating the corresponding subroutines
- Nutation using the recommended IAU model
- The tidal deformation of Earth and Moon
- Moon's interior (e.g., solid inner core) and its coupling to the Earth–Moon dynamics

Besides modeling, the overall LLR processing shall be optimized. The best strategy for the data fitting procedure needs to be explored for (highly) correlated parameters.

Finally, LLR should be prepared for a renaissance of lunar missions where transponders or new retroreflectors may be deployed on the surface of the Moon, which would enable many pure SLR stations to observe the Moon. NASA is planning to return to the Moon by 2008 with Lunar Reconnaissance Orbiter (LRO), and later with robotic landers, and then with astronauts until 2018. The primary focus of these planned missions will be lunar exploration and preparation for trips to Mars, but they will also provide opportunities for science, particularly if new reflectors are placed at more widely separated locations than the present configuration (see Fig. 8). New installations on the Moon would give stronger determinations of lunar rotation and tides. New reflectors on the Moon would provide additional accurate surface positions for cartographic control [18], would also aid navigation of surface vehicles or spacecraft at the Moon, and they also would contribute significantly to research in fundamental and gravitational physics, LLR-derived ephemeris, and lunar rotation. Moreover in the case of colocation of microwave transponders, the connection to the VLBI system may become possible which will open a wide range of further activities such as frame ties.

7 Conclusions

LLR has become a technique for measuring a variety of relativistic gravity parameters with unsurpassed precision. Sensitivity studies have been performed to estimate the order of magnitude of relativistic effects on lunar ranges and the potential capability to better determine certain relativistic features. Spectral analyses showed the typical frequencies related to each effect, indicating as well, how (highly) correlated parameters might be separated. No violations of general relativity are found in our investigations. Both the weak and strong forms of the EP are verified, while strong empirical limitations on any inverse-square law violation, time variation of G, and preferred-frame effects are also obtained.

LLR continues as an active program, and it can remain as one of the most important tools for testing Einstein's general relativity theory of gravitation if appropriate observations strategies are adopted and if the basic LLR model is further extended and improved down to the millimeter level of accuracy. The deployment of transponders on the Moon would considerably improve the performance for lunar ranging applications. Lunar science, fundamental physics, control networks for surface mapping, and navigation would benefit. Demonstration of active devices would prepare the way for very accurate ranging to Mars and other solar system bodies.

Acknowledgments

Current LLR data are collected, archived, and distributed under the auspices of the International Laser Ranging Service (ILRS). All former and current LLR data are electronically accessible through the European Data Center (EDC) in Munich, Germany, and the Crustal Dynamics Data Information Service (CDDIS) in Greenbelt, Maryland. The following Web site can be queried for further information: `http://ilrs.gsfc.nasa.gov`. We also acknowledge with thanks that the more than 35 years of LLR data used in these analyses have been obtained under the efforts of personnel at the Observatoire de la Côte d'Azur, in France, the LURE Observatory in Maui, Hawaii, and the McDonald Observatory in Texas.

A portion of the research described in this chapter was carried out at the Jet Propulsion Laboratory of the California Institute of Technology, under a contract with the National Aeronautics and Space Administration.

References

1. Bertotti B., Iess L., Tortora, P.: A test of general relativity using radio links with the Cassini spacecraft. Nature **425**, 374 (2003).
2. IAU Resolutions 2000: `http://syrte.obspm.fr/IAU_resolutions/Resol-UAI.htm`.
3. IERS Conventions 2003. IERS Technical Note No. 32, D.D. McCarthy and G. Petit (eds.), Frankfurt, BKG, 2004. Electronic version `http://www.tai.bipm.org/iers/conv2003/conv2003.html`.
4. Koch C.: Implementierung eines neuen Integrators in die LLR-Auswertesoftware und Erzeugung von Standardlösungen für die zu bestimmenden Zielparameter, unpublished diploma thesis, University of Hannover 2005.

5. Mansouri R.M. and Sexl R.U.: A test theory of Special Relativity. Gen. Rel. Grav. **8**, 497 (1977); ibid. **8**, 515 (1977); ibid. **8**, 809 (1977).
6. Müller, J.: FESG/TUM, Report about the LLR Activities. ILRS Annual Report 1999, M. Pearlman, L. Taggart (eds.), 204–208, 2000.
7. Müller, J.: FESG/TUM, Report about the LLR Activities. ILRS Annual Report 2000, M. Pearlman, M. Torrence, L. Taggart (eds.), 7–35/36, 2001.
8. Müller J. and Nordtvedt K.: Lunar laser ranging and the equivalence principle signal. Physical Review D, 58, 062001, 1998.
9. Müller, J., Nordtvedt, K., Schneider, M., Vokrouhlický, D.: Improved Determination of Relativistic Quantities from LLR. In: Proceedings of the 11th International Workshop on Laser Ranging Instrumentation, held in Deggendorf, Germany, Sept. 21–25, 1998, BKG v.10, 216–222, 1999.
10. Müller J., Nordtvedt K. and Vokrouhlický D.: Improved constraint on the α_1 PPN parameter from lunar motion. Physical Review D, 54, R5927–R5930, 1996.
11. Müller J. and Tesmer V.: Investigation of Tidal Effects in Lunar Laser Ranging. Journal of Geodesy, 76, 232–237, 2002.
12. Müller J. J.G. Williams, S.G. Turyshev, and P. Shelus.: Potential Capabilities of Lunar Laser Ranging for Geodesy and Relativity. Proceedings of the IAG General Assembly, held in Cairns, Australia, 22-26 August 2005, in print 2005, gr-qc/0509019.
13. Murphy, T.M., Jr., Strasburg, J.D., Stubbs, C.W., Adelberger, E.G., Angle, J., Nordtvedt, K., Williams, J.G., Dickey, J.O., and Gillespie, B.: "The Apache Point Observatory Lunar Laser-Ranging Operation (APOLLO)," Proceedings of 12th International Workshop on Laser, Ranging, Matera, Italy (November 2000), `http://www.astro.washington.edu/tmurphy/apollo/matera.pdf`
14. Williams J.G., Newhall X.X. and Dickey J.O.: Relativity parameters determined from lunar laser ranging. Physical Review D, 53, 6730, 1996.
15. Williams, J.G., S.G. Turyshev, and D.H. Boggs.: Progress in lunar laser ranging tests of relativistic gravity. Phys. Rev. Lett. **93**, 261101 (2004) [gr-qc/0411113].
16. Williams, J.G., S.G. Turyshev, and T.W. Murphy, Jr.: Improving LLR Tests of Gravitational Theory. (Fundamental Physics meeting, Oxnard, CA, April 2003), International Journal of Modern Physics **D 13**, 567 (2004) [gr-qc/0311021].
17. Williams, J.G., D.H. Boggs, and J.T. Ratcliff: Lunar Fluid Core and Solid-Body Tides. Abstract No. 1503 of the Lunar and Planetary Science Conference XXXVI, March 14–18, 2005a.
18. Williams, J.G., S.G. Turyshev, and D.H. Boggs.: Lunar Laser Ranging Tests of the Equivalence Principle with the Earth and Moon. In proceedings of 'Testing the Equivalence Principle on Ground and in Space', Pescara, Italy, September 20–23, 2004, C. Lämmerzahl, C.W.F. Everitt and R. Ruffini (eds.), to be published by Springer Verlag, Lect. Notes Phys., 2005, gr-qc/0507083.

Science, Technology, and Mission Design for the Laser Astrometric Test of Relativity

Slava G. Turyshev[1], Michael Shao[1], and Kenneth L. Nordtvedt, Jr.[2]

[1] Jet Propulsion Laboratory, California Institute of Technology, 4800 Oak Grove Drive, Pasadena, CA 91109, USA

[2] Northwest Analysis, 118 Sourdough Ridge Road, Bozeman, MT 59715, USA

Summary. The Laser Astrometric Test of Relativity (LATOR) is a Michelson–Morley type experiment designed to test the metric nature of gravitation – a fundamental postulate of the Einstein's general theory of relativity. The key element of LATOR is a geometric redundancy provided by the long-baseline optical interferometry and interplanetary laser ranging. By using a combination of independent time-series of gravitational deflection of light in the immediate proximity to the Sun, along with measurements of the Shapiro time delay on interplanetary scales (to a precision, respectively, better than 0.1 picoradians and 1 cm), LATOR will significantly improve our knowledge of relativistic gravity and cosmology. The primary mission objective is (1) to measure the key post-Newtonian Eddington parameter γ with accuracy of a part in 10^9. $\frac{1}{2}(1-\gamma)$ is a direct measure for presence of a new interaction in gravitational theory and, in its search, LATOR goes a factor 30,000 beyond the present best result, Cassini's 2003 test; other mission objectives include (2) first measurement of gravity's nonlinear effects on light to $\sim$0.01% accuracy, including both the traditional Eddington β parameter and also the spatial metric's second-order potential contribution (never measured before); (3) direct measurement of the solar quadrupole moment J_2 (currently unavailable) to accuracy of a part in 200 of its expected size of $\simeq 10^{-7}$; and (4) direct measurement of the "frame-dragging" effect on light due to the Sun's rotational gravitomagnetic field, to 0.1% accuracy.

LATOR's primary measurement pushes to unprecedented accuracy the search for cosmologically relevant scalar–tensor theories of gravity by looking for a remnant scalar field in today's solar system. We discuss the science objectives of the mission, its technology, mission and optical designs, as well as expected performance of this experiment. LATOR will lead to very robust advances in the tests of fundamental physics: this mission could discover a violation or extension of general relativity and/or reveal the presence of an additional long-range interaction in the physical law. There are no analogs to LATOR; it is unique and is a natural culmination of solar system gravity experiments.

1 Introduction

After almost 90 years since general theory of relativity was born, the Einstein's gravitational theory has survived every test. Such longevity, of course, does not mean that this theory is absolutely correct, but it serves to motivate more accurate tests to determine the level of accuracy at which it is violated. General theory of relativity began with its empirical success in 1915 by explaining the anomalous perihelion precession of Mercury's orbit. Shortly thereafter, Eddington's 1919 observations of star lines-of-sight during a solar eclipse confirmed the doubling of the deflection angles predicted by general relativity as compared to Newtonian-like and Equivalence Principle arguments. This test of gravitational deflection of light made general relativity an instant success.

From these beginnings, general theory of relativity has been verified at ever higher accuracy. Thus, microwave ranging to the Viking Lander on Mars yielded a ~0.2% accuracy in the tests of general relativity [80, 93, 94]. Spacecraft and planetary radar observations reached an accuracy of ~0.15% [4, 78]. The astrometric observations of quasars on the solar background performed with Very Long-Baseline Interferometry (VLBI) improved the accuracy of the tests of gravity to ~0.045% [37, 48, 81, 95]. Lunar laser ranging, a continuing legacy of the Apollo program, provided ~0.011% verification of general relativity via precision measurements of the lunar orbit [64, 66, 70–72, 111, 112, 114, 115]. Finally, the recent experiments with the Cassini spacecraft improved the accuracy of the tests to ~0.0023% [11, 43] (see Sect. 2 and Fig. 1 on page 42). As a result, today general relativity is the standard theory of gravity when astrometry and spacecraft navigation are concerned.

Considering gravitation and fundamental physics, our solar system is the laboratory that still offers many opportunities to improve the tests of relativistic gravity. A carefully designed gravitational experiment that utilizes the strongest gravity potential available in the solar system, that provided by the Sun itself, also has the advantage to conduct tests in a controlled and well-understood environment. Indeed, compared to terrestrial conditions, the Sun offers a factor of $M_\odot/M_\oplus \sim 3.3 \times 10^5$ increase in the strength of gravitational effects, the fact, that was recognized in a number of experiments proposed over the years (see discussion [109, 110]). Most of these proposals rely on sending an ensemble of ultra-stable clocks to a close proximity to the Sun, typically to distances of four solar radii [1,51,88]. An approach, alternative to sending spacecraft on a highly eccentric trajectory into the challenging near-solar environment, would be to send a laser light instead. This is due to the fact that optical technologies (i.e., long-baseline interferometry, laser ranging, etc.) have recently demonstrated a very significant progress achieving the level of maturity needed for a major improvement of the accuracy of gravitational experiments in space. The use of these technologies allows one to probe the strongest gravity in the solar system while still being separated by a safe distance from the Sun; later we will develop this idea further.

This chapter discusses the Laser Astrometric Test of Relativity (LATOR), the space-based experiment that is designed to significantly improve the tests of relativistic gravity in the solar system. The test will be performed in the solar gravity field using optical interferometry between two microspacecraft. Precise measurements of the angular position of the spacecraft will be made using a fiber-coupled optical interferometer on the ISS with a 100 m baseline. The primary objective of the LATOR mission will be to measure the gravitational deflection of light by the solar gravity to accuracy of 0.1 picoradians (prad), which corresponds to ~10 picometers (pm) on a 100 m interferometric baseline. A combination of laser ranging among the spacecraft and direct interferometric measurements will allow LATOR to measure deflection of light in the solar gravity by a factor of ~30,000 better than that had recently been accomplished with the Cassini spacecraft. In particular, LATOR will not only measure the key PPN parameter γ to unprecedented levels of accuracy of one part in 10^9, but it will also reach ability to measure the next post-Newtonian order ($\propto G^2$) of light deflection resulting from gravity's intrinsic nonlinearity. As a result, LATOR will measure values of other PPN parameters (see (1)) such as parameter δ to one part in 10^4 (never measured before), the solar quadrupole moment parameter J_2 to one part in 200, and the frame-dragging effects on light due to the solar angular momentum to a precision of one part in 10^3.

The chapter is organized as follows. Section 2 discusses the theoretical framework and science motivation for the precision gravity tests in the solar system; it also presents the science objectives for the LATOR experiment. Section 3 provides an overview for the LATOR experiment, including basic elements of the current mission and optical designs. Section 4 addresses design of the LATOR long-baseline optical interferometer, including the laser metrology system. Section 5 discusses the current design for the LATOR flight system and presents a preliminary design for LATOR optical receivers. In Sect. 6 we discuss modeling of the mission observables and address observational logic of LATOR measurements. In Sect. 7 we present major constituents of the mission's error budget and discuss the expected mission performance. Section 8 compares LATOR with other proposed gravity experiments and also discusses the next steps that will be taken in the development of LATOR.

2 Scientific Motivation

Recent remarkable progress in observational cosmology has again submitted general relativity to a test by suggesting a non-Einsteinian model of universe's evolution [76, 83, 102]. From the theoretical standpoint, the challenge is even stronger – if the gravitational field is to be quantized, general relativity will have to be modified. This is why the recent advances in the scalar–tensor extensions of gravity, which are consistent with the current inflationary model of the Big Bang, have motivated new search for a very small deviation from the

Einstein's theory, at the level of accuracy of three to five orders of magnitude below the level tested by experiment.

In this section we will consider the recent theoretical end experimental motivations for the high-accuracy gravitational tests. We will also present the scientific objectives of the LATOR experiment.

2.1 The PPN Formalism

Generalizing on a phenomenological parameterization of the gravitational metric tensor field which Eddington originally developed for a special case, a method called the parameterized post-Newtonian (PPN) metric, has been developed [62–66,107,109]. This method represents the gravity tensor's potentials for slowly moving bodies and weak interbody gravity, and it is valid for a broad class of metric theories, including general relativity as a unique case. The several parameters in the PPN metric expansion vary from theory to theory, and they are individually associated with various symmetries and invariance properties of underlying theory. Gravity experiments can be analyzed in terms of the PPN metric, and an ensemble of experiments will determine the unique value for these parameters, and hence the metric field itself.

In locally Lorentz-invariant theories, the expansion of the metric field for a single, slowly rotating gravitational source in PPN coordinates is given by

$$\begin{aligned} g_{00} &= 1 - 2\frac{GM}{c^2 r}\Big(1 - J_2\frac{R^2}{r^2}\frac{3\cos^2\theta - 1}{2}\Big) + 2\beta\Big(\frac{GM}{c^2 r}\Big)^2 + \mathcal{O}(c^{-5}), \\ g_{0i} &= 2(\gamma+1)\frac{G[\boldsymbol{J}\times\boldsymbol{r}]_i}{c^3 r^3} + \mathcal{O}(c^{-5}), \\ g_{ij} &= -\delta_{ij}\Big[1 + 2\gamma\frac{GM}{c^2 r}\Big(1 - J_2\frac{R^2}{r^2}\frac{3\cos^2\theta - 1}{2}\Big) + \frac{3}{2}\delta\Big(\frac{GM}{c^2 r}\Big)^2\Big] + \mathcal{O}(c^{-5}), \end{aligned} \tag{1}$$

where M and $\boldsymbol{J}$ being the mass and angular momentum of the Sun, J_2 being the quadrupole moment of the Sun, and R being its radius. r is the distance between the observer and the center of the Sun. β, γ, δ are the PPN parameters and in general relativity they are all equal to 1. The M/r term in the g_{00} equation is the Newtonian limit; the terms multiplied by the post-post-Newtonian parameters β, γ are post-Newtonian terms. The term multiplied by the post-Newtonian parameter δ also enters the calculation of the relativistic light deflection [68].

This PPN expansion serves as a useful framework to test relativistic gravitation in the context of the LATOR mission. In the special case, when only two PPN parameters (γ, β) are considered, these parameters have clear physical meaning. Parameter γ represents the measure of the curvature of the space–time created by a unit rest mass; parameter β is a measure of the non-linearity of the law of superposition of the gravitational fields in the theory of gravity. General relativity, which corresponds to $\gamma = \beta = 1$, is thus embedded in a two-dimensional space of theories. The Brans–Dicke is the best known theory among the alternative theories of gravity. It contains, besides the metric tensor, a scalar field and an arbitrary coupling constant ω, which yields the two PPN parameter values $\gamma = (1+\omega)/(2+\omega)$ and $\beta = 1$. More general scalar–tensor theories yield values of β different from one.

Current Limits on the PPN Parameters γ and β

The PPN formalism has proved to be a versatile method to plan gravitational experiments in the solar system and to analyze the data obtained [2, 10, 62–68, 102, 107–110]. Different experiments test different combinations of the PPN parameters (for more details, see [109, 110]). The most precise value for the PPN parameter γ is at present given by the Cassini mission as $\gamma - 1 = (2.1 \pm 2.3) \times 10^{-5}$ [11]. (Note that the Cassini result constraints the Brans–Dicke scalar coupling constant at the level of $|\omega| \geq 4.35 \times 10^4$.) The secular trend of Mercury's perihelion, when described in the PPN formalism, depends on another linear combination of the PPN parameters γ and β and the quadrupole coefficient J_2 of the solar gravity field: $\lambda_\odot = (2+2\gamma-\beta)/3 + 0.296 \times J_2 \times 10^4$. The combination of parameters $\lambda_\odot$ was obtained with Mercury ranging data as $\lambda_\odot = 0.9996 \pm 0.0006$ [77, 78]. Analysis of planetary ranging data recently yielded an independent determination of parameter γ: $\gamma - 1 = 0.0015 \pm 0.0021$; it also gave β with accuracy at the level of $\beta - 1 = -0.0010 \pm 0.0012$ [3, 4, 112]. The astrometric observations of quasars on the solar background performed with VLBI further reduced the uncertainty in the knowledge of the PPN parameter γ, resulting in the limit of $\gamma = 0.99983 \pm 0.00045$ [37, 95].

The PPN formalism has provided a useful framework for testing the violation of the strong equivalence principle (SEP) for gravitationally bound bodies. In that formalism, the ratio of passive gravitational mass M_G to inertial mass M_I of the same body is given by $M_\mathrm{G}/M_\mathrm{I} = 1 - \eta U_\mathrm{G}/(M_0 c^2)$, where M_0 is the rest mass of this body and U_G is the gravitational self-energy. The SEP violation is quantified by the parameter η, which is expressed in terms of the basic set of PPN parameters by the relation $\eta = 4\beta - \gamma - 3$. Additionally, with LLR finding that Earth and Moon fall toward the Sun at rates equal to 1.5 parts in 10^{13}, even in a conservative scenario, where a composition dependence of acceleration rates masks a gravitational self-energy dependence, η is constrained to be less than 0.0008 [3]; without such accidental cancelation the η constraint improves to 0.0003. Using the recent Cassini result [11] on the PPN γ, the parameter β was measured as $\beta - 1 = (0.9 \pm 1.1) \times 10^{-4}$ from LLR [113–115] (see Fig. 1 on page 42). The next order PPN parameter δ has not yet been measured, though its value can be inferred from other measurements.

Over the recent decade, the technology has advanced to the point that one can consider carrying out direct tests in a weak field to second order in the field strength parameter ($\propto G^2$). Although any measured anomalies in first or second-order metric gravity potentials will not determine strong-field gravity, they would signal that modifications in the strong-field domain will exist. The converse is perhaps more interesting: if to high precision no anomalies are found in the lowest order metric potentials, and this is reinforced by finding no anomalies at the next order, then it follows that any anomalies in the strong gravity environment are correspondingly quenched under all but exceptional circumstances.[1]

We shall now discuss the recent motivations for the precision gravity experiments.

2.2 Motivations for Precision Gravity Experiments

The continued inability to merge gravity with quantum mechanics and recent cosmological observations indicate that the pure tensor gravity of general relativity needs modification. The tensor–scalar theories of gravity, where the usual general relativity tensor field coexists with one or several long-range scalar fields, are believed to be the most promising extension of the theoretical foundation of modern gravitational theory. The superstring, many-dimensional Kaluza–Klein and inflationary cosmology theories have revived interest in the so-called "dilaton fields," i.e., neutral scalar fields whose background values determine the strength of the coupling constants in the effective four-dimensional theory. The importance of such theories is that they provide a possible route to the quantization of gravity and the unification of physical laws.

Although the scalar fields naturally appear in the theory, their inclusion predicts different relativistic corrections to Newtonian motions in gravitating systems. These deviations from general relativity lead to a violation of the Equivalence Principle (either weak or strong or both), modification of large-scale gravitational phenomena, and generally lead to space and time variation of physical "constants." As a result, this progress has provided new strong motivation for high-precision relativistic gravity tests.

Tensor–Scalar Extensions of General Relativity

Recent theoretical findings suggest that the present agreement between general relativity and experiment might be naturally compatible with the existence of a scalar contribution to gravity. In particular, Damour and Nordtvedt [22, 23] (see also [24, 25] for nonmetric versions of this mechanism together with [31,32] for the recent summary of a dilaton-runaway scenario) have found that a scalar–tensor theory of gravity may contain a "built-in" cosmological attractor mechanism toward general relativity. These scenarios assume that the scalar coupling parameter $\frac{1}{2}(1-\gamma)$ was of order one in the early universe (say, before inflation), and show that it then evolves to be close to, but not

[1] For example, a mechanism of a "spontaneous-scalarization" that, under certain circumstances, may exist in tensor–scalar theories [26].

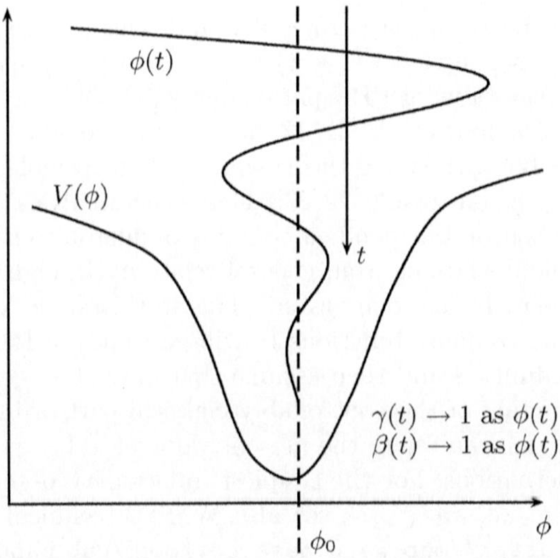

Fig. 1. Typical cosmological dynamics of a background scalar field is shown in the case when that field's coupling function to matter, $V(\phi)$, has an attracting point ϕ_0. The strength of the scalar interaction's coupling to matter is proportional to the derivative (slope) of the coupling function, and so it weakens as the attracting point is approached, and both the Eddington parameters γ and β (and all higher structure parameters as well) approach their pure tensor gravity values in this limit [22,23,27,28,31,32]. But a small residual scalar gravity should remain today because this dynamical process is not complete, and that is what LATOR experiment seeks to find.

exactly equal to, zero at the present time (Fig. 1 illustrates this mechanism in more details).

The Eddington parameter γ, whose value in general relativity is unity, is perhaps the most fundamental PPN parameter, in that $\frac{1}{2}(1-\gamma)$ is a measure, for example, of the fractional strength of the scalar gravity interaction in scalar–tensor theories of gravity [27,28]. Within perturbation theory for such theories, all other PPN parameters to all relativistic orders collapse to their general relativistic values in proportion to $\frac{1}{2}(1-\gamma)$. This is why measurement of the first-order light deflection effect at the level of accuracy comparable with the second-order contribution would provide the crucial information separating alternative scalar–tensor theories of gravity from general relativity [65] and also to probe possible ways for gravity quantization and to test modern theories of cosmological evolution [22–25, 31, 32]. Under some assumptions (see, e.g., [22, 23]) one can even estimate what is the likely order of magnitude of the left-over coupling strength at present time which, depending on the total mass density of the universe, can be given as $1-\gamma \sim 7.3 \times 10^{-7}(H_0/\Omega_0^3)^{1/2}$, where Ω_0 is the ratio of the current density to the closure density and H_0 is the Hubble constant in units of $100\,\mathrm{km\,s^{-1}\,Mpc^{-1}}$. Compared to the cosmological constant, these scalar field models are consistent with the supernovae observations for a lower matter density, $\Omega_0 \sim 0.2$, and a higher age, $(H_0 t_0) \approx 1$. If this is indeed the case, the level $(1-\gamma) \sim 10^{-6}$ to 10^{-7} would be the lower bound for the present value of PPN parameter γ [22, 23].

More recently, Damour et al. [31, 32] have estimated $\frac{1}{2}(1-\gamma)$ within the framework compatible with string theory and modern cosmology, which basically confirms the previous result [22,23]. This recent analysis discusses a scenario when a composition-independent coupling of dilaton to hadronic matter produces detectable deviations from general relativity in high-accuracy light deflection experiments in the solar system. This work assumes only some general property of the coupling functions (for large values of the field, i.e., for an "attractor at infinity") and then assume that only $(1-\gamma)$ is of the order of one at the beginning of the controllably classical part of inflation. It was shown in [32] that one can relate the present value of $\frac{1}{2}(1-\gamma)$ to the cosmological density fluctuations. For the simplest inflationary potentials (favored by WMAP mission, i.e., $m^2\chi^2$ [7], see also WMAP technical papers at the mission's website: `http://map.gsfc.nasa.gov/m_mm/pub_papers/`), Damour et al. [31,32] found that the present value of $(1-\gamma)$ could be just below 10^{-7}. In particular, within this framework $\frac{1}{2}(1-\gamma) \simeq \alpha^2_{\rm had}$, where $\alpha_{\rm had}$ is the dilaton coupling to hadronic matter, its value depends on the model taken for the inflation potential $V(\chi) \propto \chi^n$, with χ being the inflation field; the level of the expected deviations from general relativity is $\sim 0.5 \times 10^{-7}$ for $n = 2$ [32]. Note that these predictions are based on the work in scalar–tensor extensions of gravity, which are consistent with, and indeed often part of, present cosmological models.

Another example of recent theoretical progress is the Dvali–Gabadadze–Porrati (DGP) brane-world model, which explores a possibility that we live on a brane embedded in a large extra dimension, and where the strength of gravity in the bulk is substantially less than that on the brane [33]. Although such theories can lead to perfectly conventional gravity on large scales, it is also possible to choose the dynamics in such a way that new effects show up exclusively in the far infrared, providing a mechanism to explain the acceleration of the universe [76, 83]. It is interesting to note that DGP gravity and other modifications of GR hold out the possibility of having interesting and testable predictions that distinguish them from models of dynamical Dark Energy. One outcome of this work is that the physics of the accelerating universe may be deeply tied to the properties of gravity on relatively short scales, from millimeters to astronomical units [8, 34, 35].

To date general relativity and some other alternative gravitational theories are in good agreement with the experimental data collected from the relativistic celestial mechanical extremes provided by the relativistic motions in the binary millisecond pulsars. At the same time, many modern theoretical models, which include general relativity as a standard gravity theory, are faced with the problem of the unavoidable appearance of space–time singularities. It is generally suspected that the classical description, provided by general relativity, breaks down in a domain where the curvature is large, and hence, a proper understanding of such regions requires new physics. This is a reason why recently a considerable interest has been shown in the physical processes occurring in the strong gravitational field regime with relativistic pulsars, providing a promising possibility to test gravity in this qualitatively different dynamical environment. The general theoretical framework for pulsar tests of strong-field gravity was introduced in [30]; the observational data for the initial tests were obtained with PSR1534 [97]. An analysis of strong-field gravitational tests and their theoretical justification was presented in [27–29]. The recent analysis of the pulsar data resulted in $\frac{1}{2}(1-\gamma) \simeq \alpha_{\rm had}^2 \sim 4 \times 10^{-4}$ at a 3σ confidence level [47], with $\alpha_{\rm had}$ being the dilaton coupling to hadronic matter. While being a natural alternative to the weak-gravity tests, the pulsar tests of gravitation currently cannot offer the accuracy at the level that is presently available within the solar system. In fact, a carefully designed dedicated experiment that utilizes the strongest gravitational potential available in the solar system, provided by the Sun itself, offers a unique opportunity to test gravitation in a controlled and well-understood environment. Therefore, despite the relative weakness of its gravitational field, the Sun still has an advantage and offers an attractive opportunity to perform accurate tests of gravity.

The analyses discussed above not only motivate new searches for very small deviations of relativistic gravity in the solar system, but they also predict that such deviations are currently present in the range from 10^{-5} to 5×10^{-8} for $\frac{1}{2}(1-\gamma)$, i.e., for observable post-Newtonian deviations from general relativity predictions and, thus, should be easily detectable with LATOR. This would require measurement of the effects of the next post-Newtonian order ($\propto G^2$) of light deflection resulting from gravity's intrinsic nonlinearity. An ability to measure the first-order light deflection term at the accuracy comparable with the effects of the second order is of the utmost importance for gravitational theory and a major challenge for the twenty-first century fundamental physics.

Observational Motivations for Higher Accuracy Tests of Gravity

Recent astrophysical measurements of the angular structure of the cosmic microwave background (CMB) [12], the masses of large-scale structures [75], and the luminosity distances of type Ia supernovae [76, 83] have placed stringent constraints on the cosmological constant Λ and also have led to a revolutionary conclusion: the expansion of the universe is accelerating. The implication of these observations for cosmological models is that a classically evolving scalar field currently dominates the energy density of the universe. Such models have been shown to share the advantages of Λ: compatibility with the spatial flatness predicted inflation; a universe older than the standard Einstein-de Sitter model; and, combined with cold dark matter, predictions for large-scale structure formation in good agreement with data from galaxy surveys. Combined with the fact that scalar field models imprint distinctive signature on the CMB anisotropy, they remain currently viable and should be testable in the near future. This completely unexpected discovery demonstrates the importance of testing the important ideas about the nature of gravity. We are presently in the "discovery" phase of this new physics, and while there are many theoretical conjectures as to the origin of a nonzero Λ, it is essential that we exploit every available opportunity to elucidate the physics that is at the root of the observed phenomena.

There is now multiple evidence indicating that 70% of the critical density of the universe is in the form of a "negative-pressure" dark energy component; there is no understanding as to its origin and nature. The fact that the expansion of the universe is currently undergoing a period of acceleration now seems rather well tested: it is directly measured from the light curves of several hundred type Ia supernovae [76,83,98], and independently inferred from observations of CMB by the WMAP satellite [7] and other CMB experiments [41,59]. Cosmic speed-up can be accommodated within general relativity by invoking a mysterious cosmic fluid with large negative pressure, dubbed dark energy. The simplest possibility for dark energy is a cosmological constant; unfortunately, the smallest estimates for its value are 55 orders of magnitude too large (for reviews see [16, 74]). Most of the theoretical studies operate in the shadow of the cosmological constant problem, the most embarrassing hierarchy problem in physics. This fact has motivated a host of other possibilities, most of which assume $\Lambda = 0$, with the dynamical dark energy being associated with a new scalar field (see [18, 19] and references therein). However, none of these suggestions is compelling and most have serious drawbacks. Given the challenge of this problem, a number of authors considered the possibility that cosmic acceleration is not due to some kind of stuff, but rather arises from new gravitational physics (see discussion in [16–18,74]). In particular, certain extensions to general relativity in a low energy regime [15,18,20] were shown to predict an experimentally consistent universe evolution without the need for dark energy (see discussion on the interplay between theory, experiment, and observation in [6, 9]). These dynamical models are expected to explain the observed acceleration of the universe without dark energy, but may produce measurable gravitational effects on the scales of the solar system. In particular, corresponding contribution to the parameter γ in experiments conducted in the solar system are expected at the level of $1-\gamma \sim 10^{-7} - 5\times10^{-9}$, thus further motivating the relativistic gravity research. Therefore, the PPN parameter γ may be the only key parameter that holds the answer to most of the questions discussed earlier. Also an anomalous parameter δ will most likely be accompanied by a "γ mass" of the Sun, which differs from the gravitational mass of the Sun and therefore will show up as anomalous γ (see discussion in [72]).

Even in the solar system, general relativity still faces challenges. There is the long-standing problem of the size of the solar quadrupole moment and its possible effect on the relativistic perihelion precession of Mercury (see review in [109]). The interest is in the study of the behavior of the solar quadrupole moment vs. the radius and the heliographic latitudes. This solar parameter has been very often neglected in the past, because it was rather difficult to determine an accurate value. The improvement of the accuracy of our knowledge of J_2 is certainly due to the fact that, today, we are able to take into account the differential rotation with depth. In fact, the quadrupole moment plays an important role in the accurate computation of several astrophysical quantities, such as the ephemeris of the planets or general relativistic prediction for the precession of the perihelion of Mercury and minor planets such as Icarus. Finally, it is necessary to accurately know the value of the quadrupole moment to determine the shape of the Sun, i.e., to say its oblateness. Solar oblateness measurements by Dicke and others in the past gave conflicting results for J_2 (reviewed on p. 145 of [21]). A measurement of solar oblateness with the balloon-borne Solar Disk Sextant gave a quadrupole moment on the order of 2×10^{-7} [50]. Helioseismic determinations using solar oscillation data have since implied a small value for J_2, on the order of $\sim 10^{-7}$, i.e., consistent with simple uniform rotation [13, 40, 109]. However, there exist uncertainties in the helioseismic determination for depths below roughly 0.4 $R_\odot$, which might permit a rapidly rotating core. (LATOR can measure J_2 with accuracy sufficient to put this issue to rest.)

In summary, there are a number of theoretical and experimental reasons to question the validity of general relativity. Despite the success of modern gauge field theories in describing the electromagnetic, weak, and strong interactions, it is still not understood how gravity should be described at the quantum level. In theories that attempt to include gravity, new long-range forces can arise in addition to the Newtonian inverse-square law. Even at the purely classical level, and assuming the validity of the Equivalence Principle, Einstein's theory does not provide the most general way to generate the space–time metric. Regardless of whether the cosmological constant should be included, there are also important reasons to consider additional fields, especially scalar fields. The LATOR mission is designed to directly address theses challenges with an unprecedented accuracy; we shall now discuss LATOR in more details.

3 Overview of LATOR

The LATOR experiment uses the standard technique of time-of-fight laser ranging (extended to interplanetary scales) between two microspacecraft whose lines-of-sight pass close by the Sun and also a long-baseline stellar optical interferometer (placed above the Earth's atmosphere) to accurately measure deflection of light by the solar gravitational field in the extreme proximity to the Sun [101]. Figure 2 shows the general concept for the LATOR missions, including the mission-related geometry, experiment details, and required accuracies.

In this section we will consider the LATOR mission architecture in more detail.

3.1 Evolving Light Triangle

The LATOR mission architecture uses an evolving light triangle formed by laser ranging between two spacecraft (placed in ~1 AU heliocentric orbits) and a laser transceiver terminal on the International Space Station (ISS) (realized via European collaboration [106]). The objective is to measure the gravitational deflection of laser light as it passes in extreme proximity to the Sun (see Fig. 2). To that extent, the long-baseline (~100 m) fiber-coupled optical interferometer on the ISS will perform differential astrometric measurements of the laser light sources on the two spacecraft as their lines-of-sight pass behind the Sun.

As seen from the Earth, the two spacecraft will be separated by ~1°, which will be accomplished by a small maneuver immediately after their launch [101, 104]. This separation would permit differential astrometric observations to an accuracy of ~0.1 prad needed to significantly improve measurements of gravitational deflection of light in the solar gravity.

The schematic of the LATOR experiment is quite simple and is given in Fig. 2. Two spacecraft are injected into a heliocentric solar orbit on the opposite side of the Sun from the Earth. The triangle in the figure has three independent quantities and three arms are monitored with laser metrology. Each spacecraft is equipped with a laser ranging system that enables the measurement of the arms of the triangle formed by the two spacecraft and the ISS. The uniqueness of this mission comes with its geometrically redundant architecture that enables LATOR to measure the departure from Euclidean geometry ($\sim 8.48 \times 10^{-6}$ rad) caused by the solar gravity field, to a very high accuracy [101]. This departure is shown as a difference between the calculated

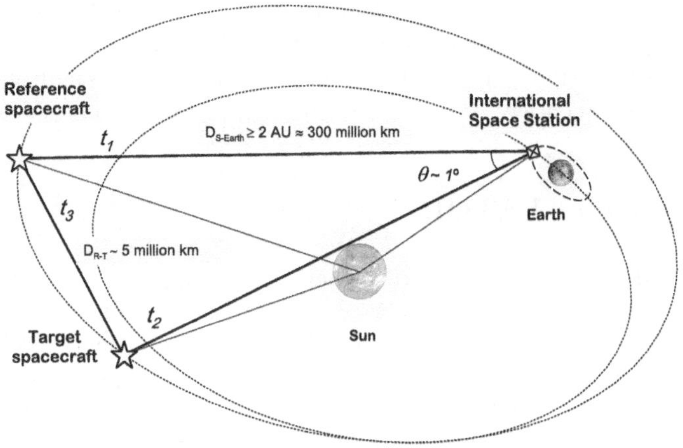

Fig. 2. The overall geometry of the LATOR experiment. The objective is to measure distances, t_1, t_2, t_3, between the three nodes of the space triangle to accuracy of 0.5 cm. To benefit from the geometric redundancy, the experiment will also measure the angle between the two spacecraft to accuracy of 0.05 prad.

Euclidean value for an angle in the triangle and its value directly measured by the interferometer. This discrepancy, which results from the curvature of the space–time around the Sun and can be computed for every alternative theory of gravity, constitutes LATOR's signal of interest. The precise measurement of this departure constitutes the primary mission objective.

A version of LATOR with a ground-based receiver was proposed in [120]. Because of atmospheric turbulence and seismic vibrations that are not common mode to the receiver optics, a very long-baseline interferometer (30 km) was proposed. This interferometer could only measure the differential light deflection to an accuracy of 0.1 μas, with a spacecraft separation of less than 1 arcmin. The shortening of the interferometric baseline (as compared to the previously studied version [90, 91, 120]) is achieved solely by going into space to avoid the atmospheric turbulence and Earth's seismic vibrations. On the space station, all vibrations can be made common mode for both ends of the interferometer by coupling them by an external laser truss. This relaxes the constraint on the separation between the spacecraft, allowing it to be as large as few degrees, as seen from the ISS. Additionally, the orbital motion of the ISS provides variability in the interferometer's baseline projection as needed to resolve the fringe ambiguity of the stable laser light detection by an interferometer.

We shall now consider the basic elements of the LATOR optical design.

3.2 General Approach in Optical Design

A single aperture of the interferometer on the ISS consists of three 30 cm diameter telescopes (see Fig. 3 for a conceptual design). One of the telescopes with a very narrow bandwidth laser line filter in front and with an InGaAs camera at its focal plane, sensitive to the 1,064 nm laser light, serves as the acquisition telescope to locate the spacecraft near the Sun.

The second telescope emits the directing beacon to the spacecraft. Both spacecraft are served out of one telescope by a pair of piezocontrolled mirrors placed on the focal plane. The properly collimated laser light (1 W) is injected into the telescope focal plane and deflected in the right direction by the piezoactuated mirrors.

The third telescope is the laser light tracking interferometer input aperture, which can track both spacecraft at the same time. To eliminate beam walk on the critical elements of this telescope, two piezoelectric X–Y–Z stages are used to move two single-mode fiber tips on a spherical surface while maintaining focus and beam position on the fibers and other optics. Dithering at a few Hz is used to make the alignment to the fibers and the subsequent tracking of the two spacecraft completely automatic. The interferometric tracking telescopes are coupled together by a network of single-mode fibers whose relative length changes are measured internally by a heterodyne metrology system to an accuracy of less than 5 pm.

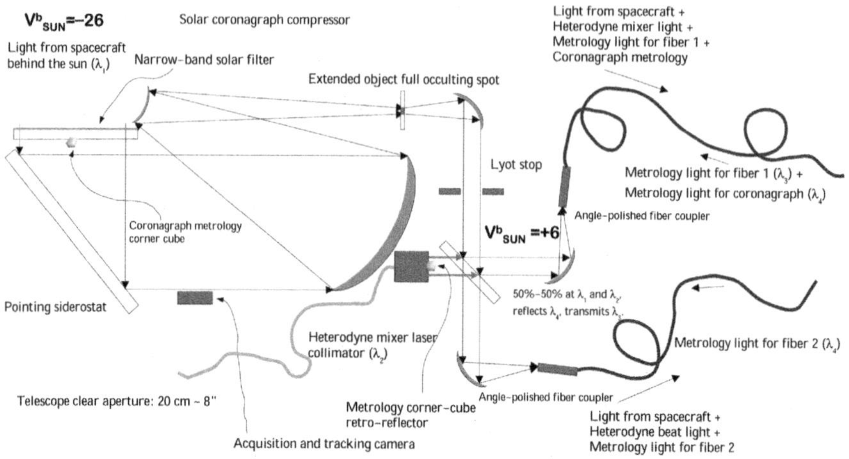

Fig. 3. Basic elements of optical design for the LATOR interferometer. The laser light (together with the solar background) is going through a full aperture ($\sim$30 cm) narrow band-pass filter with 5×10^{-5} µm bandwidth around wavelength of $\lambda = 1{,}064$ nm. The remaining light illuminates the baseline metrology corner cube and falls onto a steering flat mirror where it is reflected to an off-axis telescope with no central obscuration (needed for metrology). It then enters the solar coronograph compressor (with 1×10^{-5} suppression properties) by first going through a 1/2 plane focal plane occulter and then coming to a Lyot stop. At the Lyot stop, the background solar light is reduced by a factor of 10^6. This combination of a narrow band-pass filter and coronograph enables the solar luminosity reduction from $V = -26$ to $V = 4$ (as measured at the ISS), thus enabling the LATOR precision observations.

The spacecrafts are identical in construction and contain a relatively high-powered (1 W), stable (2 MHz h^{-1} $\sim$ 500 Hz s^{-1}), small cavity fiber-amplified laser at 1,064 nm. The power of this laser is directed to the Earth through a 20 cm aperture telescope and its phase is tracked by the interferometer. With the available power and the beam divergence, there are enough photons to track the slowly drifting phase of the laser light. There is another 0.2 W laser operating at 780 nm, the power of which is transmitted through another telescope with small aperture of 5 cm, which points toward the other spacecraft. In addition to the two transmitting telescopes, each spacecraft has two receiving telescopes. The receiving telescope, which points toward the area near the Sun, has laser line filters and a simple knife-edge coronagraph to suppress the Sun's light to one part in 10^5 of the light level of the light received from the space station. The receiving telescope that points to the other spacecraft is free of the Sun light filter and the coronagraph.

In addition to the four telescopes they carry, the spacecraft also carry a tiny (2.5 cm) telescope with a CCD camera. This telescope is used to initially point the spacecraft directly toward the Sun so that their signal may be seen at the space station. One more of these small telescopes may also be installed at right angles to the first one to determine the spacecraft attitude, using known bright stars. The receiving telescope looking toward the other spacecraft may be used for this purpose part of the time, reducing hardware complexity. Star trackers with this construction were demonstrated many years ago and they are readily available. A small RF transponder with an omni-directional antenna is also included in the instrument package to track the spacecraft while they are on their way to assume the orbital position needed for the experiment.

The LATOR experiment has a number of advantages over techniques that use radio waves to measure gravitational light deflection. Advances in optical communications technology allow low bandwidth telecommunications with the LATOR spacecraft without having to deploy high gain radio antennae needed to communicate through the solar corona. The use of the monochromatic light enables the observation of the spacecraft almost at the limb of the Sun, as seen from the ISS. The use of narrow band filters, coronagraph optics, and heterodyne detection will suppress background light to a level where the solar background is no longer the dominant noise source. In addition, the short wavelength allows much more efficient links with smaller apertures, thereby eliminating the need for a deployable antenna. Finally, the use of the ISS will allow conducting the test above the Earth's atmosphere – the major source of astrometric noise for any ground based interferometer. This fact justifies LATOR as a space mission.

3.3 Science Objectives

LATOR is a Michelson–Morley type experiment designed to test the pure tensor metric nature of gravitation – a fundamental postulate of Einstein's theory of general relativity [101]. With its focus on gravity's action on light propagation, it complements other tests that rely on the gravitational dynamics of bodies. The idea behind this experiment is to use a combination of independent time-series of highly accurate measurements of the gravitational deflection of light in the immediate proximity to the Sun along with measurements of the Shapiro time delay on the interplanetary scales (to a precision, respectively, better than 0.1 prad and 1 cm). Such a combination of observables is unique and enables LATOR to significantly improve tests of relativistic gravity.

The LATOR's primary mission objective is to measure the key post-Newtonian Eddington parameter γ with an accuracy of a part in 10^9. When the light deflection in solar gravity is concerned, the magnitude of the first-order effect as predicted by general relativity for the light ray just grazing the limb of the Sun is $\sim$1.75 arcsecond (as) (for more details see Table 1). (Note that 1 as $\simeq$ 5 µrad; when convenient, later we will use the units of radians and arcseconds interchangeably.) The effect varies inversely with the impact parameter. The second-order term is almost six orders of magnitude smaller, resulting in $\sim$3.5 microarcseconds (µas) light deflection effect,

Table 1. Comparable sizes of various light deflection effects in the solar gravity field.

Effect	Analytical form	Deflection angle (μas)	Delay (pm)
First order	$2(1+\gamma)\frac{M}{b}$	1.75 as	0.849 mm
Second order	$\left[\left(2(1+\gamma)-\beta+\frac{3}{4}\delta\right)\pi-2(1+\gamma)^2\right]\frac{M^2}{b^2}$	3.5	1,697
Frame dragging	$\pm 2(1+\gamma)\frac{J}{b^2}$	±0.7	±339
Solar quadrupole	$2(1+\gamma)J_2\frac{M}{b^3}$	0.2	97

The value of deflection angle is calculated on the limb of the Sun ($b = R_\odot$); the corresponding delay is given for a $b = 100$ m interferometric baseline proposed for LATOR.

and which falls off inversely as the square of the light ray's impact parameter [36, 38, 65, 84, 85, 101]. The relativistic frame-dragging term[2] is ±0.7 μas, and contribution of the solar quadrupole moment, J_2, is sized as 0.2 μas (using theoretical value of the solar quadrupole moment $J_2 \simeq 10^{-7}$). The small magnitudes of the effects emphasize the fact that, among the four forces of nature, gravitation is the weakest interaction; it acts at very long distances and controls the large-scale structure of the universe, thus making the precision tests of gravity a very challenging task.

If the Eddington's 1919 experiment was performed to confirm the general theory of relativity, LATOR is motivated to search for physics beyond the Einstein's theory of gravity with an unprecedented accuracy [101]. In fact, this mission is designed to address the questions of fundamental importance to modern physics. In particular, this solar system scale experiment would search for a cosmologically evolved scalar field that is predicted by modern theories of quantum gravity and cosmology, and also by superstring and brane-world models [8, 9, 33]. LATOR will also test the cosmologically motivated theories that attempt to explain the small acceleration rate of the Universe (so-called "dark energy") via modification of gravity at very large horizon or superhorizon distances.

By studying the effect of gravity on light and measuring the Eddington parameter γ, this mission will test the presently viable alternative theories of gravity, namely the scalar–tensor theories. The value of the parameter γ may hold the key to the solution of the most fundamental questions concerning the evolution of the universe. In the low energy approximation suitable for the solar system, a number of modern theories of gravity and cosmology studied as methods for gravity quantization or proposed as an explanation to the recent cosmological puzzles predict measurable contributions to the parameter γ at the level of $\frac{1}{2}(1-\gamma) \sim 10^{-6} - 10^{-8}$; detecting this deviation

[2] Gravitomagnetic frame dragging is the effect in which both the orientation and trajectory of objects in orbit around a body are altered by the gravity of the body's rotation. It was studied by Lense and Thirring in 1918.

Table 2. LATOR mission summary: science objectives.

Qualitative objectives

- To test the metric nature of the Einstein's general theory of relativity in the most intense gravitational environment available in the solar system – the extreme proximity to the Sun
- To test alternative theories of gravity and cosmology, notably scalar–tensor theories, by searching for cosmological remnants of scalar field in the solar system
- To verify the models of light propagation and motion of the gravitationally bounded systems at the second post-Newtonian order (i.e., including effects $\propto G^2$)

Quantitative objectives

- To measure the key Eddington PPN parameter γ with accuracy of one part in 10^9 – a factor of 30,000 improvement in the tests of gravitational deflection of light
- To provide direct and independent measurement of the Eddington PPN parameter β via gravity effect on light to ~0.01% accuracy
- To measure effect of the second-order gravitational deflection of light with accuracy of one part in 10^4, including first ever measurement of the post-PPN parameter δ
- To directly measure the frame-dragging effect on light (first such observation and also first direct measurement of solar spin) with accuracy of one part in 10^3
- To measure the solar quadrupole moment J_2 (using the theoretical value of $J_2 \simeq 10^{-7}$, currently unavailable) to one part in 200

is LATOR's primary objective. With the accuracy of one part in 10^9, this mission could discover a violation or extension of general relativity, and/or reveal the presence of any additional long-range interaction (see Table 2).

We now outline the basic elements of the LATOR trajectory.

3.4 Spacecraft Trajectory: A 3:2 Earth Resonant Orbit

To enable the primary objective, LATOR will place two spacecraft into a heliocentric orbit, to provide conditions for observing the spacecraft when they are behind the Sun as viewed from the ISS (see Figs. 4 and 7). With the help of the JPL Advanced Project Design Team (Team X), we recently conducted detailed mission design studies [39]. In particular, we analyzed various trajectory options for the deep-space flight segment of LATOR, using both the Orbit Determination Program (ODP) and Satellite Orbit Analysis Program (SOAP) – the two standard JPL navigation software packages.

One trajectory option would be to use a Venus flyby to place the spacecraft in a 1 year orbit (perihelion at Venus orbit ~0.73 AU and aphelion ~1.27 AU).

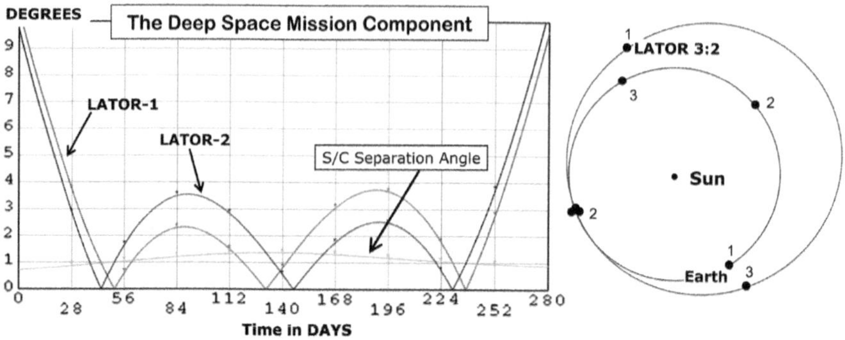

Fig. 4. *Left*: the Sun–Earth–Probe angle during the period of three occultations (*two periodic curves*) and the angular separation of the spacecraft as seen from the Earth (*lower smooth line*). Time shown is days from the moment when one of the spacecraft are at 10° distance from the Sun. *Right*: view from the North Ecliptic of the LATOR spacecraft in a 3:2 resonance. The epoch is taken near the first occultation.

One complication of this approach is that the Venus orbit is inclined about 3.4° with respect to the ecliptic and the out-of-plane position of Venus at the time of the flyby determines the orbit inclination [39]. The LATOR observations require that the spacecraft pass directly behind the Sun, i.e., with essentially no orbit inclination. To minimize the orbit inclination, the Venus' flyby would need to occur near the time of Venus nodal crossing. An approach with a type IV trajectory and a single Venus flyby requires a powered Venus flyby with about 500–900 $\mathrm{m\,s^{-1}}$. However, a type I trajectory to Venus with two Venus gravity assists would get LATOR into a desirable 1 year orbit at Earth's opposition. This option requires no velocity change, called Δv, and provides repeated opportunities for the desired science observations. (Δv is a desired spacecraft velocity change, which is typically enabled by either the onboard propulsion system or a planetary flyby.) At the same time this orbit has a short launch window ~17 days, which motivated us to look for an alternative.

An orbit with a 3:2 resonance with the Earth[3] was found to uniquely satisfy the LATOR orbital requirements [39, 101]. For this orbit, 13 months after the launch, the spacecraft are within ~10° of the Sun with first occultation occurring 15 months after launch [101]. At this point, LATOR is orbiting at a slower speed than the Earth, but as LATOR approaches its perihelion, its motion in the sky begins to reverse and the spacecraft is again occulted by the Sun 18 months after launch. As the spacecraft slows down and moves out

[3] The 3:2 resonance occurs when the Earth does three revolutions around the Sun while the spacecraft does exactly two revolutions on a 1.5 year orbit. The exact period of the orbit may vary slightly, <1%, from a 3:2 resonance depending on the time of launch.

toward aphelion, its motion in the sky reverses again, and it is occulted by the Sun for the third and final time 21 months after launch.

This entire process will again repeat itself in about 3 years after the initial occultation; however, there may be a small maneuver required to allow for more occultations. Therefore, to allow for more occultations in the future, there may be a need for an extra few tens of $\mathrm{m\,s^{-1}}$ of Δv. The energy required for launch, C_3, will vary approximately between 10.6 and $11.4\,\mathrm{km^2\,s^{-2}}$ depending on the time of launch, but it is suitable for a Delta II launch vehicle. The desirable $\sim 1°$ spacecraft separation (as seen from the Earth) is achieved by performing a $30\,\mathrm{m\,s^{-1}}$ maneuver after the launch. This results in the second spacecraft being within $\sim(0.6°–1.4°)$ separation during the entire period of three occultations by the Sun.

Figure 4 shows the trajectory and the occultations in more details. The figure on the right is the spacecraft position in the solar system showing the Earth's and LATOR's orbits (in the 3:2 resonance) relative to the Sun. The epoch of this figure shows the spacecraft passing behind the Sun as viewed from the Earth. The figure on the left shows the trajectory when the spacecraft would be within 10° of the Sun as viewed from the Earth. This period of 280 days will occur once every 3 years, provided the proper maneuvers are performed. Two similar periodic curves give the Sun–Earth–Probe (SEP) angles for the two spacecraft while the lower smooth curve gives their angular separation as seen from the Earth.

As a baseline design for the LATOR orbit,[4] both spacecraft will be launched on the same launch vehicle. Almost immediately after the launch there will be a $30\,\mathrm{m\,s^{-1}}$ maneuver that separates the two spacecraft on their 3:2 Earth resonant orbits (see Fig. 4). The sequence of events that occurs during each observation period will be initiated at the beginning of each orbit of the ISS. It is assumed that bore sighting of the spacecraft attitude with the spacecraft transmitters and receivers have already been accomplished. This sequence of operations is focused on establishing the ISS-to-spacecraft link. The interspacecraft link is assumed to be continuously established after deployment, since the spacecraft never lose line of sight with one another (for more details consult Sect. 5.5).

[4] In addition to this 3:2 Earth resonant orbit, here is also an option to have both spacecraft move in opposite directions during each of the solar conjunctions [73]. In this option, the two LATOR spacecraft move either toward or away from each other, as seen from the Earth. At the beginning of each conjunction the two craft are on the opposite sides from the Sun, moving toward each other in such a way that not only their impact parameters are equal (i.e., $p_1 = -p_2$), but also the rates of change of these impact parameters are also equal (i.e., $\dot{p}_1 = -\dot{p}_2$). This option would increase the amount of Δv, which LATOR spacecraft should carry onboard, but it significantly reduces the experiment's dependence on the accuracy of the determination of the solar impact parameter. This particular option is currently being investigated and results will be reported elsewhere.

The 3:2 Earth resonant orbit provides an almost ideal trajectory for the LATOR mission, specifically (1) it imposes no restrictions on the time of launch; (2) with a small propulsion maneuver after launch, it places the two LATOR spacecraft at the distance of less then 3.5° (or $\sim$14 $R_\odot$) for the entire duration of the experiment (or $\sim$8 months); (3) it provides three solar conjunctions even during the nominal mission lifetime of 22 months, all within a 7 month period; (4) at a cost of an extra maneuver, it offers a possibility of achieving small orbital inclinations (to enable measurements at different solar latitudes); and, finally, (5) it offers a very slow change in the Sun–Earth–Probe angle of about $\sim R_\odot$ in 4 days. Furthermore, such an orbit provides three observing sessions during the initial 21 months after the launch, with the first session starting in 15 months [101]. As such, this orbit represents a very attractive choice for LATOR. We intend to further study this 3:2 Earth resonant trajectory as the baseline option for the mission.

In Sect. 4 we will discuss the preliminary design of the LATOR interferometer on the ISS.

4 LATOR Interferometry

In this section, we describe the process of how the LATOR interferometer will be measuring angles. Since the spacecraft will carry lasers that are monochromatic sources, the interferometer can efficiently use heterodyne detection to measure the phase of the incoming signal. To this extent, we first present a simplified explanation of heterodyne interferometry proposed for the LATOR interferometer. We then describe the interferometric design of the LATOR station on the ISS.

4.1 Heterodyne Interferometry

Figures 5 and 6 show a simplified schematic of how angles are measured using a heterodyne interferometer. In Fig. 5, two siderostats are pointed at a target. Two fiducials, shown as corner cubes, define the end points of the interferometer baseline. The light from each of the two arms is interfered with stable local oscillators (LOs) and the phase difference recorded. If the LOs in each arm were phase-locked, the angles of the target with respect to the baseline normal is

$$\theta = \arcsin\left[\frac{(2\pi n + \phi_1 - \phi_2)\lambda}{2\pi b}\right], \tag{2}$$

where λ is the wavelength of the downlink laser, n is an unknown integer arising from the fringe ambiguity, and b is the baseline length. To resolve this ambiguity, multiple baselines were used in the previous mission design (this is discussed in greater detail in [120]). In reality, it is difficult to phase-lock the two LOs over the long-baseline lengths. The left graphics in Fig. 6 shows how a single LO can be used and transmitted to both siderostats, using a single mode

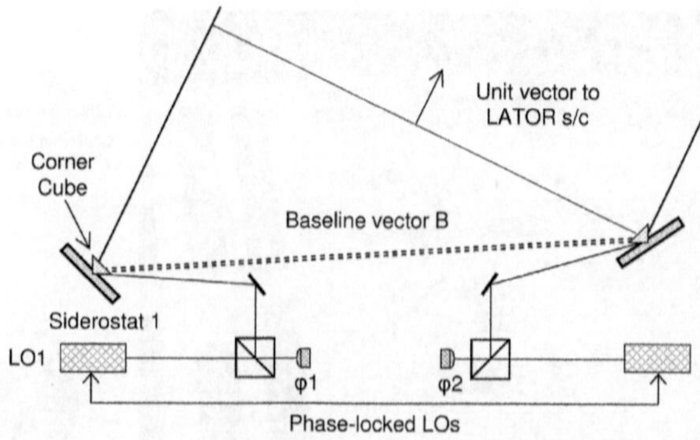

Fig. 5. Heterodyne interferometry on one spacecraft with phase-locked local oscillator.

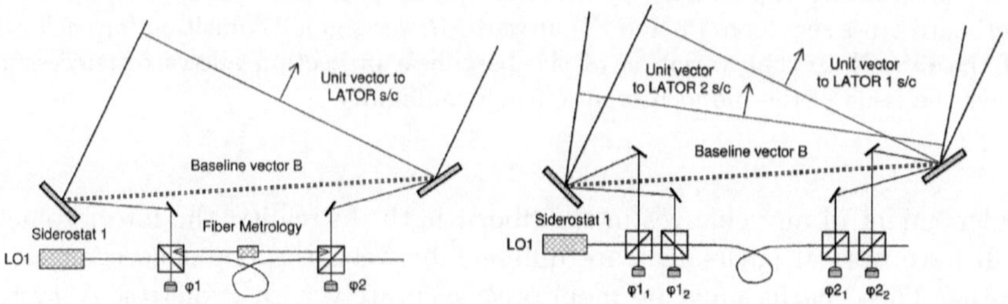

Fig. 6. *Left*: fiber-linked heterodyne interferometry and fiber metrology system. *Right*: fiber-linked heterodyne interferometry on two spacecraft.

fiber. In this configuration, a metrology system is used to monitor changes in the path length as seen by the LO as it propagates through the fiber. The metrology system measures the distance from one beam splitter to the other. In this case, the angle is given by

$$\theta = \arcsin\left[\frac{(2\pi n + \phi_1 - \phi_2 + m_1)\lambda}{2\pi b}\right], \tag{3}$$

where m_1 is the phase variations introduced by changes in the optical path of the fiber as measured by the metrology system.

Now consider the angle measurement between two spacecraft (shown on the right in Fig. 6). In this case the phase variations due to changes in the path through the fiber are common to both spacecraft. The differential angle is

$$\theta = \arcsin\left[\frac{(2\pi(n_1 - n_2) + (m_1 - m_2))\lambda}{2\pi b} + \frac{((\phi 1_1 - \phi 1_2) - (\phi 2_1 - \phi 2_2))\lambda}{2\pi b}\right]. \tag{4}$$

Since the spacecraft are monochromatic sources, the interferometer can efficiently use heterodyne detection to measure the phase of the incoming signal. Note that because of the fact that this is a differential measurement, it is

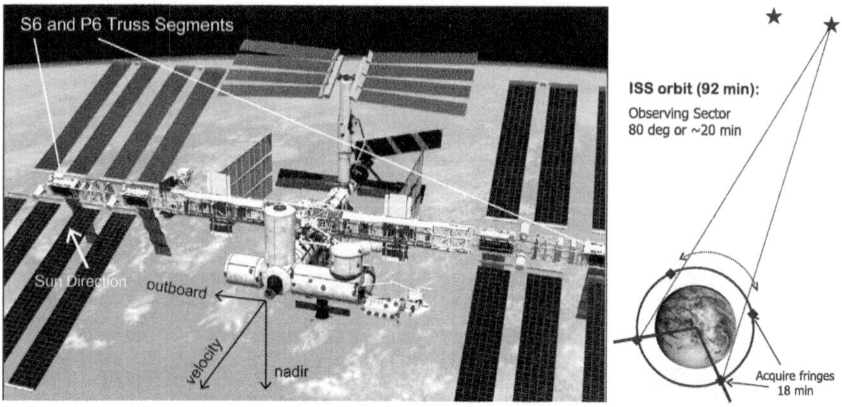

Fig. 7. *Left*: location of the LATOR interferometer on the ISS. To utilize the inherent ISS Sun-tracking capability, the LATOR optical packages will be located on the outboard truss segments P6 and S6 outward. *Right*: signal acquisition for each orbit of the ISS. Note that variation of the baseline's projection allows to successfully solve the issue of the monochromatic fringe ambiguity.

independent of any changes in the fiber length. In reality, the interferometer will have optical paths that are different between the two spacecraft signal paths. These paths must be monitored accurately with a metrology system to correct for phase changes in the optical system due to thermal variations. However, this metrology must only measure path lengths in each interferometric station and not along the entire length of the fiber.

The use of multiple interferometers is a standard solution to resolve the fringe ambiguity resulting from the interferometric detection of monochromatic light [101]. The current LATOR mission proposal is immune from the fringe ambiguity problem, as the orbit of the ISS provides enough variability (at least ∼30%) in the baseline projection (see Fig. 7 for general description of the geometry of the interferometer on the ISS and its orbit). This variablity enables one to take multiple measurements during one orbit, to uniquely resolve the baseline orientation for each ISS orbit, which successfully solves the fringe ambiguity issue for LATOR.

4.2 Long-Baseline Optical Interferometer on the ISS

The LATOR station on the ISS is used to interferometrically measure the angle between the two spacecraft and also to transmit and receive the laser ranging signals to each of the spacecraft. A block diagram of the laser station is shown in Fig. 8 and is described in more detail later. The station on the ISS is composed of a two laser beacon stations that perform communications and laser ranging to the spacecraft and two interferometer stations that collect the downlink signal for the astrometric measurement. This station also uses

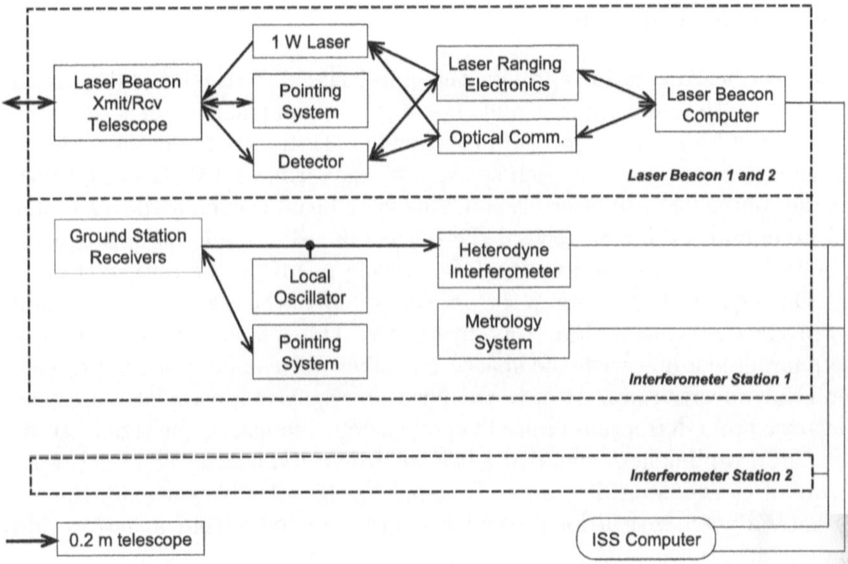

Fig. 8. The LATOR station on the ISS.

a fiber optic link to transmit the common local oscillator to the interferometer station.

General Description

The interferometer on the ISS will be formed by two optical packages (or laser beacon stations) with approximate dimensions of ($0.6 \times 0.6 \times 0.6\,\mathrm{m}^3$) for each package. The mass of each telescope assembly $\sim$120 kg. Both laser beacon stations must be physically located and integrated with the ISS infrastructure (see description of the ISS in [44]). Their location must provide a straight-line separation of $\sim$100 m between the two stations and have a clear line-of-sight (LOS) path between the two transceivers during the observation periods. Both packages must have clear LOS to both spacecraft during predefined measurement periods. Location on the ISS should maximize the inherent ISS Sun-tracking capability. Both telescope assemblies will have to be able to point toward the Sun during each observing period, which can be achieved by locating these payloads on the ISS outboard truss segments (P6 and S6 outward, see Fig. 7). In addition, a limited degree of automatic Sun-tracking capability is afforded by the α-gimbals on the ISS.

The minimum unobstructed LOS time duration between each transponder on the ISS and the transponders and their respective spacecraft will be 58 min per the 92 min orbit of the ISS. The pointing error of each transceivers to its corresponding spacecraft will be no greater than 1 μrad for control, 1 μrad for knowledge, with a stability of $0.1\,\mu\mathrm{rad\,s}^{-1}$, provided by a combination of the standard GPS link available on the ISS and μg accelerometers.

Laser Beacon Station

The laser beacon stations provide the uplink signals to the LATOR spacecraft and detect their downlink signals (see Fig. 9). The transmitter laser signal is modulated for laser ranging and to provide optical communications. Separate transmitters are used for each spacecraft, each using a 1 W laser at 1,064 nm as the source for each laser beacon. The laser beam is expanded to a diameter of 30 cm and is directed toward the spacecraft using a siderostat mirror. Fine pointing is accomplished with a fast steering mirror in the optical train.

During initial acquisition, the optical system of the laser beacon is modified to produce a beam with a 10 as divergence. This angular spread is necessary to guarantee a link with the spacecraft, albeit a weak one, in the presence of pointing uncertainties. After the acquisition sequence is complete, the beam is narrowed to a diffraction limited beam, thereby increasing the signal strength.

The downlink laser signal at 1,064 nm is detected using a 12×12 (10×10 as) array of Germanium detectors. To suppress the solar background, the signal is heterodyned with a local oscillator and detected within a narrow 1 MHz bandwidth. In the initial acquisition mode, the detection system searches over a 300 MHz bandwidth and uses a spiral search over a 30 as angular field to find the downlink signal. Upon acquisition, the search bandwidth is decreased to 1 MHz and a quad-cell subarray is used to point the siderostat and fast steering mirrors of the beacon.

Interferometer Station

The interferometer stations collect the laser signal from both spacecraft to perform the heterodyne measurements needed for the interferometric angle

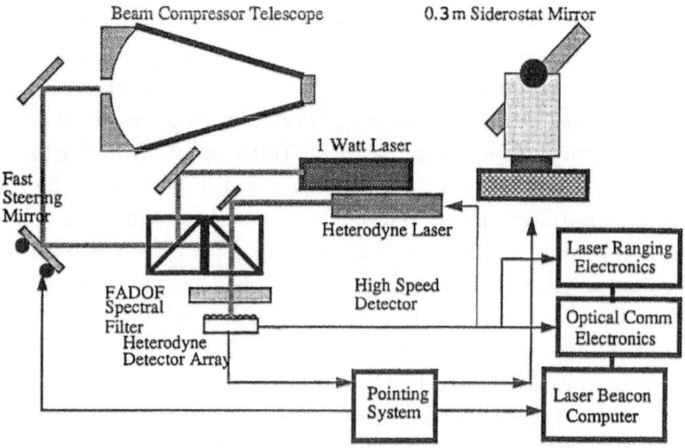

Fig. 9. The laser beacon station on the ISS.

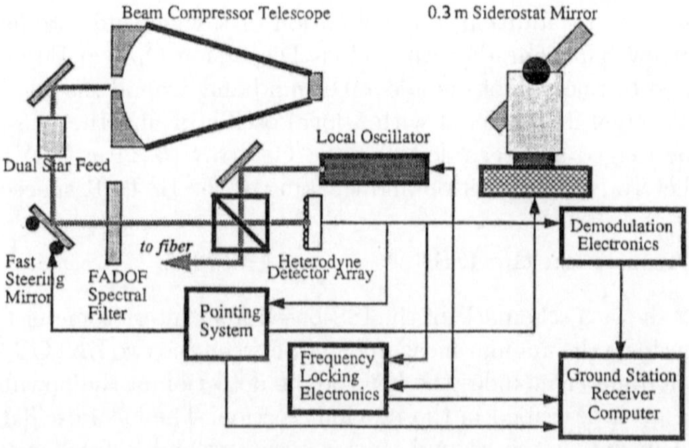

Fig. 10. Component description of the receiver on the ISS.

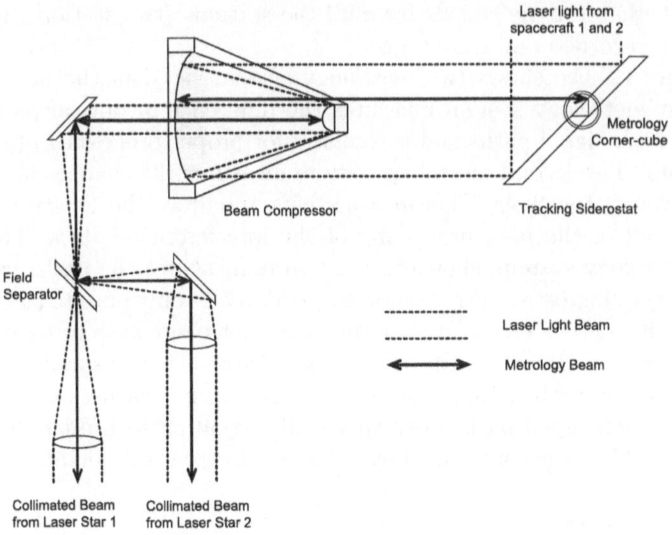

Fig. 11. Principal diagram for the dual feed optical system. The diagram describes propagation of laser starlight and metrology beams through the optical system (note metrology corner cube, beam compressor, and field separator).

measurement. There are a total of five receivers to make the four angular measurements needed to resolve fringe ambiguity.

Figure 10 shows a schematic of an interferometer station. The detection and tracking system is basically similar to the receiver arm of the laser beacon described in the previous section. Light is collected by a 0.3 m siderostat mirror and compressed with a telescope to a manageable beam size. The light from each of the spacecraft is separated using a dual feed optical system as shown in Fig. 11. A fast steering mirror is used for high bandwidth pointing of the receiver. In addition, a combination of a wideband interference filter and a narrow band Faraday Anomalous Dispersion Optical Filter (FADOF) will be used to reject light outside a 0.05 nm band around the laser line. Each spacecraft signal is interfered with a local oscillator and the phase measurement time tagged and recorded. A 6×6 Ge array (5×5 as FOV) is used to provide heterodyne acquisition and tracking of the LATOR spacecraft.

Interferometer on the ISS

Figure 12 shows a schematic of the ISS-based fiber interferometer that will be used to perform the angular measurement between the two LATOR spacecraft. The interferometer includes the heterodyne detection of the downlink signals that have been described in the previous section. The local oscillator (LO) is generated in one of the ground station receivers and is frequency locked to the laser signal from one of the spacecraft. The LO is then broadcast to the other station on the ISS through a 100 m single mode polarization preserving fiber. The heterodyne signals from all the stations (two stations, two signals each) are recorded and time tagged.

Figure 12 also shows two metrology systems used in the interferometer. The first metrology system measures the difference in optical path between the two laser signal paths and is essential for proper processing of the heterodyne data. The second metrology system measures the changes in the optical path through the fiber. This measurement monitors the length of the fiber and is used in the post processing of the interferometer data. The internal path metrology system, shown in the figure, measures the paths from corner cube on the siderostat mirror (shown as two, really only one) to the metrology beam splitter. It is essential that the laser metrology system be boresighted to the laser signal path so that the correct distance is measured. A Michelson interferometer with a frequency shift in one arm measures changes in the length of each signal path. Both spacecraft signal paths are measured simultaneously. This is accomplished by using an electro-optic cell and modulating

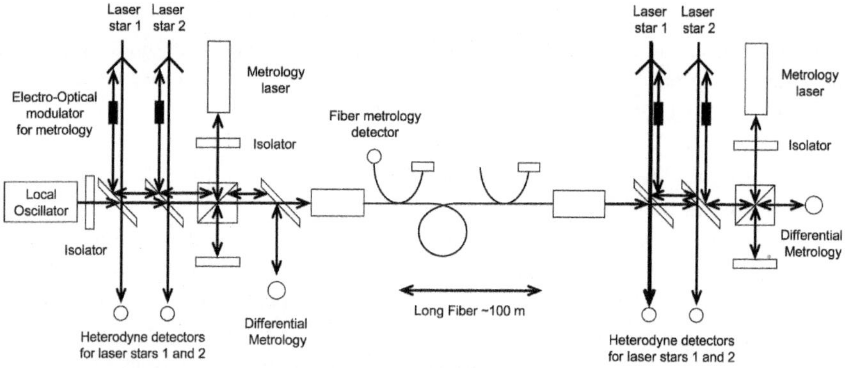

Fig. 12. Component description of the ISS-based interferometer.

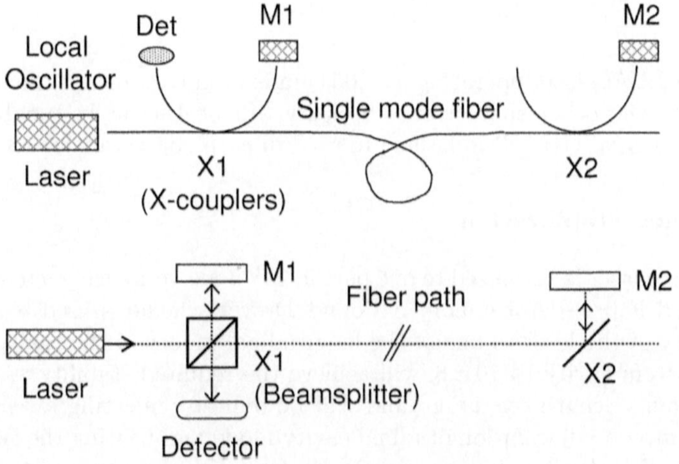

Fig. 13. Fiber metrology system.

each beam at a different frequency. A He–Ne laser is used as the light source for this metrology system. Filters at the output of the detector are then used to separate the signals corresponding to each metrology beam.

The fiber metrology system measures changes in the optical path through the fiber. This system uses local oscillator signal in a Michelson configuration. Figure 13 shows the correspondence between a standard Michelson interferometer and the fiber metrology system. The two X couplers serve as the beam splitters. Reflectors at the ends of the fiber couplers serve as the reference and signal mirrors. One of these reflectors is dithered to frequency shift the output signal. The phase measurement at the detector measures changes in the path length between points X1 and X2, if Ml-X1 and M2-X2 are held constant. This is accomplished by placing the X couplers and mirrors at each end of the fiber on a single thermally stable optical breadboard.

4.3 Laser Metrology Transceiver Subsystem

The metrology transceiver consists of the laser, frequency modulators, optics, and frequency stabilizer. The laser light is first frequency-stabilized to better than one part in 10^{10}, this is done to make the measurements. The laser light is then frequency-modulated to produce the heterodyne signal and distinguish between incoming and outgoing beams. Finally, light is collimated and injected into the beam launcher optics. The incoming metrology signal is received by the beam launcher optics and is interfered with the local laser. A cat's eye retroreflector serves as the spacecraft fiducial and is common to all three beam launchers. Below we discuss these elements in more details.

Laser

A 1 W Nd:YAG laser operating at 1,064 nm is used to transmit the metrology signals to the other spacecraft. The laser will be thermally tunable over a range of several GHz. Two lasers are used in each spacecraft for redundancy.

Frequency Stabilization

The source laser is stabilized to one part in 10^{10} long-term using a temperature controlled Fabry–Perot etalon. A Pound–Drever scheme is used to servo the frequency of the laser to one of the longitudinal modes of the cavity. Control of the ~3 cm cavity to 10 mK will achieve the required stability. Calibration of the cavity length on the ground will be done by injecting a second laser locked onto an adjacent longitudinal cavity mode and beating the two signals together. For the 3 cm cavity, the 5 GHz beat frequency must be known to 10^{-10}. Temperature control of the cavity will allow fine tuning of the laser frequencies between the spacecraft so that the heterodyne signal between two lasers lies below 2 MHz. This will require knowledge of the spacecraft relative velocity to $1\,\mathrm{m\,s^{-1}}$, which is easily achievable.

Frequency Modulators

The laser frequency is modulated to distinguish between the various transmit and receive beams used in the LATOR measurements. In addition, the relative velocity between the spacecraft can reach as high as $100\,\mathrm{m\,s^{-1}}$. This will produce a Doppler frequency of up to 200 MHz between lasers from two spacecraft. The frequency of the modulator will be tuned to slightly offset from the Doppler frequency to minimize the bandwidth at which the data needs to be recorded.

Acousto-optic modulators (AOM) with fiber-coupled input and output are used. For a single metrology channel three different frequencies are needed for the reference, and two unknown signals. One implementation is a fiber-fed modulator, which uses a bulk AOM and is insensitive to alignment errors. Other implementations for the AOMs will also be studied. These include integrated optic AOMs and multichannel Bragg cells, both of which will be capable of generating the multiple signals at much lower mass.

The metrology system will also need to phase lock the outgoing laser with the incoming laser. The AOM provides the phase modulation to the laser beam. The incoming signal and the laser output from the AOM are interfered on a high frequency detector. This signal is then used to servo the frequency of the AOM to null. This will produce a phase-locked signal whose phase error is determined by the level of the null. In reality, because of the AOM, center frequency, the interfered signal will be upshifted by a stable local oscillator and the servoing done in RF. The stability of this local oscillator is the same as the required stability of the phase-locked loop, 10^{-10} (discussed in Sect. 4.3).

Beam Launcher and Receiver Optics

In the current instrument design, the modulated laser beam is injected using a polarization preserving single mode fiber and expanded to a 0.5 cm beam. A cat's eye retroreflector is one of the several devices that can be used as the metrology fiducial and is common to the three metrology beams. The cat's eye uses two optically contacted concentric hemispheres with radius of ~10 and ~20 cm. The cat's eye is sized many times larger than the beam to minimize the effect of spherical aberration.

The beam is then expanded to a 5 cm beam using a refractive telescope. A refractive design was chosen because changes in the optical path are relatively insensitive to changes in the position and orientation of the optical elements.

To measure the path length to better ~5 pm, errors due to thermal effects on the beam launcher optics must be controlled. For example, a change in the temperature of 1 mK on a 0.5 cm beam splitter would produce a path length error of ~5 pm; consequently, an active thermal controller would be used on the beam splitters and telescope optics. Furthermore, baffles on the optics will be used to prevent external radiation from affecting the temperature of the instrument. The metrology optics will be mounted on a GrEp bench for thermal stability.

Acquisition Camera Subsystem

The acquisition camera will be used as the sensor for pointing the metrology beam. A 512×512 camera may be used to detect the position of the incoming laser beam to 0.5 as over a 1° field by interpolating the centroid of the spot to 0.1 pixel. Three cameras will be used to track each of the incoming metrology beams. The outgoing laser beam will be retroreflected from the alignment corner cube to produce a spot on the acquisition camera on which to servo the pointing gimbal. The direction of the outgoing beam is set to the position of the target spacecraft, taking into account the point-ahead angle.

Pointing Subsystem

In the current instrument design, the entire beam launcher optical assembly is gimbaled to point the metrology beam to the target spacecraft. The two-axis gimbal has a center of rotation at the center of the cat's eye retroreflector. This optical arrangement measures the distance between the optical fiducials and is not sensitive to slight misalignments to the first order. The gimbal will have a range of 1° and a pointing resolution of 0.5 as.

Laser Ranging Subsystem

The laser ranging system is used to determine the positions of the spacecraft with respect to the ISS. This is required to determine the impact parameter of the laser beam grazing the Sun as well as the coplanarity of the three spacecraft. A time of flight laser ranging system is used to triangulate the spacecraft positions. A laser transponder system on the spacecraft is used to increase the SNR of the return pulse.

Laser ranging will be performed with an accuracy of ~1 cm by integrating over a number of laser pulses. If the system were capable of instantaneously detecting delays of 100 ps (3 cm), at a 1 kHz repetition rate, it would take less than 1 s to reach the desired accuracy. Assuming this level of ranging and using baseline 100 m will result in an accuracy in the transverse direction of 1 m at LATOR's orbit, with spacecraft separated as much as 2 AU (see discussion of interspacecraft laser ranging operations in Sect. 5.1).

In Sect. 5 we turn our attention to the mission flight system.

5 LATOR Flight System

The LATOR flight system consists of two major components: the deep-space component that will be used to transmit and receive the laser signals needed to make science measurements and the interferometer on the ISS that will be used to interferometrically measure the angle between the two spacecraft and to transmit and receive the laser ranging signals to each of the spacecraft.

There are two LATOR spacecraft in the deep-space component of the mission, which will be used to transmit and receive the laser signals needed to make the science measurements. Figure 14 shows a schematic of the flight system without the solar cell array. The flight system is subdivided into the

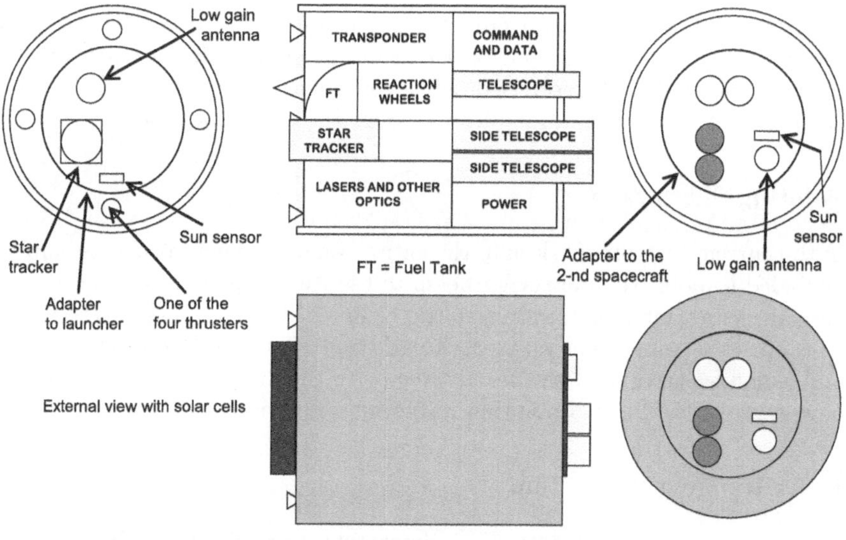

Fig. 14. LATOR spacecraft concept.

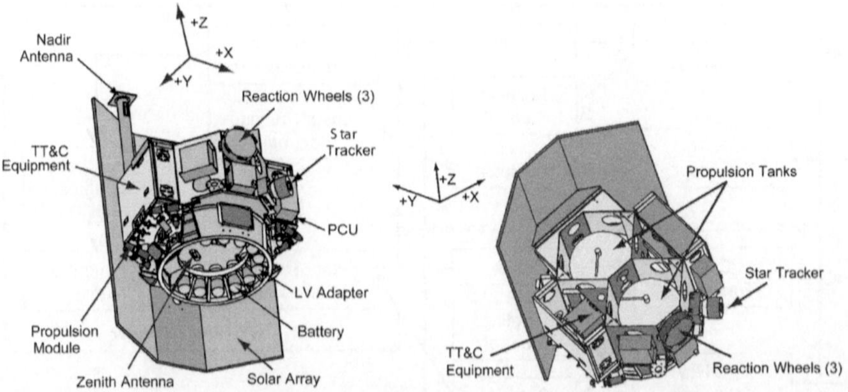

Fig. 15. A typical Spectrum Astro SA-200S/B bus. With minor modifications this configuration may be adopted for the deep-space component of the LATOR mission.

instrument payload and the spacecraft bus (note that SA200S spacecraft built by Spectrum Astro already has the needed capabilities, see Fig. 15). The instrument includes the laser ranging and communications hardware and is described in more detail in the following section. The spacecraft contains the remainder of the flight hardware, which includes solar cells, attitude control, and the spacecraft structure.

In this section we will discuss the design of these components in more detail.

5.1 LATOR Instrument

The LATOR instrument in each of the two spacecraft consists of three laser metrology transmitters and receivers that can be gimbaled to point at the other spacecraft, and a camera system to acquire the incoming laser signals and to control the pointing of the outgoing beams. In addition, the instrument contains a laser ranging transponder to determine the spacecraft position from the ground. The LATOR instrument is used to perform laser ranging between the two spacecraft; it is also used (the second set) for laser ranging and optical communications between the spacecraft and the ISS. Figure 16 shows a block diagram of the instrument subsystems, which we describe in more detail below.

ISS-to-Spacecraft Receiver and Transmitter

The ISS-to-spacecraft receiver performs the acquisition, tracking, and detection of the signals from the ISS (Fig. 17). This uplinked signal will be sent at 1,064 nm and will contain modulation both to perform laser ranging and to send control signals to the spacecraft. The signals from the ISS are detected by a telescope with a collecting aperture of 20 cm. A coronograph will be used

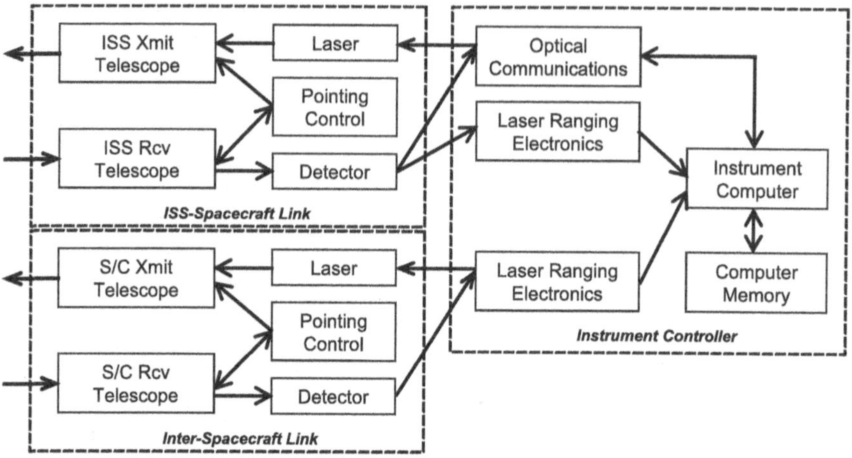

Fig. 16. LATOR instrument subsystem block diagram.

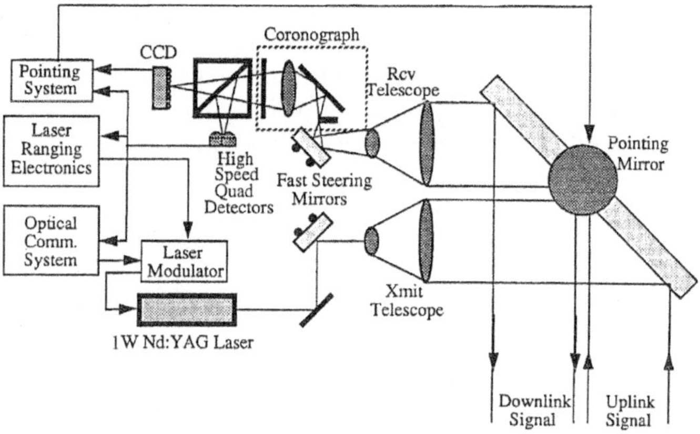

Fig. 17. Spacecraft transmitter and receiver for the ISS–spacecraft link.

to suppress stray light from the Sun. In addition, a combination of a wide-band interference filter and a narrow band FADOF filter will be used to reject light outside a 0.05 nm band around the laser line. The incoming signal is subdivided with one portion going to a high bandwidth detector and the other to an acquisition and tracking CCD array. Using a 64×64 CCD array with pixels sized to a diffraction limited spot, this array will have a 5 arcmin field of view, which is greater than the pointing knowledge of the attitude control system and the point-ahead angle (30 as). After acquisition of the ISS beacon, a 2×2 element subarray of the CCD will be used as a quad cell to control the ISS–S/C two axis steering mirror. This pointing mirror is common to both the receiver and transmitter channel to minimize misalignments between the two optical systems due to thermal variations. The pointing mirror will have

10 arcmin throw and a pointing accuracy of 0.5 as, which will enable placement of the uplink signal on the high bandwidth detector.

The ISS-to-spacecraft transmitter sends a laser signal to both the interferometer collectors and the beacon receivers. The signal will be encoded for both ranging and communication information. In particular, the transmitted signal will include the interspacecraft ranging measurements. The transmitter uses a 1 W frequency stabilized Nd:YAG laser at 1,064 nm. A 5 kHz line width is required to simplify heterodyne detection at the ground station. A 20 cm telescope is used to transmit the laser beam and a steering mirror is used for pointing. The mirror uses information from the attitude control system, the quad-cell detector in the receiver, and the point-ahead information from the instrument controller to determine the transmit direction. A fast steering mirror is used to maintain high bandwidth pointing control for both the transmitter and receiver.

We have also considered the possibility of using a common optical system for both the transmitter and receiver. Figure 18 shows a schematic of such a transmitter/receiver system. Because of the difference in the receive and transmit wavelengths, dichroic beam splitters and filters are used to minimize losses from the optics and leakage into the detectors. In this scheme a point-ahead mirror is used to maintain a constant angular offset between the received and transmitted beams. Because of the common optical elements, this system is more tolerant to misalignments than the previous configuration.

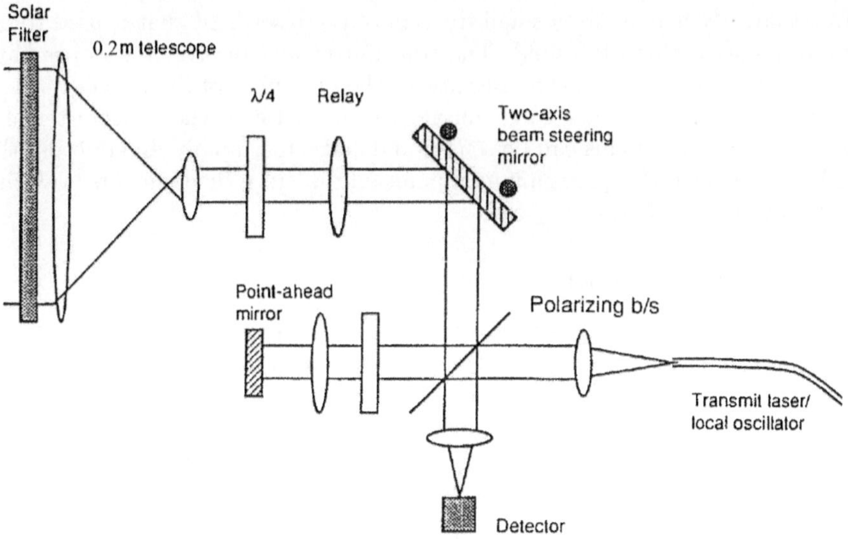

Fig. 18. ISS-to-spacecraft link with common optics: spiral scanning spatial acquisition; open loop point-ahead control with piezoactuators; fiber-coupled, frequency stabilized transmitter; pupil planes at the steering mirror and mixing apertures.

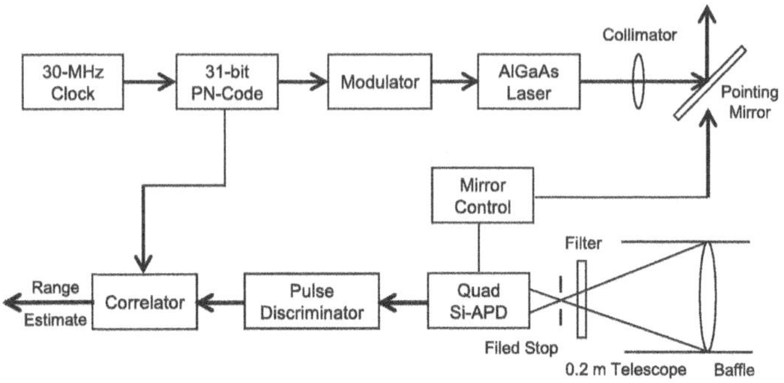

Fig. 19. The interspacecraft transmitter and receiver.

Interspacecraft Receiver and Transmitter

The interspacecraft receiver/transmitter uses two separate optical systems. The receiver detects the laser ranging signal from the other spacecraft (shown in Fig. 19). The receiver is similar in design to the ISS–spacecraft receiver subsystem. Since there is no solar background contribution, the coronograph and FADOF filter have been removed. Detection of the signal is accomplished using a CCD for acquisition and a quad cell subarray for tracking. The tracking signal is also used to control the pointing of the transmitter minor. A separate high bandwidth detector is used for detecting the laser ranging signal.

The interspacecraft transmitter sends the laser ranging signal to the other spacecraft. The transmitter uses a 780 nm laser with an output power of 0.2 W (alternatively it may use a small fraction of the laser light that is used to establish spacecraft-to-ISS link). The transmitter and receiver telescopes have an aperture of 5 cm diameter. Because of the proximity of the LATOR spacecraft, thermal drifts that cause misalignments between the transmitter and receiver optical systems can be sensed and corrected rapidly. In addition, the LATOR geometry requires minimal point ahead, since the transverse velocity between spacecraft is nearly zero.

Instrument Controller

The instrument controller subsystem contains the remainder of the instrument hardware. This includes the electronics needed for the laser ranging and optical communications as well as the computer used to control the instrument. The instrument computer will take information from the attitude control system and receiver subsystems to control the pointing of the transmit subsystems and the modulation of their laser signals.

5.2 LATOR Spacecraft

The LATOR spacecraft, like most spacecraft, will be composed of the following subsystems: thermal, structural, attitude control, power, command and data handling, telecommunications, and propulsion, which will be discussed below.

Thermal Subsystem

The basic thermal design will be similar to that of the SA-200B, with modifications to account for the variation in range. This design uses basically passive thermal control elements with electric heaters/thermostats. The thermal control flight elements are multilayer insulation, thermal surfaces, thermal conduction control, and sensors. The active elements are minimized and will be only electric heaters/thermostats. To minimize heater power thermal louvers may be used. The current design assumes that the spacecraft uses passive thermal control with heaters/thermostats, because it is basically designed for Earth orbit (alternative thermal designs that utilize active elements are also being studied).

Structural Subsystem

The current best estimate for the total dry mass is 115 kg, including a set of required modifications to the standard SA-200B bus (i.e., a small propulsion system, a 0.5 m HGA for deep-space telecom, etc.) The design calls for launching the two spacecraft on a custom carrier structure, as they should easily fit into the fairing (for instance, Delta II 2425-9.5). The total launch mass for the two spacecraft will be 552 kg. (This estimate may be further reduced, given more time to develop a point design.)

Attitude Control Subsystem

An attitude control system may be required to have pointing accuracy of 6 µrad and a pointing knowledge of 3 µrad. This may be achieved using a star tracker and Sun sensor combination to determine attitude together with reaction wheels (RWs) to control attitude. Cold-gas jets may be used to desaturate RWs. A Spectrum Astro SA-200B three-axis stabilized bus with RWs for fine pointing and thrusters for RW desaturation is a good platform [39]. For the current experiment design it is sufficient to utilize a pointing architecture with the following performance (3σ, per axis): control 6 µrad; knowledge 3 µrad; stability 0.1 µrad s^{-1}. The SA-200B readily accommodates these requirements.

Power Subsystem

The flight system will require ~50 W of power. This may be supplied by a 1 square meter GaAs solar cell array. To maintain a constant attitude with respect to the Sun, the solar cells must be deployed away from the body of the spacecraft. This will allow the cells to radiate away its heat to maintain the cells within their operating temperature range.

Telecommunications Subsystem

The telecommunications subsystem will be a hybrid that utilizes the optical communications capability of the instrument as the primary means of transmitting and receiving commands and data. In addition, a small low gain antenna for low data rate radio communications will be used for emergency purposes. This system will use a 15 W transmitter and 10 dB gain antenna. Using X band this system can operate with a 5 bit per second (bps) data rate. The design calls for an SDST X-Band transponder, operating at 15 W; X-Band SSPA; a 0.5 m HGA; two X-Band LGAs pointed opposite each other; a duplexer; two switches; and coax cabling – the standard options of present day spacecraft design.

Propulsion Subsystem

The propulsion subsystem for SA-200S bus may be used as it is. This will ensure that a minimal amount of engineering is required. The system is a blowdown monopropellant system with eight 5-N thrusters and two 32 cm tanks each with 22 kg propellant capacity. The system exists and was tested in many applications.

We shall now consider the basic elements of the LATOR optical receiver system. While we focus on the optics for the two spacecraft, the interferometer has essentially similar optical architecture.

5.3 Optical Receiver System

The LATOR 200 mm receiver optical system is a part of a proposed experiment. This system is located at each of the two separate spacecraft placed on heliocentric orbits, as shown in Fig. 2. The receiver optical system captures optical communication signals from a transmitter on the ISS, which orbits the Earth. To support the primary mission objective, this system must be able to receive the optical communication signal from the uplink system at the ISS that passes through the solar corona at the immediate proximity of the solar limb (at a distance of no more than five Airy disks).

Our recent analysis of the LATOR receiver optical system successfully satisfied all the configuration and performance requirements (shown in Table 3) [79,103,104]. We have also performed a conceptual design (see Fig. 20), which was validated with a ray-trace analysis. The ray-trace performance of the designed instrument is diffraction limited in both the APD and CCD channels over the specified field of view at 1,064 nm. The design incorporated the required field stop and Layot stop. A preliminary baffle design has been developed for controlling the stray light.

The optical housing is estimated to have very accommodating dimensions; it measures $500 \times 220 \times 250\,\mathrm{mm}^3$. The housing could be made even shorter by reducing the focal length of the primary and secondary mirrors, which

Table 3. Summary of design parameters for the LATOR spacecraft optical receiver system.

Parameters/requirements	Value/description
Aperture	200 mm, unobstructed
Wavelength	1,064 nm
Narrow band-pass filter	0.05 nm FWHM over full aperture
Focal planes	APD Data & CCD acquisition/tracking
APD field of view	Airy disk field stop (pinhole) in front of APD
APD field stop (pinhole)	Approximately 0.009 mm in diameter
APD detector size	TBD (a little larger than 0.009 mm)
CCD field of view	5 arcmin
CCD detector size	640 × 480 pixels (9.6 × 7.2 mm^2)
CCD detector pixel size	15 μm
Beam splitter ratio (APD/CCD)	90/10
Field stop	"D"-shaped at primary mirror focus
Lyot stop	Circular aperture located at telescope exit pupil

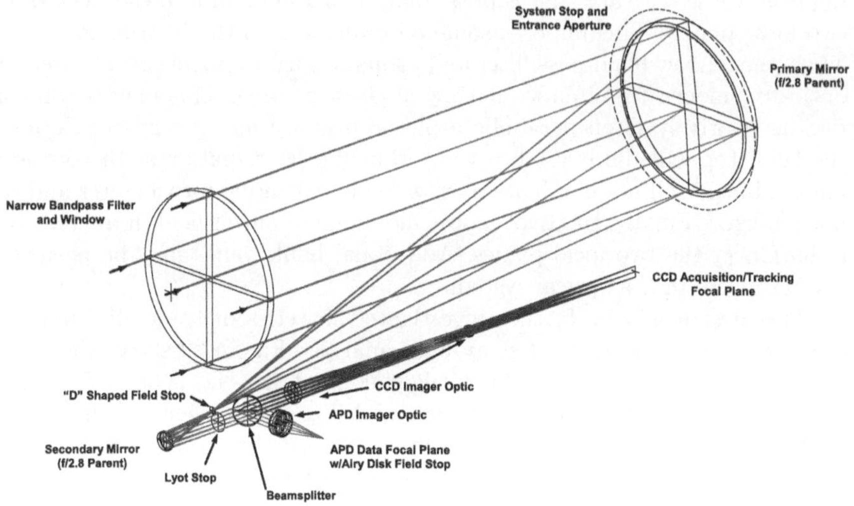

Fig. 20. Layout for LATOR optical receiver system. The following design parameters were used: (1) the primary and secondary mirrors are concave off-axis parabolas, (2) the field stop is a "D"-shaped aperture with a 5 arcmin diameter, (3) a pupil image of the primary mirrors is located at the Lyot stop, (4) the primary and secondary mirrors form an off-axis unobscured afocal 10× beam reducer, (5) the APD imager lens is an $f/3.6$ triplet, and (6) the CCD imager lens is an $f/45.5$ telephoto doublet.

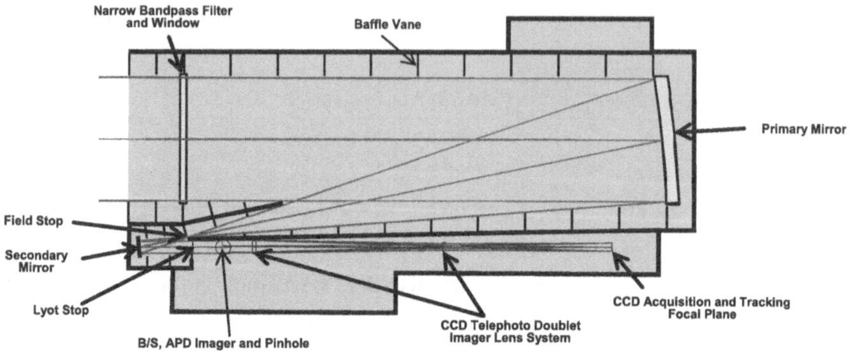

Fig. 21. Preliminary baffle design for LATOR optical receiver system.

may impose some fabrication difficulties. These design opportunities are being currently investigated.

Preliminary Baffle Design

Figure 21 shows the LATOR preliminary baffle design. The out-of-field solar radiation falls on the narrow band-pass filter and primary mirror; the scattering from these optical surfaces puts some solar radiation into the FOV of the two focal planes. This imposes some requirements on the instrument design. Thus, the narrow band-pass filter and primary mirror optical surfaces must be optically smooth to minimize narrow angle scattering. This may be difficult for the relatively steep parabolic aspheric primary mirror surface. However, the field stop will eliminate direct out-of-field solar radiation at the two focal planes, but it will not eliminate narrow angle scattering for the filter and primary mirror. Finally, the Lyot stop will eliminate out-of-field diffracted solar radiation at the two focal planes. Additional baffle vanes may be needed at several places in the optical system.

This design will be further investigated in series of trade-off studies by also focusing on the issue of stray light analysis. Figure 22 shows the design of the focal plane capping. The straight edge of the "D"-shaped CCD field stop is tangent to the limb of the Sun and it is also tangent to the edge of APD field stop. There is a 2.68 as offset between the straight edge and the concentric point for the circular edge of the CCD field stop. The results of the analysis of APD and CCD channels point spread functions (PSFs) can be found in [103, 104].

Focal Plane Mapping

Figure 22 shows the design of the focal plane capping. The straight edge of the D-shaped CCD field stop is tangent to the limb of the Sun and it is also tangent to the edge of APD field stop (pinhole). There is a 2.68 as offset between the

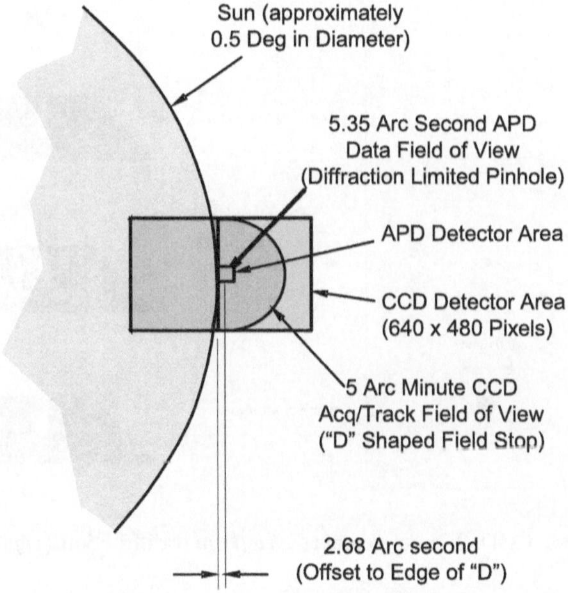

Fig. 22. LATOR focal plane mapping (the diagram not to scale).

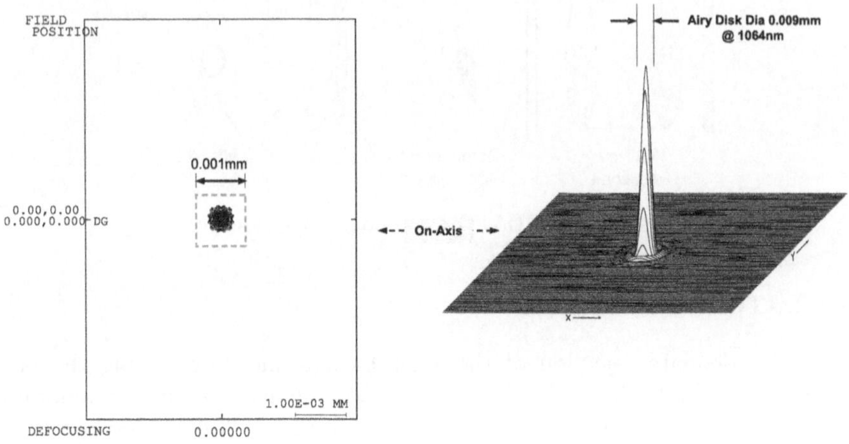

Fig. 23. APD channel geometric (*left*) and diffraction (*right*) PSF.

straight edge and the concentric point for the circular edge of the CCD field stop (D-shaped aperture). In addition, the APD field of view and the CCD field of view circular edges are concentric with each other. Depending on the spacecraft orientation and pointing ability, the D-shaped CCD field stop aperture may need to be able to be rotated to bring the straight edge into a tangent position relative to the limb of the Sun. The results of the analysis of APD and CCD channels PSFs are shown in Figs. 23 and 24.

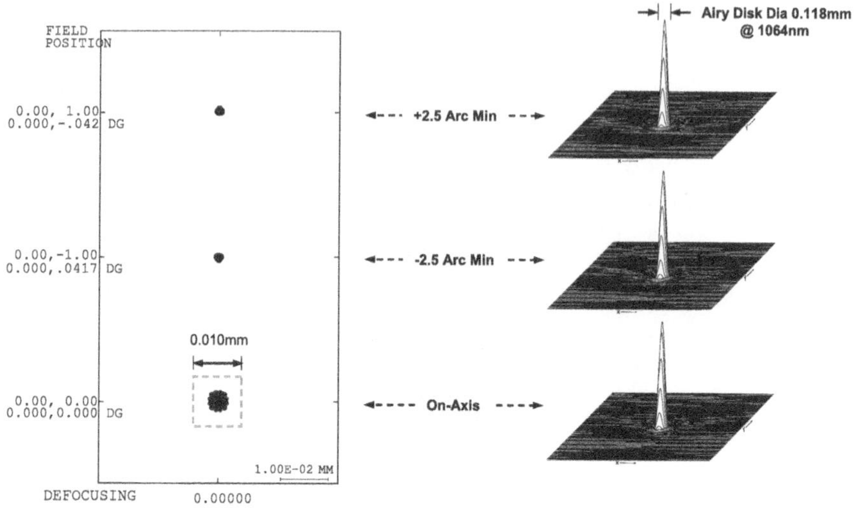

Fig. 24. CCD channel geometric (*left*) and diffraction (*right*) PSF.

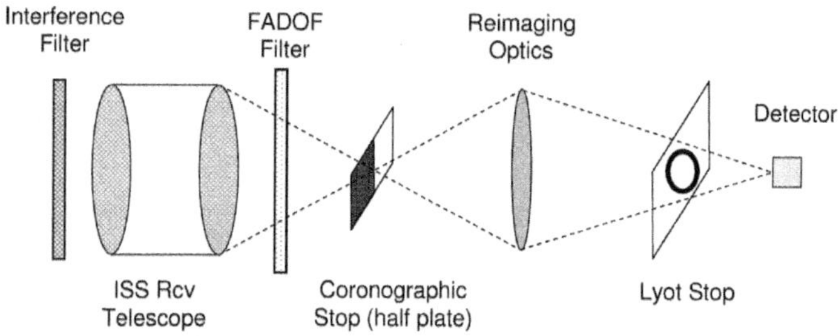

Fig. 25. LATOR coronograph system.

5.4 LATOR Coronograph

To have adequate rejection of the solar background surrounding the laser uplink from Earth, the spacecraft optical system must include a coronagraph. Figure 25 shows a schematic of the coronagraph. A 20 cm telescope forms an image on the chronographic stop. This stop consists of a knife-edge mask placed 6 arcsec beyond the solar limb. The transmitted light is then reimaged onto a Lyot stop, which transmits 88% of the incident intensity. Finally, the light is reimaged onto the tracking detector.

The results of a simulated coronograph showing the stray light rejection as a function of the distance from the solar limb is shown in Fig. 26. The solar surface has been approximated as a vertical edge extending along the entire length of a 256 × 256 array. The upper curve shows the stray light levels for an optical system without a coronograph. In this case, the flux from the solar surface has only been decreased by a factor of 100. With the

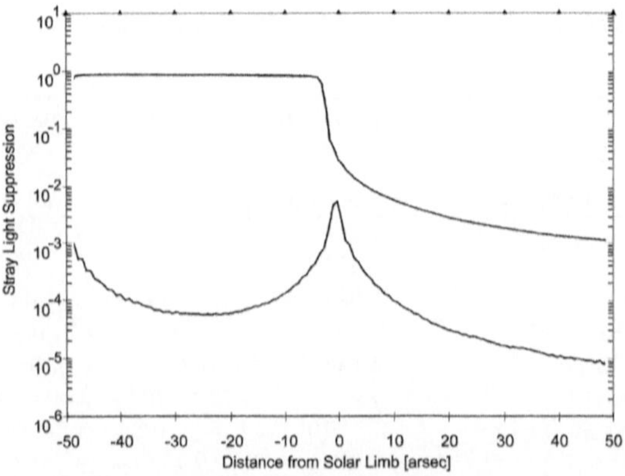

Fig. 26. Results of coronograph performance simulation.

coronograph, however, a further factor of 100 rejection can be achieved. In addition to decreasing the stray solar radiation, the coronograph will decrease the transmission of the laser signal by 78% (for a signal 12 as from limb) due to coronographic transmission and broadening of the PSF. At these levels of solar rejection, it is possible for the spectral filter to reject enough starlight to acquire the laser beacon. It is interesting to note that without the coronograph, the stray light from the Sun decreases proportionally to the distance from the limb. In contrast, with the use of the coronograph, the stray light decreases as the square of the distance from the limb.

5.5 LATOR Observing Sequence

It is important to discuss the sequence of events that will lead to the signal acquisition and that occur during each observation period (i.e., every orbit of the ISS). This sequence will be initiated at the beginning of the experiment period, after ISS emergence from the Earth's shadow (see Fig. 7). It assumes that boresighting of the spacecraft attitude with the spacecraft transmitters and receivers have already been accomplished. This sequence of operations is focused on establishing the ISS-to-spacecraft link. The interspacecraft link is assumed to be continuously established after final deployment (at $\sim 15°$ off the Sun), since the spacecraft never lose LOS with one another.

The laser beacon transmitter at the ISS is expanded to have a beam divergence of 10 as to guarantee illumination of the LATOR spacecraft (see Table 4). After reemerging from the Earth's shadow this beam is transmitted to the craft and reaches them in about 18 min. At this point, the LATOR spacecraft acquire the expanded laser beacon signal. In this mode, a signal-to-noise ratio (SNR) of 23.2 can be achieved with 60 s of integration. With an attitude knowledge of 10 as and an array field of view of 30 as no spiral search is necessary. Upon signal acquisition, the receiver mirror on the spacecraft

Table 4. Analysis of various links between ISS and spacecraft in observation and acquisition modes.

	Spacecraft-to-ISS link spacecraft=Xmit, ISS=Rcv		ISS-to-spacecraft link ISS=Xmit, spacecraft=Rcv	
Optical link parameters	Acquisition	Observation	Acquisition	Observation
Transmitter parameters				
Laser power (W)	1	1	1	1
Wavelength (μm)	1.064	1.064	1.064	1.064
Xmit telescope diameter (m)	0.2	0.2	0.3	0.3
Beam divergence (μrad)	10 as	5.32	10 as	3.55
Distance ($L = 2\,\mathrm{AU}$) (m)	3.00×10^{11}	3.00×10^{11}	3.00×10^{11}	3.00×10^{11}
Footprint diameter at L (m)	1.45×10^{7}	1.59×10^{6}	1.45×10^{7}	1.06×10^{6}
Optics efficiency	0.7	0.7	0.7	0.7
Pointing efficiency	0.9	0.9	0.9	0.9
Receiver parameters				
Rcv telescope diameter (m)	0.3	0.3	0.2	0.2
Rcv optics efficiency	0.7	0.7	0.7	0.7
Detector quantum efficiency	0.9	0.9	0.9	0.9
Power received (W)	1.70×10^{-16}	1.41×10^{-14}	7.55×10^{-17}	1.41×10^{-14}
Photon flux received ($\mathrm{ph\,s^{-1}}$)	9.10×10^{2}	7.55×10^{4}	4.04×10^{2}	7.55×10^{4}
Solar background parameters				
Solar irradiance ($\mathrm{ph\,s^{-1}\,m^{-2}\,sr^{-1}\,\mu m^{-1}}$)	4.64×10^{25}	4.64×10^{25}	4.64×10^{25}	4.64×10^{25}
Rcv detector size (μrad)	3.55	3.55	5.32	5.32
Heterodyne spectral bandwidth	300 MHz	1 MHz	–	–
Narrow band-pass filter (μm)	–	–	5×10^{-5}	5×10^{-5}
Coronograph efficiency	1×10^{-5}	1×10^{-5}	1×10^{-5}	1×10^{-5}
Solar photon flux received ($\mathrm{ph\,s^{-1}}$)	2.95×10^{2}	9.82×10^{-1}	1.30×10^{4}	1.30×10^{4}
Signal-to-noise ratio	22.5	24.7	23.2	21.4
Integration time	1 s	10 ms	60 s	10 ms

will center the signal and use only the center quad array for pointing control. Transition from acquisition to tracking should take about 1 min. Because of the weak uplink intensity, at this point, tracking of the ISS station is done at a very low bandwidth. The pointing information is fed-forward to the spacecraft transmitter pointing system and the transmitter is turned on. The signal is then retransmitted down to the ISS with a light-travel time of 18 min.

Each interferometer station and laser beacon station searches for the spacecraft laser signal. In acquisition mode, the return is heterodyned by using an expanded bandwidth of 300 MHz to assure capture of the laser frequency. In this case, the solar background is the dominant source of noise, and an SNR of 22.5 is achieved with 1 s integration. Because of the small field of view of the array, a spiral search will take 30 s to cover a 30 as field. Upon acquisition, the signal will be centered on the quad cell portion of the array and the local oscillator frequency locked to the spacecraft signal. The frequency band will

Table 5. Spacecraft-to-spacecraft link parameters.

Transmitter parameters	
Laser power (W)	0.2
Wavelength (μm)	0.780
Xmit telescope diameter (m)	0.05
Beam divergence (μrad)	15.6
Distance (r, 0.5° at 2 AU) (m)	2.61×10^9
Footprint at r (m)	4.07×10^4
Optics efficiency	0.7
Pointing efficiency	0.9
Receiver parameters	
Rcv telescope diameter (m)	0.05
Rcv optics efficiency	0.7
Detector quantum efficiency	0.9
Power at detector (W)	1.20×10^{-13}
Photon flux received (ph s^{-1})	4.70×10^5
Signal-to-noise ratio	30.1
Integration time	10 ms

then be narrowed to 1 MHz and the local oscillator frequency locked to the download laser. In this regime, the solar background is no longer the dominant noise source and an SNR of 24.7 can be achieved in only 10 ms of integration. The laser beacon transmitter will then narrow its beam to be diffraction limited (~1 as) and to point toward the LATOR spacecraft. This completes the signal acquisition phase, and the entire architecture is in-lock and transmits scientific signal. This procedure is reestablished during each 92-min orbit of the ISS.

The interspacecraft optical link budget is given in Table 5. In this case, the Sun is not contributing to the signal-to-noise analysis, one has to account for the detector's noise contribution only.

5.6 Factors Affecting SNR Analysis

In conducting the signal-to-noise analysis we pay significant attention to several important factors. In particular, we estimate what fraction of the transmitted signal power is captured by the 20 cm receiver aperture and analyze the effect of the Gaussian beam divergence (estimated at ~7 μrad) of the 30 cm transmit aperture on the ISS. Given the fact that the distance between the transmitter and receiver is on the order of 2 AU, the amount captured is about 2.3×10^{-10} of the transmitted power.

We also consider the amount of solar disk radiation scattered into the two receiver focal planes. In particular, the surface contamination, coating defects, optical roughness, and substrate defects could scatter as much as 1×10^{-4} or more (possibly 1×10^{-3}) of the solar energy incident on the receive aperture into the field of view. These issues are being considered in our current analysis. We also study the amount of the solar corona spectrum within the receive field of view that is not blocked by the narrow band-pass filter. The factors we consider is the filter's FWHM band-pass is 0.05 nm, the filter will have 4.0 optical density blocking outside the 0.05 nm filter band-pass from the X-ray region of 1,200 nm; the filter efficiency within the band-pass will be about 35%, and the detector is probably sensitive from 300 to 1,200 nm.

Additionally, we consider the amount of out-of-field solar radiation scattered into the focal plane by the optical housing. This issue needs to be investigated in a stray light analysis, which can be used to optimize the baffle design to minimize the stray light at the focal plane. Finally, we study the effectiveness of the baffle design in suppressing stray light at the focal plane. Thus, in addition to the stray light analysis, the effectiveness of the final baffle design should be verified by building an engineering model that can be tested for stray light.

Our recent conceptual design and a CODEV ray-trace analysis met all the configuration and performance requirements (shown in Table 3). The ray-trace performance of the resulted instrument is diffraction limited in both the APD and CCD channels over the specified field of view at 1,064 nm. The design incorporated the required field stop and Layot stop. A preliminary baffle design has been developed for controlling the stray light. In the near future, we plan to perform a stray light analysis, which should be performed to optimize the baffle design and calculate the amount of stray light that could be present at each of the two focal planes. This stray light analysis will take into account the optical smoothness of the band-pass filter and primary mirror surfaces. Narrow angle scattering may be a problem at the two focal planes in the filter and primary mirror are not optically very smooth and, thus, it requires a more detailed study. Finally, a rigorous signal-to-noise analysis will be performed to validate the power required to achieve a high signal-to-noise ratio in detecting received beam signal in the presence of the expected focal beam stray light predicted by the stray light analysis and the engineering model stray light tests.

In Sect. 6 we will consider the modeling of the LATOR observables and will discuss the logic of its measurements.

6 LATOR Preliminary Observational Model

The goal of measuring deflection of light in solar gravity with accuracy of one part in 10^9 requires serious consideration of systematic errors. One would have to properly identify the entire set of factors that may influence the mission accuracy at this level. Fortunately, we initiated this process aided by previous experience in the development of a number of instruments that require similar technology and a comparable level of accuracy [101], notably Space Interferometry Mission, Keck and Palomar Testbed Interferometers. This experience comes with understanding various constituents of the error budget, expertise in developing appropriate instrument models; it is also supported by the extensive verification of the expected performance with instrumental testbeds and existing flight hardware. Details of the LATOR error budget are being developed and will be published elsewhere, when fully analyzed. Recent covariance studies confirmed the expected mission performance and emphasized the significant potential of the mission [73, 79].

In this section we will discuss the LATOR observables, based on a simplified model that will be used to introduce the observational logic of this experiment. We first discuss the model for the relativistic delay of the laser signals as they transit between the nodes of the LATOR's light triangle. We then will introduce the model for differential astrometric interferometry to be implemented for the mission.

6.1 Relativistic Light–Time Model

In development of the mission's error budget we use a simple model to capture all error sources and their individual impact on the mission performance [101]. The first step into a relativistic modeling of the light path consists of determining the direction of the incoming photon as measured by an observer as a function of the barycentric coordinate position of the light source.

From a geometrical point of view, the Sun, Earth, and other planets each curve space–time in their vicinity to varying degrees. The effect of this curvature is the increase of the round-trip travel time of a laser pulse. Effects of the gravitational monopole on light propagation are the largest among those in the solar system. To the first order in gravitational constant, the one-way relativistic light–time expression was derived in heliocentric form by Shapiro [92]; in its most general form it was given by Tausner [96] and independently by Holdridge [42]. It was formulated in expanded solar system barycentric form and incorporated into JPL orbit determination software by Moyer [57, 58].

The portion of the Moyer's formulation due to the Sun and Earth is

$$(t_j - t_i) = \frac{r_{ij}^B}{c} + (1+\gamma)\left[\frac{\mu_\odot}{c^3}\ln\left(\frac{r_i^S + r_j^S + r_{ij}^S + (1+\gamma)\frac{\mu_S}{c^3}}{r_i^S + r_j^S - r_{ij}^S + (1+\gamma)\frac{\mu_\odot}{c^2}}\right)\right.$$
$$\left. + \frac{\mu_E}{c^3}\ln\left(\frac{r_i^E + r_j^E + r_{ij}^E}{r_i^E + r_j^E - r_{ij}^E}\right)\right]. \tag{5}$$

The first term on the right is the geometric travel time due to coordinate separation; the remaining two terms represent the gravitational curvature effects due to the Sun and Earth. The complete equation gives the elapsed coordinate time between two photon events, where an event is indicated by the subscript $i, j \in \{1, 3\}$ (with subscript $i = 3$ reserved for the ISS). $\mu_\odot = GM_{\mathrm{S}}$ and $\mu_E = GM_{\mathrm{E}}$ are the solar and Earth's gravitational constants correspondingly. A Latin superscript denotes the origin of a vector: B is the solar system barycenter, S is the Sun, and E is the Earth. The use of the symbols in the equation is as follows. $r_i^{\rm S} = |\mathbf{r}_i^{\rm S}|$ is the magnitude of the vector from the Sun to photon event i transmission (or reception) at coordinate time t_i. $\mathbf{r}_{ij} = \mathbf{r}_j^{\rm S} - \mathbf{r}_i^{\rm S}$ is the vector and $r_{ij} = |\mathbf{r}_{ij}|$ is the magnitude of the difference between the vector from the Sun to photon event j at time t_j and the vector from the Sun to photon event i at time t_i.

When the ray path is near the solar limb, (5) is greatly simplified taking the following standard [109] form

$$(t_j - t_i) = \frac{r_{ij}}{c} + (1+\gamma)\frac{\mu_\odot}{c^3}\ln\left(\frac{4r_ir_j}{p^2}\right). \tag{6}$$

For the shortest arm of the triangle, ℓ_{12}, the solar impact parameter is comparable to the distances involved, thus the relativistic delay in this arm is very small and, for the purposes of this chapter, it can be neglected; thus, we can write $(t_2 - t_1) \simeq r_{12}/c$, which is sufficient for our purposes here.

The obtained equations may be used to model the light paths, $\ell_{ij} = c(t_j - t_i)$, for the signals to transit between all three vortices of the LATOR triangle, as shown in Fig. 2. Indeed, one can write the following approximate expressions:

$$\ell_{3j} = r_{3j} + (1+\gamma)\frac{\mu_\odot}{c^2}\ln\left(\frac{4r_jr_3}{p_j^2}\right), \qquad \ell_{12} = r_{12}. \tag{7}$$

Another observable that will be available to LATOR is the range-rate, $\dot{\ell}_{ij}$ that essentially is a time-derivative of $\ell_{ij}(p_j(t))$. The accuracy of the range-rate time series is expected to produce even more accurate results, similar to the situation with recent Cassini experiment [5, 11]. The impact of the high-accuracy range-rate data on the final accuracy of the experiment is being investigated and results will be reported elsewhere.

A more rigorous analysis that includes effects of the order G^2v/c had been initiated at JPL. The results of this analysis and the corresponding covariance studies will be reported elsewhere. The recent covariance studies already show very interesting results [73, 79].

6.2 Interferometric Delay Model

Because a light signal propagating in a gravitational potential is retarded relative to its travel time in an field-free space, as predicted by general relativity, the computed value for the differential time of arrival of the signals at two telescopes forming an interferometer must be corrected for gravitational effects [87]. The LATOR interferometer is highly sensitive to these effects of gravity on light propagation. In this section we will derive the expression for the contribution of the relativistic gravity to the optical path difference (OPD) measured by an interferometer in solar orbit.

Let us define $\mathbf{r}_{3(i)}(t_i)$ with $i = 1, 2$ to be the barycentric positions of the two telescopes of the interferometer, such that $\mathbf{b} = \mathbf{r}_{3(2)} - \mathbf{r}_{3(1)}$ is the interferometer's baseline. By keeping only the terms to first order in gravitational constant G, the expression (5) can be used to derive the OPD $d_j = \ell_{j3}(\mathbf{r}_3 + \mathbf{b}) - \ell_{j3}(\mathbf{r}_3) = c(t_{3(2)} - t_j) - (t_{3(1)} - t_j)$ registered by the interferometer for the light received from the jth source:

$$d_j = r_{j3(2)} - r_{j3(1)} + (1+\gamma)\frac{\mu_\odot}{c^2} \ln\Big[\Big(\frac{r_j + r_{3(2)} + r_{j3(2)}}{r_j + r_{3(1)} + r_{j3(1)}}\Big)\Big(\frac{r_j + r_{3(1)} - r_{j3(1)}}{r_j + r_{3(2)} - r_{j3(2)}}\Big)\Big]. \tag{8}$$

This is the required correction to coordinate time delay due to the solar gravitational monopole to the time of arrival of the light sent by jth spacecraft and received by (i)th telescope.

Equation (8) is the differenced Shapiro time delay for the two telescopes separated by $\mathbf{b}$; it is appropriate for the most general geometry, in which $r_j \approx r_{3(i)} \approx r_{j3(i)}$. For practical purposes, however, $(r_{j3(2)} - r_{j3(1)})/r_{j3(1)}$ is a small quantity that allows further simplification of (8). Using relations $r_{3(2)} \simeq r_{3(1)} + (\mathbf{b}\cdot\hat{\mathbf{r}}_{3(1)})$, $r_{j3(2)} \simeq r_{j3(1)} + (\mathbf{b}\cdot\hat{\mathbf{r}}_{j3(1)})$, and expressing all quantities at $t_3 \equiv t_{3(1)}$, we can present the interferometric delay d_j as follows:

$$d_j \simeq \frac{(\mathbf{b}\cdot\hat{\mathbf{r}}_{j3})}{1 - (\mathbf{v}_3\cdot\hat{\mathbf{r}}_{j3})/c} + (1+\gamma)\frac{\mu_\odot}{c^2}\Big[\frac{r_j + r_3 - r_{j3}}{2r_j r_3}\frac{\mathbf{b}\cdot(\hat{\mathbf{r}}_3 + \hat{\mathbf{r}}_{j3})}{1 + (\hat{\mathbf{r}}_j\cdot\hat{\mathbf{r}}_3)} - \frac{r_j + r_3 + r_{j3}}{2r_j r_3}\frac{\mathbf{b}\cdot(\hat{\mathbf{r}}_3 - \hat{\mathbf{r}}_{j3})}{1 + (\hat{\mathbf{r}}_j\cdot\hat{\mathbf{r}}_3)}\Big], \tag{9}$$

where $\mathbf{r}_{j3} \equiv \mathbf{r}_{j3(1)}$ and $\mathbf{r}_3 \equiv \mathbf{r}_{3(1)}$, and $\mathbf{v}_3$ is the barycentric velocity of the interferometer. The obtained expression (9) is appropriate for the most general geometry; however, when the ray path is near the solar limb and also the corresponding impact parameters p_j are small compare to the barycentric distances to emitter and receiver, $p_j/r_j \ll 1$, $p_j/r_3 \ll 1$, it is further simplified, taking the form:

$$d_j \simeq \frac{bp_j}{r_3}\frac{1}{1 - (\mathbf{v}_3\cdot\hat{\mathbf{r}}_{j3})/c} + (1+\gamma)\frac{2\mu_\odot}{c^2}\frac{b}{p_j}\frac{r_j}{r_3 + r_j}. \tag{10}$$

The interferometric delay (10) is for the geometry when both transmitter and receiver are at finite and comparable distances from the Sun. Note that, if transmitter is located at a distance far greater then that of the receiver, $r_j \gg r_3$, (10) transforms to a typical expression for VLBI observations with sources being at infinity [87, 109]. Therefore, for a typical LATOR geometry with $r_j \approx r_3$, the magnitude of the relativistic delay (and corresponding angle) is approximately twice smaller when compared to the case when the light source is at infinity.

As we discussed in Sect. 4.1, the LATOR interferometer reaches its highest accuracy in differential mode by accurately measuring differential OPD, $d_{12} = d_2 - d_1$, which is given from (10) as

$$d_{12} \simeq \frac{b}{r_3}\left[\frac{p_2}{1-(\mathbf{v}_3\cdot\hat{\mathbf{r}}_{23})/c} - \frac{p_1}{1-(\mathbf{v}_3\cdot\hat{\mathbf{r}}_{13})/c}\right] + (1+\gamma)\frac{2\mu_\odot}{c^2}\left[\frac{b}{p_2}\frac{r_2}{r_3+r_2} - \frac{b}{p_1}\frac{r_1}{r_3+r_1}\right]. \quad (11)$$

Note that interferometric delay rate is another observable that will be available for LATOR. Having determined the expression for delay (11), the delay rate, $\dot{d}_{12}$, is simply given as time-derivative of the delay $d_{12}(p_j(t))$. Currently, we are investigating the impact of this observable on the experiment and will incorporate $\dot{d}_{12}$ in our further studies.

Equation (11) captures the largest terms in the model of LATOR interferometric observations. The entire LATOR model accounts for a significant number of other effects, including those due to gravitational multipoles, second-order deflection, angular momentum contribution, etc. The work to develop a complete mission model work had being initiated; the results will be reported elsewhere.

In Sect. 6.3 we will consider on a conceptual formulation of the LATOR observables.

6.3 Logic of LATOR Observations

In this section, we discuss the observational logic of the LATOR experiment. This is done to only conceptually demonstrate the features of the mission design.

The range observations (7) may be used to measure any angle between the three fiducials in the triangle. However, for observations in the solar gravity field, measuring the lengths do not give you a complete information to determine the angles, and some extra information is needed. This information is the mass of the Sun, and, at least one of the impact parameters. Nevertheless, noting that the paths $\underline{\ell}_{ij}$ correspond to the sides of the connected, but gravitationally distorted triangle, one can write $\underline{\ell}_{12}+\underline{\ell}_{23}+\underline{\ell}_{31}=0$, where $\underline{\ell}_{ij}$ is the null geodesic path for light moving between the two points i and j. This leads to an expression for the angle between the two spacecraft computed from the range measurements; using (7) this quantity may be given as

$$\sin(\angle\underline{\ell}_{31}\underline{\ell}_{32}) \simeq \sin\alpha_3 - (1+\gamma)\frac{\mu_\odot}{c^2}\frac{r_{12}}{r_{31}r_{32}}\left[\cos\alpha_1\ln\frac{4r_3r_1}{p_1^2} + \cos\alpha_2\ln\frac{4r_3r_2}{p_2^2}\right]\cot\alpha_3, \quad (12)$$

where the three Euclidian angles within the LATOR light triangle are computed from the laser ranging measurements of the three arms of the triangle, r_{12}, r_{31}, r_{32} using usual formulae $\cos\alpha_3 = (r_{32}^2+r_{31}^2-r_{12}^2)/(2r_{32}r_{31})$, $\cos\alpha_1 = (r_{31}^2+r_{12}^2-r_{32}^2)/(2r_{31}r_{12})$, and $\cos\alpha_2 = (r_{32}^2+r_{12}^2-r_{31}^2)/(2r_{32}r_{12})$, and $\sin\alpha_3$ is given as below:

$$\sin \alpha_3 = \frac{1}{2r_{31}r_{32}} \Big[2r_{12}^2(r_{31}^2 + r_{32}^2) - (r_{32}^2 - r_{31}^2)^2 - r_{12}^4 \Big]^{1/2}. \qquad (13)$$

LATOR will provide highly accurate time-series of r_{12}, r_{31}, r_{32}, which can be used to simulate measurement of $\sin(\angle \ell_{31} \ell_{32})$. For a typical LATOR configuration, (12) can be approximated as

$$\sin(\angle \ell_{31} \ell_{32}) \simeq \frac{r_{12}}{2r_3} - (1+\gamma) \frac{\mu_\odot}{c^2} \frac{r_{12}}{4r_3^2} \ln \frac{4r_3^2}{p_1 p_2} \approx 0.01745 - \frac{1}{2}(1+\gamma)\, 1.82 \times 10^{-9}, \qquad (14)$$

where we evaluated the expression for a typical set of numerical values suitable for LATOR, with one spacecraft at the solar limb: $r_1 \sim r_2 \sim r_3 \sim 1\,\mathrm{AU}$, $p_1 = R_\odot \equiv 0.2667°$, $\alpha_3 = 1°$, $p_2 = 2r_3 \sin(\angle p_1 + \alpha_3)$, and $r_{12} = 2r_3 \sin \alpha_3$.

Equation (14) has an immediate impact on the logic of the LATOR observations. In particular, it follows from this equation that the smallness of the opening angle α_3 makes it very difficult to measure γ to a sufficient accuracy using only laser ranging observables. Thus, if one desires to achieve accuracy of $\sigma_\gamma \sim 10^{-8}$ in measuring the PPN γ, one needs to know $r_{12} = 5 \times 10^6$ km to accuracy of $\Delta r_{12} \sim 2\,\mu\mathrm{m}$. This ranging accuracy is certainly achievable with modern technologies, but would rely on drag-free spacecraft and very precise clocks. Notably, in the next decade, LISA (i.e., Laser Interferometer Space Antenna) would be able to achieve the accuracy of few pm over the same distance by utilizing a complex scheme of coherent detection of light. However, this is not what LATOR is going to do. The experiment uses time-of-flight laser ranging to measure distances between the nodes of the light triangle. These observations are then used to directly compute the opening angle $\angle \ell_{31} \ell_{32}$ for further input in the astrometric observations. Below we shall discuss the interferometric component of the mission model.

Differential astrometric observations (11) will be used to obtain another measurement of the same angle $\angle \ell_{31} \ell_{32}$ between the two spacecraft. The LATOR interferometer will perform differential observations between the two sources of laser light, measuring the differential delay $d_{12} = d_2 - d_1$ to the high accuracy. Note that, for a typical LATOR configuration, $r_1 \sim r_2 \sim r_3$ and $p_2 \simeq p_1 + r_3 \sin \alpha_3$ and one can benefit from the following approximate relation:

$$\sin(\angle \ell_{31} \ell_{32}) \simeq \sin \alpha_3 \simeq \frac{p_2 - p_1}{r_3} \simeq \frac{r_{12}}{2r_3}. \qquad (15)$$

This is exactly the angle that will be measured interferometrically. However, the obtained relation (15) may be used to provide a direct geometric measurement of the opening angle $\angle \ell_{31} \ell_{32}$ to the required accuracy, which would aid the LATOR astrometric interferometry.

We will use result (15) to further simplify the expression for the differential delay (11), which for a typical LATOR geometry and circular motion of the interferometer takes the following form:

$$d_{12} \simeq b \left[\frac{r_{12}}{2r_3} - \frac{v_3}{c} \frac{p_3^2 - p_1^2}{r_3^2} + (1+\gamma) \frac{\mu_\odot}{c^2} \left(\frac{1}{p_2} - \frac{1}{p_1} \right) \right]. \qquad (16)$$

The LATOR interferometer is designed to provide highly accurate time-series of the interferometric delay d_{12}, while laser ranging will determine the light-travel times between the nodes of the triangle, $\Delta t_{ij} = t_j - t_i$ (that will be used to compute $r_{ij} = c\Delta t_{ij}$). Evaluating expression (16) for a typical LATOR configuration with one spacecraft is at the solar limb and $v_3 = 30\,\mathrm{km\,s^{-1}}$, one finds the characteristic sizes of the effects:

$$d_{12} \simeq \Big(1.745 - 4.675 \times 10^{-6} - \frac{1}{2}(1+\gamma)\; 3.352 \times 10^{-4}\Big)\ \mathrm{m}. \qquad (17)$$

Optical interferometry is a mature technology that may achieve very high accuracy in measuring the quantities involved in (17). Thus, a measurement of the delay d_{12} with uncertainty of 5 pm results in the accuracy in the parameter γ of $\Delta\gamma = 2 \times 10^{-8}$. Currently we are able to measure delays on large interferometric baselines with accuracy ~1 pm, which certainly benefit LATOR by satisfying its primary objective.

As is evident from Fig. 2, the key element of the LATOR experiment is a redundant-geometry optical truss to measure the effects of gravity on the laser signal trajectories. LATOR will generate four time series of measurements: one for the optical range of each side of the triangle, plus the angle between light signals arriving at one vertex of the light triangle. Within the context of a moving Euclidean light triangle, these measurements are redundant. From a combination of these four times series of data, the several effects of gravity on the light propagations can be precisely and separately determined. For example, the first and second-order gravity monopole deflections go as p^{-1} and p^{-2} while the solar quadrupole deflection goes as p^{-3}, with $p(t)$ being a laser signal's evolving impact parameter; the quadrupole moment's deflection has further latitude dependence if spacecraft lines-of-sight are so located.

The data will be taken over periods in which the laser light's impact parameters $p(t)$ vary from one to ten solar radii, producing time signatures in the data, which permits both the separation of the several gravitational effects and the determination of key spacecraft location coordinates needed to calibrate the deflection signals. In our mission simulations we use the complete set of LATOR observables, including range and range-rates and also delay and delay rates. These additional data types account for the temporal evolution of the entire LATOR light triangle and further improve the mission accuracy. To demonstrate our design considerations in the discussion below, we will use only equations for range and interferometric delay given by (7) and (11) correspondingly.

In Sect. 7 we will discuss the expected performance of the experiment and will present criteria for the mission design.

7 LATOR Astrometric Performance

It is convenient to present error sources in three broad categories (1) the ones that are related to mission architecture, (2) those that are external to the triangle and have an astrophysical origin, and (3) those that originated within the instrument itself. Typical mission-related errors are those that result from the uncertainties in the orbits of the spacecraft and the ISS, chosen mission design and observing scenario and, in general, those errors that result from the geometry of the experiment and affect the range and angle determination. The astrophysical errors are those that are external to the instrument and are due to various phenomena that influence both mission planning and observing scenario. Such errors are due to the gravity effects of planets and largest asteroids, unmodeled motion of the fiducial stars, optical properties of the Sun and solar corona near the limb, etc. Example of the instrumental errors include effects of the long-term laser stability, errors in pointing of the laser beams, instrumental drifts, and other systematic and random errors originating within the instrument itself. (By instrument we understand the experimental hardware situated at all three vortices of the triangle.)

We consider $\varDelta\gamma = 2 \times 10^{-8}$ to be the accuracy of determining the PPN parameter γ in a single measurement. This design accuracy drives the flow-down of mission requirements that we will discuss in this section. Thus, based on (16), such a design accuracy results in the following requirement on the accuracy of the OPD measurements:

$$\varDelta d_{12} \simeq \varDelta\gamma \, \frac{\mu_\odot}{c^2} \frac{b(p_2 - p_1)}{p_1 p_2} = 5\,\mathrm{pm} \left(\frac{\varDelta\gamma}{2 \times 10^{-8}}\right)\left(\frac{b}{100\ \mathrm{m}}\right)\left(\frac{R_\odot}{p_1}\right)\left(1 - \frac{p_1}{p_2}\right). \tag{18}$$

Therefore, in our design considerations we take $\Delta d_{12} = 5\,\mathrm{pm}$ to be the target accuracy for the interferometric measurements on the LATOR's 100 m baseline.

Here we discuss a preliminary astrometric error budget for the LATOR experiment and present design considerations that enable the desirable instrument performance. (A more detailed model to the second order in gravitational effects is available and is being used in simulations to verify the expected mission performance.) Our goal for this section is to present a comparative analysis of the accuracy that mission needs to achieve to satisfy its science requirements.

7.1 Trajectory Measurement Accuracy

In this section we will discuss the constituents of LATOR error budget for the measurement of the angular deflection light. The error budget is subdivided into three components – range and interferometer measurements, and spacecraft stability, which are described in this section.

Range Measurement

This component describes the angular errors due to uncertainties in the distance between the spacecraft and the ISS as determined by laser ranging. The angular uncertainty due to an interspacecraft ranging error, $\varDelta r_{12}$, is

$$\begin{aligned} \Delta r_{12} &\simeq \Delta\gamma \frac{\mu_\odot}{c^2} \frac{2r_3(p_2 - p_1)}{p_1 p_2} \\ &= 1 \text{ cm} \left(\frac{\Delta\gamma}{2\times 10^{-8}}\right)\left(\frac{r_3}{\text{AU}}\right)\left(\frac{R_\odot}{p_1}\right)\left(1 - \frac{p_1}{p_2}\right). \end{aligned} \tag{19}$$

Therefore, the experiment will require a spacecraft-to-spacecraft laser ranging accuracy of 1 cm, which for spacecraft separated by 1° at 2 AU distance, results in an angular error of 0.03 prad (i.e., corresponding to a delay uncertainty of $\sigma_d = 3$ pm).

The angular error due to an ISS-to-spacecraft ranging error is given by

$$\begin{aligned} \Delta r_{31} &\simeq \Delta\gamma \frac{\mu_\odot}{c^2} \frac{4r_3^2}{r_{12}} \frac{(p_2 - p_1)}{p_1 p_2} \\ &= 60 \text{ cm} \left(\frac{\Delta\gamma}{2\times 10^{-8}}\right)\left(\frac{\sin 1^\circ}{\sin\alpha_3}\right)\left(\frac{r_3}{\text{AU}}\right)\left(\frac{R_\odot}{p_1}\right)\left(1 - \frac{p_1}{p_2}\right). \end{aligned} \tag{20}$$

Therefore, we have allocated a 60 cm range uncertainty for each of the two ISS-to-spacecraft laser links, resulting in an angular uncertainty of 0.035 prad (i.e., $\sigma_\mathrm{d} = 3.5$ pm). The total error budget for the laser ranging distance measurements is 4.6 pm.

Knowledge of the Baseline

Experiment uncertainties in the ISS interferometer measurement contribute additional terms to the overall error budget. The baseline design for the instrument is a 100 m baseline with a 10 s integration time. Using (16), one obtains the requirements on the accuracy of the baseline estimation:

$$\begin{aligned} \Delta b &\simeq \Delta\gamma \frac{\mu_\odot}{c^2} \frac{2r_3}{r_{12}} \frac{b(p_2 - p_1)}{p_1 p_2} \\ &= 0.2 \text{ nm} \left(\frac{\Delta\gamma}{2\times 10^{-8}}\right)\left(\frac{\sin 1^\circ}{\sin\alpha_3}\right)\left(\frac{b}{100 \text{ m}}\right)\left(\frac{R_\odot}{p_1}\right)\left(1 - \frac{p_1}{p_2}\right). \end{aligned} \tag{21}$$

Based on theoretical predictions for narrow angle measurements, we require an angular error of 0.025 prad (i.e., $\sigma_\mathrm{d} = 2.5$ pm), limited by long-term instrument systematics.

The current requirement for systematic errors in the instrument has been set at 0.05 prad. This corresponds to measurement of the laser fringe phase to one part in 2.9×10^5 ($\lambda = 1.064\,\mu$m, $b = 100$ m). This term includes errors in the metrology and fringe detection of the interferometer, as well as the effect of photon noise.

Orbit Stability of the Spacecraft

To determine the first and second-order terms of gravitational deflection, LATOR will make a number of measurements at different spacecraft separations and various impact parameters. During the period between measurements, it is assumed that the impact parameter is known. An error in this assumption will cause an equivalent error in the computation of the deflection term. Thus, it is important to set a requirement on the knowledge of the impact parameter, which is given by

$$\Delta p_1 \simeq \Delta\gamma \frac{1}{2}\frac{p_1 p_2}{p_1 + p_2} = 5.75 \text{ m} \left(\frac{\Delta\gamma}{2\times 10^{-8}}\right)\left(\frac{p_1}{R_\odot}\right)\left(\frac{1}{1+p_1/p_2}\right). \tag{22}$$

This is a highly requirement if one conducts only a static measurement of the light deflection. In our case, the delay-rate observable and smooth motion of the spacecraft significantly reduce the sensitivity of the experiment on the absolute knowledge of the impact parameter $p_1 = p_1(t_0) + \dot{p}_j(t - t_0)$. In fact, LATOR will have a very good orbit determined by the combination of the laser ranging and conventional radiometric navigation, which will provide $\dot{p}_j$ to a high accuracy. This allows that one can solve for the initial impact parameter $p_1(t_0)$ in the numerical analysis.

Therefore, one might require that the spacecraft be stable to 4 µas, which corresponds to a drift in the transverse distance of 5.8 m and results in an angular error of 0.04 prad.

Similarly, the separation between the spacecraft should also be stable to keep the knowledge of the impact parameters. This is quantified by the allowable uncertainty in the difference between impact parameters $\delta p = p_2 - p_1$ that is given by

$$\Delta\delta p \simeq \Delta\gamma \frac{p_2}{2p_1}(p_2 - p_1) = 124.0 \text{ m} \left(\frac{\Delta\gamma}{2\times 10^{-8}}\right)\left(\frac{p_2}{p_1}\right)\left(\frac{r_3}{\text{AU}}\right)\left(\frac{\sin\alpha_3}{\sin 1^\circ}\right). \tag{23}$$

Therefore, we require that the spacecraft be stable to 0.05 mas, which results in an angular error of 0.035 prad (i.e., $\sigma_\text{d} = 3.5$ pm of corresponding delay uncertainty).

Orbit Stability of the ISS

In addition to the spacecraft error, the ISS's orbital error will also produce contribution to the angular measurement. Most of the errors on the ISS can be made common-mode; therefore, their influence on the differential astrometry with LATOR interferometer will be either negligible or it will be small and well modeled. However, there are some errors that would still produce measurable contribution to the differential delay, if not properly addressed; notably, the accuracy of the ISS orbit. The current mission design calls for an enhancement of the ISS orbit solution by utilizing GPS receivers at the location of each optical (see Sect. 5.5). This will also help to address the issue of the extended structure low-frequency vibrations of the ISS. As we mentioned earlier, the effect of these vibrations will be addressed by using µg level accelerometers, which will be integrated within both optical packages on the ISS. A combination of the GPS receivers and µg accelerometers will provide information needed to improve the ISS attitude information; this improvement will be done for each corner cube fiducial (needed for the interferometric baseline determination).

Our current error budget for the differential observations with the LATOR interferometer allocates ∼2.7 pm of error in 100 s of integration for the uncertainty in the ISS orbit, its attitude, and the extended structure vibrations.

7.2 Mission Errors

Although one could in principle setup complicated engineering models to predict all or each of the systematics, often the uncertainty of the models is too large to make them useful, despite the significant effort required. A different approach is to accept our ignorance about a nongravitational acceleration and assess to what extent these can be assumed a constant bias over the timescale of all or part of the mission. (In fact, a constant acceleration produces a linear frequency drift that can be accounted for in the data analysis by a single unknown parameter.) In fact, we will use both approaches.

In most orbit determination programs some effects, like the solar radiation pressure, are included in the set of routinely estimated parameters. Nevertheless, we want to demonstrate their influence on LATOR's navigation from the general physics standpoint. This is not only to validate our results, but also to be a model as to how to study the influence of the other physical phenomena that are not yet included in the standard navigational packages for future more demanding missions. Such missions will involve either spacecraft that will be distant or spacecraft at shorter distances where high-precision spacecraft navigation will be required.

In the current design, the LATOR experiment requires that the location of one of the spacecraft with respect to the Sun is known to be within 20 m over the duration of the each observing session or ∼92 min. The major perturbation to the spacecraft trajectory is from local spacecraft disturbances, such as gas leaks for thruster valves and solar radiation pressure. The spacecraft can be designed to eliminate spacecraft errors leaving solar radiation pressure as the major source for the position noise. Other disturbances such as solar wind, magnetic fields, cosmic rays, etc. have been identified and are at least three orders of magnitude lower than solar radiation pressure.

In this section we will discuss possible systematics generated external to the spacecraft, which might affect the LATOR's mission accuracy.

Direct Solar Radiation Pressure

There is an exchange of momentum when solar photons impact the spacecraft and are either absorbed or reflected. Models for this solar pressure effect are usually developed before a mission is launched. The models take into account various parts of the spacecraft exposed to solar radiation; they compute the acceleration directed away from the Sun as a function of spacecraft orientation and solar distance

$$a_{\mathrm{s.p.}}(r) = \frac{2 f_\odot A}{c\, m} \frac{\cos\theta(r)}{r^2}, \qquad (24)$$

where $f_{\odot} = 1,367\,\mathrm{W\,m^{-2}\,(AU)^{-2}}$ is the (effective-temperature Stefan–Boltzmann) "solar radiation constant" at 1 AU from the Sun and A is the effective size of the craft as seen by the Sun. θ is the angle between the axis of the antenna and the direction of the Sun, c is the speed of light, M is the mass of the spacecraft, and r is the distance from the Sun to the spacecraft in AU. For expected spacecraft values of $A = 1.0\,\mathrm{m}^2$ and $m = 150\,\mathrm{kg}$, gives an acceleration of $a_{\mathrm{s.p.}} \simeq 6.1 \times 10^{-8}\,\mathrm{m\,s^{-2}}$ at $r = 1\,\mathrm{AU}$ from the Sun.

This acceleration will produce an unmodeled force, which ultimately may result in the error in the radial position of the spacecraft. Over a time t this error is $\delta r = \frac{1}{2}\,\delta a\,t^2$, where δa is the unmodeled acceleration. In turn, this error would lead to a transverse position error of $\delta x = \frac{1}{4}\,\delta a\,n\,t^3$, where n is the spacecraft velocity about the Sun, $n \sim 2\times10^{-7}\,\mathrm{rad\,s^{-1}}$. If the effect of solar radiation pressure were completely unmodeled, over a period of 21 days, the transverse position error due to solar pressure would be ~18.2 km, potentially resulting in a 61 nrad astrometric error. (This is one of the reasons to consider a drag-free spacecraft for the experiment as suggested in Plowman and Hellings [79].) However, if one conducts laser ranging with position knowledge of 60 cm, the transverse position uncertainty over a period of 21 day would only be ~10 cm. Consequently, it is necessary to use the laser ranging information to predict the transverse position of the spacecraft.

The laser ranging information will be used to solve for the slowly varying changes in the solar pressure, leaving the random fluctuations of the solar pressure as the dominant source of position error. Figure 27 shows one realization of the position error of the spacecraft. In this simulation, random fluctuations correspond to 1% RMS of the total solar radiation pressure, the spacecraft wanders ~1 m in the radial direction and 8 cm in the transverse direction in a day. The use of a redundant optical truss offers an excellent

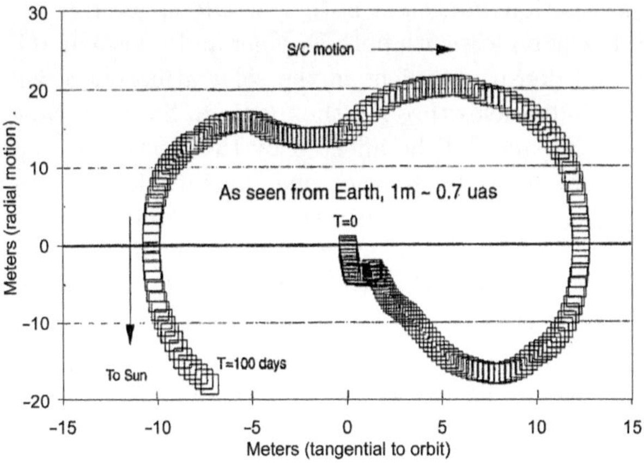

Fig. 27. Simulation of spacecraft motion due to random solar pressure fluctuations.

alternative to an ultra-precise orbit determination. This feature also makes LATOR insensitive to spacecraft buffeting from solar wind and solar radiation pressure. This is why, as opposed to other gravitational missions, there is no need for a drag-free spacecraft to enable the high accuracy of the LATOR experiment. (The drag-free option was studied in [79], resulting in several interesting design considerations that will be further investigated.)

Effect of the Solar Corona

The electron density and density gradient in the solar atmosphere influence the propagation of light and radiowaves through the medium. In fact, radiometric observables will experience effect of the electron density in the solar corona [14]; however, effect on laser ranging and astrometric observations will be significantly less. The use of optical wavelengths offer a significant advantage for the spacecraft communication in the solar system as opposed to the microwave radiation – the current navigation standard. Such a choice makes the deep-space communication effectively free from the solar corona noise. Indeed, the solar plasma effects on wave propagation decrease as λ^2 and there is a factor of 10^{10} reduction in the solar plasma optical path fluctuations by simply moving from the S-band microwave signal $\lambda = 10\,\mathrm{cm}$ ($f = 3\,\mathrm{GHz}$) to the optical wavelengths of $\lambda \sim 1\,\mu\mathrm{m}$ ($f = 300\,\mathrm{THz}$). This 10^{10} reduction of the dispersive media effects offers tremendous gain in the quality of both spacecraft navigation (increased pointing precision and timing) and deep-space communication (very high data transmission rates). LATOR will utilize design capable of rejecting background solar noise in combination with optical wavelengths for precision navigation; this combination will lead to a significant reduction of the solar corona effect, making its contribution harmless to the mission.

For the navigation purposes, both LATOR spacecraft will be equipped with X-band transponders with both Doppler and range capabilities. The electron density and density gradient in the solar atmosphere will influence the propagation of radiowaves through the medium. So, both range and Doppler observations at X-band will be affected by the electron density in the interplanetary medium and outer solar corona. This would result in the spacecraft not being able to communicate with the ground when the impact parameter will be less than $\sim 2.5R_\odot$. This is why the current mission plan includes provision that the radio-navigation will not be conducted for the solar impact parameters smaller than $2.5R_\odot$. Most of the important navigational, instrumental, and experimental information will be stored onboard until the time of clear communication link with the NASA Deep-Space Network. This is when the mission will step-up to its full potential by utilizing its optical communication capabilities from the extreme solar environment to enable high-precision navigation of the spacecraft.

Earth's Orbit Velocity

The knowledge of the Earth's orbital velocity, v_{E}, puts requirements on the correction for the stellar aberration. The error associated with the accuracy of knowledge of the Earth's motion is given as

$$\Delta v_E \simeq c\Delta\gamma \frac{\mu_\odot}{c^2} \frac{r_3^2}{p_1 p_2 (p_2 + p_1)}$$
$$= 2.15\,\mathrm{cm\,s^{-1}} \Big(\frac{\Delta\gamma}{2\times 10^{-8}}\Big)\Big(\frac{r_3}{\mathrm{AU}}\Big)^2\Big(\frac{R_\odot}{p_1}\Big)^3 \Big[\frac{1}{(1+p_2/p_1)p_2/p_1}\Big]. \quad (25)$$

Therefore, the experiment will require a knowledge of the earth's barycentric velocity of $2.2\,\mathrm{cm\,s^{-1}}$, resulting in an angular error of 0.03 prad. Note that a similar uncertainty is tolerable for the ISS's orbit about the Earth.

7.3 Astrophysical Errors

Physical phenomena of an astrophysical origin that are external to the LATOR triangle, but do not affect the mission navigation accuracy, are treated as the sources of astrophysical errors. These errors would nominally influence both the mission planning and observing scenario, they would be due to non-stationary behavior of the gravity field in the solar system (gravity effects of planets and asteroids), unmodeled motion of the fiducial stars, optical properties of the Sun and solar corona near the limb, the properties of the solar surface, etc. We will discuss these sources in some detail.

Knowledge of Solar Interior

Laser ranging between the ISS and spacecraft will be used to measure the orbits of the flight segments with ~1 cm accuracy. This implies that the solar impact parameter should be measured to 8×10^{-9}, a scaling error for the measurement of parameter γ, but an insignificant error for the other measurements.

Along with the impact parameter, other solar parameters such as its mass, angular momentum, and quadrupole moment must be also known (or will be solved for directly from the data). The LATOR instrument may actually be used to gain additional knowledge on the Sun by observing its surface with a Doppler imager. This information may than be used to study the propagation of the sound waves through the solar interior. The resulted data may be used to bootstrap the gravitational solution for the solar oblateness and the higher spherical multipoles of the solar interior. The instrumental implication of this possibility are currently being investigated and, if feasible, it may be included for the mission proposal.

Solar System Gravity

Since the solar system is not static and the spacecraft are in the orbits around the Sun, many large solar system bodies, such as the Sun itself, planets, asteroids, and even the galaxy, would have a significant effect on the measurement of γ at the eighth decimal place. Fortunately, the ephemerides for the solar system objects are known to sufficient accuracy and the motion of the solar system about the galactic center is sufficiently smooth during the 92 min of each observing session. Earth orbit crossing asteroids may cause a significant disturbance if they come within ∼10,000 km of one of the arms of the triangle. The relativity measurement may either have to be delayed or conducted with a slightly higher sampling rate if one of these are nearby.

The change in the first-order relativistic time delay due to other bodies in the solar system has to be known to $\sim 10^{-9}$ of the effect from the Sun. The final observational model would have to account for the effects due to all the major solar system bodies. A similar theoretical and algorithmic work is currently being conducted for both SIM and Gaia astrometric missions and may well be used for this mission [45, 99].

7.4 Instrument Errors

In our design considerations we address two types of instrumental errors, namely the offset and scale errors. Thus, in some cases, when a measured value has a systematic offset of a few pm, there may be instrumental errors that lead to further offset errors. There are many sources of offset (additive) errors caused by imperfect optics or imperfectly aligned optics at a pm level; there are also many sources for scale errors. We take a comfort in the fact that, for the space-based stellar interferometry, we have an ongoing technology program at JPL; not only has this program already demonstrated metrology accurate to a sub-pm level, but it also has identified a number of the error sources and developed methods to either eliminate them or to minimize their effect at the required level.

The second type of error is a scale error. For instance, to measure γ in a single measurement with accuracy of two parts in 10^8, the laser frequency also must be stable at least to 10^{-8} long term; lower accuracy would result in a scale error. The measurement strategy adopted for LATOR would require the laser stability to be only ∼1% to achieve the accuracy needed to measure the second-order gravity effect. Absolute laser frequency must be known to 10^{-9} in order for the scaling error to be negligible. Similarly, robust solutions were developed to address the effects of other known sources of scale errors.

There is a considerable effort currently underway at JPL to evaluate a number of potential errors sources for the LATOR mission, to understand their properties, and to establish methods to mitigate their contributions. (A careful strategy is needed to isolate the instrumental effects of the second order of smallness; however, our experience with SIM [55,56,100] is critical in helping us to properly capture its contribution in the instrument models.) The work is ongoing, this is why the discussion below serves for illustration purposes only. We intend to publish the corresponding analysis and simulations in the subsequent publications.

The final error would have contributions from multiple measurements of the light triangle's four attributes (to enable the redundancy) taken by range and interferometer observations at a series of times. The corresponding errors will be combined with those from the orbital position and velocity coordinate uncertainty. These issues are currently being investigated in the mission covariance analysis; the detailed results of this analysis will be reported elsewhere. However, our current understanding of the expected mission and instrumental accuracies suggests that LATOR will offer a very significant improvement compared to any other available techniques. This conclusion serves as the strongest experimental motivation to conduct the LATOR experiment.

8 Discussion

The LATOR mission aims to carry out a test of the curvature of the solar system's gravity field with an accuracy better than one part in 10^9. In spite of previous space missions exploiting radio waves for tracking the spacecraft, this mission manifests an actual breakthrough in the relativistic gravity experiments as it allows one to take full advantage of the optical techniques that recently became available. The LATOR experiment benefits from a number of advantages over techniques that use radiowaves to study the light propagation in the solar vicinity. The use of monochromatic light enables the observation of the spacecraft almost at the limb of the Sun, as seen from the ISS. The use of narrow band filters, coronagraph optics, and heterodyne detection will suppress background light to a level where the solar background is no longer the dominant noise source. The short wavelength allows much more efficient links with smaller apertures, thereby eliminating the need for a deployable antenna. Advances in optical communications technology allow low bandwidth telecommunications with the LATOR spacecraft, without having to deploy high gain radio antennae needed to communicate through the solar corona. Finally, the use of the ISS not only makes the test affordable, but also allows conducting the experiment above the Earth's atmosphere – the major source of astrometric noise for any ground based interferometer. This fact justifies the placement of LATOR's interferometer node in space.

The concept is technologically sound; the required technologies have been demonstrated as part of the international laser ranging activities and optical interferometry programs at JPL. (i.e., Space Interferometry Mission (SIM) and Keck Interferometer developments. Accuracy of 5 pm was already demonstrated in our SIM-related studies.) The LATOR concept arose from several developments at NASA and JPL that initially enabled optical astrometry and metrology, and also led to developing expertise needed for the precision gravity experiments. Technology that has become available in the last several years such as low cost microspacecraft, medium power highly efficient solid state and fiber lasers, and the development of long-range interferometric techniques make possible an unprecedented factor of 30,000 improvement in this test of general relativity. This mission is unique and is the natural next step in solar system gravity experiments that fully exploit modern technologies.

LATOR uses geometric redundancy of the optical truss to achieve a very precise determination of the interplanetary distances between the two microspacecraft and a beacon station on the ISS. The experiment takes advantage of the existing space-qualified optical technologies, leading to an outstanding performance in a reasonable mission development time. In addition, the issues of the extended structure vibrations on the ISS, interferometric fringe ambiguity, and signal acquisition on the solar backgrounds have all been analyzed, and do not compromise mission goals. The ISS is the default location for the interferometer; however, ground- and free-flying versions have also been studied. While offering programmatic benefits, these options differ in cost, reliability, and performance. The availability of the ISS (via European collaboration) makes presented concept realizable in the near future. A recent JPL Team X study [39] confirmed the feasibility of LATOR as a NASA Medium Explorer (MIDEX) class mission; the current mission concept calls for a launch as early as 2014.

8.1 LATOR vs. Other Gravity Experiments

Tests of fundamental gravitational physics feature prominently among NASA and ESA goals, missions, and programs. Prediction of possible deviation of PPN parameters from the general relativistic values provides a robust theoretical paradigm and constructive guidance for experiments that would push beyond the present empirical upper bound on the PPN parameter γ of $\gamma - 1 = (2.1 \pm 2.3) \times 10^{-5}$ obtained by recent conjunction experiments with Cassini spacecraft [11].[5] Among the future missions that will study the nature of gravity, we discuss here the missions most relevant to LATOR science:

- Configuration similar to the geometry of the Cassini conjunction experiments may be utilized for the microwave ranging between the Earth and a lander on Mars. If the lander were to be equipped with a Cassini-class dual X- and Ka-band communication system, the measurement of the PPN parameter γ is possible with the accuracy of ~one part in 10^6. However, as opposed to any scenario involving accurate ranging out to the Martian vicinity, the LATOR operations will be conducted at ~1 AU heliocentric distances (well within the asteroid belt) and, thus, will not be affected by the damaging effects of the asteroid belt [46, 69].

[5] In addition, any experiment pushing the present upper bounds on β (i.e., $\beta - 1 = (1.2 \pm 1.1) \times 10^{-4}$ from Williams et al. [114, 115] will also be of interest.

- An ambitious test of the Equivalence Principle – one of the foundations of general relativity – is proposed for the Space Test of Equivalence Principle (STEP) mission. The experiment will test the composition independence of gravitational acceleration for laboratory-sized bodies by searching for a violation of the EP with a fractional acceleration accuracy of $\Delta a/a \sim 10^{-18}$ [53, 119]. STEP will be able to test very precisely for any nonmetric, long-range interactions in physical law; however, the results of this mission will say nothing about the metric component of gravity itself. The LATOR mission is designed specifically to test the metric nature of the gravitational interaction.
- The Solar Orbit Relativity Test (SORT) mission concept proposes to use laser pulses and a drag-free spacecraft aided with a precision clock orbiting around the Sun to precisely measure γ and J_2 (solar quadrupole moment) [52,116,117]. SORT would combine a time-delay experiment (via laser signals sent from the Earth and recorded by precise clocks onboard two satellites orbiting the Sun) with a light deflection experiment (interferometric measurement on Earth of the angle between the two light signals emitted from the satellites) [52,82]. As such, SORT would attempt to measure parameter γ with accuracy of one part in 10^6. In its basic configuration, the LATOR experiment will relay on the redundant geometry formed by the three flight segments (two spacecraft and the ISS) and will depend neither on ultra-stable clocks nor on ground-based interferometry that is severely limited by the atmosphere [89].
- The ESA's BepiColombo mission will explore the planet Mercury with equipment allowing an extremely accurate tracking. This mission will conduct relativity experiments, including the study of Mercury's perihelion advance and the relativistic light propagation near the Sun. The BepiColombo mission will enable achievement of the following accuracies: $\sigma_\gamma \simeq 2 \times 10^{-6}$, $\sigma_\beta \simeq 2 \times 10^{-6}$, and $\sigma_{J_2} \simeq 2 \times 10^{-9}$ in measuring the main post-Newtonian parameters [54]. While a very impressive mission design, its expected accuracy is at least two orders of magnitude worse than that expected from LATOR. The LATOR mission is a designated relativity mission and it is designed to test solar gravity with accuracy at the level of one part in a billion.
- We stress that the future optical interferometers in space such as NASA's Space Interferometry Mission (SIM) and ESA's Gaia (formerly known as, Global Astrometric Interferometer for Astrophysics [49]) would provide improvement in measurement of relativistic parameters as a byproduct of their astrometric program. Thus, SIM will be able to reach accuracy of $\sim 10^{-6}$ in measuring PPN parameter γ. Gaia may potentially reach the accuracy of $10^{-5} - 6 \times 10^{-7}$ in measuring the γ [118]. However, both of these missions will have rather large exclusion angles and will not be able to test gravity effects on light near the Sun.
- A mission concept aiming to reach comparable accuracies in the tests of relativistic gravity in the solar system had been studied in [60] (see also references therein) and [61]. The Astrodynamical Space Test of Relativity

using Optical Devices (ASTROD) is an ambitious mission concept that utilizes three drag-free spacecraft – one near L1/L2 point, one with an inner solar orbit, and one with an outer solar orbit, ranging coherently with one another using lasers to test relativistic gravity and to detect low frequency gravitational waves. The mission may improve the accuracy of determination of the PPN parameter γ to $\sim 10^{-7}$ for mini-ASTROD and to $\sim 5\times 10^{-9}$ for a full-scale version [60]. Because of the technological and programmatic complexities, the launch of an ASTROD-like mission is not expected before 2025.

A clear advantage of the LATOR mission concept is its independence on both the drag-free spacecraft environment and ultra-precise phase-coherent laser transponding techniques. In fact, LATOR will utilize the photon-counting laser ranging methods and redundant optical truss provided by the long-baseline optical multichanneled interferometer on the ISS. The LATOR experiment is optimized for its primary science goal – to measure gravitational deflection of light in the solar gravity to one part in 10^9 (or at the level of the effects of the second post-Newtonian order of light deflection resulting from gravity's intrinsic nonlinearity). There is no major technological breakthroughs needed to satisfy the LATOR mission requirements. All the required technologies already exist and most are space-qualified as a part of our ongoing interferometry program at JPL (SIM, Terrestrial Planet Finder (TPF), and Palomar Testbed and Keck Interferometers).

Concluding this section, we point out that the recent progress in relativistic gravity research resulted in a significant tightening of the existing bounds on the PPN parameters obtained at the first post-Newtonian level of accuracy. However, this improvement is not sufficient to lead to ground-breaking tests of fundamental physical laws addressed above. This is especially true, if the cosmological attractor discovered in [23,31,32] is more robust, time variation in the fine structure constant will be confirmed in other experiments and various general relativity extensions will demonstrate feasibility of these methods for cosmology and relativistic gravity. The LATOR mission is proposed to directly address the challenges discussed above.

8.2 Conclusions and Further Considerations

Concluding, we would like to summarize the most significant results of our LATOR mission study. The most natural question is "Why is LATOR potentially orders of magnitude more sensitive and less expensive?"

First of all, there is a significant advantage in using optical wavelengths as opposed to the microwaves, the present navigational standard. This is based on the fact that solar plasma effects decrease as λ^2 and, in the case of LATOR, we gain a factor of 10^{10} reduction in the solar plasma optical path fluctuations by simply moving from $\lambda = 10\,\mathrm{cm}$ to $\lambda = 1\,\mu\mathrm{m}$. Another LATOR advantage is its independence of a drag-free technology. In addition, the use of a redundant optical truss offers an excellent alternative to an ultra-precise orbit determination. This feature also makes LATOR insensitive to spacecraft buffeting from solar wind and solar radiation pressure.

Furthermore, the use of existing technologies, laser components and spacecraft make this mission a low cost experiment. Thus, 1 W lasers with sufficient frequency stability and >10 years lifetime already developed for optical telecom and also are flight-qualified for SIM. Additionally, small optical apertures ~10–20 cm are sufficient and provide this experiment with a high signal-to-noise ratio. There is also a significant advantage in using components with no motorized moving parts. This all makes LATOR an excellent candidate for the next flight experiment in fundamental physics. Table 2 summarizes the science objectives for this mission.

LATOR is envisaged as a partnership between NASA and ESA wherein both partners are essentially equal contributors, while focusing on different mission elements: NASA provides the deep-space mission components and interferometer design, while building and servicing infrastructure on the ISS is an ESA contribution. The NASA focus is on mission management, system engineering, software management, integration (both of the payload and the mission), the launch vehicle for the deep-space component, and operations. The European focus is on interferometer components, the initial payload integration, optical assemblies, and testing of the optics in a realistic ISS environment. The proposed arrangement would provide clean interfaces between familiar mission elements.

This mission may become a twenty-first century version of the Michelson–Morley experiment in the search for a cosmologically evolved scalar field in the solar system. As such, LATOR will lead to very robust advances in the tests of fundamental physics: it could discover a violation or extension of general relativity, and/or reveal the presence of an additional long-range interaction in the physical law. With this mission testing theory to several orders of magnitude higher precision, finding a violation of general relativity or discovering a new long-range interaction could be one of this era's primary steps forward in fundamental physics. There are no analogs to the LATOR experiment; it is unique and is a natural culmination of solar system gravity experiments.

Acknowledgments

The work described here was carried out at the Jet Propulsion Laboratory, California Institute of Technology, under a contract with the National Aeronautics and Space Administration.

References

1. Anderson, J.D., Colombo, G., Friedman, L.D., and Lau, E.L., "An Arrow to the Sun," in proc. "International Symposium on Experimental Gravitation,"

Pavia, Italy, Sept 1976, ed. B. Bertotti, pp. 393–422, (Accademia Nazionale dei Lincei, Rome, 1977).

2. Anderson, J.D., Gross, M., Nordtvedt, K.L., and Turyshev, S.G., "The Solar Test of the Equivalence Principle," Astrophys. J. 459, 365 (1996).
3. Anderson, J.D., and Williams, J.G., "Long-Range Tests of the Equivalence Principle," Class. Quant. Grav. 18, 2447 (2001).
4. Anderson, J.D., Lau, E.L., Turyshev, S.G., Williams, J.G., and Nieto, M.M., "Recent Results for Solar-System Tests of General Relativity." BAAS 34, 833 (2002).
5. Anderson, J.D., Lau, E.L., Giampieri G., "Measurement of the PPN Parameter γ with Radio Signals from the Cassini Spacecraft at X- and Ka-Bands," in Proc. "The XXII Texas Symposium on Relativistic Astrophysics," Stanford University, December 13–17, 2004, eConf C041213 #0305 (2004).
6. Bean, R., Carroll, S.M., Trodden, M., "Insights into Dark Energy: Interplay Between Theory and Observation," astro-ph/0510059.
7. Bennett, C.L., Halpern, M., Hinshaw, G., Jarosik, N., Kogut, A., Limon, M., Meyer, S.S., Page, L., Spergel, D.N., Tucker, G.S., Wollack, E., Wright, E.L., Barnes, C., Greason, M.R., Hill, R.S., Komatsu, E., Nolta, M.R., Odegard, N., Peirs, H.V., Verde, L., Weiland, J.L., [i.e. WMAP Science Team], "First Year Wilkinson Microwave Anisotropy Probe (WMAP) Observations: Preliminary Maps and Basic Results," Astrophys. J. Suppl. 148, 1 (2003), astro-ph/ 0302207.
8. Bertolami, O., Páramos, J., "The Pioneer anomaly in the context of the braneworld scenario," Class. Quant. Grav. 21, 3309, (2004), gr-qc/0310101; "Astrophysical Constraints on Scalar Field Models" Phys. Rev. D71, 023521, (2004), astro-ph/0408216.
9. Bertolami, O., Páramos, J., and Turyshev, S.G., "General Theory of Relativity: Will it survive the next decade?" see this volume, page 27, gr-qc/0602016.
10. Bender, P.L., Currie, D.C., Dicke, R.H., Eckhardt, D.H., Faller, J.E., Kaula, W.M., Mulholand, J.D., Plotkin, H.H., Poultney, S.K., Silverberg, E.C., Wilkinson, D.T., Williams, J.G. and Alley, C.O., "The Lunar Laser Ranging Experiment," Science 182, 229 (1997).
11. Bertotti, B., Iess, L., Tortora, P., "A test of general relativity using radio links with the Cassini spacecraft," Nature 425, 374 (2003).
12. de Bernardis, P., Ade, P.A.R., Bock, J.J., Bond, J.R., Borrill, J., Boscaleri, A., Coble, K., Crill, B.P., De Gasperis, G., Farese, P.C., Ferreira, P.G., Ganga, K., Giacometti, M., Hivon, E., Hristov, V.V., Iacoangeli, A., Jaffe, A.H., Lange, A.E., Martinis, L., Masi, S., Mason, P.V., Mauskopf, P.D., Melchiorri, A., Miglio, L., Montroy, T., Netterfield, C.B., Pascale, E., Piacentini, F., Pogosyan, D., Prunet, S., Rao, S., Romeo, G., Ruhl, J.E., Scaramuzzi, F., Sforna, D., Vittorio, N., "A Flat Universe From High-Resolution Maps of the Cosmic Microwave Background Radiation," Nature 404, 955 (2000).
13. Brown, T.M., Christensen-Dalsgaard, J., Dziembowski, W.A., Goode, P., Gough, D.O., and Morrow, C.A., "Inferring the Sun's internal angular velocity from observed p-mode frequency splittings." Astrophys. J. 343, 526 (1989).
14. Brynjolfsson, A., "Redshift of photons penetrating a hot plasma," astro-ph/0401420.

15. Capozziello, S., Troisi, A., "PPN-limit of Fourth Order Gravity inspired by Scalar-Tensor Gravity," Phys. Rev. D72, 044022, (2005), astro-ph/0507545.
16. Carroll S.M., "The Cosmological Constant," Living Rev. Rel. 4, 1 (2001), astro-ph/0004075.
17. Carroll, S.M., Hoffman, M., and Trodden, M., "Can the dark energy equation-of-state parameter w be less than -1?" Phys. Rev. D68, 023509 (2003), astro-ph/0301273.
18. Carroll, S.M., Duvvuri, V., Trodden, M., and Turner, M., "Is Cosmic Speed-Up Due to New Gravitational Physics?" Phys. Rev. D70, 043528 (2004), astro-ph/0306438.
19. Carroll, S.M., De Felice, A., Duvvuri, V., Easson, D.A., Trodden, M., Turner, M.S., "The Cosmology of Generalized Modified Gravity Models," astro-ph/0410031.
20. Carroll, S.M., " Is Our Universe Natural?," hep-th/0512148,
21. Ciufolini, I., Wheeler, J. A., "Gravitation and Inertia." (Princeton University Press, 1995).
22. Damour, T., and Nordtvedt, K.L., Jr., "General Relativity as a Cosmological Attractor of Tensor Scalar Theories", Phys. Rev. Lett. 70, 2217 (1993a).
23. Damour, T., and Nordtvedt, K.L., Jr., "Tensor-scalar cosmological models and their relaxation toward general relativity," Phys. Rev. D48, 3436 (1993b).
24. Damour, T., Polyakov, A.M., "String Theory and Gravity," Gen. Relativ. Gravit. 26, 1171 (1994a).
25. Damour, T., and Polyakov, A.M., "The string dilaton and a least coupling principle," Nucl. Phys. B423, 532 (1994b).
26. Damour, T., Esposito-Farese, G., "Non-perturbative strong-field effects in tensor-scalar theories of gravitation," Phys. Rev. Lett. 70, 2220 (1993).
27. Damour, T., Esposito-Farese, G., " Testing gravity to second post-Newtonian order: a field-theory approach," Phys. Rev. D53, 5541 (1996a), gr-qc/9506063.
28. Damour, T., Esposito-Farese, G., "Tensor-scalar gravity and binary pulsar experiments," Phys. Rev. D54, 1474 (1996b), gr-qc/9602056.
29. Damour, T., Esposito-Farese, G., "Gravitational-wave versus binary-pulsar tests of strong-field gravity," Phys. Rev. D58, 042001 (1998), gr-qc/9803031.
30. Damour T., Taylor J.H., " Strong-field tests of relativistic gravity and binary pulsars," Phys. Rev. D45, 1840 (1992).
31. Damour, T., Piazza, F., and Veneziano, G., "Runaway dilaton and equivalence principle violations" Phys. Rev. Lett. 89, 081601, (2002a), gr-qc/0204094.
32. Damour, T., Piazza, F., and Veneziano, G., "Violations of the equivalence principle in a dilaton-runaway scenario," Phys. Rev. D66, 046007, (2002b), hep-th/0205111.
33. Dvali, G., Gabadadze, G., Porrati, M., "4D Gravity on a Brane in 5D Minkowski Space," Phys. Lett. B485, 208, (2000), hep-th/0005016.
34. Dvali, G., Gabadadze, G., Porrati, M., "On Sub-Millimeter Forces From Extra Dimensions," Mod. Phys. Lett. A15, 1717 (2000), hep-ph/0007211.
35. Dvali, G., Gruzinov, A., Zaldarriaga, M., "The Accelerated Universe and the Moon," Phys. Rev. D68, 024012 (2003), hep-ph/0212069.
36. Epstein, R., Shapiro, I.I., "Post-post-Newtonian deflection of light by the Sun," Phys. Rev. D22, 2947 (1980).
37. Eubanks, T.M. et al. "Advances in Solar System Tests of Gravity." In: Proc. of The Joint APS/AAPT 1997 Meeting, 18–21 April 1997, Washington D.C. BAAS, #K 11.05 (1997).

38. Fischbach, E., and Freeman, B.S., "Second-order contribution to the gravitational deflection of light," Phys. Rev. D22, 2950 (1980).
39. Gerber, A., et al., "LATOR 2003 Mission Analysis," JPL Advanced Project Design Team (Team X) Report #X-618 (2003).
40. Gough, D. and Toomre, J., "Seismic Observations of the Solar Interior," Ann. Rev. Astron. Astroph. 29, 627 (1991).
41. Halverson, N.W., Leitch E.M., Pryke C., Kovac, J., Carlstrom, J.E., Holzapfel, W.L., Dragovan, M., Cartwright, J.K., Mason, B.S., Padin, S., Pearson, T.J., Shepherd, M.C., Readhead, A.C.S., "DASI First Results: A Measurement of the Cosmic Microwave Background Angular Power Spectrum," Astrophys. J. 568, 38 (2002), astro-ph/0104489.
42. Holdridge, D. B., in *Supporting Research and Advanced Development*, Space Programs Summary 37–48, Jet Propulsion Laboratory Report, unpublished, Vol. III, pp. 2–4 (1967).
43. Iess L., Giampieri, G., Anderson, J.D., and Bertotti, B., "Doppler measurement of the solar gravitational deflection," Class. Quant. Grav. 16, 1487 (1999).
44. "International Space Station Evolution Data Book, Volume I. Baseline Design, Rev. A," C.A. Jorgensen, ed., Hampton, Virginia. NASA Document #NASA/SP-2000-6109/VOL1/ REV1 (October 2000), at `http://techreports.larc.nasa.gov/ltrs/PDF/2000/spec/NASA-2000-sp6109vol1rev1.pdf`
45. Klioner, S.A., this volume, (2005).
46. Konopliv, A.S., Yoder, C.F., Standish, E.M., Yuan, D.-N., Sjogren, W.L. "A Global Solution for the Mars Static and Seasonal Gravity, Mars Orientation, Phobos and Deimos Masses, and Mars Ephemeris," submitted to Icarus, (2005).
47. Lange, Ch., Camilo, F., Wex, N., Kramer, M., Backer, D.C., Lyne, A.G. and Doroshenko, O., " Precision timing measurements of PSR J1012+5307," Mon. Not. R. Astron. Soc. 326, 274 (2001).
48. Lebach, D.E., Corey, B.E., Shapiro, I.I., Ratner, M.I., Webber, J.C., Rogers, A.E.E., Davis, J.L. and Herring, T.A., "Measurement of the Solar Gravitational Deflection of Radio Waves Using Very-Long-Baseline Interferometry," Phys. Rev. Lett. 75, 1439 (1995).
49. Lindegren, L., and Perryman, M.A.C., "The GAIA Concept, in Proceedings of a Joint RGO-ESA Workshop on Future Possibilities for Astrometry in Space," Cambridge, UK, 19–21 June 1995 (ESA SP-379, September 1995), 23 (1995).
50. Lydon, T.J., Sofia, S., "A Measurement of the Shape of the Solar Disk: The Solar Quadrupole Moment, the Solar Octopole Moment, and the Advance of Perihelion of the Planet Mercury," Phys. Rev. Lett. 76, 177 (1996).
51. Maleki, L., Prestage, J., "SpaceTime Mission: Clock Test of Relativity at Four Solar Radii," in proc. of "Gyros, Clocks, Interferometers ...: Testing Relativistic Gravity in Space," Edited by C. Lämmerzahl, C.W.F. Everitt, F.W. Hehl, Lecture Notes in Physics 562, 369 (2001).
52. Melliti, T., Fridelance, F., and Samain, E., "Study of gravitational theories and of the solar quadrupole moment with the SORT experiment: Solar Orbit Relativity Test," in preparattion (2005).
53. Mester, J., Torii, R., Worden, P., Lockerbie, N., Vitale, S., and Everitt, C.W.F., "The STEP Mission: principles and baseline design," Class. Quant. Grav. 18, 2475 (2001).

54. Milani, A., Vokrouhlicky, D., Villani, D., and Rossi, A., "Testing general relativity with the BepiColombo radio science experiment," Phys. Rev. D66, 082001 (2002).
55. Milman, M., Catanzarite, J., and Turyshev, S.G., "The effect of wavenumber error on the computation of path-lenght delay in white-light interferometry," Applied Optics 41, 4884 (2002).
56. Milman, M., and Turyshev S.G., "Observational Model for Microarcsecond Astrometry with the Space Interferometry Mission," Optical Engeneering 42, 1873 (2003), physics/0301047.
57. Moyer, T. D., JPL Internal Memorandum No. 314.7–122, 1977, unpublished.
58. Moyer, T. D., "Formulation for Observed and Computed Values of Deep Space Network Data Types for Navigation" (Willey, 2003).
59. Netterfield, C.B., Ade, P.A.R., Bock, J.J., Bond, J.R., Borrill, J., Boscaleri, A., Coble, K., Contaldi, C.R., Crill, B.P., de Bernardis, P., Farese, P., Ganga, K., Giacometti, M., Hivon, E., Hristov, V.V., Iacoangeli, A., Jaffe, A.H., Jones, W.C., Lange, A.E., Martinis, L., Masi, S., Mason, P., Mauskopf, P.D., Melchiorri, A., Montroy, T., Pascale, E., Piacentini, F., Pogosyan, D., Pongetti, F., Prunet, S., Romeo, G., Ruhl, J.E., Scaramuzzi, F., [i.e. Boomerang Collaboration], "A measurement by BOOMERANG of multiple peaks in the angular power spectrum of the cosmic microwave background," Astrophys. J. 571, 604 (2002), astro-ph/0104460.
60. Ni, W.-T., "ASTROD – An Overview," Int. J. Mod. Phys. D11, 947 (2002).
61. Ni, W.-T., Shiomi, S., and Liao, A.-C. "ASTROD, ASTROD I and their gravitational-wave sensitivities," Class. Quant. Grav. 21, S641 (2004), gr-qc/0309011.
62. Nordtvedt, K.L., Jr., "Equivalence Principle for Massive Bodies. I. Phenomenology," Phys. Rev. 169, 1014 (1968a).
63. Nordtvedt, K.L., Jr., "Equivalence Principle for Massive Bodies. II. Theory," Phys. Rev. 169, 1017 (1968b).
64. Nordtvedt, K.L., Jr., "Testing Relativity with Laser Ranging to the Moon," Phys. Rev. 170, 1186 (1968c).
65. Nordtvedt, K.L., Jr., Probing Gravity to the 2nd Post-Newtonian Order and to one part in 10^7 Using the Sun, ApJ 320, 871 (1987).
66. Nordtvedt, K.L., Jr., Lunar Laser Ranging Re-examined: The Non-Null Relativistic Contribution, Phys. Rev. D43, 10 (1991).
67. Nordtvedt, K.L., Jr., The relativistic orbit observables in lunar laser ranging, Icarus 114, 51 (1995).
68. Nordtvedt, K.L., Jr., Significance of 'second-order' light propagation experiments in the solar system, Class. Quant. Grav. 13, A11 (1996).
69. Nordtvedt, K.L., Jr., Reducing Asteroid Belt Correlated Noise from Earth Mars Ranging Data, Icarus 129, 120 (1997).
70. Nordtvedt, K.L., Jr., Optimizing the observation schedule for tests of gravity in lunar laser ranging and similar experiments, Class. Quant. Grav. 15, 3363 (1998).
71. Nordtvedt, K.L., Jr., 30 years of lunar laser ranging and the gravitational interaction, Class. Quant. Grav. 16, A101 (1999).
72. Nordtvedt, K.L., Jr., Lunar Laser Ranging - A Comprehensive Probe of Post-Newtonian Gravity, (2003), gr-qc/0301024.
73. Nordtvedt, K.L., Jr., "Covariance analysis studies for LATOR mission," this volume (2005).

74. Peebles, P.J.E., and Ratra, B., The Cosmological Constant and Dark Energy, Rev. Mod. Phys. 75, 599 (2003), astro-ph/0207347.
75. Peacock, J.A., et al., A measurement of the cosmological mass density from clustering in the 2dF galaxy redshift survey, Nature 410, 169 (2001).
76. Perlmutter, S., Aldering, G., Goldhaber, G., Knop, R.A., Nugent, P., Castro, P.G., Deustua, S., Fabbro, S., Goobar, A., Groom, D.E., Hook, I.M., Kim, A.G., Kim, M.Y., Lee, J.C., Nunes, N.J., Pain, R., Pennypacker, C.R., Quimby, R., Lidman, C., Ellis, R.S., Irwin, M., McMahon, R.G., Ruiz-Lapuente, P., Walton, N., Schaefer, B., Boyle, B.J., Filippenko, A.V., Matheson, T., Fruchter, A.S., Panagia, N., Newberg, H.J.M., Couch, W.J., [i.e. Supernova Cosmology Project Collaboration], "Measurements of Omega and Lambda from 42 High-Redshift Supernovae," Astrophys. J. 517, 565 (1999), astro-ph/9812133.
77. Pitjeva, E.V., "Experimental testing of relativistic effects, variability of the gravitational constant and topography of Mercury surface from radar observations 1964–1989," Celest. Mech. Dyn. Astr. 55, 313 (1993).
78. Pitjeva, E.V., "Relativistic Effects and Solar Oblateness from Radar Observations of Planets and Spacecraft," Astronomy Letters 31, 340 (2005).
79. Plowman, J.E., Hellings, R.W., "LATOR Covariance Analysis," Class. Quant. Grav. 23, 309 (2006), gr-qc/0505064.
80. Reasenberg, R.D., Shapiro, I.I., MacNeil, P.E., Goldstein, R.B., Breidenthal, J.C., Brenkle, J.P., Cain, D.L., Kaufman, T.M., Komarek, T.A., Zygielbaum, A.I., "Viking relativity experiment: Verification of signal retardation by solar gravity," ApJ Lett. 234, L219 (1979).
81. Robertson, D.S., Carter, W.E., and Dillinger, W.H., "A new measurement of solar gravitational deflection of radio signals using VLBI," Nature 349, 768 (1991).
82. Reinhard, R., "Ten Years of Fundamental Physics in ESA's Space Science Programme." ESA Bulletin 98, (1999).
83. Riess, A.G., Filippenko, A.V., Challis, P., Clocchiatti, A., Diercks, A., Garnavich, P.M., Gilliland, R.L., Hogan, C.J., Jha, S., Kirshner, R., Leibundgut, B., Phillips, M.M., Reiss, D., Schmidt, B.P., Schommer, R.A., Smith, R.C., Spyromilio, J., Stubbs, C., Suntzeff, N.B.; Tonry, J.,"Observational Evidence from Supernovae for an Accelerating Universe and a Cosmological Constant," [i.e., Supernova Search Team Collaboration], Astron. J. 116, 1009 (1998).
84. Richter, G.W., and Matzner, R.A., "2nd-order contributions to relativistic time delay in the parametrized post-Newtonian formalism," Phys. Rev. D26, 1219 (1982a).
85. Richter, G.W., Matzner, R.A., "Second-order contributions to gravitational deflection of light in the parametrized post-Newtonian formalism. II. Photon orbits and deflections in three dimensions," Phys. Rev. D26, 2549 (1982b).
86. Richter, G.W., Matzner, R.A., "Second-order contributions to relativistic time delay in the parametrized post-Newtonian formalism," Phys. Rev. D28, 3007-3012 (1983).
87. Sovers, O.J., Fanselow, J.L., and Jacobs, C.S., "Astrometry and geodesy with radio interferometry: experiments, models, results", Rev. Mod. Phys., 70, 1393–1454 (1998).

88. Spallicci, A., “The Solar Probe mission for gravitation experiments: Orbital mechanics, thermal design, communications and instrumentation”, In proc. “48th IAF Congress,” Torino, (1997).
89. Shao, M., Colavita, M., "Potential of long-baseline infrared interferometry for narrow-angle astrometry,” A&A 262, 353 (1992).
90. Shao, M., "Prospects for Ground Based Interferometric Astrometry,” Astrophys. Space Sci. 223, 119 (1995).
91. Shao, M., Yu, G., Gürsel, Y., Hellings, R., et al. “Laser Astrometric Test of Relativity (LATOR),” JPL Internal Technical Memorandum, (1996).
92. Shapiro, I. I., “Fourth Test of General Relativity,” Phys. Rev. Lett. 13, 789–791 (1964).
93. Shapiro, I.I., Counselman, C.C., III, and King, R.W., “Verification of the Principle of Equivalence for Massive Bodies,” Phys. Rev. Lett. 36, 555 (1976).
94. Shapiro, I.I., Reasenberg, R.D., MacNeil, P.E., Goldstein, R.B., Brenkle, J.P., Cain, D.L., Komarek, T., Zygielbaum, A.I., Cuddihy, W.F., Michael, W.H., Jr., “The Viking relativity experiment,” JGR 82, 4329 (1977).
95. Shapiro, S.S., Davis, J.L., Lebach, D.E., and Gregory, J.S., "Measurement of the Solar Gravitational Deflection of RadioWaves using Geodetic Very-Long-Baseline Interferometry Data, 1979-1999,” Phys. Rev. Lett. 92, 121101 (2004).
96. Tausner, M.J., “General Relativity and Its Effects on Planetary Orbits and Intarplanetary Observations,” (Lincoln Laboratory, MIT, Cambridge, MA), Technical Report No. 425 (1966).
97. Taylor, J.H., Wolszczan, A., Damour, T., and Weisberg, J.M., “Experimental constraints on strong-field relativistic gravity,” Nature 355, 132 (1992).
98. Tonry, J.L., Schmidt, B.P., Barris, B., Candia, P., Challis, P., Clocchiatti, A., Coil, A.L., Filippenko, A.V., Garnavich, P., Hogan, C., Holland, S.T., Jha, S., Kirshner, R.P., Krisciunas, K., Leibundgut, B., Li, W., Matheson, T., Phillips, M.M., Riess, A.D., Schommer, R., Smith, R.C., Sollerman, J., Spyromilio, J., Stubbs, C.W., Suntzeff, N.B., “Cosmological Results from High-z Supernovae,” Astrophys. J. 594, 1 (2003), astro-ph/0305008.
99. Turyshev S.G., “Relativistic stellar aberration for the Space Interferometry Mission,” (2002), gr-qc/0205061.
100. Turyshev S.G., “Analytical Modeling of the White Light Fringe,” Applied Optics 42, 71 (2003), physics/0301026.
101. Turyshev, S.G., Shao, M., and Nordtvedt, K.L., Jr., “The Laser Astrometric Test of Relativity (LATOR) Mission,” Class. Quant. Grav. 21, 2773 (2004a), gr-qc/0311020.
102. Turyshev, S.G., Williams, J.G., Nordtvedt, K.L., Jr., Shao, M., and Murphy, T.W., Jr., “35 Years of Testing Relativistic Gravity: Where do we go from here?.” In Proc. “302.WE-Heraeus-Seminar: “Astrophysics, Clocks and Fundamental Constants,” 16–18 June 2003. The Physikzentrum, Bad Honnef, Germany.” Springer Verlag, Lect. Notes Phys. 648, 301 (2004), gr-qc/0311039.
103. Turyshev, S.G., Shao, M., and Nordtvedt, K.L., Jr., “Experimental Design for the LATOR Mission,” Int. J. Mod. Phys. D13, 2035 (2004b), gr-qc/0410044.
104. Turyshev, S.G., Shao, M., and Nordtvedt, K.L., Jr., “Optical Design for the Laser Astrometric Test of Relativity”, in Proc. “The XXII Texas Symposium on Relativistic Astrophysics,” Stanford University, December 13–17, 2004, eConf C041213 #0306 (2004c), gr-qc/0502113.

105. Turyshev, S.G., Shao, M., and Nordtvedt, K.L., Jr., "Mission Design for the Laser Astrometric Test Of Relativity," to be published, Advances Space. Res., (2005), gr-qc/0409111.
106. Turyshev, S.G., Dittus, H., Lämmerzahl, C., Theil, S., Ertmer, W., Rasel, E., Foerstner, R., Johann, U., Klioner, S., Soffel, M., Dachwald, B., Seboldt, W., Perlick, V., Sandford, M.C.W., Bingham, R., Kent, B., Sumner, T.J., Bertolami, O., Páramos, J., Christophe, B., Foulon, B., Touboul, P., Bouyer, P., Damour, T., Salomon, C., Reynaud, S., Brillet, A., Bondu, F., Mangin, J.-F., Samain, E., Erd, C., Grenouilleau, J.C., Izzo, D., Rathke, A., Asmar, S.W., Colavita, M., Gürsel, Y., Hemmati, H., Shao, M., Williams, J.G., Nordtvedt, K.L., Jr., Degnan, J., Plowman, J.E., Hellings, R., Murphy, T.W., Jr., "Fundamental Physics with the Laser Astrometric Test Of Relativity," In proceesings of "2005 ESLAB Symposium "Trends in Space Science and Cosmic Vision 2020," ESA/ESTEC, Noordwijk, The Netherlands, 19 April 2005, ESA Publication SP-588, 11–18 (2005), gr-qc/0506104.
107. Will, C.M., Nordtvedt, K.L., Jr., "Conservation Laws and Preferred Frames in Relativistic Gravity. I. Preferred-Frame Theories and an Extended PPN Formalism," ApJ 177, 757 (1972).
108. Will, C.M., "General Relativity at 75: How Right was Einstein?" Science 250, 770 (1990).
109. Will, C.M., "Theory and Experiment in Gravitational Physics," (Cambridge University Press, 1993).
110. Will, C.M., "The Confrontation between General Relativity and Experiment," (2005), gr-qc/0510072.
111. Williams, J.G., Newhall, X.X. and Dickey, J.O., "Relativity Parameters Determined from Lunar Laser Ranging," Phys. Rev. D53, 6730 (1996).
112. Williams, J.G., Anderson, J.D., Boggs, D.H., Lau, E.L., and Dickey, J.O., "Solar System Tests for Changing Gravity," AAS Meeting, Pasadena, CA, June 3–7, 2001, BAAS 33, 836 (2001).
113. Williams, J.G., Turyshev, S.G., and Murphy, T.W., Jr., "Improving LLR Tests of Gravitational Theory," Int. J. Mod. Phys. D13, 567, (2004a), gr-qc/0311021.
114. Williams, J.G., Turyshev, S.G., and Boggs, D.H., "Progress in Lunar Laser Ranging Tests of Relativistic Gravity," Phys. Rev. Lett. 93, 261101, (2004b), gr-qc/0411113.
115. Williams, J.G., Turyshev, S.G., and Boggs, D.H., "Lunar Laser Ranging Tests of the Equivalence Principle with the Earth and Moon." In proceedings of "Testing the Equivalence Principle on Ground and in Space," Pescara, Italy, September 20–23, 2004, C. Lämmerzahl, C.W.F. Everitt and R. Ruffini, editors. To be published by Springer Verlag, Lect. Notes Phys., (2005), gr-qc/0507083.
116. Veillet, C., et al., "Proposal in response to the ESA call for mission concepts for follow-up Horizon 2000," (1993).
117. Veillet, C., Stanford, M., "Solar Orbit Relativity Test: A mission for measuring γ through the solar gravitational deflection and delay of light." Presented at "Future Fundamental Physics Missions in Space and Enabling Technologies, Spain, April 5–7, 1994," (1994).
118. Vecchiato, A., Lattanzi, M.G., Bucciarelli, B., Costa., M., de Felice, F., Gai, M.,"Testing general relativity by micro-arcsecond global astrometry." Astron. & Astrophysics 399, 337 (2003).

119. Worden, P., Mester, J., Torii, R., "STEP error model development," Class. Quant. Grav. 18, 2543 (2001).
120. Yu, J., Shao, M., Gursel Y., and Hellings R., "LATOR Mission," SPIE Publ. 2200, 325 (1994).

LATOR's Measured Science Parameters and Mission Configuration

Kenneth Nordtvedt

Northwest Analysis, 118 Sourdough Ridge Road, Bozeman MT 59715 USA

Summary. A LATOR mission to measure the non-Euclidean features of a light triangle with sides passing close by the Sun is analyzed as a probe of General Relativity and alternative theories of gravity. It measures second post-Newtonian features of gravity as well as a novel mass parameter of the Sun — the spatial metric's "gamma mass" — while carrying out the main science goal of measuring basic PPN gamma to a part in 10^9 precision. By arranging orbits for the two LATOR spacecrafts which have their lines of sight from the near-Earth laser interferometer being about equal and opposite the Sun center during their close passings, transverse knowledge of spacecraft positions are much relaxed and allow elimination of onboard drag free systems.

1 Introduction

LATOR is a proposed mission in which four properties of a light triangle surrounding the Sun are measured. Laser ranging measures each side of the triangle in round trip propagation times that serve as surrogate lengths, while a laser interferometer measures the small angle of the triangle formed by the two arriving laser pulse streams from each of the spacecraft both located beyond the Sun from Earth. This experimental configuration is illustrated in the top panel of Fig. 1. In idealization of light propagating at fixed speed in straight Euclidean lines around a fixed triangle, the relationship between the four observables would be given by

$$\sin\psi_{Euc} = \frac{1}{2T_A T_B}\sqrt{2T_{AB}^2(T_A^2+T_B^2)-(T_A^2-T_B^2)^2-T_{AB}^4} \qquad (1)$$

But in general relativity and viable alternative theories of gravity, two relativistic effects of gravity on light propagation will alter this relationship. First, the speed of light, globally observed, decreases as the light propagates closer to a gravitating body such as the Sun, and second, this location dependence of the global speed of light results in deflection of the light paths. For

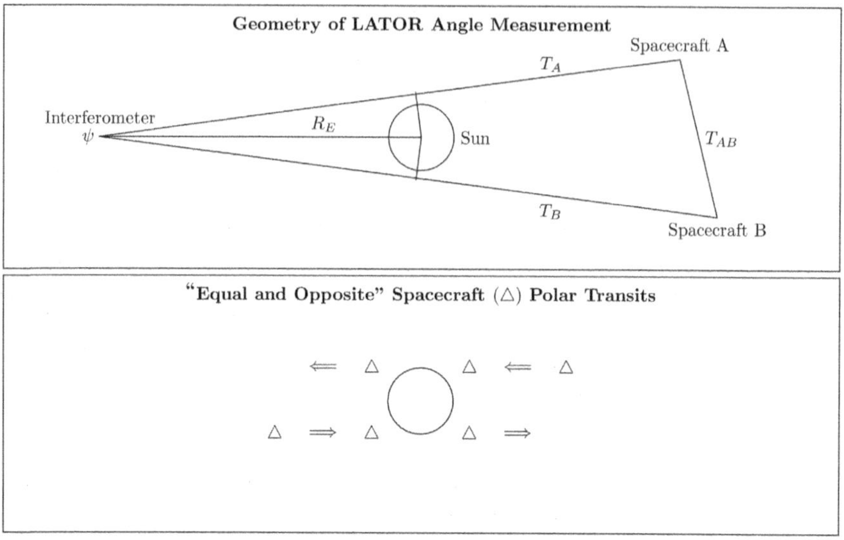

Fig. 1. The *top panel* illustrates LATOR's light triangle and its the four observables: T_A, T_B, T_{AB}, ψ. The location of the light triangle relative to the Sun's center needs also to be very precisely known to infer the two indicated impact parameters of the light paths. The *bottom panel* shows the movements of the two spacecraft lines of sight as seen from the interferometer. The shown configuration with equal-and-opposite locations relaxes the needed precision in locating the light triangle relative to the Sun. The polar passages of the spacecraft lines of sight helps to orthogonalize the mission's deflection signals due to the monopolar and quadrupolar potentials of the Sun, thereby reducing the variance of the science parameter estimates.

light triangles with two of the sides passing close by the Sun, the deflection of light by gravity becomes the more sensitive observational measure of gravity's action. The failure of (1) to correctly give the relationship between the four actual observables, T_A, T_B, T_{AB}, ψ, then becomes LATOR's probe of gravitational theory.

In Sect. 2, I show that the main scientific signal of the experiment, proposed to be measured to a part in 10^9 precision, is not simply the parameterized post-Newtonian γ coefficient, but as well involves the Sun's gravitational to inertial mass ratio. This will necessitate either an independent measurement of that ratio for the Sun or an inference of that ratio from the same underlying model of gravity theory. This brings the PPN β coefficient into that observable and calls for an improvement in the measurement precision of β from lunar laser ranging or other means.

In Sect. 3, a configurational scenario for the passage of the two spacecraft lines of sight by the Sun is discussed, which has several advantages for achieving mission goals. This scenario has the angular locations of the two spacecraft from the Sun center, as seen from the interferometer, remain close to equal but opposite throughout the data taking period(s) of the mission experiment. Doing this substantially relaxes the transverse location tracking precision needed for the two spacecraft during their transits and eliminates the need for drag-free systems included on the spacecraft. This configuration also doubles the strength of the deflection signals that correspondingly increases precision of measurement of scientific parameters. And if the spacecraft lines of sight are designed not to pass across the Sun's equator, but pass instead at equal-and-opposite high latitudes, then some of the key fit-for parameters of the scientific model acquire "partials" whose shapes more easily "separate" in the least-squares-fit process, resulting in an even greater precision of fit-for the science parameters of interest.

2 The Chief Scientific Parameter of the Mission

For a metric tensor gravitational field produced by bodies of negligible gravitational self-energies, the deflection of light rays passing close by the Sun has leading contributions

$$\begin{aligned}\Theta = 2(1+\gamma)\frac{GM_S}{c^2}\left(\frac{1}{D_A}+\frac{1}{D_B}\right)+\chi\left(\frac{GM_S}{c^2}\right)^2\left(\frac{1}{D_A^2}+\frac{1}{D_B^2}\right)\\ +(1+\gamma)J_2a_S^2\frac{GM_S}{c^2}\left(\frac{1-2\cos^2 L_A}{D_A^3}+\frac{1-2\cos^2 L_B}{D_B^3}\right)+\ldots \qquad (2)\end{aligned}$$

with D_A, D_B being light path distances of closest approach, and these expressions neglecting small corrections for light trajectories' finite endpoints being about 1 AU from the Sun. The first-order deflection term includes the PPN coefficient γ, which appears in the spatial metric field potential. γ equals precisely one in general relativity but generally deviates from this value in alternative theories of gravity. Its very precise measurement at the part in 10^9 is the chief scientific goal of LATOR. The next contributions to the light deflection scale as the inverse square of light impact parameters and receive contributions from both the nonlinear properties of the metric gravity field – including nonlinearities in the spatial metric potentials – and from quadratic corrections to the first-order deflection. The third shown contribution is due to the Sun's Newtonian quadrupole moment contributing a corrective deflection that depends on the directional location of the passing light rays with respect to the Sun's equator; it scales as the inverse third power of impact parameter. M_S is the Sun's mass. An experiment aiming for a part in 10^9 measurement of PPN γ requires knowledge of the solar mass parameter GM_S to five parts in 10^{10}. The accumulated observations of planetary motion in the solar system have achieved this level of knowledge for only the solar gravitational mass – $GM(G)_S$. For experiments of the past, that was sufficient.

But at the part in 10^9 precision, account must be taken that the Sun's gravitational self-energy, about four parts in 10^6 of the whole solar mass, could produce a solar gravitational mass that differs from that body's inertial mass [1].

$$M(G)_S = M(I)_S + (4\beta - \gamma - 3)\frac{U_S}{c^2} \quad (3)$$

U_S being the Sun's gravitational self-energy and β being the other first post-Newtonian order coefficient for metric gravity ($\beta = 1$ in general relativity). But the solar mass parameter, which appears in the spatial metric, will also generally differ from the Sun's inertial and gravitational masses; this depends on the structure of the nonlinear gravitational potentials in the spatial metric [3]:

$$M(\gamma)_S = M(I)_S + \xi\frac{U_s}{c^2} \quad (4)$$

with parameter ξ summarizing the effect of compactifying the higher-order spatial metric potential contributions to the Sun's coupling strength due to its internal gravitational binding. This parameter probes novel second post-Newtonian order aspects of gravity; it has value zero in general relativity but not generally in alternative theories. The first-order metric tensor potentials then read [2]

$$\begin{aligned} g_{00} &= 1 - 2\frac{GM(G)}{c^2 R} \\ -g_{ij} &= \delta_{ij}\left(1 + 2\gamma\frac{GM(\gamma)}{c^2 R}\right) \end{aligned} \quad (5)$$

and the speed of light function that determines the ranging times and deflections and that follows from the null geodesic principle for light, $g_{\mu\nu}dx^\mu dx^\nu = 0$, then becomes in leading order

$$c(\boldsymbol{R}) = c\left(1 - \frac{G(M(G) + \gamma M(\gamma))}{c^2 R}\right) \quad (6)$$

The effect of considering the Sun as a realistic body with substantial gravitational self-energy then results in the first-order deflection term in (2) being proportional to the combination $G\left(M(G)_S + \gamma M(\gamma)_S\right)$; its measurement does not yield simply γ.

It has been shown, however, that if the metric gravitational field not only fulfills local Lorentz-invariance, but also fulfills "extended Lorentz-invariance,"[1] then the unknown solar $GM(\gamma)$ will be given by more familiar mass parameters of the Sun, the gravitational and inertial masses. The required relationship is [2]

[1] The local Lorenz-invariance violating PPN coefficients α_1 and α_2 have been experimentally found to be zero to respective precisions of a few times 10^{-5} and 10^{-7} suggesting that this invariance is not destroyed by cosmic boundary conditions. Generalized Brans–Dicke type scalar–tensor theories automatically fulfill extended Lorentz-invariance

$$GM(G) + \gamma GM(\gamma) = (1+\gamma)GM(I) \tag{7}$$

After pulling out the well-measured gravitational mass parameter of the Sun, $GM(G)_S$, it is seen that the science parameter to be measured by LATOR and which scales the leading order light deflection signal is the combination

$$\gamma^* = (1+\gamma)\left(\frac{M(I)}{M(G)}\right)_S - 1 \tag{8}$$

The Sun's gravitational to inertial mass ratio has not yet been measured to precision of five parts in 10^{10}. With a slight improvement in our knowledge of PPN β, however, this ratio can be inferred to that accuracy. By using metric theory's expression for a body's gravitational to inertial mass ratio (3) and with the Sun's gravitational binding energy fraction being about $4 \cdot 10^{-6}$, knowledge of β is needed to $3 \cdot 10^{-5}$ precision. That is about a factor of 3 better than lunar laser ranging presently measures this parameter, but the new generation (APOLLO)lunar ranging program should improve the β measurement to the needed precision by the time the LATOR mission is carried out.

The Sun does contain a type of energy that has not been appreciably present in other bodies for which their gravitational to inertial mass ratios have been measured – the energy of free electromagnetic radiation. In the Sun, this fractional contribution to the body's mass reaches about $1.6 \cdot 10^{-7}$. So a direct measurement of the Sun's gravitational to inertial mass ratio to $5 \cdot 10^{-10}$ precision would be desirable. This could be achieved by transponded laser ranging between Earth and Mars [4]. An anomalous gravitational to inertial mass ratio of the Sun will produce a polarization of the inner planet orbits in the direction of Jupiter if the Sun experiences an anomalous acceleration toward Jupiter. Measuring these polarization amplitudes to several centimeters precision would yield the desired precision in direct measurement of the Sun's gravitational to inertial mass ratio as needed in (8).

The nonlinearity in the spatial components of the tensor metric field is also tested by LATOR through the direct rescaling of the PPN γ due to the distant matter in our part of the universe, which creates the cosmologically local gravity potential inhomogeneities. If U_C/c^2 is considered as the dimensionless background Newtonian potential deviation from cosmic average of distant matter outside of the solar system, then by considering an enlarged portion of the universe as the system under consideration, the two types of nonlinear potentials in the spatial metric components become renormalization or rescaling contributions to the linear gravitational potentials within the solar system;

$$\frac{G^2}{c^4}\sum_{ij}\frac{m_i m_j}{r_i r_j} \longrightarrow 2\frac{U_C}{c^2}\,G\sum_i \frac{m_i}{r_i} \tag{9}$$

$$\frac{G^2}{c^4}\sum_{ij}\frac{m_i m_j}{r_i r_{ij}} \longrightarrow \frac{U_C}{c^2}\,G\sum_i \frac{m_i}{r_i} \tag{10}$$

Then after rescaling Newtonian G and the spatial coordinates within the solar system,

$$G \longrightarrow G\left(1-(4\beta-\gamma-3)\frac{U_C}{c^2}\right) \tag{11}$$

$$r \longrightarrow r\left(1+\gamma\frac{U_C}{c^2}\right) \tag{12}$$

a rescaled PPN γ results. With our galaxy's contribution to U_C/c^2 being of order 10^{-6} and the nearest cluster of galaxies contributing of order 10^{-5} to U_C/c^2, a LATOR mission can test the spatial metric's nonlinearity at the part in 10^3 or 10^4 level by means of its γ measurement. These nondirect methods of testing the spatial metric's nonlinearity in the LATOR mission are generally more precise than its direct measurement from the deflections proportional to χ in (10).

3 LATOR Mission Configuration

No matter how precise the three sides and one angle of LATOR's light triangle are measured, the relationship between those four observables cannot be used to test theory unless that light triangle can also be located with respect to the Sun's gravitational potential center with sufficient precision. In the expression for light angle deflection (2), the distances of closest approach of the light rays to the Sun's center must apparently be known to five parts in 10^{10} if the theory-dependent coefficient of the leading monopolar deflection is to be inferred to a part in 10^9. For the rays passing closest by the Sun that amounts to a less than 35 cm precision in impact parameter, or about twice that in transverse location of spacecraft. Over the whole data collection period that requirement relaxes somewhat as the line of sight passes the Sun at larger distances; but nevertheless a transverse (to lines of sight), nonmodelable acceleration of spacecraft greater than about $10^{-10}\,\mathrm{cm\,s^{-2}}$ seems unacceptable. This would seem to require drag-free or drag-measuring systems onboard the spacecraft with very high performance near-zero frequency.

However, a mission configuration in which the two spacecraft lines of sight are about equal-and-opposite with respect to the Sun center relaxes the precision with which the transverse location of the light triangle is needed. This can be seen by expressing the leading monopolar deflection in terms of the angle measured by the interferometer, the distance to Sun's center from the interferometer, and a common mode angle ψ_C of the two spacecraft angular locations (See Fig. 1.)

$$\begin{aligned}\Theta_\gamma &= 2(1+\gamma)\frac{GM_S}{c^2}\left(\frac{1}{D_A}+\frac{1}{D_B}\right)\\ &= 4\frac{GM_s}{c^2R_E}\left(\frac{1}{\sin(\psi/2+\psi_C)}+\frac{1}{\sin(\psi/2-\psi_C)}\right)+\ \ldots\end{aligned} \tag{13}$$

with

$$D_a = R_e \sin(\psi/2 - \psi_C) \tag{14}$$
$$D_b = R_e \sin(\psi/2 + \psi_C) \tag{15}$$

The distance from Earth to Sun will be known to part in 10^9 precision, and the interferometer angle is very well measured in the mission. The common mode angle then acts as the variable that transversely locates the light triangle with respect to the Sun. If the partial of this monopolar deflection signal is taken with respect to the common mode angle,

$$\frac{\partial\Theta_\gamma}{\partial\psi_C} \cong \frac{8GM_S R_E}{c^2 D^2}\frac{D_A - D_B}{D} \leq 3.6\ 10^{-3}\left(\frac{D_A - D_B}{D}\right) \tag{16}$$

this sensitivity is diminished for near-equal impact parameters, $D_A \simeq D_B$. This directly translates into a less stringent requirement in the knowledge of this coordinate. Since it is the total deflection of the two light paths of the triangle that contributes to the measured angle at the interferometer, in this equal-and-opposite mission configuration, the error to first order in inferring the contribution of one light path to that total angle is compensated for by an opposite error from the other light path when there is uncertainty in the light triangle's location, i.e., ψ_C. One way to then treat the time-varying common mode angle $\psi_C(t)$ is to include both its value and time rate of change at some fiducial time, $\psi_C(t_o)$ and $d\psi_C(t_o)/dt$, as two fit-for parameters in the least-squares-fit of the mission's entire data set of observations, and to include the modelable gravitational dynamics of this variable routinely as part of the model. The accumulated error in the common mode angle due to the unmodeled, nongravitational accelerations on the spacecraft will not, over the life of the data taking phase of the mission, disturb the deflection observable due to the suppression of sensitivity indicated in (16).

Another positive feature of the equal-and-opposite configuration of spacecraft passages by the Sun is that the total deflection of the two light paths are additive and double the sensitivity (mission performance) of the observed interferometer angles to the scientific parameters of interest and being estimated in the least-squares-fit of the data.

There is the additional question of whether the two spacecraft trajectories should pass by the Sun such that their lines of sight basically traverse the solar equator, or should they pass above and below that line toward the solar poles? Although the explicit nonlinear parameter χ in (2) may or may not need to be a fit-or parameter along with the chief scientific parameter γ^*, there is no question that the Sun's quadrupole moment parameter J_2 will have to be simultaneously fit to the data. LATOR will measure this parameter to higher precision than it is known from other means, and that higher precision will be necessary so that the estimate for γ^* is not biased by its uncertainty. In a multiparameter fit of data, the precision with which any parameter is determined is proportional to the norm of that parameter's orthogonalized partial–the

Table 1. Standard deviations for science parameter estimations (based on 100 independent light triangle measurements).

	Three-parameter fit			Two-parameter fit	
Z/R_S	$10^9 \cdot \delta\gamma$	$10^8 \cdot \delta J_2$	$10^3 \cdot \delta\chi$	$10^9 \cdot \delta\gamma$	$10^8 \cdot \delta J_2$
0	23	33	22	3.7	4.1
0.25	28	49	29	3.8	4.7
0.50	15	79	26	4.8	8.2
0.75	9	15	3	3.6	13.3
1	11	6	5	1.9	2.6

orthogonalized partial being that portion of the partial, considered as a data vector with component values for each observation, which is orthogonal to all the other partial vectors for the other fit-for parameters of the model that are adjusted to best-fit the data. The latitude dependence of the partial for the quadrupole parameter J_2 has a zero and sign change at $\pm 45°$ of latitude, so passages of the spacecraft near the poles produce a J_2 partial, which tends to have minimal projection onto the partial for γ^*, consequently improving the precision for the estimation of this latter parameter. The high latitude passages shown in the lower panel of Fig. 1 also have the spacecraft spending more time close to the Sun's center, permitting more high quality data to be accumulated and thereby enhancing the magnitude of the parameter partials. Table 1 shows the results of some simulation studies in which least-squares-fit for the triplet of parameters, γ^*, J_2, χ, or just the doublet, γ^*, J_2, were estimated for spacecraft passages at various out-of-ecliptic distances measured in units of the solar radius. High latitude passages are seen to improve estimation precision for the science parameters. The table values are based on there being 100 independent measurements of the four light triangle observables over the mission; these results can be rescaled as $10/N^{1/2}$ for different assumed numbers N of total mission measurements.

References

1. K. Nordtvedt: *Phys. Rev.* **169**, 1017 (1968).
2. K. Nordtvedt: *Astrophys. J.* **297**, 390 (1985).
3. K. Nordtvedt: *Astrophys. J.* **407**, (1993)
4. J.D Anderson, M. Gross, K. Nordtvedt, and S.G. Turyshev: *Astrophys. J.* **459**, 365 (1996).

OPTIS: High-Precision Tests of Special and General Relativity in Space

Claus Lämmerzahl[1], Hansjörg Dittus[1], Achim Peters[2], Silvia Scheithauer[3], and Stephan Schiller[4]

[1] ZARM, University of Bremen, Am Fallturm, 28359 Bremen, Germany
[2] Humboldt–University Berlin, Berlin, Germany
[3] Max-Planck-Institut für Astronomie, Königstuhl 17, 69117 Heidelberg, Germany
[4] Institute for Experimental Physics, Heinrich–Heine University Düsseldorf, 40225 Düsseldorf, Germany

Summary. The OPTIS mission is a satellite equipped with a variety of clocks and laser ranging and tracking facilities for performing improved tests of the foundations as well as predictions of special and general relativity. This mission makes advantage of the space conditions of large differences in the velocity and the gravitational potential. Here, we report on recent progress made in the studies of the behavior of the resonator in the field of a gravity gradient.

1 Introduction

It has been outlined in the survey articles on pages 26 and 75 in this volume that Einstein's general relativity (GR) is mainly based on the Einstein equivalence principle, i.e., on the universality of free fall, the universality of the gravitational redshift (UGR), and the local validity of Lorentz invariance (LI). The proposed OPTIS mission aims at an improvement of the complete test of LI and of UGR by three orders of magnitude compared with the present ground experiments. Therefore, together with the test of UFF by MICROSCOPE and STEP, we will have a complete tests of the foundations of GR.

2 Science Objectives

The mission OPTIS is a collection of clocks in an highly elliptic Earth bound orbit. Laser tracking devices and a laser link to the Earth completes the scientific hardware components. As a consequence, many issues related to clocks and to the orbit can be measured with high precision. Therefore, OPTIS aims at improving tests of the *foundations of special and general relativity* by up to three orders of magnitude. The scientific basis for this has been outlined in detail in [5]. The science objectives are listed in Table 1.

Table 1. The scientific objectives of OPTIS.

	Test	Method	Present accuracy	OPTIS accuracy
1	Isotropy of speed of light	Cavity–cavity comparison	$1.5 \cdot 10^{-9}$	10^{-12}
2	Constancy of speed of light	Cavity–clock comparison	$7 \cdot 10^{-7}$	10^{-8}
3	Time dilation – Doppler effect	Laser link	$2 \cdot 10^{-7}$	10^{-9}
4	Universality of gravitational redshift I	Cavity–clock comparison	$1.7 \cdot 10^{-2}$	10^{-4}
5	Universality of gravitational redshift II	Clock–clock comparison	$2.5 \cdot 10^{-5}$	10^{-7}
6	Absolute gravitational redshift	Time transfer	$1.4 \cdot 10^{-4}$	10^{-8}
7	Lense–Thirring effect	Laser tracking	0.3	10^{-3}
8	Einstein perigee advance	Laser tracking	$3 \cdot 10^{-3}$	$6 \cdot 10^{-4}$
9	Test of Newton potential	Laser tracking	10^{-5}	10^{-12}

Except for the Universality of Free Fall, which will be tested by MICROSCOPE and STEP, OPTIS represents a complete test of the foundations of the metric theories of gravity. Furthermore, relativistic orbital effects as predicted by Einstein's theory of gravity will be tested.

3 Mission Design

To have a good test of the UGR, the clocks should move through a large gravitational potential difference. Therefore, a highly elliptical orbit is preferable. For tests of special relativity, also a large variation of the velocity is needed. This again can be obtained from a highly elliptical orbit. From tracking and ranging, highly precise orbital data can be obtained, which are of use for the exploration of the structure of the gravitational field of the Earth, the Sun and the Moon.

The mission scenario is shown below in Fig. 1.

4 Mission Technology

For this mission, technologies are required, which have been used recently, to carry through the most precise tests of special relativity. The precision of these tests can be further increased under space conditions thanks to longer integration times, larger changes in the orbital velocity, and larger differences of the gravitational potential. Furthermore, very precise laser tracking and linking of satellites is a well-established technique and will provide, in combination with the active drag-free control system, very accurate orbit data.

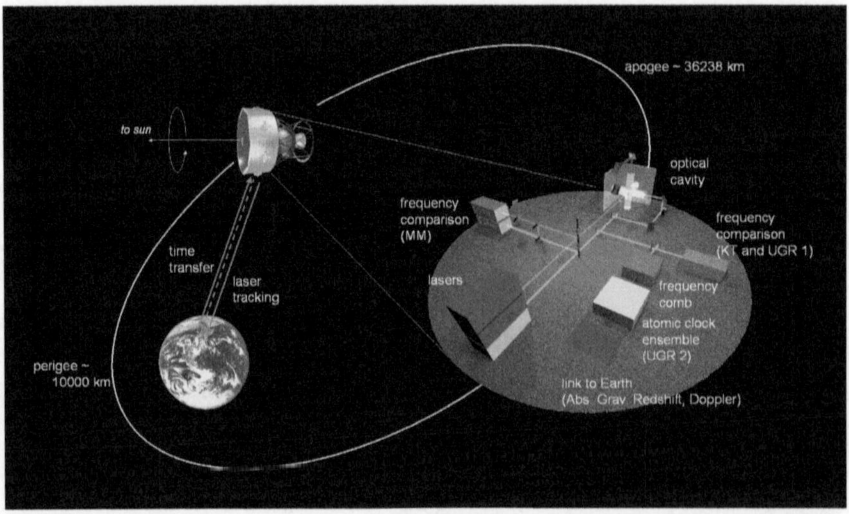

Fig. 1. The mission scenario (apogee and perigee heights measured from Earth's surface).

The core technologies for OPTIS are mentioned below:

1. Optical resonators
2. Highly stabilized lasers
3. Capacitive gravitational reference sensors
4. Drag-free control
5. Ion clocks
6. Frequency combs
7. Laser tracking systems

These technologies are also key technologies for other future missions.

5 The Deformation of the Resonator in a Tidal Gravitational Field

Although many of the disturbances acting on a resonator can be minimized by means of an appropriate satellite control system, some intrinsic disturbances cannot be eliminated as a matter of principle and distort the resonator shape leading to a systematic frequency shift. In particular, the tidal gravitational force,[1] which acts through every extended body, cannot be eliminated by choosing an appropriate frame and, thus, will induce distortions on the resonator.

[1] In space and engineering sciences, the tidal gravitational force is often referred to as "gravity gradient."

We give a rough estimate of the expected effect of the tidal gravitational force on a freely moving cube of length L. If the position of the cube is at a distance R from the center of the Earth, then the difference of the Earth's acceleration on the top and bottom of the cube is $\Delta a = (\partial^2 U/\partial r^2)L$, where U is the Earth's Newtonian potential $U = GM_\oplus/R$. For an orbit with $R = 10,000$ km and a typical resonator length of $L = 5$ cm we have $\Delta a \approx GM_\oplus/R^3 L \approx 2 \cdot 10^{-8}\,\mathrm{m\,s^{-2}}$. In a rough estimate, we assume that a force $F = m\Delta a$ to act on the top surface of the cube. Now Hook's law of elasticity

$$\frac{F}{A} = E\frac{\Delta L}{L} \tag{1}$$

gives the change of the length ΔL of the cube due to a force F acting on the area A. In our case $F = m\Delta a = \rho L^3 \Delta a$ and we get

$$\frac{\Delta L}{L} = \frac{\rho L}{E}\Delta a \approx 10^{-17} \tag{2}$$

assuming an elasticity modulus of $E = 90\,\mathrm{GPa}$ and a density of 2,350 $\mathrm{kg\,m^{-3}}$, which is typical for Zerodur.

In the OPTIS mission, the science goal for the measurement of the isotropy of the speed of light is better than $\Delta c/c = 10^{-18}$ [5]. This can only be achieved if the resonator has a length stability of $\Delta L/L = 10^{-18}$ [5]. As one can see from our estimates, the tidal gravitational force will lead to systematic deformations, which are one order of magnitude larger than the expected accuracy. Therefore, the effect has to be investigated carefully by including the tidal gravitational force into the equations of elasticity, calculating the resulting resonator shape, and then subtract the effect.

5.1 Analytical Solution for an elastic Cylinder in a Tidal Gravitational Force Field

Simplified Model of the Resonator

The problem of an optical resonator flying on a geodetic Earth orbit can be simplified by treating the problem in a body-fixed coordinate system. We also consider, for simplicity, the body to be a homogeneous and isotropic cylinder of radius R and height $2L$. The body coordinates are (r, φ, z) with the origin being at the center-of-mass of the cylinder. The z-axis coincides with the symmetry axis of the cylinder, see Fig. 2. The only force present is the volume force due to the tidal gravitational force $\boldsymbol{K} = -\rho\nabla U$, which will be modeled as gradient of a spherically symmetric Earth acceleration field.

To calculate the elastic deformations of the cylinder, the equations of elasticity have to be solved including the influence of the tidal gravitational force.

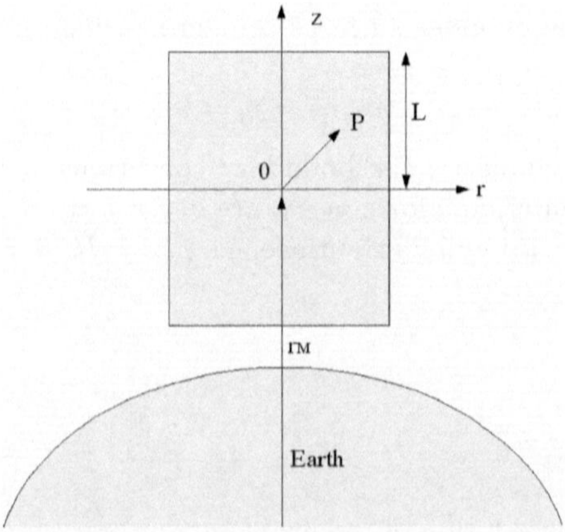

Fig. 2. Simplified model of an optical resonator on a geodetic Earth orbit.

The equilibrium equation of elasticity describes the equilibrium state of a homogeneous isotropic body when a volume force $\boldsymbol{K}$ is acting [4]

$$\Delta\boldsymbol{\xi} + \frac{1}{1-2\nu}\nabla(\nabla\cdot\boldsymbol{\xi}) + \frac{1}{\rho}\boldsymbol{K} = 0 \tag{3}$$

where $\boldsymbol{\xi}$ is the displacement vector. For vanishing volume forces $\boldsymbol{K} = 0$, we get the homogeneous equation of elasticity. The general solution of (3) is a superposition of a homogeneous and a particular solution

$$\boldsymbol{\xi} = \boldsymbol{\xi}^{\mathrm{h}} + \boldsymbol{\xi}^{\mathrm{p}} . \tag{4}$$

Assuming a spherical Earth potential, $U(r) = GM_\oplus/r$, where $GM_\oplus$ is the gravitational constant times the mass of Earth, the potential acting at an arbitrary point P inside the cylinder can be calculated via Taylor expansion

$$U(\boldsymbol{r}_M + \boldsymbol{r}) = U(\boldsymbol{r}_M) + \frac{\partial U(\boldsymbol{r}_M)}{\partial r_i} r_i + \frac{1}{2}\frac{\partial^2 U(\boldsymbol{r}_M)}{\partial r_i \partial r_j} r_i r_j \tag{5}$$

$$= U(\boldsymbol{r}_M) + \nabla U(\boldsymbol{r}_M)\boldsymbol{r} + \frac{GM_\oplus}{2r_M^3}(r^2 - 2z^2) \tag{6}$$

where $\boldsymbol{r}_M$ is the vector from the center-of-mass of the Earth to the center-of-mass of the cylinder and $\boldsymbol{r}$ is the vector from the cylinder center-of-mass to point P. The linear term in the Taylor expansion vanishes as this equation is valid in the freely falling reference frame of the cylinder.

Beside the axial symmetry, we also have the following symmetries for the r and z components of the displacement vector for reflection at the $z = 0$ plane: $\xi_z(r, -z) = -\xi_z(r, z)$ and $\xi_r(r, -z) = \xi_r(r, z)$.

The boundary conditions for the solution are given through the condition of weightlessness in space, that means no external forces are present and, thus, the forces $\boldsymbol{p}$ at the cylinder surfaces are zero, which gives us the boundary conditions

$$\sigma_{ij} n_j = p_i = 0 \; . \tag{7}$$

Concerning axis symmetry, the boundary conditions at the top and bottom surface of the cylinder, i.e., for $z = \pm L$ are $\sigma_{rz}(r, z = \pm L)$ and $\sigma_{zz}(r, z = \pm L)$ and at the superficies cylinder surface, i.e., $r = R$, $0 = \sigma_{rr}(r = R, z)$ and $0 = \sigma_{zr}(r = R, z)$.

The Solution

Particular Solution

To find a particular solution of the problem, one can assume that the solution of the equilibrium equation can be written as gradient of a scalar ψ [4] $\boldsymbol{\xi}^{\mathrm{p}} = \nabla\psi$. Inserting this approach into (3) and using the potential U calculated in (5), we get for the r and z components of the displacement (with $\mu = \lambda\frac{1-2\nu}{2\nu}$)

$$\begin{aligned} \xi^{\mathrm{p}}_r &= -\frac{1-2\nu}{2(1-\nu)\mu}\rho\frac{GM_\oplus}{2r_M^3}\left(-\frac{r^3}{4} + 2cr\right) = \frac{\nu}{\lambda(1-\nu)}\gamma\left(-\frac{r^3}{4} + 2cr\right) \\ \xi^{\mathrm{p}}_z &= -\frac{1-2\nu}{2(1-\nu)\mu}\rho\frac{GM_\oplus}{2r_M^3}\left(\frac{2}{3}z^3 + 2dz\right) = \frac{\nu}{\lambda(1-\nu)}\gamma\left(\frac{2}{3}z^3 + 2dz\right) , \end{aligned} \tag{8}$$

where we substituted $\gamma := -GM/(2r_M^3)\rho$. Herein, c and d are arbitrary constants.

Using the well-known relations between displacements and stresses σ_{ij} for cylindrical coordinates [2, 4], we obtain the stress components out of the displacements

$$\begin{aligned} \sigma^{\mathrm{p}}_{rr} &= \gamma\left(\frac{(2\nu-3)}{4(1-\nu)}r^2 + \frac{2(c+\nu d)}{1-\nu} + \frac{2\nu}{1-\nu}z^2\right) \\ \sigma^{\mathrm{p}}_{zz} &= \gamma\left(\frac{4c\nu}{1-\nu} - \frac{\nu}{1-\nu}r^2 + 2(z^2+d)\right) \\ \sigma^{\mathrm{p}}_{rz} &= 0 \; . \end{aligned} \tag{9}$$

Homogeneous Solution

Derivation of the Boundary Conditions. Now we are looking for a homogeneous solution, i.e., a displacement vector that satisfies the homogeneous equation of elasticity

$$\Delta\boldsymbol{\xi}^{\mathrm{h}} + \frac{1}{1-2\nu}\nabla(\nabla\cdot\boldsymbol{\xi}^{\mathrm{h}}) = 0 \; . \tag{10}$$

The boundary conditions the homogeneous solution part has to fulfill can be derived from the boundary conditions (7) of the complete solution.

Then the boundary conditions at the cylinder top and bottom surface, i.e., $z = \pm L$ are

$$\sigma^{\mathrm{h}}_{zz}(r, z = \pm L) = -\sigma^{\mathrm{p}}_{zz}(r, z = \pm L) = -\gamma\left(\frac{4c\nu}{1-\nu} - \frac{\nu}{1-\nu}r^2 + 2(L^2 + d)\right)$$
$$\sigma^{\mathrm{h}}_{rz}(r, z = L) = -\sigma^{\mathrm{p}}_{rz}(r, z = L) = 0 \,. \qquad (11)$$

For the superficies surface, i.e., $r = R$ we get

$$\sigma^{\mathrm{h}}_{rr}(r = R, z) = -\sigma^{\mathrm{p}}_{rr}(r = R, z)$$
$$= -\gamma\left(\frac{(2\nu-3)}{4(1-\nu)}R^2 + \frac{2(c+\nu d)}{1-\nu} + \frac{2\nu}{1-\nu}z^2\right)$$
$$\sigma^{\mathrm{h}}_{zr}(r = R, z) = -\sigma^{\mathrm{p}}_{zr}(r = R, z) = 0 \quad . \qquad (12)$$

General Ansatz for the Homogeneous Solution. Love [3] showed that the deformations in an elastic axis-symmetric body can be expressed in terms of the so-called Love function χ

$$\xi^{\mathrm{h}}_r = -\frac{1+\nu}{E}\frac{\partial^2\chi}{\partial r\partial z}$$
$$\xi^{\mathrm{h}}_z = \frac{1+\nu}{E}\left((1-2\nu)\nabla^2\chi + \frac{\partial^2\chi}{\partial r^2} + \frac{1}{r}\frac{\partial\chi}{\partial r}\right) \qquad (13)$$

where ξ_r and ξ_z are the displacement components.

The corresponding stress components are

$$\sigma^{\mathrm{h}}_{rr} = \frac{\partial}{\partial z}\left(\nu\nabla^2\chi - \frac{\partial^2\chi}{\partial r^2}\right) \qquad \sigma^{\mathrm{h}}_{rz} = \frac{\partial}{\partial r}\left((1-\nu)\nabla^2\chi - \frac{\partial^2\chi}{\partial z^2}\right)$$
$$\sigma^{\mathrm{h}}_{zz} = \frac{\partial}{\partial z}\left((2-\nu)\nabla^2\chi - \frac{\partial^2\chi}{\partial z^2}\right) \qquad \sigma^{\mathrm{h}}_{\phi\phi} = \frac{\partial}{\partial r}\left(\nu\nabla^2\chi - \frac{1}{r}\frac{\partial\chi}{\partial r}\right) \,. \qquad (14)$$

The Love function χ necessarily fulfills the biharmonic equation

$$\nabla^2\nabla^2\chi = 0 \,. \qquad (15)$$

The major obstacle is to find an adequate approach for the Love function fulfilling all boundary conditions. By modifying the so-called Papkovich–Neuber approach [1], one can make an ansatz for the Love function suggested in [6]

$$\chi = B_0 z^3 + \sum_{j=1}^{\infty}\left(A_j\frac{\sinh(\lambda_j z)}{\sinh(\lambda_j L)} + B_j z\frac{\cosh(\lambda_j z)}{\sinh(\lambda_j L)}\right)\frac{J_0(\lambda_j r)}{\lambda_j^2}$$
$$+D_0 r^2 z + \sum_{n=1}^{\infty}\left(C_n\frac{I_0(k_n r)}{I_1(k_n R)} + D_n r\frac{I_1(k_n r)}{I_1(k_n R)}\right)\frac{\sin(k_n z)}{k_n^2} \,. \qquad (16)$$

Herein, J_0 are the Bessel functions of first kind and order zero and I_1 are the modified Bessel functions of order one. Furthermore, $\zeta_j = \lambda_j R$ are the zeros of the Bessel functions of order one, $J_1(\zeta_j) = 0$, and $k_n = \frac{n\pi}{L}$ where n is an integer number.

By inserting the Love function approach into (14), one obtains the equations for the stress components that have to fulfill the boundary conditions (11) and (12). Thus, one can determine the unknown coefficients. The Love function approach was chosen in such a way that arbitrary boundary conditions at the cylinder boundaries can be fulfilled. For the determination of the coefficients, the properties of Bessel functions, trigonometric functions as well as Fourier and Dini series have to be used. As the derivation is somewhat bulky, we just give the results here and refer to [7] for the detailed calculations.

The coefficients in (16) are $(n, j = 1 \ldots \infty)$:

$$A_j \lambda_j = -B_j \left(2\nu + \lambda_j L \coth(\lambda_j L)\right) \tag{17}$$

$$C_n k_n = -D_n \left((2 - 2\nu) + k_n R \frac{I_0(k_n R)}{I_1(k_n R)} \right) \tag{18}$$

$$\begin{aligned} D_n &= -\left(k_n R \left(\frac{I_0(k_n R)^2}{I_1(k_n R)^2} - 1 \right) - \frac{2 - 2\nu}{k_n R} \right)^{-1} \\ &\quad \times \left(\frac{4\hat{c}_2 (-1)^n}{k_n^2} + \sum_{m=1}^{\infty} B_m (-1)^n \frac{4\lambda_m k_n^2}{L(k_n^2 + \lambda_m^2)^2} J_0(\lambda_m R) \right) \end{aligned} \tag{19}$$

$$\begin{aligned} B_j &= \left(J_0(\lambda_j R) \left(\coth(\lambda_j L) + \frac{\lambda_j L}{\sinh^2(\lambda_j L)} \right) \right)^{-1} \\ &\quad \times \left(\frac{2}{\lambda_j^2} \hat{c}_2 - \sum_{k=1}^{\infty} (-1)^k D_k k_k \left(\frac{4\lambda_j^2}{R(k_k^2 + \lambda_j^2)^2} \right) \right) \end{aligned} \tag{20}$$

$$B_0 = \frac{1}{6} \left(-\hat{c}_0 - \hat{c}_1 - \mathcal{S} \right) - D_0 \tag{21}$$

$$D_0 = \frac{1}{2(1+\nu)} \left((1-\nu)\hat{c}_1 - \nu \left(\hat{c}_0 + \mathcal{S} \right) + \mathcal{R} \right) \tag{22}$$

with

$$\hat{c}_0 = \gamma \left(\frac{4c\nu}{1-\nu} + 2(L^2 + d) \right) \quad \hat{c}_1 = \gamma \left(\frac{(2\nu - 3)}{4(1-\nu)} R^2 + \frac{2(c + \nu d)}{1-\nu} \right) \quad \hat{c}_2 = \gamma \frac{2\nu}{1-\nu} \tag{23}$$

and

$$\mathcal{S} = \mathcal{R} + \mathcal{Z} \tag{24}$$

$$\mathcal{R} = \sum_{j=1}^{\infty} B_j \left(\frac{(1 - \lambda_j L \coth(\lambda_j L))}{\sinh(\lambda_j L)} \right) J_0(\lambda_j R)$$
$$+ \sum_{n=1}^{\infty} D_n \left(k_n R \left(\frac{I_0(k_n R)^2}{I_1(k_n R)^2} - 1 \right) - \frac{2 - 2\nu}{k_n R} \right) \tag{25}$$

$$\mathcal{Z} = \sum_{j=1}^{\infty} B_j \left(\coth(\lambda_j L) + \frac{\lambda_j L}{\sinh^2(\lambda_j L)} \right)$$
$$+ \sum_{n=1}^{\infty} (-1)^n D_n \left(2 - k_n R \frac{I_0(k_n R)}{I_1(k_n R)} \right) \frac{1}{I_1(k_n R)} \tag{26}$$

Having found all unknown coefficients in (16), we can, following (13), directly calculate the homogeneous displacement. By adding the homogeneous and the particular solution from (8), we get the total displacement of the cylinder. Note that the solution forms an infinite system of equations whose convergence can be shown. So it can be approximately solved by reducing it to a finite system, i.e., by expanding the sums only to $n = N$ and $j = J$. Then, we have a system of $N + J$ equations. By increasing the values of N and J, one can improve the accuracy of the solution and find their limits.

Comparison Between Analytical and FEM Solution

Table 2 shows the results from numerical evaluation of the infinite system of equations for different orders $N = J$. Although the infinite series converge very quickly, an expansion to higher orders $N = J$ still gives an improvement of accuracy.

Figure 3 shows the resulting total displacement field from the analytical solution of (4). The infinite sums in the analytical solution were expanded to $N = J = 1{,}700$.

Now we can compare our analytical solution for our simplified problem with the result obtained with help of a finite element code to verify its applicability. The finite element analysis was done with the commercial FEM code ANSYS. The cylinder model of radius $R = 1$ and length $L = 2R$ was

Table 2. Comparison of the analytical solution at point $r = 1 = R$, $z = 2 = L$ for different expansion orders $N = J$ of the infinite sums in equation system (19).

$N = J$	ξ_r	ξ_z
100	$-6.471 \cdot 10^{-15}$	$1.637 \cdot 10^{-13}$
1,000	$-6.233 \cdot 10^{-15}$	$1.629 \cdot 10^{-13}$
1,700	$-6.218 \cdot 10^{-15}$	$1.628 \cdot 10^{-13}$

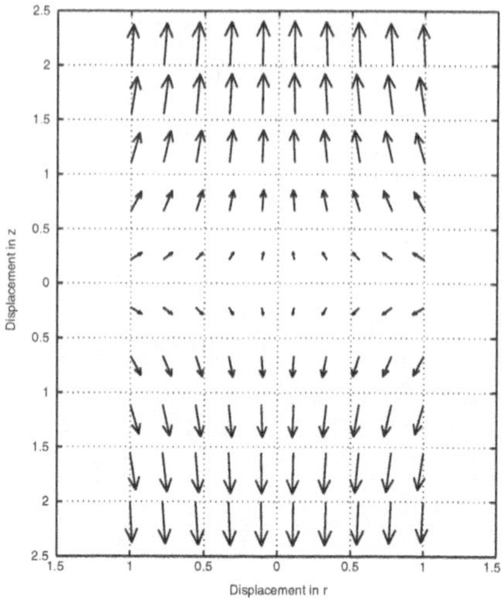

Fig. 3. Displacement field of a cylinder under influence of a spherical tidal gravitational force field. The displacements are plotted over the body coordinates r and z. The cylinder boundaries are at $z = \pm L = \pm 2$ and $r = R = 1$.

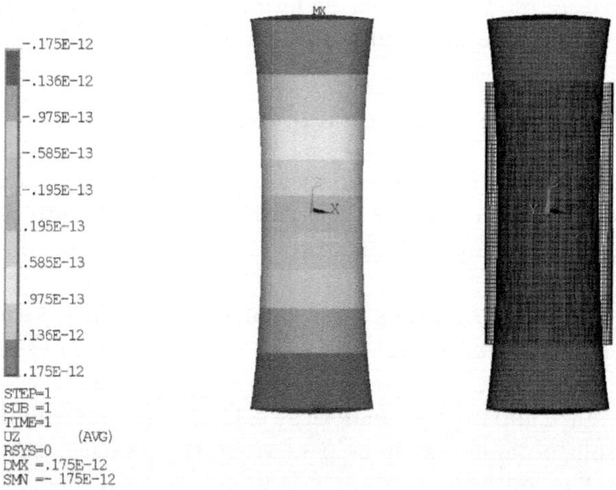

Fig. 4. Finite element solution: deformation of the cylinder under influence of a spherical tidal gravitational force field. The deformation is scaled by a factor of $6 \cdot 10^{13}$. *Right*: deformed cylinder shape and original finite element mesh. *Left*: the scale shows the z displacements.

divided into approximately 110,000 hexahedron elements. Hexahedron elements allow the creation of a structured finite element mesh that ensures a high relative accuracy of the finite element solution. Figure 4 shows the deformation of the cylinder in the tidal gravitational force field as result of the FEM calculation.

Table 3. Displacements ξ_r and ξ_z in r- and z-direction.

		Analytical		Numerical	
r	z	ξ_r	ξ_z	ξ_r	ξ_z
1	2	$-6.218 \cdot 10^{-15}$	$1.628 \cdot 10^{-13}$	$-6.202 \cdot 10^{-15}$	$1.625 \cdot 10^{-13}$
1	0	$-3.491 \cdot 10^{-14}$	0	$-3.486 \cdot 10^{-14}$	$-1.66 \cdot 10^{-23}$
$0.49507\sqrt{2}$	1	$-2.059 \cdot 10^{-14}$	$1.151 \cdot 10^{-13}$	$-2.062 \cdot 10^{-14}$	$1.150 \cdot 10^{-13}$
$0.11888\sqrt{2}$	1	$-5.208 \cdot 10^{-15}$	$1.183 \cdot 10^{-13}$	$-5.236 \cdot 10^{-15}$	$1.182 \cdot 10^{-13}$

Comparison between analytical and FEM solution. The cylinder boundaries are $r = R = 1$ and $z = \pm L = \pm 2$. In the analytical solution, the infinite series were expanded to $N = J = 1,700$. The constant $\gamma := -\frac{GM_\oplus}{2r_M^3}\rho \approx -1.4 \cdot 10^{-3}$ which corresponds to a 7,000 km Earth orbit.

Table 3 contains a quantitative comparison between the displacements from the analytical and FEM solution for special points of the cylinder. Analytical as well as FEM solutions are completely symmetric, i.e., the norms of the displacements of two opposing points of the cylinder are equal. Please note that we have chosen a very fine mesh for the finite element solution to get high accuracy.

Obviously, the FEM solution and the analytical solution agree very well. The small differences are based on the nature of the FEM analysis. The elements in which the cylinder is divided cannot be chosen to be infinite small, they are "finite" and thus the FEM solution is a kind of summation over all elements of finite size instead of an integration where the limit to infinitesimal small element size can be performed. Furthermore, during the FEM analysis, at least three points must be fixed to prevent the cylinder from rigid rotations. In the current FEM analysis, the center-of-mass as well as four of the next nodes were fixed so as to assure the perfect symmetry of the FEM solution.

5.2 The OPTIS Resonator Under a Tidal Gravitational Force

As we have estimated analytically that the influence of the gravity gradient on a resonator in space will be larger than the science requirement for the OPTIS mission, it has to be investigated numerically. Therefore, we carried out structural analysis within the FEM program ANSYS.

We modeled the OPTIS resonator as a cube with side length 6 cm, having three drillings, one in each body axis direction. The drillings are closed by circular mirrors at each end. In contrast to the Earth experiments, all three axis are equal, thus laser beams can be coupled into the resonator in three axis directions. We used the material properties of Corning ULE (ultra-low expansion) glass, which means elasticity modulus $E = 67.6\,\text{GPa}$, Possion number $\nu = 0.17$, density $\rho = 2,210\,\text{kg}\,\text{m}^{-3}$.

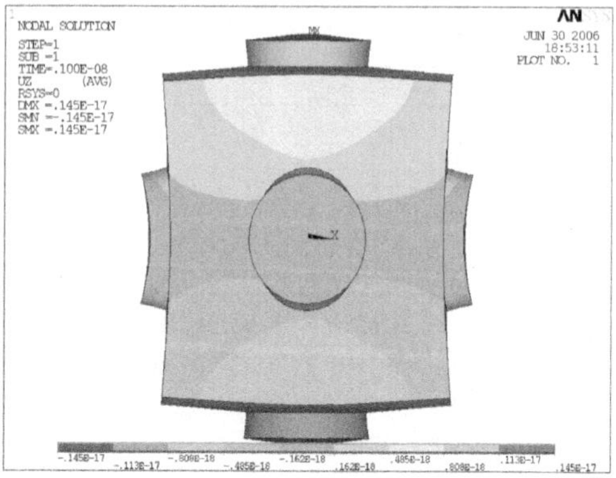

Fig. 5. Deformation of the OPTIS resonator under the influence of a gravity-gradient force. The color bar gives the values of the z displacements in meter.

Figure 5 shows the result of the FEM analysis giving the deformation of the resonator under the influence of a gravity-gradient force field acting in z-direction of the form

$$K_x^0 = 2x\gamma$$

$$K_y^0 = 2y\gamma$$

$$K_z^0 = -4z\gamma \tag{27}$$

where $\gamma := -\frac{GM\oplus/}{2r_M^3}\rho$. For the numerical calculations, the constant is $\gamma \approx\approx -1.4 \cdot 10^{-3}$ which corresponds to a 7,000 km Earth orbit.

One can see that the originally cubical resonator is squeezed in x- and y-direction and is pulled out in z-direction.

As the resonator is rotating around the Earth, it is subject to a changing gravity-gradient field. Therefore, the deformations of the resonator on his orbits around the Earth have to be investigated. In addition, the resonator is rotating around its body z-axis approximately once per 10 s.

Thus, the resulting changing deformation due to the gravity gradient will be visible in the measurement signal of the Michelson–Morley experiment and has to be subtracted.

In the following, we show plots for the relative displacements dx, dy, dz between two mirror midpoints on the resonator axes. The FEM calculations were carried out for half an orbital rotation of the resonator within the inertial $x-z$ plane around the Earth (that means a rotation around the inertial y-axis with angle $\phi_y = 0 \dots \pi$) and a simultaneous rotation of the resonator around its body z-axis with angle $\phi_z = 0 \dots 2\pi$.

Figures 6–9 show the displacements of the mirrors $\Delta x, \Delta y, \Delta z$ for the two opposing mirrors on each resonator axis as well as the relative displacements

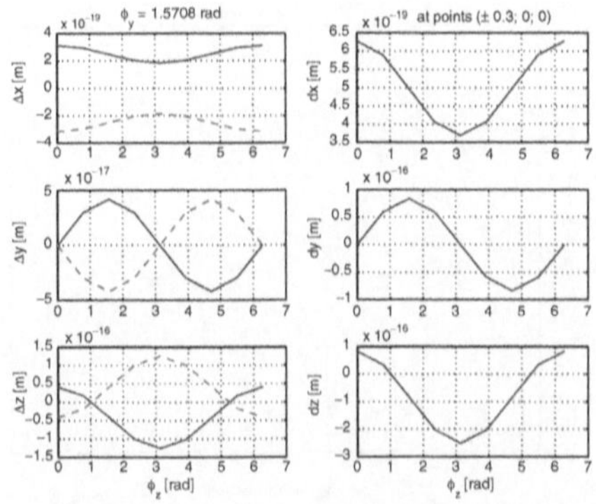

(a) Mirrors on the x-axis

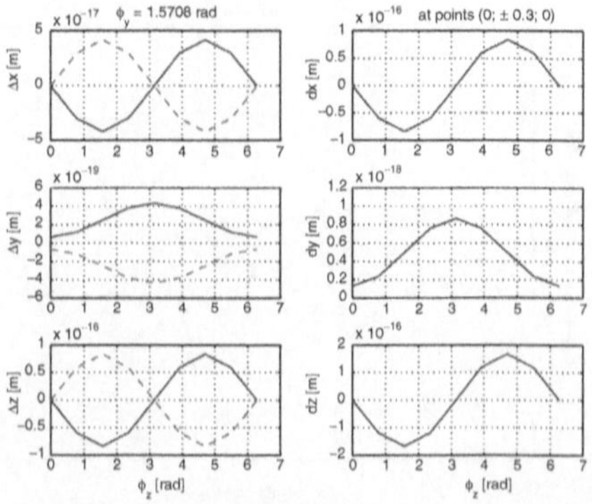

(b) Mirrors on the y-axis

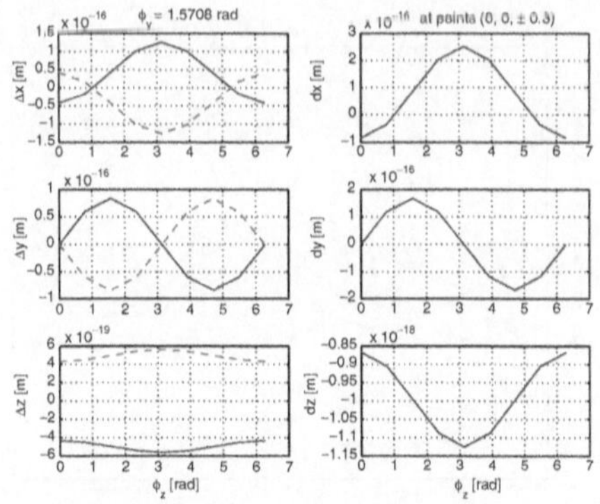

(c) Mirrors on the z-axis

Fig. 6. OPTIS resonator with gravity gradient at point $\phi_y = \pi/2$ on its orbit around the Earth. *Left side* of each subfigure: displacements at the mirror midpoints, *dotted line* represents mirror on negative original coordinate point and *full line* represents mirror on positive original coordinate point; *Right side*: relative displacements between two opposing mirrors.

dx, dy, dz on these axes. The data show one complete rotation of the resonator around its body z-axis when the resonator is at point $\phi_y = \pi/2$ on it way around the Earth, which means when the gravity-gradient force is acting completely in the body x-direction.

Figure 7–9 shows contour plots as two-dimensional representation of the three-dimensional data. In all plots, the data for $\phi_y = \phi_z = 0$, that means at the status of deformation where the laser beams are coupled into the resonator, have already been subtracted. That means the numbers on the contour lines give the relative displacements of the mirrors with respect to this starting state.

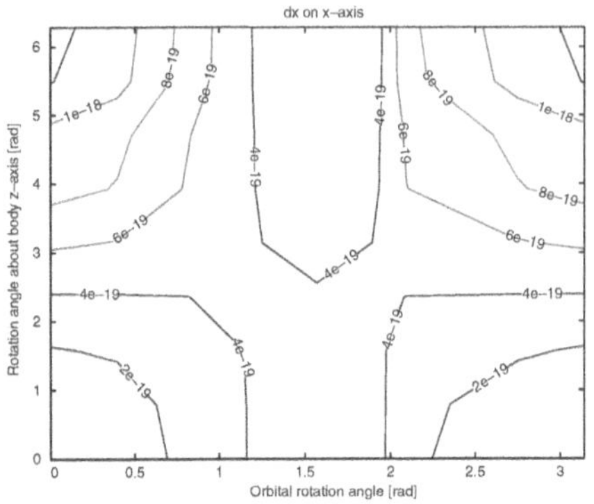

(a) Contour plots of the relative mirror displacements dx on the x-axis

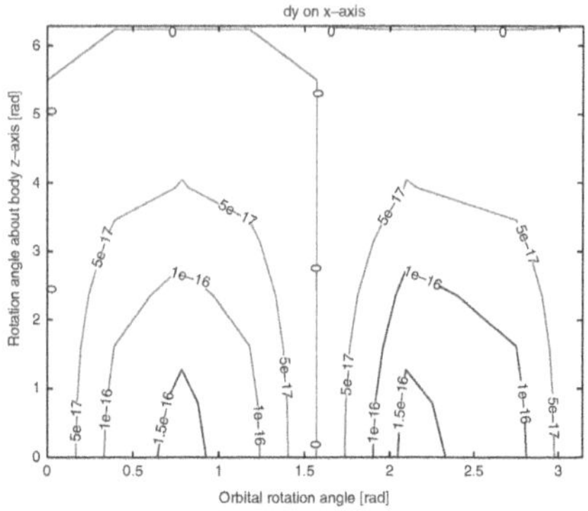

(b) Contour plots of the relative mirror displacements dy on the x-axis

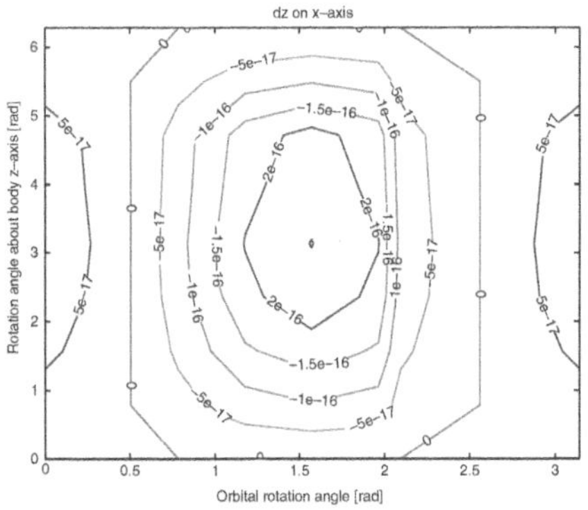

(c) Contour plots of the relative mirror displacements dz on the x-axis

Fig. 7. OPTIS resonator with gravity gradient. The data correspond to an orbital rotation around the inertial y-axis for half an orbit with angle ϕ_y and a rotation of the resonator around its body z-axis with angle ϕ_z.

6 OPTIS Resonator Under Thermal Gradient

To analyze the influence of thermal gradients on the deformation of the OPTIS resonator, a thermal analysis within the FEM program ANSYS was carried out. Figure 10 shows the deformation of the resonator assuming a thermal gradient of 10^{-9} K between the upper and lower surface in z-direction. We used the material properties of Corning ULE (ultra-low expansion) glass, which has so far the lowest thermal expansion coefficient of $\alpha \approx 10^{-9}\,\mathrm{K}^{-1}$.

The relative deformations between the mirrors on the x, y, and z-axis resulting from this temperature gradient are $dx = dy = dz = 2.99 \cdot 10^{-20}\,\mathrm{m}$. This is still within the science requirement $dL/L \leq 10^{-18}$, which means for a resonator of $L = 6\,\mathrm{cm}$ $dL \leq 6 \cdot 10^{-20}$.

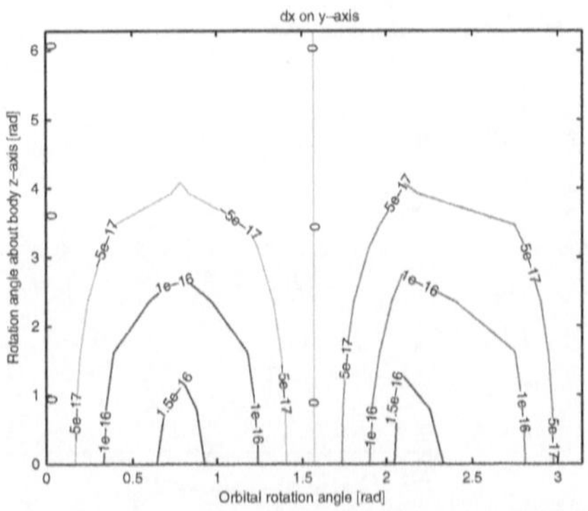

(a) Contour plots of the relative mirror displacements dx on the y-axis

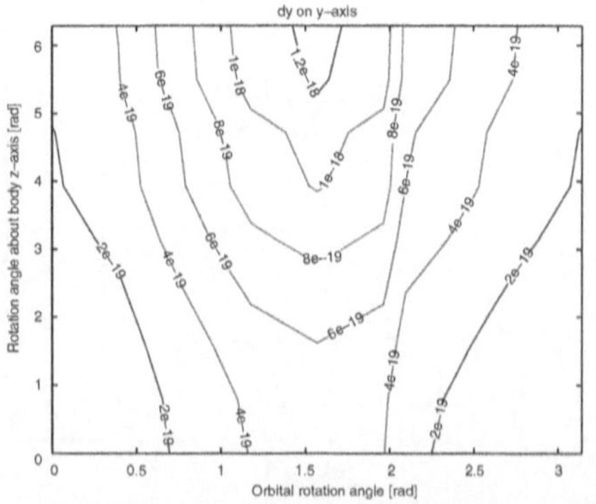

(b) Contour plots of the relative mirror displacements dy on the y-axis

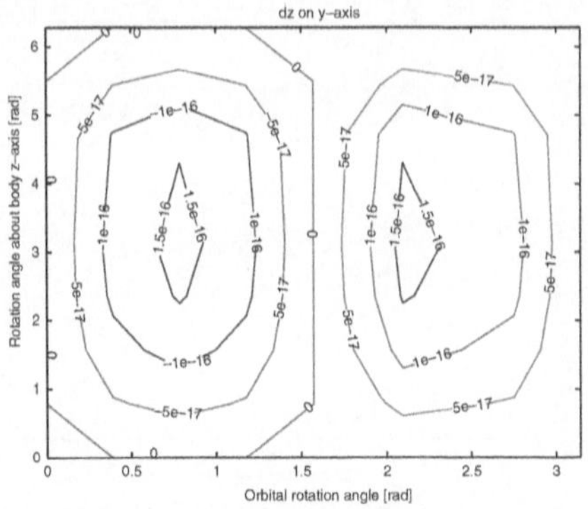

(c) Contour plots of the relative mirror displacements dz on the y-axis

Fig. 8. OPTIS resonator with gravity gradient. The data correspond to an orbital rotation around the inertial y-axis for half an orbit with angle ϕ_y and a rotation of the resonator around its body z-axis with angle ϕ_z.

Enlarging the temperature gradient by one order of magnitude increases the displacements by one order of magnitude and thus the science requirement cannot be fulfilled. Using a material with an higher thermal expansion coefficient, e.g., Zerodur with $\alpha \approx 10^{-7}\,\mathrm{K}^{-1}$, enlarges the deformations by two orders of magnitude.

Thus, we can conclude that for the OPTIS resonator made of ULE the temperature gradient must be smaller than $10^{-9}\,\mathrm{K}$ over its length of 6 cm, which corresponds to a requirement of smaller than $10^{-7}\,\mathrm{K\,m}^{-1}$, to fulfill the science requirements. Furthermore, this means that the temperature stability between the endpoints of the resonator must be better than $10^{-9}\,\mathrm{K}$ on a timescale (L^2/χ_T) that is determined by the temperature conductivity χ_T of the material.

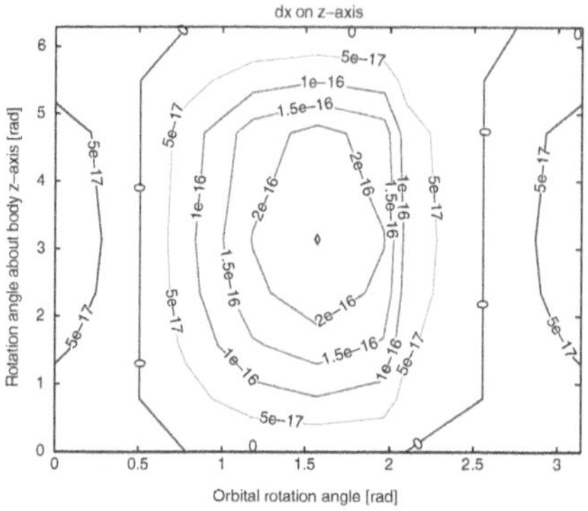

(a) Contour plots of the relative mirror displacements dx on the z-axis

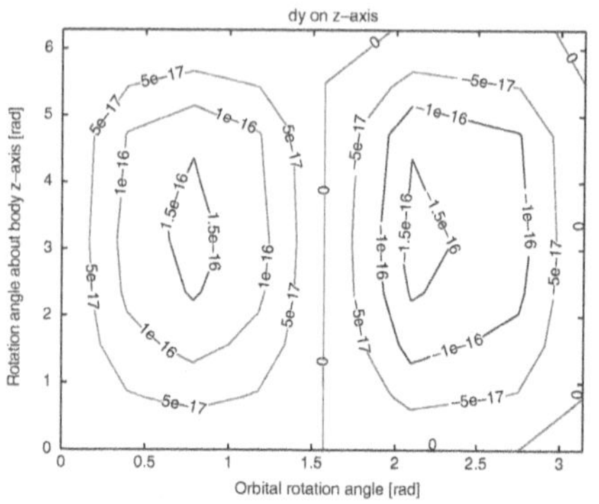

(b) Contour plots of the relative mirror displacements dy on the z-axis

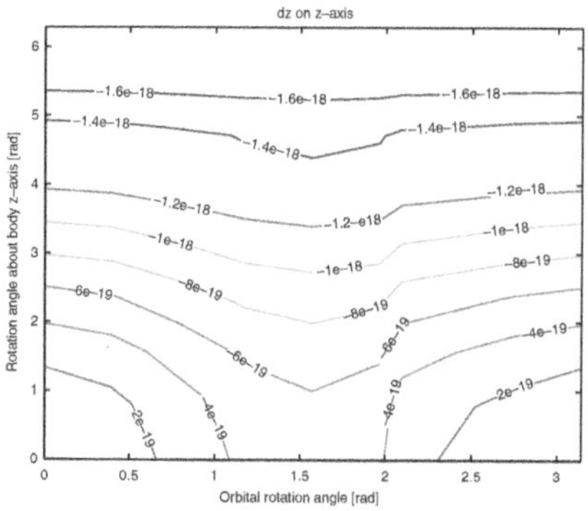

(c) Contour plots of the relative mirror displacements dz on the z-axis

Fig. 9. OPTIS resonator with gravity gradient. The data correspond to an orbital rotation around the inertial y-axis for half an orbit with angle ϕ_y and a rotation of the resonator around its body z-axis with angle ϕ_z

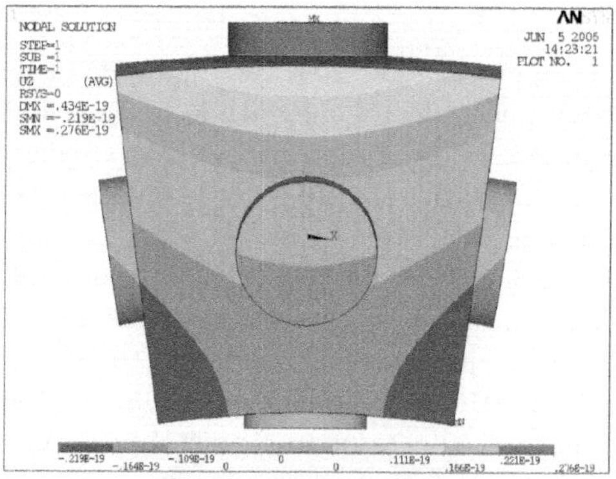

Fig. 10. OPTIS resonator under the influence of a linear temperature-gradient field of 10^{-9} K in z-axis direction. The color bar gives the values of the z displacements in meter. The deformation is scaled by a factor of $3 \cdot 10^{17}$.

Acknowledgments

We thank I. Ciufolini, L. Iorio, and G. Schäfer for their support and various discussions. Financial support from the German Research Foundation DFG and the German Aerospace Agency DLR is gratefully acknowledged.

References

1. A. D. Kovalenko. *Thermoelasticity – Basic Theory and Applications.* Wolters-Noordhoff Publishing Groningen, 1969.
2. H. Leipholz. *Theory of Elasticity.* Noordhoff International Publishing, Leyden, 1974.
3. A.E.H. Love. *A Treatise on the Mathematical Theory of Elasticity.* 4th edn. Cambridge University Press, Cambridge, 1927.
4. A.I. Lurje. *Rumliche Probleme der Elastizittstheorie.* Akademie-Verlag, Berlin, 1963.
5. C. Lämmerzahl, I. Ciufolini, H. Dittus, L. Iorio, r H. Mlle, A. Peters, E. Samain, S. Scheithauer, and S. Schiller. OPTIS – An Einstein Mission for Improved Tests of Special and General Relativity. *General Relativity and Gravitation*, 36(10), 2004.
6. V.V. Meleshko. Equilibrium of an elastic finite cylinder: Filon's problem revisted. *Journal of Engineering Mathematics*, 16:355–376, 2003.
7. S. Scheithauer and C. Lämmerzahl. Analytical Solution for the Deformation of a Cylinder under Tidal Gravitational Forces. Class. Quantum Grav. 23:7273, 2006.

Testing Relativistic Gravity to One Part per Billion

Wei-Tou Ni, Antonio Pulido Patón, and Yan Xia

Purple Mountain Observatory, No. 2, Beijing W. Rd., Nanjing, 210008 China
wtni@pmo.ac.cn

Summary. In 1859, Le Verrier discovered the mercury perihelion advance anomaly. This is a 10^{-7}-part deviation from Newtonian gravity. This together with new theoretical ideas prompted the formulation of general relativity. In a similar way, will a test of relativistic gravity to 10^{-7}–10^{-9} prompt a new theory of gravity and probe significantly into the microscopic origin of gravity? We envisage that the time is ripe to test relativistic gravity to 10^{-9} level of accuracy (1 ppb) within the next 25 years. Deep-space pulse and interferometric ranging holds such a key. This will ensure a revival of interest both in experimental and theoretical aspects of gravitation, and may lead to answer some profound questions about the origin of gravity and the cosmos.

1 Introduction

The discovery that the expansion of our Universe is accelerating and the cosmological constant is nonvanishing put general relativity into an empirically tight position. The correct semiclassical or microscopic gravitation theory is needed – the cosmological constant needs to be explained and the microscopic origin of gravity needs to be probed.

As in other fields of Physics, we advocate the interaction approach: experiments $\longleftrightarrow$ theory. We are now in an unique position in time epoch to discern a test of relativistic gravity to one part per billion (i.e., 10^{-9}) within the next 25 years. Since Le Verrier [1], the precision of tests of relativistic gravity has been improved by 3–4 orders of magnitude [2]. Recently, in 2003, Bertotti, Iess, and Tortora reported a measurement of the relativistic Shapiro time delay from the Cassini spacecraft to give the relativistic-gravity parameter γ the value 1.000021 ± 0.000023 of general relativity – a 1.5-order improvement over previous results [3]; in 2004, Ciufolini and Pavlis [4] reported a measurement of the Lense–Thirring effect on the LAGEOS and LAGEOS2 satellites to be 0.99 ± 0.10 times the value predicted by general relativity. In April 2004, Gravity Probe B was launched and has recently finished the experiment; for measuring the Lense–Thirring effect, a preliminary analysis

Table 1. Aimed accuracy of PPN light deflection/light retardation parameter γ for various ongoing/proposed experiments.

Mission	Ongoing/proposed experiment	Aimed accuracy of γ
GP-B [5]	Geodetic precession	$1 \cdot 10^{-5}$
Bepi-Colombo [10]	Retardation	$2 \cdot 10^{-6}$
GAIA [11]	Deflection	$1 \cdot 10^{-5} - 2 \cdot 10^{-7}$
ASTROD I [7]	Retardation	$1 \cdot 10^{-7}$
LATOR [9]	Deflection	$1 \cdot 10^{-8}$
ASTROD [8]	Retardation	$1 \cdot 10^{-9}$

The types of experiments (deflection, retardation, or geodetic precession) are given in the second row.

has given an accuracy of about 1% level; for the light deflection/light retardation parameter γ of relativistic-gravity. Gravity Probe B will reach $1 \cdot 10^{-5}$ level [5].

Deep-space laser ranging holds the key to test relativistic gravity to one part per billion (i.e., 10^{-9}) [6]. Pulse laser ranging with drag-free spacecraft will reach 10^{-7} in testing relativistic gravity. One such mission under study is ASTROD I (Astrodynamical Space Test of Relativity using Optical Devices I) [7]. Interferometric laser ranging with drag-free spacecraft will reach 10^{-9}; ASTROD is aiming at this level of accuracy [8]. The astrometric mission LATOR is aiming at an accuracy of 10^{-8} by using laser light for measuring angles [9]. Table 1 compiles these aims together with those of Bepi-Colombo mission [10] to Mercury and GAIA (Global Astrometric Interferometer for Astrophysics) mission [11] for global astrometry.

2 ASTROD I

ASTROD I with one spacecraft interferometric ranging and pulse ranging with ground stations is the first step for a full ASTROD mission [7]. Spacecraft orbit can be designed for a 2012/2013/2015 launch (separated by about a synchronous year (584 days) of Venus), using two encounters with Venus to swing the spacecraft to the other side of the Sun in 370 days and having the second opposition around 720 days, to conduct Shapiro time-delay measurements and various tests of relativistic gravity. The spacecraft is three-axis stabilized, contains a three-axis drag-free proof (test) mass, and is to follow this proof mass using microthrusters. The drag-free performance requirement is $10^{-13}\,\mathrm{m\,s^{-2}\,Hz^{-1/2}}$ between 0.1 and 1 mHz (three-axis) and would give an error comparable to 10 ps timing error in about 1 year. A $50 \times 50 \times 35\ \mathrm{mm}^3$ rectangular parallelepiped proof mass made from Au–Pt alloy of low magnetic susceptibility ($<5 \cdot 10^{-5}$) is planned to be used. Six-degree-of-freedom capacitive sensing for the proof mass will be implemented. The laser ranging is between a fiducial point in the spacecraft and a fiducial point in the ground laser station. Assuming a 10 ps timing accuracy and $10^{-13}\,\mathrm{m\,s^{-2}\,Hz^{-1/2}}$ (at $f \approx 0.1$ mHz) accelerometer noise, our simulation for 400 days (350–750 days after launch) of the accuracy for determining the relativistic parameters γ and β (relativistic nonlinearity parameter), and the solar quadrupole parameter J_2 gives 10^{-7}, 10^{-7}, and $3.8 \cdot 10^{-9}$, respectively, for their uncertainties [12].

3 ASTROD

A baseline implementation of the Astrodynamical Space Test of Relativity using Optical Devices (ASTROD) is to have two spacecraft in separate solar orbits carrying a payload of one proof mass, two telescopes, two 1–2 W lasers, a clock, and a drag-free system, together with a similar L1/L2 spacecraft [6,8]. The three spacecraft range coherently with one another using lasers to map the solar-system gravity, test relativistic gravity, and detect gravitational waves. In 2.5 years, the inner spacecraft goes three orbits, the outer spacecraft goes two orbits, and the earth goes 2.5 orbits to the other side of the Sun to conduct a Shapiro time-delay experiment efficiently. The spacecraft configuration after 700 days from launch is shown in Fig. 1. With technological development and with the implementation of ASTROD I, the drag-free requirement and the timing accuracy can be improved for ASTROD. ASTROD aims at a tenfold improvement with respect to LISA's accelerometer noise performance,

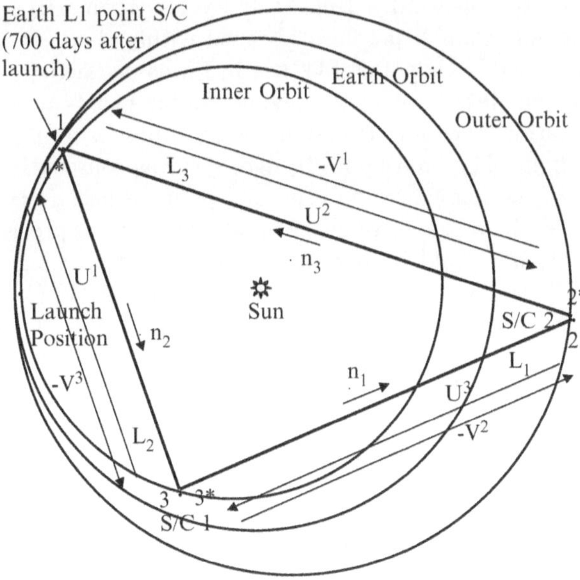

Fig. 1. A schematic ASTROD configuration (baseline ASTROD after 700 days of launch).

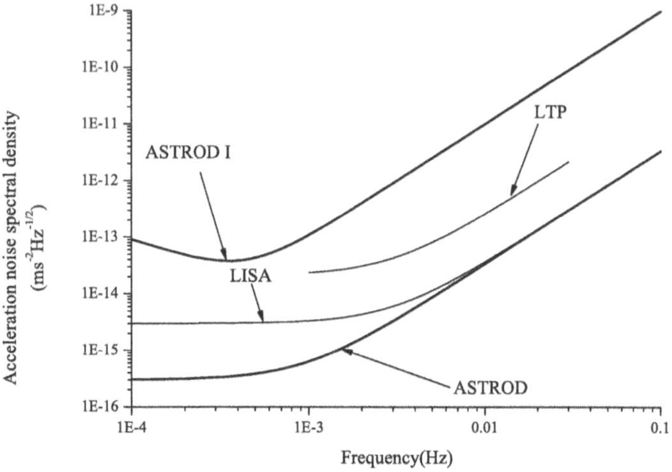

Fig. 2. Acceleration noise spectral density requirements for ASTROD and ASTROD I. The LISA Technology Package (LTP) and LISA requirements are illustrated in the figure for comparison.

i.e., about $3 \cdot 10^{-16}\,\mathrm{m\,s^{-2}\,Hz^{-1/2}}$ at 0.1 mHz [13]. The spectral density of the residual target acceleration noise curve of ASTROD is shown in Fig. 2 together with those of ASTROD I, the LTP and LISA. At present, the prototype of the OCA timer is fully operational having a precision less than 3 ps, a linearity error of 1 ps rms, and a time stability less than 0.01 ps over 1,000 s with dead time less than 10 μs. For a mission within the next 10–20 years, a timing accuracy better than 1 ps (300 μm in terms of ranging) can be anticipated. In coherent interferometric ranging, timing events need to be generated by modulation/encoding technique or by superposing timing pulses on the CW laser light. The interference fringes serve as consecutive time marks. With timing events aggregated to a normal point using an orbit model, the precision can reach 30 μm in range. The effective range precision for parameter determination could be better, reaching 3–10 μm using orbit models. Since ASTROD range is typically of the order of 1–2 AU (1.5–$3 \cdot 10^{11}$ m), a range precision of 3 μm would give a fractional precision of distance determination of 10^{-17}. Therefore, the desired clock accuracy/stability should be 10^{-17} over 1,000 s travel time. Optical clocks with this accuracy/stability are under research development. This development would facilitate ASTROD to use optical clocks [6]. The present range precision for radio tracking is a couple of meters. The improvement of ASTROD would be five orders of magnitude, allowing testing relativistic gravity in the solar system to the 1 ppb (part per billion) realm. Simulation also gave the same results.

4 Outlook

Owing to Geodesy missions [14] and LISA (Laser Interferometer Space Antenna) Program [15], improvements in many technologies critical to the tests of gravity are under development – the highly accurate deep-space navigation, high-precision frequency standards, precise pointing and attitude control, and the drag-free technologies. These will provide a technological platform for performing relativistic gravity experiments. Technology is ripe for deep-space laser ranging to test relativistic gravity to 1 ppb. As precision is increased by orders of magnitude, we are in a position to explore deeper into the origin of gravitation. The current and coming generations are holding such promises.

Acknowledgments

We thank the National Natural Science Foundation (Grant No. 10475114) and the Foundation of Minor Planets for funding this work.

References

1. U.J.J. Le Verrier. Theorie du mouvement de Mercure. *Ann. Observ. Imp. Paris* (Mrń.) **5**, 1 (1859).
2. W.-T. Ni. Empirical foundations of the relativistic gravity. *Int. J. Mod. Phys.* **D 14**, 901 (2005); and references therein.
3. B. Bertotti, L. Iess, and P. Tortora. A test of general relativity using radio links with the Cassini spacecraft. *Nature* **425**, 374 (2003).
4. I. Ciufolini and E.C. Pavlis. A confirmation of the general relativistic prediction of the Lense–Thirring effect. *Nature* **431**, 958 (2004). N. Ashby. General relativity - Frame-dragging confirmed. *Nature* **431**, 918 (2004).
5. GP-B (Gravity Probe B) `http://einstein.stanford.edu/`; C.W.F. Everitt, Seminar, Stanford University, June 2005.
6. W.-T. Ni, et al. Testing relativistic gravity to 1 ppb and mapping the solar-system gravity using laser ranging and optical clocks. CD-ROM Proceedings of the First ESA Optical Clock Workshop, June 7-10, 2005, ESTEC, Noordwijk, The Netherlands (2005).
7. W.-T. Ni, et al. *J. Korean Phys. Soc.* **45**, S118 (2004).
8. W.-T. Ni, S. Shiomi, and A.-C. Liao ASTROD, ASTROD I and their gravitational-wave sensitivities. *Class. Quantum Grav.* **21**, S641 (2004); and references therein.
9. S.G. Turyshev, M. Shao, and K. Nordtvedt. The laser astrometric test of relativity mission. *Class. Quantum Grav.* **21**, 2773 (2004).
10. A. Milani, D. Vokrouhlický, D. Villani, C. Bonanno, and A. Rossi. Testing general relativity with the BepiColombo radio science experiment. *Phys. Rev.* **D 66**, 082001 (2004).
11. `http://www.esa.int/esaSC/120377_index_0_m.html`; A. Vecchiato, et al. *Astron. Astrophys.* **399**, 337 (2003).

12. Y. Xia, et al. Orbit Design and Orbit Simulation for ASTROD I. PMO report.
13. A. Pulido Patón and W.-T. Ni. The low-frequency sensitivity to gravitational waves for ASTROD. *Gen. Rel. Grav.* **39** (2007), in press.
14. A. Albertella, F. Migliaccio and F. Sansó. *Celestial Mech. Dyn. Astron.* **83** 1 (2002); and references therein.
15. LISA study team 2000 LISA (Laser Interferometer Space Antenna): A Cornerstone Mission for the Observation of Gravitational Waves *ESA System and Technology Study Report* ESA-SCI 11.

Exploring the Pioneer Anomaly: Concept Considerations for a Deep-Space Gravity Probe Based on Laser-Controlled Free-Flying Reference Masses

Ulrich Johann[1], Hansjörg Dittus[2], and Claus Lämmerzahl[2]

[1] EADS Astrium GmbH, 88039 Friedrichshafen, Germany
[2] Centre of Applied Space Technology and Microgravity (ZARM)
University of Bremen, Am Fallturm, 28359 Bremen, Germany

Summary. The analysis of their radiometric tracking data has consistently indicated a small, anomalous, Doppler frequency drift at heliocentric distances of 20–70 AU. The drift is a blueshift, uniformly changing with a rate of $\sim (5.99 \pm 0.01) \times 10^{-9}\,\mathrm{Hz\,s^{-1}}$, which can be interpreted as a constant acceleration of each spacecraft of $a_{\mathrm{Pioneer}} = (8.74 \pm 1.33) \times 10^{-10}\ \mathrm{m\,s^{-2}}$ toward the Sun. This signal has become known as the Pioneer anomaly. The inability to explain the anomalous behavior of the Pioneer spacecraft with conventional physics and the search for "new physics" motivated by the search for quantum gravity or local effects of dark matter and/or dark energy emphasizes the need for a new experiment to explore the detected signal. Only a dedicated experiment could ultimately determine the nature of the found signal. We discuss the Pioneer anomaly and present the next steps toward an understanding of its origin. We specifically focus on the development of a mission to explore the Pioneer Anomaly in a dedicated experiment conducted in deep space.

1 Introduction

The exploration of the solar system's frontiers – the region between 25 and 50 astronomical units (AU) from the Sun – is a most ambitious and exciting technological challenge. The scientific goals for possible deep-space missions are well-recognized and include studies of the gas and dust distributions, exploration of the heliopause and the space beyond, measurements of the magnetic fields and particle fluxes, studies of the Oort Cloud and Kuiper Belt Objects, encounters with distant bodies, and investigation of the dynamical background of the solar system by studying the effects of various forces that influence the trajectory of the spacecraft. We are most interested in this last goal.

Our interest comes from navigating the Pioneer 10 and 11 spacecraft that yielded an exceptionally good acceleration sensitivity. Surprisingly, the accuracies of their orbit reconstruction were limited by a small, anomalous, Doppler frequency drift that can be interpreted as a constant acceleration of the spacecraft of $a_{\text{Pioneer}} = (8.74 \pm 1.33) \cdot 10^{-10}\ \text{m}\,\text{s}^{-2}$ toward the Sun (see [1]). This is called the *Pioneer anomaly*.

The nature of this anomaly remains a mystery, with possible explanations ranging from nominal sources of onboard systematics to exotic gravity extensions on solar system scales. Although the most obvious cause would be that there is a systematic origin to the effect, the limited data analyzed do not unambiguously support any of the suggested mechanisms [1]. The inability either to explain the anomaly or to test it with other spacecraft has contributed to a growing discussion about its origin.

Recently there was a significant interest in developing a dedicated mission to study the detected signal. Previous extensive efforts have included formulating theoretical mechanisms to explain the anomaly and analyzing existing solar system data, including both planetary and spacecraft data. Analysis of the capabilities of spacecraft currently in operation or in design demonstrated their inability to fulfill an independent verification of the anomaly. These efforts led to the conclusion that only a dedicated experiment could ultimately determine the nature of the anomalous signal.

In this contribution, we first discuss the Pioneer missions and the detected anomaly. We also review mechanisms proposed to explain the Pioneer anomaly, both with conventional and "new" physics. Then we outline the program of experimental tests, focusing on a dedicated mission concept to explore the Pioneer anomaly.

2 The Pioneer Missions and the Anomaly

The Pioneer 10/11 missions, launched on 2 March 1972 (Pioneer 10) and 5 April 1973 (Pioneer 11), respectively, were the first spacecraft to explore the outer solar system [1]. After Jupiter and (for Pioneer 11) Saturn encounters, the craft followed escape hyperbolic orbits near the plane of the ecliptic to opposite sides of the solar system (see Fig. 1). Pioneer 10 eventually became the first human-made object to leave the solar system. The last telemetry was obtained from Pioneer 10 on 27 April 2002 when the craft was 80 AU from the Sun. (The last signal from Pioneer 10 was received on 23 January 2003.)

The Pioneers were excellent craft with which to perform precise celestial mechanics experiments. This was due to a combination of many factors,

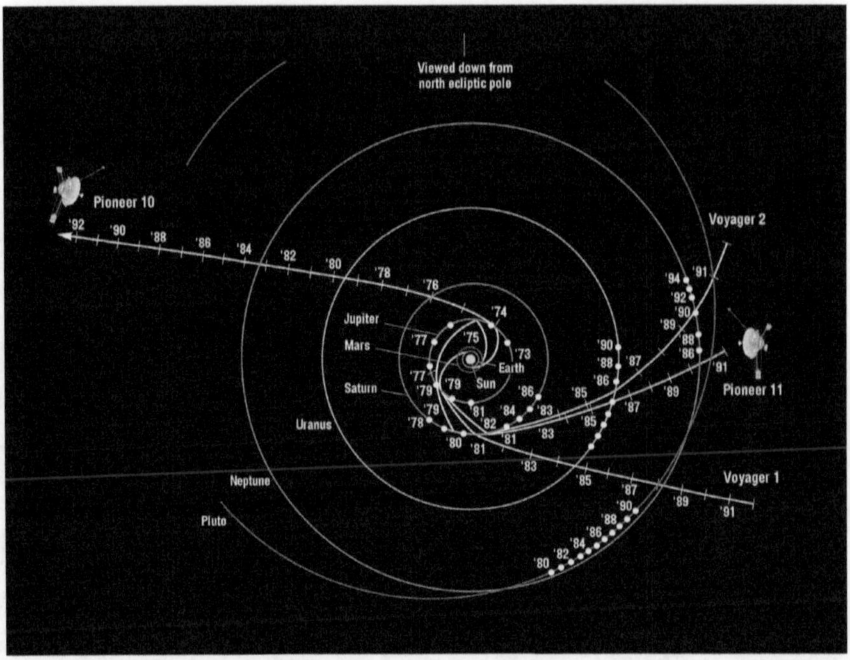

Fig. 1. Ecliptic pole view of Pioneer and Voyager trajectories. Pioneer 11 is traveling approximately in the direction of the Sun's orbital motion about the galactic center. The galactic center is approximately in the direction of the top of the figure.

including their attitude control (spin-stabilized, with a minimum number of attitude correction maneuvers using thrusters), power design (the Plutonium-238 powered heat-source RTGs – Radioisotope Thermoelectric Generators – being on extended booms aided the stability of the craft and also reduced the effects due to heating), and precise Doppler tracking (with the accuracy of postfit Doppler residuals at the level of mHz). The result was the most precise navigation in deep space to date. (See Fig. 2 for a design drawing of the spacecraft.)

By 1980, when Pioneer 10 passed a distance of ~20 AU from the Sun, the acceleration contribution from solar-radiation pressure on the craft (directed away from the Sun) had decreased to less than $4 \times 10^{-10}\,\mathrm{m\,s^{-2}}$. This meant that small effects could unambiguously be determined from the data, and the anomalous acceleration began to be seen. A detailed study of the anomaly began in 1994, using data starting in 1987. By then the external systematics (like solar-radiation pressure) were limited and the existence of the anomaly in the Pioneers' data became clearly evident [4, 82, 84].

In Sect. 2.1, we shall review our current knowledge of the Pioneer anomaly.

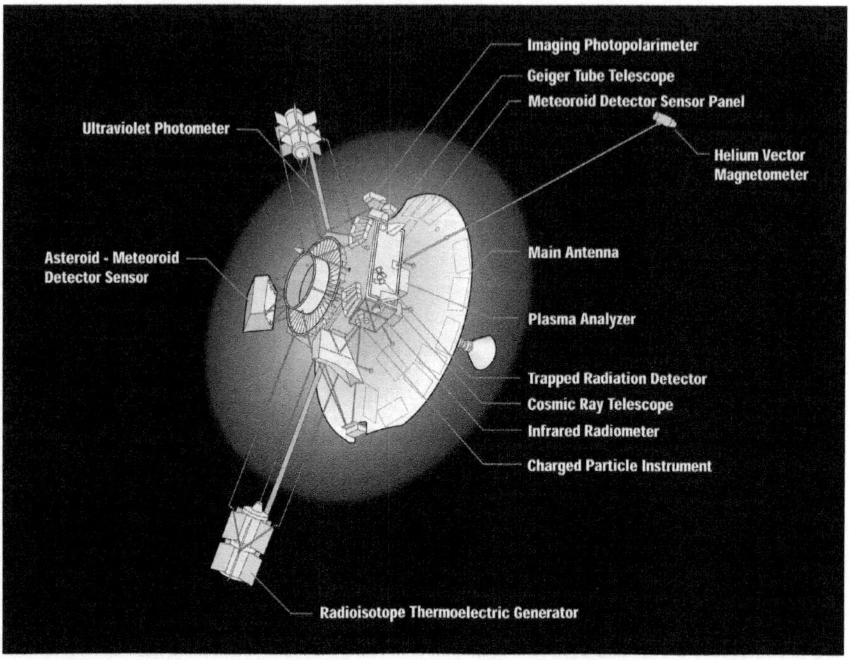

Fig. 2. The Pioneer spacecraft.

2.1 A Summary of the Pioneer Anomaly

As discussed above, the analysis of the Pioneer 10 and 11 data [1] demonstrated the presence of an anomalous, Doppler frequency blueshift drift, uniformly changing with a rate of [84]

$$\frac{d\nu_{\mathrm{obs}}}{dt} \sim (5.99 \pm 0.01) \times 10^{-9}\mathrm{Hz\,s^{-1}}. \quad (1)$$

To understand the phenomenology of the effect, consider ν_{obs}, the frequency of the retransmitted signal observed by a DSN antenna, and ν_{model}, the predicted frequency of that signal. The observed, two-way (round-trip) anomalous effect can be expressed to first order in v/c as

$$\nu_{\mathrm{obs}}(t) - \nu_{\mathrm{model}}(t) = -2\dot{\nu}_P\, t, \quad (2)$$

with ν_{model} being the modeled frequency change due to conventional forces accounted for in the spacecraft's motion (for more details see [1]). This motion is outward from the Sun and hence it produces a redshift.

After accounting for the gravitational and other large forces included in standard orbit determination programs this translates to

$$\nu_{\mathrm{obs}}(t) - \nu_{\mathrm{model}}(t) = -\nu_0 \frac{2a_P}{c} t\,. \quad (3)$$

Here ν_0 is the reference frequency [1].

Furthermore, after accounting for all *known* (not modeled) sources of systematic error (discussed in [1]), the conclusion remained that there was an anomalous sunward constant acceleration signal of

$$a_{\text{Pioneer}} = (8.74 \pm 1.33) \cdot 10^{-10}\,\text{m}\,\text{s}^{-2}. \tag{4}$$

We have already included the sign showing that a_P is inward using the DSN convention (see [1] for more information).

Realizing the potential significance of the result, all *known* sources of a possible systematic origin for the detected anomaly were specifically addressed. We emphasize *known* because one might naturally expect that there is a systematic origin of the effect, perhaps generated by the spacecraft themselves from excessive heat or propulsion gas leaks. However, a convincing explanation for the anomalous behavior of the spacecraft is still unavailable (for more details, see [2, 4]).

For the most detailed analysis of the Pioneer anomaly to date, Anderson et al. [1] used the following Pioneer 10/11 Doppler data [84]:

1. *Pioneer 10*: the data used were obtained between 3 January 1987 and 22 July 1998. This interval covers heliocentric distances ~40 – 70.5 AU. This data set had 20,055 data points obtained over the 11.5 years.
2. *Pioneer 11*: the data used were obtained between 5 January 1987 and 1 October 1990. This interval covers heliocentric distances ~22.42–31.7 AU. This data set had 10,616 data points obtained over the 3.75 years.

By now, several studies of the Pioneer Doppler navigational data have demonstrated that the anomaly is unambiguously present in the Pioneer 10 and 11 data. These studies were performed with three independent and different navigational computer programs [2–4]:

1. Various versions of JPL's Orbit Determination Program (ODP) code developed in 1980–2005
2. A version of The Aerospace Corporation's CHASPM (latest version of POEAS, see [1]) code extended for deep-space navigation
3. A third code written by C. Markwardt [3], of the Goddard Space Flight Center (GFSC). He analyzed Pioneer 10 data obtained from the National Space Science Data Center (NSSDC, with more information at `http://nssdc.gsfc.nasa.gov/`), for the time period 1987–1994

All analyses of the Pioneer 10 and 11 radiometric data [2–4, 82, 84] have established the following basic properties of the Pioneer anomaly:

1. *Direction.* Within the 10 dbm bandwidth of the Pioneer high-gain antenna, the anomaly behaves as a constant acceleration of the spacecraft directed toward the Sun.
2. *Distance.* It is unclear how far out the anomaly goes, but the Pioneer 10 data support the presence of the anomaly at distances up to ~70 AU from

the Sun. In addition, the Pioneer 11 Doppler data show the presence of the anomalous constant frequency drift as close in as ∼20 AU.
3. *Constancy.* Both temporal and spatial variations of the anomaly's magnitude are less then 3.4% for each spacecraft.

All these information have been used as guidance in investigating applicability of attempts to explain the Pioneer anomaly with both conventional and "new" physical mechanisms. We will briefly review these attempts later. We will use the same information later for discussing our proposal to explore the Pioneer anomaly in a dedicated deep-space experiment.

2.2 Attempts at Explanations

Conventional Physics Mechanisms

There were many attempts to explain the anomaly with a conventional physics mechanism that are either not strong enough to support the anomaly or else exhibit significant temporal and/or spatial variations contradicting the known properties of the anomaly. The observable effect is seen in the Doppler frequency of the Pioneer navigation data. As such it may have one of the following two flavors – either it is due to physical forces that are not yet properly modeled, or it is due to a blueshift/time–acceleration (for instance, due to a uniform drift of the frequency standards) of the radio signal. Consequently, attempts of explanation along these standard lines of conventional physics have addressed both possibilities [4–7].

Unknown Solar System Objects

The most straightforward way to generate a putative real physical force is the gravitational attraction due to not precisely known mass distribution in the outer solar system. Such a distribution could be due to the Kuiper belt objects or dust. The known density distribution for the Kuiper belt has been studied in [4,21], but these distributions found to be incompatible with the discovered properties of the anomaly. Even worse, these distributions cannot circumvent the constraint from the undisturbed orbits of Mars and Jupiter [7]. Hence a gravitational attraction by the Kuiper belt gravity can to a large extent be ruled out.

Dust

Interplanetary dust leads to (a) a gravitational acceleration, (b) an additional drag force (resistance), and (c) to a frequency shift of the radio signals proportional to the distance. The analysis of data from the inner parts of the solar system taken by the Pioneer 10 and 11 dust detectors strongly favors a spherical distribution of dust over a disk [8]. Ulysses and Galileo measurements in the inner solar system find very few dust grains in the $10^{-18} - 10^{-12}$ kg range [9, 10]. Infrared observations rule out more than 0.3 Earth's mass from Kuiper Belt dust in the trans-Neptunian region [11–13]. Furthermore, the density varies greatly within the Kuiper belt, precluding any constant acceleration.

The density of dust is not large enough to produce a gravitational acceleration on the order of the Pioneer anomaly [14–17]. The resistance caused by the interplanetary dust [18, 19] is too small [20, 22] to provide support for the anomaly. Also the dust induced frequency shift is too small to account for the frequency drift.

Accelerating Sun

Another attempt suggests that our Sun, by some reason, is accelerating in the direction perpendicular to the plain of ecliptic. In this case, the Pioneer Anomaly would be just a projection of this acceleration onto the orbits of the spacecraft [23] that are slightly inclined with respect to the plane. One of the proposed mechanisms, an asymmetric neutrino emission from the Sun, was unable to provide enough power to explain the solar acceleration. In addition, the planetary orbits would show secular perturbations at an already detectable level.

Spin–Rotation Coupling

The spin–rotation coupling on the circularly polarized radio signal when it interacts with the rotation of the Pioneer spacecraft and the Earth leads to a constant acceleration, which, however, is too small to explain the Pioneer data [24]. Furthermore, the helicity–rotation coupling has already been phenomenologically incorporated in the analysis of Doppler data.

Local Effect of Expansion of the Universe

Motivated by the numerical coincidence $a_{\text{Pioneer}} \approx H_0 c$ where H_0 is the Hubble constant and c the speed of light there are many attempts to explain the anomaly in terms of the expansion of the Universe. The ways in which the cosmic expansion might be responsible for the Pioneer Anomaly vary considerably between the approaches. In [1], it has been shown that such a mechanism would produce an opposite sign for the effect. Furthermore, it is known [2, 4] that the very presence of the Pioneer anomalous acceleration when acting on planets contradicts the accurately known motion of the inner planets of our solar system. This motivated focusing on the effect of cosmic acceleration on the radiocommunication signal rather than on the spacecraft themselves. In [25,26], it has been argued that the cosmic expansion influences the measurement process via a change in the frequency of the traveling electromagnetic signals. However, in [27], it has been shown that taking all effects of the cosmic expansion on the frequency as well as on the Pioneer motion into account, the resulting acceleration is $-vH$ and, thus, has the correct sign but is too small by a factor v/c. Similar results have been found in [29, 61].

In general, an expansion of the universe may influence (a) the motion of the Pioneers, (b) the size of the solar system (i.e., the planetary orbits), and (c) the propagation of signals. The first step is to calculate the gravitational field of a central (nonrotating or rotating) body in an expanding universe. For the nonrotating case, this has been done first by McVittie [30] using an expansion scheme. The use of matching conditions led to the famous Einstein–Straus vacuole and variations of it [31–35]. The vacuole is unstable [36] but on such a huge timescale [35] so that the instability is of no relevance; practically Einstein–Straus vacuoles can stay for a huge time. No axial symmetric vacuoles have been found hitherto. The proof that there are no cylindrical symmetric vacuoles [37, 38] does not apply to axial symmetry. For physical reasons, the vacuole seems not to be a viable model for the physics in the solar system embedded into cosmology.

Another approach considers a mass in a cosmology with perfect fluid [39–42] and pressure [43]. Gautreau also discusses the motion of planets in that metric and obtains an outspiraling of the orbits. However, the assumption that the cosmic fluid is also present inside the solar system might go too far.

The Schwarzschild–de Sitter solutions, see, e.g., [44], describe a spherically symmetric mass in a universe with cosmological constant, which, due to the cosmological constant, is expanding. This has been generalized to rotating masses leading to the Kerr–de Sitter solution, e.g., [45]. The trajectories in these space–times have been discussed in [46] and [47]. Also rotating mass configurations have been embedded into a cosmological context [48, 49]. (The opposite situation, the rotation of voids, has been considered in [50], for example.)

General statements regarding the size of the solar system using different methods have been obtained by McVittie [30], Dicke and Peebles [51], Gautreau [42], Anderson [52], Cooperstock et al. [28], Gautreau [41], and Bonnor [53–55]. Very interesting is the result in [52] where it has been found that the expansion couples to escape orbits while it does not couple to bound orbits.

Of some interest are considerations that connect these local physics in the expanding universe with the presence of a cosmological constant Λ [56] or, in a dynamical but more speculative context, with quintessence [57].

If the anomaly finds its origin in standard physics, such an explanation will be important for solar system physics, astrophysics, and also for advanced high-accuracy navigation. However, there is also a possibility for discovering new physics.

Possibility for New Physics?

The apparent difficulty to explain the anomaly within standard physics became a motivation to look for "new physics." These attempts in general did not produce a viable mechanism for the anomaly. Some of these proposals are the following.

Extensions of General Relativity

An inverse time dependence for the gravitational constant G [58] produces effects similar to that of an expanding universe. So did a length or momentum scale-dependent cosmological term in the gravitational action functional [59, 60]. The anomalous acceleration could be explained in the frame of a quasimetric theory of relativity [61]. The possible influence of the cosmological constant on the notion of inertial systems leading to an additional acceleration has been discussed in [56]. In addition, there were ideas to invoke a model for superstrong interaction of photons or massive bodies with the graviton background [62]. A five-dimensional cosmological model with a variable extra-dimensional scale factor in a static external space [63] was also proposed. There is also an attempt to explain the anomaly in the framework of a nonsymmetric gravitational theory [64].

Gravity Modifications

One approach, called MOND (Modified Newtonian Dynamics), induces a long-range modification of gravity [65–67], which was proposed to "explain" the rotation curves of galaxies. It has been pointed out that a scalar field with a suitable potential can account for a constant acceleration as experienced by the spacecraft [68]. Recently, a modification of the gravitational field equations for a metric gravitational field by introducing a general linear relation between the Einstein tensor and the energy–momentum tensor has been claimed to account for the Pioneer anomaly [69].

Dark Matter

Various distributions of dark matter in the inner and outer solar system have been proposed to explain the Pioneer anomaly, e.g., dark matter distributed in a form of a disk in outer solar system of a density of $\sim 4 \cdot 10^{-16}\,\mathrm{kg\,m^{-3}}$, which yields the wanted effect. However, it would have to be a special variety of dark matter that was not seen in other nongravitational processes. Also dark matter in the form of a spherical halo of a degenerate gas of heavy neutrinos around the Sun [70] mirror matter [71] has been discussed.

String Theory and Higher-Dimensional Models

Though in general scalar fields cannot explain the anomalous acceleration, a nonuniformly coupled scalar field could have the wanted effect [72]. Though brane-world models with large extra dimensions may offer richer phenomenology than the one by the standard scalar–tensor theories, it seems difficult to find a convincing explanation for the Pioneer Anomaly [73].

Further Ideas

Some further ideas are Yukawa-like or higher-order corrections to the Newtonian potential [1]; a scalar–tensor extension to the standard gravitational model [74]; Newtonian gravity as a long wavelength excitation of a scalar condensate inducing electroweak symmetry breaking [75]; interaction of the spacecraft with a long-range scalar field, unrelated to gravity, determined by an external source term proportional to the Newtonian potential [76]. In addition, there were suggestions based on the flavor oscillations of neutrinos in the Brans–Dicke theory of gravity [77]; a theory of conformal gravity with dynamical mass generation, including the Higgs scalar [78]. Although these models are quite interesting, they may not be accepted as a viable explanation at their present development stage.

3 Attempts of an Independent Confirmation

Attempts to verify the anomaly using other existing spacecraft have been not successful. This is because the Voyager, Galileo, Ulysses, and Cassini spacecraft navigation data all have their own individual difficulties for use in an independent test of the anomaly. In addition, many of the deep-space missions that are currently being considered either may not provide the needed navigational accuracy and trajectory stability sensitive to accelerations of under $10^{-10}\,\mathrm{m\,c^{-2}}$ (e.g., NASA New Horizons mission), or else they will have significant onboard systematics that mask the anomaly (e.g., JIMO – Jupiter Icy Moons Orbiter).

To enable a clean test of the anomaly, there is also a requirement to have an escape hyperbolic trajectory. This makes a number of other missions (i.e., LISA, the Laser Interferometric Space Antenna; STEP, Satellite Test of Equivalence Principle; etc.) less able to directly test the anomalous acceleration. Although these missions all have excellent scientific goals and technologies, nevertheless, their orbits lend them a less advantageous position to conduct a precise test of the detected anomaly.

A number of alternative ground-based verifications of the anomaly have also been considered, for example, using Very Long Baseline Interferometry (VLBI) astrometric observations. However, the trajectories of spacecraft like the Pioneers, with small proper motions in the sky, make it presently impossible to use VLBI in accurately isolating an anomalous sunward acceleration of the size of a_{Pioneer}.

To summarize, the origin of this anomaly remains unclear. Therefore, it is highly recommended to develop a dedicated deep-space experiment to explore the Pioneer Anomaly with an accuracy for acceleration resolution at the level of $10^{-12}\,\mathrm{m\,s^{-2}}$ in the extremely low frequency (or nearly dc) range.

Advantages of a dedicated concept include demonstration of new technologies and capabilities, especially in developing technologies for a low disturbance craft, advanced thermal design, formation-flying, accurate navigation, attitude control, etc., all of which are important for future deep-space missions. It also enables synergies with other science and technology, namely solar system studies (including plasma, dust distributions), Kuiper belt, gravitational waves, heliopause, etc. The goal here would be to explore the anomaly at the $10^{-12}\,\mathrm{m\,s^{-2}}$ level in the near dc frequency range and to develop technologies important for future deep-space navigation and attitude control.

In the following sections, we discuss the work needed to experimentally find the origin of the Pioneer anomaly.

4 A Deep-Space Gravity Explorer Mission to Explore the Pioneer Anomaly

Our experience with studying the Pioneer spacecraft and our current understanding of the Pioneer Anomaly lead to the following set of science objectives and technological goals for a dedicated mission to explore it.

4.1 Scientific Objectives

The main science goals of such a new Deep-Space Gravity Probe are the following:

1. Confirmation of the anomalous acceleration. This is the minimum aim for any such mission.
2. Test Newtonian gravity at large distances, or, more general.
3. Explore the gravitational field(s) on the scale of the solar system. This may give new information on, e.g., a hypothetical Yukawa part of the gravitational potential or of influences of dark matter and dark energy on the scale of the solar system.
4. Test the universality of free fall at large distances.
5. Test Newtonian dynamics for small gravitational acceleration. Since the MOND ansatz can be used to simulate the galactic rotation curves and, thus, gives an effect which usually is ascribed to dark matter, it is a model that should be taken seriously and should be confronted with a dedicated experiment.
6. Improvement of the limits on the extremely low-frequency gravitational radiation. This is an add-on to the previous goals that can be obtained just through the precise navigation and by a particular analysis of the data.

In particular, any new mission should be able to provide the following:

(a) Confirmation of the Pioneer anomaly with an improvement of the accuracy by a factor of 10^3.
(b) Improvement of spatial, temporal, and directional resolution of the anomalous acceleration. These points are extremely important. The directional resolution may help in identifying the origin of the anomaly: if the acceleration points toward the Sun, then the origin is gravitational; if it points toward the Earth, then the anomaly has something to do with the signal exchange; if the acceleration is along the spin-axis of the spacecraft, then the anomaly is most likely due to a systematics on the spacecraft; and if, finally, the acceleration is along the velocity vector, then an external nongravitational force should be responsible to the acceleration. The spatial and temporal resolution gives important information about when the anomaly starts and whether it turns on during the last flyby where the spacecraft is turned into an escape orbit.
(c) Identification and measurement of all possible disturbing and competing effects.
(d) Monitoring the state of motion of the spacecraft continuously.
(e) Study of the deep-space environment in the outer solar system.

4.2 Mission Design

The main science goal, the measurement of an anomalous acceleration, requires the information about the motion of a spacecraft unperturbed by nongravitational forces like solar wind or drag by dust. For an unambiguous verification of the anomalous acceleration and its precise characterization, the influence of nongravitational, nearly constant (dc) acceleration on a reference mass or probe has to be determined and carefully discriminated from the well-modeled gravitational motion in free fall to a bias accuracy of around $10^{-12}\,\mathrm{m\,s^{-2}}$ in all three coordinates. Several novel mission and payload concepts for such a Deep-Space Gravity Probe have been proposed and elaborated by the authors recently [79–81]. Promising concepts have been identified, which are based on a two-step measurement process. They involve a standard radio science link of a nearly classically operated noisy primary spacecraft with respect to an Earth reference combined with radio or laser ranging between the primary spacecraft and one or more freely moving reference masses, spheres, covered with corner-cube retroreflectors. The resulting distance is then combined with the distance between the Earth and the primary spacecraft, determined with radiometric methods (see Fig. 3). This combination is insensitive to disturbances acting on the primary spacecraft and will give the solely gravitationally determined motion of the reference mass. One main feature is the shielding of the reference mass from solar influences through the position of the primary spacecraft being on the line between the reference mass and the Sun.

The (nearly) free-falling reference masses are shielded or well modeled with respect to spacecraft and space environment interaction. Consequently, they are either kept in free space in formation flying in the vicinity of the

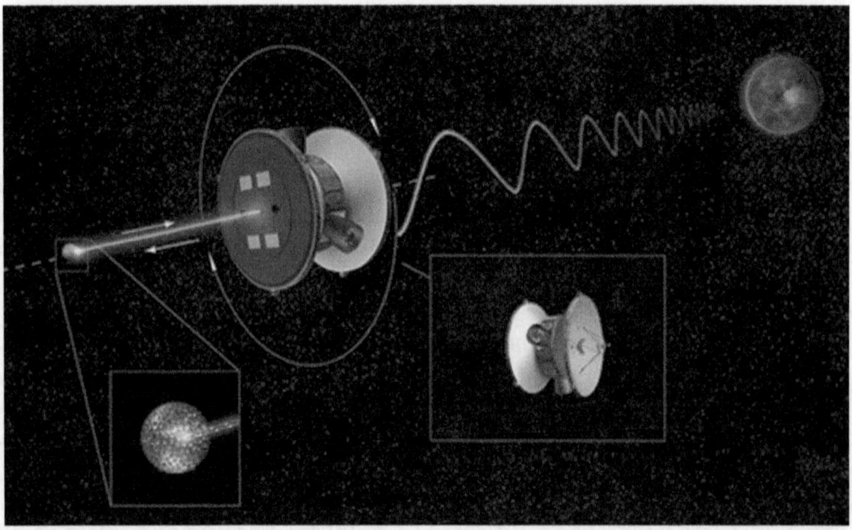

Fig. 3. A drawing for the measurement concept chosen of DSGP. The formation-flying approach relies on actively controlled spacecraft and a set of passive test masses. The reference mass (probe) is shielded from solar-radiation pressure by shadowing. The sphere is tracked by a laser radar and a star sensor payload on the spacecraft and allowed to drift in a range of an environmentally quiet distance between about 100 and 1,000 m at 1 AU (10,000 m at distances larger than 10 AU). The trajectory of the spacecraft will be determined based on the standard methods of radio tracking. The spacecraft is actively chasing the sphere with very low thrust maneuvers with period of several days, depending on Sun distance and design parameters. The concept establishes a flexible craft-to-reference mass formation. The laser radiation pressure and the thermal radiation pressure from the spacecraft are sufficiently small. The main objective is to accurately determine the heliocentric motion of the test mass by utilizing the two-step tracking needed for common-mode noise rejection purposes.

spacecraft at sufficient distance to minimize any spacecraft-to-reference mass interaction, or, alternatively, they are placed inside the spacecraft and shielded from space environment, but then subject to disturbing interaction with the spacecraft itself. In the latter case, the reference mass can be operated also as a test mass of an accelerometer, which is coupled to the spacecraft, while measuring any nongravitational acceleration of the test mass itself. In any case, a nearly constant (dc-bias) nongravitational acceleration must be monitored, removed, or modeled to the quoted level of accuracy. The requirement of operation in the dc regime or at extremely low frequency (days or weeks in terms of period) is a distinct technical challenge as compared to similar devices to measure or shield very small accelerations in other space missions (GOCE, MICROSCOPE, LISA Pathfinder, LISA) operating at a low, but non-dc frequency band (hours in terms of period). This requirement is the main reason to consider formation-flying reference masses external to the spacecraft, despite their added operational complexity. The same requirement, however, also imposes operational and design constraints of similar complexity in concepts, where the reference mass is placed inside the spacecraft.

Not only will this design allow for the most accurate orbit determination ever, but it will also lead to the development of optical navigation, communication, and accurate formation-flying technologies. This mission benefits from improvements in low-frequency accelerometers, ultra-stable oscillators, precision star trackers, dust detectors, spectrometers, and real-time autonomous attitude control.

A realization of the formation flight concept requires the main spacecraft to have a precision star-tracker and an accelerometer capable of precise navigation to a level less than $\sim 10^{-12}\,\mathrm{m\,s^{-2}}$ in the low-frequency acceleration regime. Mounted on the front would be a container holding a probe–a spherical test mass covered with corner cubes. Once the configuration is on its solar system escape trajectory and will undergo no further navigation maneuvers, and is at a heliocentric distance of $\sim$5–20 AU, the reference mass would be released from the primary spacecraft. The reference mass will be passively laser-ranged from the primary spacecraft with the latter having enough fuel to maneuver with respect to the probe, if needed. The distance from the Earth to the primary would be determined either with standard radiometric methods operating at Ka-band or with optical communication. Note that any dynamical noise at the primary would be a common-mode contribution to the Earth–spacecraft and spacecraft–probe distances. This design satisfies the primary objective, which would be accomplished by the two-staged accurate navigation of the reference mass with sensitivity down to the $10^{-12}\,\mathrm{m\,s^{-2}}$ level in the dc of extremely low frequency bandwidth.

Since the small forces affecting the motion of a craft from four possible directions, all having entirely different characters (i.e., sunward, Earth-pointed, and along the velocity vector or spin-axis [83, 84]), it is clear that an antenna with a highly pointed radiation pattern and star sensors will create even better conditions for resolving the true direction of the anomaly, when compared to standard navigation techniques. On a craft with these additional capabilities, all onboard systematics will become a common-mode factor contributing to all the attitude sensors and antennas. The combination of all the attitude measurements will enable one to clearly separate the effects of the onboard systematics referenced to the direction toward the Sun.

An additional wish is to have a fast orbital transfer to the outer regions of the solar system beyond 20 AU to a hyperbolic escape trajectory.

To summarize, the mission design is characterized by the following:

- Range of heliocentric distances of interest 25–45 AU
- Hyperbolic escape trajectory beyond 15 AU
- Fast orbit transfer with a velocity of larger than 5 AU year^{-1}
- Radio-tracked spacecraft

- An additional reference mass to be shielded from the Sun and laser tracked by the spacecraft

Most of the technology is readily available and could lead to rapid mission design and components fabrication.

4.3 Technological Challenges

To meet the above science goals, it is necessary to develop the following:

1. Methods for precise spacecraft navigation and attitude control (needed for all future interplanetary missions)
2. Drag-free technologies operating at extremely low frequencies (needed also for, e.g., next generation of gravitational wave missions)
3. Fast orbit transfer for deep-space access, namely propulsion concepts (including solar sails) and power management at large heliocentric distances (including the use of RTGs)
4. Advanced onboard environmental sensors may give additional and independent information of the dust content in the outer solar system from which the drag can be calculated and the additional gravitational acceleration due to the dust estimated

The experience gained from the Pioneer spacecraft leads to an approach to spacecraft and mission design that responds directly to the set of objectives and goals presented above. In particular, this experience translates in the following design requirements for the new mission, which are characterized by the following.

Navigation and Attitude Control

- Spin-stabilized spacecraft
- Three-dimensional acceleration sensitivity $\sim 10^{-12}\,\mathrm{m\,s^{-2}}$, in very low frequency or dc range
- Propulsion system with precisely calibrated thrusters, propellant lines, and fuel gauges with real-time control
- X- and Ka-band with significant dual-band tracking
- Data types: Doppler, range, ΔDOR, and VLBI

Drag-Free Technologies

The dc drag-free technology is central to this mission. Such dc drag-free technologies may be based, e.g., on atomic interferometry or on capacitive methods (as employed for the MICROSCOPE mission) with time-dependent coupling constant. Here we suggest a concept using laser ranging of an external freely flying test mass. This will be outlined in Sect. 4.4.

Thermal Design

To minimize the number of maneuvers needed to keep the spacecraft in line between the test mass and the Sun, one should keep thermal effects as small as possible. What needs to be considered is as follows:

- Entire spacecraft is heat-balanced and heat-symmetric
- Active control of all heat dissipation within and outward
- Knowledge of three-dimensional vector of thermal recoil force
- Optical surfaces with understood ageing properties
- Onboard power – the use of RTGs
- Must provide thermal and inertial balance and stability

Fast Orbit Transfer

To enable fast orbital transfer to distances greater than 20 AU, hyperbolic escape trajectories enabled by, e.g., solar sail propulsion technology (see the article of Dachwald et al. on page 379 in this volume) are considered as an attractive candidate. Among other options are standard chemical rocket and nuclear electric propulsion, as was successfully demonstrated recently. The proposed combination of a formation-flying flight system aided by solar sail propulsion for fast trajectory transfer leads to a technology combination that will benefit many astronomy and fundamental physics missions in the future.

4.4 Technological Developments for a dc Drag-Free Control

The Scheme

Laser ranging lends itself as a suitable tool to monitor the distance between a reference mass and fiducial points located in the spacecraft, because it can be made to be essentially without bias and drift and can provide the necessary resolution with a minimum of equipment based upon proven technology. The reference mass then would be a passive sphere with well-defined surface and electromagnetic properties, equipped with equally distributed corner-cube retroreflectors, similar to the LAGEOS satellites in Earth orbit, but much smaller (approximately 20 cm diameter). Laser metrology to read out the relative position of a reference mass inside an inertial sensor/accelerometer is presently employed for the LISA Pathfinder payload. A nonpolarizing heterodyne interferometer with differential wavefront sensing for attitude measurement is complementing here the capacitive read out system at much improved accuracy along the sensitive measurement axis and for lateral attitude angles. It provides a resolution of $10\,\mathrm{pm\,Hz^{-1/2}}$ and $10\,\mathrm{nrad\,Hz^{-1/2}}$, respectively, within the band 10^{-3}–10^{-1} Hz. Augmented by a molecular frequency reference and a thermally stable reference arm, sufficient stability may be reached for much lower or quasi-dc frequencies. A different and much simpler laser ranging technique can be employed in the case of external, formation-flying reference masses. The following discussion will focus on the latter scenario.

The present hypothesis is that the Pioneer Anomaly is a constant or very smooth effect, featuring a constant anomalous acceleration either in the direction of the Sun, the Earth line of sight, or opposite to the reference masses velocity vector in the solar system coordinate frame. Which case applies cannot be discriminated on the basis of available data. The effect is constant in the sense that it is independent of time and orbital position or velocity at least during cruising phases. As such, any deviation in Doppler – interpreted as a corresponding velocity change – should accumulate linear in time. In terms of distance ranging, the effect hence should be quadratic in time. It is obvious that the required range rate or ranging precision for the measurement scales inversely with the measurement intervals (or integration time). In fact, for cruising intervals of 10 years or more, periodic short measurements about every month should be perfectly sufficient. Even outliers in form of sudden accelerations caused, e.g., by micrometeorites can be discriminated and tolerated. Nevertheless, a certain higher frequency measurement capability (hours) may be desirable for reliable chasing of the reference mass by the spacecraft. The ranging budget is required to achieve a 1% accuracy goal ($\delta a_{\mathrm{RM}} = 8 \cdot 10^{-12}\,\mathrm{m\,s^{-2}}$) of the reference mass acceleration a_{RM} measurement. Hence, the measurements on both legs, the Earth–spacecraft radio link and the spacecraft–reference mass laser link, respectively, must be accurate enough to unambiguously discriminate the Pioneer Anomaly effective contribution from all known other effects influencing the trajectory. In addition, the modeling accuracy for the acceleration of the RM by all known gravitational and nongravitational effects must be within the δa_{RM} budget. As a consequence, the reference mass should be purely passive with well-defined surface and electrical properties. The Pioneer Anomaly effect accumulates to a Doppler shift, velocity deviation, or distance deviation relative to the modeled parameter values over one day or one month, respectively, as shown in Table 1 together with the accuracy to be provided by the ranging. These figures of course apply to the complete two-step process, comprising the radio link to Earth and the local laser ranger and are relative values. It is evident that – even if factorized – only moderate requirements are imposed for the laser ranger in terms of measurement precision. This fact has the important consequence that the tracking can be accomplished with rather simple sensor equipment on the spacecraft in addition to the anyway existing radio link to Earth.

Because the spacecraft is on a presumably close to Sun-radial outbound trajectory, the ranging accuracy along the Earth line of sight by far dominates the measurement accuracy and the lateral coordinates can be determined with less precision. Nevertheless, VLBI techniques for the radio link to Earth allow μ arcsec precision if desired. The orientation of the laser ranging vector relative to the Earth line of sight can be easily determined to within a sufficient few arcsec accuracy, e.g., by a simple star tracker. The capabilities of current

Table 1. Single bin measurement accuracy required to resolve a (smooth) Pioneer acceleration with sufficient precision of 1% as function of measurement interval.

Measurement period	Relative velocity	Relative distance	Relative Doppler Ka	Required ranging accuracy
1 day	$70\,\mu m\,s^{-1}$	6 m ($25\,cm\,h^{-1}$)	7 mHz	6 cm $0.7\,\mu m\,s^{-1}$ 0.07 mHz
1 month	$2{,}100\,\mu m\,s^{-1}$	5,300 m	200 mHz	50 m

"Relative" refers here to the trajectory without the PA effect. The values are for the total measurement (Earth–spacecraft–reference mass). A monthly period is considered to be sufficient, greatly relaxing sensor requirements. A daily period may however be desirable for active tracking of the free-fall reference mass. Please note that no absolute ranging accuracy to that precision is required for the Earth, spacecraft link, but a constant bias can be tolerated.

Table 2. Present DSN radio link capabilities for spacecraft tracking Earth–spacecraft (DSN Handbook [4]).

Parameter	Measures	Accuracy ($1\,\sigma$)
Doppler	Range rate	$0.03\,mm\,s^{-1}$
Range	Range	~1–2 m
Angle	Lateral angular position (right ascension, declination)	0.01° ($170\,\mu rad$)
DDOR (VLBI)	Lateral angular position (right ascension, declination)	$0.14\,\mu°$ (2.4 nrad)

radio ranging in Deep-Space Network communication links are summarized in Table 2.

In principle, the following ranging technologies can be combined in the proposed two-step process, see Table 3.

The ranging requirements between spacecraft and formation-flying reference mass (or several masses) are constrained further by the necessity to chase the latter while actively maneuvering the spacecraft. The impact on spacecraft operations shall however be kept at a minimum, in particular in scenarios, where the experiment is a passenger on a mission with different objectives. Obviously, the frequency of (very low Δv) correction maneuvers and the allowance of letting the reference mass drifting far away, while stressing the local ranging requirements, are conjugate. A further requirement indirectly related is the necessity to shield or model any reference disturbance at sufficient precision. Among the various disturbance sources, here the dominant effects of solar light pressure and gravity or thermal interaction with the spacecraft itself define constraints. A sphere of 25 cm diameter, 5 kg equipped with retroreflecting corner-cube experiences at 1 AU a light pressure acceleration of about $5{\cdot}10^{-8}\,m\,s^{-2}$, a figure about 100 times the Pioneer effect. To achieve the

Table 3. Principle ranging technologies applicable for the proposed two-step process and rendering the measurement insensitive to noisy spacecraft effects. This chapter focuses on the first option.

Earth–spacecraft link	Spacecraft–reference mass (RM) link
Classical bidirectional biwavelength radio science	Laser ranging + star tracker of reference mass with corner cubes (in shadow)
Classical bidirectional biwavelength radio science	Radio tracking of passive radar reflector (in shadow)
Classical bidirectional biwavelength radio science	Active transponder on reference mass (in shadow)
Classical bidirectional biwavelength radio science	Radar reflector tracking in main communication beam (in light)
Laser ranging to earth station (ground or orbit)	Everything above

1% measurement accuracy, a modeling of the induced acceleration to within 10^{-4} accuracy is required, which is a challenging task, considering ageing surface properties, etc. At 10 AU, still an accuracy of 1% is necessary. Hence, flying the reference mass in the shadow of the spacecraft would be an advantage at least in early parts of the trajectory. Assuming a spacecraft carrying an opaque 2.5 m antenna, it will cast a core shadow (umbra) about 260 and 2,600 m for 1 and 10 AU, respectively. Equipped with a dedicated Sun shield of say 10 m diameter, the shielded range could be extended to 1 and 10 km, respectively. On close range, a 500 kg spacecraft would pull the reference mass equivalent to the desired acceleration accuracy at about 65 m distance. In case, the gravity interaction can be modeled to 1% accuracy, a lower limit of only 6.5 m could be allowed, but then thermal and electrostatic interaction with the spacecraft become important. Figure 4 illustrates the geometry of the formation-flying mission.

The frequency of maneuvers to keep the reference mass in the range interval allowed by above constraints, is then set primarily by the differential light pressure acceleration. For a typical 2.5 m diameter spacecraft, a range walk of 180 and 1.8 m day^{-1} would occur by this effect at 1 and 10 AU, respectively. Hence, at 1 AU a daily maneuver imposing a Δv of about 10 mm s^{-1} in Sun direction would be required or alternatively, a continuous thrust of about 25 μN. At 10 AU that reduces to a monthly maneuver imposing a Δv of about 3 mm s^{-1}. Beyond 10 AU, the spacecraft can be allowed to drift away in sunlight until the tracking capabilities are exhausted. The reference mass can be centered in the spacecraft shadow by very low thrust lateral maneuvers of the spacecraft. It is important to note that these positioning maneuvers are not comparable with the complex and risky orbit correction maneuvers in classical missions, as only very low thrust authority is employed. A possible alternative operation mode would be to let the reference mass and the spacecraft drift apart independently for long intervals (month), to reacquire the target

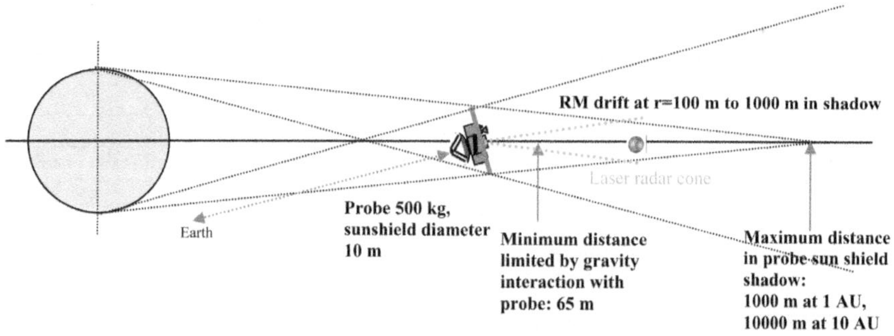

Fig. 4. Geometry of formation-flying reference mass shielded from solar-radiation pressure by shadowing. The sphere is tracked by a laser radar and a star sensor payload on the spacecraft and allowed to drift in a range between about 100 and 1,000 m at 1 AU (10,000 m $>$ 10 AU). The spacecraft is actively chasing the sphere with very low thrust maneuvers with period of several days, depending on Sun distance and design parameters. The laser radiation pressure and the thermal radiation pressure from the spacecraft are sufficiently small.

with the star tracker and to actively steer the spacecraft close within few m followed by a very simple short range calibration. Obviously, the shadowing would be lost in that mode.

Suitable Technologies for Laser Ranging to Reference Mass

Numerous laser ranging techniques exist for many applications on ground and in space and for a large range of requirements. For the envisaged application, the selection depends on the mission scenario, in particular, whether it is based upon close range continuous tracking or large range drifts. It is further driven by the least impact on spacecraft operations (mass, power, operation modes, required actuation, or AOCS maneuvers) and by simplicity, robustness, and space heritage.

A (nonexhaustive) list of candidate measurement principles is:

- Pulsed time of flight (TOF) laser radar for ranging and star tracker for relative attitude
- Frequency chirp coherent laser radar (FMCW)
- Continuous tracking with coherent laser heterodyne interferometer
- Calibrated measurement of sphere diameter (interferometric fringe contrast)
- Triangulation using three star trackers
- Intensity ranging

Another important requirement is the capability to periodically acquire the spacecraft following inactive intervals. Obviously, the system robustness demands that the reference mass may not be lost. The periodic operation modes are then:

1. Warm up
2. Calibration (option)
3. Target acquisition
4. Target tracking (option)
5. Range, range rate, and angular measurement
6. Stand-by

In the following, the pulsed laser radar based upon existing technology is shown to be perfectly suitable even for the large range scenarios. A detailed trade, however, and final selection have to be incorporated into a complete system level study for the experiment. It is assumed that the ranging done by a laser radar is supported by a directional measurement using a state-of-the-art simple star tracker. The star tracker adds also significant robustness to the system, because the target can be "seen" in its field of view and in front of the star field, illuminated either by the laser or by the Sun.

One interesting option for a very simple system is a uniform (white) sphere, illuminated by a defocused laser beam, which can be broadband, but must be calibrated in power. The backscattered laser light is detected by the star tracker, who also receives a power calibration from the transmission (e.g., fiber link) and locates the sphere relative to the beam axis. For laser intensity calibrated to 10^{-3} relative accuracy and an intensity flat beam lobe within the angular accuracy of the star tracker (~50 µrad), a range resolution of $\delta R = 2.5 \cdot 10^{-4} R$ can be achieved. That corresponds to 25 mm at $R = 100$ m. A mode for initial calibration and recalibration, compensating for aging target surface properties would have to be incorporated however.

A TOF radar would require a target (sphere or disk) packed with corner cubes to support the link budget. Acquisition strategies typically employ scanning laser beams (spiral or rectangular patterns), defocused beams followed by reorientation and refocussing or combinations thereof. Also the star tracker signal can be used to actively point the focused laser beam to an acquired target, if both boresights are aligned. In that case, a scanning or defocusing of the beam can be avoided, provided the target is illuminated.

The received power for a defocused (nondiffraction limited) transmitted beam reflected off a corner-cube carrying sphere can be expressed as

$$P_r = \frac{D_t^2}{\left(D_t + \frac{R\lambda}{d_T}\right)^2} n\rho \frac{d_T^2}{d_b^2} P_t \tag{5}$$

where, in the example considered here, $D_t = 0.1$ m is the transmit/receive telescope diameter, R is the range, $\lambda = 1\,\mu$m is the wavelength, $d_T = 0.02$ m is the diameter of the reflector corner cubes, $n = 19$ is the number of illuminated cubes with an average efficiency of $\rho = 0.5$, $d_b = 10$ m is the diameter of the laser illuminated area at the location of the sphere, and P_t is the transmitted power.

For a defocused beam to 10 m diameter at 10 km distance, the received power then is

$$P_r = 1 \cdot 10^{-6} P_t \tag{6}$$

A suitable transmitted power of 100 mW hence provides a reception signature of 100 nW, which is by far sufficient for localization. A focussed beam (about diffraction limited) provides

$$P_r = 2.6 \cdot 10^{-3} P_t \tag{7}$$

And hence 260 µW. This value again is by far sufficient for ranging accuracy of about 10 mm in a TOF laser radar (see later). The example is in fact representing an overdesigned system, illustrating the capabilities.

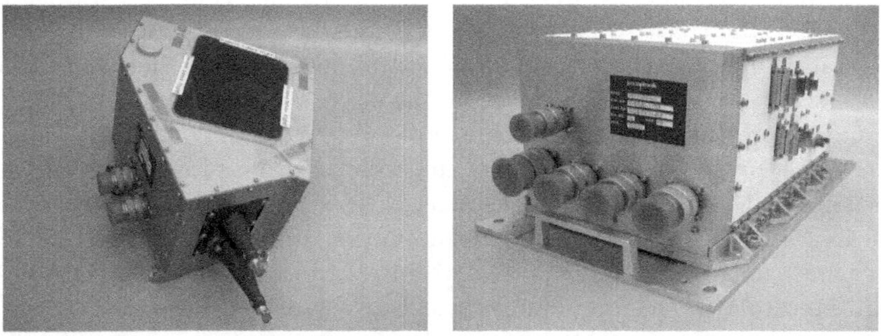

Fig. 5. Space demonstrated laser range finder developed by Jena Optronik, Astrium GmbH, and Riegel GmbH for Space Shuttle rendezvous and docking.

Table 4. Performance and budgets of the Jena Optronik laser range finder [87]. Custom-specific modification should lead to significantly reduced mass and power budgets.

Field of view	Up to 40° × 40°
Measurement parameter	Azimuth α, elevation β, roll R, pitch P, yaw Y, time, range r
Accuracy (700–3 m)	
LOS (noise + bias)	<0.1°
LOS bias	<0.1°
Power	35 W minimum 70 W maximum
Temperature	
Operational	35–65°C
Nonoperational	55–70°C
Mechanical size	
Optical head (w/o fiber connectors)	270 × 287 × 196 mm^3
Electronic box	315 × 224 × 176 mm^3
Mass	
Optical head	6.1 kg
Electronic box	8.2 kg

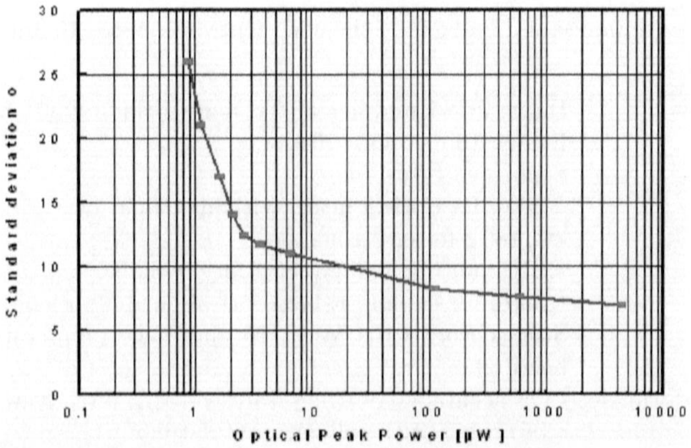

Fig. 6. Ranging performance measured as function of received optical power. A resolution of 10 mm is obtained for 10 μW (Jena Optronik).

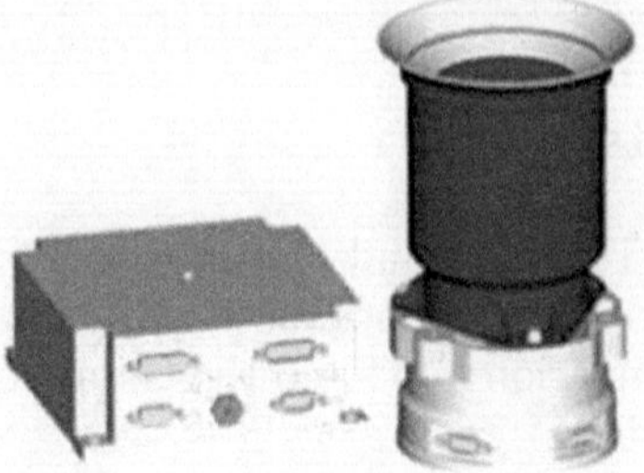

Fig. 7. Autonomous star tracker Astro 10 (Jena Optronik).

Existing Suitable Sensors

A space-qualified TOF laser radar that could be employed with few (descoping) modifications has been developed by Jena Optronik together with EADS Astrium GmbH and Riegel GmbH. It has flown successfully on several Space Shuttle missions where it served as a rendezvous and docking sensor. Figure 5 illustrates the device and Table 4 summarizes main parameters [85]. The scanning mechanism would be omitted and replaced by a refocussing mechanism in the foreseen application, further reducing power and mass budgets.

A suitable star tracker is also available from Jena Optronik, see Fig. 6, although other devices exist, which may be even more compact, lighter, and less power consuming (e.g., the ATC of the Technical University of Denmark [86]). Figure 7 and Table 5 present main design and performance figures for the Astro 10 autonomous star tracker [87].

Table 5. Performance and budgets of the Jena Optronik Astro 10 autonomous star tracker [87].

Dimensions	Head: ∅ 185 mm, height 242 mm (including 30° baffle) E-box: $150 \times 145 \times 75\,\text{mm}^3$ Separated box design
Mass	<960 g for optical head (without baffle) <1,180 g for electronic box <510 g for 30° baffle (380 g for 40° baffle) <350 g for cabling E-box/optical head (1 m length)
Power	Star sensor: <10.0 W at 20° interface temperature optical head
Sensor performance	LOS accuracy (BOL): ≤5 arcsec (3 μ), pitch/yaw ≤35 arcsec (3 μ), roll, slew rate at $0.6°\,\text{s}^{-1}$
Operating modes	Boot, stand-by, initial acquisition, attitude lock-in, high-accuracy attitude, simulation
Data interface	RS 422, alternatively MIL 1553 B
Input voltage range	22–35 V

Acknowledgments

We like to thank S. Turyshev and the Pioneer Exploration Collaboration for support, information, and various discussions. Financial support from the German Research Foundation DFG and the German Aerospace Agency DLR is gratefully acknowledged.

References

1. J.D. Anderson, P.A. Laing, E.L. Lau, A.S. Liu, M.M. Nieto, and S.G. Turyshev. Study of the anomalous acceleration of Pioneer 10 and 11. *Phys. Rev.*, D 65:082004, 2002.
2. J.D Anderson, P.A. Laing, E.L. Lau, A.S. Liu, M.M. Nieto, and S.G. Turyshev. Indication, from Pioneer 10/11, Galileo, and Ulysses Data, of an Apparent Anomalous, Weak, Long–Range Acceleration. *Phys. Rev. Lett.*, 81:2858, 1998.
3. C.B. Markwardt. Independent confirmation of the Pioneer 10 anomalous acceleration. gr-qc/0208046.
4. J.D. Anderson, E.L. Lau, S.G. Turyshev, P.A. Laing, and M.M. Nieto. Search for a standard explanation of the Pioneer anomaly. *Mod. Phys. Lett.*, A 17:875, 2002.
5. J.D. Anderson, M.M. Nieto, and S.G. Turyshev. A mission to test the Pioneer anomaly. *Int. J. Mod. Phys.*, D 11:1545, 2002.
6. M.M. Nieto and S.G. Turyshev. Finding the origin of the pioneer anomaly. *Class. Quantum Grav.* **21**, 4005 (2004).
7. L.K. Scheffer. Conventional forces can explain the anomalous acceleration of Pioneer 10. *Phys. Rev.*, D 67:084021, 2004.
8. D.H. Humes. Results of Pioneer 10 and 11 meteoroid experiments: Interplanetary and Near Saturn. *J. Geophys. Res.* **85**, 5481 (1980).

9. D.A. Gurnett, J.A. Ansher, W.S. Kurth, and L.J. Granroth. Micron-sized dust particles detected in the outer solar system by the Voyager 1 and 2 plasma wave instruments. *Geophys. Res. Lett.* **24**, 3125 (1997).
10. M. Landgraf, K. Augustsson, E. Grun, and A.S. Gustafson. Deflection of the local interstellar dust flow by solar radiation pressure. *Science* **286**, 2319 (1999).
11. G.E. Backman, A. Dasgupta, and R.E. Stencel. Model of a Kuiper belt small–grain population and resulting far–infrared emission. *Astrophys. J.* **450**, L35 (1995).
12. S.A. Stern. Signatures of collisions in the Kuiper Disk. *Astron. Astrophys.* **310**, 999 (1996).
13. V.L. Teplitz, S.A. Stern, J.D. Anderson, D. Rosenbaum, R.J. Scalise, and P. Wentzler. Infrared Kuiper belt constraints. *Astrophys. J.* **516**, 425 (1999).
14. R. Malhotra. The origin of Plutos orbit – implications for the Solar system beyond Neptune. *Astron. J.* **110**, 420 (1995).
15. R. Malhotra. The phase space structure near Neptune resonances in the Kuiper Belt. *Astron. J.* **111**, 504 (1996).
16. A.P. Boss and S.J. Peale. *Icarus* **27**, 119 (1976).
17. A.S. Liu, J.D. Anderson, and E. Lau. Proc. AGU (Fall Meeting, San Francisco, 16-18 December 1996), paper No. SH22B-05.
18. P. Ingersoll, T.V. Johnson, J. Kargel, R. Kirk, D.I.N. Didon, J. Perchoux, and E. Courtens. Preprint Université de Montpellier (1999).
19. N. Didon, J. Perchoux, and E. Courtens. Universite de Montpellier preprint (1999).
20. D.E. Backman A. and Dasgupta. Model of a Kuiper belt small grain population and resulting far-infrared emission. *Astroph. J.* **450**, L35 (1995).
21. M.M. Nieto. Analytic gravitational–force calculations for models of the kuiper belt, with application to the Pioneer anomaly. *Phys. Rev.*, D 72:083004, 2005.
22. M.M. Nieto, S. Turyshev, and J.D. Anderson. Directly measured limit on the interplanetary matter density from Pioneer 10 and 11. *Phys. Lett. B*, 613:11, 2005.
23. D. Bini, C. Cherubini, and B. Mashhoon. Vacuum c-metric and the gravitational stark effect. *Phys. Rev.*, D 70:044020, 2004.
24. J.D. Anderson and B. Mashhoon. Pioneer anomaly and the helicity–rotation coupling. *Phys. Lett.*, A 315:199, 2003.
25. A.F. Ranada. The Pioneer riddle, the quantum vacuum and the acceleration of light. *Europhys. Lett.*, 63:653, 2002.
26. J.L. Rosales. The Pioneer's acceleration anomaly and Hubble's constant. gr-qc/0212019.
27. C. Lämmerzahl and H. Dittus. Doppler tracking in the expanding universe, 2005. preprint, ZARM, University of Bremen.
28. F.I. Cooperstock, V. Faraoni, and D.N. Vollik. The influence of the cosmological expansion on local systems. *Astroph. J.*, 503:61, 1998.
29. S. Klioner and M.H. Soffel. Refining the relativistic model for Gaia: cosmological effects in the BCRS. **in:** C. Turon, K.S. O'Flaherty, M.A.C. Perryman (eds.). *Proc. of the Symposium "The three-dimensional Universe with Gaia"*, ESA SP-576 (2005), p. 305.
30. G.C. MacVittie. The mass–particle in an expanding universe. *Mon. Not. Roy. Astr. Soc.* **93**, 325 (1933).

31. A. Einstein and E.G. Strauss. The Influence of the Expansion of Space on the Gravitation Fields Surrounding the Individual Stars. *Rev. Mod. Phys.* **17**, 120 (1945).
32. A. Einstein and E.G. Strauss. Corrections and Additional Remarks to our Paper: The Influence of the Expansion of Space on the Gravitation Fields Surrounding the Individual Stars. *Rev. Mod. Phys.* **18**, 148 (1946).
33. E. Schücking. Das Schwarzschildsche Linienelement und die Expansion des Weltalls. *Z. Physik*, 137:595, 1954.
34. R. Balbinot, R. Bergamini, and A. Comastri. Solution of the Einstein–Straus problem with a Λ term. *Phys. Rev.*, 38:2415, 1988.
35. W.B. Bonnor. A generalization of the Einstein–Straus vacuole. *Class. Quantum Grav.*, 17:2739, 2000.
36. A. Krasiński. *Inhomogenous Cosmological Models.* Cambridge University Press, Cambridge, 1997.
37. J.M.M. Senovilla and R. Vera. Impossibility of the cylindrically symmetric Einstein–Straus model. *Phys. Rev. Lett.*, 78:2284, 1997.
38. B.C. Nolan and L.V. Nolan. On isotropic cylindrical symmetric stallar models. *Class. Quantum Grav.*, 21:3693, 2004.
39. J. Eisenstadt. Spherical mass emmersed in a cosmologicasl universe: A class of solutions. *Phys. Rev.* **D 11**, 2021 (1975).
40. J. Eisenstadt. Spherical mass emmersed in a cosmologicasl universe: A class of solutions II. *Phys. Rev.* **D 12**, 1573 (1976).
41. R. Gautreau. Curvature coordinates in cosmology. *Phys. Rev.* **D 29**, 186 (1984).
42. R. Gautreau. Imbedding a Schwarzschild mass solution into cosmology. *Phys. Rev.* **D 29**, 198 (1984).
43. C. Bona and J. Stela. "Swiss cheese" models with pressure. *Phys. Rev.* **36**, 2915 (1987).
44. W. Rindler. *Relativity.* Oxford University Press, Oxford, 2001.
45. B. Carter. **In** C. DeWitt and B.D. DeWitt (eds.). *Black Holes.* (Gordon and Breach, New York 1973).
46. Z. Stuchlik and S. Hledik. Some properties of the Schwarzschildde Sitter and Schwarzschildanti-de Sitter spacetimes. *Phys. Rev.*, D 60:044006, 1999.
47. Z. Stuchlik and P. Slany. Equatorial circular orbits in the Kerr–de Sitter spacetimes. *Phys. Rev.*, D 69:064001, 2004.
48. C. Klein. Rotational perturbations and frame dragging in a Friedmann universe. *Class. Quantum Grav.* **10**, 1619 (1993).
49. C. Klein. Second-order effects of rotational perturbations of a Friedmann universe. *Class. Quantum Grav.* **11**, 1539 (1994).
50. T. Doležel, J. Bičák and N. Deruelle. Slowly rotating voids in cosmology. *Class. Quantum Grav.* **17**, 2719 (2000).
51. R.H. Dicke and J.E. Peebles. Evolution of the solar system and the expansion of the universe. *Phys. Rev. Lett.*, 12:435, 1964.
52. J.L. Anderson. Multiparticle dynamics in an expanding universe. *Phys. Rev. Lett.*, 75:3602, 1995.
53. W.B. Bonnor. The cosmic expansion and local dynamics. *Mon. Not. Roy. Astron. Soc.* **282**, 1467 (1996).
54. W.B. Bonnor. The hydrogen atom in an expanding universe. *Class. Quantum Grav.*, 16:1313, 1999.
55. W.B. Bonnor. A generalization of the Einstein–Straus vacuole. *Class. Quantum Grav.*, 17:2739, 2000.

56. L. Nottale. The Pioneer anomalous acceleration: a measurement of the cosmological constant at the scale of the solar system. gr-qc/0307042.
57. J.P. Mbelek. General relativity and quintessence explain the Pioneer anomaly. gr-qc/0402088.
58. B.G. Sidharth. Effects of varying *G*. *Il Nuovo Cim.* **B 115**, 151 (2000).
59. G. Modanese. Effect of a scale-dependent cosmological term on the motion of small test particles in a Schwarzschild background. *Nucl. Phys.*, B 556:397, 1999.
60. J.L. Rosales and J.L. Sanchez-Gomez. The "Pioneer effect" as a manifestation of the cosmic expansion in the solar system. arXiv:gr-qc/9810085.
61. D. Østvang. An explanation of the 'Pioneer effect' based on quasi–metric relativity. *Class. Quantum Grav.*, 19:4132, 2002.
62. M.A. Ivanov. Possible Manifestations of the Graviton Background. *Gen. Rel. Grav.* **33**, 479 (2001).
63. W.B. Belayev. Cosmological model in 5D, stationarity, yes or no. gr-qc/9903016.
64. J.W. Moffat. Modified gravitational theory and the Pioneer 10 and 11 spacecraft anomalous accelerartion. gr-qc:0405076.
65. M. Milgrom. MOND - A pedagogical review. *Acta Phys. Pol.* **B 32**, 3613 (2001).
66. M. Milgrom. MOND - theoretical aspects. *New Astr. Rev.*, 46:741, 2002.
67. J.D. Bekenstein. Relativistic gravitation theory for the MOND paradigm. astro-ph/0403694.
68. J.W. Moffat. Scalar–tensor–vector gravity theory. gr-qc/0506021.
69. M.-T. Jaekel and S. Reynaud. Gravity tests in the Solar system and the Pioneer anomaly. gr-qc/0410148.
70. M. Munyaneza and R.D. Viollier. Heavy neutrino dark matter in the solar system. astro-ph/9910566
71. R. Foot and R.R. Volkas. A Mirror World Explanation for the Pioneer Spacecraft Anomalies?' *Phys. Lett.* **B 517**, 13 (2001).
72. O. Bertolami and J. Paramos. The Pioneer anomaly in a bimetric theory of gravity on the brane. *Class. Quant. Grav.* **21**, 3309 (2004).
73. A. Aguirre. Alternatives to Dark Matter (?). astro-ph/0310572.
74. S. Calchi Novati, S. Capozziello, and G. Lambiase. *Grav. Cosmol.* **6**, 173 (2000).
75. M. Consoli and F. Siringo. Newtonian gravity from Higgs condensates. hep-ph/9910372.
76. J.P. Mbelek and M. Lachieze-Rey. Long-range acceleration induced by a scalar field external to gravity and the indication from Pioneer 10/11, Galileo and Ulysses Data. gr-qc/9910105.
77. S. Capozziello, S. De Martino, S. De Siena, and F. Illuminati. *Mod. Phys. Lett.* **A 16**, 693 (2001).
78. J. Wood and W. Moreau. Solutions of Conformal Gravity with Dynamical Mass Generation in the Solar System. gr-qc/0102056.
79. U. Johann and R. Förstner. Enigma', 2003. Unsolicited Proposal to ESA/ESTEC, Fund. Physics and Advanced Concepts.
80. U. Johann and R. Förstner. Enigma, `http://www.Zarm.uni-bremen.de/Pioneer`. Presentation on the first international Pioneer Anomaly Workshop, ZARM, Bremen, May 2004.
81. H. Dittus and the Pioneer Explorer Collaboration. A mission to explore the Pioneer anomaly. In Á. Giménez et al, editor, *Trends in Space Science and Cosmic Vision 2020*, page 3. ESA, Noordwijk, 2005; gr-qc/0506139.

82. M.M. Nieto, S.G. Turyshev, and J.D. Anderson. The Pioneer Anomaly: The Data, its Meaning, and a Future Test, **in:** AIP Conf. Proc. **758**, 113 (2005) [gr-qc/0411077].
83. M.M. Nieto and S.G. Turyshev. Finding the origin of the Pioneer anomaly. *Class. Quantum Grav.* **21**, 4005 (2004).
84. S.G. Turyshev, M.M. Nieto, and J.D. Anderson. A Route to Understanding of the Pioneer Anomaly. **in:** The XXII Texas Symposium on Relativistic Astrophysics, Stanford University, Dec. 13-17, 2004, edited by P. Chen, E. Bloom, G. Madejski, and V. Petrosian. (2005), [gr-qc/0503021].
85. DSN 810-005, Rev. E, 203.
86. `http://oersted.dtu.dk`
87. `http://www.jena-optronik.de` (April 2004).

Pioneer Anomaly: What Can We Learn from LISA?

Denis Defrère[1] and Andreas Rathke[2]

[1] Faculty of Applied Sciences, University of Liege, chemin des Chevreuils, 1 Bât. B52/3 Sart Tilman, 4000 Liege, Belgium
[2] EADS Astrium GmbH, Dept. AED41, 88039 Friedrichshafen, Germany

Summary. The Doppler tracking data from two deep-space spacecraft, Pioneer 10 and 11, show an anomalous blueshift, which has been dubbed the "Pioneer anomaly". The effect is most commonly interpreted as a real deceleration of the spacecraft – an interpretation that faces serious challenges from planetary ephemerides. The Pioneer anomaly could as well indicate an unknown effect on the radio signal itself. Several authors have made suggestions how such a blueshift could be related to cosmology. We consider this interpretation of the Pioneer anomaly and study the impact of an anomalous blueshift on the Laser Interferometer Space Antenna (LISA), a planned joint ESA–NASA mission aiming at the detection of gravitational waves. The relative frequency shift (proportional to the light travel time) for the LISA arm length is estimated to 10^{-16}, which is much bigger than the expected amplitude of gravitational waves. The anomalous blueshift enters the LISA signal in two ways, as a small term folded with the gravitational-wave signal, and as larger term at low frequencies. A detailed analysis shows that both contributions remain undetectable and do not impair the gravitational-wave detection. This suggests that the Pioneer anomaly will have to be tested in the outer solar system regardless if the effect is caused by an anomalous blueshift or by a real force.

1 Introduction

The Laser Interferometer Space Antenna (LISA) is a joint ESA–NASA mission to be launched after 2012 that will detect gravitational waves (GWs) in a frequency range between 10^{-4} and 1 Hz and study their sources [1]. LISA will consist of three spacecraft forming a roughly equilateral triangle of 5×10^9 m baseline placed on an orbit similar to that of the Earth. The spacecraft will exchange phase-coherent laser signals with each other to conduct picometer interferometry to measure passing GWs through the modulation in the light travel time between the spacecraft that the waves cause.

In this study, we consider the impact of an anomalous blueshift, which is homogeneous in the light travel time and isotropic, on LISA. The motivation to consider such an effect comes from the Doppler tracking data of the Pioneer

10 and 11 deep-space probes. Both spacecraft show a deviation between their orbit reconstruction and their Doppler tracking signal [2,3]. The discrepancy, which has become known as the Pioneer anomaly, can correspond either to a small constant deceleration of the spacecraft of roughly $9 \times 10^{-10}\,\mathrm{m\,s^{-2}}$, or to an anomalous blueshift of the radio signal at a rate of $6 \times 10^{-9}\,\mathrm{Hz\,s^{-1}}$. Since no unambiguous conventional mechanism, like small onboard forces, to explain the anomaly has been identified, there is a growing number of studies, which consider an explanation in terms of a novel physical effect (see [3–5] for an overview of the theoretical models). It has been realised that it is difficult to explain the Pioneer anomaly by a real force which satisfies all constraints from planetary ephemerides [3,5,6]. Hence an explanation in terms of an anomalous blueshift seems particularly attractive.

In view of the increasing interest in an experimental verification of the Pioneer anomaly [7], it is a logical step to consider if such a verification might be possible with a space mission that is already planned. Unfortunately, the current and upcoming exploration missions are hardly suited for a verification of the Pioneer anomaly [5,8]. LISA is the first upcoming high-precision fundamental physics mission that might be sensitive to the anomaly. Already in an early discussion, Scheffer [9] expressed the expectation that the Pioneer anomaly, if not due to a spacecraft-specific conventional reason, should be detectable in data from LISA. In a proposal to ESA's "Cosmic Vision 2015–2025 Call for Themes" the question was raised again if LISA could be a suitable testbed for a verification of the Pioneer anomaly – in particular if the effect were due to an anomalous blueshift [10]. Even more important might be the opposite question: if the Pioneer anomaly is indeed a novel physical effect could it impair the performance of LISA? In this case it would be of crucial importance to ensure that the LISA science goals can be achieved despite of the presence of the anomaly. The present study addresses both of these questions and comes to the conclusion that LISA is neither sufficiently sensitive to the Pioneer anomaly to detect it nor impeded in its mission goals by the potential presence of the anomaly.

The layout of our considerations is as follows. Section 2 gives an overview of the Pioneer anomaly and its possible relevance for LISA. In Sect. 2.1, we review the observational evidence for the Pioneer anomaly, and briefly review the models that have been put forward to explain the anomaly. In Sect. 2.2, we discuss which models of the Pioneer anomaly are relevant for observations with LISA and derive a first order of magnitude estimate for the maximal effect to expect on the interferometric signal of LISA. We also find the generic response function of LISA in the presence of an anomalous blueshift. Section 3 discusses the effect of the blueshift in the frequency domain. The frequency domain method has been discarded for the actual evaluation of LISA interferometric data because time-delay interferometry (TDI) achieves a far superior cancellation of the laser phase noise (cf. [11]). However, the frequency domain method has the advantage that it gives direct physical insight into the impact of an anomalous blueshift on the interferometer. Section 3.1 briefly reviews the structure of the interferometric signal of LISA and its Fourier transform. It is followed by the analysis of the impact of the blueshift, which is split into two parts. First Sect. 3.2 discusses the effect of the anomalous blueshift in the sensitivity band of LISA. Then the detectability of the blueshift at very low frequencies outside of the sensitivity band of LISA is considered in Sect. 3.3. In both cases, no measurable impact of the anomaly is found. Section 4 reconsiders the effect of the anomaly in the framework of TDI, the current baseline method for LISA. Section 4.1 discusses the signature of the anomaly on first-generation TDI observables for the idealised case of fixed arm length. It is found that the symmetry of TDI observables leads to an exact cancellation of the effect of the anomalous blueshift in the case of fixed interferometer arms. Section 4.2 generalises these considerations to the realistic case of moving spacecraft. Also in this setting, the effect of the anomalous blueshift would remain below the detection threshold of LISA. Section 5 summarises our results and discusses their implications for options to verify and characterise the Pioneer anomaly.

2 The Pioneer Anomaly and LISA

2.1 The Characteristics of the Pioneer Anomaly

The Pioneer 10 and 11 spacecraft, launched on 2 March 1972 and 5 April 1973, respectively, were the first to explore the outer solar system (see [12] for an overview of the Pioneer 10 and 11 missions). Since its Jupiter gravity assisted on 4 December 1973, Pioneer 10 is on a hyperbolic coast. Pioneer 11 used a Saturn swingby on 1 September 1979 to reach a hyperbola, in approximately opposite direction to Pioneer 10. Already before the swingby a discrepancy between the Doppler signal from Pioneer 10 and its orbit integration was observed, which was originally ascribed to fuel leaks and a mis-modelling in the solar radiation pressure model (cf. [13]). This interpretation became more and more untenable after the swingbys due to the decrease of the solar radiation pressure, inversely proportional to the square of the heliocentric distance, and the quiet state of the spacecraft, with very little thruster activity. Moreover an anomaly of the same magnitude became apparent in the Pioneer 11 data [14].

The anomaly on both probes has been subject to three independent analyses with different orbit determination programs [3, 15]. The result of the investigations is that an anomalous Doppler blueshift is present in the data from both spacecraft of approximately $6 \times 10^{-9}\,\mathrm{Hz\,s^{-1}}$ corresponding to an apparent deceleration of the spacecraft of approximately $9 \times 10^{-10}\,\mathrm{m\,s^{-2}}$. From the Doppler data, it is not possible to distinguish between an anomalous frequency shift of the radio signal and a real deceleration of the spacecraft (see later). The principle investigators of the anomaly have conducted a thorough investigation of possible biases and concluded that no conventional effect is likely to have caused the anomaly [3]. Meanwhile, there exists an ample body of literature discussing various aspects of possible systematic effects, without definitive conclusion [9,16–20]. For various reasons, all other deep-space probes have lower navigational accuracy [3, 5, 8]. Hence to date the effect could not be verified with another spacecraft.

The inability to explain the anomalous acceleration of the Pioneer spacecraft with conventional physics has contributed to the growing discussion about its origin. The possibility that it could come from a new physical effect is now being seriously considered. In particular, the coincidence in magnitude of the Pioneer anomaly and the Hubble acceleration has stirred the suggestion that the Pioneer anomaly could be related to the cosmological expansion.

One of the obstacles for attempting an explanation of the Pioneer anomaly in terms of new physics is that a modification of gravity, large enough to explain the Pioneer anomaly, is likely to run into contradiction with the planetary ephemerides. This is readily illustrated by adding a term corresponding to the Pioneer anomaly to the Newtonian potential of the Sun,

$$V(r) = -\frac{\mu_\odot}{r} - a^* r\,, \tag{1}$$

($\mu_\odot$ is the reduced mass of the Sun, r is the heliocentric distance and $a^* \approx 9 \times 10^{-10}\,\mathrm{m\,s^{-2}}$ is the anomalous acceleration) and considering the orbit of Neptune. At 30 AU, the Pioneer anomaly is visible in the Doppler data of both Pioneer 10 and 11. The influence of an additional radial acceleration of $9 \times 10^{-10}\mathrm{m\,s^{-2}}$ on Neptune is conveniently parameterised by a change of the effective reduced solar mass $\mu_\odot$, felt by the planet (cf. [21]). The value resulting for the anomaly, $\Delta\mu_\odot = a^* r_{♆}^2 \approx 1.4 \times 10^{-4}\,\mu_\odot$, is nearly two orders of magnitude beyond the current observational constraint of $\Delta\mu_\odot = (-1.9 \pm 1.8) \times 10^{-6}\,\mu_\odot$ [22]. Similarly, the Pioneer 11 data contradict the Uranus ephemerides by more than one order of magnitude. Thus, the Pioneer anomaly can hardly be ascribed to a gravitational force since this would indicate a considerable violation of the weak equivalence principle. In particular, planetary constraints rule out an explanation in terms of a long-range Yukawa force [3, 23]. Hence, more subtle explanations are to be attempted.

One line of reasoning is to consider an effect on the radio signal rather than a force on the spacecraft. Already the principle investigators have considered several phenomenological models of accelerating time [3]. The main purpose of these models was to investigate the possibility of a systematic drift of atomic clocks. Most of the phenomenological models failed the cross-check with tracking data from other spacecraft. Only a time acceleration restricted to the signal propagation itself yielded a good fit to all spacecraft data, although this model is still statistically disfavoured to a real deceleration of the spacecraft. The time acceleration of this model is indistinguishable from a runtime/travel distance-dependent blueshift of the radio signal.

To first order in v/c, the anomalous Doppler drift is related to the anomalous acceleration as

$$\frac{1}{\nu}\frac{d\nu}{dt} = -\frac{a^*}{c}\,, \tag{2}$$

where ν is the emitter frequency of the signal, v is the spacecraft velocity and c is the velocity of light (cf. [3]). Note that a^* is negative since it indicates a deceleration. At first sight this coincidence in phenomenology between an anomalous deceleration and an anomalous blueshift is surprising. It gets explained if one considers that the anomaly was only thoroughly investigated for the part of the Pioneer trajectories through the outer solar system: here the back-reaction of the spacecraft's orbit to a small perturbing force can be neglected and an anomalous acceleration can be treated linearly to high accuracy [5].[1]

Several theoretical models have been put forward that implement an anomalous blueshift by very distinct mechanisms [24–30]. The works [24, 25] consider the anomaly as a kinematical effect of the cosmological expansion. The anomaly arises from the fact that the coordinate system, in which local measurements are carried out, is not a synchronous one. The studies [26, 27] consider an adiabatic effect of the cosmic expansion on the phase of light viewed as the phase of a quantum state. While [26] considers a closed path Berry phase, [27] drops the closed path requirement and considers an open-path Berry phase. In [28–30], the anomaly arises from a time dependence of the local metric which leads to an effective time acceleration.

All of the above models to explain the blueshift of the Radio signals transponded by the Pioneers have to be considered as incomplete. This is most obvious for the model of [24,25], where only a Robertson–Walker metric is considered and the influence of the gravitational field of the Sun is completely neglected. This seems too much of a simplification considering the predominant opinion that the local Schwarzschild geometry of the solar system remains practically unaffected by the cosmological expansion (see the contribution of Lämmerzahl in this volume). The problem is ameliorated a bit for the quantum effect considered in [26,27] because in this case one could argue that the adiabatic evolution of quantum states is governed by a different metric than the non-adiabatic dynamics of large bodies. Also the definition of the open-path Berry phase in [27] does not seem to be compatible with the general discussion of the open-path Berry phase in [31]. In the models of Ranada [28–30], the embedding problem does not seem to spoil the model because both the cosmic and the local metric are treated as perturbations of a locally flat metric and can (at least formally) be superimposed linearly. However, the model of Ranada [28, 29] suffers from the introduction of two ad-hoc coupling parameters between the electromagnetic and the gravitational field [32]. Furthermore, the models of Ranada [28–30] lack a relativistic derivation of the background potential from the cosmological parameters. Despite of the deficiencies of the current models, the idea that the Pioneer anomaly

[1] This simple observation illustrates the need for the analysis of the full Doppler data of Pioneer 10 and 11 because from data further inward in the solar system a discrimination between a real force and a blueshift might be possible through the presence or absence of a change of the orbital parameters due to the anomaly.

is caused by a blueshift of light is attractive because it automatically satisfies all constraints from planetary ephemerides.

2.2 Relevance for LISA

Among the proposed explanations of the Pioneer anomaly, most would have no significance for LISA. For example, this is the case for all models based on systematics generated onboard the Pioneer spacecraft. Generally, if the anomaly corresponds to a real acceleration on the Pioneers, the anomaly should have no influence on LISA. This can be concluded from the fact that the LISA orbit is practically identical to the Earth's orbit. For the Earth itself an anomalous acceleration of the magnitude of the Pioneers would lead to an orbital perturbation, which is beyond current observational limits (cf. [21]). Hence only a considerable violation of the weak equivalence principle (e.g. between bodies of different mass) could result in an anomalous acceleration on LISA but not on the Earth. On the other hand an anomalous blueshift of light could be highly relevant for LISA, since the mission is supposed to detect GWs through small frequency shifts. The blueshift for light travelling along an arm of LISA is found by integrating (2) in time,

$$\frac{\Delta\nu^*}{\nu_0} = -\frac{a^*}{c} T \,, \tag{3}$$

where T is the light travel time and ν_0 is the laser frequency. For the LISA values, $\nu_0 \approx 3 \times 10^{14}$ Hz and $T \approx 17$ s, one finds $\Delta\nu^* \simeq 1.5 \times 10^{-2}$ Hz. Although the absolute blueshift is very small compared with the nominal frequency, it might nevertheless be within the reach of LISA. Indeed, the corresponding relative change of the frequency is $\Delta\nu^*/\nu \simeq 10^{-16}$ and the expected value for the weakest GWs, that will be detectable by LISA, is about 10^{-23} [1]. The frequency shift due to the anomaly is therefore seven orders of magnitude bigger than the lowest signal to be detectable by LISA. The ability to measure the contribution of the anomalous blueshift will, however, depend on the sensitivity of LISA at the frequencies where the anomaly is present.

For a comprehensive analysis of the impact of the anomalous blueshift on LISA, one has to take into consideration the change of the light travel time by passing GWs. In linear order in the GW strain h the rate of change of the light travel time caused by a plane wave is proportional to the projection of the difference of the GW strains at the point of reception and the point of emission onto the light travel direction [33, 34],

$$\frac{d}{dt}\Delta T = \frac{1}{2}(1+\beta)\big(h(t) - h[t + (1-\beta)T]\big) \,. \tag{4}$$

Here T is the unperturbed one-way light travel time, ΔT is the change of light travel time and β is the cosine of the angle between the light travel direction and the normal of the wave front of the GW. The time t is the time measured by a clock at the point of reception.

Writing (3) for the modified light travel time $T + \Delta T$ and using (4) to express ΔT by the linear term of a Taylor expansion, one obtains the frequency shift for the combined effect of the anomalous blueshift and GWs up to linear order in a^* and h for a one-way signal

$$\frac{\nu_1 - \nu_0}{\nu_0} = -\frac{a^*}{c}T + \frac{1}{2}(1+\beta)\left[1 - \frac{a^*}{c}T\right](h(t) - h[t + (1-\beta)T])\,, \qquad (5)$$

where ν_1 is the frequency at reception. Equation (5) is generic for any model of a homogeneous isotropic blueshift or time-acceleration. In particular it holds for the models considered in [3–30].

Depending on the model there might arise one subtlety, which has not been addressed up to now. In the same way as the electromagnetic waves are blueshifted an analogous blueshift might arise for the GWs. For example, this is the case in the models of Ranada [28–30], where the anomalous blueshift originates from a time-dependent term in the g_{00} component of the metric caused by a homogeneous cosmological background potential. This additional term leads to a modified dispersion relation for all types of waves. The anomalous blueshift of GWs could be investigated by a parameter estimation of the dispersion relation via matched filtering of GW signals detected by LISA. The method would be analogous to the search for a graviton mass in GW signals (cf. [35–37]). In the present study, we restrict ourselves to the possible manifestations of the Pioneer anomaly in electromagnetic waves because the occurrence of a blueshift of GWs is model dependent and hence would hardly allow a generic statement about the LISA's capability to verify the Pioneer anomaly.

Rather than the one-way response function of (5), the two-way response function of a signal transponded back to its emitter is the relevant observable for LISA. It is found analogous to (5) as

$$\frac{\nu_2 - \nu_0}{\nu_0} = -\frac{a^*}{c}T + \frac{1}{2}(1+\beta)\left[1 - \frac{a^*}{2c}T\right]h(t) - \beta\left[1 - \frac{a^*}{2c}T\right]h(t + (1-\beta)T/2)$$
$$-\frac{1}{2}(1-\beta)\left[1 - \frac{a^*}{2c}T\right]h(t+T), \qquad (6)$$

where T now denotes the unperturbed *two-way* light time and ν_2 is the frequency at reception [33,34]. The anomalous blueshift contributes to the frequency shift by two types of terms. On the one hand, it arises proportional to the unperturbed light travel time. If the light travel time is time dependent $T(t)$, as will be the case for LISA, the influence of the anomalous blueshift arises at the different frequencies contained in $T(t)$ and at null frequency anyway. On the other hand, the anomalous frequency shift appears as a cross-term with the GW strain. This effect is hence suppressed by the smallness of the GW strain but still several orders of magnitude larger than terms quadratic in the GW strain. Both manifestations of the anomalous blueshift will be investigated in the following.

3 Frequency Domain Analysis

In Sect. 2, we have discussed how the Pioneer anomaly could find its explanation in a blueshift of light and we found the generic Doppler response function to a plane GW in the presence of an anomalous blueshift. This Doppler response function describes the influence of the anomalous blueshift on GW detection by an interferometer arm of LISA. In the following, we analyse this signal both *inside* and *outside* the sensitivity band of the LISA. This is done first through an analysis in the frequency domain [38] and afterwards in the framework of TDI [11,39].

The noise cancelation algorithm in the frequency domain [38,40] is now considered as obsolete for the LISA data analysis and has been superseded by TDI, which achieves a far superior cancellation of laser phase noise in the signal than the frequency domain algorithm. For our purposes, the analysis in the frequency domain has the considerable advantage that it allows an intuitive understanding of the influence of the anomalous blueshift. In this method, the magnitude of various contributions to the signal can be easily compared and have a direct interpretation in terms of phase shifts of an idealised signal.

This convenient interpretation is partially lost in TDI, in which combinations of signals are formed following an algebraic method to cancel the dominant noise source of the interferometer. A more physical picture can in some part be regained by interpreting the TDI combinations as synthesised interferometers. Nevertheless, an investigation of the impact of an anomalous blueshift purely in terms of TDI might miss important effects of the blueshift, which might be cancelled by the specific symmetries of the TDI observables. On the other hand, the possibility exists that TDI combinations become particularly sensitive to the blueshift on behalf of their symmetries. Hence, it is important to investigate if TDI remains unimpaired by an anomalous blueshift and if TDI is capable of detecting a potential anomaly.

Our analysis of the anomalous blueshift in the frequency domain is based on the method of Giampieri [38]. We amend the original discussion by the consideration of additional noise sources, such as acceleration noise, which were not addressed in [38] and we update the values of laser and shot noise to match the current expectations for LISA (cf. [1,11]).

3.1 The Two-Way Doppler Signals

For simplicity, we assumed in our analysis that each laser has the same fundamental frequency ν, whereas in a realistic LISA configuration the frequency of the lasers may differ to each other by several hundreds of MHz. As a further simplification, we start our discussion by assuming constant and exactly known (but unequal) lengths of the interferometer arms. This assumption will be dropped later.

The basic interferometer configuration is displayed in the Fig. 1. The distances between pairs of spacecraft are L_1, L_2 and L_3, with L_i corresponding

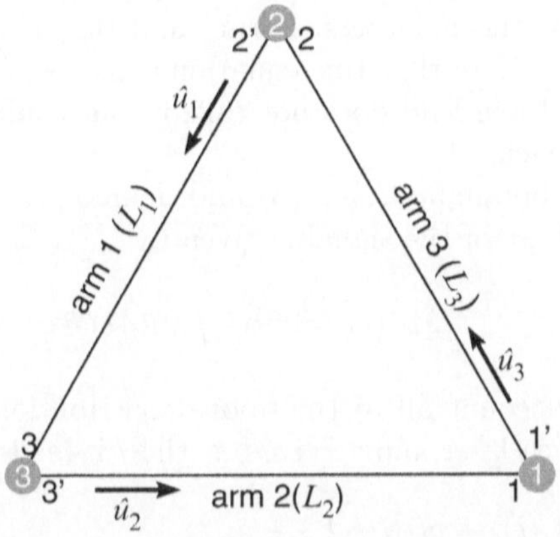

Fig. 1. Geometry of the LISA formation.

to the interferometer arm opposite to spacecraft i. The optical benches of each spacecraft are labelled by a number, which corresponds to that of the host spacecraft. An apostrophe allows to distinguish the two optical benches of the same spacecraft. In addition, a unit vector $\hat{u}_i$ is assigned to each arm, with i being the label of the opposite spacecraft. The orientation of the three unit vectors are such that $\hat{u}_1 L_1 + \hat{u}_2 L_2 + \hat{u}_3 L_3 = 0$.

The phase of the signal received from a distant spacecraft of the LISA constellation is the sum of the following contributions:

1. The phase $2\pi\nu l_i(t)$ due to the runtime of the signal, where $l_i(t)$ is the one-way light time for the signal along the ith interferometer arm. It changes due to the slow relative velocities between the spacecraft and on shorter timescales due to GWs.
2. The laser phase noise, $p_i(t)$, is the phase noise of the ith laser, so that the phase of the ith laser is $P_i = 2\pi\nu t + p_i(t)$.
3. The shot noise. Its effect is immediate at the time of reception, so that the response of the Doppler measurement at the ith laser is simply given by $n_i(t)$.
4. The acceleration noise. The phase variation $\Delta\varphi$ of a signal depends on the path length x through

$$\Delta\varphi = \frac{2\pi\nu}{c}\Delta x\,. \tag{7}$$

Therefore, the residual acceleration $\boldsymbol{a}_i(t)$ of the optical bench of the ith spacecraft appears in the second derivative of the phase of the signal. Obviously, the residual accelerations at the two spacecraft, both at transmission and reception, have to be taken into account according to the following expression

$$\Delta\varphi_i = \frac{2\pi\nu}{c}\int\!\!\int \left[\hat{u}_j \cdot \boldsymbol{a}_i(t) - \hat{u}_j \cdot \boldsymbol{a}_{k'}(t - l_j)\right] dt^2\,, \tag{8}$$

where i and k' are the end lasers of arm j and $\Delta\varphi_i$ is the phase variation at the photodiode i. Note that this equation considers only the acceleration along the optical axis and does not take into account a possible turning of the optical bench.

5. The anomalous blueshift. Using (3) and defining $\alpha^* \equiv -a^*/c$ its contribution to the phase of the signal is given by

$$\Delta\varphi_i^*(t) = 2\pi\alpha^* \int \Delta l_i(t)dt\,. \tag{9}$$

By taking into account all of the above contributions, the phase of the signal sent by the k'th laser and received at the ith reads

$$\begin{aligned}\varphi_i(t) &= 2\pi\nu\,(t - l_j(t)) + p_{k'}(t - l_j) + n_i(t)\\ &\quad + 2\frac{\pi\nu}{c}\int\!\!\int [\hat{u}_j \cdot \boldsymbol{a}_i(t) - \hat{u}_j \cdot \boldsymbol{a}_{k'}(t - l_j)]dt^2 + 2\pi\alpha^* \int l_j(t)dt\,. \end{aligned}\tag{10}$$

At the reception, the incoming signal is beaten with the signal P_i of the local laser to give the beat signal

$$\sin(\varphi_i(t)) + \sin(P_i(t)) = 2\sin\left[\frac{\varphi_i + P_i}{2}\right]\cos\left[\frac{\varphi_i - P_i}{2}\right]. \tag{11}$$

The high-frequency sine term is too fast to be read and is not used in the data analysis. Therefore, on the jth arm, the phase of the beat signal read in the spacecraft photodiode is given by

$$\begin{aligned}s_i(t) &= \varphi_i(t) - P_i(t)\\ &= -2\pi\nu l_j(t) + p_{k'}(t - l_j) - p_i(t) + 2\pi\alpha^* \int l_j(t)dt + n_{k'}(t)\\ &\quad + 2\frac{\pi\nu}{c}\int\!\!\int [\hat{u}_j \cdot \boldsymbol{a}_i(t) - \hat{u}_j \cdot \boldsymbol{a}_{k'}(t - l_j)]\,dt^2\,, \end{aligned}\tag{12}$$

where we have dropped the factor 1/2 from the argument of the cosine in (11). Furthermore, the two lasers on each spacecraft are tied to each other in phase by the exchange of a two-way reference signal between them.

The two-way Doppler signal is then formed by the combination of the phase measurements from two photodiodes on the same arm (cf. [38]),

$$\begin{aligned}z_i(t) &= s_i(t) + s_{k'}(t - l_j)\\ &= p_i(t - 2l_j) - p_i(t) - 4\pi\nu l_j(t) + 4\pi\alpha^* \int l_j(t)dt + n_{k'}(t) + n_i(t - l_j)\\ &\quad + 2\frac{\pi\nu}{c}\int\!\!\int [\hat{u}_j \cdot \boldsymbol{a}_i(t) - 2\hat{u}_j \cdot \boldsymbol{a}_{k'}(t - l_j) + \hat{u}_j \cdot \boldsymbol{a}_i(t - 2l_j)]\,dt^2\,. \end{aligned}\tag{13}$$

To obtain $z_i(t)$, $s_{k'}(t)$ is sent to the ith laser to be beaten with $s_i(t)$. Here, the beat signal is filtered to preserve the GW contribution, i.e. by reading the cosine term in the expression of a beat (analogous to (11)).

3.2 Inside the Sensitivity Band

In (13), the light travel time as a function of reception time $l_j(t)$ includes both the orbital motion of the spacecraft and the GWs. We write explicitly the contribution of the GWs by now considering $l_j(t)$ in as the nominal arm length in the undisturbed spacetime and adding the disturbance by the GW as a separate term. Using the Doppler response function (5), the effect of a GW, transverse to the LISA plane, i.e. $\beta \equiv 0$, and with appropriate polarisation, on the two-way Doppler signal is given by

$$\frac{\Delta\nu}{\nu} = \frac{1}{2}\epsilon(1+\alpha^* l_j)\left[h(t) - h(t-2l_j)\right], \tag{14}$$

where $\Delta\nu$ is the difference between the frequency of the signal sent and received at the central spacecraft and h is the GW strain amplitude. The ϵ can take any value between -1 and 1, depending on the orientation of the arm with respect to the polarisation of the GW. Particularly, for an angle of 60° between the arms of LISA, one can have $\epsilon = 1$ for one arm and $\epsilon = -1/2$ for the other (cf. [1] p. 102 for the general expressions). The GW adds a contribution to the signal (13),

$$\Delta\varphi_{\mathrm{gw}} = \epsilon\,\pi\nu \int (1+\alpha^*)[h(t) - h(t-2l_j)]dt\,. \tag{15}$$

To estimate the importance of each term in (13) and (14), it is useful to compute the power spectral density of $z_i(t)$. To begin, we restrict our study to the sensitivity band of LISA, i.e. from 10^{-4} to 1 Hz. In a first estimate we can drop the two terms, $-4\pi\nu l_j(t) + 4\pi\alpha^* \int l_j(t)dt$ because the orbital motion has little impact at the frequencies of the sensitivity band. To compute the power spectral density, we consider the Fourier transform of $z_i(t)$

$$\begin{aligned} z_i(f) &= p_i(f)\left(e^{4\pi i f l_j} - 1\right) + n_i(f)\left(1 + e^{2\pi i f l_j}\right) + \nu a_i(f)\frac{e^{4\pi i f l_j} + 2e^{2\pi i f l_j} + 1}{2\pi c f^2} \\ &+ \epsilon\nu h(f)\frac{e^{4\pi i f l_j} - 1}{2if} + \epsilon\nu\alpha^* l_j\, h(f)\frac{e^{4\pi i f l_j} - 1}{2if}, \end{aligned} \tag{16}$$

where we have assumed that the shot noise and acceleration spectra for both optical benches are the same, $n_i(f) = n_{k'}(f)$ and $a_i(f) = a_{k'}(f)$. Furthermore, we have assumed the maximum value for the direction cosine between $\boldsymbol{a}_{i,k'}$ and $\hat{u}_j$. The Fourier transform (16) supposes that the observing time is infinite. In practice, LISA is expected to operate in data-taking intervals of $T \sim 10{,}000$ s and thus, (16) only gives an estimate of the true spectrum. We will return to the effect of finite observation time below.

From (16), it can be read off immediately that the effect of the anomalous blueshift would be undetectable. The blueshift enters the spectrum folded with the GW strain h. Hence its effect will be 15 orders of magnitude below the GW signal. This corresponds to a spectral power at least ten orders of magnitude below the secondary noises, shot noise and acceleration noise (cf. [1,11] for the estimated noise spectra for LISA). Currently, no procedure exists to cancel the shot noise in the LISA signal. Hence the anomalous blueshift would be overwhelmed by the secondary noises and remain unnoticed.

This conclusion has, however, to be reconsidered taking into account that the data-taking periods of LISA are limited in length. This leads to the leakage of spectral power to other frequencies. In particular, the low-frequency terms neglected in the two-way signal $z_i(t)$ (13) have now to be addressed. For typical integration times, $T \sim 10{,}000\,\mathrm{s}$ (cf. [11]), the arm length rate of change is nearly constant. Its magnitude depends on the position of the spacecraft along its orbit [41]. The relative velocity, $\boldsymbol{v}$ can reach up to $13\,\mathrm{m\,s^{-1}}$.[2] In the approximation of constant relative velocity, the Fourier transform of $l_j(t) = l_{j,0} + (v_j/c)t$ is given by

$$\widetilde{F.T.}[l_j(t)] = \int_0^T \left(l_{j,0} + \frac{v_j}{c}t\right) e^{2\pi i f t} dt$$
$$= \frac{v_j}{c} e^{i\pi T f} \frac{\pi T f \cos(\pi T f) - \sin(\pi T f)}{2\pi^2 i f^2}$$
$$+ \left[l_{j,0} + \frac{v_j}{2c}T\right] e^{i\pi T f} \frac{\sin(\pi T f)}{\pi f}, \quad (17)$$

where $cl_{j,0} \approx 5 \times 10^9\,\mathrm{m}$ is the initial light time between the spacecraft. In (17), the constant term due to the arm length, $l_{j,0}$, is dominant. This term will remain present even after the application of the laser noise cancellation algorithm in the frequency domain (cf. [38]). Using (17), the Fourier transform of the two-way Doppler signal becomes

$$z_i(f) = p_i(f)\left[e^{4\pi i f l_j} - 1\right] + n_i(f)\left[1 + e^{2\pi i f l_j}\right]$$
$$+\nu a_i(f)\left[e^{4\pi i f l_j} + 2e^{2\pi i f l_j} + 1\right] + 4\pi\nu l_j(f) - \frac{2\alpha^*}{if} l_j(f). \quad (18)$$

In this expression, we have dropped the contribution of GWs because above it was found irrelevant for the discussion of the blueshift (see [38] for a discussion of the GW signal in terms of the frequency domain algorithm). The contributions to the amplitude power spectrum corresponding to (18) are shown in Fig. 2. The signal of the anomalous blueshift is higher than the secondary noise sources but below the laser phase noise. The nominal term from l_j, i.e. the orbital motion is much higher than the laser noise. Hence this term would have to be removed by a preprocessing method before the laser noise cancellation algorithm could be applied (see later). The signal of the anomalous blueshift is below the laser phase noise but above the secondary noise sources.

The further processing of the signal is distinct for the search for GWs and for the search for an anomalous blueshift. In the search for GWs the spectral

[2] See, however, [42] for recent suggestion of a modified orbit, which could reduce the relative velocity between the spacecraft by a factor of 6.

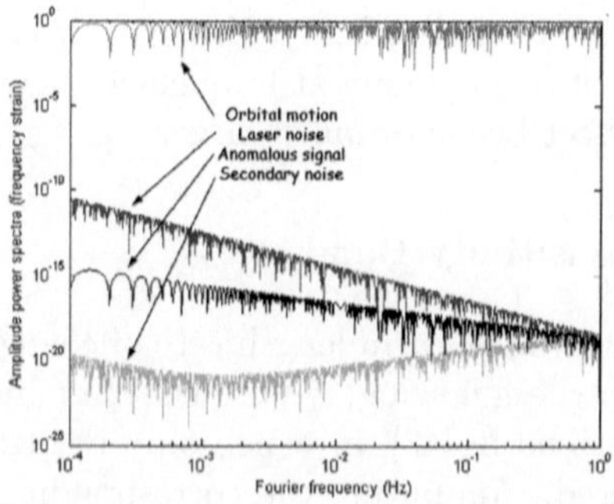

Fig. 2. Amplitude power spectra contributing to the two-way Doppler signal in the sensitivity band.

leakage is unwanted. Hence a suitable approach is pre-multiplying the time-domain data sets by a window function before taking Fourier transform [11]. With this preprocessing, the laser noise cancellation can be performed and GWs could be detected. On the other hand, in the search for an anomalous blueshift the spectral leakage has to be preserved. However, before the cancellation of the laser noise could be attempted, one has to generate a signal in which the laser noise is the dominant disturbance. Hence the nominal orbital term needs to be removed from the signal. The natural approach to this task would be to pre-process the data with information on the orbits acquired from a different source, e.g. ground tracking of the spacecraft.

The crucial question is how accurately we would need to determine the arm length of LISA to sufficiently remove the nominal orbital motion term. According to Fig. 2, a cancellation of about 10^{15} orders of magnitude would have to be performed. We suppose that the arm length l_j is known up to a factor k, i.e. the real arm length l_j differs from the assumed arm length $l_{j,\mathrm{measured}}$ by the length kl_j, $l_{j,\mathrm{measured}} = (1+k)l_j$. Then the Fourier transform of the one-way Doppler signal becomes

$$z_i(f) = p_i(f)\left[e^{4\pi i f l_j} - 1\right] + n_i(f)\left[1 + e^{2\pi i f l_j}\right] + a_i(f)\left[e^{4\pi i f l_j} + 2e^{2\pi i f l_j} + 1\right] \\ +4(1+k)\pi\nu l_j(f) - (1+k)\frac{2\alpha^*}{if} l_j(f). \tag{19}$$

Therefore, after the removal of the nominal orbital motion term from the knowledge of $l_{j,\mathrm{measured}}$, the term $4k\pi\nu l_j(f)$ remains which has to be sufficiently low to detect the anomaly. However, at 10^{-4} Hz, this requirement corresponds to a knowledge of the arm length of about $10^{-15} \times l_j = 5 \times 10^{-6}$ m, which is far beyond the experimental capabilities of LISA.

In conclusion, an anomalous blueshift of the magnitude of the Pioneer anomaly would remain undetected in the sensitivity band of LISA. Furthermore, it would not affect LISA's capability to observe GWs. The blueshift remains unimportant because it is peaked around zero frequency and thus far away from the sensitivity band of LISA. However, it is still worth considering the potential impact of the anomaly at frequencies below the sensitivity band of LISA where the effect becomes much larger.

3.3 Outside the Sensitivity Band

For long integration times, the arm length rate of change cannot be treated as constant anymore. In the following, we use the simplified analytical model of the LISA orbits described in [43], in which only the Kepler problem for each spacecraft is considered. Computing the corresponding power spectrum, one can plot the two-way Doppler signal, outside the sensitivity band of LISA.

Unfortunately, the noise spectra for LISA at frequencies below the measurement band have not yet been fully investigated (cf. [45]). For our purposes, we use an extrapolation of the noise spectra obtained in [45]. The acceleration noises might become considerably higher if a suspension mode for low frequencies would be implemented along the optical axes of the interferometer. However, the results of [44] indicate that the best performance of LISA is obtained if the drag-free mode along the sensitivity axes is maintained also for low frequencies. Hence an extrapolation of the noise spectra given in [45] should give a reasonable impression of the actual performance to expect from LISA at low frequencies.

The result is displayed in Fig. 3. The term due to the anomalous blueshift is of the same order of magnitude as the secondary noise but remains still below the laser phase noise. However, for the integration time required to reach these frequencies, the laser noise cancellation algorithm can no longer be implemented because the arm length changes from the orbital motion of LISA are so big that the algorithm becomes ineffective [38, 40]. Thus, the presence of an anomalous blueshift cannot be revealed at low frequencies either.

To summarise, the anomalous blueshift would have an amplitude several orders of magnitude higher than the weakest GWs detectable, on the LISA's arms. However, this "large" impact of the anomaly comes from the constant

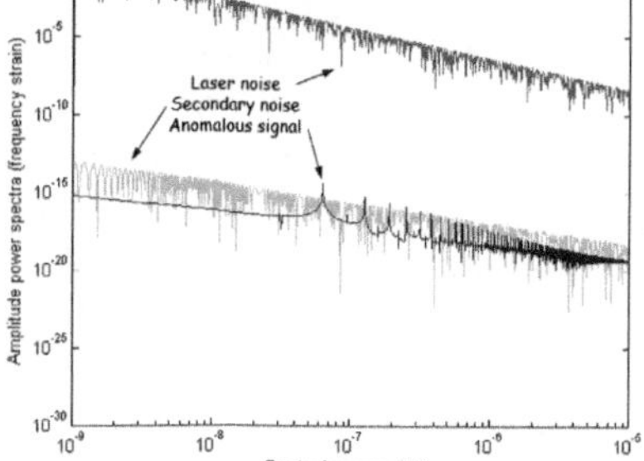

Fig. 3. Amplitude power spectra contributing to the two-way Doppler signal outside the sensitivity band.

part of the arm lengths. Hence it is located at null Fourier frequency while relevant GWs for LISA are expected at Fourier frequencies between 10^{-4} and 1 Hz. In the sensitivity band of LISA, the effect of the anomaly is well below all the instrumental noises and hence is neither detectable nor does it have an impact on the GW detection. With a finite observation time, power of the constant contribution of the anomaly can leak in the sensitivity band. The analysis of the spectral leakage of the anomalous blueshift would, however, require a knowledge of the arm length that would have to be far more precise than it is achievable. Below the sensitivity band, we found that the contribution of the anomalous blueshift should be just above the secondary noise sources but still below the laser phase noise. On these timescales, the arm lengths change much more than it is allowed to remove efficiently the laser phase noise. Therefore, we can conclude that the Pioneer anomaly has no impact on the GW detection and cannot be detected with the frequency domain method.

4 Time-Delay Interferometry

TDI is a noise cancellation method for unequal-arm interferometers that is performed in the time domain [11,39,46,47] (see also the contribution by Tinto in this volume). The basic principle of TDI consists in combining appropriate one-way Doppler signals to remove the laser phase noise (actually, it also cancels the acceleration noise of the optical benches). TDI had originally been developed as a purely algebraic method, whereas its data combinations have a physical interpretation as virtual measurements of a synthesised interferometer [48]. The major question to be addressed here is how the Pioneer anomaly affects the TDI combinations. Since the frequency domain study showed that the effect of the anomalous blueshift is negligible when folded with the GW strain, we can restrict our attention on the anomalous blueshift occurring at low frequencies.

4.1 Linear Data Combinations

In principle, there is an unlimited number of TDI observables corresponding to more and more complicated synthesised interferometers. For applications to LISA, the number of beams to combine is usually limited to eight in the limit of a static interferometer. For this maximum number of beams, there are ten linear combinations, which cancel the laser noises from all the spacecraft.

These TDI combinations cancel the laser phase noise of an interferometer at rest with unequal but constant arm lengths and are commonly dubbed *first generation TDI*. For the nominal operation mode of LISA, the unequal-arm Michelson interferometer, three independent possible combinations exist, which are called X, Y and Z. In the following, we only consider the X combination,

$$
\begin{aligned}
X = &\, y_{32,322} - y_{23,233} + y_{31,22} - y_{21,33} + y_{23,2} - y_{32,3} + y_{21} - y_{31} \\
&+ \frac{1}{2}(-z_{21,2233} + z_{21,33} + z_{21,22} - z_{21}) \\
&+ \frac{1}{2}(+z_{31,2233} - z_{31,33} - z_{31,22} + z_{31}) \, .
\end{aligned}
\tag{20}
$$

The Doppler data to be analysed are now called $y_{ij} = \Delta\nu/\nu$, where $\Delta\nu$ is the frequency deviation from the centre frequency ν. The subscripts label the transmitting and receiving spacecraft. The convention is that y_{ij} is the beam transmitted from spacecraft i and received at spacecraft j. Internal metrology signals to correct for optical bench motions are denoted by z_{ij} , with the same labelling convention. These will, however, play no role in our considerations because their travel times are too short to show an appreciable anomalous blueshift. They are hence omitted from now on. Delay of laser data streams is indicated by commas in the subscripts: $y_{31,23} = y_{31}(t - l_2 - l_3) = y_{31,32}$, etc. (l_i is the light time on the i-arm).

The Y and Z combinations are obtained from the X combination by cyclic relabelling of the spacecraft. Hence our results hold for all three of the unequal-arm Michelson combinations. It is easy to verify by direct substitution of the laser noise contribution that the combination (20) does not contain any laser noise. In the unequal-arm Michelson combinations, each one-way signal occurs twice, at two different times; one term is added and the other subtracted. As a consequence, the Pioneer anomaly component, given by (3),

$$
y^*_{ij} = -\frac{a^*}{c} l_i(t) = \alpha^* l_i(t) \, ,
\tag{21}
$$

disappears. Even the spectral leakage of the data has no impact because the terms, which contain the anomaly are all cancelled *exactly*. The same property holds for the other combinations of the data present in the literature (see [39] for a description of the other combinations).

In the combinations called α, β, γ and ζ, which represent synthetic Sagnac interferometers, the contribution of the anomaly would not be cancelled if the frequency shift would depend on the direction of the light beam with respect to the Sun. To obtain such a direction-dependent anomaly, which does not decay significantly over tens of AUs, one would, however, have to resort to exotic ideas like a topological defect located in the Sun. Such a model is hard to envisage and no such effect has been suggested as an explanation of the Pioneer anomaly. Hence we do not further consider this possibility.

4.2 Effect of the Orbital Motion on Time-Delay Interferometry

The first-generation TDI observables, as presented above, have been formulated in the limit that LISA is fixed in space. However, each year, LISA will accomplish a complete rotation around its centre and the symmetries provided by a fixed interferometer will be broken. Because of this loss of symmetry, the contribution of the anomalous blueshift, arising on each arm, would not be cancelled completely anymore in the TDI combinations. Moreover, the laser phase noise does not cancel exactly, either. More complicated TDI combinations have been developed to overcome this problem [49–51]. In addition to the rotation, there occurs a flexing of the arms of the detector, which is caused by the orbital motion and the perturbations of the planets. The interaction of these two effects with the anomaly is considered in the following.

The Effects of Rigid Rotation

For the discussion of rigid rotation, a more subtle notation for the Doppler signals is required because the light travel times will now depend on the direction of the signal with respect to the rotation. While in Sect. 3, L_3 was the length of the arm between the first and the second spacecraft, we denote now by L_{12} the length travelled by the signal sent from spacecraft 1 and received at spacecraft 2. The length travelled by the signal sent from spacecraft 2 and received at spacecraft 1 is called L_{21}. As illustrated in Fig. 4, the interferometer is rotating in the clockwise direction if viewed from the celestial pole. The spacecraft move while the signals are travelling along the arms. If we define the length of the arm between spacecraft 1 and 2 to be $\mathcal{L}_{12}$ in the limit of no rotation, then the actual distance travelled by the signal from spacecraft 1 to spacecraft 2 will be $L_{12} < \mathcal{L}_{12}$. In the same manner, the signal from spacecraft 2 to spacecraft 1 will have to reach spacecraft 1 in its motion and will therefore travel a distance $L_{21} > \mathcal{L}_{12}$. Hence also the magnitude of

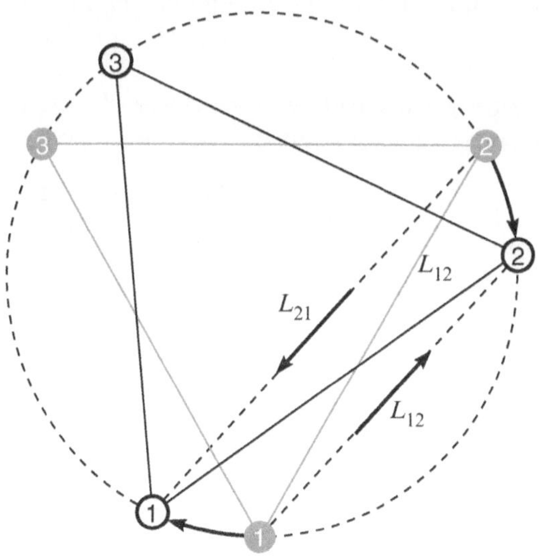

Fig. 4. The rotation of the interferometer breaks the direction symmetry in the arm lengths.

an anomalous blueshift on an arm would depend on the direction, in which the signal has travelled. Then, if the signals, which have travelled on the same arm but in opposite direction, are subtracted, a residual contribution of the anomalous blueshift would remain.

The direction dependence of the light travel times has different effects on the individual TDI combinations. For the unequal-arm length interferometric combinations $X(t)$, $Y(t)$ and $Z(t)$, the contribution of the anomaly still cancels exactly. Indeed, if we take the $X(t)$ combination (the reasoning is the same for $Y(t)$ and $Z(t)$), we see that the one-way signals appear twice for a given direction with opposite signs in the combination but that they are delayed by different times. For a rigid rotation the relation $L_{ij}(t+\tau) = L_{ij}$ holds and the contribution of the anomaly is cancelled. The (P, Q, R), (E, F, G) and (U, V, W) have the same property so that one reaches the same conclusion for these combinations.

For the Sagnac combinations (α, β, γ and ζ), the structure of the signal is different (cf. [39]). In these observables, two signals from each arm, running in opposite direction, are combined. For instance, the signal ζ reads,

$$\zeta = y_{32} - y_{23,3} + y_{13,3} - y_{31,1} + y_{21,1} - y_{12,2} \, . \tag{22}$$

Hence the effect of the anomaly is not totally removed. The anomalous component in (22) reads

$$\zeta^* = \alpha^*(l_{12} - l_{13} + l_{23} - l_{21} + l_{31} - l_{32}) \, . \tag{23}$$

The terms of this equation can be grouped into two parts: $\Delta l_- \equiv l_{12}+l_{23}+l_{31}$, which is the total time around the interferometer in the counterclockwise direction and $\Delta l_+ \equiv l_{13} + l_{21} + l_{32}$, which is the total time in the clockwise direction. Even for a perfectly rigid triangle, the times of flight are not equal. Since the LISA constellation rotates in clockwise direction, we always have $\Delta l_- < \Delta l_+$. The corresponding Sagnac time shift is given by, cf. [49],

$$\Delta l_- - \Delta l_+ = \Delta l_{\mathrm{Sagnac}} = \frac{4A\Omega}{c^2} \approx \frac{2\pi\sqrt{3}L^2}{c^2 T} \, . \tag{24}$$

Here Ω is the angular velocity of the rotation, A is the area enclosed by the light path, T is the period of rotation and L is a typical arm length. For the LISA orbit ($T = 1\,\mathrm{year}$ and $L = 5 \times 10^9\,\mathrm{m}$), the Sagnac effect has the magnitude $\Delta l_{\mathrm{Sagnac}} = -10^{-4}\,\mathrm{s}$. Therefore, the residual effect of the anomalous blueshift on the combination ζ would be

$$\zeta^* = \alpha^*(\Delta l_- - \Delta l_+) \simeq 3 \times 10^{-22} \, . \tag{25}$$

The same result is obtained for the other Sagnac combinations α, β, γ. The effect of the anomalous blueshift is to add a constant frequency shift in the Sagnac combinations. The amplitude of this additional Doppler shift would be comparable to the weakest GWs detectable by LISA. However, the optimal sensitivity of LISA occurs in a Fourier frequency range far from the zero Fourier frequency, where the constant residual contribution of the anomaly has its impact. Hence, again the effect would not be detectable.

As mentioned above the rigid rotation induces a Sagnac effect on the noises as well. To maintain noise cancellation up to linear order in the rotational velocity for all observables *modified TDI* has been introduced, in which each Doppler signal from a specific arm enters twice travelling in *the same* direction. In modified TDI, the anomalous blueshift is cancelled at linear order in the rotational velocity and the effect of the anomaly would become even smaller.

The Effects of Flexing

As we have seen in Sect. 3.3, the arm lengths of LISA do not remain constant due to the orbital motion and the perturbations of the orbits by the planets. Unlike the rigid rotation, the flexing of the arms does not preserve the continuous symmetry, $L_{ij}(t+\tau) = L_{ij}(t)$. For example, the X combination (20) becomes for time varying runtimes (cf. [49])

$$\begin{aligned} X = y_{12}\left[t - l_{31} - l_{13}^{(1)} - l_{21}^{(2)}\right] - y_{13}\left[t - l_{21} - l_{12}^{(1)} - l_{31}^{(2)}\right] + y_{21}\left[t - l_{31} - l_{13}^{(1)}\right] \\ -y_{31}\left[t - l_{21} - l_{12}^{(1)}\right] + y_{13}(t - l_{31}) - y_{12}(t - l_{21}) + y_{31}(t) - y_{21}(t)\,, \quad (26) \end{aligned}$$

where $l_{21} \equiv l_{21}(t)$, $l_{31} \equiv l_{31}(t)$, $l_{12}^{(1)} \equiv l_{12}(t - l_{21})$, $l_{13}^{(1)} \equiv l_{13}(t - l_{31})$, $l_{21}^{(2)} \equiv l_{21}(t - l_{31} - l_{13}^{(1)})$ and $l_{31}^{(2)} \equiv l_{31}(t - l_{21} - l_{12}^{(1)})$. Using (21), the contribution of the anomalous blueshift in the X combination is given by

$$\begin{aligned} X^* &= \alpha^*\left[l_{12}\left(t - l_{31} - l_{13}^{(1)} - l_{21}^{(2)}\right) - l_{13}\left(t - l_{21} - l_{12}^{(1)} - l_{31}^{(2)}\right) + l_{21}\left(t - l_{31} - l_{13}^{(1)}\right)\right] \\ &\quad + \alpha^*\left[-l_{31}\left(t - l_{21} - l_{12}^{(1)}\right) + l_{13}\left(t - l_{31}\right) - l_{12}(t - l_{21}) + l_{31}(t) - l_{21}(t)\right] \\ &= \alpha^*\left[l_{12}^{(3)} - l_{13}^{(3)} + l_{21}^{(2)} - l_{31}^{(2)} + l_{13}^{(1)} - l_{12}^{(1)} + l_{31} - l_{21}\right], \quad (27) \end{aligned}$$

where $l_{12}^{(3)} \equiv l_{12}(t - l_{31} - l_{13}^{(1)} - l_{21}^{(2)})$ and $l_{13}^{(3)} \equiv l_{13}(t - l_{21} - l_{12}^{(1)} - l_{31}^{(2)})$. The arm lengths can be estimated by their first-order changes $l_{ij}^{(n)} = l_{ij}(t) - nV_{ij}l$, where V_{ij} is the rate of change of the arm's light travel time in seconds per second and l is a typical one-way light time [49]. Then we find from (27)

$$X^* = 4\alpha^*(V_{13} - V_{12})l = 6 \times 10^{-24}\,, \quad (28)$$

where, in accordance with [49], we have used $V_{13} - V_{12} \lesssim 10\,(\mathrm{m\,s^{-1}})/c$. This relation determines the maximum effect of the blueshift due to the flexing induced by the orbital motion. The Doppler shift stays below the optimal sensitivity of LISA, which is about 10^{-23}. Results at the same order of magnitude are obtained for all TDI combinations. Hence we conclude that the effect of the anomalous blueshift due the flexing of the interferometer arms will not be detectable.

The flexing of the arms inhibits laser noise cancellation to first order in the velocities for first-generation TDI or modified TDI. To achieve the noise cancellation up to first order, inclusive, another set of observables has been designed. The so-called *second-generation TDI* achieves the cancellation by having not only a signal for each arm entering twice in the same direction, but also having each term linear in the change rate of the length of an arm entering twice. Hence second-generation TDI also cancels the anomalous blueshift from terms linear in the change rate of the arms.

We found that TDI observables, especially those of second-generation TDI, are particularly insensitive to the anomalous blueshift. This result has a simple geometric justification. The anomalous blueshift of sizable magnitude arises proportional to arm length differences. However, TDI is based on combining signals in a way that yields overall light times of zero (cf. [48]). Hence the very principle of the TDI algorithm leads to an automatic cancellation of the anomalous blueshift in the signal.

5 Summary and Conclusions

The Pioneer anomaly is attracting a growing interest in the scientific community. Hence a verification of the effect beyond the Pioneer data would be highly desirable. Here we studied in which way the LISA mission could contribute to a test of the Pioneer anomaly. Because of its Earth-like orbit, LISA would most likely not experience an anomalous force since this would require a strong violation of the weak equivalence principle. On the other hand, if one interprets the Pioneer anomaly as an anomalous blueshift of light, this effect would also affect the LISA interferometer.

Several models in the literature consider the Pioneer anomaly as a homogeneous and isotropic blueshift originating from the cosmic expansion through various mechanisms [24–30]. All of these distinct models lead to a common Doppler response function for LISA up to linear order in the anomalous blueshift and in the GW strain. We derived this Doppler response function as an extension of the well known two-point response to GWs. We found that the blueshift arises on the one hand as a cross-term with the GW signal and on the other hand as a low-frequency bias depending on the interferometer arm length. The cross-term with the GW signal is much larger than a possible second-order GW term but still too small to be detectable by LISA.

The low-frequency term was found to induce a relative frequency shift of 10^{-16}, which is several orders of magnitude larger than the weakest measurable GW strain of 10^{-23}. The implications of this number are, however, not immediate because the anomalous blueshift arises at zero frequency, whereas the LISA sensitivity lies between 10^{-4} and 1 Hz.

Consequently, we investigated the power spectral density of the anomalous frequency shift, which arises from the orbital motion of the LISA satellites. We considered both short times and timescales, which comprise a considerable fraction of the orbital period of LISA. These results were compared with the noise spectra of LISA. Unfortunately, due to its low-frequency nature, the anomalous blueshift is always overwhelmed by some noise source of the LISA interferometer. Hence an anomalous blueshift would remain undetectable.

This conclusion is then reconsidered in the framework of TDI, the current baseline method for laser phase noise cancellation in LISA. The dedicated discussion of TDI is necessary because the complicated signals synthesised in TDI could produce a by-chance amplification of a homogeneous isotropic signal. Our results show on the contrary that TDI cancels the blueshift in all data combinations to a high degree due to the inherent symmetries of the TDI observables. Only from the rotation and flexion of the interferometer, a residual contribution of the anomaly would arise. This effect would, however, be below the detection threshold of LISA. Hence an anomalous blueshift will not be recognisable in TDI and will not degrade the performance of TDI for the detection of GWs either.

In the present study, we have focused our attention on models of the anomaly that predict a homogeneous and isotropic blueshift. At first glance, this seems a bit restrictive, since also models, in which the Pioneer anomaly arises from a central force, can lead to a considerable blueshift of light. An example is the model of [52], which introduces separate momentum-dependent gravitational couplings for the scalar and the conformal sector of the Einstein equations. In this model, about half of the Pioneer anomaly is due an anomalous blueshift. However, the blueshift induced into the LISA signal by a central force would be proportional to the difference in light travel time between the way back and forth in a two-way signal, whereas a homogeneous and isotropic blueshift is proportional to the two-way light travel time. Thus, the blueshift from a central force, which is supposed to explain the Pioneer anomaly, would in general have much less effect on the LISA signal than a homogeneous and isotropic blueshift.

In conclusion, LISA cannot be used to test the Pioneer anomaly and one will have to look for other options to verify if the Pioneer anomaly could be a novel physical effect. Considering the blueshift interpretation of the anomaly, missions for a test of general relativity by interferometry like LATOR [53] (see also the contribution by Turyshev in this volume) or ASTROD [54] (see also the contribution by Wei-Tou Ni in this volume) might be sensitive to this effect. However, also these missions would face the problem that the anomalous blueshift becomes significant only at low frequencies, i.e. for large changes of the light travel time in the interferometer.

More promising – and probably mandatory if the Pioneer anomaly represents a force and not a blueshift – would be a test in the outer solar system by radio-tracking of a deep-space vehicle with very well know onboard systematics [7]. Preferably this would be a dedicated mission to explore the anomaly, although a planetary exploration spacecraft, which has been designed with the secondary goal to test the Pioneer anomaly could already gain considerable insights [5].[3] The analysis of the full archive of Pioneer 10 and 11 Doppler data, which is currently being initiated, might further help to identify mission scenarios that are especially suited for a test of the anomaly.

Acknowledgements

The authors would like to thank the organisers of the 359th WE-Heraeus-Seminar "Lasers, Clocks and Drag-Free: Key Technologies for Future High-Precision Test of General Relativity" for the realisation of this stimulating conference. The authors acknowledge support by the Advanced Concepts Team of the European Space Agency where the major part of this work was carried out. This work has much benefitted for discussions with O. Jennrich and D. Izzo.

References

1. K. Danzmann *et al.* [LISA Study Team], *Laser Interferometer Space Antenna: A cornerstone mission for the observation of Gravitational waves*, System and technology study report, ESA-SCI(2000)11, July 2000.
2. J.D. Anderson, P.A. Laing, E.L. Lau, A.S. Liu, M.M. Nieto and S.G. Turyshev, "Indication, from Pioneer 10/11, Galileo, and Ulysses data, of an apparent anomalous, weak, long-range acceleration," Phys. Rev. Lett. **81** (1998) 2858 [arXiv:gr-qc/9808081].
3. J.D. Anderson, P.A. Laing, E.L. Lau, A.S. Liu, M.M. Nieto and S.G. Turyshev, "Study of the anomalous acceleration of Pioneer 10 and 11," Phys. Rev. D **65** (2002) 082004 [arXiv:gr-qc/0104064].
4. O. Bertolami and J. Paramos, "The Pioneer anomaly in a bimetric theory of gravity on the brane," Class. Quant. Grav. **21** (2004) 3309 [arXiv:gr-qc/0310101].
5. D. Izzo and A. Rathke, "Options for a non-dedicated test of the Pioneer anomaly," arXiv:astro-ph/0504634.
6. R.H. Sanders, "A tensor-vector-scalar framework for modified dynamics and cosmic dark matter," arXiv:astro-ph/0502222.
7. H. Dittus *et al.* [Pioneer Collaboration], "A mission to explore the Pioneer anomaly," to appear in the proceedings of the 39th ESLAB Symposium: Trends in Space Science and Cosmic Vision 2020, Noordwijk, The Netherlands, 19–21 Apr 2005, arXiv:gr-qc/0506139.
8. M.M. Nieto and S.G. Turyshev, "Finding the origin of the Pioneer anomaly," Class. Quant. Grav. **21** (2004) 4005 [arXiv:gr-qc/0308017].
9. L.K. Scheffer, "Conventional forces can explain the anomalous acceleration of Pioneer 10," Phys. Rev. D **67** (2003) 084021 [arXiv:gr-qc/0107092].

[3] See [55, 56] for an example implementation of a Pioneer anomaly test on a Pluto exploration mission.

10. J.L. Rosales, "LISA mission and the Pioneer anomaly", proposal in response to the Call for Themes for Cosmic Vision 2015–2025 of the European Space Agency (May 2004).
11. M. Tinto and J.W. Armstrong, "Cancellation of laser noise in an unequal-arm interferometer detector of gravitational radiation," Phys. Rev. D **59** (1999) 102003.
12. L. Lasher and J. Dyer, "Pioneer Missions," in: P. Murdin (ed.), *Encyclopedia of Astronomy and Astrophysics* (Institute of Physics Publishing, Bristol, UK, 2002), `http://eaa.iop.org/abstract/0333750888/2187`.
13. G.W. Null, "Gravity field of Jupiter and its satellites from Pioneer 10 and Pioneer 11 tracking data," AJ **81** (1976) 1153.
14. M.M. Nieto and J. D. Anderson, "Using early data to illuminate the Pioneer anomaly," arXiv:gr-qc/0507052.
15. C.B. Markwardt, "Independent confirmation of the Pioneer 10 anomalous acceleration," arXiv:gr-qc/0208046.
16. J.I. Katz, "Comment on "Indication, from Pioneer 10/11, Galileo, and Ulysses data, of an apparent anomalous, weak, long-range acceleration"," Phys. Rev. Lett. **83** (1999) 1892 [arXiv:gr-qc/9809070].
17. E.M. Murphy, "A prosaic explanation for the anomalous accelerations seen in distant spacecraft," Phys. Rev. Lett. **83** (1999) 1890 [arXiv:gr-qc/9810015].
18. J.D. Anderson, P. A. Laing, E.L. Lau, M.M. Nieto and S.G. Turyshev, "The search for a standard explanation of the Pioneer anomaly," Mod. Phys. Lett. A **17** (2002) 875 [arXiv:gr-qc/0107022].
19. B. Mashhoon, "Modification of the Doppler effect due to the helicity-rotation coupling," Phys. Lett. A **306** (2002) 66 [arXiv:gr-qc/0209079].
20. J.D. Anderson and B. Mashhoon, "Pioneer anomaly and the helicity-rotation coupling," Phys. Lett. A **315** (2003) 199 [arXiv:gr-qc/0306024].
21. C. Talmadge, J.P. Berthias, R.W. Hellings and E.M. Standish, "Model independent constraints on possible modifications of Newtonian gravity," Phys. Rev. Lett. **61**, 1159 (1988).
22. J.D. Anderson, E.L. Lau, T.P. Krisher, D.A. Dicus, D. C. Rosenbaum and V. L. Teplitz, "Improved bounds on nonluminous matter in Solar orbit," Astrophys. J. **448** (1998) 885 [arXiv:hep-ph/9503368].
23. S. Reynaud and M.T. Jaekel, "Testing the Newton law at long distances," Int. J. Mod. Phys. A **20** (2005) 2294 [arXiv:gr-qc/0501038].
24. J.L. Rosales and J.L. Sanchez-Gomez, "The 'Pioneer effect' as a manifestation of the cosmic expansion in the solar system," arXiv:gr-qc/9810085.
25. J.L. Rosales, "The Pioneer effect: a cosmological Foucault's experiment," arXiv:gr-qc/0212019.
26. J.L. Rosales, "The Pioneer's anomalous Doppler drift as a Berry phase," arXiv:gr-qc/0401014.
27. J.L. Rosales, "The Pioneer anomaly: The measure of a topological phase defect of light in cosmology," arXiv:quant-ph/0501041.
28. A.F. Ranada, "The Pioneer riddle, the quantum vacuum and the acceleration of light," Europhys. Lett. **63** (2003) 653 [arXiv:gr-qc/0211052].
29. A.F. Ranada, "The light speed and the interplay of the quantum vacuum, the gravitation of all the universe and the fourth Heisenberg relation," Int. J. Mod. Phys. D **12** (2003) 1755.
30. A.F. Ranada, "The Pioneer anomaly as acceleration of the clocks," Found. Phys. **34** (2005) 1955 [arXiv:gr-qc/0410084].

31. A.K. Pati, "Adiabatic Berry phase and Hannay angle for open paths," Annals Phys. **270** (1998) 178 [arXiv:quant-ph/9804057].
32. N. Jafari and A. Shariati, "Comments on the quantum vacuum and the light acceleration," arXiv:gr-qc/0409113.
33. F.B. Estabrook and H. Wahlquist, "Response of Doppler spacecraft tracking to gravitational radiation," Gen. Rel. Grav. **6** (1975) 439.
34. R.W. Hellings, "Spacecraft Doppler gravity wave detection. 1. Theory," Phys. Rev. D **23**, 832 (1981).
35. C.M. Will, "Bounding the mass of the graviton using gravitational-wave observations of inspiralling compact binaries," Phys. Rev. D **57** (1998) 2061 [arXiv: gr-qc/9709011].
36. C.M. Will and N. Yunes, "Testing alternative theories of gravity using LISA," Class. Quant. Grav. **21** (2004) 4367 [arXiv:gr-qc/0403100].
37. E. Berti, A. Buonanno and C.M. Will, "Estimating spinning binary parameters and testing alternative theories of gravity with LISA," Phys. Rev. D **71** (2005) 084025 [arXiv:gr-qc/0411129].
38. Giacomo Giampieri, Ronald W. Hellings, Massimo Tinto, James E. Faller, "Algorithms for unequal-arm Michelson interferometers," Optics Communications **123** (1996) 669.
39. Sanjeev V. Dhurandhar and Massimo Tinto, "Time-delay interferometry," Living Rev. Relativity **8**, (2005), 4. URL (cited on 15 August 2005): `http://www.livingreviews.org/lrr-2005-4` [arXiv:gr-qc/0409034].
40. G. Giampieri, R.W. Hellings, L. Maleki, M. Tinto, K. Danzmann, J. Hough, D. Robertson, "Heterodyne laser tracking at high Doppler rates", Optics Communications **124** (1996) 313.
41. W.M. Folkner, T.H. Sweetser, M.A. Vincent, F. Hechler and P.L. Bender, "LISA orbit selection and stability," Class. Quant. Grav. **14** (1997) 1405.
42. K.R. Nayak, S. Koshti, S. V. Dhurandhar and J. Y. Vinet, "Reducing the flexing of the arms of LISA," arXiv:gr-qc/0507105.
43. S.V. Dhurandhar, K. Rajesh Nayak, S. Koshti and J. Y. Vinet, "Fundamentals of the LISA stable flight formation," Class. Quant. Grav. **22** (2005) 481 [arXiv:gr-qc/0410093].
44. A. Hammesfahr *et al.*, *LISA, Study of the Laser Interferometer Space Antenna, final technical report,* ESA Industrial Study at Phase A, ESTEC Contract no. 13631/99/NL/MS, DSS Report No. LI-RP-DS-009 (Dornier Satellitensysteme GmbH, Friedrichshafen, Germany).
45. P.L. Bender, "LISA sensitivity below 0.1 mHz," Class. Quant. Grav. **20** (2003) S301.
46. F.B. Estabrook, M. Tinto and J.W. Armstrong, "Time delay analysis of LISA gravitational wave data: Elimination of spacecraft motion effects," Phys. Rev. D **62** (2000) 042002.
47. M. Tinto, F. B. Estabrook and J. W. Armstrong, "Time-delay interferometry for LISA," Phys. Rev. D **65** (2002) 082003.
48. M. Vallisneri, "Geometric time delay interferometry," arXiv:gr-qc/0504145.
49. N.J. Cornish and R.W. Hellings, "The effects of orbital motion on LISA time delay interferometry," Class. Quant. Grav. **20** (2003) 4851 [arXiv:gr-qc/0306096].
50. D.A. Shaddock, M. Tinto, F.B. Estabrook and J.W. Armstrong, "Data combinations accounting for LISA spacecraft motion," Phys. Rev. D **68** (2003) 061303 [arXiv:gr-qc/0307080].

51. M. Tinto, F.B. Estabrook and J. W. Armstrong, "Time delay interferometry with moving spacecraft arrays," Phys. Rev. D **69** (2004) 082001 [arXiv: gr-qc/0310017].
52. M.T. Jaekel and S. Reynaud, "Post-Einsteinian tests of linearized gravitation," Class. Quant. Grav. **22** (2005) 2135 [arXiv:gr-qc/0502007].
53. S.G. Turyshev *et al.* [LATOR Collaboration], "Fundamental physics with the laser astrometric test of relativity," to appear in the proceedings of the 39th ESLAB Symposium: Trends in Space Science and Cosmic Vision 2020, Noordwijk, The Netherlands, 19–21 Apr 2005, arXiv:gr-qc/0506104.
54. W.T. Ni, "ASTROD: An overview," Int. J. Mod. Phys. D **11** (2002) 947.
55. A. Rathke, "Testing for the Pioneer anomaly on a Pluto exploration mission," in: A. Morseli, P. Picozza and M. Ricci (eds.), *Proceedings of the third international workshop on Frontier Science 2004. Physics and Astrophysics in Space* (INFN, Frascati, Italy 2004) [arXiv:astro-ph/0409373].
56. T. Bondo, R. Walker, A. Willig, A. Rathke, D. Izzo and M. Ayre, "Preliminary Design of an Advanced Mission to Pluto". to appear in the proceedings of the 24th International Symposium on Space Technology and Science, Miyazaki, Japan, June 2004, `http://www.esa.int/gsp/ACT/doc/ACT-RPR-4200-ISTS2004.pdf`.

Index

Recently Published in the ASSL series

Volume 349: *Lasers, Clocks and Drag-Free Control: Exploration of Relativistic Gravity in Space*, edited by Hansjörg Dittus, Claus Lämmerzahl, Salva Turyshev. Hardbound ISBN: 978-3-540-34376-9, September 2007

Volume 348: *The Paraboloidal Reflector Antenna in Radio Astronomy and Communication – Theory and Practice,* by Jacob W.M. Baars. Hardbound 978-0-387-69733-8, July 2007

Volume 347: *The Sun and Space Weather,* by Arnold Hanslmeier. Hardbound 978-1-4020-5603-1, June 2007

Volume 346: *Exploring the Secrets of the Aurora,* by Syun-Ichi Akasofu. Hardbound 978-0-387-45094-0, July 2007

Volume 345: *Canonical Perturbation Theories – Degenerate Systems and Resonance*, by Sylvio Ferraz-Mello. Hardbound 978-0-387-38900-4, January 2007

Volume 344: *Space Weather: Research Toward Applications in Europe*, edited by Jean Lilensten. Hardbound 1-4020-5445-9, January 2007

Volume 343: *Organizations and Strategies in Astronomy: Volume 7*, edited by A. Heck. Hardbound 1-4020-5300-2, December 2006

Volume 342: *The Astrophysics of Emission Line Stars*, by Tomokazu Kogure, Kam-Ching Leung. Hardbound ISBN: 0-387-34500-0, June 2007

Volume 341: *Plasma Astrophysics, Part II: Reconnection and Flares*, by Boris V. Somov. Hardbound ISBN: 0-387-34948-0, November 2006

Volume 340: *Plasma Astrophysics, Part I: Fundamentals and Practice*, by Boris V. Somov. Hardbound ISBN 0-387-34916-9, September 2006

Volume 339: *Cosmic Ray Interactions, Propagation, and Acceleration in Space Plasmas*, by Lev Dorman. Hardbound ISBN 1-4020-5100-X, August 2006

Volume 338: *Solar Journey: The Significance of Our Galactic Environment for the Heliosphere and the Earth*, edited by Priscilla C. Frisch. Hardbound ISBN 1-4020-4397-0, September 2006

Volume 337: *Astrophysical Disks*, edited by A. M. Fridman, M. Y. Marov, I. G. Kovalenko. Hardbound ISBN 1-4020-4347-3, June 2006

Volume 336: *Scientific Detectors for Astronomy 2005*, edited by J. E. Beletic, J. W. Beletic, P. Amico. Hardbound ISBN 1-4020-4329-5, December 2005

Volume 335: *Organizations and Strategies in Astronomy 6*, edited by A. Heck. Hardbound ISBN 1-4020-4055-5, November 2005

Volume 334: *The New Astronomy: Opening the Electromagnetic Window and Expanding our View of Planet Earth*, edited by W. Orchiston. Hardbound ISBN 1-4020-3723-6, October 2005

For other titles see www.springer.com/astronomy

GPSR Compliance

The European Union's (EU) General Product Safety Regulation (GPSR) is a set of rules that requires consumer products to be safe and our obligations to ensure this.

If you have any concerns about our products, you can contact us on ProductSafety@springernature.com

In case Publisher is established outside the EU, the EU authorized representative is:

Springer Nature Customer Service Center GmbH
Europaplatz 3
69115 Heidelberg, Germany

Batch number: 10761593

Printed by Printforce, the Netherlands